Numerical Analysis

THIRD EDITION

Timothy Sauer
George Mason University

Director, Portfolio Management: Deirdre Lynch
Executive Editor: Jeff Weidenaar
Editorial Assistant: Jennifer Snyder
Content Producer: Tara Corpuz
Managing Producer: Scott Disanno
Producer: Jean Choe
Product Marketing Manager: Yvonne Vannatta
Field Marketing Manager: Evan St. Cyr
Marketing Assistant: Jon Bryant
Senior Author Support/Technology Specialist: Joe Vetere
Manager, Rights and Permissions: Gina Cheselka
Manufacturing Buyer: Carol Melville, LSC Communications
Cover Image: Gyn9037/ Shutterstock
Text and Cover Design, Illustrations, Production Coordination, Composition:
    Integra Software Services Pvt. Ltd

Photo Credits: Page 1 Zsolt Biczo/ Shutterstock; Page 26 Polonio Video/ Shutterstock; Page 41 DEA PICTURE LIBRARY /
Getty Images; Page 74 Redswept /Shutterstock; Page 144 Rosenfeld Images, Ltd./Photo Researchers, Inc.; Page 196
dolgachov/ 123RF; Page 253 wklzzz / 123RF; Page 293 UPPA/Photoshot; Page 366 Paul Springett 04/Alamy Stock Photo;
Page 394 iStock/Getty Images Plus; Page 453 xPACIFICA / Alamy; Page 489 Picture Alliance/Photoshot; Page 518 Chris
Rout/Alamy Stock Photo; Pages 528 & 534 Toni Angermayer/Photo Researchers, Inc.; Page 556 Jinx Photography
Brands/Alamy Stock Photo; Page 593 Astronoman /Shutterstock.

Text Credits: Page 50 J. H. Wilkinson, The perfidious polynomial, In ed. by Gene H. Golub. Studies in Numerical Analysis.
Mathematical Association of America, 24 (1984); Page 153 & Page 188 "Author-created using the software from MATLAB.
The MathWorks, Inc., Natick, Massachusetts, USA, http://www.mathworks.com."; Page 454 Von Neumann, John (1951).
"Various techniques used in connection with random digits." In A. S. Householder, G. E. Forsythe, and H. H. Germond,
eds., Proceedings of Symposium on "Monte Carlo Method" held June-July 1949 in Los Angeles. Journal of Research of the
National Bureau of Standards, Applied Mathematics Series, no. 12, pp 36–38 (Washington, D.C.: USGPO, 1951) Summary
written by George E. Forsythe. Reprinted in von Neumann, John von Neumann Collected Works, ed. A. H. Taub, vol. 5
(New York: Macmillan, 1963) Vol. V, pp 768–770; Page 622 Author-created using the software from MATLAB. The
MathWorks, Inc., Natick, Massachusetts, USA, http://www.mathworks.com.; Page 623 Author-created using the software
from MATLAB. The MathWorks, Inc., Natick, Massachusetts, USA, http://www.mathworks.com.

Library of Congress Cataloging-in-Publication Data

Names: Sauer, Tim, author.
Title: Numerical analysis / Timothy Sauer, George Mason University.
Description: Third edition. | Hoboken : Pearson, [2019] | Includes
    bibliographical references and index.
Identifiers: LCCN 2017028491| ISBN 9780134696454 (alk. paper) |
    ISBN 013469645X (alk. paper)
Subjects: LCSH: Numerical analysis. | Mathematical analysis.
Classification: LCC QA297 .S348 2019 | DDC 518–dc23
LC record available at https://lccn.loc.gov/2017028491

16 2020

ISBN 10:   0-13-469645-X
ISBN 13: 978-0-13-469645-4

Contents

Preface

Numerical Analysis is a text for students of engineering, science, mathematics, and computer science who have completed elementary calculus and matrix algebra. The primary goal is to construct and explore algorithms for solving science and engineering problems. The not-so-secret secondary mission is to help the reader locate these algorithms in a landscape of some potent and far-reaching principles. These unifying principles, taken together, constitute a dynamic field of current research and development in modern numerical and computational science.

The discipline of numerical analysis is jam-packed with useful ideas. Textbooks run the risk of presenting the subject as a bag of neat but unrelated tricks. For a deep understanding, readers need to learn much more than how to code Newton's Method, Runge–Kutta, and the Fast Fourier Transform. They must absorb the big principles, the ones that permeate numerical analysis and integrate its competing concerns of accuracy and efficiency.

The notions of *convergence, complexity, conditioning, compression*, and *orthogonality* are among the most important of the big ideas. Any approximation method worth its salt must converge to the correct answer as more computational resources are devoted to it, and the complexity of a method is a measure of its use of these resources. The conditioning of a problem, or susceptibility to error magnification, is fundamental to knowing how it can be attacked. Many of the newest applications of numerical analysis strive to realize data in a shorter or compressed way. Finally, orthogonality is crucial for efficiency in many algorithms, and is irreplaceable where conditioning is an issue or compression is a goal.

In this book, the roles of these five concepts in modern numerical analysis are emphasized in short thematic elements labeled ***Spotlight***. They comment on the topic at hand and make informal connections to other expressions of the same concept elsewhere in the book. We hope that highlighting the five concepts in such an explicit way functions as a Greek chorus, accentuating what is really crucial about the theory on the page.

Although it is common knowledge that the ideas of numerical analysis are vital to the practice of modern science and engineering, it never hurts to be obvious. The feature entitled ***Reality Check*** provide concrete examples of the way numerical methods lead to solutions of important scientific and technological problems. These extended applications were chosen to be timely and close to everyday experience. Although it is impossible (and probably undesirable) to present the full details of the problems, the Reality Checks attempt to go deeply enough to show how a technique or algorithm can leverage a small amount of mathematics into a great payoff in technological design and function. The Reality Checks were popular as a source of student projects in previous editions, and they have been extended and amplified in this edition.

NEW TO THIS EDITION

Features of the third edition include:

- Short URLs in the side margin of the text (235 of them in all) take students directly to relevant content that supports their use of the textbook. Specifically:
  - **MATLAB Code:** Longer instances of MATLAB code are available for students in *.m format. The homepage for all of the instances of MATLAB code is `bit.ly/2yupqhx`.

○ **Solutions to Selected Exercises**: This text used to be supported by a Student Solutions Manual that was available for purchase separately. In this edition we are providing students with access solutions to selected exercises online *at no extra charge*. The homepage for the selected solutions is `bit.ly/2PG6q69`.

○ **Additional Examples**: Each section of the third edition is enhanced with extra new examples, designed to reinforce the text exposition and to ease the reader's transition to active solution of exercises and computer problems. The full worked-out details of these examples, more than one hundred in total, are available online. Some of the solutions are in video format (created by the author). The homepage for the solutions to Additional Examples is `bit.ly/2PG6q69`.

○ NOTE: The homepage for *all* web content supporting the text is `bit.ly/2yN3AEX`.

- More detailed discussion of several key concepts has been added in this edition, including theory of polynomial interpolation, multi-step differential equation solvers, boundary value problems, and the singular value decomposition, among others.

- The Reality Check on audio compression in Chapter 11 has been refurbished and simplified, and other MATLAB codes have been added and updated throughout the text.

- Several dozen new exercises and computer problems have been added to the third edition.

TECHNOLOGY

The software package MATLAB is used both for exposition of algorithms and as a suggested platform for student assignments and projects. The amount of MATLAB code provided in the text is carefully modulated, due to the fact that too much tends to be counterproductive. More MATLAB code is found in the early chapters, allowing the reader to gain proficiency in a gradual manner. Where more elaborate code is provided (in the study of interpolation, and ordinary and partial differential equations, for example), the expectation is for the reader to use what is given as a jumping-off point to exploit and extend.

It is not essential that any particular computational platform be used with this textbook, but the growing presence of MATLAB in engineering and science departments shows that a common language can smooth over many potholes. With MATLAB, all of the interface problems—data input/output, plotting, and so on—are solved in one fell swoop. Data structure issues (for example those that arise when studying sparse matrix methods) are standardized by relying on appropriate commands. MATLAB has facilities for audio and image file input and output. Differential equations simulations are simple to realize due to the animation commands built into MATLAB. These goals can all be achieved in other ways. But it is helpful to have one package that will run on almost all operating systems and simplify the details so that students can focus on the real mathematical issues. Appendix B is a MATLAB tutorial that can be used as a first introduction to students, or as a reference for those already familiar.

SUPPLEMENTS

The Instructor's Solutions Manual contains detailed solutions to the odd-numbered exercises, and answers to the even-numbered exercises. The manual also shows how to use MATLAB software as an aid to solving the types of problems that are presented in the Exercises and Computer Problems.

DESIGNING THE COURSE

Numerical Analysis is structured to move from foundational, elementary ideas at the outset to more sophisticated concepts later in the presentation. Chapter 0 provides fundamental building blocks for later use. Some instructors like to start at the beginning; others (including the author) prefer to start at Chapter 1 and fold in topics from Chapter 0 when required. Chapters 1 and 2 cover equation-solving in its various forms. Chapters 3 and 4 primarily treat the fitting of data, interpolation and least squares methods. In chapters 5–8, we return to the classical numerical analysis areas of continuous mathematics: numerical differentiation and integration, and the solution of ordinary and partial differential equations with initial and boundary conditions.

Chapter 9 develops random numbers in order to provide complementary methods to Chapters 5–8: the Monte-Carlo alternative to the standard numerical integration schemes and the counterpoint of stochastic differential equations are necessary when uncertainty is present in the model.

Compression is a core topic of numerical analysis, even though it often hides in plain sight in interpolation, least squares, and Fourier analysis. Modern compression techniques are featured in Chapters 10 and 11. In the former, the Fast Fourier Transform is treated as a device to carry out trigonometric interpolation, both in the exact and least squares sense. Links to audio compression are emphasized, and fully carried out in Chapter 11 on the Discrete Cosine Transform, the standard workhorse for modern audio and image compression. Chapter 12 on eigenvalues and singular values is also written to emphasize its connections to data compression, which are growing in importance in contemporary applications. Chapter 13 provides a short introduction to optimization techniques.

Numerical Analysis can also be used for a one-semester course with judicious choice of topics. Chapters 0–3 are fundamental for any course in the area. Separate one-semester tracks can be designed as follows:

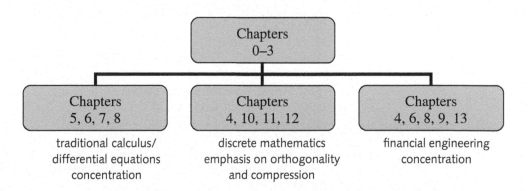

ACKNOWLEDGMENTS

The third edition owes a debt to many people, including the students of many classes who have read and commented on earlier versions. In addition, Paul Lorczak was essential in helping me avoid embarrassing blunders. The resourceful staff at Pearson, including Jeff Weidenaar, Jenn Snyder, Yvonne Vannatta, and Tara Corpuz, made the production of the third edition almost enjoyable. Finally, thanks are due to the helpful readers from other universities for their encouragement of this project and indispensable advice for improvement of earlier versions:

- Eugene Allgower, Colorado State University
- Constantin Bacuta, University of Delaware
- Michele Benzi, Emory University
- Jerry Bona, University of Illinois at Chicago
- George Davis, Georgia State University
- Chris Danforth, University of Vermont
- Alberto Delgado, Illinois State University
- Robert Dillon. Washington State University
- Qiang Du, Columbia University
- Ahmet Duran, University of Michigan
- Gregory Goeckel, Presbyterian College
- Herman Gollwitzer, Drexel University
- Weimin Han, University of Iowa *
- Don Hardcastle, Baylor University
- David R. Hill, Temple University
- Alberto Jimenez, California Polytechnic State University *
- Hideaki Kaneko, Old Dominion University
- Ashwani Kapila, Rensselaer Polytechnic Institute *
- Daniel Kaplan, Macalester College
- Fritz Keinert, Iowa State University
- Akhtar A. Khan, Rochester Institute of Technology
- Lucia M. Kimball, Bentley College
- Colleen M. Kirk, California Polytechnic State University
- Seppo Korpela, Ohio State University
- William Layton, University of Pittsburgh
- Brenton LeMesurier, College of Charleston
- Melvin Leok, University of California, San Diego
- Doron Levy, University of Maryland
- Bo Li, University of California, San Diego *
- Jianguo Liu, University of North Texas *
- Mark Lyon, University of New Hampshire *
- Shankar Mahalingam, University of Alabama, Huntsville
- Amnon Meir, Southern Methodist University
- Peter Monk, University of Delaware
- Joseph E. Pasciak, Texas A&M University
- Jeff Parker, Harvard University
- Jacek Polewczak, California State University
- Jorge Rebaza, Missouri State University
- Jeffrey Scroggs, North Carolina State University
- David Stewart, University of Iowa *
- David Stowell, Brigham Young University *
- Sergei Suslov, Arizona State University
- Daniel Szyld, Temple University
- Ahlam Tannouri, Morgan State University

- Janos Turi, University of Texas, Dallas *
- Jin Wang, Old Dominion University
- Bruno Welfert, Arizona State University
- Nathaniel Whitaker, University of Massachusetts

* Contributed to the current edition

CHAPTER

0

Fundamentals

This introductory chapter provides basic building blocks necessary for the construction and understanding of the algorithms of the book. They include fundamental ideas of introductory calculus and function evaluation, the details of machine arithmetic as it is carried out on modern computers, and discussion of the loss of significant digits resulting from poorly designed calculations.

After discussing efficient methods for evaluating polynomials, we study the binary number system, the representation of floating point numbers, and the common protocols used for rounding. The effects of the small rounding errors on computations are magnified in ill-conditioned problems. The battle to limit these pernicious effects is a recurring theme throughout the rest of the chapters.

The goal of this book is to present and discuss methods of solving mathematical problems with computers. The most fundamental operations of arithmetic are addition and multiplication. These are also the operations needed to evaluate a polynomial $P(x)$ at a particular value x. It is no coincidence that polynomials are the basic building blocks for many computational techniques we will construct.

Because of this, it is important to know how to evaluate a polynomial. The reader probably already knows how and may consider spending time on such an easy problem slightly ridiculous! But the more basic an operation is, the more we stand to gain by doing it right. Therefore we will think about how to implement polynomial evaluation as efficiently as possible.

0.1 EVALUATING A POLYNOMIAL

What is the best way to evaluate

$$P(x) = 2x^4 + 3x^3 - 3x^2 + 5x - 1,$$

say, at $x = 1/2$? Assume that the coefficients of the polynomial and the number 1/2 are stored in memory, and try to minimize the number of additions and multiplications required to get $P(1/2)$. To simplify matters, we will not count time spent storing and fetching numbers to and from memory.

METHOD 1   The first and most straightforward approach is

$$P\left(\frac{1}{2}\right) = 2 * \frac{1}{2} * \frac{1}{2} * \frac{1}{2} * \frac{1}{2} + 3 * \frac{1}{2} * \frac{1}{2} * \frac{1}{2} - 3 * \frac{1}{2} * \frac{1}{2} + 5 * \frac{1}{2} - 1 = \frac{5}{4}. \quad (0.1)$$

The number of multiplications required is 10, together with 4 additions. Two of the additions are actually subtractions, but because subtraction can be viewed as adding a negative stored number, we will not worry about the difference.

There surely is a better way than (0.1). Effort is being duplicated—operations can be saved by eliminating the repeated multiplication by the input $1/2$. A better strategy is to first compute $(1/2)^4$, storing partial products as we go. That leads to the following method:

METHOD 2   Find the powers of the input number $x = 1/2$ first, and store them for future use:

$$\frac{1}{2} * \frac{1}{2} = \left(\frac{1}{2}\right)^2$$
$$\left(\frac{1}{2}\right)^2 * \frac{1}{2} = \left(\frac{1}{2}\right)^3$$
$$\left(\frac{1}{2}\right)^3 * \frac{1}{2} = \left(\frac{1}{2}\right)^4.$$

Now we can add up the terms:

$$P\left(\frac{1}{2}\right) = 2 * \left(\frac{1}{2}\right)^4 + 3 * \left(\frac{1}{2}\right)^3 - 3 * \left(\frac{1}{2}\right)^2 + 5 * \frac{1}{2} - 1 = \frac{5}{4}.$$

There are now 3 multiplications of $1/2$, along with 4 other multiplications. Counting up, we have reduced to 7 multiplications, with the same 4 additions. Is the reduction from 14 to 11 operations a significant improvement? If there is only one evaluation to be done, then probably not. Whether Method 1 or Method 2 is used, the answer will be available before you can lift your fingers from the computer keyboard. However, suppose the polynomial needs to be evaluated at different inputs x several times per second. Then the difference may be crucial to getting the information when it is needed.

Is this the best we can do for a degree 4 polynomial? It may be hard to imagine that we can eliminate three more operations, but we can. The best elementary method is the following one:

METHOD 3   (Nested Multiplication) Rewrite the polynomial so that it can be evaluated from the inside out:

$$\begin{aligned} P(x) &= -1 + x(5 - 3x + 3x^2 + 2x^3) \\ &= -1 + x(5 + x(-3 + 3x + 2x^2)) \\ &= -1 + x(5 + x(-3 + x(3 + 2x))) \\ &= -1 + x * (5 + x * (-3 + x * (3 + x * 2))). \quad (0.2) \end{aligned}$$

Here the polynomial is written backwards, and powers of x are factored out of the rest of the polynomial. Once you can see to write it this way—no computation is required to do the rewriting—the coefficients are unchanged. Now evaluate from the inside out:

$$\text{multiply } \frac{1}{2} * 2, \quad \text{add } + 3 \rightarrow 4$$

$$\text{multiply } \frac{1}{2} * 4, \quad \text{add } - 3 \rightarrow -1$$

$$\text{multiply } \frac{1}{2} * -1, \quad \text{add } + 5 \rightarrow \frac{9}{2}$$

$$\text{multiply } \frac{1}{2} * \frac{9}{2}, \quad \text{add } - 1 \rightarrow \frac{5}{4}. \tag{0.3}$$

This method, called **nested multiplication** or **Horner's method**, evaluates the polynomial in 4 multiplications and 4 additions. A general degree d polynomial can be evaluated in d multiplications and d additions. Nested multiplication is closely related to synthetic division of polynomial arithmetic.

The example of polynomial evaluation is characteristic of the entire topic of computational methods for scientific computing. First, computers are very fast at doing very simple things. Second, it is important to do even simple tasks as efficiently as possible, since they may be executed many times. Third, the best way may not be the obvious way. Over the last half-century, the fields of numerical analysis and scientific computing, hand in hand with computer hardware technology, have developed efficient solution techniques to attack common problems.

While the standard form for a polynomial $c_1 + c_2 x + c_3 x^2 + c_4 x^3 + c_5 x^4$ can be written in nested form as

$$c_1 + x(c_2 + x(c_3 + x(c_4 + x(c_5)))), \tag{0.4}$$

some applications require a more general form. In particular, interpolation calculations in Chapter 3 will require the form

$$c_1 + (x - r_1)(c_2 + (x - r_2)(c_3 + (x - r_3)(c_4 + (x - r_4)(c_5)))), \tag{0.5}$$

where we call r_1, r_2, r_3, and r_4 the **base points**. Note that setting $r_1 = r_2 = r_3 = r_4 = 0$ in (0.5) recovers the original nested form (0.4).

The following MATLAB code implements the general form of nested multiplication (compare with (0.3)):

MATLAB code
shown here can be found
at bit.ly/20tdR4b

```
%Program 0.1 Nested multiplication
%Evaluates polynomial from nested form using Horner's Method
%Input: degree d of polynomial,
%       array of d+1 coefficients c (constant term first),
%       x-coordinate x at which to evaluate, and
%       array of d base points b, if needed
%Output: value y of polynomial at x
function y=nest(d,c,x,b)
if nargin<4, b=zeros(d,1); end
y=c(d+1);
for i=d:-1:1
  y = y.*(x-b(i))+c(i);
end
```

Running this MATLAB function is a matter of substituting the input data, which consist of the degree, coefficients, evaluation points, and base points. For example, polynomial (0.2) can be evaluated at $x = 1/2$ by the MATLAB command

```
>> nest(4,[-1 5 -3 3 2],1/2,[0 0 0 0])

ans =

    1.2500
```

as we found earlier by hand. The file `nest.m`, as the rest of the MATLAB code shown in this book, must be accessible from the MATLAB path (or in the current directory) when executing the command.

If the `nest` command is to be used with all base points 0 as in (0.2), the abbreviated form

```
>> nest(4,[-1 5 -3 3 2],1/2)
```

may be used with the same result. This is due to the `nargin` statement in `nest.m`. If the number of input arguments is less than 4, the base points are automatically set to zero.

Because of MATLAB's seamless treatment of vector notation, the `nest` command can evaluate an array of x values at once. The following code is illustrative:

```
>> nest(4,[-1 5 -3 3 2],[-2 -1 0 1 2])

ans =

   -15    -10    -1     6     53
```

Finally, the degree 3 interpolating polynomial

$$P(x) = 1 + x\left(\frac{1}{2} + (x-2)\left(\frac{1}{2} + (x-3)\left(-\frac{1}{2}\right)\right)\right)$$

from Chapter 3 has base points $r_1 = 0, r_2 = 2, r_3 = 3$. It can be evaluated at $x = 1$ by

```
>> nest(3,[1 1/2 1/2 -1/2],1,[0 2 3])

ans =

    0
```

► **EXAMPLE 0.1** Find an efficient method for evaluating the polynomial $P(x) = 4x^5 + 7x^8 - 3x^{11} + 2x^{14}$.

Some rewriting of the polynomial may help reduce the computational effort required for evaluation. The idea is to factor x^5 from each term and write as a polynomial in the quantity x^3:

$$P(x) = x^5(4 + 7x^3 - 3x^6 + 2x^9)$$
$$= x^5 * (4 + x^3 * (7 + x^3 * (-3 + x^3 * (2)))).$$

For each input x, we need to calculate $x * x = x^2$, $x * x^2 = x^3$, and $x^2 * x^3 = x^5$ first. These three multiplications, combined with the multiplication of x^5, and the three multiplications and three additions from the degree 3 polynomial in the quantity x^3 give the total operation count of 7 multiplies and 3 adds per evaluation. ◄

▶ **ADDITIONAL EXAMPLES**

1. Use nested multiplication to evaluate the polynomial
   $P(x) = x^6 - 2x^5 + 3x^4 - 4x^3 + 5x^2 - 6x + 7$ at $x = 2$.
2. Rewrite the polynomial $P(x) = 3x^{18} - 5x^{15} + 4x^{12} + 2x^6 - x^3 + 4$ in nested form. How many additions and how many multiplications are required for each input x?

   ⊑ **Solutions** for Additional Examples can be found at `bit.ly/2NOoKZ4`

0.1 Exercises

⊑ **Solutions** for Exercises numbered in blue can be found at `bit.ly/2CSG91u`

1. Rewrite the following polynomials in nested form. Evaluate with and without nested form at $x = 1/3$.

   (a) $P(x) = 6x^4 + x^3 + 5x^2 + x + 1$
   (b) $P(x) = -3x^4 + 4x^3 + 5x^2 - 5x + 1$
   (c) $P(x) = 2x^4 + x^3 - x^2 + 1$

2. Rewrite the following polynomials in nested form and evaluate at $x = -1/2$:

   (a) $P(x) = 6x^3 - 2x^2 - 3x + 7$
   (b) $P(x) = 8x^5 - x^4 - 3x^3 + x^2 - 3x + 1$
   (c) $P(x) = 4x^6 - 2x^4 - 2x + 4$

3. Evaluate $P(x) = x^6 - 4x^4 + 2x^2 + 1$ at $x = 1/2$ by considering $P(x)$ as a polynomial in x^2 and using nested multiplication.

4. Evaluate the nested polynomial with base points $P(x) = 1 + x(1/2 + (x - 2)(1/2 + (x - 3)(-1/2)))$ at (a) $x = 5$ and (b) $x = -1$.

5. Evaluate the nested polynomial with base points $P(x) = 4 + x(4 + (x - 1)(1 + (x - 2)(3 + (x - 3)(2))))$ at (a) $x = 1/2$ and (b) $x = -1/2$.

6. Explain how to evaluate the polynomial for a given input x, using as few operations as possible. How many multiplications and how many additions are required?
   (a) $P(x) = a_0 + a_5x^5 + a_{10}x^{10} + a_{15}x^{15}$
   (b) $P(x) = a_7x^7 + a_{12}x^{12} + a_{17}x^{17} + a_{22}x^{22} + a_{27}x^{27}$.

7. How many additions and multiplications are required to evaluate a degree n polynomial with base points, using the general nested multiplication algorithm?

0.1 Computer Problems

⊑ **Solutions** for Computer Problems numbered in blue can be found at `bit.ly/2NMwtqv`

1. Use the function `nest` to evaluate $P(x) = 1 + x + \cdots + x^{50}$ at $x = 1.00001$. (Use the MATLAB `ones` command to save typing.) Find the error of the computation by comparing with the equivalent expression $Q(x) = (x^{51} - 1)/(x - 1)$.

2. Use `nest.m` to evaluate $P(x) = 1 - x + x^2 - x^3 + \cdots + x^{98} - x^{99}$ at $x = 1.00001$. Find a simpler, equivalent expression, and use it to estimate the error of the nested multiplication.

0.2 BINARY NUMBERS

In preparation for the detailed study of computer arithmetic in the next section, we need to understand the binary number system. Decimal numbers are converted from base 10 to base 2 in order to store numbers on a computer and to simplify computer operations like addition and multiplication. To give output in decimal notation, the process is reversed. In this section, we discuss ways to convert between decimal and binary numbers.

Binary numbers are expressed as

$$\ldots b_2 b_1 b_0 . b_{-1} b_{-2} \ldots ,$$

where each binary digit, or **bit**, is 0 or 1. The base 10 equivalent to the number is

$$\ldots b_2 2^2 + b_1 2^1 + b_0 2^0 + b_{-1} 2^{-1} + b_{-2} 2^{-2} \ldots .$$

For example, the decimal number 4 is expressed as $(100.)_2$ in base 2, and 3/4 is represented as $(0.11)_2$.

0.2.1 Decimal to binary

The decimal number 53 will be represented as $(53)_{10}$ to emphasize that it is to be interpreted as base 10. To convert to binary, it is simplest to break the number into integer and fractional parts and convert each part separately. For the number $(53.7)_{10} = (53)_{10} + (0.7)_{10}$, we will convert each part to binary and combine the results.

Integer part. Convert decimal integers to binary by dividing by 2 successively and recording the remainders. The remainders, 0 or 1, are recorded by starting at the decimal point (or more accurately, **radix**) and moving away (to the left). For $(53)_{10}$, we would have

$$
\begin{aligned}
53 \div 2 &= 26 \text{ R } 1 \\
26 \div 2 &= 13 \text{ R } 0 \\
13 \div 2 &= 6 \text{ R } 1 \\
6 \div 2 &= 3 \text{ R } 0 \\
3 \div 2 &= 1 \text{ R } 1 \\
1 \div 2 &= 0 \text{ R } 1.
\end{aligned}
$$

Therefore, the base 10 number 53 can be written in bits as 110101, denoted as $(53)_{10} = (110101.)_2$. Checking the result, we have $110101 = 2^5 + 2^4 + 2^2 + 2^0 = 32 + 16 + 4 + 1 = 53$.

Fractional part. Convert $(0.7)_{10}$ to binary by reversing the preceding steps. Multiply by 2 successively and record the integer parts, moving away from the decimal point to the right.

$$
\begin{aligned}
.7 \times 2 &= .4 + 1 \\
.4 \times 2 &= .8 + 0 \\
.8 \times 2 &= .6 + 1 \\
.6 \times 2 &= .2 + 1 \\
.2 \times 2 &= .4 + 0 \\
.4 \times 2 &= .8 + 0 \\
&\vdots
\end{aligned}
$$

Notice that the process repeats after four steps and will repeat indefinitely exactly the same way. Therefore,

$$(0.7)_{10} = (.1011001100110\ldots)_2 = (.1\overline{0110})_2,$$

where overbar notation is used to denote infinitely repeated bits. Putting the two parts together, we conclude that

$$(53.7)_{10} = (110101.1\overline{0110})_2.$$

0.2.2 Binary to decimal

To convert a binary number to decimal, it is again best to separate into integer and fractional parts.

Integer part. Simply add up powers of 2 as we did before. The binary number $(10101)_2$ is simply $1 \cdot 2^4 + 0 \cdot 2^3 + 1 \cdot 2^2 + 0 \cdot 2^1 + 1 \cdot 2^0 = (21)_{10}$.

Fractional part. If the fractional part is finite (a terminating base 2 expansion), proceed the same way. For example,

$$(.1011)_2 = \frac{1}{2} + \frac{1}{8} + \frac{1}{16} = \left(\frac{11}{16}\right)_{10}.$$

The only complication arises when the fractional part is not a finite base 2 expansion. Converting an infinitely repeating binary expansion to a decimal fraction can be done in several ways. Perhaps the simplest way is to use the shift property of multiplication by 2.

For example, suppose $x = (0.\overline{1011})_2$ is to be converted to decimal. Multiply x by 2^4, which shifts 4 places to the left in binary. Then subtract the original x:

$$2^4 x = 1011.\overline{1011}$$
$$x = 0000.\overline{1011}.$$

Subtracting yields

$$(2^4 - 1)x = (1011)_2 = (11)_{10}.$$

Then solve for x to find $x = (.\overline{1011})_2 = 11/15$ in base 10.

As another example, assume that the fractional part does not immediately repeat, as in $x = .10\overline{101}$. Multiplying by 2^2 shifts to $y = 2^2 x = 10.\overline{101}$. The fractional part of y, call it $z = .\overline{101}$, is calculated as before:

$$2^3 z = 101.\overline{101}$$
$$z = 000.\overline{101}.$$

Therefore, $7z = 5$, and $y = 2 + 5/7$, $x = 2^{-2}y = 19/28$ in base 10. It is a good exercise to check this result by converting 19/28 to binary and comparing to the original x.

Binary numbers are the building blocks of machine computations, but they turn out to be long and unwieldy for humans to interpret. It is useful to use base 16 at times just to present numbers more easily. **Hexadecimal numbers** are represented by the 16 numerals $0, 1, 2, \ldots, 9, A, B, C, D, E, F$. Each hex number can be represented by 4 bits. Thus $(1)_{16} = (0001)_2$, $(8)_{16} = (1000)_2$, and $(F)_{16} = (1111)_2 = (15)_{10}$. In the next section, MATLAB's format hex for representing machine numbers will be described.

► **ADDITIONAL EXAMPLES**

*1. Convert the decimal number 98.6 to binary.

2. Convert the repeating binary number $0.1\overline{000111}$ to a base 10 fraction.

⌘ **Solutions** for Additional Examples can be found at `bit.ly/2RX2gsd`
(* example with video solution)

0.2 Exercises

⌘ **Solutions**
for Exercises
numbered in blue
can be found at
`bit.ly/2RVTZVl`

1. Find the binary representation of the base 10 integers. (a) 64 (b) 17 (c) 79 (d) 227

2. Find the binary representation of the base 10 numbers. (a) 1/8 (b) 7/8 (c) 35/16 (d) 31/64

3. Convert the following base 10 numbers to binary. Use overbar notation for nonterminating binary numbers. (a) 10.5 (b) 1/3 (c) 5/7 (d) 12.8 (e) 55.4 (f) 0.1

4. Convert the following base 10 numbers to binary. (a) 11.25 (b) 2/3 (c) 3/5 (d) 3.2 (e) 30.6 (f) 99.9

5. Find the first 15 bits in the binary representation of π.

6. Find the first 15 bits in the binary representation of e.

7. Convert the following binary numbers to base 10: (a) 1010101 (b) 1011.101 (c) $10111.\overline{01}$ (d) $110.\overline{10}$ (e) $10.\overline{110}$ (f) $110.1\overline{101}$ (g) $10.010\overline{1101}$ (h) $111.\overline{1}$

8. Convert the following binary numbers to base 10: (a) 11011 (b) 110111.001 (c) $111.\overline{001}$ (d) $1010.\overline{01}$ (e) $10111.1\overline{0101}$ (f) $1111.010\overline{001}$

0.3 FLOATING POINT REPRESENTATION OF REAL NUMBERS

There are several models for computer arithmetic of floating point numbers. The models in modern use are based on the IEEE 754 Floating Point Standard. The Institute of Electrical and Electronics Engineers (IEEE) takes an active interest in establishing standards for the industry. Their floating point arithmetic format has become the common standard for single precision and double precision arithmetic throughout the computer industry.

Rounding errors are inevitable when finite-precision computer memory locations are used to represent real, infinite precision numbers. Although we would hope that small errors made during a long calculation have only a minor effect on the answer, this turns out to be wishful thinking in many cases. **Simple algorithms, such as Gaussian elimination or methods for solving differential equations, can magnify microscopic errors to macroscopic size**. In fact, a main theme of this book is to help the reader to recognize when a calculation is at risk of being unreliable due to magnification of the small errors made by digital computers and to know how to avoid or minimize the risk.

0.3.1 Floating point formats

The IEEE standard consists of a set of binary representations of real numbers. A **floating point number** consists of three parts: the **sign** ($+$ or $-$), a **mantissa**, which contains the string of significant bits, and an **exponent**. The three parts are stored together in a single computer **word**.

There are three commonly used levels of precision for floating point numbers: single precision, double precision, and extended precision, also known as long-double precision. The number of bits allocated for each floating point number in the three formats is 32, 64, and 80, respectively. The bits are divided among the parts as follows:

precision	sign	exponent	mantissa
single	1	8	23
double	1	11	52
long double	1	15	64

All three types of precision work essentially the same way. The form of a **normalized** IEEE floating point number is

$$\pm 1.bbb\ldots b \times 2^p, \tag{0.6}$$

where each of the N b's is 0 or 1, and p is an M-bit binary number representing the exponent. Normalization means that, as shown in (0.6), the leading (leftmost) bit must be 1.

When a binary number is stored as a normalized floating point number, it is "left-justified," meaning that the leftmost 1 is shifted just to the left of the radix point. The shift is compensated by a change in the exponent. For example, the decimal number 9, which is 1001 in binary, would be stored as

$$+1.001 \times 2^3,$$

because a shift of 3 bits, or multiplication by 2^3, is necessary to move the leftmost one to the correct position.

For concreteness, we will specialize to the double precision format for most of the discussion. The double precision format, common in C compilers, python, and MAT-LAB, uses exponent length $M = 11$ and mantissa length $N = 52$. Single and long double precision are handled in the same way, but with different choices for M and N as specified above.

The double precision number 1 is

$$+1.\boxed{0000000000000000000000000000000000000000000000000000} \times 2^0,$$

where we have boxed the 52 bits of the mantissa. The next floating point number greater than 1 is

$$+1.\boxed{0000000000000000000000000000000000000000000000000001} \times 2^0,$$

or $1 + 2^{-52}$.

DEFINITION 0.1 The number **machine epsilon**, denoted ϵ_{mach}, is the distance between 1 and the smallest floating point number greater than 1. For the IEEE double precision floating point standard,

$$\epsilon_{mach} = 2^{-52}.$$ ◻

The decimal number $9.4 = (1001.\overline{0110})_2$ is left-justified as

$$+1.\boxed{0010110011001100110011001100110011001100110011001100}110\ldots \times 2^3,$$

where we have boxed the first 52 bits of the mantissa. A new question arises: How do we fit the infinite binary number representing 9.4 in a finite number of bits?

We must truncate the number in some way, and in so doing we necessarily make a small error. One method, called **chopping**, is to simply throw away the bits that fall off the end—that is, those beyond the 52nd bit to the right of the decimal point. This protocol is simple, but it is biased in that it always moves the result toward zero.

The alternative method is **rounding**. In base 10, numbers are customarily rounded up if the next digit is 5 or higher, and rounded down otherwise. In binary, this corresponds to rounding up if the bit is 1. Specifically, the important bit in the double precision format is the 53rd bit to the right of the radix point, the first one lying outside of the box. The default rounding technique, implemented by the IEEE standard, is to add 1 to bit 52 (round up) if bit 53 is 1, and to do nothing (round down) to bit 52 if bit 53 is 0, with one exception: If the bits following bit 52 are 10000..., exactly halfway between up and down, we round up or round down according to which choice makes the final bit 52 equal to 0. (Here we are dealing with the mantissa only, since the sign does not play a role.)

Why is there the strange exceptional case? Except for this case, the rule means rounding to the normalized floating point number closest to the original number—hence its name, the Rounding to Nearest Rule. The error made in rounding will be equally likely to be up or down. Therefore, the exceptional case, the case where there are two equally distant floating point numbers to round to, should be decided in a way that doesn't prefer up or down systematically. This is to try to avoid the possibility of an unwanted slow drift in long calculations due simply to a biased rounding. The choice to make the final bit 52 equal to 0 in the case of a tie is somewhat arbitrary, but at least it does not display a preference up or down. Problem 8 sheds some light on why the arbitrary choice of 0 is made in case of a tie.

IEEE Rounding to Nearest Rule

For double precision, if the 53rd bit to the right of the binary point is 0, then round down (truncate after the 52nd bit). If the 53rd bit is 1, then round up (add 1 to the 52 bit), unless all known bits to the right of the 1 are 0's, in which case 1 is added to bit 52 if and only if bit 52 is 1.

For the number 9.4 discussed previously, the 53rd bit to the right of the binary point is a 1 and is followed by other nonzero bits. The Rounding to Nearest Rule says to round up, or add 1 to bit 52. Therefore, the floating point number that represents 9.4 is

$$+1.\boxed{0010110011001100110011001100110011001100110011001101} \times 2^3. \qquad (0.7)$$

DEFINITION 0.2 Denote the IEEE double precision floating point number associated to x, using the Rounding to Nearest Rule, by $\mathbf{fl(x)}$. ⬚

Representation of floating point number

To represent a real number as a double precision floating point number, convert the number to binary, and carry out two steps:

1. **Justify.** Shift radix point to the right of the leftmost 1, and compensate with the exponent.

2. **Round.** Apply a rounding rule, such as the IEEE Rounding to Nearest Rule, to reduce the mantissa to 52 bits.

To find $\mathrm{fl}(1/6)$, note that $1/6$ is equal to $0.0\overline{01} = 0.001010101\ldots$ in binary.

1. **Justify.** The radix point is moved three places to the right, to obtain the justified number

$+1. \boxed{0101010101010101010101010101010101010101010101010101} 0101\ldots \times 2^{-3}$

2. **Round.** Bit 53 of the justified number is 0, so round down.

$$fl(1/6) = +1. \boxed{0101010101010101010101010101010101010101010101010101} \times 2^{-3}$$

To find $fl(11.3)$, note that 11.3 is equal to $1011.0\overline{1001}$ in binary.

1. **Justify.** The radix point is moved three places to the left, to obtain the justified number

$+1. \boxed{0110100110011001100110011001100110011001100110011001} 1001\ldots \times 2^3$

2. **Round.** Bit 53 of the justified number is 1, so round up, which means adding 1 to bit 52. Notice that the addition causes carrying to bit 51.

$$fl(11.3) = +1. \boxed{0110100110011001100110011001100110011001100110011010} \times 2^3$$

In computer arithmetic, the real number x is replaced with the string of bits $fl(x)$. According to this definition, $fl(9.4)$ is the number in the binary representation (0.7). We arrived at the floating point representation by discarding the infinite tail $\overline{1100} \times 2^{-52} \times 2^3 = .\overline{0110} \times 2^{-51} \times 2^3 = .4 \times 2^{-48}$ from the right end of the number and then adding $2^{-52} \times 2^3 = 2^{-49}$ in the rounding step. Therefore,

$$\begin{aligned} fl(9.4) &= 9.4 + 2^{-49} - 0.4 \times 2^{-48} \\ &= 9.4 + (1 - 0.8)2^{-49} \\ &= 9.4 + 0.2 \times 2^{-49}. \end{aligned} \tag{0.8}$$

In other words, a computer using double precision representation and the Rounding to Nearest Rule makes an error of 0.2×2^{-49} when storing 9.4. We call 0.2×2^{-49} the **rounding error**.

The important message is that the floating point number representing 9.4 is not equal to 9.4, although it is very close. To quantify that closeness, we use the standard definition of error.

DEFINITION 0.3   Let x_c be a computed version of the exact quantity x. Then

$$\textbf{absolute error} = |x_c - x|,$$

and

$$\textbf{relative error} = \frac{|x_c - x|}{|x|},$$

if the latter quantity exists. ⬜

Relative rounding error

In the IEEE machine arithmetic model, the relative rounding error of $fl(x)$ is no more than one-half machine epsilon:

$$\frac{|fl(x) - x|}{|x|} \leq \frac{1}{2}\epsilon_{mach}. \tag{0.9}$$

In the case of the number $x = 9.4$, we worked out the rounding error in (0.8), which must satisfy (0.9):

$$\frac{|fl(9.4) - 9.4|}{9.4} = \frac{0.2 \times 2^{-49}}{9.4} = \frac{8}{47} \times 2^{-52} < \frac{1}{2}\epsilon_{mach}.$$

▶ **EXAMPLE 0.2**   Find the double precision representation fl(x) and rounding error for $x = 0.4$.

Since $(0.4)_{10} = (.\overline{0110})_2$, left-justifying the binary number results in

$$0.4 = 1.100\overline{110} \times 2^{-2}$$
$$= +1.\boxed{1001100110011001100110011001100110011001100110011001}$$
$$100110\ldots \times 2^{-2}.$$

Therefore, according to the rounding rule, fl(0.4) is

$$+1.\boxed{1001100110011001100110011001100110011001100110011010} \times 2^{-2}.$$

Here, 1 has been added to bit 52, which caused bit 51 also to change, due to carrying in the binary addition.

Analyzing carefully, we discarded $2^{-53} \times 2^{-2} + .\overline{0110} \times 2^{-54} \times 2^{-2}$ in the truncation and added $2^{-52} \times 2^{-2}$ by rounding up. Therefore,

$$\text{fl}(0.4) = 0.4 - 2^{-55} - 0.4 \times 2^{-56} + 2^{-54}$$
$$= 0.4 + 2^{-54}(-1/2 - 0.1 + 1)$$
$$= 0.4 + 2^{-54}(.4)$$
$$= 0.4 + 0.1 \times 2^{-52}.$$

Notice that the relative error in rounding for 0.4 is $0.1/0.4 \times \epsilon_{\text{mach}} = 1/4 \times \epsilon_{\text{mach}}$, obeying (0.9).   ◀

0.3.2 Machine representation

So far, we have described a floating point representation in the abstract. Here are a few more details about how this representation is implemented on a computer. Again, in this section we will discuss the double precision format; the other formats are very similar.

Each double precision floating point number is assigned an 8-byte word, or 64 bits, to store its three parts. Each such word has the form

$$\boxed{se_1e_2\ldots e_{11}b_1b_2\ldots b_{52}}, \tag{0.10}$$

where the sign is stored, followed by 11 bits representing the exponent and the 52 bits following the decimal point, representing the mantissa. The sign bit s is 0 for a positive number and 1 for a negative number. The 11 bits representing the exponent come from the positive binary integer resulting from adding $2^{10} - 1 = 1023$ to the exponent, at least for exponents between -1022 and 1023. This covers values of $e_1\ldots e_{11}$ from 1 to 2046, leaving 0 and 2047 for special purposes, which we will return to later.

The number 1023 is called the **exponent bias** of the double precision format. It is used to convert both positive and negative exponents to positive binary numbers for storage in the exponent bits. For single and long-double precision, the exponent bias values are 127 and 16383, respectively.

MATLAB's `format hex` consists simply of expressing the 64 bits of the machine number (0.10) as 16 successive hexadecimal, or base 16, numbers. Thus, the first 3 hex numerals represent the sign and exponent combined, while the last 13 contain the mantissa.

For example, the number 1, or

$$1 = +1.\boxed{0000000000000000000000000000000000000000000000000000} \times 2^0,$$

has double precision machine number form

$$\boxed{0}\;\boxed{01111111111}\;\boxed{0000000000000000000000000000000000000000000000000000}$$

once the usual 1023 is added to the exponent. The first three hex digits correspond to

$$001111111111 = 3FF,$$

so the `format hex` representation of the floating point number 1 will be $3FF0000000000000$. You can check this by typing `format hex` into MATLAB and entering the number 1.

▶ **EXAMPLE 0.3**   Find the hex machine number representation of the real number 9.4.

From (0.7), we find that the sign is $s = 0$, the exponent is 3, and the 52 bits of the mantissa after the decimal point are

$$\boxed{0010}\;\boxed{1100}\;\boxed{1100}\;\boxed{1100}\;\boxed{1100}\;\boxed{1100}\;\boxed{1100}\;\boxed{1100}\;\boxed{1100}\;\boxed{1100}\;\boxed{1100}\;\boxed{1100}\;\boxed{1101}$$
$$\rightarrow (2CCCCCCCCCCCD)_{16}.$$

Adding 1023 to the exponent gives $1026 = 2^{10} + 2$, or $(10000000010)_2$. The sign and exponent combination is $(010000000010)_2 = (402)_{16}$, making the hex format $4022CCCCCCCCCCCD$.   ◀

Now we return to the special exponent values 0 and 2047. The latter, 2047, is used to represent ∞ if the mantissa bit string is all zeros and NaN, which stands for Not a Number, otherwise. Since 2047 is represented by eleven 1 bits, or $e_1 e_2 \ldots e_{11} = (111\ 1111\ 1111)_2$, the first twelve bits of `Inf` and `-Inf` are $\boxed{0111}\;\boxed{1111}\;\boxed{1111}$ and $\boxed{1111}\;\boxed{1111}\;\boxed{1111}$, respectively, and the remaining 52 bits (the mantissa) are zero. The machine number `NaN` also begins $\boxed{1111}\;\boxed{1111}\;\boxed{1111}$ but has a nonzero mantissa. In summary,

machine number	example	hex format
+Inf	1/0	7FF0000000000000
-Inf	−1/0	FFF0000000000000
NaN	0/0	FFFxxxxxxxxxxxxx

where the x's denote bits that are not all zero.

The special exponent 0, meaning $e_1 e_2 \ldots e_{11} = (000\ 0000\ 0000)_2$, also denotes a departure from the standard floating point form. In this case the machine number is interpreted as the non-normalized floating point number

$$\pm 0.\boxed{b_1 b_2 \ldots b_{52}} \times 2^{-1022}. \tag{0.11}$$

That is, *in this case only, the left-most bit is no longer assumed to be* 1. These non-normalized numbers are called **subnormal** floating point numbers. They extend the range of very small numbers by a few more orders of magnitude. Therefore, $2^{-52} \times 2^{-1022} = 2^{-1074}$ is the smallest nonzero representable number in double precision. Its machine word is

$$\boxed{0}\;\boxed{00000000000}\;\boxed{0000000000000000000000000000000000000000000000000001}.$$

Be sure to understand the difference between the smallest representable number 2^{-1074} and $\epsilon_{mach} = 2^{-52}$. Many numbers below ϵ_{mach} are machine representable, even though adding them to 1 may have no effect. On the other hand, double precision numbers below 2^{-1074} cannot be represented at all.

The subnormal numbers include the most important number 0. In fact, the subnormal representation includes two different floating point numbers, $+0$ and -0, that are treated in computations as the same real number. The machine representation of $+0$ has sign bit $s = 0$, exponent bits $e_1 \ldots e_{11} = 00000000000$, and mantissa 52 zeros; in short, all 64 bits are zero. The hex format for $+0$ is 0000000000000000. For the number -0, all is exactly the same, except for the sign bit $s = 1$. The hex format for -0 is 8000000000000000.

The term **overflow** refers to the condition when the result of an arithmetic operation is too large to be stored as a regular floating point number. For double precision floating point numbers, this means the exponent p in (0.6) is greater than 1023. Most computer languages will convert an overflow condition to machine number +Inf, -Inf, or NaN.

The term **underflow** refers to the condition when the result is too small to be represented. For double precision, this occurs for numbers less than 2^{-1074}. In most cases, an underflow will be set to zero. In both overflow and underflow situations, all significant digits are lost.

0.3.3 Addition of floating point numbers

Machine addition consists of lining up the decimal points of the two numbers to be added, adding them, and then storing the result again as a floating point number. The addition itself can be done in higher precision (with more than 52 bits) since it takes place in a register dedicated just to that purpose. Following the addition, the result must be rounded back to 52 bits beyond the binary point for storage as a machine number.

For example, adding 1 to 2^{-53} would appear as follows:

$$1.\boxed{00\ldots0} \times 2^0 + 1.\boxed{00\ldots0} \times 2^{-53}$$

$$= 1.\boxed{0000000000000000000000000000000000000000000000000000} \times 2^0$$

$$+ \; 0.\boxed{0000000000000000000000000000000000000000000000000000}1 \times 2^0$$

$$= 1.\boxed{0000000000000000000000000000000000000000000000000000}1 \times 2^0$$

This is saved as $1. \times 2^0 = 1$, according to the rounding rule. Therefore, $1 + 2^{-53}$ is equal to 1 in double precision IEEE arithmetic. Note that 2^{-53} is the largest floating point number with this property; anything larger added to 1 would result in a sum greater than 1 under computer arithmetic.

The fact that $\epsilon_{mach} = 2^{-52}$ does not mean that numbers smaller than ϵ_{mach} are negligible in the IEEE model. As long as they are representable in the model, computations with numbers of this size are just as accurate, assuming that they are not added or subtracted to numbers of unit size.

It is important to realize that computer arithmetic, because of the truncation and rounding that it carries out, can sometimes give surprising results. For example, if a double precision computer with IEEE rounding to nearest is asked to store 9.4, then subtract 9, and then subtract 0.4, the result will be something other than zero! What happens is the following: First, 9.4 is stored as $9.4 + 0.2 \times 2^{-49}$, as shown previously. When 9 is subtracted (note that 9 can be represented with no error), the result is $0.4 + 0.2 \times 2^{-49}$. Now, asking the computer to subtract 0.4 results in subtracting (as we found in Example 0.2) the machine number $\mathrm{fl}(0.4) = 0.4 + 0.1 \times 2^{-52}$, which will leave

$$0.2 \times 2^{-49} - 0.1 \times 2^{-52} = .1 \times 2^{-52}(2^4 - 1) = 3 \times 2^{-53}$$

instead of zero. This is a small number, on the order of ϵ_{mach}, but it is not zero. Since MATLAB's basic data type is the IEEE double precision number, we can illustrate this finding in a MATLAB session:

```
>> format long
>> x=9.4

x =

   9.40000000000000

>> y=x-9
y =

   0.40000000000000

>> z=y-0.4

z =

    3.330669073875470e-16

>> 3*2^(-53)

ans =

    3.330669073875470e-16
```

► **EXAMPLE 0.4**   Find the double precision floating point sum $(1 + 3 \times 2^{-53}) - 1$.

Of course, in real arithmetic the answer is 3×2^{-53}. However, floating point arithmetic may differ. Note that $3 \times 2^{-53} = 2^{-52} + 2^{-53}$. The first addition is

$$1.\boxed{00\ldots0} \times 2^0 + 1.\boxed{10\ldots0} \times 2^{-52}$$

$$= 1.\boxed{0000000000000000000000000000000000000000000000000000} \times 2^0$$

$$+ 0.\boxed{0000000000000000000000000000000000000000000000000001}1 \times 2^0$$

$$= 1.\boxed{0000000000000000000000000000000000000000000000000001}1 \times 2^0.$$

This is again the exceptional case for the rounding rule. Since bit 52 in the sum is 1, we must round up, which means adding 1 to bit 52. After carrying, we get

$$+ 1.\boxed{0000000000000000000000000000000000000000000000000010} \times 2^0,$$

which is the representation of $1 + 2^{-51}$. Therefore, after subtracting 1, the result will be 2^{-51}, which is equal to $2\epsilon_{mach} = 4 \times 2^{-53}$. Once again, note the difference between computer arithmetic and exact arithmetic. Check this result by using MATLAB.   ◄

Calculations in MATLAB, or in any compiler performing floating point calculation under the IEEE standard, follow the precise rules described in this section. Although floating point calculation can give surprising results because it differs from exact arithmetic, it is always predictable. The Rounding to Nearest Rule is the typical default rounding, although, if desired, it is possible to change to other rounding rules by using compiler flags. The comparison of results from different rounding protocols is sometimes useful as an informal way to assess the stability of a calculation.

It may be surprising that small rounding errors alone, of relative size ϵ_{mach}, are capable of derailing meaningful calculations. One mechanism for this is introduced in the next section. More generally, the study of error magnification and conditioning is a recurring theme in Chapters 1, 2, and beyond.

▶ **ADDITIONAL EXAMPLES**

*1. Determine the double-precision floating point number fl(20.1) and find its machine number representation.

2. Calculate $(2 + (2^{-51} + 2^{-52})) - 2$ in double precision floating point.

⌨ **Solutions** for Additional Examples can be found at bit.ly/2OzcIrC
(* example with video solution)

0.3 Exercises

⌨ **Solutions** for Exercises numbered in blue can be found at bit.ly/2yKx7PS

1. Convert the following base 10 numbers to binary and express each as a floating point number fl(x) by using the Rounding to Nearest Rule: (a) 1/4 (b) 1/3 (c) 2/3 (d) 0.9

2. Convert the following base 10 numbers to binary and express each as a floating point number fl(x) by using the Rounding to Nearest Rule: (a) 9.5 (b) 9.6 (c) 100.2 (d) 44/7

3. For which positive integers k can the number $5 + 2^{-k}$ be represented exactly (with no rounding error) in double precision floating point arithmetic?

4. Find the largest integer k for which $fl(19 + 2^{-k}) > fl(19)$ in double precision floating point arithmetic.

5. Do the following sums by hand in IEEE double precision computer arithmetic, using the Rounding to Nearest Rule. (Check your answers, using MATLAB.)

   (a) $(1 + (2^{-51} + 2^{-53})) - 1$
   (b) $(1 + (2^{-51} + 2^{-52} + 2^{-53})) - 1$

6. Do the following sums by hand in IEEE double precision computer arithmetic, using the Rounding to Nearest Rule:

   (a) $(1 + (2^{-51} + 2^{-52} + 2^{-54})) - 1$
   (b) $(1 + (2^{-51} + 2^{-52} + 2^{-60})) - 1$

7. Write each of the given numbers in MATLAB's format hex. Show your work. Then check your answers with MATLAB. (a) 8 (b) 21 (c) 1/8 (d) fl(1/3) (e) fl(2/3) (f) fl(0.1) (g) fl(−0.1) (h) fl(−0.2)

8. Is $1/3 + 2/3$ exactly equal to 1 in double precision floating point arithmetic, using the IEEE Rounding to Nearest Rule? You will need to use fl(1/3) and fl(2/3) from Exercise 1. Does this help explain why the rule is expressed as it is? Would the sum be the same if chopping after bit 52 were used instead of IEEE rounding?

9. (a) Explain why you can determine machine epsilon on a computer using IEEE double precision and the IEEE Rounding to Nearest Rule by calculating $(7/3 - 4/3) - 1$. (b) Does $(4/3 - 1/3) - 1$ also give ϵ_{mach}? Explain by converting to floating point numbers and carrying out the machine arithmetic.

10. Decide whether $1 + x > 1$ in double precision floating point arithmetic, with Rounding to Nearest. (a) $x = 2^{-53}$ (b) $x = 2^{-53} + 2^{-60}$

11. Does the associative law hold for IEEE computer addition?

12. Find the IEEE double precision representation $\mathrm{fl}(x)$, and find the exact difference $\mathrm{fl}(x) - x$ for the given real numbers. Check that the relative rounding error is no more than $\epsilon_{\mathrm{mach}}/2$. (a) $x = 1/3$ (b) $x = 3.3$ (c) $x = 9/7$

13. There are 64 double precision floating point numbers whose 64-bit machine representations have exactly one nonzero bit. Find the (a) largest (b) second-largest (c) smallest of these numbers.

14. Do the following operations by hand in IEEE double precision computer arithmetic, using the Rounding to Nearest Rule. (Check your answers, using MATLAB.) (a) $(4.3 - 3.3) - 1$ (b) $(4.4 - 3.4) - 1$ (c) $(4.9 - 3.9) - 1$

15. Do the following operations by hand in IEEE double precision computer arithmetic, using the Rounding to Nearest Rule. (a) $(8.3 - 7.3) - 1$ (b) $(8.4 - 7.4) - 1$ (c) $(8.8 - 7.8) - 1$

16. Find the IEEE double precision representation $\mathrm{fl}(x)$, and find the exact difference $\mathrm{fl}(x) - x$ for the given real numbers. Check that the relative rounding error is no more than $\epsilon_{\mathrm{mach}}/2$. (a) $x = 2.75$ (b) $x = 2.7$ (c) $x = 10/3$

0.4 LOSS OF SIGNIFICANCE

An advantage of knowing the details of computer arithmetic is that we are therefore in a better position to understand potential pitfalls in computer calculations. One major problem that arises in many forms is the loss of significant digits that results from subtracting nearly equal numbers. In its simplest form, this is an obvious statement. Assume that through considerable effort, as part of a long calculation, we have determined two numbers correct to seven significant digits, and now need to subtract them:

$$
\begin{array}{r}
123.4567 \\
- \ 123.4566 \\
\hline
000.0001
\end{array}
$$

The subtraction problem began with two input numbers that we knew to seven-digit accuracy, and ended with a result that has only one-digit accuracy. Although this example is quite straightforward, there are other examples of loss of significance that are more subtle, and in many cases this can be avoided by restructuring the calculation.

▶ **EXAMPLE 0.5** Calculate $\sqrt{9.01} - 3$ on a three-decimal-digit computer.

This example is still fairly simple and is presented only for illustrative purposes. Instead of using a computer with a 52-bit mantissa, as in double precision IEEE standard format, we assume that we are using a three-decimal-digit computer. Using a three-digit computer means that storing each intermediate calculation along the way implies storing into a floating point number with a three-digit mantissa. The problem data (the 9.01 and 3.00) are given to three-digit accuracy. Since we are going to use a three-digit computer, being optimistic, we might hope to get an answer that is good to three digits. (Of course, we can't expect more than this because we only carry along three digits during the calculation.) Checking on a hand calculator, we see that the correct answer is approximately $0.0016662 = 1.6662 \times 10^{-3}$. How many correct digits do we get with the three-digit computer?

None, as it turns out. Since $\sqrt{9.01} \approx 3.0016662$, when we store this intermediate result to three significant digits we get 3.00. Subtracting 3.00, we get a final answer of 0.00. No significant digits in our answer are correct.

Surprisingly, there is a way to save this computation, even on a three-digit computer. What is causing the loss of significance is the fact that we are explicitly subtracting nearly equal numbers, $\sqrt{9.01}$ and 3. We can avoid this problem by using algebra to rewrite the expression:

$$
\sqrt{9.01} - 3 = \frac{(\sqrt{9.01} - 3)(\sqrt{9.01} + 3)}{\sqrt{9.01} + 3}
$$

$$
= \frac{9.01 - 3^2}{\sqrt{9.01} + 3}
$$

$$
= \frac{0.01}{3.00 + 3} = \frac{.01}{6} = 0.00167 \approx 1.67 \times 10^{-3}.
$$

Here, we have rounded the last digit of the mantissa up to 7 since the next digit is 6. Notice that we got all three digits correct this way, at least the three digits that the correct answer rounds to. The lesson is that it is important to find ways to avoid subtracting nearly equal numbers in calculations, if possible.   ◀

The method that worked in the preceding example was essentially a trick. Multiplying by the "conjugate expression" is one trick that can help restructure the calculation. Often, specific identities can be used, as with trigonometric expressions. For example, calculation of $1 - \cos x$ when x is close to zero is subject to loss of significance. Let's compare the calculation of the expressions

$$
E_1 = \frac{1 - \cos x}{\sin^2 x} \quad \text{and} \quad E_2 = \frac{1}{1 + \cos x}
$$

for a range of input numbers x. We arrived at E_2 by multiplying the numerator and denominator of E_1 by $1 + \cos x$, and using the trig identity $\sin^2 x + \cos^2 x = 1$. In infinite precision, the two expressions are equal. Using the double precision of MATLAB computations, we get the following table:

x	E_1	E_2
1.00000000000000	0.64922320520476	0.64922320520476
0.10000000000000	0.50125208628858	0.50125208628857
0.01000000000000	0.50001250020848	0.50001250020834
0.00100000000000	0.50000012499219	0.50000012500002
0.00010000000000	0.49999999862793	0.50000000125000
0.00001000000000	0.50000004138685	0.50000000001250
0.00000100000000	0.50004445029134	0.50000000000013
0.00000010000000	0.49960036108132	0.50000000000000
0.00000001000000	0.00000000000000	0.50000000000000
0.00000000100000	0.00000000000000	0.50000000000000
0.00000000010000	0.00000000000000	0.50000000000000
0.00000000001000	0.00000000000000	0.50000000000000
0.00000000000100	0.00000000000000	0.50000000000000

The right column E_2 is correct up to the digits shown. The E_1 computation, due to the subtraction of nearly equal numbers, is having major problems below $x = 10^{-5}$ and has no correct significant digits for inputs $x = 10^{-8}$ and below.

The expression E_1 already has several incorrect digits for $x = 10^{-4}$ and gets worse as x decreases. The equivalent expression E_2 does not subtract nearly equal numbers and has no such problems.

The quadratic formula is often subject to loss of significance. Again, it is easy to avoid as long as you know it is there and how to restructure the expression.

▶ **EXAMPLE 0.6**   Find both roots of the quadratic equation $x^2 + 9^{12}x = 3$.

Try this one in double precision arithmetic, for example, using MATLAB. Neither one will give the right answer unless you are aware of loss of significance and know how to counteract it. The problem is to find both roots, let's say, with four-digit accuracy. So far it looks like an easy problem. The roots of a quadratic equation of form $ax^2 + bx + c = 0$ are given by the quadratic formula

$$x = \frac{-b \pm \sqrt{b^2 - 4ac}}{2a}. \tag{0.12}$$

For our problem, this translates to

$$x = \frac{-9^{12} \pm \sqrt{9^{24} + 4(3)}}{2}.$$

Using the minus sign gives the root

$$x_1 = -2.824 \times 10^{11},$$

correct to four significant digits. For the plus sign root

$$x_2 = \frac{-9^{12} + \sqrt{9^{24} + 4(3)}}{2},$$

MATLAB calculates 0. Although the correct answer is close to 0, the answer has no correct significant digits—even though the numbers defining the problem were specified exactly (essentially with infinitely many correct digits) and despite the fact that MATLAB computes with approximately 16 significant digits (an interpretation of the fact that the machine epsilon of MATLAB is $2^{-52} \approx 2.2 \times 10^{-16}$). How do we explain the total failure to get accurate digits for x_2?

The answer is loss of significance. It is clear that 9^{12} and $\sqrt{9^{24} + 4(3)}$ are nearly equal, relatively speaking. More precisely, as stored floating point numbers, their mantissas not only start off similarly, but also are actually identical. When they are subtracted, as directed by the quadratic formula, of course the result is zero.

Can this calculation be saved? We must fix the loss of significance problem. The correct way to compute x_2 is by restructuring the quadratic formula:

$$\begin{aligned}
x_2 &= \frac{-b + \sqrt{b^2 - 4ac}}{2a} \\[2mm]
&= \frac{(-b + \sqrt{b^2 - 4ac})(b + \sqrt{b^2 - 4ac})}{2a(b + \sqrt{b^2 - 4ac})} \\[2mm]
&= \frac{-4ac}{2a(b + \sqrt{b^2 - 4ac})} \\[2mm]
&= \frac{-2c}{b + \sqrt{b^2 - 4ac}}.
\end{aligned}$$

Substituting a, b, c for our example yields, according to MATLAB, $x_2 = 1.062 \times 10^{-11}$, which is correct to four significant digits of accuracy, as required. ◄

This example shows us that the quadratic formula (0.12) must be used with care in cases where a and/or c are small compared with b. More precisely, if $4|ac| \ll b^2$, then b and $\sqrt{b^2 - 4ac}$ are nearly equal in magnitude, and one of the roots is subject to loss of significance. If b is positive in this situation, then the two roots should be calculated as

$$x_1 = -\frac{b + \sqrt{b^2 - 4ac}}{2a} \quad \text{and} \quad x_2 = -\frac{2c}{(b + \sqrt{b^2 - 4ac})}. \qquad (0.13)$$

Note that neither formula suffers from subtracting nearly equal numbers. On the other hand, if b is negative and $4|ac| \ll b^2$, then the two roots are best calculated as

$$x_1 = \frac{-b + \sqrt{b^2 - 4ac}}{2a} \quad \text{and} \quad x_2 = \frac{2c}{(-b + \sqrt{b^2 - 4ac})}. \qquad (0.14)$$

▶ **ADDITIONAL EXAMPLES**

1. Define $f(x) = x^2 - x\sqrt{x^2 + 9}$. Calculate $f(8^{12})$ correct to 3 significant digits.
2. Calculate both roots of $3x^2 - 9^{14}x + 100 = 0$ correct to 3 significant digits.

Solutions for Additional Examples can be found at bit.ly/2yK4h21

0.4 Exercises

Solutions for Exercises numbered in blue can be found at bit.ly/2AeCVms

1. Identify for which values of x there is subtraction of nearly equal numbers, and find an alternate form that avoids the problem.

   (a) $\dfrac{1 - \sec x}{\tan^2 x}$ (b) $\dfrac{1 - (1 - x)^3}{x}$ (c) $\dfrac{1}{1 + x} - \dfrac{1}{1 - x}$

2. Find the roots of the equation $x^2 + 3x - 8^{-14} = 0$ with three-digit accuracy.
3. Explain how to most accurately compute the two roots of the equation $x^2 + bx - 10^{-12} = 0$, where b is a number greater than 100.
4. Evaluate the quantity $x\sqrt{x^2 + 17} - x^2$ where $x = 9^{10}$, correct to at least 3 decimal places.
5. Evaluate the quantity $\sqrt{16x^4 - x^2} - 4x^2$ where $x = 8^{12}$, correct to at least 3 decimal places.
6. Prove formula (0.14).

0.4 Computer Problems

Solutions for Computer Problems numbered in blue can be found at bit.ly/2CSnlzy

1. Calculate the expressions that follow in double precision arithmetic (using MATLAB, for example) for $x = 10^{-1}, \ldots, 10^{-14}$. Then, using an alternative form of the expression that doesn't suffer from subtracting nearly equal numbers, repeat the calculation and make a table of results. Report the number of correct digits in the original expression for each x.

   (a) $\dfrac{1 - \sec x}{\tan^2 x}$ (b) $\dfrac{1 - (1 - x)^3}{x}$

2. Find the smallest value of p for which the expression calculated in double precision arithmetic at $x = 10^{-p}$ has no correct significant digits. (Hint: First find the limit of the expression as $x \to 0$.)

$$\text{(a)} \quad \frac{\tan x - x}{x^3} \qquad \text{(b)} \quad \frac{e^x + \cos x - \sin x - 2}{x^3}$$

3. Evaluate the quantity $a + \sqrt{a^2 + b^2}$ to four correct significant digits, where $a = -12345678987654321$ and $b = 123$.

4. Evaluate the quantity $\sqrt{c^2 + d} - c$ to four correct significant digits, where $c = 246886422468$ and $d = 13579$.

5. Consider a right triangle whose legs are of length 3344556600 and 1.2222222. How much longer is the hypotenuse than the longer leg? Give your answer with at least four correct digits.

0.5 REVIEW OF CALCULUS

Some important basic facts from calculus will be necessary later. The Intermediate Value Theorem and the Mean Value Theorem are important for solving equations in Chapter 1. Taylor's Theorem is important for understanding interpolation in Chapter 3 and becomes of paramount importance for solving differential equations in Chapters 6, 7, and 8.

The graph of a continuous function has no gaps. For example, if the function is positive for one x-value and negative for another, it must pass through zero somewhere. This fact is basic for getting equation solvers to work in the next chapter. The first theorem, illustrated in Figure 0.1(a), generalizes this notion.

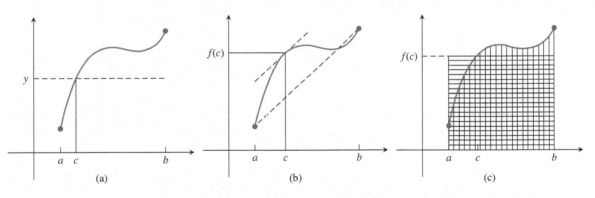

Figure 0.1 Three important theorems from calculus. There exist numbers c between a and b such that: (a) $f(c) = y$, for any given y between $f(a)$ and $f(b)$, by Theorem 0.4, the Intermediate Value Theorem (b) the instantaneous slope of f at c equals $(f(b) - f(a))/(b - a)$ by Theorem 0.6, the Mean Value Theorem (c) the vertically shaded region is equal in area to the horizontally shaded region, by Theorem 0.9, the Mean Value Theorem for Integrals, shown in the special case $g(x) = 1$.

THEOREM 0.4 (Intermediate Value Theorem) Let f be a continuous function on the interval $[a, b]$. Then f realizes every value between $f(a)$ and $f(b)$. More precisely, if y is a number between $f(a)$ and $f(b)$, then there exists a number c with $a \le c \le b$ such that $f(c) = y$. ∎

▶ **EXAMPLE 0.7**   Show that $f(x) = x^2 - 3$ on the interval $[1, 3]$ must take on the values 0 and 1.

Because $f(1) = -2$ and $f(3) = 6$, all values between -2 and 6, including 0 and 1, must be taken on by f. For example, setting $c = \sqrt{3}$, note that $f(c) = f(\sqrt{3}) = 0$, and secondly, $f(2) = 1$.   ◀

THEOREM 0.5   (Continuous Limits) Let f be a continuous function in a neighborhood of x_0, and assume $\lim_{n \to \infty} x_n = x_0$. Then

$$\lim_{n \to \infty} f(x_n) = f\left(\lim_{n \to \infty} x_n\right) = f(x_0).$$ ■

In other words, limits may be brought inside continuous functions.

THEOREM 0.6   (Mean Value Theorem) Let f be a continuously differentiable function on the interval $[a, b]$. Then there exists a number c between a and b such that $f'(c) = (f(b) - f(a))/(b - a)$.   ■

▶ **EXAMPLE 0.8**   Apply the Mean Value Theorem to $f(x) = x^2 - 3$ on the interval $[1, 3]$.

The content of the theorem is that because $f(1) = -2$ and $f(3) = 6$, there must exist a number c in the interval $(1, 3)$ satisfying $f'(c) = (6 - (-2))/(3 - 1) = 4$. It is easy to find such a c. Since $f'(x) = 2x$, the correct $c = 2$.   ◀

The next statement is a special case of the Mean Value Theorem.

THEOREM 0.7   (Rolle's Theorem) Let f be a continuously differentiable function on the interval $[a, b]$, and assume that $f(a) = f(b)$. Then there exists a number c between a and b such that $f'(c) = 0$.   ■

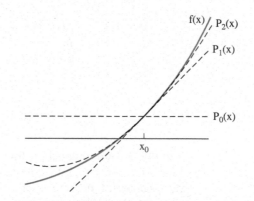

Figure 0.2 Taylor's Theorem with Remainder. The function $f(x)$, denoted by the solid curve, is approximated successively better near x_0 by the degree 0 Taylor polynomial (horizontal dashed line), the degree 1 Taylor polynomial (slanted dashed line), and the degree 2 Taylor polynomial (dashed parabola). The difference between $f(x)$ and its approximation at x is the Taylor remainder.

Taylor approximation underlies many simple computational techniques that we will study. If a function f is known well at a point x_0, then a lot of information about f at nearby points can be learned. If the function is continuous, then for points x near x_0, the function value $f(x)$ will be approximated reasonably well by $f(x_0)$. However, if

$f'(x_0) > 0$, then f has greater values for nearby points to the right, and lesser values for points to the left, since the slope near x_0 is approximately given by the derivative. The line through $(x_0, f(x_0))$ with slope $f'(x_0)$, shown in Figure 0.2, is the Taylor approximation of degree 1. Further small corrections can be extracted from higher derivatives, and give the higher degree Taylor approximations. Taylor's Theorem uses the entire set of derivatives at x_0 to give a full accounting of the function values in a small neighborhood of x_0.

THEOREM 0.8   (Taylor's Theorem with Remainder) Let x and x_0 be real numbers, and let f be $k + 1$ times continuously differentiable on the interval between x and x_0. Then there exists a number c between x and x_0 such that

$$f(x) = f(x_0) + f'(x_0)(x - x_0) + \frac{f''(x_0)}{2!}(x - x_0)^2 + \frac{f'''(x_0)}{3!}(x - x_0)^3 + \cdots$$

$$+ \frac{f^{(k)}(x_0)}{k!}(x - x_0)^k + \frac{f^{(k+1)}(c)}{(k+1)!}(x - x_0)^{k+1}.$$

∎

The polynomial part of the result, the terms up to degree k in $x - x_0$, is called the **degree k Taylor polynomial** for f centered at x_0. The final term is called the **Taylor remainder**. To the extent that the Taylor remainder term is small, Taylor's Theorem gives a way to approximate a general, smooth function with a polynomial. This is very convenient in solving problems with a computer, which, as mentioned earlier, can evaluate polynomials very efficiently.

▶ **EXAMPLE 0.9**   Find the degree 4 Taylor polynomial $P_4(x)$ for $f(x) = \sin x$ centered at the point $x_0 = 0$. Estimate the maximum possible error when using $P_4(x)$ to estimate $\sin x$ for $|x| \le 0.0001$.

     The polynomial is easily calculated to be $P_4(x) = x - x^3/6$. Note that the degree 4 term is absent, since its coefficient is zero. The remainder term is

$$\frac{x^5}{120} \cos c,$$

which in absolute value cannot be larger than $|x|^5/120$. For $|x| \le 0.0001$, the remainder is at most $10^{-20}/120$ and will be invisible when, for example, $x - x^3/6$ is used in double precision to approximate $\sin 0.0001$. Check this by computing both in MATLAB. ◀

     Finally, the integral version of the Mean Value Theorem is illustrated in Figure 0.1(c).

THEOREM 0.9   (Mean Value Theorem for Integrals) Let f be a continuous function on the interval $[a, b]$, and let g be an integrable function that does not change sign on $[a, b]$. Then there exists a number c between a and b such that

$$\int_a^b f(x)g(x)\, dx = f(c) \int_a^b g(x)\, dx.$$

∎

► **ADDITIONAL EXAMPLES**

1. Find c satisfying the Mean Value Theorem for $f(x) = \ln x$ on the interval $[1, 2]$.
2. Find the Taylor polynomial of degree 4 about the point $x = 0$ for $f(x) = e^{-x}$.

⌐□ **Solutions** for Additional Examples can be found at bit.ly/2AhOPMl

0.5 Exercises

⌐□ **Solutions**
for Exercises
numbered in blue
can be found at
bit.ly/2Cpe7tK

1. Use the Intermediate Value Theorem to prove that $f(c) = 0$ for some $0 < c < 1$.
   (a) $f(x) = x^3 - 4x + 1$ (b) $f(x) = 5\cos\pi x - 4$ (c) $f(x) = 8x^4 - 8x^2 + 1$

2. Find c satisfying the Mean Value Theorem for $f(x)$ on the interval $[0, 1]$. (a) $f(x) = e^x$
   (b) $f(x) = x^2$ (c) $f(x) = 1/(x + 1)$

3. Find c satisfying the Mean Value Theorem for Integrals with $f(x), g(x)$ in the interval
   $[0, 1]$. (a) $f(x) = x, g(x) = x$ (b) $f(x) = x^2, g(x) = x$ (c) $f(x) = x, g(x) = e^x$

4. Find the Taylor polynomial of degree 2 about the point $x = 0$ for the following functions:
   (a) $f(x) = e^{x^2}$ (b) $f(x) = \cos 5x$ (c) $f(x) = 1/(x + 1)$

5. Find the Taylor polynomial of degree 5 about the point $x = 0$ for the following functions:
   (a) $f(x) = e^{x^2}$ (b) $f(x) = \cos 2x$ (c) $f(x) = \ln(1 + x)$ (d) $f(x) = \sin^2 x$

6. (a) Find the Taylor polynomial of degree 4 for $f(x) = x^{-2}$ about the point $x = 1$.

   (b) Use the result of (a) to approximate $f(0.9)$ and $f(1.1)$.

   (c) Use the Taylor remainder to find an error formula for the Taylor polynomial. Give error bounds for each of the two approximations made in part (b). Which of the two approximations in part (b) do you expect to be closer to the correct value?

   (d) Use a calculator to compare the actual error in each case with your error bound from part (c).

7. Carry out Exercise 6 (a)–(d) for $f(x) = \ln x$.

8. (a) Find the degree 5 Taylor polynomial $P(x)$ centered at $x = 0$ for $f(x) = \cos x$. (b) Find an upper bound for the error in approximating $f(x) = \cos x$ for x in $-\pi/4, \pi/4$ by $P(x)$.

9. A common approximation for $\sqrt{1 + x}$ is $1 + \frac{1}{2}x$, when x is small. Use the degree 1 Taylor polynomial of $f(x) = \sqrt{1 + x}$ with remainder to determine a formula of form $\sqrt{1 + x} = 1 + \frac{1}{2}x \pm E$. Evaluate E for the case of approximating $\sqrt{1.02}$. Use a calculator to compare the actual error to your error bound E.

Software and Further Reading

The IEEE standard for floating point computation is published as IEEE Standard 754 [1985]. Goldberg [1991] and Stallings [2003] discuss floating point arithmetic in great detail, and Overton [2001] emphasizes the IEEE 754 standard. The texts Wilkinson [1994] and Knuth [1981] had great influence on the development of both hardware and software.

There are several software packages that specialize in general-purpose scientific computing, the bulk of it done in floating point arithmetic. Netlib (http://www.netlib.org) is a collection of free software maintained by AT&T Bell Laboratories, the University of Tennessee, and Oak Ridge National Laboratory. The collection consists of high-quality programs available in Fortran, C, and Java. The comments in the code are meant to be sufficiently instructive for the user to operate the program.

The Numerical Algorithms Group (NAG) (http://www.nag.co.uk) markets a library containing over 1400 user-callable subroutines for solving general applied math problems. The programs are available in Fortran and C and are callable from Java programs. NAG includes libraries for shared memory and distributed memory computing.

The computing environments Mathematica, Maple, and MATLAB have grown to encompass many of the same computational methods previously described and have built-in editing and graphical interfaces. Mathematica (http://www.wolframresearch.com) and Maple (www.maplesoft.com) came to prominence due to novel symbolic computing engines. MATLAB has grown to serve many science and engineering applications through "toolboxes," which leverage the basic high-quality software into divers directions.

In this text, we frequently illustrate basic algorithms with MATLAB implementations. The MATLAB code given is meant to be instructional only. Quite often, speed and reliability are sacrificed for clarity and readability. Readers who are new to MATLAB should begin with the tutorial in Appendix B; they will soon be doing their own implementations.

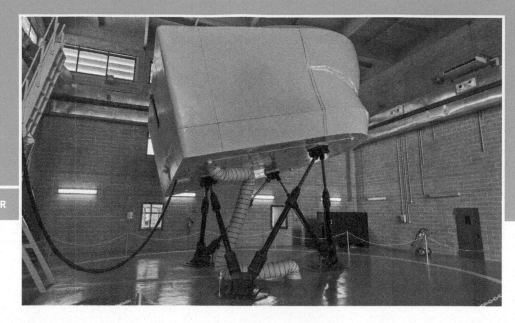

Solving Equations

A recently excavated cuneiform tablet shows that the Babylonians calculated the square root of 2 correctly to within five decimal places. Their technique is unknown, but in this chapter we introduce iterative methods that they may have used and that are still used by modern calculators to find square roots.

The Stewart platform, a six-degree-of-freedom robot that can be located with extreme precision, was originally developed by Eric Gough of Dunlop Tire Corporation in the 1950s to test airplane tires. Today its applications range from flight simulators, which are

often of considerable mass, to medical and surgical applications, where precision is very important. Solving the forward kinematics problem requires determining the position and orientation of the platform, given the strut lengths.

Reality Check 📍 Reality Check 1 on page 70 uses the methods developed in this chapter to solve the forward kinematics of a planar version of the Stewart platform.

E quation solving is one of the most basic problems in scientific computing. This chapter introduces a number of iterative methods for locating solutions x of the equation $f(x) = 0$. These methods are of great practical importance. In addition, they illustrate the central roles of convergence and complexity in scientific computing.

Why is it necessary to know more than one method for solving equations? Often, the choice of method will depend on the cost of evaluating the function f and perhaps its derivative. If $f(x) = e^x - \sin x$, it may take less than one-millionth of a second to determine $f(x)$, and its derivative is available if needed. If $f(x)$ denotes the freezing temperature of an ethylene glycol solution under x atmospheres of pressure, each function evaluation may require considerable time in a well-equipped laboratory, and determining the derivative may be infeasible.

In addition to introducing methods such as the Bisection Method, Fixed-Point Iteration, and Newton's Method, we will analyze their rates of convergence and discuss their computational complexity. Later, more sophisticated equation solvers are presented, including Brent's Method, that combines the best properties of several solvers.

1.1 THE BISECTION METHOD

How do you look up a name in an unfamiliar phone book? To look up "Smith," you might begin by opening the book at your best guess, say, the letter Q. Next you may turn a sheaf of pages and end up at the letter U. Now you have "bracketed" the name Smith and need to hone in on it by using smaller and smaller brackets that eventually converge to the name. The Bisection Method represents this type of reasoning, done as efficiently as possible.

1.1.1 Bracketing a root

DEFINITION 1.1   The function $f(x)$ has a **root** at $x = r$ if $f(r) = 0$.   ☐

The first step to solving an equation is to verify that a root exists. One way to ensure this is to bracket the root: to find an interval $[a, b]$ on the real line for which one of the pair $\{f(a), f(b)\}$ is positive and the other is negative. This can be expressed as $f(a)f(b) < 0$. If f is a continuous function, then there will be a root: an r between a and b for which $f(r) = 0$. This fact is summarized in the following corollary of the Intermediate Value Theorem 0.4:

THEOREM 1.2   Let f be a continuous function on $[a, b]$, satisfying $f(a)f(b) < 0$. Then f has a root between a and b, that is, there exists a number r satisfying $a < r < b$ and $f(r) = 0$.   ■

In Figure 1.1, $f(0)f(1) = (-1)(1) < 0$. There is a root just to the left of 0.7. How can we refine our first guess of the root's location to more decimal places?

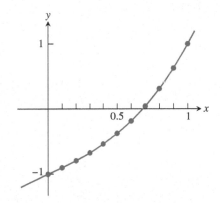

Figure 1.1 A plot of $f(x) = x^3 + x - 1$. The function has a root between 0.6 and 0.7.

We'll take a cue from the way our eye finds a solution when given a plot of a function. It is unlikely that we start at the left end of the interval and move to the right, stopping at the root. Perhaps a better model of what happens is that the eye first decides the general location, such as whether the root is toward the left or the right of the interval. It then follows that up by deciding more precisely just how far right or left the root lies and gradually improves its accuracy, just like looking up a name in the phone book. This general approach is made quite specific in the Bisection Method, shown in Figure 1.2.

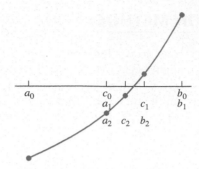

Figure 1.2 The Bisection Method. On the first step, the sign of $f(c_0)$ is checked. Since $f(c_0)f(b_0) < 0$, set $a_1 = c_0, b_1 = b_0$, and the interval is replaced by the right half $[a_1, b_1]$. On the second step, the subinterval is replaced by its left half $[a_2, b_2]$.

Bisection Method

Given initial interval $[a, b]$ such that $f(a)f(b) < 0$
while $(b - a)/2 > \text{TOL}$
    $c = (a + b)/2$
    if $f(c) = 0$, **stop, end**
    if $f(a)f(c) < 0$
        $b = c$
    else
        $a = c$
    end
end
The final interval $[a, b]$ contains a root.
The approximate root is $(a + b)/2$.

Check the value of the function at the midpoint $c = (a + b)/2$ of the interval. Since $f(a)$ and $f(b)$ have opposite signs, either $f(c) = 0$ (in which case we have found a root and are done), or the sign of $f(c)$ is opposite the sign of either $f(a)$ or $f(b)$. If $f(c)f(a) < 0$, for example, we are assured a solution in the interval $[a, c]$, whose length is half that of the original interval $[a, b]$. If instead $f(c)f(b) < 0$, we can say the same of the interval $[c, b]$. In either case, one step reduces the problem to finding a root on an interval of one-half the original size. This step can be repeated to locate the function more and more accurately.

A solution is bracketed by the new interval at each step, reducing the uncertainty in the location of the solution as the interval becomes smaller. An entire plot of the function f is not needed. We have reduced the work of function evaluation to only what is necessary.

▶ **EXAMPLE 1.1**   Find a root of the function $f(x) = x^3 + x - 1$ by using the Bisection Method on the interval $[0, 1]$.

As noted, $f(a_0)f(b_0) = (-1)(1) < 0$, so a root exists in the interval. The interval midpoint is $c_0 = 1/2$. The first step consists of evaluating $f(1/2) = -3/8 < 0$ and choosing the new interval $[a_1, b_1] = [1/2, 1]$, since $f(1/2)f(1) < 0$. The second step consists of evaluating $f(c_1) = f(3/4) = 11/64 > 0$, leading to the new interval $[a_2, b_2] = [1/2, 3/4]$. Continuing in this way yields the following intervals:

i	a_i	$f(a_i)$	c_i	$f(c_i)$	b_i	$f(b_i)$
0	0.0000	−	0.5000	−	1.0000	+
1	0.5000	−	0.7500	+	1.0000	+
2	0.5000	−	0.6250	−	0.7500	+
3	0.6250	−	0.6875	+	0.7500	+
4	0.6250	−	0.6562	−	0.6875	+
5	0.6562	−	0.6719	−	0.6875	+
6	0.6719	−	0.6797	−	0.6875	+
7	0.6797	−	0.6836	+	0.6875	+
8	0.6797	−	0.6816	−	0.6836	+
9	0.6816	−	0.6826	+	0.6836	+

We conclude from the table that the solution is bracketed between $a_9 \approx 0.6816$ and $c_9 \approx 0.6826$. The midpoint of that interval $c_{10} \approx 0.6821$ is our best guess for the root.

Although the problem was to find a root, what we have actually found is an interval $[0.6816, 0.6826]$ that contains a root; in other words, the root is $r = 0.6821 \pm 0.0005$. We will have to be satisfied with an approximation. Of course, the approximation can be improved, if needed, by completing more steps of the Bisection Method. ◄

At each step of the Bisection Method, we compute the midpoint $c_i = (a_i + b_i)/2$ of the current interval $[a_i, b_i]$, calculate $f(c_i)$, and compare signs. If $f(c_i)f(a_i) < 0$, we set $a_{i+1} = a_i$ and $b_{i+1} = c_i$. If, instead, $f(c_i)f(a_i) > 0$, we set $a_{i+1} = c_i$ and $b_{i+1} = b_i$. Each step requires one new evaluation of the function f and bisects the interval containing a root, reducing its length by a factor of 2. After n steps of calculating c and $f(c)$, we have done $n + 2$ function evaluations, and our best estimate of the solution is the midpoint of the latest interval. The algorithm can be written in the following MATLAB code:

MATLAB code
shown here can be found
at bit.ly/2Esx9SA

```
%Program 1.1 Bisection Method
%Computes approximate solution of f(x)=0
%Input: function handle f; a,b such that f(a)*f(b)<0,
%         and tolerance tol
%Output: Approximate solution xc
function xc=bisect(f,a,b,tol)
if sign(f(a))*sign(f(b)) >= 0
  error('f(a)f(b)<0 not satisfied!') %ceases execution
end
fa=f(a);
fb=f(b);
while (b-a)/2>tol
  c=(a+b)/2;
  fc=f(c);
  if fc == 0            %c is a solution, done
```

```
      break
   end
   if sign(fc)*sign(fa)<0    %a and c make the new interval
      b=c;fb=fc;
   else                      %c and b make the new interval
      a=c;fa=fc;
   end
end
xc=(a+b)/2;                  %new midpoint is best estimate
```

To use `bisect.m`, first define a MATLAB function by:

```
>> f=@(x) x^3+x-1;
```

This command actually defines a "function handle" f, which can be used as input for other MATLAB functions. See Appendix B for more details on MATLAB functions and function handles. Then the command

```
» xc=bisect(f,0,1,0.00005)
```

returns a solution correct to a tolerance of 0.00005.

1.1.2 How accurate and how fast?

If $[a, b]$ is the starting interval, then after n bisection steps, the interval $[a_n, b_n]$ has length $(b - a)/2^n$. Choosing the midpoint $x_c = (a_n + b_n)/2$ gives a best estimate of the solution r, which is within half the interval length of the true solution. Summarizing, after n steps of the Bisection Method, we find that

$$\text{Solution error} = |x_c - r| < \frac{b - a}{2^{n+1}} \tag{1.1}$$

and

$$\text{Function evaluations} = n + 2. \tag{1.2}$$

A good way to assess the efficiency of the Bisection Method is to ask how much accuracy can be bought per function evaluation. Each step, or each function evaluation, cuts the uncertainty in the root by a factor of two.

DEFINITION 1.3   A solution is **correct within p decimal places** if the error is less than 0.5×10^{-p}.   ❏

▶ **EXAMPLE 1.2**   Use the Bisection Method to find a root of $f(x) = \cos x - x$ in the interval $[0, 1]$ to within six correct places.

First we decide how many steps of bisection are required. According to (1.1), the error after n steps is $(b - a)/2^{n+1} = 1/2^{n+1}$. From the definition of p decimal places, we require that

$$\frac{1}{2^{n+1}} < 0.5 \times 10^{-6}$$

$$n > \frac{6}{\log_{10} 2} \approx \frac{6}{0.301} = 19.9.$$

Therefore, $n = 20$ steps will be needed. Proceeding with the Bisection Method, the following table is produced:

k	a_k	$f(a_k)$	c_k	$f(c_k)$	b_k	$f(b_k)$
0	0.000000	+	0.500000	+	1.000000	−
1	0.500000	+	0.750000	−	1.000000	−
2	0.500000	+	0.625000	+	0.750000	−
3	0.625000	+	0.687500	+	0.750000	−
4	0.687500	+	0.718750	+	0.750000	−
5	0.718750	+	0.734375	+	0.750000	−
6	0.734375	+	0.742188	−	0.750000	−
7	0.734375	+	0.738281	+	0.742188	−
8	0.738281	+	0.740234	−	0.742188	−
9	0.738281	+	0.739258	−	0.740234	−
10	0.738281	+	0.738770	+	0.739258	−
11	0.738769	+	0.739014	+	0.739258	−
12	0.739013	+	0.739136	−	0.739258	−
13	0.739013	+	0.739075	+	0.739136	−
14	0.739074	+	0.739105	−	0.739136	−
15	0.739074	+	0.739090	−	0.739105	−
16	0.739074	+	0.739082	+	0.739090	−
17	0.739082	+	0.739086	−	0.739090	−
18	0.739082	+	0.739084	+	0.739086	−
19	0.739084	+	0.739085	−	0.739086	−
20	0.739084	+	0.739085	−	0.739085	−

The approximate root to six correct places is 0.739085. ◄

For the Bisection Method, the question of how many steps to run is a simple one—just choose the desired precision and find the number of necessary steps, as in (1.1). We will see that more high-powered algorithms are often less predictable and have no analogue to (1.1). In those cases, we will need to establish definite "stopping criteria" that govern the circumstances under which the algorithm terminates. Even for the Bisection Method, the finite precision of computer arithmetic will put a limit on the number of possible correct digits. We will look into this issue further in Section 1.3.

► **ADDITIONAL EXAMPLES**

1. Apply two steps of the Bisection Method on the interval $[1, 2]$ to find the approximate root of $f(x) = 2x^3 - x - 7$.

2. Use the `bisect.m` code to find the solution of $e^x = 3$ correct to six decimal places.

⊡ **Solutions** for Additional Examples can be found at `bit.ly/2P4Qhdw`

1.1 Exercises

⊡ **Solutions** for Exercises numbered in blue can be found at `bit.ly/2Eucbme`

1. Use the Intermediate Value Theorem to find an interval of length one that contains a root of the equation. (a) $x^3 = 9$ (b) $3x^3 + x^2 = x + 5$ (c) $\cos^2 x + 6 = x$

2. Use the Intermediate Value Theorem to find an interval of length one that contains a root of the equation. (a) $x^5 + x = 1$ (b) $\sin x = 6x + 5$ (c) $\ln x + x^2 = 3$

3. Consider the equations in Exercise 1. Apply two steps of the Bisection Method to find an approximate root within 1/8 of the true root.

4. Consider the equations in Exercise 2. Apply two steps of the Bisection Method to find an approximate root within 1/8 of the true root.

5. Consider the equation $x^4 = x^3 + 10$.

(a) Find an interval $[a, b]$ of length one inside which the equation has a solution.

(b) Starting with $[a, b]$, how many steps of the Bisection Method are required to calculate the solution within 10^{-10}? Answer with an integer.

6. Suppose that the Bisection Method with starting interval $[-2, 1]$ is used to find a root of the function $f(x) = 1/x$. Does the method converge to a real number? Is it the root?

1.1 Computer Problems

Solutions for Computer Problems numbered in blue can be found at bit.ly/2NOVeSK

1. Use the Bisection Method to find the root to six correct decimal places. (a) $x^3 = 9$ (b) $3x^3 + x^2 = x + 5$ (c) $\cos^2 x + 6 = x$

2. Use the Bisection Method to find the root to eight correct decimal places. (a) $x^5 + x = 1$ (b) $\sin x = 6x + 5$ (c) $\ln x + x^2 = 3$

3. Use the Bisection Method to locate all solutions of the following equations. Sketch the function by using MATLAB's plot command and identify three intervals of length one that contain a root. Then find the roots to six correct decimal places. (a) $2x^3 - 6x - 1 = 0$ (b) $e^{x-2} + x^3 - x = 0$ (c) $1 + 5x - 6x^3 - e^{2x} = 0$

4. Calculate the square roots of the following numbers to eight correct decimal places by using the Bisection Method to solve $x^2 - A = 0$, where A is (a) 2 (b) 3 (c) 5. State your starting interval and the number of steps needed.

5. Calculate the cube roots of the following numbers to eight correct decimal places by using the Bisection Method to solve $x^3 - A = 0$, where A is (a) 2 (b) 3 (c) 5. State your starting interval and the number of steps needed.

6. Use the Bisection Method to calculate the solution of $\cos x = \sin x$ in the interval $[0, 1]$ within six correct decimal places.

7. Use the Bisection Method to find the two real numbers x, within six correct decimal places, that make the determinant of the matrix

$$A = \begin{bmatrix} 1 & 2 & 3 & x \\ 4 & 5 & x & 6 \\ 7 & x & 8 & 9 \\ x & 10 & 11 & 12 \end{bmatrix}$$

equal to 1000. For each solution you find, test it by computing the corresponding determinant and reporting how many correct decimal places (after the decimal point) the determinant has when your solution x is used. (In Section 1.2, we will call this the "backward error" associated with the approximate solution.) You may use the MATLAB command det to compute the determinants.

8. The **Hilbert matrix** is the $n \times n$ matrix whose ijth entry is $1/(i + j - 1)$. Let A denote the 5×5 Hilbert matrix. Its largest eigenvalue is about 1.567. Use the Bisection Method to decide how to change the upper left entry A_{11} to make the largest eigenvalue of A equal to π. Determine A_{11} within six correct decimal places. You may use the MATLAB commands hilb, pi, eig, and max to simplify your task.

9. Find the height reached by 1 cubic meter of water stored in a spherical tank of radius 1 meter. Give your answer ± 1 mm. (Hint: First note that the sphere will be less than half full. The volume of the bottom H meters of a hemisphere of radius R is $\pi H^2(R - 1/3H)$.)

10. A planet orbiting the sun traverses an ellipse. The eccentricity e of the ellipse is the distance between the center of the ellipse and either of its foci divided by the length of the semimajor axis. The perihelion is the nearest point of the orbit to the sun. *Kepler's equation $M = E - e \sin E$* relates the *eccentric anomaly E*, the true angular distance (in radians) from perihelion, to the *mean anomaly M*, the fictitious angular distance from

perihelion if it were on a circular orbit with the same period as the ellipse. (a) Assume $e = 0.1$. Use the Bisection Method to find the eccentric anomalies E when $M = \pi/6$ and $M = \pi/2$. Begin by finding a starting interval and explain why it works. (b) How do the answers to (a) change if the eccentricity is changed to $e = 0.2$?

1.2 FIXED-POINT ITERATION

Use a calculator or computer to apply the cos function repeatedly to an arbitrary starting number. That is, apply the cos function to the starting number, then apply cos to the result, then to the new result, and so forth. (If you use a calculator, be sure it is in radian mode.) Continue until the digits no longer change. The resulting sequence of numbers converges to 0.7390851332, at least to the first 10 decimal places. In this section, our goal is to explain why this calculation, an instance of Fixed-Point Iteration (FPI), converges. While we do this, most of the major issues of algorithm convergence will come under discussion.

1.2.1 Fixed points of a function

The sequence of numbers produced by iterating the cosine function appears to converge to a number r. Subsequent applications of cosine do not change the number. For this input, the output of the cosine function is equal to the input, or $\cos r = r$.

DEFINITION 1.4 The real number r is a **fixed point** of the function g if $g(r) = r$. ❏

The number $r = 0.7390851332$ is an approximate fixed point for the function $g(x) = \cos x$. The function $g(x) = x^3$ has three fixed points, $r = -1, 0$, and 1.

We used the Bisection Method in Example 1.2 to solve the equation $\cos x - x = 0$. The fixed-point equation $\cos x = x$ is the same problem from a different point of view. When the output equals the input, that number is a fixed point of $\cos x$, and simultaneously a solution of the equation $\cos x - x = 0$.

Once the equation is written as $g(x) = x$, Fixed-Point Iteration proceeds by starting with an initial guess x_0 and iterating the function g.

Fixed-Point Iteration

$$x_0 = \text{initial guess}$$
$$x_{i+1} = g(x_i) \text{ for } i = 0, 1, 2, \ldots$$

Therefore,

$$x_1 = g(x_0)$$
$$x_2 = g(x_1)$$
$$x_3 = g(x_2)$$
$$\vdots$$

and so forth. The sequence x_i may or may not converge as the number of steps goes to infinity. However, if g is continuous and the x_i converge, say, to a number r, then r is a fixed point. In fact, Theorem 0.5 implies that

$$g(r) = g\left(\lim_{i \to \infty} x_i\right) = \lim_{i \to \infty} g(x_i) = \lim_{i \to \infty} x_{i+1} = r. \tag{1.3}$$

The Fixed-Point Iteration algorithm applied to a function g is easily written in MATLAB code:

⌨ **MATLAB code**
shown here can be found
at bit.ly/2QWFwah

```
%Program 1.2 Fixed-Point Iteration
%Computes approximate solution of g(x)=x
%Input: function handle g, starting guess x0,
%        number of iteration steps k
%Output: Approximate solution xc
function xc=fpi(g, x0, k)
x(1)=x0;
for i=1:k
  x(i+1)=g(x(i));
end
xc=x(k+1);
```

After defining a MATLAB function by

```
>>   g=@(x) cos(x)
```

the code of Program 1.2 can be called as

```
>>   xc=fpi(g,0,10)
```

to run 10 steps of Fixed-Point Iteration with initial guess 0.

Fixed-Point Iteration solves the fixed-point problem $g(x) = x$, but we are primarily interested in solving equations. Can every equation $f(x) = 0$ be turned into a fixed-point problem $g(x) = x$? Yes, and in many different ways. For example, the root-finding equation of Example 1.1,

$$x^3 + x - 1 = 0, \tag{1.4}$$

can be rewritten as

$$x = 1 - x^3, \tag{1.5}$$

and we may define $g(x) = 1 - x^3$. Alternatively, the x^3 term in (1.4) can be isolated to yield

$$x = \sqrt[3]{1 - x}, \tag{1.6}$$

where $g(x) = \sqrt[3]{1 - x}$. As a third and not very obvious approach, we might add $2x^3$ to both sides of (1.4) to get

$$3x^3 + x = 1 + 2x^3$$
$$(3x^2 + 1)x = 1 + 2x^3$$
$$x = \frac{1 + 2x^3}{1 + 3x^2} \tag{1.7}$$

and define $g(x) = (1 + 2x^3)/(1 + 3x^2)$.

Next, we demonstrate Fixed-Point Iteration for the preceding three choices of $g(x)$. The underlying equation to be solved is $x^3 + x - 1 = 0$. First we consider the form $x = g(x) = 1 - x^3$. The starting point $x_0 = 0.5$ is chosen somewhat arbitrarily. Applying FPI gives the following result:

i	x_i
0	0.50000000
1	0.87500000
2	0.33007813
3	0.96403747
4	0.10405419
5	0.99887338
6	0.00337606
7	0.99999996
8	0.00000012
9	1.00000000
10	0.00000000
11	1.00000000
12	0.00000000

Instead of converging, the iteration tends to alternate between the numbers 0 and 1. Neither is a fixed point, since $g(0) = 1$ and $g(1) = 0$. The Fixed-Point Iteration fails. With the Bisection Method, we know that if f is continuous and $f(a)f(b) < 0$ on the original interval, we must see convergence to the root. This is not so for FPI.

The second choice is $g(x) = \sqrt[3]{1 - x}$. We will keep the same initial guess, $x_0 = 0.5$.

i	x_i		i	x_i
0	0.50000000		13	0.68454401
1	0.79370053		14	0.68073737
2	0.59088011		15	0.68346460
3	0.74236393		16	0.68151292
4	0.63631020		17	0.68291073
5	0.71380081		18	0.68191019
6	0.65900615		19	0.68262667
7	0.69863261		20	0.68211376
8	0.67044850		21	0.68248102
9	0.69072912		22	0.68221809
10	0.67625892		23	0.68240635
11	0.68664554		24	0.68227157
12	0.67922234		25	0.68236807

This time FPI is successful. The iterates are apparently converging to a number near 0.6823.

Finally, let's use the rearrangement $x = g(x) = (1 + 2x^3)/(1 + 3x^2)$. As in the previous case, there is convergence, but in a much more striking way.

i	x_i
0	0.50000000
1	0.71428571
2	0.68317972
3	0.68232842
4	0.68232780
5	0.68232780
6	0.68232780
7	0.68232780

Here we have four correct digits after four iterations of Fixed-Point Iteration, and many more correct digits soon after. Compared with the previous attempts, this is an astonishing result. Our next goal is to try to explain the differences between the three outcomes.

1.2.2 Geometry of Fixed-Point Iteration

In the previous section, we found three different ways to rewrite the equation $x^3 + x - 1 = 0$ as a fixed-point problem, with varying results. To find out why the FPI method converges in some situations and not in others, it is helpful to look at the geometry of the method.

Figure 1.3 shows the three different $g(x)$ discussed before, along with an illustration of the first few steps of FPI in each case. The fixed point r is the same for each $g(x)$. It is represented by the point where the graphs $y = g(x)$ and $y = x$ intersect. Each step of FPI can be sketched by drawing line segments (1) **vertically to the function** and then (2) **horizontally to the diagonal** line $y = x$. The vertical and horizontal arrows in Figure 1.3 follow the steps made by FPI. The vertical arrow moving from the x-value to the function g represents $x_i \rightarrow g(x_i)$. The horizontal arrow represents turning the output $g(x_i)$ on the y-axis and transforming it into the same number x_{i+1} on the x-axis, ready to be input into g in the next step. This is done by drawing the horizontal line segment from the output height $g(x_i)$ across to the diagonal line $y = x$. This geometric illustration of a Fixed-Point Iteration is called a **cobweb diagram**.

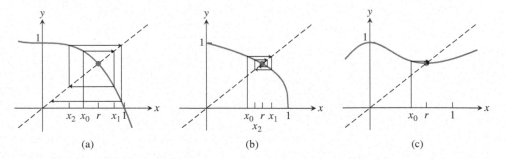

Figure 1.3 Geometric view of FPI. The fixed point is the intersection of $g(x)$ and the diagonal line. Three examples of $g(x)$ are shown together with the first few steps of FPI. (a) $g(x) = 1 - x^3$ (b) $g(x) = (1 - x)^{1/3}$ (c) $g(x) = (1 + 2x^3)/(1 + 3x^2)$.

In Figure 1.3(a), the path starts at $x_0 = 0.5$, and moves up to the function and horizontal to the point $(0.875, 0.875)$ on the diagonal, which is (x_1, x_1). Next, x_1 should be substituted into $g(x)$. This is done the same way it was done for x_0, by moving vertically to the function. This yields $x_2 \approx 0.3300$, and after moving horizontally to move the y-value to an x-value, we continue the same way to get $x_3, x_4, \ldots$. As we saw earlier, the result of FPI for this $g(x)$ is not successful—the iterates eventually tend toward alternating between 0 and 1, neither of which are fixed points.

Fixed-Point Iteration is more successful in Figure 1.3(b). Although the $g(x)$ here looks roughly similar to the $g(x)$ in part (a), there is a significant difference, which we will clarify in the next section. You may want to speculate on what the difference is. What makes FPI spiral in toward the fixed point in (b), and spiral out away from the fixed point in (a)? Figure 1.3(c) shows an example of very fast convergence. Does this picture help with your speculation? If you guessed that it has something to do with the slope of $g(x)$ near the fixed point, you are correct.

1.2.3 Linear convergence of Fixed-Point Iteration

The convergence properties of FPI can be easily explained by a careful look at the algorithm in the simplest possible situation. Figure 1.4 shows Fixed-Point Iteration for two linear functions $g_1(x) = -\frac{3}{2}x + \frac{5}{2}$ and $g_2(x) = -\frac{1}{2}x + \frac{3}{2}$. In each case, the fixed point is $x = 1$, but $|g_1'(1)| = \left|-\frac{3}{2}\right| > 1$ while $|g_2'(1)| = \left|-\frac{1}{2}\right| < 1$. Following the vertical and horizontal arrows that describe FPI, we see the reason for the difference. Because the slope of g_1 at the fixed point is greater than one, the vertical segments, the ones that represent the change from x_n to x_{n+1}, are increasing in length as FPI proceeds. As a result, the iteration "spirals out" from the fixed point $x = 1$, even if the initial guess x_0 was quite near. For g_2, the situation is reversed: The slope of g_2 is less than one, the vertical segments decrease in length, and FPI "spirals in" toward the solution. Thus, $|g'(r)|$ makes the crucial difference between divergence and convergence.

That's the geometric view. In terms of equations, it helps to write $g_1(x)$ and $g_2(x)$ in terms of $x - r$, where $r = 1$ is the fixed point:

$$g_1(x) = -\tfrac{3}{2}(x - 1) + 1$$
$$g_1(x) - 1 = -\tfrac{3}{2}(x - 1)$$
$$x_{i+1} - 1 = -\tfrac{3}{2}(x_i - 1). \qquad (1.8)$$

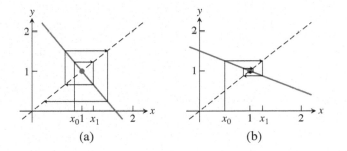

Figure 1.4 Cobweb diagram for linear functions. (a) If the linear function has slope greater than one in absolute value, nearby guesses move farther from the fixed point as FPI progresses, leading to failure of the method. (b) For slope less than one in absolute value, the reverse happens, and the fixed point is found.

If we view $e_i = |r - x_i|$ as the error at step i (meaning the distance from the best guess at step n to the fixed point), we see from (1.8) that $e_{i+1} = 3e_i/2$, implying that errors increase at each step by a factor of approximately $3/2$. This is divergence.

Repeating the preceding algebra for g_2, we have

$$g_2(x) = -\tfrac{1}{2}(x - 1) + 1$$
$$g_2(x) - 1 = -\tfrac{1}{2}(x - 1)$$
$$x_{i+1} - 1 = -\tfrac{1}{2}(x_i - 1).$$

The result is $e_{i+1} = e_i/2$, implying that the error, the distance to the fixed point, is multiplied by $1/2$ on each step. The error decreases to zero as the number of steps increases. This is convergence of a particular type.

DEFINITION 1.5   Let e_i denote the error at step i of an iterative method. If

$$\lim_{i \to \infty} \frac{e_{i+1}}{e_i} = S < 1,$$

the method is said to obey **linear convergence** with rate S. ❑

Fixed-Point Iteration for g_2 is linearly convergent to the root $r = 1$ with rate $S = 1/2$. Although the previous discussion was simplified because g_1 and g_2 are linear, the same reasoning applies to a general continuously differentiable function $g(x)$ with fixed point $g(r) = r$, as shown in the next theorem.

THEOREM 1.6 Assume that g is continuously differentiable, that $g(r) = r$, and that $S = |g'(r)| < 1$. Then Fixed-Point Iteration converges linearly with rate S to the fixed point r for initial guesses sufficiently close to r. ∎

Proof. Let x_i denote the iterate at step i. The next iterate is $x_{i+1} = g(x_i)$. Since $g(r) = r$,

$$x_{i+1} - r = g(x_i) - g(r)$$
$$= g'(c_i)(x_i - r) \tag{1.9}$$

for some c_i between x_i and r, according to the Mean Value Theorem. Defining $e_i = |x_i - r|$, (1.9) can be written as

$$e_{i+1} = |g'(c_i)|e_i. \tag{1.10}$$

If $S = |g'(r)|$ is less than one, then by the continuity of g', there is a small neighborhood around r for which $|g'(x)| < (S + 1)/2$, slightly larger than S, but still less than one. If x_i happens to lie in this neighborhood, then c_i does, too (it is trapped between x_i and r), and so

$$e_{i+1} \leq \frac{S + 1}{2}e_i.$$

Thus, the error decreases by a factor of $(S + 1)/2$ or better on this and every future step. That means $\lim_{i\to\infty} x_i = r$, and taking the limit of (1.10) yields

$$\lim_{i\to\infty}\frac{e_{i+1}}{e_i} = \lim_{i\to\infty}|g'(c_i)| = |g'(r)| = S. \qquad \square$$

According to Theorem 1.6, the approximate error relationship

$$e_{i+1} \approx Se_i \tag{1.11}$$

holds in the limit as convergence is approached, where $S = |g'(r)|$. See Exercise 25 for a variant of this theorem.

DEFINITION 1.7 An iterative method is called **locally convergent** to r if the method converges to r for initial guesses sufficiently close to r. ❏

In other words, the method is locally convergent to the root r if there exists a neighborhood $(r - \epsilon, r + \epsilon)$, where $\epsilon > 0$, such that convergence to r follows from all initial guesses from the neighborhood. The conclusion of Theorem 1.6 is that Fixed-Point Iteration is locally convergent if $|g'(r)| < 1$.

Theorem 1.6 explains what happened in the previous Fixed-Point Iteration runs for $f(x) = x^3 + x - 1 = 0$. We know the root $r \approx 0.6823$. For $g(x) = 1 - x^3$, the derivative is $g'(x) = -3x^2$. Near the root r, FPI behaves as $e_{i+1} \approx Se_i$, where $S = |g'(r)| = |-3(0.6823)^2| \approx 1.3966 > 1$, so errors increase, and there can be no convergence. This error relationship between e_{i+1} and e_i is only guaranteed to hold near r, but it does mean that no convergence to r can occur.

For the second choice, $g(x) = \sqrt[3]{1-x}$, the derivative is $g'(x) = 1/3(1-x)^{-2/3}$ (-1), and $S = |(1-0.6823)^{-2/3}/3| \approx 0.716 < 1$. Theorem 1.6 implies convergence, agreeing with our previous calculation.

For the third choice, $g(x) = (1+2x^3)/(1+3x^2)$,

$$g'(x) = \frac{6x^2(1+3x^2) - (1+2x^3)6x}{(1+3x^2)^2}$$

$$= \frac{6x(x^3+x-1)}{(1+3x^2)^2},$$

and $S = |g'(r)| = 0$. This is as small as S can get, leading to the very fast convergence seen in Figure 1.3(c).

▶ **EXAMPLE 1.3** Explain why the Fixed-Point Iteration $g(x) = \cos x$ converges.

This is the explanation promised early in the chapter. Applying the cosine button repeatedly corresponds to FPI with $g(x) = \cos x$. According to Theorem 1.6, the solution $r \approx 0.74$ attracts nearby guesses because $g'(r) = -\sin r \approx -\sin 0.74 \approx -0.67$ is less than 1 in absolute value. ◀

▶ **EXAMPLE 1.4** Use Fixed-Point Iteration to find a root of $\cos x = \sin x$.

The simplest way to convert the equation to a fixed-point problem is to add x to each side of the equation. We can rewrite the problem as

$$x + \cos x - \sin x = x$$

and define

$$g(x) = x + \cos x - \sin x. \tag{1.12}$$

The result of applying the Fixed-Point Iteration method to this $g(x)$ is shown in the table.

i	x_i	$g(x_i)$	$e_i = |x_i - r|$	e_i/e_{i-1}
0	0.0000000	1.0000000	0.7853982	
1	1.0000000	0.6988313	0.2146018	0.273
2	0.6988313	0.8211025	0.0865669	0.403
3	0.8211025	0.7706197	0.0357043	0.412
4	0.7706197	0.7915189	0.0147785	0.414
5	0.7915189	0.7828629	0.0061207	0.414
6	0.7828629	0.7864483	0.0025353	0.414
7	0.7864483	0.7849632	0.0010501	0.414
8	0.7849632	0.7855783	0.0004350	0.414
9	0.7855783	0.7853235	0.0001801	0.414
10	0.7853235	0.7854291	0.0000747	0.415
11	0.7854291	0.7853854	0.0000309	0.414
12	0.7853854	0.7854035	0.0000128	0.414
13	0.7854035	0.7853960	0.0000053	0.414
14	0.7853960	0.7853991	0.0000022	0.415
15	0.7853991	0.7853978	0.0000009	0.409
16	0.7853978	0.7853983	0.0000004	0.444
17	0.7853983	0.7853981	0.0000001	0.250
18	0.7853981	0.7853982	0.0000001	1.000
19	0.7853982	0.7853982	0.0000000	

There are several interesting things to notice in the table. First, the iteration appears to converge to 0.7853982. Since $\cos \pi/4 = \sqrt{2}/2 = \sin \pi/4$, the true solution to the equation $\cos x - \sin x = 0$ is $r = \pi/4 \approx 0.7853982$. The fourth column is the "error column." It shows the absolute value of the difference between the best guess x_i at step i and the actual fixed point r. This difference becomes small near the bottom of the table, indicating convergence toward a fixed point.

Notice the pattern in the error column. The errors seem to decrease by a constant factor, each error being somewhat less than half the previous error. To be more precise, the ratio between successive errors is shown in the final column. In most of the table, we are seeing the ratio e_{k+1}/e_k of successive errors to approach a constant number, about 0.414. In other words, we are seeing the linear convergence relation

$$e_i \approx 0.414 e_{i-1}. \tag{1.13}$$

This is exactly what is expected, since Theorem 1.6 implies that

$$S = |g'(r)| = |1 - \sin r - \cos r| = \left| 1 - \frac{\sqrt{2}}{2} - \frac{\sqrt{2}}{2} \right| = |1 - \sqrt{2}| \approx 0.414. \quad \blacktriangleleft$$

The careful reader will notice a discrepancy toward the end of the table. We have used only seven correct digits for the correct fixed point r in computing the errors e_i. As a result, the relative accuracy of the e_i is poor as the e_i near 10^{-8}, and the ratios e_i/e_{i-1} become inaccurate. This problem would disappear if we used a much more accurate value for r.

▶ **EXAMPLE 1.5**   Find the fixed points of $g(x) = 2.8x - x^2$.

The function $g(x) = 2.8x - x^2$ has two fixed points 0 and 1.8, which can be determined by solving $g(x) = x$ by hand, or alternatively, by noting where the graphs of $y = g(x)$ and $y = x$ intersect. Figure 1.5 shows a cobweb diagram for FPI with initial guess $x = 0.1$. For this example, the iterates

$$x_0 = 0.1000$$
$$x_1 = 0.2700$$
$$x_2 = 0.6831$$
$$x_3 = 1.4461$$
$$x_4 = 1.9579,$$

and so on, can be read as the intersections along the diagonal.

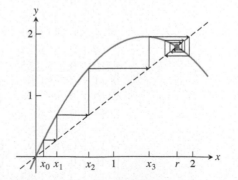

Figure 1.5 Cobweb diagram for Fixed-Point Iteration. Example 1.5 has two fixed points, 0 and 1.8. An iteration with starting guess 0.1 is shown. Only 1.8 will be converged to by FPI.

Even though the initial point $x_0 = 0.1$ is near the fixed point 0, FPI moves toward the other fixed point $x = 1.8$ and converges there. The difference between the two fixed points is that the slope of g at $x = 1.8$, given by $g'(1.8) = -0.8$, is smaller than one in absolute value. On the other hand, the slope of g at the other fixed point $x = 0$, the one that repels points, is $g'(0) = 2.8$, which is larger than one in absolute value. ◄

Theorem 1.6 is useful *a posteriori*—at the end of the FPI calculation, we know the root and can calculate the step-by-step errors. The theorem helps explain why the rate of convergence S turned out as it did. It would be much more useful to have that information before the calculation starts. In some cases, we are able to do this, as the next example shows.

▶ **EXAMPLE 1.6**   Calculate $\sqrt{2}$ by using FPI.

An ancient method for determining square roots can be expressed as an FPI. Suppose we want to find the first 10 digits of $\sqrt{2}$. Start with the initial guess $x_0 = 1$. This guess is obviously too low; therefore, $2/1 = 2$ is too high. In fact, any initial guess $0 < x_0 < 2$, together with $2/x_0$, form a bracket for $\sqrt{2}$. Because of that, it is reasonable to average the two to get a better guess:

$$x_1 = \frac{1 + \frac{2}{1}}{2} = \frac{3}{2}.$$

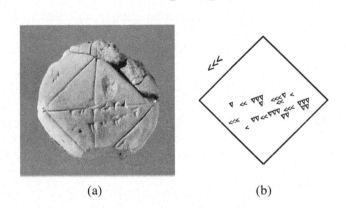

(a)   (b)

Figure 1.6 Ancient calculation of $\sqrt{2}$. (a) Tablet YBC7289 (b) Schematic of tablet. The Babylonians calculated in base 60, but used some base 10 notation. The < denotes 10, and the ∇ denotes 1. In the upper left is 30, the length of the side. Along the middle are 1, 24, 51, and 10, which represents the square root of 2 to five correct decimal places (see Spotlight on page 42). Below, the numbers 42, 25, and 35 represent $30\sqrt{2}$ in base 60.

Now repeat. Although $3/2$ is closer, it is too large to be $\sqrt{2}$, and $2/(3/2) = 4/3$ is too small. As before, average to get

$$x_2 = \frac{\frac{3}{2} + \frac{4}{3}}{2} = \frac{17}{12} = 1.41\overline{6},$$

which is even closer to $\sqrt{2}$. Once again, x_2 and $2/x_2$ bracket $\sqrt{2}$.

The next step yields

$$x_3 = \frac{\frac{17}{12} + \frac{24}{17}}{2} = \frac{577}{408} \approx 1.414215686.$$

Check with a calculator to see that this guess agrees with $\sqrt{2}$ within 3×10^{-6}. The FPI we are executing is

$$x_{i+1} = \frac{x_i + \frac{2}{x_i}}{2}. \tag{1.14}$$

Note that $\sqrt{2}$ is a fixed point of the iteration.

Convergence   The ingenious method of Example 1.6 converges to $\sqrt{2}$ within five decimal places after only three steps. This simple method is one of the oldest in the history of mathematics. The cuneiform tablet YBC7289 shown in Figure 1.6(a) was discovered near Baghdad in 1962, dating from around 1750 B.C. It contains the base 60 approximation $(1)(24)(51)(10)$ for the side length of a square of area 2. In base 10, this is

$$1 + \frac{24}{60} + \frac{51}{60^2} + \frac{10}{60^3} = 1.41421296.$$

The Babylonians' method of calculation is not known, but some speculate it is the computation of Example 1.6, in their customary base 60. In any case, this method appears in Book 1 of *Metrica*, written by Heron of Alexandria in the first century A.D., to calculate $\sqrt{720}$.

Before finishing the calculation, let's decide whether it will converge. According to Theorem 1.6, we need $S < 1$. For this iteration, $g(x) = 1/2(x + 2/x)$ and $g'(x) = 1/2(1 - 2/x^2)$. Evaluating at the fixed point yields

$$g'(\sqrt{2}) = \frac{1}{2}\left(1 - \frac{2}{(\sqrt{2})^2}\right) = 0, \tag{1.15}$$

so $S = 0$. We conclude that the FPI will converge, and very fast.

Exercise 18 asks whether this method, now often referred to as the Mechanic's Rule, will be successful in finding the square root of an arbitrary positive number. ◄

1.2.4 Stopping criteria

Unlike the case of bisection, the number of steps required for FPI to converge within a given tolerance is rarely predictable beforehand. In the absence of an error formula like (1.1) for the Bisection Method, a decision must be made about terminating the algorithm, called a **stopping criterion**.

For a set tolerance, TOL, we may ask for an absolute error stopping criterion

$$|x_{i+1} - x_i| < \text{TOL} \tag{1.16}$$

or, in case the solution is not too near zero, the relative error stopping criterion

$$\frac{|x_{i+1} - x_i|}{|x_{i+1}|} < \text{TOL}. \tag{1.17}$$

A hybrid absolute/relative stopping criterion such as

$$\frac{|x_{i+1} - x_i|}{\max(|x_{i+1}|, \theta)} < \text{TOL} \tag{1.18}$$

for some $\theta > 0$ is often useful in cases where the solution is near 0. In addition, good FPI code sets a limit on the maximum number of steps in case convergence fails. The issue of stopping criteria is important, and will be revisited in a more sophisticated way when we study forward and backward error in Section 1.3.

The Bisection Method is guaranteed to converge linearly. Fixed-Point Iteration is only locally convergent, and when it converges it is linearly convergent. Both methods require one function evaluation per step. The bisection cuts uncertainty by $1/2$ for each step, compared with approximately $S = |g'(r)|$ for FPI. Therefore, Fixed-Point Iteration may be faster or slower than bisection, depending on whether S is smaller or larger than $1/2$. In Section 1.4, we study Newton's Method, a particularly refined version of FPI, where S is designed to be zero.

▶ **ADDITIONAL EXAMPLES**

*1. (a) Show that $-1, 1$, and 2 are fixed points of $g(x) = \dfrac{2 + x^3 - 7x}{2x - 6}$. (b) To which of $-1, 1$, and 2 will Fixed-Point Iteration converge? Will the convergence be faster or slower than the Bisection Method?

2. Use the `fpi.m` code to find the three real roots of the equation $x^5 + 4x^2 = \sin x + 4x^4 + 1$ correct to six decimal places.

⌑ **Solutions** for Additional Examples can be found at `bit.ly/2CRvZ1c` (* example with video solution)

1.2 Exercises

⌑ **Solutions** for Exercises numbered in blue can be found at `bit.ly/2NLs3Qs`

1. Find all fixed points of the following $g(x)$.
   (a) $\dfrac{3}{x}$ (b) $x^2 - 2x + 2$ (c) $x^2 - 4x + 2$

2. Find all fixed points of the following $g(x)$.
   (a) $\dfrac{x + 6}{3x - 2}$ (b) $\dfrac{8 + 2x}{2 + x^2}$ (c) x^5

3. Show that $1, 2$, and 3 are fixed points of the following $g(x)$.
   (a) $\dfrac{x^3 + x - 6}{6x - 10}$ (b) $\dfrac{6 + 6x^2 - x^3}{11}$

4. Show that $-1, 0$, and 1 are fixed points of the following $g(x)$.
   (a) $\dfrac{4x}{x^2 + 3}$ (b) $\dfrac{x^2 - 5x}{x^2 + x - 6}$

5. For which of the following $g(x)$ is $r = \sqrt{3}$ a fixed point?
   (a) $g(x) = \dfrac{x}{\sqrt{3}}$ (b) $g(x) = \dfrac{2x}{3} + \dfrac{1}{x}$ (c) $g(x) = x^2 - x$ (d) $g(x) = 1 + \dfrac{2}{x + 1}$

6. For which of the following $g(x)$ is $r = \sqrt{5}$ a fixed point?
   (a) $g(x) = \dfrac{5 + 7x}{x + 7}$ (b) $g(x) = \dfrac{10}{3x} + \dfrac{x}{3}$ (c) $g(x) = x^2 - 5$ (d) $g(x) = 1 + \dfrac{4}{x + 1}$

7. Use Theorem 1.6 to determine whether Fixed-Point Iteration of $g(x)$ is locally convergent to the given fixed point r. (a) $g(x) = (2x - 1)^{1/3}, r = 1$ (b) $g(x) = (x^3 + 1)/2, r = 1$ (c) $g(x) = \sin x + x, r = 0$

8. Use Theorem 1.6 to determine whether Fixed-Point Iteration of $g(x)$ is locally convergent to the given fixed point r. (a) $g(x) = (2x - 1)/x^2, r = 1$ (b) $g(x) = \cos x + \pi + 1, r = \pi$ (c) $g(x) = e^{2x} - 1, r = 0$

9. Find each fixed point and decide whether Fixed-Point Iteration is locally convergent to it. (a) $g(x) = \frac{1}{2}x^2 + \frac{1}{2}x$ (b) $g(x) = x^2 - \frac{1}{4}x + \frac{3}{8}$

10. Find each fixed point and decide whether Fixed-Point Iteration is locally convergent to it.
    (a) $g(x) = x^2 - \frac{3}{2}x + \frac{3}{2}$ (b) $g(x) = x^2 + \frac{1}{2}x - \frac{1}{2}$

11. Express each equation as a fixed-point problem $x = g(x)$ in three different ways.
    (a) $x^3 - x + e^x = 0$ (b) $3x^{-2} + 9x^3 = x^2$

12. Consider the Fixed-Point Iteration $x \to g(x) = x^2 - 0.24$. (a) Do you expect Fixed-Point Iteration to calculate the root -0.2, say, to 10 or to correct decimal places, faster or slower than the Bisection Method? (b) Find the other fixed point. Will FPI converge to it?

13. (a) Find all fixed points of $g(x) = 0.39 - x^2$. (b) To which of the fixed-points is Fixed-Point Iteration locally convergent? (c) Does FPI converge to this fixed point faster or slower than the Bisection Method?

14. Which of the following three Fixed-Point Iterations converge to $\sqrt{2}$? Rank the ones that converge from fastest to slowest.

    (A) $x \longrightarrow \dfrac{1}{2}x + \dfrac{1}{x}$ (B) $x \longrightarrow \dfrac{2}{3}x + \dfrac{2}{3x}$ (C) $x \longrightarrow \dfrac{3}{4}x + \dfrac{1}{2x}$

15. Which of the following three Fixed-Point Iterations converge to $\sqrt{5}$? Rank the ones that converge from fastest to slowest.

    (A) $x \longrightarrow \dfrac{4}{5}x + \dfrac{1}{x}$ (B) $x \longrightarrow \dfrac{x}{2} + \dfrac{5}{2x}$ (C) $x \longrightarrow \dfrac{x+5}{x+1}$

16. Which of the following three Fixed-Point Iterations converge to the cube root of 4? Rank the ones that converge from fastest to slowest.

    (A) $g(x) = \dfrac{2}{\sqrt{x}}$ (B) $g(x) = \dfrac{3x}{4} + \dfrac{1}{x^2}$ (C) $g(x) = \dfrac{2}{3}x + \dfrac{4}{3x^2}$

17. Check that $1/2$ and -1 are roots of $f(x) = 2x^2 + x - 1 = 0$. Isolate the x^2 term and solve for x to find two candidates for $g(x)$. Which of the roots will be found by the two Fixed-Point Iterations?

18. Prove that the method of Example 1.6 will calculate the square root of any positive number.

19. Explore the idea of Example 1.6 for cube roots. If x is a guess that is smaller than $A^{1/3}$, then A/x^2 will be larger than $A^{1/3}$, so that the average of the two will be a better approximation than x. Suggest a Fixed-Point Iteration on the basis of this fact, and use Theorem 1.6 to decide whether it will converge to the cube root of A.

20. Improve the cube root algorithm of Exercise 19 by reweighting the average. Setting $g(x) = wx + (1 - w)A/x^2$ for some fixed number $0 < w < 1$, what is the best choice for w?

21. Consider Fixed-Point Iteration applied to $g(x) = 1 - 5x + \frac{15}{2}x^2 - \frac{5}{2}x^3$. (a) Show that $1 - \sqrt{3/5}$, 1, and $1 + \sqrt{3/5}$ are fixed points. (b) Show that none of the three fixed points is locally convergent. (Computer Problem 7 investigates this example further.)

22. Show that the initial guesses 0, 1, and 2 lead to a fixed point in Exercise 21. What happens to other initial guesses close to those numbers?

23. Assume that $g(x)$ is continuously differentiable and that the Fixed-Point Iteration $g(x)$ has exactly three fixed points, $r_1 < r_2 < r_3$. Assume also that $|g'(r_1)| = 0.5$ and $|g'(r_3)| = 0.5$. What range of values is possible for $g'(r_2)$ under these assumptions? To which of the fixed points will FPI converge?

24. Assume that g is a continuously differentiable function and that the Fixed-Point Iteration $g(x)$ has exactly three fixed points, -3, 1, and 2. Assume that $g'(-3) = 2.4$ and that FPI started sufficiently near the fixed point 2 converges to 2. Find $g'(1)$.

25. Prove the variant of Theorem 1.6: If g is continuously differentiable and $0 < g'(x) \le B < 1$ on an interval $[a, b]$ containing the fixed point r, then FPI converges to r from any initial guess in $[a, b]$.

26. Prove that a continuously differentiable function $g(x)$ satisfying $|g'(x)| < 1$ on a closed interval cannot have two fixed points on that interval.

27. Consider Fixed-Point Iteration with $0 \le g'(x) \le B < 1$. (a) Show that $x = 0$ is the only fixed point. (b) Show that if $0 < x_0 < 1$, then $x_0 > x_1 > x_2 \ldots > 0$. (c) Show that FPI converges to $r = 0$, while $g'(0) = 1$. (Hint: Use the fact that every bounded monotonic sequence converges to a limit.)

28. Consider Fixed-Point Iteration with $g(x) = x + x^3$. (a) Show that $x = 0$ is the only fixed point. (b) Show that if $0 < x_0 < 1$, then $x_0 < x_1 < x_2 < \ldots$. (c) Show that FPI fails to converge to a fixed point, while $g'(0) = 1$. Together with Exercise 27, this shows that FPI may converge to a fixed point r or diverge from r when $|g'(r)| = 1$.

29. Consider the equation $x^3 + x - 2 = 0$, with root $r = 1$. Add the term cx to both sides and divide by c to obtain $g(x)$. (a) For what c is FPI locally convergent to $r = 1$? (b) For what c will FPI converge fastest?

30. Assume that Fixed-Point Iteration is applied to a twice continuously differentiable function $g(x)$ and that $g'(r) = 0$ for a fixed point r. Show that if FPI converges to r, then the error obeys $\lim_{i \to \infty} (e_{i+1})/e_i^2 = M$, where $M = |g''(r)|/2$.

31. Define Fixed-Point Iteration on the equation $x^2 + x = 5/16$ by isolating the x term. Find both fixed points, and determine which initial guesses lead to each fixed point under iteration.
    (Hint: Plot $g(x)$, and draw cobweb diagrams.)

32. Find the set of all initial guesses for which the Fixed-Point Iteration $x \to 4/9 - x^2$ converges to a fixed point.

33. Let $g(x) = a + bx + cx^2$ for constants a, b, and c. (a) Specify one set of constants a, b, and c for which $x = 0$ is a fixed-point of $x = g(x)$ and Fixed-Point Iteration is locally convergent to 0. (b) Specify one set of constants a, b, and c for which $x = 0$ is a fixed-point of $x = g(x)$ but Fixed-Point Iteration is not locally convergent to 0.

1.2 Computer Problems

Solutions for Computer Problems numbered in blue can be found at
`bit.ly/2CRlpHC`

1. Apply Fixed-Point Iteration to find the solution of each equation to eight correct decimal places. (a) $x^3 = 2x + 2$ (b) $e^x + x = 7$ (c) $e^x + \sin x = 4$.

2. Apply Fixed-Point Iteration to find the solution of each equation to eight correct decimal places. (a) $x^5 + x = 1$ (b) $\sin x = 6x + 5$ (c) $\ln x + x^2 = 3$

3. Calculate the square roots of the following numbers to eight correct decimal places by using Fixed-Point Iteration as in Example 1.6: (a) 3 (b) 5. State your initial guess and the number of steps needed.

4. Calculate the cube roots of the following numbers to eight correct decimal places, by using Fixed-Point Iteration with $g(x) = (2x + A/x^2)/3$, where A is (a) 2 (b) 3 (c) 5. State your initial guess and the number of steps needed.

5. Example 1.3 showed that $g(x) = \cos x$ is a convergent FPI. Is the same true for $g(x) = \cos^2 x$? Find the fixed point to six correct decimal places, and report the number of FPI steps needed. Discuss local convergence, using Theorem 1.6.

6. Derive three different $g(x)$ for finding roots to six correct decimal places of the following $f(x) = 0$ by Fixed-Point Iteration. Run FPI for each $g(x)$ and report results, convergence or divergence. Each equation $f(x) = 0$ has three roots. Derive more $g(x)$ if necessary until all roots are found by FPI. For each convergent run, determine the value of S from the errors e_{i+1}/e_i, and compare with S determined from calculus as in (1.11).
   (a) $f(x) = 2x^3 - 6x - 1$ (b) $f(x) = e^{x-2} + x^3 - x$ (c) $f(x) = 1 + 5x - 6x^3 - e^{2x}$

7. Exercise 21 considered Fixed-Point Iteration applied to $g(x) = 1 - 5x + \frac{15}{2}x^2 - \frac{5}{2}x^3 = x$. Find initial guesses for which FPI (a) cycles endlessly through numbers in the interval

(0, 1) (b) the same as (a), but the interval is (1, 2) (c) diverges to infinity. Cases (a) and (b) are examples of chaotic dynamics. In all three cases, FPI is unsuccessful.

1.3 LIMITS OF ACCURACY

One of the goals of numerical analysis is to compute answers within a specified level of accuracy. Working in double precision means that we store and operate on numbers that are kept to 52-bit accuracy, about 16 decimal digits.

Can answers always be computed to 16 correct significant digits? In Chapter 0, it was shown that, with a naive algorithm for computing roots of a quadratic equation, it was possible to lose some or all significant digits. An improved algorithm eliminated the problem. In this section, we will see something new—a calculation that a double-precision computer cannot make to anywhere near 16 correct digits, even with the best algorithm.

1.3.1 Forward and backward error

The first example shows that, in some cases, pencil and paper can still outperform a computer.

► **EXAMPLE 1.7** Use the Bisection Method to find the root of $f(x) = x^3 - 2x^2 + \frac{4}{3}x - \frac{8}{27}$ to within six correct significant digits.

Note that $f(0)f(1) = (-8/27)(1/27) < 0$, so the Intermediate Value Theorem guarantees a solution in [0, 1]. According to Example 1.2, 20 bisection steps should be sufficient for six correct places.

In fact, it is easy to check without a computer that $r = 2/3 = 0.666666666\ldots$ is a root:

$$f(2/3) = \frac{8}{27} - 2\left(\frac{4}{9}\right) + \left(\frac{4}{3}\right)\left(\frac{2}{3}\right) - \frac{8}{27} = 0.$$

How many of these digits can the Bisection Method obtain?

i	a_i	$f(a_i)$	c_i	$f(c_i)$	b_i	$f(b_i)$
0	0.0000000	−	0.5000000	−	1.0000000	+
1	0.5000000	−	0.7500000	+	1.0000000	+
2	0.5000000	−	0.6250000	−	0.7500000	+
3	0.6250000	−	0.6875000	+	0.7500000	+
4	0.6250000	−	0.6562500	−	0.6875000	+
5	0.6562500	−	0.6718750	+	0.6875000	+
6	0.6562500	−	0.6640625	−	0.6718750	+
7	0.6640625	−	0.6679688	+	0.6718750	+
8	0.6640625	−	0.6660156	−	0.6679688	+
9	0.6660156	−	0.6669922	+	0.6679688	+
10	0.6660156	−	0.6665039	−	0.6669922	+
11	0.6665039	−	0.6667480	+	0.6669922	+
12	0.6665039	−	0.6666260	−	0.6667480	+
13	0.6666260	−	0.6666870	+	0.6667480	+
14	0.6666260	−	0.6666565	−	0.666687	+
15	0.6666565	−	0.6666718	+	0.6666870	+
16	0.6666565	−	0.6666641	0	0.6666718	+

Surprisingly, the Bisection Method stops after 16 steps, when it computes $f(0.6666641) = 0$. This is a serious failure if we care about six or more digits of precision. Figure 1.7 shows the difficulty. As far as IEEE double precision is concerned, there are many floating point numbers within 10^{-5} of the correct root $r = 2/3$ that are evaluated to machine zero, and therefore have an equal right to be called the root! To make matters worse, although the function f is monotonically increasing, part (b) of the figure shows that even the sign of the double precision value of f is often wrong.

Figure 1.7 shows that the problem lies not with the Bisection Method, but with the inability of double precision arithmetic to compute the function f accurately enough near the root. Any other solution method that relies on this computer arithmetic is bound to fail. For this example, 16-digit precision cannot even check whether a candidate solution is correct to six places. ◄

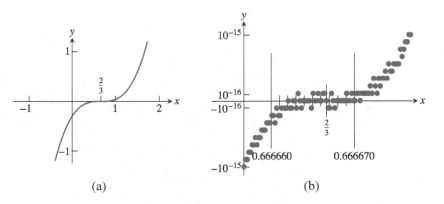

(a)   (b)

Figure 1.7 The shape of a function near a multiple root. (a) Plot of $f(x) = x^3 - 2x^2 + 4/3x - 8/27$. (b) Magnification of (a), near the root $r = 2/3$. There are many floating point numbers within 10^{-5} of 2/3 that are roots as far as the computer is concerned. We know from calculus that 2/3 is the only root.

To convince you that it's not the fault of the Bisection Method, we apply MATLAB's most high-powered multipurpose rootfinder, fzero.m. We will discuss its details later in this chapter; for now, we just need to feed it the function and a starting guess. It has no better luck:

```
>> fzero('x.^3-2*x.^2+4*x/3-8/27',1)

ans =

    0.66666250845989
```

The reason that all methods fail to get more than five correct digits for this example is clear from Figure 1.7. The only information any method has is the function, computed in double precision. If the computer arithmetic is showing the function to be zero at a nonroot, there is no way the method can recover. Another way to state the difficulty is to say that an approximate solution can be as close as possible to a solution as far as the y-axis is concerned, but not so close on the x-axis.

These observations motivate some key definitions.

DEFINITION 1.8 Assume that f is a function and that r is a root, meaning that it satisfies $f(r) = 0$. Assume that x_a is an approximation to r. For the root-finding problem, the **backward error** of the approximation x_a is $|f(x_a)|$ and the **forward error** is $|r - x_a|$. ❑

The usage of "backward" and "forward" may need some explanation. Our viewpoint considers the process of finding a solution as central. The problem is the input, and the solution is the output:

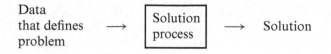

In this chapter, the "problem" is an equation in one variable, and the "solution process" is an algorithm that solves equations:

Equation $\longrightarrow$ | Equation solver | $\longrightarrow$ Solution

Backward error is on the left or input (problem data) side. It is the amount we would need to change the problem (the function f) to make the equation balance with the output approximation x_a. This amount is $|f(x_a)|$. Forward error is the error on the right or output (problem solution) side. It is the amount we would need to change the approximate solution to make it correct, which is $|r - x_a|$.

The difficulty with Example 1.7 is that, according to Figure 1.7, the backward error is near $\epsilon_{mach} \approx 2.2 \times 10^{-16}$, while forward error is approximately 10^{-5}. Double precision numbers cannot be computed reliably below a relative error of machine epsilon. Since the backward error cannot be decreased further with reliability, neither can the forward error.

Example 1.7 is rather special because the function has a triple root at $r = 2/3$. Note that

$$f(x) = x^3 - 2x^2 + \frac{4}{3}x - \frac{8}{27} = \left(x - \frac{2}{3}\right)^3 .$$

This is an example of a multiple root.

DEFINITION 1.9   Assume that r is a root of the differentiable function f; that is, assume that $f(r) = 0$. Then if $0 = f(r) = f'(r) = f''(r) = \cdots = f^{(m-1)}(r)$, but $f^{(m)}(r) \neq 0$, we say that f has a **root** of **multiplicity** m at r. We say that f has a **multiple root** at r if the multiplicity is greater than one. The root is called **simple** if the multiplicity is one.   ❐

For example, $f(x) = x^2$ has a multiplicity two, or double, root at $r = 0$, because f(0) = 0, $f'(0) = 2(0) = 0$, but $f''(0) = 2 \neq 0$. Likewise, $f(x) = x^3$ has a multiplicity three, or triple, root at $r = 0$, and $f(x) = x^m$ has a multiplicity m root there. Example 1.7 has a multiplicity three, or triple, root at $r = 2/3$.

Because the graph of the function is relatively flat near a multiple root, a great disparity exists between backward and forward errors for nearby approximate solutions. The backward error, measured in the vertical direction, is often much smaller than the forward error, measured in the horizontal direction.

▶ **EXAMPLE 1.8**   The function $f(x) = \sin x - x$ has a triple root at $r = 0$. Find the forward and backward error of the approximate root $x_c = 0.001$.

The root at 0 has multiplicity three because

$$f(0) = \sin 0 - 0 = 0$$
$$f'(0) = \cos 0 - 1 = 0$$
$$f''(0) = -\sin 0 - 0 = 0$$
$$f'''(0) = -\cos 0 = -1.$$

The forward error is $\text{FE} = |r - x_a| = 10^{-3}$. The backward error is the constant that would need to be added to $f(x)$ to make x_a a root, namely $\text{BE} = |f(x_a)| = |\sin(0.001) - 0.001| \approx 1.6667 \times 10^{-10}$.   ◀

The subject of backward and forward error is relevant to stopping criteria for equation solvers. The goal is to find the root r satisfying $f(r) = 0$. Suppose our algorithm produces an approximate solution x_a. How do we decide whether it is good enough?

Two possibilities come to mind: (1) to make $|x_a - r|$ small and (2) to make $|f(x_a)|$ small. In case $x_a = r$, there is no decision to be made—both ways of looking at it are the same. However, we are rarely lucky enough to be in this situation. In the more typical case, approaches (1) and (2) are different and correspond to forward and backward error.

Whether forward or backward error is more appropriate depends on the circumstances surrounding the problem. If we are using the Bisection Method, both errors are easily observable. For an approximate root x_a, we can find the backward error by evaluating $f(x_a)$, and the forward error can be no more than half the length of the current interval. For FPI, our choices are more limited, since we have no bracketing interval. As before, the backward error is known as $f(x_a)$, but to know the forward error would require knowing the true root, which we are trying to find.

Stopping criteria for equation-solving methods can be based on either forward or backward error. There are other stopping criteria that may be relevant, such as a limit on computation time. The context of the problem must guide our choice.

Functions are flat in the vicinity of a multiple root, since the derivative f' is zero there. Because of this, we can expect some trouble in isolating a multiple root, as we have demonstrated. But multiplicity is only the tip of the iceberg. Similar difficulties can arise where no multiple roots are in sight, as shown in the next section.

1.3.2 The Wilkinson polynomial

A famous example with simple roots that are hard to determine numerically is discussed in Wilkinson [1994]. The **Wilkinson polynomial** is

$$W(x) = (x - 1)(x - 2) \cdots (x - 20), \tag{1.19}$$

which, when multiplied out, is

$$
\begin{aligned}
W(x) = {} & x^{20} - 210x^{19} + 20615x^{18} - 1256850x^{17} + 53327946x^{16} - 1672280820x^{15} \\
& + 40171771630x^{14} - 756111184500x^{13} + 11310276995381x^{12} \\
& - 135585182899530x^{11} + 1307535010540395x^{10} - 10142299865511450x^{9}
\end{aligned}
$$

$$+\ 63030812099294896x^8 - 311333643161390640x^7$$
$$+\ 1206647803780373360x^6 - 3599979517947607200x^5$$
$$+\ 8037811822645051776x^4 - 12870931245150988800x^3$$
$$+\ 13803759753640704000x^2 - 8752948036761600000x$$
$$+\ 2432902008176640000. \tag{1.20}$$

The roots are the integers from 1 to 20. However, when $W(x)$ is defined according to its unfactored form (1.20), its evaluation suffers from cancellation of nearly equal, large numbers. To see the effect on root-finding, define the MATLAB m-file wilkpoly.m by typing in the nonfactored form (1.20), or obtaining it from the textbook website.

MATLAB code shown here can be found at bit.ly/2NM8Tdl

Again we will try MATLAB's fzero. To make it as easy as possible, we feed it an actual root $x = 16$ as a starting guess:

```
>> fzero(@wilkpoly,16)

ans =

   16.01468030580458
```

The surprising result is that MATLAB's double precision arithmetic could not get the second decimal place correct, even for the simple root $r = 16$. It is not due to a deficiency of the algorithm—both fzero and Bisection Method have the same problem, as do Fixed-Point Iteration and any other floating point method. Referring to his work with this polynomial, Wilkinson wrote in 1984: "Speaking for myself I regard it as the most traumatic experience in my career as a numerical analyst." The roots of $W(x)$ are clear: the integers $x = 1, \ldots, 20$. To Wilkinson, the surprise had to do with the huge error magnification in the roots caused by small relative errors in storing the coefficients, which we have just seen in action.

The difficulty of getting accurate roots of the Wilkinson polynomial disappears if factored form (1.19) is used instead of (1.20). Of course, if the polynomial is factored before we start, there is no need to compute roots.

1.3.3 Sensitivity of root-finding

The Wilkinson polynomial and Example 1.7 with the triple root cause difficulties for similar reasons—small floating point errors in the equation translate into large errors in the root. A problem is called **sensitive** if small errors in the input, in this case the equation to be solved, lead to large errors in the output, or solution. In this section, we will quantify sensitivity and introduce the concepts of error magnification factor and condition number.

To understand what causes this magnification of error, we will establish a formula predicting how far a root moves when the equation is changed. Assume that the problem is to find a root r of $f(x) = 0$, but that a small change $\epsilon g(x)$ is made to the input, where ϵ is small. Let Δr be the corresponding change in the root, so that

$$f(r + \Delta r) + \epsilon g(r + \Delta r) = 0.$$

Expanding f and g in degree-one Taylor polynomials implies that

$$f(r) + (\Delta r)f'(r) + \epsilon g(r) + \epsilon(\Delta r)g'(r) + O((\Delta r)^2) = 0,$$

where we use the "big O" notation $O((\Delta r)^2)$ to stand for terms involving $(\Delta r)^2$ and higher powers of Δr. For small Δr, the $O((\Delta r)^2)$ terms can be neglected to get

$$(\Delta r)(f'(r) + \epsilon g'(r)) \approx -f(r) - \epsilon g(r) = -\epsilon g(r)$$

or

$$\Delta r \approx \frac{-\epsilon g(r)}{f'(r) + \epsilon g'(r)} \approx -\epsilon \frac{g(r)}{f'(r)},$$

assuming that ϵ is small compared with $f'(r)$, and in particular, that $f'(r) \neq 0$.

Sensitivity Formula for Roots

Assume that r is a root of $f(x)$ and $r + \Delta r$ is a root of $f(x) + \epsilon g(x)$. Then

$$\Delta r \approx -\frac{\epsilon g(r)}{f'(r)} \tag{1.21}$$

if $\epsilon \ll f'(r)$.

▶ **EXAMPLE 1.9**   Estimate the largest root of $P(x) = (x - 1)(x - 2)(x - 3)(x - 4)(x - 5)(x - 6) - 10^{-6}x^7$.

Set $f(x) = (x - 1)(x - 2)(x - 3)(x - 4)(x - 5)(x - 6), \epsilon = -10^{-6}$ and $g(x) = x^7$. Without the $\epsilon g(x)$ term, the largest root is $r = 6$. The question is, how far does the root move when we add the extra term?

The Sensitivity Formula yields

$$\Delta r \approx -\frac{\epsilon 6^7}{5!} = -2332.8\epsilon,$$

meaning that input errors of relative size ϵ in $f(x)$ are magnified by a factor of over 2000 into the output root. We estimate the largest root of $P(x)$ to be $r + \Delta r = 6 - 2332.8\epsilon = 6.0023328$. Using fzero on $P(x)$, we get the correct value 6.0023268.   ◀

The estimate in Example 1.9 is good enough to tell us how errors propagate in the root-finding problem. An error in the sixth digit of the problem data caused an error in the third digit of the answer, meaning that three decimal digits were lost due to the factor of 2332.8. It is useful to have a name for this factor. For a general algorithm that produces an approximation x_c, we define its

$$\textbf{error magnification factor} = \frac{\text{relative forward error}}{\text{relative backward error}}.$$

The forward error is the change in the solution that would make x_a correct, which for root-finding problems is $|x_a - r|$. The backward error is a change in input that makes x_c the correct solution. There is a wider variety of choices, depending on what sensitivity we want to investigate. Changing the constant term by $|f(x_a)|$ is the choice that was used earlier in this section, corresponding to $g(x) = 1$ in the Sensitivity Formula (1.21). More generally, any change in the input data can be used as the backward error, such as the choice $g(x) = x^7$ in Example 1.9. The error magnification factor for root-finding is

$$\text{error magnification factor} = \left| \frac{\Delta r / r}{\epsilon g(r)/g(r)} \right| = \left| \frac{-\epsilon g(r)/(rf'(r))}{\epsilon} \right| = \frac{|g(r)|}{|rf'(r)|}, \quad (1.22)$$

which in Example 1.9 is $6^7/(5!6) = 388.8$.

▶ **EXAMPLE 1.10**   Use the Sensitivity Formula for Roots to investigate the effect of changes in the x^{15} term of the Wilkinson polynomial on the root $r = 16$. Find the error magnification factor for this problem.

Define the perturbed function $W_\epsilon(x) = W(x) + \epsilon g(x)$, where $g(x) = -1,672,280,820x^{15}$. Note that $W'(16) = 15!4!$ (see Exercise 7). Using (1.21), the change in the root can be approximated by

$$\Delta r \approx \frac{16^{15}1,672,280,820\epsilon}{15!4!} \approx 6.1432 \times 10^{13}\epsilon. \quad (1.23)$$

Practically speaking, we know from Chapter 0 that a relative error on the order of machine epsilon must be assumed for every stored number. A relative change in the x^{15} term of machine epsilon ϵ_{mach} will cause the root $r = 16$ to move by

$$\Delta r \approx (6.1432 \times 10^{13})(\pm 2.22 \times 10^{-16}) \approx \pm 0.0136$$

to $r + \Delta r \approx 16.0136$, not far from what was observed on page 50. Of course, many other powers of x in the Wilkinson polynomial are making their own contributions, so the complete picture is complicated. However, the Sensitivity Formula allows us to see the mechanism for the huge magnification of error.

Finally, the error magnification factor is computed from (1.22) as

$$\frac{|g(r)|}{|rf'(r)|} = \frac{16^{15}1,672,280,820}{15!4!16} \approx 3.8 \times 10^{12}. \qquad ◀$$

The significance of the error magnification factor is that it tells us how many of the 16 digits of operating precision are lost from input to output. For a problem with error magnification factor of 10^{12}, we expect to lose 12 of the 16 and have about four correct significant digits left in the root, which is the case for the Wilkinson approximation $x_c = 16.014\ldots$.

SPOTLIGHT ON

Conditioning   This is the first appearance of the concept of condition number, a measure of error magnification. Numerical analysis is the study of algorithms, which take data defining the problem as input and deliver an answer as output. Condition number refers to the part of this magnification that is inherent in the theoretical problem itself, irrespective of the particular algorithm used to solve it.

It is important to note that the error magnification factor measures only magnification due to the problem. Along with conditioning, there is a parallel concept, stability, that refers to the magnification of small input errors due to the algorithm, not the problem itself. An algorithm is called stable if it always provides an approximate solution with small backward error. If the problem is well-conditioned and the algorithm is stable, we can expect both small backward and forward error.

The preceding error magnification examples show the sensitivity of root-finding to a particular input change. The problem may be more or less sensitive, depending on how the input change is designed. The **condition number** of a problem is defined to be the maximum error magnification over all input changes, or at least all changes of a prescribed type. A problem with high condition number is called **ill-conditioned**, and a problem with a condition number near 1 is called **well-conditioned**. We will return to this concept when we discuss matrix problems in Chapter 2.

▶ **ADDITIONAL EXAMPLES**

1. Find the multiplicity of the root $r = 0$ of $f(x) = 6x - 6\sin x - x^3$.

2. Use the MATLAB command `fzero` with initial guess 0.001 to approximate the root of $f(x) = 6x - 6\sin x - x^3$. Compute the forward and backward errors of the approximate root.

Solutions for Additional Examples can be found at `bit.ly/2CT0mEI`

1.3 Exercises

Solutions for Exercises numbered in blue can be found at bit.ly/2pXrd9W

1. Find the forward and backward error for the following functions, where the root is 3/4 and the approximate root is $x_a = 0.74$: (a) $f(x) = 4x - 3$ (b) $f(x) = (4x - 3)^2$ (c) $f(x) = (4x - 3)^3$ (d) $f(x) = (4x - 3)^{1/3}$

2. Find the forward and backward error for the following functions, where the root is 1/3 and the approximate root is $x_a = 0.3333$: (a) $f(x) = 3x - 1$ (b) $f(x) = (3x - 1)^2$ (c) $f(x) = (3x - 1)^3$ (d) $f(x) = (3x - 1)^{1/3}$

3. (a) Find the multiplicity of the root $r = 0$ of $f(x) = 1 - \cos x$. (b) Find the forward and backward errors of the approximate root $x_a = 0.0001$.

4. (a) Find the multiplicity of the root $r = 0$ of $f(x) = x^2 \sin x^2$. (b) Find the forward and backward errors of the approximate root $x_a = 0.01$.

5. Find the relation between forward and backward error for finding the root of the linear function $f(x) = ax - b$.

6. Let n be a positive integer. The equation defining the nth root of a positive number A is $x^n - A = 0$. (a) Find the multiplicity of the root. (b) Show that, for an approximate nth root with small forward error, the backward error is approximately $nA^{(n-1)/n}$ times the forward error.

7. Let $W(x)$ be the Wilkinson polynomial. (a) Prove that $W'(16) = 15!4!$. (b) Find an analogous formula for $W'(j)$, where j is an integer between 1 and 20.

8. Let $f(x) = x^n - ax^{n-1}$, and set $g(x) = x^n$. (a) Use the Sensitivity Formula to give a prediction for the nonzero root of $f_\epsilon(x) = x^n - ax^{n-1} + \epsilon x^n$ for small ϵ. (b) Find the nonzero root and compare with the prediction.

1.3 Computer Problems

Solutions for Computer Problems numbered in blue can be found at bit.ly/2Aig4GW

1. Let $f(x) = \sin x - x$. (a) Find the multiplicity of the root $r = 0$. (b) Use MATLAB's `fzero` command with initial guess $x = 0.1$ to locate a root. What are the forward and backward errors of `fzero`'s response?

2. Carry out Computer Problem 1 for $f(x) = \sin x^3 - x^3$.

3. (a) Use `fzero` to find the root of $f(x) = 2x \cos x - 2x + \sin x^3$ on $[-0.1, 0.2]$. Report the forward and backward errors. (b) Run the Bisection Method with initial interval $[-0.1, 0.2]$ to find as many correct digits as possible, and report your conclusion.

4. (a) Use (1.21) to approximate the root near 3 of $f_\epsilon(x) = (1 + \epsilon)x^3 - 3x^2 + x - 3$ for a constant ϵ. (b) Setting $\epsilon = 10^{-3}$, find the actual root and compare with part (a).

5. Use (1.21) to approximate the root of $f(x) = (x - 1)(x - 2)(x - 3)(x - 4) - 10^{-6}x^6$ near $r = 4$. Find the error magnification factor. Use `fzero` to check your approximation.

6. Use the MATLAB command `fzero` to find the root of the Wilkinson polynomial near $x = 15$ with a relative change of $\epsilon = 2 \times 10^{-15}$ in the x^{15} coefficient, making the coefficient slightly more negative. Compare with the prediction made by (1.21).

1.4 NEWTON'S METHOD

Newton's Method, also called the Newton–Raphson Method, usually converges much faster than the linearly convergent methods we have seen previously. The geometric picture of Newton's Method is shown in Figure 1.8. To find a root of $f(x) = 0$, a starting guess x_0 is given, and the tangent line to the function f at x_0 is drawn. The tangent line will approximately follow the function down to the x-axis toward the root. The intersection point of the line with the x-axis is an approximate root, but probably not exact if f curves. Therefore, this step is iterated.

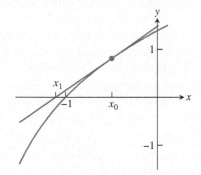

Figure 1.8 One step of Newton's Method. Starting with x_0, the tangent line to the curve $y = f(x)$ is drawn. The intersection point with the x-axis is x_1, the next approximation to the root.

From the geometric picture, we can develop an algebraic formula for Newton's Method. The tangent line at x_0 has slope given by the derivative $f'(x_0)$. One point on the tangent line is $(x_0, f(x_0))$. The point-slope formula for the equation of a line is $y - f(x_0) = f'(x_0)(x - x_0)$, so that looking for the intersection point of the tangent line with the x-axis is the same as substituting $y = 0$ in the line:

$$f'(x_0)(x - x_0) = 0 - f(x_0)$$
$$x - x_0 = -\frac{f(x_0)}{f'(x_0)}$$
$$x = x_0 - \frac{f(x_0)}{f'(x_0)}.$$

Solving for x gives an approximation for the root, which we call x_1. Next, the entire process is repeated, beginning with x_1, to produce x_2, and so on, yielding the following iterative formula:

Newton's Method

$$x_0 = \text{initial guess}$$

$$x_{i+1} = x_i - \frac{f(x_i)}{f'(x_i)} \quad \text{for } i = 0, 1, 2, \ldots.$$

▶ **EXAMPLE 1.11**   Find the Newton's Method formula for the equation $x^3 + x - 1 = 0$.

Since $f'(x) = 3x^2 + 1$, the formula is given by

$$x_{i+1} = x_i - \frac{x_i^3 + x_i - 1}{3x_i^2 + 1}$$

$$= \frac{2x_i^3 + 1}{3x_i^2 + 1}.$$

Iterating this formula from initial guess $x_0 = -0.7$ yields

$$x_1 = \frac{2x_0^3 + 1}{3x_0^2 + 1} = \frac{2(-0.7)^3 + 1}{3(-0.7)^2 + 1} \approx 0.1271$$

$$x_2 = \frac{2x_1^3 + 1}{3x_1^2 + 1} \approx 0.9577.$$

These steps are shown geometrically in Figure 1.9. Further steps are given in the following table:

i	x_i	$e_i = \lvert x_i - r \rvert$	e_i / e_{i-1}^2
0	-0.70000000	1.38232780	
1	0.12712551	0.55520230	0.2906
2	0.95767812	0.27535032	0.8933
3	0.73482779	0.05249999	0.6924
4	0.68459177	0.00226397	0.8214
5	0.68233217	0.00000437	0.8527
6	0.68232780	0.00000000	0.8541
7	0.68232780	0.00000000	

After only six steps, the root is known to eight correct digits. There is a bit more we can say about the error and how fast it becomes small. Note in the table that once convergence starts to take hold, the number of correct places in x_i approximately doubles on each iteration. This is characteristic of "quadratically convergent" methods, as we shall see next.

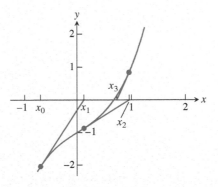

Figure 1.9 Three steps of Newton's Method. Illustration of Example 1.11. Starting with $x_0 = -0.7$, the Newton's Method iterates are plotted along with the tangent lines. The method appears to be converging to the root.

◀

1.4.1 Quadratic convergence of Newton's Method

The convergence in Example 1.11 is qualitatively faster than the linear convergence we have seen for the Bisection Method and Fixed-Point Iteration. A new definition is needed.

DEFINITION 1.10 Let e_i denote the error after step i of an iterative method. The iteration is **quadratically convergent** if

$$M = \lim_{i \to \infty} \frac{e_{i+1}}{e_i^2} < \infty.$$

$\square$

THEOREM 1.11 Let f be twice continuously differentiable and $f(r) = 0$. If $f'(r) \neq 0$, then Newton's Method is locally and quadratically convergent to r. The error e_i at step i satisfies

$$\lim_{i \to \infty} \frac{e_{i+1}}{e_i^2} = M,$$

where

$$M = \frac{f''(r)}{2f'(r)}.$$

■

Proof. To prove local convergence, note that Newton's Method is a particular form of Fixed-Point Iteration, where

$$g(x) = x - \frac{f(x)}{f'(x)},$$

with derivative

$$g'(x) = 1 - \frac{f'(x)^2 - f(x)f''(x)}{f'(x)^2} = \frac{f(x)f''(x)}{f'(x)^2}.$$

Since $g'(r) = 0$, Newton's Method is locally convergent according to Theorem 1.6.

To prove quadratic convergence, we derive Newton's Method a second way, this time keeping a close eye on the error at each step. By error, we mean the difference between the correct root and the current best guess.

Taylor's formula in Theorem 0.8 tells us the difference between the values of a function at a given point and another nearby point. For the two points, we will use the root r and the current guess x_i after i steps, and we will stop and take a remainder after two terms:

$$f(r) = f(x_i) + (r - x_i)f'(x_i) + \frac{(r - x_i)^2}{2}f''(c_i).$$

Here, c_i is between x_i and r. Because r is the root, we have

$$0 = f(x_i) + (r - x_i)f'(x_i) + \frac{(r - x_i)^2}{2}f''(c_i)$$

$$-\frac{f(x_i)}{f'(x_i)} = r - x_i + \frac{(r - x_i)^2}{2}\frac{f''(c_i)}{f'(x_i)},$$

assuming that $f'(x_i) \neq 0$. With some rearranging, we can compare the next Newton iterate with the root:

$$x_i - \frac{f(x_i)}{f'(x_i)} - r = \frac{(r - x_i)^2}{2} \frac{f''(c_i)}{f'(x_i)}$$

$$x_{i+1} - r = e_i^2 \frac{f''(c_i)}{2f'(x_i)}$$

$$e_{i+1} = e_i^2 \left| \frac{f''(c_i)}{2f'(x_i)} \right|. \tag{1.24}$$

In this equation, we have defined the error at step i to be $e_i = |x_i - r|$. Since c_i lies between r and x_i, it converges to r just as x_i does, and

$$\lim_{i \to \infty} \frac{e_{i+1}}{e_i^2} = \left| \frac{f''(r)}{2f'(r)} \right|,$$

the definition of quadratic convergence. ❑

The error formula (1.24) we have developed can be viewed as

$$e_{i+1} \approx M e_i^2, \tag{1.25}$$

where $M = |f''(r)/2f'(r)|$, under the assumption that $f'(r) \neq 0$. The approximation gets better as Newton's Method converges, since the guesses x_i move toward r, and because c_i is caught between x_i and r. This error formula should be compared with $e_{i+1} \approx S e_i$ for the linearly convergent methods, where $S = |g'(r)|$ for FPI and $S = 1/2$ for bisection.

Although the value of S is critical for linearly convergent methods, the value of M is less critical, because the formula involves the square of the previous error. Once the error gets significantly below 1, squaring will cause a further decrease; and as long as M is not too large, the error according to (1.25) will decrease as well.

Returning to Example 1.11, we can analyze the output table to demonstrate this error rate. The right column shows the ratio e_i/e_{i-1}^2, which, according to the Newton's Method error formula (1.25), should tend toward M as convergence to the root takes place. For $f(x) = x^3 + x - 1$, the derivatives are $f'(x) = 3x^2 + 1$ and $f''(x) = 6x$; evaluating at $x_c \approx 0.6823$ yields $M \approx 0.85$, which agrees with the error ratio in the right column of the table.

With our new understanding of Newton's Method, we can more fully explain the square root calculator of Example 1.6. Let a be a positive number, and consider finding roots of $f(x) = x^2 - a$ by Newton's Method. The iteration is

$$x_{i+1} = x_i - \frac{f(x_i)}{f'(x_i)} = x_i - \frac{x_i^2 - a}{2x_i}$$

$$= \frac{x_i^2 + a}{2x_i} = \frac{x_i + \frac{a}{x_i}}{2}, \tag{1.26}$$

which is the method from Example 1.6, for arbitrary a.

To study its convergence, evaluate the derivatives at the root $\sqrt{a}$:

$$f'(\sqrt{a}) = 2\sqrt{a}$$
$$f''(\sqrt{a}) = 2. \tag{1.27}$$

Newton is quadratically convergent, since $f'(\sqrt{a}) = 2\sqrt{a} \neq 0$, and the convergence rate is

$$e_{i+1} \approx M e_i^2, \tag{1.28}$$

where $M = 2/(2 \cdot 2\sqrt{a}) = 1/(2\sqrt{a})$.

1.4.2 Linear convergence of Newton's Method

Theorem 1.11 does not say that Newton's Method always converges quadratically. Recall that we needed to divide by $f'(r)$ for the quadratic convergence argument to make sense. This assumption turns out to be crucial. The following example shows an instance where Newton's Method does not converge quadratically:

▶ **EXAMPLE 1.12** Use Newton's Method to find a root of $f(x) = x^2$.

This may seem like a trivial problem, since we know there is one root: $r = 0$. But often it is instructive to apply a new method to an example we understand thoroughly. The Newton's Method formula is

$$
\begin{aligned}
x_{i+1} &= x_i - \frac{f(x_i)}{f'(x_i)} \\
&= x_i - \frac{x_i^2}{2x_i} \\
&= \frac{x_i}{2}.
\end{aligned}
$$

The surprising result is that Newton's Method simplifies to dividing by two. Since the root is $r = 0$, we have the following table of Newton iterates for initial guess $x_0 = 1$:

i	x_i	$e_i = |x_i - r|$	e_i/e_{i-1}
0	1.000	1.000	
1	0.500	0.500	0.500
2	0.250	0.250	0.500
3	0.125	0.125	0.500
⋮	⋮	⋮	⋮

Newton's Method does converge to the root $r = 0$. The error formula is $e_{i+1} = e_i/2$, so the convergence is linear with convergence proportionality constant $S = 1/2$. ◀

A similar result exists for x^m for any positive integer m, as the next example shows.

▶ **EXAMPLE 1.13** Use Newton's Method to find a root of $f(x) = x^m$.

The Newton formula is

$$
\begin{aligned}
x_{i+1} &= x_i - \frac{x_i^m}{mx_i^{m-1}} \\
&= \frac{m-1}{m}x_i.
\end{aligned}
$$

SPOTLIGHT ON

Convergence   Equations (1.28) and (1.29) express the two different rates of convergence to the root r possible in Newton's Method. At a simple root, $f'(r) \neq 0$, and the convergence is quadratic, or fast convergence, which obeys (1.28). At a multiple root, $f'(r) = 0$, and the convergence is linear and obeys (1.29). In the latter case of linear convergence, the slower rate puts Newton's Method in the same category as bisection and FPI.

Again, the only root is $r = 0$, so defining $e_i = |x_i - r| = x_i$ yields

$$e_{i+1} = Se_i,$$

where $S = (m - 1)/m$.   ◀

This is an example of the general behavior of Newton's Method at multiple roots. Note that Definition 1.9 of multiple root is equivalent to $f(r) = f'(r) = 0$, exactly the case where we could not make our derivation of the Newton's Method error formula work. There is a separate error formula for multiple roots. The pattern that we saw for multiple roots of monomials is representative of the general case, as summarized in Theorem 1.12.

THEOREM 1.12   Assume that the $(m + 1)$-times continuously differentiable function f on $[a, b]$ has a multiplicity m root at r. Then Newton's Method is locally convergent to r, and the error e_i at step i satisfies

$$\lim_{i \to \infty} \frac{e_{i+1}}{e_i} = S, \qquad (1.29)$$

where $S = (m - 1)/m$.   ■

▶ **EXAMPLE 1.14**   Find the multiplicity of the root $r = 0$ of $f(x) = \sin x + x^2 \cos x - x^2 - x$, and estimate the number of steps of Newton's Method required to converge within six correct places (use $x_0 = 1$).

It is easy to check that

$$f(x) = \sin x + x^2 \cos x - x^2 - x$$
$$f'(x) = \cos x + 2x \cos x - x^2 \sin x - 2x - 1$$
$$f''(x) = -\sin x + 2 \cos x - 4x \sin x - x^2 \cos x - 2$$

and that each evaluates to 0 at $r = 0$. The third derivative,

$$f'''(x) = -\cos x - 6 \sin x - 6x \cos x + x^2 \sin x, \qquad (1.30)$$

satisfies $f'''(0) = -1$, so the root $r = 0$ is a triple root, meaning that the multiplicity is $m = 3$. By Theorem 1.12, Newton should converge linearly with $e_{i+1} \approx 2e_i/3$.

Using starting guess $x_0 = 1$, we have $e_0 = 1$. Near convergence, the error will decrease by 2/3 on each step. Therefore, a rough approximation to the number of steps needed to get the error within six decimal places, or smaller than 0.5×10^{-6}, can be found by solving

$$\left(\frac{2}{3}\right)^n < 0.5 \times 10^{-6}$$

$$n > \frac{\log_{10}(0.5) - 6}{\log_{10}(2/3)} \approx 35.78. \qquad (1.31)$$

Approximately 36 steps will be needed. The first 20 steps are shown in the table.

i	x_i	$e_i = |x_i - r|$	e_i/e_{i-1}
1	1.00000000000000	1.00000000000000	
2	0.72159023986075	0.72159023986075	0.72159023986075
3	0.52137095182040	0.52137095182040	0.72253049309677
4	0.37530830859076	0.37530830859076	0.71984890466250
5	0.26836349052713	0.26836349052713	0.71504809348561
6	0.19026161369924	0.19026161369924	0.70896981301561
7	0.13361250532619	0.13361250532619	0.70225676492686
8	0.09292528672517	0.09292528672517	0.69548345417455
9	0.06403926677734	0.06403926677734	0.68914790617474
10	0.04377806216009	0.04377806216009	0.68361279513559
11	0.02972805552423	0.02972805552423	0.67906284694649
12	0.02008168373777	0.02008168373777	0.67551285759009
13	0.01351212730417	0.01351212730417	0.67285828621786
14	0.00906579564330	0.00906579564330	0.67093770205249
15	0.00607029292263	0.00607029292263	0.66958192766231
16	0.00405885109627	0.00405885109627	0.66864171927113
17	0.00271130367793	0.00271130367793	0.66799781850081
18	0.00180995966250	0.00180995966250	0.66756065624029
19	0.00120772384467	0.00120772384467	0.66726561353325
20	0.00080563307149	0.00080563307149	0.66706728946460

Note the convergence of the error ratio in the right column to the predicted 2/3. ◄

If the multiplicity of a root is known in advance, convergence of Newton's Method can be improved with a small modification.

THEOREM 1.13 If f is $(m + 1)$-times continuously differentiable on $[a, b]$, which contains a root r of multiplicity $m > 1$, then **Modified Newton's Method**

$$x_{i+1} = x_i - \frac{mf(x_i)}{f'(x_i)} \tag{1.32}$$

converges locally and quadratically to r. ∎

Returning to Example 1.14, we can apply Modified Newton's Method to achieve quadratic convergence. After five steps, convergence to the root $r = 0$ has taken place to about eight digits of accuracy:

i	x_i
0	1.00000000000000
1	0.16477071958224
2	0.01620733771144
3	0.00024654143774
4	0.00000006072272
5	−0.00000000633250

There are several points to note in the table. First, the quadratic convergence to the approximate root is observable, as the number of correct places in the approximation more or less doubles at each step, up to Step 4. Steps 6, 7, ... are identical to Step 5. The reason Newton's Method lacks convergence to machine precision is familiar to us from Section 1.3. We know that 0 is a multiple root. While the backward error is driven near ϵ_{mach} by Newton's Method, the forward error, equal to x_i, is several orders of magnitude larger.

Newton's Method, like FPI, may not converge to a root. The next example shows just one of its possible nonconvergent behaviors.

▶ **EXAMPLE 1.15**   Apply Newton's Method to $f(x) = 4x^4 - 6x^2 - 11/4$ with starting guess $x_0 = 1/2$.

This function has roots, since it is continuous, negative at $x = 0$, and goes to positive infinity for large positive and large negative x. However, no root will be found for the starting guess $x_0 = 1/2$, as shown in Figure 1.10. The Newton formula is

$$x_{i+1} = x_i - \frac{4x_i^4 - 6x_i^2 - \frac{11}{4}}{16x_i^3 - 12x_i}. \tag{1.33}$$

Substitution gives $x_1 = -1/2$, and then $x_2 = 1/2$ again. Newton's Method alternates on this example between the two nonroots $1/2$ and $-1/2$, and fails to find a root.

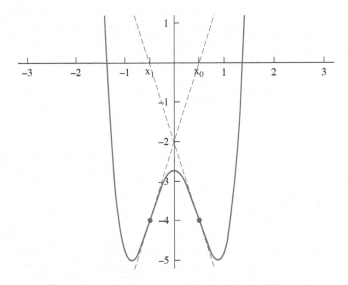

Figure 1.10 Failure of Newton's Method in Example 1.15. The iteration alternates between 1/2 and −1/2, and does not converge to a root.   ◀

Newton's Method can fail in other ways. Obviously, if $f'(x_i) = 0$ at any iteration step, the method cannot continue. There are other examples where the iteration diverges to infinity (see Exercise 6) or mimics a random number generator (see Computer Problem 13). Although not every initial guess leads to convergence to a root, Theorems 1.11 and 1.12 guarantee a neighborhood of initial guesses surrounding each root for which convergence to that root is assured.

▶ **ADDITIONAL EXAMPLES**

*1. Investigate the convergence of Newton's Method applied to the roots 1 and −1 of $f(x) = x^3 - 2x^2 + 2 - 1/x$. Use Theorems 1.11 and 1.12 to approximately express the error e_{i+1} in terms of e_i during convergence.

2. Adapt the fpi.m code from Section 1.2 to calculate all three roots of the equation $x^5 + 4x^2 = \sin x + 4x^4 + 1$ by Newton's Method.

🖳 **Solutions** for Additional Examples can be found at bit.ly/2yoJCkz (* example with video solution)

1.4 Exercises

Solutions
for Exercises
numbered in blue
can be found at
bit.ly/2PEwgHG

1. Apply two steps of Newton's Method with initial guess $x_0 = 0$. (a) $x^3 + x - 2 = 0$
   (b) $x^4 - x^2 + x - 1 = 0$ (c) $x^2 - x - 1 = 0$

2. Apply two steps of Newton's Method with initial guess $x_0 = 1$. (a) $x^3 + x^2 - 1 = 0$
   (b) $x^2 + 1/(x + 1) - 3x = 0$ (c) $5x - 10 = 0$

3. Use Theorem 1.11 or 1.12 to estimate the error e_{i+1} in terms of the previous error e_i as
   Newton's Method converges to the given roots. Is the convergence linear or quadratic?
   (a) $x^5 - 2x^4 + 2x^2 - x = 0$; $r = -1, r = 0, r = 1$ (b) $2x^4 - 5x^3 + 3x^2 + x - 1 = 0$;
   $r = -1/2, r = 1$

4. Estimate e_{i+1} as in Exercise 3. (a) $32x^3 - 32x^2 - 6x + 9 = 0$; $r = -1/2, r = 3/4$
   (b) $x^3 - x^2 - 5x - 3 = 0$; $r = -1, r = 3$

5. Consider the equation $8x^4 - 12x^3 + 6x^2 - x = 0$. For each of the two solutions $x = 0$
   and $x = 1/2$, decide which will converge faster (say, to eight-place accuracy), the Bisection
   Method or Newton's Method, without running the calculation.

6. Sketch a function f and initial guess for which Newton's Method diverges.

7. Let $f(x) = x^4 - 7x^3 + 18x^2 - 20x + 8$. Does Newton's Method converge quadratically
   to the root $r = 2$? Find $\lim_{i \to \infty} e_{i+1}/e_i$, where e_i denotes the error at step i.

8. Prove that Newton's Method applied to $f(x) = ax + b$ converges in one step.

9. Show that applying Newton's Method to $f(x) = x^2 - A$ produces the iteration of
   Example 1.6.

10. Find the Fixed-Point Iteration produced by applying Newton's Method to
    $f(x) = x^3 - A$. See Exercise 1.2.10.

11. Use Newton's Method to produce a quadratically convergent method for calculating the
    nth root of a positive number A, where n is a positive integer. Prove quadratic
    convergence.

12. Suppose Newton's Method is applied to the function $f(x) = 1/x$. If the initial guess is
    $x_0 = 1$, find x_{50}.

13. (a) The function $f(x) = x^3 - 4x$ has a root at $r = 2$. If the error $e_i = x_i - r$ after four
    steps of Newton's Method is $e_4 = 10^{-6}$, estimate e_5. (b) Apply the same question as (a) to
    the root $r = 0$. (Caution: The usual formula is not useful.)

14. Let $g(x) = x - f(x)/f'(x)$ denote the Newton's Method iteration for the function f.
    Define $h(x) = g(g(x))$ to be the result of two successive steps of Newton's Method. Then
    $h'(x) = g'(g(x))g'(x)$ according to the Chain Rule of calculus. (a) Assume that c is a fixed
    point of h, but not of g, as in Example 1.15. Show that if c is an inflection point of $f(x)$,
    that is, $f''(x) = 0$, then the fixed point iteration h is locally convergent to c. It follows that
    for initial guesses near c, Newton's Method itself does not converge to a root of f, but
    tends toward the oscillating sequence $\{c, g(c)\}$ (b) Verify that the stable oscillation
    described in (a) actually occurs in Example 1.15. Computer Problem 14 elaborates on this
    example.

1.4 Computer Problems

Solutions for
Computer Problems
numbered in blue can
be found at
bit.ly/2CSApou

1. Each equation has one root. Use Newton's Method to approximate the root to eight
   correct decimal places. (a) $x^3 = 2x + 2$ (b) $e^x + x = 7$ (c) $e^x + \sin x = 4$

2. Each equation has one real root. Use Newton's Method to approximate the root to eight
   correct decimal places. (a) $x^5 + x = 1$ (b) $\sin x = 6x + 5$ (c) $\ln x + x^2 = 3$

3. Apply Newton's Method to find the only root to as much accuracy as possible,
   and find the root's multiplicity. Then use Modified Newton's Method to converge

to the root quadratically. Report the forward and backward errors of the best approximation obtained from each method. (a) $f(x) = 27x^3 + 54x^2 + 36x + 8$ (b) $f(x) = 36x^4 - 12x^3 + 37x^2 - 12x + 1$

4. Carry out the steps of Computer Problem 3 for (a) $f(x) = 2e^{x-1} - x^2 - 1$ (b) $f(x) = \ln(3 - x) + x - 2$.

5. A silo composed of a right circular cylinder of height 10 m surmounted by a hemispherical dome contains 400 m^3 of volume. Find the base radius of the silo to four correct decimal places.

6. A 10-cm-high cone contains 60 cm^3 of ice cream, including a hemispherical scoop on top. Find the radius of the scoop to four correct decimal places.

7. Consider the function $f(x) = e^{\sin^3 x} + x^6 - 2x^4 - x^3 - 1$ on the interval $[-2, 2]$. Plot the function on the interval, and find all three roots to six correct decimal places. Determine which roots converge quadratically, and find the multiplicity of the roots that converge linearly.

8. Carry out the steps of Computer Problem 7 for the function $f(x) = 94\cos^3 x - 24\cos x + 177\sin^2 x - 108\sin^4 x - 72\cos^3 x \sin^2 x - 65$ on the interval $[0, 3]$.

9. Apply Newton's Method to find both roots of the function $f(x) = 14xe^{x-2} - 12e^{x-2} - 7x^3 + 20x^2 - 26x + 12$ on the interval $[0, 3]$. For each root, print out the sequence of iterates, the errors e_i, and the relevant error ratio e_{i+1}/e_i^2 or e_{i+1}/e_i that converges to a nonzero limit. Match the limit with the expected value M from Theorem 1.11 or S from Theorem 1.12.

10. Set $f(x) = 54x^6 + 45x^5 - 102x^4 - 69x^3 + 35x^2 + 16x - 4$. Plot the function on the interval $[-2, 2]$, and use Newton's Method to find all five roots in the interval. Determine for which roots Newton converges linearly and for which the convergence is quadratic.

11. The ideal gas law for a gas at low temperature and pressure is $PV = nRT$, where P is pressure (in atm), V is volume (in L), T is temperature (in K), n is the number of moles of the gas, and $R = 0.0820578$ is the molar gas constant. The van der Waals equation

$$\left(P + \frac{n^2 a}{V^2}\right)(V - nb) = nRT$$

covers the nonideal case where these assumptions do not hold. Use the ideal gas law to compute an initial guess, followed by Newton's Method applied to the van der Waals equation to find the volume of one mole of oxygen at 320 K and a pressure of 15 atm. For oxygen, $a = 1.36$ L^2-atm/mole^2 and $b = 0.003183$ L/mole. State your initial guess and solution with three significant digits.

12. Use the data from Computer Problem 11 to find the volume of 1 mole of benzene vapor at 700 K under a pressure of 20 atm. For benzene, $a = 18.0$ L^2-atm/mole^2 and $b = 0.1154$ L/mole.

13. (a) Find the root of the function $f(x) = (1 - 3/(4x))^{1/3}$. (b) Apply Newton's Method using an initial guess near the root, and plot the first 50 iterates. This is another way Newton's Method can fail, by producing a chaotic trajectory. (c) Why are Theorems 1.11 and 1.12 not applicable?

14. (a) Fix real numbers $a, b > 0$ and plot the graph of $f(x) = a^2 x^4 - 6abx^2 - 11b^2$ for your chosen values. Do not use $a = 2, b = 1/2$, since that case already appears in Example 1.15. (b) Apply Newton's Method to find both the negative root and the positive root of $f(x)$. Then find intervals of positive initial guesses $[d_1, d_2]$, where $d_2 > d_1$, for which Newton's Method: (c) converges to the positive root, (d) converges to the negative root,

(e) is defined, but does not converge to any root. Your intervals should not contain any initial guess where $f'(x) = 0$, at which Newton's Method is not defined.

15. Solve Computer Problem 1.1.9 using Newton's Method.

16. Solve Computer Problem 1.1.10 using Newton's Method.

17. Consider the national population growth model $P(t) = (P(0) + \frac{m}{r})e^{rt} - \frac{m}{r}$, where m and r are the immigration rate and intrinsic growth rate, respectively, and time t is measured in years. (a) From 1990 to 2000, the U.S. population increased from 248.7 million to 281.4 million, and the immigration rate was $m = 0.977$ million per year. Use Newton's Method to find the intrinsic growth rate r during the decade, according to the model. (b) The immigration rate from 2000 to 2010 was $m = 1.030$ million per year, and the population in 2010 was 308.7 million. Find the intrinsic growth rate r during the 2000–2010 decade.

18. A crucial quantity in pipeline design is the pressure drop due to friction under turbulent flow. The pressure drop per unit length is described by the *Darcy number* f, a unitless quantity that satisfies the empirical *Colebrook equation*

$$\frac{1}{\sqrt{f}} = -2\log_{10}\left[\frac{\epsilon}{3.7D} + \frac{2.51}{R\sqrt{f}}\right]$$

where D is the inside pipe diameter, ϵ is the roughness height of the pipe interior, and R is the Reynolds number of the flow. (Flows in pipes are considered turbulent when $R > 4000$ or so.) (a) For $D = 0.3$ m, $\epsilon = 0.0002$ m, and $R = 10^5$, use Newton's Method to calculate the Darcy number f. (b) Fix D and ϵ as in (a), and calculate the Darcy number for several Reynolds numbers R between 10^4 and 10^8. Make a plot of the Darcy number versus Reynolds number, using a log axis for the latter.

1.5 ROOT-FINDING WITHOUT DERIVATIVES

Apart from multiple roots, Newton's Method converges at a faster rate than the bisection and FPI methods. It achieves this faster rate because it uses more information—in particular, information about the tangent line of the function, which comes from the function's derivative. In some circumstances, the derivative may not be available.

The Secant Method is a good substitute for Newton's Method in this case. It replaces the tangent line with an approximation called the secant line, and converges almost as quickly. Variants of the Secant Method replace the line with an approximating parabola, whose axis is either vertical (Muller's Method) or horizontal (Inverse Quadratic Interpolation). The section ends with the description of Brent's Method, a hybrid method which combines the best features of iterative and bracketing methods.

1.5.1 Secant Method and variants

The Secant Method is similar to the Newton's Method, but replaces the derivative by a difference quotient. Geometrically, the tangent line is replaced with a line through the two last known guesses. The intersection point of the "secant line" is the new guess.

An approximation for the derivative at the current guess x_i is the difference quotient

$$\frac{f(x_i) - f(x_{i-1})}{x_i - x_{i-1}}.$$

A straight replacement of this approximation for $f'(x_i)$ in Newton's Method yields the Secant Method.

Secant Method

$$x_0, x_1 = \text{initial guesses}$$

$$x_{i+1} = x_i - \frac{f(x_i)(x_i - x_{i-1})}{f(x_i) - f(x_{i-1})} \text{ for } i = 1, 2, 3, \ldots.$$

Unlike Fixed-Point Iteration and Newton's Method, two starting guesses are needed to begin the Secant Method.

It can be shown that under the assumption that the Secant Method converges to r and $f'(r) \neq 0$, the approximate error relationship

$$e_{i+1} \approx \left| \frac{f''(r)}{2f'(r)} \right| e_i e_{i-1}$$

holds and that this implies that

$$e_{i+1} \approx \left| \frac{f''(r)}{2f'(r)} \right|^{\alpha-1} e_i^{\alpha},$$

where $\alpha = (1 + \sqrt{5})/2 \approx 1.62$. (See Exercise 6.) The convergence of the Secant Method to simple roots is called **superlinear**, meaning that it lies between linearly and quadratically convergent methods.

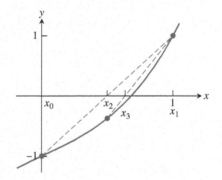

Figure 1.11 Two steps of the Secant Method. Illustration of Example 1.16. Starting with $x_0 = 0$ and $x_1 = 1$, the Secant Method iterates are plotted along with the secant lines.

▶ **EXAMPLE 1.16**   Apply the Secant Method with starting guesses $x_0 = 0, x_1 = 1$ to find the root of $f(x) = x^3 + x - 1$.

The formula gives

$$x_{i+1} = x_i - \frac{(x_i^3 + x_i - 1)(x_i - x_{i-1})}{x_i^3 + x_i - (x_{i-1}^3 + x_{i-1})}. \tag{1.34}$$

Starting with $x_0 = 0$ and $x_1 = 1$, we compute

$$x_2 = 1 - \frac{(1)(1 - 0)}{1 + 1 - 0} = \frac{1}{2}$$

$$x_3 = \frac{1}{2} - \frac{-\frac{3}{8}(1/2 - 1)}{-\frac{3}{8} - 1} = \frac{7}{11},$$

as shown in Figure 1.11. Further iterates form the following table:

i	x_i
0	0.00000000000000
1	1.00000000000000
2	0.50000000000000
3	0.63636363636364
4	0.69005235602094
5	0.68202041964819
6	0.68232578140989
7	0.68232780435903
8	0.68232780382802
9	0.68232780382802

◄

There are three generalizations of the Secant Method that are also important. The **Method of False Position**, or **Regula Falsi**, is similar to the Bisection Method, but where the midpoint is replaced by a Secant Method–like approximation. Given an interval $[a, b]$ that brackets a root (assume that $f(a)f(b) < 0$), define the next point

$$c = a - \frac{f(a)(a - b)}{f(a) - f(b)} = \frac{bf(a) - af(b)}{f(a) - f(b)}$$

as in the Secant Method, but unlike the Secant Method, the new point is guaranteed to lie in $[a, b]$, since the points $(a, f(a))$ and $(b, f(b))$ lie on separate sides of the x-axis. The new interval, either $[a, c]$ or $[c, b]$, is chosen according to whether $f(a)f(c) < 0$ or $f(c)f(b) < 0$, respectively, and still brackets a root.

Method of False Position

Given interval $[a, b]$ such that $f(a)f(b) < 0$
for $i = 1, 2, 3, \ldots$
    $c = \dfrac{bf(a) - af(b)}{f(a) - f(b)}$
    if $f(c) = 0$, **stop**, **end**
    if $f(a)f(c) < 0$
        $b = c$
    else
        $a = c$
    end
end

The Method of False Position at first appears to be an improvement on both the Bisection Method and the Secant Method, taking the best properties of each. However, while the Bisection Method guarantees cutting the uncertainty by 1/2 on each step, False Position makes no such promise, and for some examples can converge very slowly.

► **EXAMPLE 1.17**   Apply the Method of False Position on initial interval $[-1, 1]$ to find the root $r = 0$ of $f(x) = x^3 - 2x^2 + \frac{3}{2}x$.

Given $x_0 = -1, x_1 = 1$ as the initial bracketing interval, we compute the new point

$$x_2 = \frac{x_1 f(x_0) - x_0 f(x_1)}{f(x_0) - f(x_1)} = \frac{1(-9/2) - (-1)1/2}{-9/2 - 1/2} = \frac{4}{5}.$$

Since $f(-1)f(4/5) < 0$, the new bracketing interval is $[x_0, x_2] = [-1, 0.8]$. This completes the first step. Note that the uncertainty in the solution has decreased by far less than a factor of $1/2$. As Figure 1.12(b) shows, further steps continue to make slow progress toward the root at $x = 0$.

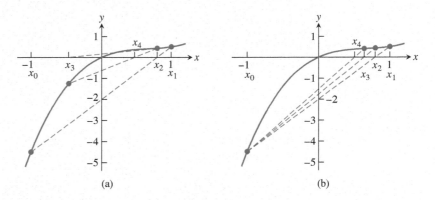

(a)   (b)

Figure 1.12 Slow convergence in Example 1.17. Both the (a) Secant Method and (b) Method of False Position converge slowly to the root $r = 0$.

Muller's Method is a generalization of the Secant Method in a different direction. Instead of intersecting the line through two previous points with the x-axis, we use three previous points x_0, x_1, x_2, draw the parabola $y = p(x)$ through them, and intersect the parabola with the x-axis. The parabola will generally intersect in 0 or 2 points. If there are two intersection points, the one nearest to the last point x_2 is chosen to be x_3. It is a simple matter of the quadratic formula to determine the two possibilities. If the parabola misses the x-axis, there are complex number solutions. This enables software that can handle complex arithmetic to locate complex roots. We will not pursue this idea further, although there are several sources in the literature that follow this direction.

Inverse Quadratic Interpolation (IQI) is a similar generalization of the Secant Method to parabolas. However, the parabola is of form $x = p(y)$ instead of $y = p(x)$, as in Muller's Method. One problem is solved immediately: This parabola will intersect the x-axis in a single point, so there is no ambiguity in finding x_{i+3} from the three previous guesses, x_i, x_{i+1}, and x_{i+2}.

The second-degree polynomial $x = P(y)$ that passes through the three points $(a, A), (b, B), (c, C)$ is

$$P(y) = a\frac{(y - B)(y - C)}{(A - B)(A - C)} + b\frac{(y - A)(y - C)}{(B - A)(B - C)} + c\frac{(y - A)(y - B)}{(C - A)(C - B)}. \qquad (1.35)$$

This is an example of Lagrange interpolation, one of the topics of Chapter 3. For now, it is enough to notice that $P(A) = a$, $P(B) = b$, and $P(C) = c$. Substituting $y = 0$ gives a formula for the intersection point of the parabola with the x-axis. After some rearrangement and substitution, we have

$$P(0) = c - \frac{r(r - q)(c - b) + (1 - r)s(c - a)}{(q - 1)(r - 1)(s - 1)}, \qquad (1.36)$$

where $q = f(a)/f(b), r = f(c)/f(b)$, and $s = f(c)/f(a)$.

For IQI, after setting $a = x_i, b = x_{i+1}, c = x_{i+2}$, and $A = f(x_i), B = f(x_{i+1})$, $C = f(x_{i+2})$, the next guess $x_{i+3} = P(0)$ is

$$x_{i+3} = x_{i+2} - \frac{r(r - q)(x_{i+2} - x_{i+1}) + (1 - r)s(x_{i+2} - x_i)}{(q - 1)(r - 1)(s - 1)}, \qquad (1.37)$$

where $q = f(x_i)/f(x_{i+1}), r = f(x_{i+2})/f(x_{i+1})$, and $s = f(x_{i+2})/f(x_i)$. Given three initial guesses, the IQI method proceeds by iterating (1.37), using the new guess x_{i+3} to replace the oldest guess x_i. An alternative implementation of IQI uses the new guess to replace one of the previous three guesses with largest backward error.

Figure 1.13 compares the geometry of Muller's Method with Inverse Quadratic Interpolation. Both methods converge faster than the Secant Method due to the higher-order interpolation. We will study interpolation in more detail in Chapter 3. The concepts of the Secant Method and its generalizations, along with the Bisection Method, are key ingredients of Brent's Method, the subject of the next section.

1.5.2 Brent's Method

Brent's Method [Brent, 1973] is a hybrid method—it uses parts of solving techniques introduced earlier to develop a new approach that retains the most useful properties of each. It is most desirable to combine the property of guaranteed convergence, from the Bisection Method, with the property of fast convergence from the more sophisticated methods. It was originally proposed by Dekker and Van Wijngaarden in the 1960s.

The method is applied to a continuous function f and an interval bounded by a and b, where $f(a)f(b) < 0$. Brent's Method keeps track of a current point x_i that is best in the sense of backward error, and a bracket $[a_i, b_i]$ of the root. Roughly speaking, the Inverse Quadratic Interpolation method is attempted, and the result is used to replace one of x_i, a_i, b_i if (1) the backward error improves and (2) the bracketing interval is cut at least in half. If not, the Secant Method is attempted with the same goal. If it fails as well, a Bisection Method step is taken, guaranteeing that the uncertainty is cut at least in half.

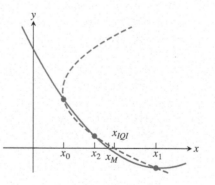

Figure 1.13 Comparison of Muller's Method step with Inverse Quadratic Iteration step. The former is determined by an interpolating parabola $y = p(x)$; the latter, by an interpolating parabola $x = p(y)$.

MATLAB's command `fzero` implements a version of Brent's Method, along with a preprocessing step, to discover a good initial bracketing interval if one is not provided by the user. The stopping criterion is of a mixed forward/backward error type. The algorithm terminates when the change from x_i to the new point x_{i+1} is less than $2\epsilon_{mach} \max(1, x_i)$, or when the backward error $|f(x_i)|$ achieves machine zero.

The preprocessing step is not triggered if the user provides an initial bracketing interval. The following use of the command enters the function $f(x) = x^3 + x - 1$ and the initial bracketing interval [0, 1] and asks MATLAB to display partial results on each iteration:

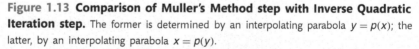

```
>> f=@(x) x^3+x-1;
>> fzero(f,[0 1],optimset('Display','iter'))
```

```
    Func-count         x                    f(x)           Procedure
        1               0                    -1             initial
        2               1                     1             initial
        3              0.5                  -0.375          bisection
        4           0.636364              -0.105935         interpolation
        5           0.684910               0.00620153       interpolation
        6           0.682225              -0.000246683      interpolation
        7           0.682328              -5.43508e-007     interpolation
        8           0.682328               1.50102e-013     interpolation
        9           0.682328                 0              interpolation
Zero found in the interval: [0, 1].

ans=

    0.68232780382802
```

Alternatively, the command

```
>> fzero(f,1)
```

looks for a root of $f(x)$ near $x = 1$ by first locating a bracketing interval and then applying Brent's Method.

► **ADDITIONAL EXAMPLES**

1. Apply two steps of the Secant Method on the interval with initial guesses $x_0 = 1$ and $x_1 = 2$ to find the approximate root of $f(x) = 2x^3 - x - 7$.

2. Write a MATLAB program that uses the Secant Method to find both roots of $f(x) = 8x^6 - 12x^5 + 6x^4 - 17x^3 + 24x^2 - 12x + 2$. Is the Secant Method superlinearly convergent to both roots?

Solutions for Additional Examples can be found at bit.ly/2Cr3rKT

1.5 Exercises

Solutions for Exercises numbered in blue can be found at bit.ly/2CQaDS4

1. Apply two steps of the Secant Method to the following equations with initial guesses $x_0 = 1$ and $x_1 = 2$. (a) $x^3 = 2x + 2$ (b) $e^x + x = 7$ (c) $e^x + \sin x = 4$

2. Apply two steps of the Method of False Position with initial bracket $[1, 2]$ to the equations of Exercise 1.

3. Apply two steps of Inverse Quadratic Interpolation to the equations of Exercise 1. Use initial guesses $x_0 = 1, x_1 = 2$, and $x_2 = 0$, and update by retaining the three most recent iterates.

4. A commercial fisher wants to set the net at a water depth where the temperature is 10 degrees C. By dropping a line with a thermometer attached, she finds that the temperature is 8 degrees at a depth of 9 meters, and 15 degrees at a depth of 5 meters. Use the Secant Method to determine a best estimate for the depth at which the temperature is 10.

5. Derive equation (1.36) by substituting $y = 0$ into (1.35).

6. If the Secant Method converges to r, $f'(r) \neq 0$, and $f''(r) \neq 0$, then the approximate error relationship $e_{i+1} \approx |f''(r)/(2f'(r))|e_i e_{i-1}$ can be shown to hold. Prove that if in addition $\lim_{i \to \infty} e_{i+1}/e_i^\alpha$ exists and is nonzero for some $\alpha > 0$, then $\alpha = (1 + \sqrt{5})/2$ and $e_{i+1} \approx |(f''(r)/2f'(r))|^{\alpha-1}e_i^\alpha$.

7. Consider the following four methods for calculating $2^{1/4}$, the fourth root of 2. (a) Rank them for speed of convergence, from fastest to slowest. Be sure to give reasons for your ranking.

(A) Bisection Method applied to $f(x) = x^4 - 2$

(B) Secant Method applied to $f(x) = x^4 - 2$

(C) Fixed-Point Iteration applied to $g(x) = \dfrac{x}{2} + \dfrac{1}{x^3}$

(D) Fixed-Point Iteration applied to $g(x) = \dfrac{2x}{3} + \dfrac{2}{3x^3}$

(b) Are there any methods that will converge faster than all above suggestions?

1.5 Computer Problems

Solutions for Computer Problems numbered in blue can be found at bit.ly/2pYPHzJ

1. Use the Secant Method to find the (single) solution of each equation in Exercise 1.

2. Use the Method of False Position to find the solution of each equation in Exercise 1.

3. Use Inverse Quadratic Interpolation to find the solution of each equation in Exercise 1.

4. Set $f(x) = 54x^6 + 45x^5 - 102x^4 - 69x^3 + 35x^2 + 16x - 4$. Plot the function on the interval $[-2, 2]$, and use the Secant Method to find all five roots in the interval. To which of the roots is the convergence linear, and to which is it superlinear?

5. In Exercise 1.1.6, you were asked what the outcome of the Bisection Method would be for $f(x) = 1/x$ on the interval $[-2, 1]$. Now compare that result with applying fzero to the problem.

6. What happens if fzero is asked to find the root of $f(x) = x^2$ near 1 (do not use a bracketing interval)? Explain the result. (b) Apply the same question to $f(x) = 1 + \cos x$ near -1.

Reality Check **1** *Kinematics of the Stewart Platform*

A Stewart platform consists of six variable length struts, or prismatic joints, supporting a payload. Prismatic joints operate by changing the length of the strut, usually pneumatically or hydraulically. As a six-degree-of-freedom robot, the Stewart platform can be placed at any point and inclination in three-dimensional space that is within its reach.

To simplify matters, the project concerns a two-dimensional version of the Stewart platform. It will model a manipulator composed of a triangular platform in a fixed plane controlled by three struts, as shown in Figure 1.14. The inner triangle represents

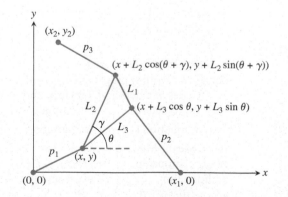

Figure 1.14 Schematic of planar Stewart platform. The forward kinematics problem is to use the lengths p_1, p_2, p_3 to determine the unknowns x, y, θ.

the planar Stewart platform whose dimensions are defined by the three lengths L_1, L_2, and L_3. Let γ denote the angle across from side L_1. The position of the platform is controlled by the three numbers p_1, p_2, and p_3, the variable lengths of the three struts.

Finding the position of the platform, given the three strut lengths, is called the forward, or direct, kinematics problem for this manipulator. Namely, the problem is to compute (x, y) and θ for each given p_1, p_2, p_3. Since there are three degrees of freedom, it is natural to expect three numbers to specify the position. For motion planning, it is important to solve this problem as fast as possible, often in real time. Unfortunately, no closed-form solution of the planar Stewart platform forward kinematics problem is known.

The best current methods involve reducing the geometry of Figure 1.14 to a single equation and solving it by using one of the solvers explained in this chapter. Your job is to complete the derivation of this equation and write code to carry out its solution.

Simple trigonometry applied to Figure 1.14 implies the following three equations:

$$
\begin{aligned}
p_1^2 &= x^2 + y^2 \\
p_2^2 &= (x + A_2)^2 + (y + B_2)^2 \\
p_3^2 &= (x + A_3)^2 + (y + B_3)^2.
\end{aligned}
\tag{1.38}
$$

In these equations,

$$
\begin{aligned}
A_2 &= L_3 \cos\theta - x_1 \\
B_2 &= L_3 \sin\theta \\
A_3 &= L_2 \cos(\theta + \gamma) - x_2 = L_2[\cos\theta \cos\gamma - \sin\theta \sin\gamma] - x_2 \\
B_3 &= L_2 \sin(\theta + \gamma) - y_2 = L_2[\cos\theta \sin\gamma + \sin\theta \cos\gamma] - y_2.
\end{aligned}
$$

Note that (1.38) solves the inverse kinematics problem of the planar Stewart platform, which is to find p_1, p_2, p_3, given x, y, θ. Your goal is to solve the forward problem, namely, to find x, y, θ, given p_1, p_2, p_3.

Multiplying out the last two equations of (1.38) and using the first yields

$$
\begin{aligned}
p_2^2 &= x^2 + y^2 + 2A_2 x + 2B_2 y + A_2^2 + B_2^2 = p_1^2 + 2A_2 x + 2B_2 y + A_2^2 + B_2^2 \\
p_3^2 &= x^2 + y^2 + 2A_3 x + 2B_3 y + A_3^2 + B_3^2 = p_1^2 + 2A_3 x + 2B_3 y + A_3^2 + B_3^2,
\end{aligned}
$$

which can be solved for x and y as

$$
\begin{aligned}
x &= \frac{N_1}{D} = \frac{B_3(p_2^2 - p_1^2 - A_2^2 - B_2^2) - B_2(p_3^2 - p_1^2 - A_3^2 - B_3^2)}{2(A_2 B_3 - B_2 A_3)} \\
y &= \frac{N_2}{D} = \frac{-A_3(p_2^2 - p_1^2 - A_2^2 - B_2^2) + A_2(p_3^2 - p_1^2 - A_3^2 - B_3^2)}{2(A_2 B_3 - B_2 A_3)},
\end{aligned}
\tag{1.39}
$$

as long as $D = 2(A_2 B_3 - B_2 A_3) \neq 0$.

Substituting these expressions for x and y into the first equation of (1.38), and multiplying through by D^2, yields one equation, namely,

$$
f = N_1^2 + N_2^2 - p_1^2 D^2 = 0
\tag{1.40}
$$

in the single unknown θ. (Recall that $p_1, p_2, p_3, L_1, L_2, L_3, \gamma, x_1, x_2, y_2$ are known.) If the roots of $f(\theta)$ can be found, the corresponding x- and y- values follow immediately from (1.39).

Note that $f(\theta)$ is a polynomial in $\sin\theta$ and $\cos\theta$, so, given any root θ, there are other roots $\theta + 2\pi k$ that are equivalent for the platform. For that reason, we can restrict attention to θ in $[-\pi, \pi]$. It can be shown that $f(\theta)$ has at most six roots in that interval.

Suggested activities:

1. Write a MATLAB function file for $f(\theta)$. The parameters $L_1, L_2, L_3, \gamma, x_1, x_2, y_2$ are fixed constants, and the strut lengths p_1, p_2, p_3 will be known for a given pose. Check Appendix B.5 if you are new to MATLAB function files. Here, for free, are the first and last lines:

```
function out=f(theta)
    ⋮
    ⋮
out=N1^2+N2^2-p1^2*D^2;
```

To test your code, set the parameters $L_1 = 2, L_2 = L_3 = \sqrt{2}, \gamma = \pi/2, p_1 = p_2 = p_3 = \sqrt{5}$ from Figure 1.15. Then, substituting $\theta = -\pi/4$ or $\theta = \pi/4$, corresponding to Figures 1.15(a, b), respectively, should make $f(\theta) = 0$.

2. Plot $f(\theta)$ on $[-\pi, \pi]$. You may use the @ symbol as described in Appendix B.5 to assign a function handle to your function file in the plotting command. You may also need to precede arithmetic operations with the "." character to vectorize the operations, as explained in Appendix B.2. As a check of your work, there should be roots at $\pm\pi/4$.

3. Reproduce Figure 1.15. The MATLAB commands

```
>> plot([u1 u2 u3 u1],[v1 v2 v3 v1],'r'); hold on
>> plot([0 x1 x2],[0 0 y2],'bo')
```

will plot a red triangle with vertices (u1,v1),(u2,v2),(u3,v3) and place small circles at the strut anchor points (0,0),(x1,0),(x2,y2). In addition, draw the struts.

4. Solve the forward kinematics problem for the planar Stewart platform specified by $x_1 = 5, (x_2, y_2) = (0, 6), L_1 = L_3 = 3, L_2 = 3\sqrt{2}, \gamma = \pi/4, p_1 = p_2 = 5, p_3 = 3$. Begin by plotting $f(\theta)$. Use an equation solver to find all four poses, and plot them. Check your answers by verifying that p_1, p_2, p_3 are the lengths of the struts in your plot.

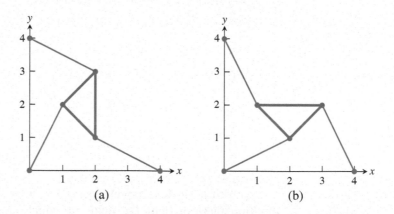

(a)　　　　　　(b)

Figure 1.15 Two poses of the planar Stewart platform with identical arm lengths. Each pose corresponds to a solution of (1.38) with strut lengths $p_1 = p_2 = p_3 = \sqrt{5}$. The shape of the triangle is defined by $L_1 = 2, L_2 = L_3 = \sqrt{2}, \gamma = \pi/2$.

5. Change strut length to $p_2 = 7$ and re-solve the problem. For these parameters, there are six poses.

6. Find a strut length p_2, with the rest of the parameters as in Step 4, for which there are only two poses.

7. Calculate the intervals in p_2, with the rest of the parameters as in Step 4, for which there are 0, 2, 4, and 6 poses, respectively.

8. Derive or look up the equations representing the forward kinematics of the three-dimensional, six-degrees-of-freedom Stewart platform. Write a MATLAB program and demonstrate its use to solve the forward kinematics. See Merlet [2000] for a good introduction to prismatic robot arms and platforms. ○

Software and Further Reading

There are many algorithms for locating solutions of nonlinear equations. The slow, but always convergent, algorithms like the Bisection Method contrast with routines with faster convergence, but without guarantees of convergence, including Newton's Method and variants. Equation solvers can also be divided into two groups, depending on whether or not derivative information is needed from the equation. The Bisection Method, the Secant Method, and Inverse Quadratic Interpolation are examples of methods that need only a black box providing a function value for a given input, while Newton's Method requires derivatives. Brent's Method is a hybrid that combines the best aspects of slow and fast algorithms and does not require derivative calculations. For this reason, it is heavily used as a general-purpose equation solver and is included in many comprehensive software packages.

MATLAB's `fzero` command implements Brent's Method and needs only an initial interval or one initial guess as input. The NAG routine c05adc and `netlib` FORTRAN program fzero.f both rely on this basic approach. The MATLAB `roots` command finds all roots of a polynomial with an entirely different approach, computing all eigenvalues of the companion matrix, constructed to have eigenvalues identical to all roots of the polynomial.

Other often-cited algorithms are based on Muller's Method and Laguerre's Method, which, under the right conditions, is cubically convergent. For more details, consult the classic texts on equation solving by Traub [1964], Ostrowski [1966], and Householder [1970].

2

Systems of Equations

Physical laws govern every engineered structure, from skyscrapers and bridges to diving boards and medical devices. Static and dynamic loads cause materials to deform, or bend. Mathematical models of bending are basic tools in the structural engineer's workbench. The degree to which a structure bends under a load depends on the stiffness of the material, as measured by its Young's modulus. The competition between stress and stiffness is modeled by a differential equation, which, after discretization, is reduced to a system of linear equations for solution.

To increase accuracy, a fine discretization is used, making the system of linear equations large and usually sparse. Gaussian elimination methods are efficient for moderately sized matrices, but special iterative algorithms are necessary for large, sparse systems.

Reality Check ⦿ Reality Check 2 on page 107 studies solution methods applicable to the Euler–Bernoulli model for pinned and cantilever beams.

I n the previous chapter, we studied methods for solving a single equation in a single variable. In this chapter, we consider the problem of solving several simultaneous equations in several variables. Most of our attention will be paid to the case where the number of equations and the number of unknown variables are the same.

Gaussian elimination is the workhorse for reasonably sized systems of linear equations. The chapter begins with the development of efficient and stable versions of this well-known technique. Later in the chapter our attention shifts to iterative methods, required for very large systems. Finally, we develop methods for systems of nonlinear equations.

2.1 GAUSSIAN ELIMINATION

Consider the system

$$x + y = 3$$
$$3x - 4y = 2. \qquad (2.1)$$

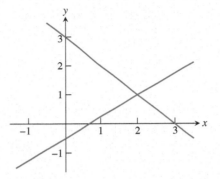

Figure 2.1 Geometric solution of a system of equations. Each equation of (2.1) corresponds to a line in the plane. The intersection point is the solution.

A system of two equations in two unknowns can be considered in terms either of algebra or of geometry. From the geometric point of view, each linear equation represents a line in the xy-plane, as shown in Figure 2.1. The point $x = 2, y = 1$ at which the lines intersect satisfies both equations and is the solution we are looking for.

The geometric view is very helpful for visualizing solutions of systems, but for computing the solution with a great deal of accuracy we return to algebra. The method known as Gaussian elimination is an efficient way to solve n equations in n unknowns. In the next few sections, we will explore implementations of Gaussian elimination that work best for typical problems.

2.1.1 Naive Gaussian elimination

We begin by describing the simplest form of Gaussian elimination. In fact, it is so simple that it is not guaranteed to proceed to completion, let alone find an accurate solution. The modifications that will be needed to improve the "naive" method will be introduced beginning in the next section.

Three useful operations can be applied to a linear system of equations that yield an equivalent system, meaning one that has the same solutions. These operations are as follows:

(1) Swap one equation for another.
(2) Add or subtract a multiple of one equation from another.
(3) Multiply an equation by a nonzero constant.

For equation (2.1), we can subtract 3 times the first equation from the second equation to eliminate the x variable from the second equation. Subtracting $3 \cdot [x + y = 3]$ from the second equation leaves us with the system

$$x + y = 3$$
$$-7y = -7. \tag{2.2}$$

Starting with the bottom equation, we can "backsolve" our way to a full solution, as in

$$-7y = -7 \longrightarrow y = 1$$

and

$$x + y = 3 \longrightarrow x + (1) = 3 \longrightarrow x = 2.$$

Therefore, the solution of (2.1) is $(x, y) = (2, 1)$.

The same elimination work can be done in the absence of variables by writing the system in so-called tableau form:

$$
\begin{bmatrix} 1 & 1 & | & 3 \\ 3 & -4 & | & 2 \end{bmatrix} \xrightarrow[\text{from row 2}]{\text{subtract } 3 \times \text{row 1}} \begin{bmatrix} 1 & 1 & | & 3 \\ 0 & -7 & | & -7 \end{bmatrix}. \tag{2.3}
$$

The advantage of the tableau form is that the variables are hidden during elimination. When the square array on the left of the tableau is "triangular," we can backsolve for the solution, starting at the bottom.

▶ **EXAMPLE 2.1**   Apply Gaussian elimination in tableau form for the system of three equations in three unknowns:

$$
\begin{aligned}
x + 2y - z &= 3 \\
2x + y - 2z &= 3 \\
-3x + y + z &= -6.
\end{aligned} \tag{2.4}
$$

This is written in tableau form as

$$
\begin{bmatrix} 1 & 2 & -1 & | & 3 \\ 2 & 1 & -2 & | & 3 \\ -3 & 1 & 1 & | & -6 \end{bmatrix}. \tag{2.5}
$$

Two steps are needed to eliminate column 1:

$$
\begin{bmatrix} 1 & 2 & -1 & | & 3 \\ 2 & 1 & -2 & | & 3 \\ -3 & 1 & 1 & | & -6 \end{bmatrix} \xrightarrow[\text{from row 2}]{\text{subtract } 2 \times \text{row 1}} \begin{bmatrix} 1 & 2 & -1 & | & 3 \\ 0 & -3 & 0 & | & -3 \\ -3 & 1 & 1 & | & -6 \end{bmatrix}
$$

$$
\xrightarrow[\text{from row 3}]{\text{subtract } -3 \times \text{row 1}} \begin{bmatrix} 1 & 2 & -1 & | & 3 \\ 0 & -3 & 0 & | & -3 \\ 0 & 7 & -2 & | & 3 \end{bmatrix}
$$

and one more step to eliminate column 2:

$$
\begin{bmatrix} 1 & 2 & -1 & | & 3 \\ 0 & -3 & 0 & | & -3 \\ 0 & 7 & -2 & | & 3 \end{bmatrix} \xrightarrow[\text{from row 3}]{\text{subtract } -\frac{7}{3} \times \text{row 2}} \begin{bmatrix} 1 & 2 & -1 & | & 3 \\ 0 & -3 & 0 & | & -3 \\ 0 & 0 & -2 & | & -4 \end{bmatrix}
$$

Returning to the equations

$$
\begin{aligned}
x + 2y - z &= 3 \\
-3y &= -3 \\
-2z &= -4,
\end{aligned} \tag{2.6}
$$

we can solve for the variables

$$
\begin{aligned}
x &= 3 - 2y + z \\
-3y &= -3 \\
-2z &= -4
\end{aligned} \tag{2.7}
$$

and solve for z, y, x in that order. The latter part is called **back substitution**, or **back-solving** because, after elimination, the equations are readily solved from the bottom up. The solution is $x = 3, y = 1, z = 2$. ◀

2.1.2 Operation counts

In this section, we do an approximate operation count for the two parts of Gaussian elimination: the elimination step and the back-substitution step. In order to do this, it will help to write out for the general case the operations that were carried out in the preceding two examples. To begin, recall two facts about sums of integers.

LEMMA 2.1 For any positive integer n, (a) $1 + 2 + 3 + 4 + \cdots + n = n(n+1)/2$ and (b) $1^2 + 2^2 + 3^2 + 4^2 + \cdots + n^2 = n(n+1)(2n+1)/6$. ■

The general form of the tableau for n equations in n unknowns is

$$
\left[
\begin{array}{cccc|c}
a_{11} & a_{12} & \cdots & a_{1n} & b_1 \\
a_{21} & a_{22} & \cdots & a_{2n} & b_2 \\
\vdots & \vdots & \cdots & \vdots & \vdots \\
a_{n1} & a_{n2} & \cdots & a_{nn} & b_n
\end{array}
\right].
$$

To carry out the elimination step, we need to put zeros in the lower triangle, using the allowed row operations.

We can write the elimination step as the loop

```
for j = 1 : n-1
  eliminate column j
end
```

where, by "eliminate column j," we mean "use row operations to put a zero in each location below the main diagonal, which are the locations $a_{j+1,j}, a_{j+2,j}, \ldots, a_{nj}$." For example, to carry out elimination on column 1, we need to put zeros in $a_{21}, \ldots, a_{n1}$. This can be written as the following loop within the former loop:

```
for j = 1 : n-1
  for i = j+1 : n
    eliminate entry a(i,j)
  end
end
```

It remains to fill in the inner step of the double loop, to apply a row operation that sets the a_{ij} entry to zero. For example, the first entry to be eliminated is the a_{21} entry. To accomplish this, we subtract a_{21}/a_{11} times row 1 from row 2, assuming that $a_{11} \neq 0$. That is, the first two rows change from

$$
\begin{array}{cccc|c}
a_{11} & a_{12} & \cdots & a_{1n} & b_1 \\
a_{21} & a_{22} & \cdots & a_{2n} & b_2
\end{array}
$$

to

$$
\begin{array}{cccc|c}
a_{11} & a_{12} & \cdots & a_{1n} & b_1 \\
0 & a_{22} - \dfrac{a_{21}}{a_{11}}a_{12} & \cdots & a_{2n} - \dfrac{a_{21}}{a_{11}}a_{1n} & b_2 - \dfrac{a_{21}}{a_{11}}b_1
\end{array}.
$$

Accounting for the operations, this requires one division (to find the multiplier a_{21}/a_{11}), plus n multiplications and n additions. The row operation used to eliminate entry a_{i1} of the first column, namely,

$$
\begin{array}{cccc|c}
a_{11} & a_{12} & \cdots & a_{1n} & b_1 \\
\vdots & \vdots & \cdots & \vdots & \vdots \\
0 & a_{i2} - \dfrac{a_{i1}}{a_{11}}a_{12} & \cdots & a_{in} - \dfrac{a_{i1}}{a_{11}}a_{1n} & b_i - \dfrac{a_{i1}}{a_{11}}b_1
\end{array}
$$

requires similar operations.

The procedure just described works as long as the number a_{11} is nonzero. This number and the other numbers a_{ii} that are eventually divisors in Gaussian elimination are called **pivots**. A zero pivot will cause the algorithm to halt, as we have explained it so far. This issue will be ignored for now and taken up more carefully in Section 2.4.

Returning to the operation count, note that eliminating each entry a_{i1} in the first column uses one division, n multiplications, and n addition/subtractions, or $2n + 1$ operations when counted together. Putting zeros into the first column requires a repeat of these $2n + 1$ operations a total of $n - 1$ times.

After the first column is eliminated, the pivot a_{22} is used to eliminate the second column in the same way and the remaining columns after that. For example, the row operation used to eliminate entry a_{ij} is

$$
\begin{array}{cccccccc}
0 & 0 & a_{jj} & a_{j,j+1} & & \cdots & a_{jn} & \Big| \quad b_j \\[4pt]
\vdots & \vdots & \vdots & \vdots & & \cdots & \vdots & \Big| \quad \vdots \\[4pt]
0 & 0 & 0 & a_{i,j+1} - \dfrac{a_{ij}}{a_{jj}} a_{j,j+1} & \cdots & a_{in} - \dfrac{a_{ij}}{a_{jj}} a_{jn} & \Big| \quad b_i - \dfrac{a_{ij}}{a_{jj}} b_j.
\end{array}
$$

In our notation, a_{22}, for example, refers to the revised number in that position after the elimination of column 1, which is not the original a_{22}. The row operation to eliminate a_{ij} requires one division, $n - j + 1$ multiplications, and $n - j + 1$ addition/subtractions.

Inserting this step into the same double loop results in

```
for j = 1 : n-1
  if abs(a(j,j))<eps; error('zero pivot encountered'); end
  for i = j+1 : n
    mult = a(i,j)/a(j,j);
    for k = j+1:n
      a(i,k) = a(i,k) - mult*a(j,k);
    end
    b(i) = b(i) - mult*b(j);
  end
end
```

Two comments on this code fragment are called for: First, asking the index k to move from j to n will put a zero in the a_{ij} location; however, moving from $j + 1$ to n is the most efficient coding. The latter will not place a zero in the a_{ij} entry, which was the entry we are trying to eliminate! Although this seems to be a mistake, note that we will never return to this entry in the remainder of the Gaussian elimination or back-substitution process, so actually putting a zero there represents a wasted step from the point of view of efficiency. Second, we ask the code to shut down, using MATLAB's error command, if a zero pivot is encountered. As mentioned, this possibility will be considered more seriously when row exchanges are discussed in Section 2.4.

We can make a total count of operations for the elimination step of Gaussian elimination. The elimination of each a_{ij} requires the following number of operations, including divisions, multiplication, and addition/subtractions:

$$
\begin{bmatrix}
0 & & & & & \\
2n + 1 & 0 & & & & \\
2n + 1 & 2(n - 1) + 1 & 0 & & & \\
2n + 1 & 2(n - 1) + 1 & 2(n - 2) + 1 & 0 & & \\
\vdots & \vdots & \vdots & & \ddots & \ddots \\
\vdots & \vdots & \vdots & & & \\
2n + 1 & 2(n - 1) + 1 & 2(n - 2) + 1 & \cdots & 2(3) + 1 & 0 \\
2n + 1 & 2(n - 1) + 1 & 2(n - 2) + 1 & \cdots & 2(3) + 1 & 2(2) + 1 & 0
\end{bmatrix}.
$$

It is convenient to add up the operations in reverse order of how they are applied. Starting on the right, we total up the operations as

$$\sum_{j=1}^{n-1}\sum_{i=1}^{j} 2(j+1) + 1 = \sum_{j=1}^{n-1} 2j(j+1) + j$$

$$= 2\sum_{j=1}^{n-1} j^2 + 3\sum_{j=1}^{n-1} j = 2\frac{(n-1)n(2n-1)}{6} + 3\frac{(n-1)n}{2}$$

$$= (n-1)n\left[\frac{2n-1}{3} + \frac{3}{2}\right] = \frac{n(n-1)(4n+7)}{6}$$

$$= \frac{2}{3}n^3 + \frac{1}{2}n^2 - \frac{7}{6}n,$$

where Lemma 2.1 has been applied.

Operation count for the elimination step of Gaussian elimination

The elimination step for a system of n equations in n variables can be completed in $\frac{2}{3}n^3 + \frac{1}{2}n^2 - \frac{7}{6}n$ operations.

Normally, the exact operation count is less important than order-of-magnitude estimates, since the details of implementation on various computer processors differ. The main point is that the number of operations is approximately proportional to the execution time of the algorithm. We will commonly make the approximation of $\frac{2}{3}n^3$ operations for elimination, which is a reasonably accurate approximation when n is large.

After the elimination is completed, the tableau is upper triangular:

$$\begin{bmatrix} a_{11} & a_{12} & \dots & a_{1n} & | & b_1 \\ 0 & a_{22} & \dots & a_{2n} & | & b_2 \\ \vdots & \vdots & \ddots & \vdots & | & \vdots \\ 0 & 0 & \dots & a_{nn} & | & b_n \end{bmatrix}.$$

In equation form,

$$a_{11}x_1 + a_{12}x_2 + \dots + a_{1n}x_n = b_1$$
$$a_{22}x_2 + \dots + a_{2n}x_n = b_2$$
$$\vdots$$
$$a_{nn}x_n = b_n, \tag{2.8}$$

where, again, the a_{ij} refer to the revised, not original, entries. To complete the computation of the solution x, we must carry out the back-substitution step, which is simply a rewriting of (2.8):

$$x_1 = \frac{b_1 - a_{12}x_2 - \dots - a_{1n}x_n}{a_{11}}$$

$$x_2 = \frac{b_2 - a_{23}x_3 - \dots - a_{2n}x_n}{a_{22}}$$

$$\vdots$$

$$x_n = \frac{b_n}{a_{nn}}. \tag{2.9}$$

Complexity   The operation count shows that direct solution of n equations in n unknowns by Gaussian elimination is an $O(n^3)$ process. This is a useful fact for estimating time required for solving large systems. For example, to estimate the time needed to solve a system of $n = 500$ equations on a particular computer, we could get a fair guess by solving a system of $n = 50$ equations and then scaling the elapsed time by $10^3 = 1000$.

Because of the triangular shape of the nonzero coefficients of the equations, we start at the bottom and work our way up to the top equation. In this way, the required x_i's are known when they are needed to compute the next one. Counting operations yields

$$1 + 3 + 5 + \cdots + (2n - 1) = \sum_{i=1}^{n} 2i - 1 = 2\sum_{i=1}^{n} i - \sum_{i=1}^{n} 1 = 2\frac{n(n + 1)}{2} - n = n^2.$$

In MATLAB syntax, the back-substitution step is

```
for i = n : -1 : 1
  for j = i+1 : n
    b(i) = b(i) - a(i,j)*x(j);
  end
  x(i) = b(i)/a(i,i);
end
```

Operation count for the back-substitution step of Gaussian elimination

The back-substitution step for a triangular system of n equations in n variables can be completed in n^2 operations.

The two operation counts, taken together, show that Gaussian elimination is made up of two unequal parts: the relatively expensive elimination step and the relatively cheap back-substitution step. If we ignore the lower order terms in the expressions for the number of multiplication/divisions, we find that elimination takes on the order of $2n^3/3$ operations and that back substitution takes on the order of n^2.

We will often use the shorthand terminology of "big-O" to mean "on the order of," saying that elimination is an $O(n^3)$ algorithm and that back substitution is $O(n^2)$.

This usage implies that the emphasis is on large n, where lower powers of n become negligible by comparison. For example, if $n = 100$, only about 1 percent or so of the calculation time of Gaussian elimination goes into the back-substitution step. Overall, Gaussian elimination takes $2n^3/3 + n^2 \approx 2n^3/3$ operations. In other words, for large n, the lower order terms in the complexity count will not have a large effect on the estimate for running time of the algorithm and can be ignored if only an estimated time is required.

▶ **EXAMPLE 2.2**   Estimate the time required to carry out back substitution on a system of 500 equations in 500 unknowns, on a computer where elimination takes 1 second.

Since we have just established that elimination is far more time consuming than back substitution, the answer will be a fraction of a second. Using the approximate number $2(500)^3/3$ for the number of multiply/divide operations for the elimination step, and $(500)^2$ for the back-substitution step, we estimate the time for back substitution to be

$$\frac{(500)^2}{2(500)^3/3} = \frac{3}{2(500)} = 0.003 \text{ sec.}$$

◀

The example shows two points: (1) Smaller powers of n in operation counts can often be safely neglected, and (2) the two parts of Gaussian elimination can be very unequal in running time—the total computation time is 1.003 seconds, almost all of which would be taken by the elimination step. The next example shows a third point. While the back-substitution time may sometimes be negligible, it may factor into an important calculation.

▶ **EXAMPLE 2.3** On a particular computer, back substitution of a 5000 × 5000 triangular matrix takes 0.1 seconds. Estimate the time needed to solve a general system of 3000 equations in 3000 unknowns by Gaussian elimination.

The computer can carry out $(5000)^2$ operations in 0.1 seconds, or $(5000)^2(10) = 2.5 \times 10^8$ operations/second. Solving a general (nontriangular) system requires about $2(3000)^3/3$ operations, which can be done in approximately

$$\frac{2(3000)^3/3}{(5000)^2(10)} \approx 72 \text{ sec.}$$

◀

▶ **ADDITIONAL EXAMPLES**

1. Put the system $x + 2y - z = 3, -3x + y + z = -6, 2x + z = 8$ into tableau form and solve by Gaussian elimination.

2. Assume that a computer can solve 200 upper-triangular matrix problems of 3000 variables in 3000 unknowns per second. Estimate how long it would take to solve one, not necessarily upper-triangular, problem of 5000 equations in 5000 unknowns.

📟 **Solutions** for Additional Examples can be found at `bit.ly/2P9UNHx`

2.1 Exercises

📟 **Solutions** for Exercises numbered in blue can be found at `bit.ly/2yuprSD`

1. Use Gaussian elimination to solve the systems:

(a) $\begin{array}{l} 2x - 3y = 2 \\ 5x - 6y = 8 \end{array}$ (b) $\begin{array}{l} x + 2y = -1 \\ 2x + 3y = 1 \end{array}$ (c) $\begin{array}{l} -x + y = 2 \\ 3x + 4y = 15 \end{array}$

2. Use Gaussian elimination to solve the systems:

(a) $\begin{array}{l} 2x - 2y - z = -2 \\ 4x + y - 2z = 1 \\ -2x + y - z = -3 \end{array}$ (b) $\begin{array}{l} x + 2y - z = 2 \\ 3y + z = 4 \\ 2x - y + z = 2 \end{array}$ (c) $\begin{array}{l} 2x + y - 4z = -7 \\ x - y + z = -2 \\ -x + 3y - 2z = 6 \end{array}$

3. Solve by back substitution:

(a) $\begin{array}{l} 3x - 4y + 5z = 2 \\ 3y - 4z = -1 \\ 5z = 5 \end{array}$ (b) $\begin{array}{l} x - 2y + z = 2 \\ 4y - 3z = 1 \\ -3z = 3 \end{array}$

4. Solve the tableau form

(a) $\left[\begin{array}{ccc|c} 3 & -4 & -2 & 3 \\ 6 & -6 & 1 & 2 \\ -3 & 8 & 2 & -1 \end{array} \right]$ (b) $\left[\begin{array}{ccc|c} 2 & 1 & -1 & 2 \\ 6 & 2 & -2 & 8 \\ 4 & 6 & -3 & 5 \end{array} \right]$

5. Use the approximate operation count $2n^3/3$ for Gaussian elimination to estimate how much longer it takes to solve n equations in n unknowns if n is tripled.

6. Assume that your computer completes a 5000 equation back substitution in 0.005 seconds. Use the approximate operation counts n^2 for back substitution and $2n^3/3$ for elimination to estimate how long it will take to do a complete Gaussian elimination of this size. Round your answer to the nearest second.

7. Assume that a given computer requires 0.002 seconds to complete back substitution on a 4000×4000 upper triangular matrix equation. Estimate the time needed to solve a general system of 9000 equations in 9000 unknowns. Round your answer to the nearest second.

8. If a system of 3000 equations in 3000 unknowns can be solved by Gaussian elimination in 5 seconds on a given computer, how many back substitutions of the same size can be done per second?

2.1 Computer Problems

📠 **Solutions** for Computer Problems numbered in blue can be found at bit.ly/2PGVNjs

1. Put together the code fragments in this section to create a MATLAB program for "naive" Gaussian elimination (meaning no row exchanges allowed). Use it to solve the systems of Exercise 2.

2. Let H denote the $n \times n$ Hilbert matrix, whose (i, j) entry is $1/(i + j - 1)$. Use the MATLAB program from Computer Problem 1 to solve $Hx = b$, where b is the vector of all ones, for (a) $n = 2$ (b) $n = 5$ (c) $n = 10$.

2.2 THE LU FACTORIZATION

Carrying the idea of tableau form one step farther brings us to the matrix form of a system of equations. Matrix form will save time in the long run by simplifying the algorithms and their analysis.

2.2.1 Matrix form of Gaussian elimination

The system (2.1) can be written as $Ax = b$ in matrix form, or

$$\begin{bmatrix} 1 & 1 \\ 3 & -4 \end{bmatrix} \begin{bmatrix} x_1 \\ x_2 \end{bmatrix} = \begin{bmatrix} 3 \\ 2 \end{bmatrix}. \tag{2.10}$$

We will usually denote the **coefficient matrix** by A and the **right-hand-side** vector as b. In the matrix form of the systems of equations, we interpret x as a column vector and Ax as matrix-vector multiplication. We want to find x such that the vector Ax is equal to the vector b. Of course, this is equivalent to having Ax and b agree in all components, which is exactly what is required by the original system (2.1).

The advantage of writing systems of equations in matrix form is that we can use matrix operations, like matrix multiplication, to keep track of the steps of Gaussian elimination. The LU factorization is a matrix representation of Gaussian elimination. It consists of writing the coefficient matrix A as a product of a lower triangular matrix L and an upper triangular matrix U. The LU factorization is the Gaussian elimination version of a long tradition in science and engineering—breaking down a complicated object into simpler parts.

DEFINITION 2.2 An $m \times n$ matrix L is **lower triangular** if its entries satisfy $l_{ij} = 0$ for $i < j$. An $m \times n$ matrix U is **upper triangular** if its entries satisfy $u_{ij} = 0$ for $i > j$. ❑

► **EXAMPLE 2.4**   Find the LU factorization for the matrix A in (2.10).

The elimination steps are the same as for the tableau form seen earlier:

$$\begin{bmatrix} 1 & 1 \\ 3 & -4 \end{bmatrix} \xrightarrow[\text{from row 2}]{\text{subtract } 3 \times \text{row 1}} \begin{bmatrix} 1 & 1 \\ 0 & -7 \end{bmatrix} = U. \qquad (2.11)$$

The difference is that now we store the multiplier 3 used in the elimination step. Note that we have defined U to be the upper triangular matrix showing the result of Gaussian elimination. Define L to be the 2×2 lower triangular matrix with 1's on the main diagonal and the multiplier 3 in the (2,1) location:

$$\begin{bmatrix} 1 & 0 \\ 3 & 1 \end{bmatrix}.$$

Then check that

$$LU = \begin{bmatrix} 1 & 0 \\ 3 & 1 \end{bmatrix}\begin{bmatrix} 1 & 1 \\ 0 & -7 \end{bmatrix} = \begin{bmatrix} 1 & 1 \\ 3 & -4 \end{bmatrix} = A. \qquad (2.12)$$

◄

We will discuss the reason this works soon, but first we demonstrate the steps with a 3×3 example.

► **EXAMPLE 2.5**   Find the LU factorization of

$$A = \begin{bmatrix} 1 & 2 & -1 \\ 2 & 1 & -2 \\ -3 & 1 & 1 \end{bmatrix}. \qquad (2.13)$$

This matrix is the matrix of coefficients of system (2.4). The elimination steps proceed as before:

$$\begin{bmatrix} 1 & 2 & -1 \\ 2 & 1 & -2 \\ -3 & 1 & 1 \end{bmatrix} \xrightarrow[\text{from row 2}]{\text{subtract } 2 \times \text{row 1}} \begin{bmatrix} 1 & 2 & -1 \\ 0 & -3 & 0 \\ -3 & 1 & 1 \end{bmatrix}$$

$$\xrightarrow[\text{from row 3}]{\text{subtract } -3 \times \text{row 1}} \begin{bmatrix} 1 & 2 & -1 \\ 0 & -3 & 0 \\ 0 & 7 & -2 \end{bmatrix}$$

$$\xrightarrow[\text{from row 3}]{\text{subtract } -\frac{7}{3} \times \text{row 2}} \begin{bmatrix} 1 & 2 & -1 \\ 0 & -3 & 0 \\ 0 & 0 & -2 \end{bmatrix} = U.$$

The lower triangular L matrix is formed, as in the previous example, by putting 1's on the main diagonal and the multipliers in the lower triangle—in the specific places they were used for elimination. That is,

$$L = \begin{bmatrix} 1 & 0 & 0 \\ 2 & 1 & 0 \\ -3 & -\frac{7}{3} & 1 \end{bmatrix}. \qquad (2.14)$$

Notice that, for example, 2 is the (2,1) entry of L, because it was the multiplier used to eliminate the (2,1) entry of A. Now check that

$$\begin{bmatrix} 1 & 0 & 0 \\ 2 & 1 & 0 \\ -3 & -\frac{7}{3} & 1 \end{bmatrix} \begin{bmatrix} 1 & 2 & -1 \\ 0 & -3 & 0 \\ 0 & 0 & -2 \end{bmatrix} = \begin{bmatrix} 1 & 2 & -1 \\ 2 & 1 & -2 \\ -3 & 1 & 1 \end{bmatrix} = A. \qquad (2.15)$$

◄

The reason that this procedure gives the LU factorization follows from three facts about lower triangular matrices.

FACT 1 Let $L_{ij}(-c)$ denote the lower triangular matrix whose only nonzero entries are 1's on the main diagonal and $-c$ in the (i, j) position. Then $A \longrightarrow L_{ij}(-c)A$ represents the row operation "subtracting c times row j from row i."

For example, multiplication by $L_{21}(-c)$ yields

$$A = \begin{bmatrix} a_{11} & a_{12} & a_{13} \\ a_{21} & a_{22} & a_{23} \\ a_{31} & a_{32} & a_{33} \end{bmatrix} \longrightarrow \begin{bmatrix} 1 & 0 & 0 \\ -c & 1 & 0 \\ 0 & 0 & 1 \end{bmatrix} \begin{bmatrix} a_{11} & a_{12} & a_{13} \\ a_{21} & a_{22} & a_{23} \\ a_{31} & a_{32} & a_{33} \end{bmatrix}$$

$$= \begin{bmatrix} a_{11} & a_{12} & a_{13} \\ a_{21} - ca_{11} & a_{22} - ca_{12} & a_{23} - ca_{13} \\ a_{31} & a_{32} & a_{33} \end{bmatrix}.$$

❐

FACT 2 $L_{ij}(-c)^{-1} = L_{ij}(c)$.

For example,

$$\begin{bmatrix} 1 & 0 & 0 \\ -c & 1 & 0 \\ 0 & 0 & 1 \end{bmatrix}^{-1} = \begin{bmatrix} 1 & 0 & 0 \\ c & 1 & 0 \\ 0 & 0 & 1 \end{bmatrix}.$$

Using Facts 1 and 2, we can understand the LU factorization of Example 2.4. Since the elimination step can be represented by

$$L_{21}(-3)A = \begin{bmatrix} 1 & 0 \\ -3 & 1 \end{bmatrix} \begin{bmatrix} 1 & 1 \\ 3 & -4 \end{bmatrix} = \begin{bmatrix} 1 & 1 \\ 0 & -7 \end{bmatrix},$$

we can multiply both sides on the left by $L_{21}(-3)^{-1}$ to get

$$A = \begin{bmatrix} 1 & 1 \\ 3 & -4 \end{bmatrix} = \begin{bmatrix} 1 & 0 \\ 3 & 1 \end{bmatrix} \begin{bmatrix} 1 & 1 \\ 0 & -7 \end{bmatrix},$$

which is the LU factorization of A.

❐

To handle $n \times n$ matrices for $n > 2$, we need one more fact.

FACT 3 The following matrix product equation holds.

$$\begin{bmatrix} 1 & & \\ c_1 & 1 & \\ & & 1 \end{bmatrix} \begin{bmatrix} 1 & & \\ & 1 & \\ c_2 & & 1 \end{bmatrix} \begin{bmatrix} 1 & & \\ & 1 & \\ & c_3 & 1 \end{bmatrix} = \begin{bmatrix} 1 & & \\ c_1 & 1 & \\ c_2 & c_3 & 1 \end{bmatrix}.$$

This fact allows us to collect the inverse L_{ij}'s into one matrix, which becomes the L of the LU factorization. For Example 2.5, this amounts to

$$\begin{bmatrix} 1 & & \\ & 1 & \\ & \frac{7}{3} & 1 \end{bmatrix} \begin{bmatrix} 1 & & \\ & 1 & \\ 3 & & 1 \end{bmatrix} \begin{bmatrix} 1 & & \\ -2 & 1 & \\ & & 1 \end{bmatrix} \begin{bmatrix} 1 & 2 & -1 \\ 2 & 1 & -2 \\ -3 & 1 & 1 \end{bmatrix} = \begin{bmatrix} 1 & 2 & -1 \\ 0 & -3 & 0 \\ 0 & 0 & -2 \end{bmatrix} = U$$

$$A = \begin{bmatrix} 1 & & \\ 2 & 1 & \\ & & 1 \end{bmatrix} \begin{bmatrix} 1 & & \\ & 1 & \\ -3 & & 1 \end{bmatrix} \begin{bmatrix} 1 & & \\ & 1 & \\ & -\frac{7}{3} & 1 \end{bmatrix} \begin{bmatrix} 1 & 2 & -1 \\ 0 & -3 & 0 \\ 0 & 0 & -2 \end{bmatrix}$$

$$= \begin{bmatrix} 1 & & \\ 2 & 1 & \\ -3 & -\frac{7}{3} & 1 \end{bmatrix} \begin{bmatrix} 1 & 2 & -1 \\ 0 & -3 & 0 \\ 0 & 0 & -2 \end{bmatrix} = LU. \quad (2.16)$$

□

2.2.2 Back substitution with the LU factorization

Now that we have expressed the elimination step of Gaussian elimination as a matrix product LU, how do we translate the back-substitution step? More importantly, how do we actually get the solution x?

Once L and U are known, the problem $Ax = b$ can be written as $LUx = b$. Define a new "auxiliary" vector $c = Ux$. Then back substitution is a two-step procedure:

(a) Solve $Lc = b$ for c.
(b) Solve $Ux = c$ for x.

Both steps are straightforward since L and U are triangular matrices. We demonstrate with the two examples used earlier.

► **EXAMPLE 2.6**   Solve system (2.10), using the LU factorization (2.12).

The system has LU factorization

$$\begin{bmatrix} 1 & 1 \\ 3 & -4 \end{bmatrix} - LU - \begin{bmatrix} 1 & 0 \\ 3 & 1 \end{bmatrix} \begin{bmatrix} 1 & 1 \\ 0 & -7 \end{bmatrix}$$

from (2.12), and the right-hand side is $b = [3, 2]$. Step (a) is

$$\begin{bmatrix} 1 & 0 \\ 3 & 1 \end{bmatrix} \begin{bmatrix} c_1 \\ c_2 \end{bmatrix} = \begin{bmatrix} 3 \\ 2 \end{bmatrix},$$

which corresponds to the system

$$c_1 + 0c_2 = 3$$
$$3c_1 + c_2 = 2.$$

Starting at the top, the solutions are $c_1 = 3, c_2 = -7$.
Step (b) is

$$\begin{bmatrix} 1 & 1 \\ 0 & -7 \end{bmatrix} \begin{bmatrix} x_1 \\ x_2 \end{bmatrix} = \begin{bmatrix} 3 \\ -7 \end{bmatrix},$$

which corresponds to the system

$$x_1 + x_2 = 3$$
$$-7x_2 = -7.$$

Starting at the bottom, the solutions are $x_2 = 1, x_1 = 2$. This agrees with the "classical" Gaussian elimination computation done earlier. ◄

► **EXAMPLE 2.7**   Solve system (2.4), using the LU factorization (2.15).

The system has LU factorization

$$\begin{bmatrix} 1 & 2 & -1 \\ 2 & 1 & -2 \\ -3 & 1 & 1 \end{bmatrix} = LU = \begin{bmatrix} 1 & 0 & 0 \\ 2 & 1 & 0 \\ -3 & -\frac{7}{3} & 1 \end{bmatrix} \begin{bmatrix} 1 & 2 & -1 \\ 0 & -3 & 0 \\ 0 & 0 & -2 \end{bmatrix}$$

from (2.15), and $b = (3, 3, -6)$. The $Lc = b$ step is

$$\begin{bmatrix} 1 & 0 & 0 \\ 2 & 1 & 0 \\ -3 & -\frac{7}{3} & 1 \end{bmatrix} \begin{bmatrix} c_1 \\ c_2 \\ c_3 \end{bmatrix} = \begin{bmatrix} 3 \\ 3 \\ -6 \end{bmatrix},$$

which corresponds to the system

$$c_1 = 3$$
$$2c_1 + c_2 = 3$$
$$-3c_1 - \frac{7}{3}c_2 + c_3 = -6.$$

Starting at the top, the solutions are $c_1 = 3, c_2 = -3, c_3 = -4$.
The $Ux = c$ step is

$$\begin{bmatrix} 1 & 2 & -1 \\ 0 & -3 & 0 \\ 0 & 0 & -2 \end{bmatrix} \begin{bmatrix} x_1 \\ x_2 \\ x_3 \end{bmatrix} = \begin{bmatrix} 3 \\ -3 \\ -4 \end{bmatrix},$$

which corresponds to the system

$$x_1 + 2x_2 - x_3 = 3$$
$$-3x_2 = -3$$
$$-2x_3 = -4,$$

and is solved from the bottom up to give $x = [3, 1, 2]$. ◄

2.2.3 Complexity of the LU factorization

Now that we have learned the "how" of the LU factorization, here are a few words about "why." Classical Gaussian elimination involves both A and b in the elimination step of the computation. This is by far the most expensive part of the process, as we have seen. Now, suppose that we need to solve a number of different problems with the same A and different b. That is, we are presented with the set of problems

$$Ax = b_1$$
$$Ax = b_2$$
$$\vdots$$
$$Ax = b_k$$

with various right-hand side vectors b_i. Classical Gaussian elimination will require approximately $2kn^3/3$ operations, where A is an $n \times n$ matrix, since we must start over at the beginning for each problem. With the LU approach, on the other hand, the right-hand-side b doesn't enter the calculations until the elimination (the $A = LU$ factorization) is finished. By insulating the calculations involving A from b, we can solve the previous set of equations with only one elimination, followed by two back substitutions ($Lc = b, Ux = c$) for each new b. The approximate number of operations with the LU approach is, therefore, $2n^3/3 + 2kn^2$. When n^2 is small compared with n^3 (i.e., when n is large), this is a significant difference.

Even when $k = 1$, there is no extra computational work done by the $A = LU$ approach, compared with classical Gaussian elimination. Although there appears to be an extra back substitution that was not part of classical Gaussian elimination, these "extra" calculations exactly replace the calculations that were saved during elimination because the right-hand-side b was absent.

SPOTLIGHT ON

Complexity   The main reason for the LU factorization approach to Gaussian elimination is the ubiquity of problems of form $Ax = b_1, Ax = b_2, \ldots$. Often, A is a so-called structure matrix, depending only on the design of a mechanical or dynamic system, and b corresponds to a "loading vector." In structural engineering, the loading vector gives the applied forces at various points on the structure. The solution x then corresponds to the stresses on the structure induced by that particular combination of loadings. Repeated solution of $Ax = b$ for various b's would be used to test potential structural designs. Reality Check 2 carries out this analysis for the loading of a beam.

If all b_i were available at the outset, we could solve all k problems simultaneously in the same number of operations. But in typical applications, we are asked to solve some of the $Ax = b_i$ problems before other b_i's are available. The LU approach allows efficient handling of all present and future problems that involve the same coefficient matrix A.

▶ **EXAMPLE 2.8**   Assume that it takes one second to factorize the 3000×3000 matrix A into $A = LU$. How many problems $Ax = b_1, \ldots, Ax = b_k$ can be solved in the next second?

The two back substitutions for each b_i require a total of $2n^2$ operations. Therefore, the approximate number of b_i that can be handled per second is

$$\frac{\frac{2n^3}{3}}{2n^2} = \frac{n}{3} = 1000. \qquad \blacktriangleleft$$

The LU factorization is a significant step forward in our quest to run Gaussian elimination efficiently. Unfortunately, not every matrix allows such a factorization.

▶ **EXAMPLE 2.9**   Prove that $A = \begin{bmatrix} 0 & 1 \\ 1 & 1 \end{bmatrix}$ does not have an LU factorization.

The factorization must have the form

$$\begin{bmatrix} 0 & 1 \\ 1 & 1 \end{bmatrix} = \begin{bmatrix} 1 & 0 \\ a & 1 \end{bmatrix} \begin{bmatrix} b & c \\ 0 & d \end{bmatrix} = \begin{bmatrix} b & c \\ ab & ac + d \end{bmatrix}.$$

Equating coefficients yields $b = 0$ and $ab = 1$, a contradiction. $\blacktriangleleft$

The fact that not all matrices have an LU factorization means that more work is required before we can declare the LU factorization a general Gaussian elimination algorithm. The related problem of swamping is described in the next section. In Section 2.4, the $PA = LU$ factorization is introduced, which will overcome both problems.

▶ **ADDITIONAL EXAMPLES**

*1 Solve

$$\begin{bmatrix} 2 & 4 & -2 \\ 1 & -2 & 1 \\ 4 & -4 & 8 \end{bmatrix} \begin{bmatrix} x_1 \\ x_2 \\ x_3 \end{bmatrix} = \begin{bmatrix} 6 \\ 3 \\ 0 \end{bmatrix}$$

using the $A = LU$ factorization.

2. Assume that a computer can carry out a LU-factorization of a 5000 × 5000 matrix in 1 second. How long will it take to solve 100 problems $Ax = b$, with the same 3000 × 3000 matrix A and 100 different b?

💾 Solutions for Additional Examples can be found at bit.ly/2NOHveI
(* example with video solution)

2.2 Exercises

💾 **Solutions** for Exercises numbered in blue can be found at bit.ly/2yIpPMj

1. Find the LU factorization of the given matrices. Check by matrix multiplication.

(a) $\begin{bmatrix} 1 & 2 \\ 3 & 4 \end{bmatrix}$ (b) $\begin{bmatrix} 1 & 3 \\ 2 & 2 \end{bmatrix}$ (c) $\begin{bmatrix} 3 & -4 \\ -5 & 2 \end{bmatrix}$

2. Find the LU factorization of the given matrices. Check by matrix multiplication.

(a) $\begin{bmatrix} 3 & 1 & 2 \\ 6 & 3 & 4 \\ 3 & 1 & 5 \end{bmatrix}$ (b) $\begin{bmatrix} 4 & 2 & 0 \\ 4 & 4 & 2 \\ 2 & 2 & 3 \end{bmatrix}$ (c) $\begin{bmatrix} 1 & -1 & 1 & 2 \\ 0 & 2 & 1 & 0 \\ 1 & 3 & 4 & 4 \\ 0 & 2 & 1 & -1 \end{bmatrix}$

3. Solve the system by finding the LU factorization and then carrying out the two-step back substitution.

(a) $\begin{bmatrix} 3 & 7 \\ 6 & 1 \end{bmatrix} \begin{bmatrix} x_1 \\ x_2 \end{bmatrix} = \begin{bmatrix} 1 \\ -11 \end{bmatrix}$ (b) $\begin{bmatrix} 2 & 3 \\ 4 & 7 \end{bmatrix} \begin{bmatrix} x_1 \\ x_2 \end{bmatrix} = \begin{bmatrix} 1 \\ 3 \end{bmatrix}$

4. Solve the system by finding the LU factorization and then carrying out the two-step back substitution.

(a) $\begin{bmatrix} 3 & 1 & 2 \\ 6 & 3 & 4 \\ 3 & 1 & 5 \end{bmatrix} \begin{bmatrix} x_1 \\ x_2 \\ x_3 \end{bmatrix} = \begin{bmatrix} 0 \\ 1 \\ 3 \end{bmatrix}$ (b) $\begin{bmatrix} 4 & 2 & 0 \\ 4 & 4 & 2 \\ 2 & 2 & 3 \end{bmatrix} \begin{bmatrix} x_1 \\ x_2 \\ x_3 \end{bmatrix} = \begin{bmatrix} 2 \\ 4 \\ 6 \end{bmatrix}$

5. Solve the equation $Ax = b$, where

$$A = \begin{bmatrix} 1 & 0 & 0 & 0 \\ 0 & 1 & 0 & 0 \\ 1 & 3 & 1 & 0 \\ 4 & 1 & 2 & 1 \end{bmatrix} \begin{bmatrix} 2 & 1 & 0 & 0 \\ 0 & 1 & 2 & 0 \\ 0 & 0 & -1 & 1 \\ 0 & 0 & 0 & 1 \end{bmatrix} \text{ and } b = \begin{bmatrix} 1 \\ 1 \\ 2 \\ 0 \end{bmatrix}.$$

6. Given the 1000 × 1000 matrix A, your computer can solve the 500 problems $Ax = b_1, \ldots, Ax = b_{500}$ in exactly one minute, using $A = LU$ factorization methods. How much of the minute was the computer working on the $A = LU$ factorization? Round your answer to the nearest second.

7. Assume that your computer can solve 1000 problems of type $Ux = c$, where U is an upper-triangular 500×500 matrix, per second. Estimate how long it will take to solve a full 5000×5000 matrix problem $Ax = b$. Answer in minutes and seconds.

8. Assume that your computer can solve a 2000×2000 linear system $Ax = b$ in 0.1 second. Estimate the time required to solve 100 systems of 8000 equations in 8000 unknowns with the same coefficient matrix, using the LU factorization method.

9. Let A be an $n \times n$ matrix. Assume that your computer can solve 100 problems $Ax = b_1, \ldots, Ax = b_{100}$ by the LU method in the same amount of time it takes to solve the first problem $Ax = b_0$. Estimate n.

2.2 Computer Problems

Solutions for Computer Problems numbered in blue can be found at bit.ly/2ywGxiL

1. Use the code fragments for Gaussian elimination in the previous section to write a MATLAB script to take a matrix A as input and output L and U. No row exchanges are allowed—the program should be designed to shut down if it encounters a zero pivot. Check your program by factoring the matrices in Exercise 2.

2. Add two-step back substitution to your script from Computer Problem 1, and use it to solve the systems in Exercise 4.

2.3 SOURCES OF ERROR

There are two major potential sources of error in Gaussian elimination as we have described it so far. The concept of ill-conditioning concerns the sensitivity of the solution to the input data. We will discuss condition number, using the concepts of backward and forward error from Chapter 1. Very little can be done to avoid errors in computing the solution of ill-conditioned matrix equations, so it is important to try to recognize and avoid ill-conditioned matrices when possible. The second source of error is swamping, which can be avoided in the large majority of problems by a simple fix called partial pivoting, the subject of Section 2.4.

The concept of vector and matrix norms are introduced next to measure the size of errors, which are now vectors. We will give the main emphasis to the so-called infinity norm.

2.3.1 Error magnification and condition number

In Chapter 1, we found that some equation-solving problems show a great difference between backward and forward error. The same is true for systems of linear equations. In order to quantify the errors, we begin with a definition of the infinity norm of a vector.

DEFINITION 2.3 The **infinity norm**, or **maximum norm**, of the vector $x = (x_1, \ldots, x_n)$ is $||x||_\infty = \max |x_i|, i = 1, \ldots, n$, that is, the maximum of the absolute values of the components of x. ❑

The backward and forward errors are defined in analogy with Definition 1.8. Backward error represents differences in the input, or problem data side, and forward error represents differences in the output, solution side of the algorithm.

DEFINITION 2.4 Let x_a be an approximate solution of the linear system $Ax = b$. The **residual** is the vector $r = b - Ax_a$. The **backward error** is the norm of the residual $||b - Ax_a||_\infty$, and the **forward error** is $||x - x_a||_\infty$. ❑

▶ **EXAMPLE 2.10** Find the backward and forward errors for the approximate solution $x_a = [1, 1]$ of the system

$$\begin{bmatrix} 1 & 1 \\ 3 & -4 \end{bmatrix} \begin{bmatrix} x_1 \\ x_2 \end{bmatrix} = \begin{bmatrix} 3 \\ 2 \end{bmatrix}.$$

The correct solution is $x = [2, 1]$. In the infinity norm, the backward error is

$$||b - Ax_a||_\infty = \left\| \begin{bmatrix} 3 \\ 2 \end{bmatrix} - \begin{bmatrix} 1 & 1 \\ 3 & -4 \end{bmatrix} \begin{bmatrix} 1 \\ 1 \end{bmatrix} \right\|_\infty$$
$$= \left\| \begin{bmatrix} 1 \\ 3 \end{bmatrix} \right\|_\infty = 3,$$

and the forward error is

$$||x - x_a||_\infty = \left\| \begin{bmatrix} 2 \\ 1 \end{bmatrix} - \begin{bmatrix} 1 \\ 1 \end{bmatrix} \right\|_\infty = \left\| \begin{bmatrix} 1 \\ 0 \end{bmatrix} \right\|_\infty = 1. \qquad ◄$$

In other cases, the backward and forward errors can be of different orders of magnitude.

▶ **EXAMPLE 2.11** Find the forward and backward errors for the approximate solution $[-1, 3.0001]$ of the system

$$x_1 + x_2 = 2$$
$$1.0001x_1 + x_2 = 2.0001. \qquad (2.17)$$

First, find the exact solution $[x_1, x_2]$. Gaussian elimination consists of the steps

$$\begin{bmatrix} 1 & 1 & | & 2 \\ 1.0001 & 1 & | & 2.0001 \end{bmatrix} \xrightarrow[\text{from row 2}]{\text{subtract } 1.0001 \times \text{ row 1}} \begin{bmatrix} 1 & 1 & | & 2 \\ 0 & -0.0001 & | & -0.0001 \end{bmatrix}.$$

Solving the resulting equations

$$x_1 + x_2 = 2$$
$$-0.0001x_2 = -0.0001$$

yields the solution $[x_1, x_2] = [1, 1]$.

The backward error is the infinity norm of the vector

$$b - Ax_a = \begin{bmatrix} 2 \\ 2.0001 \end{bmatrix} - \begin{bmatrix} 1 & 1 \\ 1.0001 & 1 \end{bmatrix} \begin{bmatrix} -1 \\ 3.0001 \end{bmatrix}$$
$$= \begin{bmatrix} 2 \\ 2.0001 \end{bmatrix} - \begin{bmatrix} 2.0001 \\ 2 \end{bmatrix} = \begin{bmatrix} -0.0001 \\ 0.0001 \end{bmatrix},$$

which is 0.0001. The forward error is the infinity norm of the difference

$$x - x_a = \begin{bmatrix} 1 \\ 1 \end{bmatrix} - \begin{bmatrix} -1 \\ 3.0001 \end{bmatrix} = \begin{bmatrix} 2 \\ -2.0001 \end{bmatrix},$$

which is 2.0001. ◄

Figure 2.2 helps to clarify how there can be a small backward error and large forward error at the same time. Even though the "approximate root" $(-1, 3.0001)$ is

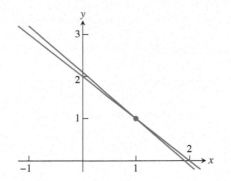

Figure 2.2 **The geometry behind Example 2.11.** System (2.17) is represented by the lines $x_2 = 2 - x_1$ and $x_2 = 2.0001 - 1.0001x_1$, which intersect at (1,1). The point (-1, 3.0001) nearly misses lying on both lines and being a solution. The differences between the lines is exaggerated in the figure—they are actually much closer.

relatively far from the exact root $(1, 1)$, it nearly lies on both lines. This is possible because the two lines are almost parallel. If the lines are far from parallel, the forward and backward errors will be closer in magnitude.

Denote the residual by $r = b - Ax_a$. The **relative backward error** of system $Ax = b$ is defined to be

$$\frac{||r||_\infty}{||b||_\infty},$$

and the **relative forward error** is

$$\frac{||x - x_a||_\infty}{||x||_\infty}.$$

SPOTLIGHT ON

> **Conditioning**   Condition number is a theme that runs throughout numerical analysis. In the discussions of the Wilkinson polynomial in Chapter 1, we found how to compute the error magnification factor for root-finding, given small perturbations of an equation $f(x) = 0$. For matrix equations $Ax = b$, there is a similar error magnification factor, and the maximum possible factor is given by $cond(A) = ||A|| \, ||A^{-1}||$.

The **error magnification factor** for $Ax = b$ is the ratio of the two, or

$$\text{error magnification factor} = \frac{\text{relative forward error}}{\text{relative backward error}} = \frac{\dfrac{||x - x_a||_\infty}{||x||_\infty}}{\dfrac{||r||_\infty}{||b||_\infty}}. \qquad (2.18)$$

For system (2.17), the relative backward error is

$$\frac{0.0001}{2.0001} \approx 0.00005 = 0.005\%,$$

and the relative forward error is

$$\frac{2.0001}{1} = 2.0001 \approx 200\%.$$

The error magnification factor is $2.0001/(0.0001/2.0001) = 40004.0001$.

In Chapter 1, we defined the concept of condition number to be the maximum error magnification over a prescribed range of input errors. The "prescribed range" depends on the context. Now we will be more precise about it for the current context of systems of linear equations. For a fixed matrix A, consider solving $Ax = b$ for various vectors b. In this context, b is the input and the solution x is the output. A small change in input is a small change in b, which has an error magnification factor. We therefore make the following definition:

DEFINITION 2.5 The **condition number** of a square matrix A, **cond**(A), is the maximum possible error magnification factor for solving $Ax = b$, over all right-hand sides b. ❐

Surprisingly, there is a compact formula for the condition number of a square matrix. Analogous to the norm of a vector, define the **matrix norm** of an $n \times n$ matrix A as

$$||A||_\infty = \text{maximum absolute row sum}, \tag{2.19}$$

that is, total the absolute values of each row, and assign the maximum of these n numbers to be the norm of A.

THEOREM 2.6 The condition number of the $n \times n$ matrix A is

$$\text{cond}(A) = ||A|| \cdot ||A^{-1}||. \qquad ■$$

Theorem 2.6, proved below, allows us to calculate the condition number of the coefficient matrix in Example 2.11. The norm of

$$A = \begin{bmatrix} 1 & 1 \\ 1.0001 & 1 \end{bmatrix}$$

is $||A|| = 2.0001$, according to (2.19). The inverse of A is

$$A^{-1} = \begin{bmatrix} -10000 & 10000 \\ 10001 & -10000 \end{bmatrix},$$

which has norm $||A^{-1}|| = 20001$. The condition number of A is

$$\text{cond}(A) = (2.0001)(20001) = 40004.0001.$$

This is exactly the error magnification we found in Example 2.11, which evidently achieves the worst case, defining the condition number. The error magnification factor for any other b in this system will be less than or equal to 40004.0001. Exercise 3 asks for the computation of some of the other error magnification factors.

The significance of the condition number is the same as in Chapter 1. Error magnification factors of the magnitude cond(A) are possible. In floating point arithmetic, the relative backward error cannot be expected to be less than ϵ_{mach}, since storing the entries of b already causes errors of that size. According to (2.18), relative forward errors of size $\epsilon_{\text{mach}} \cdot \text{cond}(A)$ are possible in solving $Ax = b$. In other words, if cond$(A) \approx 10^k$, we should prepare to lose k digits of accuracy in computing x.

In Example 2.11, cond$(A) \approx 4 \times 10^4$, so in double precision we should expect about $16 - 4 = 12$ correct digits in the solution x. We can test this by introducing MATLAB's best general-purpose linear equation solver: \.

In MATLAB, the backslash command x = A\b solves the linear system by using an advanced version of the *LU* factorization that we will explore in Section 2.4. For now, we will use it as an example of what we can expect from the best possible algorithm operating in floating point arithmetic. The following MATLAB commands deliver the computer solution x_a of Example 2.10:

```
>> A = [1 1;1.0001 1]; b=[2;2.0001];
>> xa = A\b
xa =
   1.00000000000222
   0.99999999999778
```

Compared with the correct solution $x = [1, 1]$, the computed solution has about 11 correct digits, close to the prediction from the condition number.

The Hilbert matrix H, with entries $H_{ij} = 1/(i + j - 1)$, is notorious for its large condition number.

▶ **EXAMPLE 2.12** Let H denote the $n \times n$ Hilbert matrix. Use MATLAB's \ to compute the solution of $Hx = b$, where $b = H \cdot [1, \ldots, 1]^T$, for $n = 6$ and 10.

The right-hand side b is chosen to make the correct solution the vector of n ones, for ease of checking the forward error. MATLAB finds the condition number (in the infinity norm) and computes the solution:

```
>> n=6;H=hilb(n);
>> cond(H,inf)
ans =
     2.907027900294064e+007
>> b=H*ones(n,1);
>> xa=H\b
xa =
   0.99999999999923
   1.00000000002184
   0.99999999985267
   1.00000000038240
   0.99999999957855
   1.00000000016588
```

The condition number of about 10^7 predicts $16 - 7 = 9$ correct digits in the worst case; there are about 9 correct in the computed solution. Now repeat with $n = 10$:

```
>> n=10;H=hilb(n);
>> cond(H,inf)
ans =
     3.535371683074594e+013
>> b=H*ones(n,1);
>> xa=H\b
xa =
   0.99999999875463
   1.00000010746631
   0.99999771299818
   1.00002077769598
   0.99990094548472
   1.00027218303745
   0.99955359665722
   1.00043125589482
   0.99977366058043
   1.00004976229297
```

Since the condition number is 10^{13}, only $16 - 13 = 3$ correct digits appear in the solution.

For n slightly larger than 10, the condition number of the Hilbert matrix is larger than 10^{16}, and no correct digits can be guaranteed in the computed x_a. ◀

Even excellent software may have no defense against an ill-conditioned problem. Increased precision helps; in extended precision, $\epsilon_{mach} = 2^{-64} \approx 5.42 \times 10^{-20}$, and we start with 20 digits instead of 16. However, the condition number of the Hilbert matrix grows fast enough with n to eventually disarm any reasonable finite precision.

Fortunately, the large condition numbers of the Hilbert matrix are unusual. Well-conditioned linear systems of n equations in n unknowns are routinely solved in double precision for $n = 10^4$ and larger. However, it is important to know that ill-conditioned problems exist, and that the condition number is useful for diagnosing that possibility. See Computer Problems 1–4 for more examples of error magnification and condition numbers.

The infinity vector norm was used in this section as a simple way to assign a length to a vector. It is an example of a **vector norm** $||x||$, which satisfies three properties:

(i) $||x|| \geq 0$ with equality if and only if $x = [0, \ldots, 0]$
(ii) for each scalar α and vector x, $||\alpha x|| = |\alpha| \cdot ||x||$
(iii) for vectors x, y, $||x + y|| \leq ||x|| + ||y||$.

In addition, $||A||_\infty$ is an example of a **matrix norm**, which satisfies three similar properties:

(i) $||A|| \geq 0$ with equality if and only if $A = 0$
(ii) for each scalar α and matrix A, $||\alpha A|| = |\alpha| \cdot ||A||$
(iii) for matrices A, B, $||A + B|| \leq ||A|| + ||B||$.

As a different example, the vector **1-norm** of the vector $x = [x_1, \ldots, x_n]$ is $||x||_1 = |x_1| + \cdots + |x_n|$. The matrix 1-norm of the $n \times n$ matrix A is $||A||_1 = $ maximum absolute column sum—that is, the maximum of the 1-norms of the column vectors. See Exercises 9 and 10 for verification that these definitions define norms.

The error magnification factor, condition number, and matrix norm just discussed can be defined for any vector and matrix norm. We will restrict our attention to matrix norms that are **operator norms**, meaning that they can be defined in terms of a particular vector norm as

$$||A|| = \max \frac{||Ax||}{||x||},$$

where the maximum is taken over all nonzero vectors x. Then, by definition, the matrix norm is consistent with the associated vector norm, in the sense that

$$||Ax|| \leq ||A|| \cdot ||x|| \tag{2.20}$$

for any matrix A and vector x. See Exercises 10 and 11 for verification that the norm $||A||_\infty$ defined by (2.20) is not only a matrix norm, but also the operator norm for the infinity vector norm.

This fact allows us to prove the aforementioned simple expression for cond(A). The proof works for the infinity norm and any other operator norm.

Proof of Theorem 2.6. We use the equalities $A(x - x_a) = r$ and $Ax = b$. By consistency property (2.20),

$$||x - x_a|| \leq ||A^{-1}|| \cdot ||r||$$

and

$$\frac{1}{||b||} \geq \frac{1}{||A|| \, ||x||}.$$

Putting the two inequalities together yields

$$\frac{||x - x_a||}{||x||} \leq \frac{||A||}{||b||} ||A^{-1}|| \cdot ||r||,$$

showing that $||A|| \, ||A^{-1}||$ is an upper bound for all error magnification factors. Second, we can show that the quantity is always attainable. Choose x such that $||A|| = ||Ax||/||x||$ and r such that $||A^{-1}|| = ||A^{-1}r||/||r||$, both possible by the definition of operator matrix norm. Set $x_a = x - A^{-1}r$ so that $x - x_a = A^{-1}r$. Then it remains to check the equality

$$\frac{||x - x_a||}{||x||} = \frac{||A^{-1}r||}{||x||} = \frac{||A^{-1}|| \, ||r|| \, ||A||}{||Ax||}$$

for this particular choice of x and r.

2.3.2 Swamping

A second significant source of error in classical Gaussian elimination is much easier to fix. We demonstrate swamping with the next example.

▶ **EXAMPLE 2.13**   Consider the system of equations

$$10^{-20}x_1 + x_2 = 1$$
$$x_1 + 2x_2 = 4.$$

We will solve the system three times: once with complete accuracy, second where we mimic a computer following IEEE double precision arithmetic, and once more where we exchange the order of the equations first.

1. **Exact solution.** In tableau form, Gaussian elimination proceeds as

$$\begin{bmatrix} 10^{-20} & 1 & | & 1 \\ 1 & 2 & | & 4 \end{bmatrix} \xrightarrow[\text{from row 2}]{\text{subtract } 10^{20} \times \text{ row 1}} \begin{bmatrix} 10^{-20} & 1 & | & 1 \\ 0 & 2 - 10^{20} & | & 4 - 10^{20} \end{bmatrix}.$$

The bottom equation is

$$(2 - 10^{20})x_2 = 4 - 10^{20} \longrightarrow x_2 = \frac{4 - 10^{20}}{2 - 10^{20}},$$

and the top equation yields

$$10^{-20}x_1 + \frac{4 - 10^{20}}{2 - 10^{20}} = 1$$

$$x_1 = 10^{20}\left(1 - \frac{4 - 10^{20}}{2 - 10^{20}}\right)$$

$$x_1 = \frac{-2 \times 10^{20}}{2 - 10^{20}}.$$

The exact solution is

$$[x_1, x_2] = \left[\frac{2 \times 10^{20}}{10^{20} - 2}, \frac{4 - 10^{20}}{2 - 10^{20}}\right] \approx [2, 1].$$

2. **IEEE double precision.** The computer version of Gaussian elimination proceeds slightly differently:

$$\begin{bmatrix} 10^{-20} & 1 & | & 1 \\ 1 & 2 & | & 4 \end{bmatrix} \xrightarrow[\text{from row 2}]{\text{subtract } 10^{20} \times \text{ row 1}} \begin{bmatrix} 10^{-20} & 1 & | & 1 \\ 0 & 2 - 10^{20} & | & 4 - 10^{20} \end{bmatrix}.$$

In IEEE double precision, $2 - 10^{20}$ is the same as -10^{20}, due to rounding. Similarly, $4 - 10^{20}$ is stored as -10^{20}. Now the bottom equation is

$$-10^{20} x_2 = -10^{20} \longrightarrow x_2 = 1.$$

The machine arithmetic version of the top equation becomes

$$10^{-20} x_1 + 1 = 1,$$

so $x_1 = 0$. The computed solution is exactly

$$[x_1, x_2] = [0, 1].$$

This solution has large relative error compared with the exact solution.

3. **IEEE double precision, after row exchange.** We repeat the computer version of Gaussian elimination, after changing the order of the two equations:

$$\begin{bmatrix} 1 & 2 & | & 4 \\ 10^{-20} & 1 & | & 1 \end{bmatrix} \xrightarrow[\text{from row 2}]{\text{subtract } 10^{-20} \times \text{ row 1}}$$

$$\longrightarrow \begin{bmatrix} 1 & 2 & | & 4 \\ 0 & 1 - 2 \times 10^{-20} & | & 1 - 4 \times 10^{-20} \end{bmatrix}.$$

In IEEE double precision, $1 - 2 \times 10^{-20}$ is stored as 1 and $1 - 4 \times 10^{-20}$ is stored as 1. The equations are now

$$\begin{aligned} x_1 + 2x_2 &= 4 \\ x_2 &= 1, \end{aligned}$$

which yield the computed solution $x_1 = 2$ and $x_2 = 1$. Of course, this is not the exact answer, but it is correct up to approximately 16 digits, which is the most we can ask from a computation that uses 52-bit floating point numbers.

The difference between the last two calculations is significant. Version 3 gave us an acceptable solution, while version 2 did not. An analysis of what went wrong with version 2 leads to considering the multiplier 10^{20} that was used for the elimination step. The effect of subtracting 10^{20} times the top equation from the bottom equation was to overpower, or "swamp," the bottom equation. While there were originally two independent equations, or sources of information, after the elimination step in version 2, there are essentially two copies of the top equation. Since the bottom equation has disappeared, for all practical purposes, we cannot expect the computed solution to satisfy the bottom equation; and it does not.

Version 3, on the other hand, completes elimination without swamping, because the multiplier is 10^{-20}. After elimination, the original two equations are still largely existent, slightly changed into triangular form. The result is an approximate solution that is much more accurate. ◄

The moral of Example 2.13 is that multipliers in Gaussian elimination should be kept as small as possible to avoid swamping. Fortunately, there is a simple modification to naive Gaussian elimination that forces the absolute value of multipliers to be no larger than 1. This new protocol, which involves judicious row exchanges in the tableau, is called partial pivoting, the topic of the next section.

▶ **ADDITIONAL EXAMPLES**

1. Find the determinant and the condition number (in the infinity norm) of the matrix $\begin{bmatrix} 811802 & 810901 \\ 810901 & 810001 \end{bmatrix}$.

2. The solution of the system $\begin{bmatrix} 2 & 4.01 \\ 3 & 6 \end{bmatrix} \begin{bmatrix} x_1 \\ x_2 \end{bmatrix} = \begin{bmatrix} 6.01 \\ 9 \end{bmatrix}$ is $[1, 1]$. (a) Find the relative forward and backward errors and error magnification (in the infinity norm) for the approximate solution $[21, -9]$. (b) Find the condition number of the coefficient matrix.

▭ **Solutions** for Additional Examples can be found at `bit.ly/2NP6q1I`

2.3 Exercises

▭ **Solutions** for Exercises numbered in blue can be found at `bit.ly/2QTocD5`

1. Find the norm $\|A\|_\infty$ of each of the following matrices:

   (a) $A = \begin{bmatrix} 1 & 2 \\ 3 & 4 \end{bmatrix}$ (b) $A = \begin{bmatrix} 1 & 5 & 1 \\ -1 & 2 & -3 \\ 1 & -7 & 0 \end{bmatrix}$.

2. Find the (infinity norm) condition number of

   (a) $A = \begin{bmatrix} 1 & 2 \\ 3 & 4 \end{bmatrix}$ (b) $A = \begin{bmatrix} 1 & 2.01 \\ 3 & 6 \end{bmatrix}$ (c) $A = \begin{bmatrix} 6 & 3 \\ 4 & 2 \end{bmatrix}$.

3. Find the forward and backward errors, and the error magnification factor (in the infinity norm) for the following approximate solutions x_a of the system in Example 2.11: (a) $[-1, 3]$ (b) $[0, 2]$ (c) $[2, 2]$ (d) $[-2, 4]$ (e) $[-2, 4.0001]$.

4. Find the forward and backward errors and error magnification factor for the following approximate solutions of the system $x_1 + 2x_2 = 1, 2x_1 + 4.01x_2 = 2$: (a) $[-1, 1]$ (b) $[3, -1]$ (c) $[2, -1/2]$.

5. Find the relative forward and backward errors and error magnification factor for the following approximate solutions of the system $x_1 - 2x_2 = 3, 3x_1 - 4x_2 = 7$: (a) $[-2, -4]$ (b) $[-2, -3]$ (c) $[0, -2]$ (d) $[-1, -1]$ (e) What is the condition number of the coefficient matrix?

6. Find the relative forward and backward errors and error magnification factor for the following approximate solutions of the system $x_1 + 2x_2 = 3, 2x_1 + 4.01x_2 = 6.01$: (a) $[-10, 6]$ (b) $[-100, 52]$ (c) $[-600, 301]$ (d) $[-599, 301]$ (e) What is the condition number of the coefficient matrix?

7. Find the norm $\|H\|_\infty$ of the 5×5 Hilbert matrix.

8. (a) Find the condition number of the coefficient matrix in the system $\begin{bmatrix} 1 & 1 \\ 1+\delta & 1 \end{bmatrix} \begin{bmatrix} x_1 \\ x_2 \end{bmatrix} = \begin{bmatrix} 2 \\ 2+\delta \end{bmatrix}$ as a function of $\delta > 0$. (b) Find the error magnification factor for the approximate root $x_a = [-1, 3 + \delta]$.

9. (a) Find the condition number (in the infinity norm) of the matrix $A = \begin{bmatrix} 0 & 1 & 0 \\ 0 & 0 & 1 \\ 1 & 0 & 0 \end{bmatrix}$.

   (b) Let D be an $n \times n$ diagonal matrix with diagonal entries $d_1, d_2, \ldots, d_n$. Express the condition number (in the infinity norm) of D in terms of the d_i.

10. (a) Find the (infinity norm) condition number of the matrix $A = \begin{bmatrix} 1 & 2 \\ 2 & 4.001 \end{bmatrix}$.

    (b) Let $b = \begin{bmatrix} 3 \\ 6.001 \end{bmatrix}$ and let $x = \begin{bmatrix} 1 \\ 1 \end{bmatrix}$ denote the exact solution of $Ax = b$. Find the relative forward error, relative backward error, and error magnification factor of the approximate solution $x_a = \begin{bmatrix} -6000 \\ 3001 \end{bmatrix}$.

    (c) Show that for any $\delta > 0$, the error magnification factor of the approximate solution $x_a = \begin{bmatrix} 1 - 6001\delta \\ 1 + 3000\delta \end{bmatrix}$ is equal to the condition number of A.

11. (a) Prove that the infinity norm $||x||_\infty$ is a vector norm. (b) Prove that the 1-norm $||x||_1$ is a vector norm.

12. (a) Prove that the infinity norm $||A||_\infty$ is a matrix norm. (b) Prove that the 1-norm $||A||_1$ is a matrix norm.

13. Prove that the matrix infinity norm is the operator norm of the vector infinity norm.

14. Prove that the matrix 1-norm is the operator norm of the vector 1-norm.

15. For the matrices in Exercise 1, find a vector x satisfying $||A||_\infty = ||Ax||_\infty / ||x||_\infty$.

16. For the matrices in Exercise 1, find a vector x satisfying $||A||_1 = ||Ax||_1 / ||x||_1$.

17. Find the LU factorization of

$$A = \begin{bmatrix} 10 & 20 & 1 \\ 1 & 1.99 & 6 \\ 0 & 50 & 1 \end{bmatrix}.$$

    What is the largest magnitude multiplier l_{ij} needed?

18. (a) Show that the system of equations $\begin{bmatrix} 811802 & 810901 \\ 810901 & 810001 \end{bmatrix} \begin{bmatrix} x_1 \\ x_2 \end{bmatrix} = \begin{bmatrix} 901 \\ 900 \end{bmatrix}$ has solution $[1, -1]$. (b) Solve the system in double precision arithmetic using Gaussian elimination (in tableau form, or any other form). How many decimal places are correct in your answer? Explain, using the concept of condition number.

2.3 Computer Problems

Solutions for Computer Problems numbered in blue can be found at
bit.ly/2RYIgoX

1. For the $n \times n$ matrix with entries $A_{ij} = 5/(i + 2j - 1)$, set $x = [1, \dots, 1]^T$ and $b = Ax$. Use the MATLAB program from Computer Problem 2.1.1 or MATLAB's backslash command to compute x_c, the double precision computed solution. Find the infinity norm of the forward error and the error magnification factor of the problem $Ax = b$, and compare it with the condition number of A: (a) $n = 6$ (b) $n = 10$.

2. Carry out Computer Problem 1 for the matrix with entries $A_{ij} = 1/(|i - j| + 1)$.

3. Let A be the $n \times n$ matrix with entries $A_{ij} = |i - j| + 1$. Define $x = [1, \dots, 1]^T$ and $b = Ax$. For $n = 100, 200, 300, 400$, and 500, use the MATLAB program from Computer Problem 2.1.1 or MATLAB's backslash command to compute x_c, the double precision computed solution. Calculate the infinity norm of the forward error for each solution. Find the five error magnification factors of the problems $Ax = b$, and compare with the corresponding condition numbers.

4. Carry out the steps of Computer Problem 3 for the matrix with entries $A_{ij} = \sqrt{(i - j)^2 + n/10}$.

5. For what values of n does the solution in Computer Problem 1 have no correct significant digits?

6. Use the MATLAB program from Computer Problem 2.1.1 to carry out double precision implementations of versions 2 and 3 of Example 2.13, and compare with the theoretical results found in the text.

2.4 THE PA = LU FACTORIZATION

The form of Gaussian elimination considered so far is often called "naive," because of two serious difficulties: encountering a zero pivot and swamping. For a nonsingular matrix, both can be avoided with an improved algorithm. The key to this improvement is an efficient protocol for exchanging rows of the coefficient matrix, called partial pivoting.

2.4.1 Partial pivoting

At the start of classical Gaussian elimination of n equations in n unknowns, the first step is to use the diagonal element a_{11} as a pivot to eliminate the first column. The **partial pivoting** protocol consists of comparing numbers before carrying out each elimination step. The largest entry of the first column is located, and its row is swapped with the pivot row, in this case the top row.

In other words, at the start of Gaussian elimination, partial pivoting asks that we select the pth row, where

$$|a_{p1}| \geq |a_{i1}| \tag{2.21}$$

for all $1 \leq i \leq n$, and exchange rows 1 and p. Next, elimination of column 1 proceeds as usual, using the "new" version of a_{11} as the pivot. The multiplier used to eliminate a_{i1} will be

$$m_{i1} = \frac{a_{i1}}{a_{11}}$$

and $|m_{i1}| \leq 1$.

The same check is applied to every choice of pivot during the algorithm. When deciding on the second pivot, we start with the current a_{22} and check all entries directly below. We select the row p such that

$$|a_{p2}| \geq |a_{i2}|$$

for all $2 \leq i \leq n$, and if $p \neq 2$, rows 2 and p are exchanged. Row 1 is never involved in this step. If $|a_{22}|$ is already the largest, no row exchange is made.

The protocol applies to each column during elimination. Before eliminating column k, the p with $k \leq p \leq n$ and largest $|a_{pk}|$ is located, and rows p and k are exchanged if necessary before continuing with the elimination. Note that using partial pivoting ensures that all multipliers, or entries of L, will be no greater than 1 in absolute value. With this minor change in the implementation of Gaussian elimination, the problem of swamping illustrated in Example 2.13 is completely avoided.

▶ **EXAMPLE 2.14** Apply Gaussian elimination with partial pivoting to solve the system (2.1).

The equations can be written in tableau form as

$$\begin{bmatrix} 1 & 1 & | & 3 \\ 3 & -4 & | & 2 \end{bmatrix}.$$

According to partial pivoting, we compare $|a_{11}| = 1$ with all entries below it, in this case the single entry $a_{21} = 3$. Since $|a_{21}| > |a_{11}|$, we must exchange rows 1 and 2. The new tableau is

$$\begin{bmatrix} 3 & -4 & | & 2 \\ 1 & 1 & | & 3 \end{bmatrix} \xrightarrow[\text{from row 2}]{\text{subtract } \frac{1}{3} \times \text{row 1}} \begin{bmatrix} 3 & -4 & | & 2 \\ 0 & \frac{7}{3} & | & \frac{7}{3} \end{bmatrix}.$$

After back substitution, the solution is $x_2 = 1$ and then $x_1 = 2$, as we found earlier. When we solved this system the first time, the multiplier was 3, but under partial pivoting this would never occur. ◄

▶ **EXAMPLE 2.15** Apply Gaussian elimination with partial pivoting to solve the system

$$x_1 - x_2 + 3x_3 = -3$$
$$-x_1 - 2x_3 = 1$$
$$2x_1 + 2x_2 + 4x_3 = 0.$$

This example is written in tableau form as

$$\begin{bmatrix} 1 & -1 & 3 & | & -3 \\ -1 & 0 & -2 & | & 1 \\ 2 & 2 & 4 & | & 0 \end{bmatrix}.$$

Under partial pivoting we compare $|a_{11}| = 1$ with $|a_{21}| = 1$ and $|a_{31}| = 2$, and choose a_{31} for the new pivot. This is achieved through an exchange of rows 1 and 3:

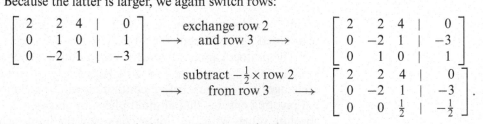

Before eliminating column 2 we must compare the current $|a_{22}|$ with the current $|a_{32}|$. Because the latter is larger, we again switch rows:

$$\begin{bmatrix} 2 & 2 & 4 & | & 0 \\ 0 & 1 & 0 & | & 1 \\ 0 & -2 & 1 & | & -3 \end{bmatrix} \xrightarrow[\text{and row 3}]{\text{exchange row 2}} \begin{bmatrix} 2 & 2 & 4 & | & 0 \\ 0 & -2 & 1 & | & -3 \\ 0 & 1 & 0 & | & 1 \end{bmatrix}$$

$$\xrightarrow[\text{from row 3}]{\text{subtract } -\frac{1}{2} \times \text{row 2}} \begin{bmatrix} 2 & 2 & 4 & | & 0 \\ 0 & -2 & 1 & | & -3 \\ 0 & 0 & \frac{1}{2} & | & -\frac{1}{2} \end{bmatrix}.$$

Note that all three multipliers are less than 1 in absolute value.

The equations are now simple to solve. From

$$\frac{1}{2}x_3 = -\frac{1}{2}$$

$$-2x_2 + x_3 = -3$$

$$2x_1 + 2x_2 + 4x_3 = 0,$$

we find that $x = [1, 1, -1]$. ◄

Notice that partial pivoting also solves the problem of zero pivots. When a potential zero pivot is encountered, for example, if $a_{11} = 0$, it is immediately exchanged for a nonzero pivot somewhere in its column. If there is no such nonzero entry at or below the diagonal entry, then the matrix is singular and Gaussian elimination will fail to provide a solution anyway.

2.4.2 Permutation matrices

Before showing how row exchanges can be used with the LU factorization approach to Gaussian elimination, we will discuss the fundamental properties of permutation matrices.

DEFINITION 2.7 A **permutation matrix** is an $n \times n$ matrix consisting of all zeros, except for a single 1 in every row and column. ◻

Equivalently, a permutation matrix P is created by applying arbitrary row exchanges to the $n \times n$ identity matrix (or arbitrary column exchanges). For example,

$$\begin{bmatrix} 1 & 0 \\ 0 & 1 \end{bmatrix}, \begin{bmatrix} 0 & 1 \\ 1 & 0 \end{bmatrix}$$

are the only 2×2 permutation matrices, and

$$\begin{bmatrix} 1 & 0 & 0 \\ 0 & 1 & 0 \\ 0 & 0 & 1 \end{bmatrix}, \begin{bmatrix} 0 & 1 & 0 \\ 1 & 0 & 0 \\ 0 & 0 & 1 \end{bmatrix}, \begin{bmatrix} 1 & 0 & 0 \\ 0 & 0 & 1 \\ 0 & 1 & 0 \end{bmatrix},$$

$$\begin{bmatrix} 0 & 0 & 1 \\ 0 & 1 & 0 \\ 1 & 0 & 0 \end{bmatrix}, \begin{bmatrix} 0 & 0 & 1 \\ 1 & 0 & 0 \\ 0 & 1 & 0 \end{bmatrix}, \begin{bmatrix} 0 & 1 & 0 \\ 0 & 0 & 1 \\ 1 & 0 & 0 \end{bmatrix}$$

are the six 3×3 permutation matrices.

The next theorem tells us at a glance what action a permutation matrix causes when multiplied on the left of another matrix.

THEOREM 2.8 **Fundamental Theorem of Permutation Matrices.** Let P be the $n \times n$ permutation matrix formed by a particular set of row exchanges applied to the identity matrix. Then, for any $n \times n$ matrix A, PA is the matrix obtained by applying exactly the same set of row exchanges to A. ∎

For example, the permutation matrix

$$\begin{bmatrix} 1 & 0 & 0 \\ 0 & 0 & 1 \\ 0 & 1 & 0 \end{bmatrix}$$

is formed by exchanging rows 2 and 3 of the identity matrix. Multiplying an arbitrary matrix on the left with P has the effect of exchanging rows 2 and 3:

$$\begin{bmatrix} 1 & 0 & 0 \\ 0 & 0 & 1 \\ 0 & 1 & 0 \end{bmatrix} \begin{bmatrix} a & b & c \\ d & e & f \\ g & h & i \end{bmatrix} = \begin{bmatrix} a & b & c \\ g & h & i \\ d & e & f \end{bmatrix}.$$

A good way to remember Theorem 2.8 is to imagine multiplying P times the identity matrix I:

$$\begin{bmatrix} 1 & 0 & 0 \\ 0 & 0 & 1 \\ 0 & 1 & 0 \end{bmatrix} \begin{bmatrix} 1 & 0 & 0 \\ 0 & 1 & 0 \\ 0 & 0 & 1 \end{bmatrix} = \begin{bmatrix} 1 & 0 & 0 \\ 0 & 0 & 1 \\ 0 & 1 & 0 \end{bmatrix}.$$

There are two different ways to view this equality: first, as multiplication by the identity matrix (so we get the permutation matrix on the right); second, as the permutation matrix acting on the rows of the identity matrix. The content of Theorem 2.8 is that the row exchanges caused by multiplication by P are exactly the ones involved in the construction of P.

2.4.3 PA = LU factorization

In this section, we put together everything we know about Gaussian elimination into the PA = LU factorization. This is the matrix formulation of elimination with partial pivoting. The PA = LU factorization is the established workhorse for solving systems of linear equations.

As its name implies, the PA = LU factorization is simply the LU factorization of a row-exchanged version of A. Under partial pivoting, the rows that need exchanging are not known at the outset, so we must be careful about fitting the row exchange information into the factorization. In particular, we need to keep track of previous multipliers when a row exchange is made. We begin with an example.

▶ **EXAMPLE 2.16**   Find the PA = LU factorization of the matrix

$$A = \begin{bmatrix} 2 & 1 & 5 \\ 4 & 4 & -4 \\ 1 & 3 & 1 \end{bmatrix}.$$

First, rows 1 and 2 need to be exchanged, according to partial pivoting:

$$P = \begin{bmatrix} 0 & 1 & 0 \\ 1 & 0 & 0 \\ 0 & 0 & 1 \end{bmatrix}$$

$$\begin{bmatrix} 2 & 1 & 5 \\ 4 & 4 & -4 \\ 1 & 3 & 1 \end{bmatrix} \longrightarrow \text{exchange rows 1 and 2} \longrightarrow \begin{bmatrix} 4 & 4 & -4 \\ 2 & 1 & 5 \\ 1 & 3 & 1 \end{bmatrix}.$$

We will use the permutation matrix P to keep track of the cumulative permutation of rows that have been done along the way. Now we perform two row operations, namely,

$$\begin{array}{c} \text{subtract } \frac{1}{2} \times \text{row 1} \\ \longrightarrow \quad \text{from row 2} \end{array} \longrightarrow \begin{bmatrix} 4 & 4 & -4 \\ \frac{1}{2} & -1 & 7 \\ 1 & 3 & 1 \end{bmatrix} \quad \begin{array}{c} \text{subtract } \frac{1}{4} \times \text{row 1} \\ \longrightarrow \quad \text{from row 3} \end{array} \longrightarrow \begin{bmatrix} 4 & 4 & -4 \\ \frac{1}{2} & -1 & 7 \\ \frac{1}{4} & 2 & 2 \end{bmatrix},$$

to eliminate the first column. We have done something new—instead of putting only a zero in the eliminated position, we have made the zero a storage location. Inside the zero at the (i, j) position, we store the multiplier m_{ij} that we used to eliminate that position. We do this for a reason. This is the mechanism by which the multipliers will stay with their row, in case future row exchanges are made.

Next we must make a comparison to choose the second pivot. Since $|a_{22}| = 1 < 2 = |a_{32}|$, a row exchange is required before eliminating the second column. Notice that the previous multipliers move along with the row exchange:

$$P = \begin{bmatrix} 0 & 1 & 0 \\ 0 & 0 & 1 \\ 1 & 0 & 0 \end{bmatrix}$$

$\longrightarrow$ exchange rows 2 and 3 $\longrightarrow$ $\begin{bmatrix} 4 & 4 & -4 \\ \boxed{\tfrac{1}{4}} & 2 & 2 \\ \boxed{\tfrac{1}{2}} & -1 & 7 \end{bmatrix}$

Finally, the elimination ends with one more row operation:

subtract $-\tfrac{1}{2} \times$ row 2
$\longrightarrow$ from row 3 $\longrightarrow$ $\begin{bmatrix} 4 & 4 & -4 \\ \boxed{\tfrac{1}{4}} & 2 & 2 \\ \boxed{\tfrac{1}{2}} & \boxed{-\tfrac{1}{2}} & 8 \end{bmatrix}.$

This is the finished elimination. Now we can read off the PA = LU factorization:

$$\begin{bmatrix} 0 & 1 & 0 \\ 0 & 0 & 1 \\ 1 & 0 & 0 \end{bmatrix} \begin{bmatrix} 2 & 1 & 5 \\ 4 & 4 & -4 \\ 1 & 3 & 1 \end{bmatrix} = \begin{bmatrix} 1 & 0 & 0 \\ \tfrac{1}{4} & 1 & 0 \\ \tfrac{1}{2} & -\tfrac{1}{2} & 1 \end{bmatrix} \begin{bmatrix} 4 & 4 & -4 \\ 0 & 2 & 2 \\ 0 & 0 & 8 \end{bmatrix} \qquad (2.22)$$

$\qquad\qquad P \qquad\qquad\quad A \qquad\qquad\qquad L \qquad\qquad\qquad U$

The entries of L are sitting inside the zeros in the lower triangle of the matrix (below the main diagonal), and U comes from the upper triangle. The final (cumulative) permutation matrix serves as P. ◄

Using the PA = LU factorization to solve a system of equations $Ax = b$ is just a slight variant of the $A = LU$ version. Multiply through the equation $Ax = b$ by P on the left, and then proceed as before:

$$\begin{aligned} PAx &= Pb \\ LUx &= Pb. \end{aligned} \qquad (2.23)$$

Solve

$$\begin{aligned} &1.\ Lc = Pb \text{ for } c. \\ &2.\ Ux = c \text{ for } x. \end{aligned} \qquad (2.24)$$

The important point, as mentioned earlier, is that the expensive part of the calculation, determining PA = LU, can be done without knowing b. Since the resulting LU factorization is of PA, a row-permuted version of the equation coefficients, it is necessary to permute the right-hand-side vector b in precisely the same way before proceeding with the back-substitution stage. That is achieved by using Pb in the first step of back substitution. The value of the matrix formulation of Gaussian elimination is apparent: All of the bookkeeping details of elimination and pivoting are automatic and contained in the matrix equations.

► **EXAMPLE 2.17** Use the PA = LU factorization to solve the system $Ax = b$, where

$$A = \begin{bmatrix} 2 & 1 & 5 \\ 4 & 4 & -4 \\ 1 & 3 & 1 \end{bmatrix}, \quad b = \begin{bmatrix} 5 \\ 0 \\ 6 \end{bmatrix}.$$

The PA = LU factorization is known from (2.22). It remains to complete the two back substitutions.

1. $Lc = Pb$:

$$\begin{bmatrix} 1 & 0 & 0 \\ \frac{1}{4} & 1 & 0 \\ \frac{1}{2} & -\frac{1}{2} & 1 \end{bmatrix} \begin{bmatrix} c_1 \\ c_2 \\ c_3 \end{bmatrix} = \begin{bmatrix} 0 & 1 & 0 \\ 0 & 0 & 1 \\ 1 & 0 & 0 \end{bmatrix} \begin{bmatrix} 5 \\ 0 \\ 6 \end{bmatrix} = \begin{bmatrix} 0 \\ 6 \\ 5 \end{bmatrix}.$$

Starting at the top, we have

$$c_1 = 0$$
$$\frac{1}{4}(0) + c_2 = 6 \Rightarrow c_2 = 6$$
$$\frac{1}{2}(0) - \frac{1}{2}(6) + c_3 = 5 \Rightarrow c_3 = 8.$$

2. $Ux = c$:

$$\begin{bmatrix} 4 & 4 & -4 \\ 0 & 2 & 2 \\ 0 & 0 & 8 \end{bmatrix} \begin{bmatrix} x_1 \\ x_2 \\ x_3 \end{bmatrix} = \begin{bmatrix} 0 \\ 6 \\ 8 \end{bmatrix}$$

Starting at the bottom,

$$8x_3 = 8 \Rightarrow x_3 = 1$$
$$2x_2 + 2(1) = 6 \Rightarrow x_2 = 2$$
$$4x_1 + 4(2) - 4(1) = 0 \Rightarrow x_1 = -1. \tag{2.25}$$

Therefore, the solution is $x = [-1, 2, 1]$. ◄

▶ **EXAMPLE 2.18** Solve the system $2x_1 + 3x_2 = 4$, $3x_1 + 2x_2 = 1$ using the $PA = LU$ factorization with partial pivoting.

In matrix form, this is the equation

$$\begin{bmatrix} 2 & 3 \\ 3 & 2 \end{bmatrix} \begin{bmatrix} x_1 \\ x_2 \end{bmatrix} = \begin{bmatrix} 4 \\ 1 \end{bmatrix}.$$

We begin by ignoring the right-hand-side b. According to partial pivoting, rows 1 and 2 must be exchanged (because $a_{21} > a_{11}$). The elimination step is

$$A = \begin{bmatrix} 2 & 3 \\ 3 & 2 \end{bmatrix} \xrightarrow[\text{exchange rows 1 and 2}]{P = \begin{bmatrix} 0 & 1 \\ 1 & 0 \end{bmatrix}} \begin{bmatrix} 3 & 2 \\ 2 & 3 \end{bmatrix}$$

$$\xrightarrow[\substack{\text{subtract } \frac{2}{3} \times \text{ row 1} \\ \text{from row 2}}]{} \begin{bmatrix} 3 & 2 \\ \boxed{\frac{2}{3}} & \frac{5}{3} \end{bmatrix}.$$

Therefore, the $PA = LU$ factorization is

$$\underbrace{\begin{bmatrix} 0 & 1 \\ 1 & 0 \end{bmatrix}}_{P} \underbrace{\begin{bmatrix} 2 & 3 \\ 3 & 2 \end{bmatrix}}_{A} = \underbrace{\begin{bmatrix} 1 & 0 \\ \frac{2}{3} & 1 \end{bmatrix}}_{L} \underbrace{\begin{bmatrix} 3 & 2 \\ 0 & \frac{5}{3} \end{bmatrix}}_{U}.$$

The first back substitution $Lc = Pb$ is

$$\begin{bmatrix} 1 & 0 \\ \frac{2}{3} & 1 \end{bmatrix} \begin{bmatrix} c_1 \\ c_2 \end{bmatrix} = \begin{bmatrix} 0 & 1 \\ 1 & 0 \end{bmatrix} \begin{bmatrix} 4 \\ 1 \end{bmatrix} = \begin{bmatrix} 1 \\ 4 \end{bmatrix}.$$

Starting at the top, we have

$$c_1 = 1$$

$$\frac{2}{3}(1) + c_2 = 4 \Rightarrow c_2 = \frac{10}{3}.$$

The second back substitution $Ux = c$ is

$$\begin{bmatrix} 3 & 2 \\ 0 & \frac{5}{3} \end{bmatrix} \begin{bmatrix} x_1 \\ x_2 \end{bmatrix} = \begin{bmatrix} 1 \\ \frac{10}{3} \end{bmatrix}.$$

Starting at the bottom, we have

$$\frac{5}{3}x_2 = \frac{10}{3} \Rightarrow x_2 = 2$$

$$3x_1 + 2(2) = 1 \Rightarrow x_1 = -1. \tag{2.26}$$

Therefore, the solution is $x = [-1, 2]$. ◄

Every $n \times n$ matrix has a PA = LU factorization. We simply follow the partial pivoting rule, and if the resulting pivot is zero, it means that all entries that need to be eliminated are already zero, so the column is done.

All of the techniques described so far are implemented in MATLAB. The most sophisticated form of Gaussian elimination we have discussed is the PA = LU factorization. MATLAB's lu command accepts a square coefficient matrix A and returns P, L, and U. The following MATLAB script defines the matrix of Example 2.16 and computes its factorization:

```
>> A=[2 1 5; 4 4 -4; 1 3 1];
>> [L,U,P]=lu(A)

L=

    1.0000         0         0
    0.2500    1.0000         0
    0.5000   -0.5000    1.0000

U=

    4    4   -4
    0    2    2
    0    0    8

P=

    0    1    0
    0    0    1
    1    0    0
```

▶ **ADDITIONAL EXAMPLES**

1. Find the PA $=$ LU factorization of the matrix $\begin{bmatrix} 1 & 2 \\ 3 & 4 \end{bmatrix}$.

*2. Solve

$$\begin{bmatrix} -1 & 1 & -2 \\ 2 & 3 & 1 \\ 4 & 8 & -4 \end{bmatrix} \begin{bmatrix} x_1 \\ x_2 \\ x_3 \end{bmatrix} = \begin{bmatrix} -5 \\ 2 \\ -4 \end{bmatrix}$$

using the PA $=$ LU factorization with partial pivoting.

⊑⊡ **Solutions** for Additional Examples can be found at `bit.ly/2CQMJpx`
(* example with video solution)

2.4 Exercises

⊑⊡ **Solutions**
for Exercises
numbered in blue
can be found at
`bit.ly/2OttZCx`

1. Find the PA $=$ LU factorization (using partial pivoting) of the following matrices:

(a) $\begin{bmatrix} 1 & 3 \\ 2 & 3 \end{bmatrix}$ (b) $\begin{bmatrix} 2 & 4 \\ 1 & 3 \end{bmatrix}$ (c) $\begin{bmatrix} 1 & 5 \\ 5 & 12 \end{bmatrix}$ (d) $\begin{bmatrix} 0 & 1 \\ 1 & 0 \end{bmatrix}$

2. Find the PA $=$ LU factorization (using partial pivoting) of the following matrices:

(a) $\begin{bmatrix} 1 & 1 & 0 \\ 2 & 1 & -1 \\ -1 & 1 & -1 \end{bmatrix}$ (b) $\begin{bmatrix} 0 & 1 & 3 \\ 2 & 1 & 1 \\ -1 & -1 & 2 \end{bmatrix}$ (c) $\begin{bmatrix} 1 & 2 & -3 \\ 2 & 4 & 2 \\ -1 & 0 & 3 \end{bmatrix}$ (d) $\begin{bmatrix} 0 & 1 & 0 \\ 1 & 0 & 2 \\ -2 & 1 & 0 \end{bmatrix}$

3. Solve the system by finding the PA $=$ LU factorization and then carrying out the two-step back substitution.

(a) $\begin{bmatrix} 3 & 7 \\ 6 & 1 \end{bmatrix} \begin{bmatrix} x_1 \\ x_2 \end{bmatrix} = \begin{bmatrix} 1 \\ -11 \end{bmatrix}$ (b) $\begin{bmatrix} 3 & 1 & 2 \\ 6 & 3 & 4 \\ 3 & 1 & 5 \end{bmatrix} \begin{bmatrix} x_1 \\ x_2 \\ x_3 \end{bmatrix} = \begin{bmatrix} 0 \\ 1 \\ 3 \end{bmatrix}$

4. Solve the system by finding the PA $=$ LU factorization and then carrying out the two-step back substitution.

(a) $\begin{bmatrix} 4 & 2 & 0 \\ 4 & 4 & 2 \\ 2 & 2 & 3 \end{bmatrix} \begin{bmatrix} x_1 \\ x_2 \\ x_3 \end{bmatrix} = \begin{bmatrix} 2 \\ 4 \\ 6 \end{bmatrix}$ (b) $\begin{bmatrix} -1 & 0 & 1 \\ 2 & 1 & 1 \\ -1 & 2 & 0 \end{bmatrix} \begin{bmatrix} x_1 \\ x_2 \\ x_3 \end{bmatrix} = \begin{bmatrix} -2 \\ 17 \\ 3 \end{bmatrix}$

5. Write down a 5×5 matrix P such that multiplication of another matrix by P on the left causes rows 2 and 5 to be exchanged.

6. (a) Write down the 4×4 matrix P such that multiplying a matrix on the left by P causes the second and fourth rows of the matrix to be exchanged. (b) What is the effect of multiplying on the *right* by P? Demonstrate with an example.

7. Change four entries of the leftmost matrix to make the matrix equation correct:

$$\begin{bmatrix} 0 & 0 & 0 & 0 \\ 0 & 0 & 0 & 0 \\ 0 & 0 & 0 & 0 \\ 0 & 0 & 0 & 0 \end{bmatrix} \begin{bmatrix} 1 & 2 & 3 & 4 \\ 3 & 4 & 5 & 6 \\ 5 & 6 & 7 & 8 \\ 7 & 8 & 9 & 0 \end{bmatrix} = \begin{bmatrix} 5 & 6 & 7 & 8 \\ 3 & 4 & 5 & 6 \\ 7 & 8 & 9 & 0 \\ 1 & 2 & 3 & 4 \end{bmatrix}.$$

8. Find the PA $=$ LU factorization of the matrix A in Exercise 2.3.15. What is the largest multiplier l_{ij} needed?

9. (a) Find the PA = LU factorization of $A = \begin{bmatrix} 1 & 0 & 0 & 1 \\ -1 & 1 & 0 & 1 \\ -1 & -1 & 1 & 1 \\ -1 & -1 & -1 & 1 \end{bmatrix}$. (b) Let A be the $n \times n$ matrix of the same form as in (a). Describe the entries of each matrix of its PA = LU factorization.

10. (a) Assume that A is an $n \times n$ matrix with entries $|a_{ij}| \leq 1$ for $1 \leq i, j \leq n$. Prove that the matrix U in its PA = LU factorization satisfies $|u_{ij}| \leq 2^{n-1}$ for all $1 \leq i, j \leq n$. See Exercise 9(b). (b) Formulate and prove an analogous fact for an arbitrary $n \times n$ matrix A.

Reality Check 2 The Euler–Bernoulli Beam

The Euler–Bernoulli beam is a fundamental model for a material bending under stress. Discretization converts the differential equation model into a system of linear equations. The smaller the discretization size, the larger is the resulting system of equations. This example will provide us an interesting case study of the roles of system size and ill-conditioning in scientific computation.

The vertical displacement of the beam is represented by a function $y(x)$, where $0 \leq x \leq L$ along the beam of length L. We will use MKS units in the calculation: meters, kilograms, seconds. The displacement $y(x)$ satisfies the Euler–Bernoulli equation

$$EIy'''' = f(x) \tag{2.27}$$

where E, the Young's modulus of the material, and I, the area moment of inertia, are constant along the beam. The right-hand-side $f(x)$ is the applied load, including the weight of the beam, in force per unit length.

Techniques for discretizing derivatives are found in Chapter 5, where it will be shown that a reasonable approximation for the fourth derivative is

$$y''''(x) \approx \frac{y(x - 2h) - 4y(x - h) + 6y(x) - 4y(x + h) + y(x + 2h)}{h^4} \tag{2.28}$$

for a small increment h. The discretization error of this approximation is proportional to h^2 (see Exercise 5.1.21). Our strategy will be to consider the beam as the union of many segments of length h, and to apply the discretized version of the differential equation on each segment.

For a positive integer n, set $h = L/n$. Consider the evenly spaced grid $0 = x_0 < x_1 < \ldots < x_n = L$, where $h = x_i - x_{i-1}$ for $i = 1, \ldots, n$. Replacing the differential equation (2.27) with the difference approximation (2.28) to get the system of linear equations for the displacements $y_i = y(x_i)$ yields

$$y_{i-2} - 4y_{i-1} + 6y_i - 4y_{i+1} + y_{i+2} = \frac{h^4}{EI} f(x_i). \tag{2.29}$$

We will develop n equations in the n unknowns $y_1, \ldots, y_n$. The coefficient matrix, or structure matrix, will have coefficients from the left-hand side of this equation. However, notice that we must alter the equations near the ends of the beam to take the boundary conditions into account.

A diving board is a beam with one end clamped at the support, and the opposite end free. This is called the **clamped-free** beam or sometimes the **cantilever** beam. The boundary conditions for the clamped (left) end and free (right) end are

$$y(0) = y'(0) = y''(L) = y'''(L) = 0.$$

In particular, $y_0 = 0$. Note that finding y_1, however, presents us with a problem, since applying the approximation (2.29) to the differential equation (2.27) at x_1 results in

$$y_{-1} - 4y_0 + 6y_1 - 4y_2 + y_3 = \frac{h^4}{EI} f(x_1), \qquad (2.30)$$

and y_{-1} is not defined. Instead, we must use an alternate derivative approximation at the point x_1 near the clamped end. Exercise 5.1.22(a) derives the approximation

$$y''''(x_1) \approx \frac{16y(x_1) - 9y(x_1 + h) + \frac{8}{3}y(x_1 + 2h) - \frac{1}{4}y(x_1 + 3h)}{h^4} \qquad (2.31)$$

which is valid when $y(x_0) = y'(x_0) = 0$.

Calling the approximation "valid," for now, means that the discretization error of the approximation is proportional to h^2, the same as for equation (2.28). In theory, this means that the error in approximating the derivative in this way will decrease toward zero in the limit of small h. This concept will be the focal point of the discussion of numerical differentiation in Chapter 5. The result for us is that we can use approximation (2.31) to take the endpoint condition into account for $i = 1$, yielding

$$16y_1 - 9y_2 + \frac{8}{3}y_3 - \frac{1}{4}y_4 = \frac{h^4}{EI} f(x_1).$$

The free right end of the beam requires a little more work because we must compute y_i all the way to the end of the beam. Again, we need alternative derivative approximations at the last two points x_{n-1} and x_n. Exercise 5.1.22 gives the approximations

$$y''''(x_{n-1}) \approx \frac{-28y_n + 72y_{n-1} - 60y_{n-2} + 16y_{n-3}}{17h^4} \qquad (2.32)$$

$$y''''(x_n) \approx \frac{72y_n - 156y_{n-1} + 96y_{n-2} - 12y_{n-3}}{17h^4} \qquad (2.33)$$

which are valid under the assumption $y''(x_n) = y'''(x_n) = 0$.

Now we can write down the system of n equations in n unknowns for the diving board. This matrix equation summarizes our approximate versions of the original differential equation (2.27) at each point $x_1, \ldots, x_n$, accurate within terms of order h^2:

$$\begin{bmatrix} 16 & -9 & \frac{8}{3} & -\frac{1}{4} & & & & & \\ -4 & 6 & -4 & 1 & & & & & \\ 1 & -4 & 6 & -4 & 1 & & & & \\ & 1 & -4 & 6 & -4 & & 1 & & \\ & & \ddots & \ddots & \ddots & \ddots & \ddots & & \\ & & & 1 & -4 & 6 & -4 & 1 & \\ & & & & 1 & -4 & 6 & -4 & 1 \\ & & & & & \frac{16}{17} & -\frac{60}{17} & \frac{72}{17} & -\frac{28}{17} \\ & & & & & -\frac{12}{17} & \frac{96}{17} & -\frac{156}{17} & \frac{72}{17} \end{bmatrix} \begin{bmatrix} y_1 \\ y_2 \\ \vdots \\ \vdots \\ \vdots \\ y_{n-1} \\ y_n \end{bmatrix} = \frac{h^4}{EI} \begin{bmatrix} f(x_1) \\ f(x_2) \\ \vdots \\ \vdots \\ \vdots \\ f(x_{n-1}) \\ f(x_n) \end{bmatrix}. \qquad (2.34)$$

The structure matrix A in (2.34) is a **banded matrix**, meaning that all entries sufficiently far from the main diagonal are zero. Specifically, the matrix entries $a_{ij} = 0$, except for $|i - j| \le 3$. The **bandwidth** of this banded matrix is 7, since $i - j$ takes on 7 values for nonzero a_{ij}.

Finally, we are ready to model the clamped-free beam. Let us consider a solid wood diving board composed of Douglas fir. Assume that the diving board is $L = 2$ meters long, 30 cm wide, and 3 cm thick. The density of Douglas fir is approximately 480 kg/m^3. One Newton of force is 1 kg-m/sec^2, and the Young's modulus of this wood is approximately $E = 1.3 \times 10^{10}$ Pascals, or Newton/m^2. The area moment of inertia I around the center of mass of a beam is $wd^3/12$, where w is the width and d the thickness of the beam.

You will begin by calculating the displacement of the beam with no payload, so that $f(x)$ represents only the weight of the beam itself, in units of force per meter. Therefore $f(x)$ is the mass per meter $480wd$ times the downward acceleration of gravity $-g = -9.81$ m/sec^2, or the constant $f(x) = f = -480wdg$. The reader should check that the units match on both sides of (2.27). There is a closed-form solution of (2.27) in the case f is constant, so that the result of your computation can be checked for accuracy.

Following the check of your code for the unloaded beam, you will model two further cases. In the first, a sinusoidal load (or "pile") will be added to the beam. In this case, there is again a known closed-form solution, but the derivative approximations are not exact, so you will be able to monitor the error of your modeling as a function of the grid size h, and see the effect of conditioning problems for large n. Later, you will put a diver on the beam.

Suggested activities:

1. Write a MATLAB program to define the structure matrix A in (2.34). Then, using the MATLAB \ command or code of your own design, solve the system for the displacements y_i using $n = 10$ grid steps.

2. Plot the solution from Step 1 against the correct solution
   $y(x) = (f/24EI)x^2(x^2 - 4Lx + 6L^2)$, where $f = f(x)$ is the constant defined above. Check the error at the end of the beam, $x = L$ meters. In this simple case the derivative approximations are exact, so your error should be near machine roundoff.

3. Rerun the calculation in Step 1 for $n = 10 \cdot 2^k$, where $k = 1, \ldots, 11$. Make a table of the errors at $x = L$ for each n. For which n is the error smallest? Why does the error begin to increase with n after a certain point? You may want to make an accompanying table of the condition number of A as a function of n to help answer the last question. To carry out this step for large k, you may need to ask MATLAB to store the matrix A as a sparse matrix to avoid running out of memory. To do this, just initialize A with the command A=sparse(n,n), and proceed as before. We will discuss sparse matrices in more detail in the next section.

4. Add a sinusoidal pile to the beam. This means adding a function of form
   $s(x) = -pg \sin \frac{\pi}{L} x$ to the force term $f(x)$. Prove that the solution

$$y(x) = \frac{f}{24EI}x^2(x^2 - 4Lx + 6L^2) - \frac{pgL}{EI\pi}\left(\frac{L^3}{\pi^3}\sin\frac{\pi}{L}x - \frac{x^3}{6} + \frac{L}{2}x^2 - \frac{L^2}{\pi^2}x\right)$$

satisfies the Euler–Bernoulli beam equation and the clamped-free boundary conditions.

5. Rerun the calculation as in Step 3 for the sinusoidal load. (Be sure to include the weight of the beam itself.) Set $p = 100$ kg/m and plot your computed solutions against the correct solution. Answer the questions from Step 3, and in addition the following one: Is the error at $x = L$ proportional to h^2 as claimed above? You may want to plot

the error versus h on a log–log graph to investigate this question. Does the condition number come into play?

6. Now remove the sinusoidal load and add a 70 kg diver to the beam, balancing on the last 20 cm of the beam. You must add a force per unit length of $-g$ times $70/0.2$ kg/m to $f(x_i)$ for all $1.8 \leq x_i \leq 2$, and solve the problem again with the optimal value of n found in Step 5. Plot the solution and find the deflection of the diving board at the free end.

7. If we also fix the free end of the diving board, we have a "clamped-clamped" beam, obeying identical boundary conditions at each end: $y(0) = y'(0) = y(L) = y'(L) = 0$. This version is used to model the sag in a structure, like a bridge. Begin with the slightly different evenly spaced grid $0 = x_0 < x_1 < \ldots < x_n < x_{n+1} = L$, where $h = x_i - x_{i-1}$ for $i = 1, \ldots, n$, and find the system of n equations in n unknowns that determine $y_1, \ldots, y_n$. (It should be similar to the clamped-free version, except that the last two rows of the coefficient matrix A should be the first two rows reversed.) Solve for a sinusoidal load and answer the questions of Step 5 for the center $x = L/2$ of the beam. The exact solution for the clamped-clamped beam under a sinusoidal load is

$$y(x) = \frac{f}{24EI}x^2(L-x)^2 - \frac{pgL^2}{\pi^4 EI}\left(L^2 \sin\frac{\pi}{L}x + \pi x(x-L)\right).$$

8. Ideas for further exploration: If the width of the diving board is doubled, how does the displacement of the diver change? Does it change more or less than if the thickness is doubled? (Both beams have the same mass.) How does the maximum displacement change if the cross-section is circular or annular with the same area as the rectangle? (The area moment of inertia for a circular cross-section of radius r is $I = \pi r^4/4$, and for an annular cross-section with inner radius r_1 and outer radius r_2 is $I = \pi(r_2^4 - r_1^4)/4$.) Find out the area moment of inertia for I-beams, for example. The Young's modulus for different materials are also tabulated and available. For example, the density of steel is about 7850 kg/m^3 and its Young's modulus is about 2×10^{11} Pascals.

The Euler–Bernoulli beam is a relatively simple, classical model. More recent models, such as the Timoshenko beam, take into account more exotic bending, where the beam cross-section may not be perpendicular to the beam's main axis.

2.5 ITERATIVE METHODS

Gaussian elimination is a finite sequence of $O(n^3)$ floating point operations that result in a solution. For that reason, Gaussian elimination is called a **direct** method for solving systems of linear equations. Direct methods, in theory, give the exact solution within a finite number of steps. (Of course, when carried out by a computer using limited precision, the resulting solution will be only approximate. As we saw earlier, the loss of precision is quantified by the condition number.) Direct methods stand in contrast to the root-finding methods described in Chapter 1, which are iterative in form.

So-called **iterative** methods also can be applied to solving systems of linear equations. Similar to Fixed-Point Iteration, the methods begin with an initial guess and refine the guess at each step, converging to the solution vector.

2.5.1 Jacobi Method

The Jacobi Method is a form of fixed-point iteration for a system of equations. In FPI the first step is to rewrite the equations, solving for the unknown. The first step of the Jacobi Method is to do this in the following standardized way: Solve the ith equation for the ith unknown. Then, iterate as in Fixed-Point Iteration, starting with an initial guess.

▶ **EXAMPLE 2.19**

Apply the Jacobi Method to the system $3u + v = 5, u + 2v = 5$.

Begin by solving the first equation for u and the second equation for v. We will use the initial guess $(u_0, v_0) = (0, 0)$. We have

$$u = \frac{5 - v}{3}$$

$$v = \frac{5 - u}{2}. \tag{2.35}$$

The two equations are iterated:

$$\begin{bmatrix} u_0 \\ v_0 \end{bmatrix} = \begin{bmatrix} 0 \\ 0 \end{bmatrix}$$

$$\begin{bmatrix} u_1 \\ v_1 \end{bmatrix} = \begin{bmatrix} \frac{5-v_0}{3} \\ \frac{5-u_0}{2} \end{bmatrix} = \begin{bmatrix} \frac{5-0}{3} \\ \frac{5-0}{2} \end{bmatrix} = \begin{bmatrix} \frac{5}{3} \\ \frac{5}{2} \end{bmatrix}$$

$$\begin{bmatrix} u_2 \\ v_2 \end{bmatrix} = \begin{bmatrix} \frac{5-v_1}{3} \\ \frac{5-u_1}{2} \end{bmatrix} = \begin{bmatrix} \frac{5-5/2}{3} \\ \frac{5-5/3}{2} \end{bmatrix} = \begin{bmatrix} \frac{5}{6} \\ \frac{5}{3} \end{bmatrix}$$

$$\begin{bmatrix} u_3 \\ v_3 \end{bmatrix} = \begin{bmatrix} \frac{5-5/3}{3} \\ \frac{5-5/6}{2} \end{bmatrix} = \begin{bmatrix} \frac{10}{9} \\ \frac{25}{12} \end{bmatrix}. \tag{2.36}$$

Further steps of Jacobi show convergence toward the solution, which is $[1, 2]$. ◀

Now suppose that the equations are given in the reverse order.

▶ **EXAMPLE 2.20**

Apply the Jacobi Method to the system $u + 2v = 5, 3u + v = 5$.

Solve the first equation for the first variable u and the second equation for v. We begin with

$$u = 5 - 2v$$

$$v = 5 - 3u. \tag{2.37}$$

The two equations are iterated as before, but the results are quite different:

$$\begin{bmatrix} u_0 \\ v_0 \end{bmatrix} = \begin{bmatrix} 0 \\ 0 \end{bmatrix}$$

$$\begin{bmatrix} u_1 \\ v_1 \end{bmatrix} = \begin{bmatrix} 5 - 2v_0 \\ 5 - 3u_0 \end{bmatrix} = \begin{bmatrix} 5 \\ 5 \end{bmatrix}$$

$$\begin{bmatrix} u_2 \\ v_2 \end{bmatrix} = \begin{bmatrix} 5 - 2v_1 \\ 5 - 3u_1 \end{bmatrix} = \begin{bmatrix} -5 \\ -10 \end{bmatrix}$$

$$\begin{bmatrix} u_3 \\ v_3 \end{bmatrix} = \begin{bmatrix} 5 - 2(-10) \\ 5 - 3(-5) \end{bmatrix} = \begin{bmatrix} 25 \\ 20 \end{bmatrix}. \tag{2.38}$$

In this case the Jacobi Method fails, as the iteration diverges. ◀

Since the Jacobi Method does not always succeed, it is helpful to know conditions under which it does work. One important condition is given in the following definition:

DEFINITION 2.9   The $n \times n$ matrix $A = (a_{ij})$ is **strictly diagonally dominant** if, for each $1 \leq i \leq n$, $|a_{ii}| > \sum_{j \neq i} |a_{ij}|$. In other words, each main diagonal entry dominates its row in the sense that it is greater in magnitude than the sum of magnitudes of the remainder of the entries in its row.   ☐

THEOREM 2.10   If the $n \times n$ matrix A is strictly diagonally dominant, then (1) A is a nonsingular matrix, and (2) for every vector b and every starting guess, the Jacobi Method applied to $Ax = b$ converges to the (unique) solution.   ∎

Theorem 2.10 says that, if A is strictly diagonally dominant, then the Jacobi Method applied to the equation $Ax = b$ converges to a solution for each starting guess. The proof of this fact is given in Section 2.5.3. In Example 2.19, the coefficient matrix is at first

$$A = \begin{bmatrix} 3 & 1 \\ 1 & 2 \end{bmatrix},$$

which is strictly diagonally dominant because $3 > 1$ and $2 > 1$. Convergence is guaranteed in this case. On the other hand, in Example 2.20, Jacobi is applied to the matrix

$$A = \begin{bmatrix} 1 & 2 \\ 3 & 1 \end{bmatrix},$$

which is not diagonally dominant, and no such guarantee exists. Note that strict diagonal dominance is only a sufficient condition. The Jacobi Method may still converge in its absence.

▶ **EXAMPLE 2.21**   Determine whether the matrices

$$A = \begin{bmatrix} 3 & 1 & -1 \\ 2 & -5 & 2 \\ 1 & 6 & 8 \end{bmatrix} \quad \text{and} \quad B = \begin{bmatrix} 3 & 2 & 6 \\ 1 & 8 & 1 \\ 9 & 2 & -2 \end{bmatrix}$$

are strictly diagonally dominant.

The matrix A is diagonally dominant because $|3| > |1| + |-1|, |-5| > |2| + |2|$, and $|8| > |1| + |6|$. B is not, because, for example, $|3| > |2| + |6|$ is not true. However, if the first and third rows of B are exchanged, then B is strictly diagonally dominant and Jacobi is guaranteed to converge.   ◀

The Jacobi Method is a form of fixed-point iteration. Let D denote the main diagonal of A, L denote the lower triangle of A (entries below the main diagonal), and U denote the upper triangle (entries above the main diagonal). Then $A = L + D + U$, and the equation to be solved is $Lx + Dx + Ux = b$. Note that this use of L and U differs from the use in the LU factorization, since all diagonal entries of this L and U are zero. The system of equations $Ax = b$ can be rearranged in a fixed-point iteration of form:

$$Ax = b$$
$$(D + L + U)x = b$$
$$Dx = b - (L + U)x$$
$$x = D^{-1}(b - (L + U)x). \tag{2.39}$$

Since D is a diagonal matrix, its inverse is the matrix of reciprocals of the diagonal entries of A. The Jacobi Method is just the fixed-point iteration of (2.39):

Jacobi Method

$$x_0 = \text{initial vector}$$
$$x_{k+1} = D^{-1}(b - (L + U)x_k) \quad \text{for} \quad k = 0, 1, 2, \ldots. \qquad (2.40)$$

For Example 2.19,

$$\begin{bmatrix} 3 & 1 \\ 1 & 2 \end{bmatrix} \begin{bmatrix} u \\ v \end{bmatrix} = \begin{bmatrix} 5 \\ 5 \end{bmatrix},$$

the fixed-point iteration (2.40) with $x_k = \begin{bmatrix} u_k \\ v_k \end{bmatrix}$ is

$$\begin{bmatrix} u_{k+1} \\ v_{k+1} \end{bmatrix} = D^{-1}(b - (L + U)x_k)$$
$$= \begin{bmatrix} 1/3 & 0 \\ 0 & 1/2 \end{bmatrix} \left(\begin{bmatrix} 5 \\ 5 \end{bmatrix} - \begin{bmatrix} 0 & 1 \\ 1 & 0 \end{bmatrix} \begin{bmatrix} u_k \\ v_k \end{bmatrix} \right)$$
$$= \begin{bmatrix} (5 - v_k)/3 \\ (5 - u_k)/2 \end{bmatrix},$$

which agrees with our original version.

2.5.2 Gauss–Seidel Method and SOR

Closely related to the Jacobi Method is an iteration called the **Gauss–Seidel** Method. The only difference between Gauss–Seidel and Jacobi is that in the former, the most recently updated values of the unknowns are used at each step, even if the updating occurs in the current step. Returning to Example 2.19, we see that Gauss–Seidel looks like this:

$$\begin{bmatrix} u_0 \\ v_0 \end{bmatrix} = \begin{bmatrix} 0 \\ 0 \end{bmatrix}$$
$$\begin{bmatrix} u_1 \\ v_1 \end{bmatrix} = \begin{bmatrix} \frac{5 - v_0}{3} \\ \frac{5 - u_1}{2} \end{bmatrix} = \begin{bmatrix} \frac{5 - 0}{3} \\ \frac{5 - 5/3}{2} \end{bmatrix} = \begin{bmatrix} \frac{5}{3} \\ \frac{5}{3} \end{bmatrix}$$
$$\begin{bmatrix} u_2 \\ v_2 \end{bmatrix} = \begin{bmatrix} \frac{5 - v_1}{3} \\ \frac{5 - u_2}{2} \end{bmatrix} = \begin{bmatrix} \frac{5 - 5/3}{3} \\ \frac{5 - 10/9}{2} \end{bmatrix} = \begin{bmatrix} \frac{10}{9} \\ \frac{35}{18} \end{bmatrix}$$
$$\begin{bmatrix} u_3 \\ v_3 \end{bmatrix} = \begin{bmatrix} \frac{5 - v_2}{3} \\ \frac{5 - u_3}{2} \end{bmatrix} = \begin{bmatrix} \frac{5 - 35/18}{3} \\ \frac{5 - 55/54}{2} \end{bmatrix} = \begin{bmatrix} \frac{55}{54} \\ \frac{215}{108} \end{bmatrix}. \qquad (2.41)$$

Note the difference between Gauss–Seidel and Jacobi: The definition of v_1 uses u_1, not u_0. We see the approach to the solution $[1, 2]$ as with the Jacobi Method, but somewhat more accurately at the same number of steps. Gauss–Seidel often converges faster than Jacobi if the method is convergent. Theorem 2.11 verifies that the Gauss–Seidel Method, like Jacobi, converges to the solution as long as the coefficient matrix is strictly diagonally dominant.

Gauss–Seidel can be written in matrix form and identified as a fixed-point iteration where we isolate the equation $(L + D + U)x = b$ as

$$(L + D)x_{k+1} = -Ux_k + b.$$

Note that the usage of newly determined entries of x_{k+1} is accommodated by including the lower triangle of A into the left-hand side. Rearranging the equation gives the Gauss–Seidel Method.

Gauss–Seidel Method

$$x_0 = \text{initial vector}$$
$$x_{k+1} = D^{-1}(b - Ux_k - Lx_{k+1}) \quad \text{for} \ \ k = 0, 1, 2, \ldots.$$

▶ **EXAMPLE 2.22**   Apply the Gauss–Seidel Method to the system

$$\begin{bmatrix} 3 & 1 & -1 \\ 2 & 4 & 1 \\ -1 & 2 & 5 \end{bmatrix} \begin{bmatrix} u \\ v \\ w \end{bmatrix} = \begin{bmatrix} 4 \\ 1 \\ 1 \end{bmatrix}.$$

The Gauss–Seidel iteration is

$$u_{k+1} = \frac{4 - v_k + w_k}{3}$$
$$v_{k+1} = \frac{1 - 2u_{k+1} - w_k}{4}$$
$$w_{k+1} = \frac{1 + u_{k+1} - 2v_{k+1}}{5}.$$

Starting with $x_0 = [u_0, v_0, w_0] = [0, 0, 0]$, we calculate

$$\begin{bmatrix} u_1 \\ v_1 \\ w_1 \end{bmatrix} = \begin{bmatrix} \frac{4-0-0}{3} = \frac{4}{3} \\ \frac{1-8/3-0}{4} = -\frac{5}{12} \\ \frac{1+4/3+5/6}{5} = \frac{19}{30} \end{bmatrix} \approx \begin{bmatrix} 1.3333 \\ -0.4167 \\ 0.6333 \end{bmatrix}$$

and

$$\begin{bmatrix} u_2 \\ v_2 \\ w_2 \end{bmatrix} = \begin{bmatrix} \frac{101}{60} \\ -\frac{3}{4} \\ \frac{251}{300} \end{bmatrix} \approx \begin{bmatrix} 1.6833 \\ -0.7500 \\ 0.8367 \end{bmatrix}.$$

The system is strictly diagonally dominant, and therefore the iteration will converge to the solution $[2, -1, 1]$. ◀

The method called **Successive Over-Relaxation (SOR)** takes the Gauss–Seidel direction toward the solution and "overshoots" to try to speed convergence. Let ω be a real number, and define each component of the new guess x_{k+1} as a weighted average of ω times the Gauss–Seidel formula and $1 - \omega$ times the current guess x_k. The number ω is called the **relaxation parameter**, and $\omega > 1$ is referred to as **over-relaxation**.

▶ **EXAMPLE 2.23**   Apply SOR with $\omega = 1.25$ to the system of Example 2.22.

Successive Over-Relaxation yields

$$u_{k+1} = (1 - \omega)u_k + \omega \frac{4 - v_k + w_k}{3}$$
$$v_{k+1} = (1 - \omega)v_k + \omega \frac{1 - 2u_{k+1} - w_k}{4}$$
$$w_{k+1} = (1 - \omega)w_k + \omega \frac{1 + u_{k+1} - 2v_{k+1}}{5}.$$

Starting with $[u_0, v_0, w_0] = [0, 0, 0]$, we calculate

$$\begin{bmatrix} u_1 \\ v_1 \\ w_1 \end{bmatrix} \approx \begin{bmatrix} 1.6667 \\ -0.7292 \\ 1.0312 \end{bmatrix}$$

and

$$\begin{bmatrix} u_2 \\ v_2 \\ w_2 \end{bmatrix} \approx \begin{bmatrix} 1.9835 \\ -1.0672 \\ 1.0216 \end{bmatrix}.$$

In this example, the SOR iteration converges faster than Jacobi and Gauss–Seidel to the solution $[2, -1, 1]$. ◀

Just as with Jacobi and Gauss–Seidel, an alternative derivation of SOR follows from treating the system as a fixed-point problem. The problem $Ax = b$ can be written $(L + D + U)x = b$, and, upon multiplication by ω and rearranging,

$$(\omega L + \omega D + \omega U)x = \omega b$$
$$(\omega L + D)x = \omega b - \omega U x + (1 - \omega)Dx$$
$$x = (\omega L + D)^{-1}[(1 - \omega)Dx - \omega U x] + \omega(D + \omega L)^{-1}b.$$

Successive Over-Relaxation (SOR)

$x_0 = $ initial vector
$x_{k+1} = (\omega L + D)^{-1}[(1 - \omega)Dx_k - \omega U x_k] + \omega(D + \omega L)^{-1}b$ **for** $k = 0, 1, 2, \ldots$.

SOR with $\omega = 1$ is exactly Gauss–Seidel. The parameter ω can also be allowed to be less than 1, in a method called Successive Under-Relaxation.

▶ **EXAMPLE 2.24** Compare Jacobi, Gauss–Seidel, and SOR on the system of six equations in six unknowns:

$$\begin{bmatrix} 3 & -1 & 0 & 0 & 0 & \frac{1}{2} \\ -1 & 3 & -1 & 0 & \frac{1}{2} & 0 \\ 0 & -1 & 3 & -1 & 0 & 0 \\ 0 & 0 & -1 & 3 & -1 & 0 \\ 0 & \frac{1}{2} & 0 & -1 & 3 & -1 \\ \frac{1}{2} & 0 & 0 & 0 & -1 & 3 \end{bmatrix} \begin{bmatrix} u_1 \\ u_2 \\ u_3 \\ u_4 \\ u_5 \\ u_6 \end{bmatrix} = \begin{bmatrix} \frac{5}{2} \\ \frac{3}{2} \\ 1 \\ 1 \\ \frac{3}{2} \\ \frac{5}{2} \end{bmatrix}. \qquad (2.42)$$

The solution is $x = [1, 1, 1, 1, 1, 1]$. The approximate solution vectors x_6, after running six steps of each of the three methods, are shown in the following table:

Jacobi	Gauss–Seidel	SOR
0.9879	0.9950	0.9989
0.9846	0.9946	0.9993
0.9674	0.9969	1.0004
0.9674	0.9996	1.0009
0.9846	1.0016	1.0009
0.9879	1.0013	1.0004

The parameter ω for Successive Over-Relaxation was set at 1.1. SOR appears to be superior for this problem. ◄

Figure 2.3 compares the infinity norm error in Example 2.24 after six iterations for various ω. Although there is no general theory describing the best choice of ω, clearly there is a best choice in this case. See Ortega [1972] for discussion of the optimal ω in some common special cases.

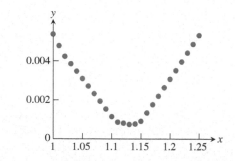

Figure 2.3 Infinity norm error after six steps of SOR in Example 2.24, as a function of over-relaxation parameter ω**.** Gauss–Seidel corresponds to $\omega = 1$. Minimum error occurs for $\omega \approx 1.13$.

2.5.3 Convergence of iterative methods

In this section we prove that the Jacobi and Gauss–Seidel Methods converge for strictly diagonally dominant matrices. This is the content of Theorems 2.10 and 2.11.

The Jacobi Method is written as

$$x_{k+1} = -D^{-1}(L + U)x_k + D^{-1}b. \tag{2.43}$$

Theorem A.7 of Appendix A governs convergence of such an iteration. According to this theorem, we need to know that the spectral radius $\rho(D^{-1}(L + U)) < 1$ in order to guarantee convergence of the Jacobi Method. This is exactly what strict diagonal dominance implies, as shown next.

Proof of Theorem 2.10. Let $R = L + U$ denote the nondiagonal part of the matrix. To check $\rho(D^{-1}R) < 1$, let λ be an eigenvalue of $D^{-1}R$ with corresponding eigenvector v. Choose this v so that $\|v\|_\infty = 1$, so that for some $1 \le m \le n$, the component $v_m = 1$ and all other components are no larger than 1. (This can be achieved by starting with any eigenvector and dividing by the largest component. Any constant multiple of an eigenvector is again an eigenvector with the same eigenvalue.) The definition of eigenvalue means that $D^{-1}Rv = \lambda v$, or $Rv = \lambda Dv$.

Since $r_{mm} = 0$, taking absolute values of the mth component of this vector equation implies

$$|r_{m1}v_1 + r_{m2}v_2 + \cdots + r_{m,m-1}v_{m-1} + r_{m,m+1}v_{m+1} + \cdots + r_{mn}v_n|$$
$$= |\lambda d_{mm}v_m| = |\lambda||d_{mm}|.$$

Since all $|v_i| \le 1$, the left-hand side is at most $\sum_{j\ne m}|r_{mj}|$, which, according to the strict diagonal dominance hypothesis, is less than $|d_{mm}|$. This implies that $|\lambda||d_{mm}| < |d_{mm}|$, which in turn forces $|\lambda| < 1$. Since λ was an arbitrary eigenvalue, we have shown $\rho(D^{-1}R) < 1$, as desired. Now Theorem A.7 from Appendix A implies that Jacobi converges to a solution of $Ax = b$. Finally, since $Ax = b$ has a solution for arbitrary b, A is a nonsingular matrix.

Putting the Gauss–Seidel Method into the form of (2.43) yields

$$x_{k+1} = -(L + D)^{-1}U x_k + (L + D)^{-1}b.$$

It then becomes clear that convergence of Gauss–Seidel follows if the spectral radius of the matrix

$$(L + D)^{-1}U \tag{2.44}$$

is less than one. The next theorem shows that strict diagonal dominance implies that this requirement is imposed on the eigenvalues.

THEOREM 2.11 If the $n \times n$ matrix A is strictly diagonally dominant, then (1) A is a nonsingular matrix, and (2) for every vector b and every starting guess, the Gauss–Seidel Method applied to $Ax = b$ converges to a solution. ∎

Proof. Let λ be an eigenvalue of (2.44), with corresponding eigenvector v. Choose the eigenvector so that $v_m = 1$ and all other components are smaller in magnitude, as in the preceding proof. Note that the entries of L are the a_{ij} for $i > j$, and the entries of U are the a_{ij} for $i < j$. Then viewing row m of the eigenvalue equation of (2.44),

$$\lambda(D + L)v = Uv,$$

yields a string of inequalities similar to the previous proof:

$$|\lambda|\left(\sum_{i>m}|a_{mi}|\right) < |\lambda|\left(|a_{mm}| - \sum_{i<m}|a_{mi}|\right)$$

$$\leq |\lambda|\left(|a_{mm}| - \left|\sum_{i<m}a_{mi}v_i\right|\right)$$

$$\leq |\lambda|\left|a_{mm} + \sum_{i<m}a_{ml}v_l\right|$$

$$= \left|\sum_{i>m}a_{mi}v_i\right|$$

$$\leq \sum_{i>m}|a_{mi}|.$$

It follows that $|\lambda| < 1$, which finishes the proof. □

2.5.4 Sparse matrix computations

Direct methods based on Gaussian elimination provide the user a finite number of steps that terminate in the solution. What is the reason for pursuing iterative methods, which are only approximate and may require several steps for convergence?

There are two major reasons for using iterative methods like Gauss–Seidel. Both reasons stem from the fact that one step of an iterative method requires only a fraction of the floating point operations of a full LU factorization. As we established earlier in the chapter, Gaussian elimination for an $n \times n$ matrix costs on the order of n^3

operations. A single step of Jacobi's Method, for example, requires about n^2 multiplications (one for each matrix entry) and about the same number of additions. The question is how many steps will be needed for convergence within the user's tolerance.

One particular circumstance that argues for an iterative technique is when a good approximation to the solution is already known. For example, suppose that a solution to $Ax = b$ is known, after which A and/or b change by a small amount. We could imagine a dynamic problem where A and b are remeasured constantly as they change, and an accurate updated solution x is constantly required. If the solution to the previous problem is used as a starting guess for the new but similar problem, fast convergence of Jacobi or Gauss–Seidel can be expected.

Suppose the b in problem (2.42) is changed slightly from the original $b = [2.5, 1.5, 1, 1, 1.5, 2.5]$ to a new $b = [2.2, 1.6, 0.9, 1.3, 1.4, 2.45]$. We can check that the true solution of the system is changed from $[1, 1, 1, 1, 1, 1]$ to $[0.9, 1, 1, 1.1, 1, 1]$. Assume that we have in memory the sixth step of the Gauss–Seidel iteration x_6 from the preceding table, to use as a starting guess. Continuing Gauss–Seidel with the new b and with the helpful starting guess x_6 yields a good approximation in only one additional step. The next two steps are as follows:

x_7	x_8
0.8980	0.8994
0.9980	0.9889
0.9659	0.9927
1.0892	1.0966
0.9971	1.0005
0.9993	1.0003

This technique is often called **polishing**, because the method begins with an approximate solution, which could be the solution from a previous, related problem, and then merely refines the approximate solution to make it more accurate. Polishing is common in real-time applications where the same problem needs to be re-solved repeatedly with data that is updated as time passes. If the system is large and time is short, it may be impossible to run an entire Gaussian elimination or even a back substitution in the allotted time. If the solution hasn't changed too much, a few steps of a relatively cheap iterative method might keep sufficient accuracy as the solution moves through time.

The second major reason to use iterative methods is to solve sparse systems of equations. A coefficient matrix is called **sparse** if many of the matrix entries are known to be zero. Often, of the n^2 eligible entries in a sparse matrix, only $O(n)$ of them are nonzero. A **full** matrix is the opposite, where few entries may be assumed to be zero. Gaussian elimination applied to a sparse matrix usually causes **fill-in**, where the coefficient matrix changes from sparse to full due to the necessary row operations. For this reason, the efficiency of Gaussian elimination and its PA = LU implementation become questionable for sparse matrices, leaving iterative methods as a feasible alternative.

Example 2.24 can be extended to a sparse matrix as follows:

▶ **EXAMPLE 2.25** Use the Jacobi Method to solve the 100,000-equation version of Example 2.24.

Let n be an even integer, and consider the $n \times n$ matrix A with 3 on the main diagonal, -1 on the super- and subdiagonal, and $1/2$ in the $(i, n + 1 - i)$ position for all $i = 1, \ldots, n$, except for $i = n/2$ and $n/2 + 1$. For $n = 12$,

$$
A = \begin{bmatrix}
3 & -1 & 0 & 0 & 0 & 0 & 0 & 0 & 0 & 0 & 0 & \frac{1}{2} \\
-1 & 3 & -1 & 0 & 0 & 0 & 0 & 0 & 0 & 0 & \frac{1}{2} & 0 \\
0 & -1 & 3 & -1 & 0 & 0 & 0 & 0 & 0 & \frac{1}{2} & 0 & 0 \\
0 & 0 & -1 & 3 & -1 & 0 & 0 & 0 & \frac{1}{2} & 0 & 0 & 0 \\
0 & 0 & 0 & -1 & 3 & -1 & 0 & \frac{1}{2} & 0 & 0 & 0 & 0 \\
0 & 0 & 0 & 0 & -1 & 3 & -1 & 0 & 0 & 0 & 0 & 0 \\
0 & 0 & 0 & 0 & 0 & -1 & 3 & -1 & 0 & 0 & 0 & 0 \\
0 & 0 & 0 & 0 & \frac{1}{2} & 0 & -1 & 3 & -1 & 0 & 0 & 0 \\
0 & 0 & 0 & \frac{1}{2} & 0 & 0 & 0 & -1 & 3 & -1 & 0 & 0 \\
0 & 0 & \frac{1}{2} & 0 & 0 & 0 & 0 & 0 & -1 & 3 & -1 & 0 \\
0 & \frac{1}{2} & 0 & 0 & 0 & 0 & 0 & 0 & 0 & -1 & 3 & -1 \\
\frac{1}{2} & 0 & 0 & 0 & 0 & 0 & 0 & 0 & 0 & 0 & -1 & 3
\end{bmatrix}. \tag{2.45}
$$

Define the vector $b = (2.5, 1.5, \ldots, 1.5, 1.0, 1.0, 1.5, \ldots, 1.5, 2.5)$, where there are $n - 4$ repetitions of 1.5 and 2 repetitions of 1.0. Note that if $n = 6$, A and b define the system of Example 2.24. The solution of the system for general n is $[1, \ldots, 1]$. No row of A has more than 4 nonzero entries. Since fewer than $4n$ of the n^2 potential entries are nonzero, we may call the matrix A sparse.

If we want to solve this system of equations for $n = 100,000$ or more, what are the options? Treating the coefficient matrix A as a full matrix means storing $n^2 = 10^{10}$ entries, each as a floating point double precision number requiring 8 bytes of storage. Note that 8×10^{10} bytes is approximately 80 gigabytes. Depending on your computational setup, it may be impossible to fit the entire n^2 entries into RAM.

Not only is size an enemy, but so is time. The number of operations required by Gaussian elimination will be on the order of $n^3 \approx 10^{15}$. If your machine runs on the order of a few GHz (10^9 cycles per second), an upper bound on the number of floating point operations per second is around 10^8. Therefore, $10^{15}/10^8 = 10^7$ is a reasonable guess at the number of seconds required for Gaussian elimination. There are 3×10^7 seconds in a year. Although this is back-of-the-envelope accounting, it is clear that Gaussian elimination for this problem is not an overnight computation.

On the other hand, one step of an iterative method will require approximately $2 \times 4n = 800,000$ operations, two for each nonzero matrix entry. We could do 100 steps of Jacobi iteration and still finish with fewer than 10^8 operations, which should take roughly a second or less on a modern PC. For the system just defined, with $n = 100,000$, the following Jacobi code jacobi.m needs only 50 steps to converge from a starting guess of $(0, \ldots, 0)$ to the solution $(1, \ldots, 1)$ within six correct decimal places. The 50 steps require less than 1 second on a typical PC.

MATLAB code shown here can be found at bit.ly/2RVqQJR

```
% Program 2.1 Sparse matrix setup
% Input: n = size of system
% Outputs: sparse matrix a, r.h.s. b
function [a,b] = sparsesetup(n)
e = ones(n,1); n2=n/2;
a = spdiags([-e 3*e -e],-1:1,n,n);    % Entries of a
c=spdiags([e/2],0,n,n);c=fliplr(c);a=a+c;
a(n2+1,n2) = -1; a(n2,n2+1) = -1;     % Fix up 2 entries
b=zeros(n,1);                         % Entries of r.h.s. b
b(1)=2.5;b(n)=2.5;b(2:n-1)=1.5;b(n2:n2+1)=1;
```

MATLAB code shown here can be found at bit.ly/2RWwBqz

```
% Program 2.2 Jacobi Method
% Inputs: full or sparse matrix a, r.h.s. b,
%         number of Jacobi iterations, k
% Output: solution x
function x = jacobi(a,b,k)
```

```
n=length(b);       % find n
d=diag(a);         % extract diagonal of a
r=a-diag(d);       % r is the remainder
x=zeros(n,1);      % initialize vector x
for j=1:k          % loop for Jacobi iteration
  x = (b-r*x)./d;
end                % End of Jacobi iteration loop
```

Note a few interesting aspects of the preceding code. The program spars-esetup.m uses MATLAB's spdiags command, which defines the matrix A as a sparse data structure. Essentially, this means that the matrix is represented by a set of triples (i, j, d), where d is the real number entry in position (i, j) of the matrix. Memory is not reserved for the entire n^2 potential entries, but only on an as-needed basis. The spdiags command takes the columns of a matrix and places them along the main diagonal, or a specified sub- or super-diagonal below or above the main diagonal.

MATLAB's matrix manipulation commands are designed to work seamlessly with the sparse matrix data structure. For example, an alternative to the preceding code would be to use MATLAB's lu command to solve the system directly. However, for that example, even though A is sparse, the upper-triangular matrix U that follows from Gaussian elimination suffers from fill-in during the process. For example, the upper-triangular U from Gaussian elimination for size $n = 12$ of the preceding matrix A is

$$\begin{bmatrix} 3 & -1.0 & 0 & 0 & 0 & 0 & 0 & 0 & 0 & 0 & 0 & 0.500 \\ 0 & 2.7 & -1.0 & 0 & 0 & 0 & 0 & 0 & 0 & 0 & 0.500 & 0.165 \\ 0 & 0 & 2.6 & -1.0 & 0 & 0 & 0 & 0 & 0 & 0.500 & 0.187 & 0.062 \\ 0 & 0 & 0 & 2.6 & -1.000 & 0 & 0 & 0 & 0.500 & 0.191 & 0.071 & 0.024 \\ 0 & 0 & 0 & 0 & 2.618 & -1.000 & 0 & 0.500 & 0.191 & 0.073 & 0.027 & 0.009 \\ 0 & 0 & 0 & 0 & 0 & 2.618 & -1.000 & 0.191 & 0.073 & 0.028 & 0.010 & 0.004 \\ 0 & 0 & 0 & 0 & 0 & 0 & 2.618 & -0.927 & 0.028 & 0.011 & 0.004 & 0.001 \\ 0 & 0 & 0 & 0 & 0 & 0 & 0 & 2.562 & -1.032 & -0.012 & -0.005 & -0.001 \\ 0 & 0 & 0 & 0 & 0 & 0 & 0 & 0 & 2.473 & -1.047 & -0.018 & -0.006 \\ 0 & 0 & 0 & 0 & 0 & 0 & 0 & 0 & 0 & 2.445 & -1.049 & -0.016 \\ 0 & 0 & 0 & 0 & 0 & 0 & 0 & 0 & 0 & 0 & 2.440 & -1.044 \\ 0 & 0 & 0 & 0 & 0 & 0 & 0 & 0 & 0 & 0 & 0 & 2.458 \end{bmatrix}$$

Since U turns out to be a relatively full matrix, the memory restrictions previously mentioned again become a limitation. A significant fraction of the n^2 memory locations will be necessary to store U on the way to completing the solution process. It is more efficient, by several orders of magnitude in execution time and storage, to solve this large sparse system by an iterative method. ◄

► **ADDITIONAL EXAMPLES**

1. Rearrange the equations $-x + 4y + z = 2, x - y + 3z = 8, 2x - z = 4$ to be strictly diagonally dominant, and apply two steps of the Gauss-Seidel method with initial guess $[0, 0, 0]$ to approximate the solution.

2. Adapt the sparsesetup.m and jacobi.m codes to apply the Jacobi method to the 100×100 tridiagonal system

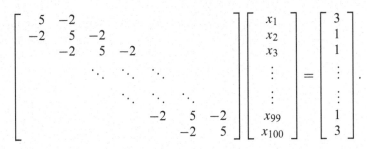

$$
\begin{bmatrix}
5 & -2 & & & & & \\
-2 & 5 & -2 & & & & \\
& -2 & 5 & -2 & & & \\
& & \ddots & \ddots & \ddots & & \\
& & & \ddots & \ddots & \ddots & \\
& & & & -2 & 5 & -2 \\
& & & & & -2 & 5
\end{bmatrix}
\begin{bmatrix}
x_1 \\ x_2 \\ x_3 \\ \vdots \\ \vdots \\ x_{99} \\ x_{100}
\end{bmatrix}
=
\begin{bmatrix}
3 \\ 1 \\ 1 \\ \vdots \\ \vdots \\ 1 \\ 3
\end{bmatrix}.
$$

Plot the solution error as a function of the number of Jacobi iteration steps.

⊡ **Solutions** for Additional Examples can be found at bit.ly/2yKacEh

2.5 Exercises

⊡ **Solutions** for Exercises numbered in blue can be found at bit.ly/2EpdsLu

1. Compute the first two steps of the Jacobi and the Gauss–Seidel Methods with starting vector $[0, \dots, 0]$.

(a) $\begin{bmatrix} 3 & -1 \\ -1 & 2 \end{bmatrix} \begin{bmatrix} u \\ v \end{bmatrix} = \begin{bmatrix} 5 \\ 4 \end{bmatrix}$
(b) $\begin{bmatrix} 2 & -1 & 0 \\ -1 & 2 & -1 \\ 0 & -1 & 2 \end{bmatrix} \begin{bmatrix} u \\ v \\ w \end{bmatrix} = \begin{bmatrix} 0 \\ 2 \\ 0 \end{bmatrix}$

(c) $\begin{bmatrix} 3 & 1 & 1 \\ 1 & 3 & 1 \\ 1 & 1 & 3 \end{bmatrix} \begin{bmatrix} u \\ v \\ w \end{bmatrix} = \begin{bmatrix} 6 \\ 3 \\ 5 \end{bmatrix}$

2. Rearrange the equations to form a strictly diagonally dominant system. Apply two steps of the Jacobi and Gauss–Seidel Methods from starting vector $[0, \dots, 0]$.

(a) $\begin{matrix} u + 3v = -1 \\ 5u + 4v = 6 \end{matrix}$
(b) $\begin{matrix} u - 8v - 2w = 1 \\ u + v + 5w = 4 \\ 3u - v + w = -2 \end{matrix}$
(c) $\begin{matrix} u + 4v = 5 \\ v + 2w = 2 \\ 4u + 3w = 0 \end{matrix}$

3. Apply two steps of SOR to the systems in Exercise 1. Use starting vector $[0, \dots, 0]$ and $\omega = 1.5$.

4. Apply two steps of SOR to the systems in Exercise 2 after rearranging. Use starting vector $[0, \dots, 0]$ and $\omega = 1$ and 1.2.

5. Let λ be an eigenvalue of an $n \times n$ matrix A. (a) Prove the Gershgorin Circle Theorem: There is a diagonal entry A_{mm} such that $|A_{mm} - \lambda| \le \sum_{j \ne m} |A_{mj}|$. (Hint: Begin with an eigenvector v such that $||v||_\infty = 1$, as in the proof of Theorem 2.10.) (b) Prove that a strictly diagonally dominant matrix cannot have a zero eigenvalue. This is an alternative proof of part (1) of Theorem 2.10.

2.5 Computer Problems

⊡ **Solutions** for Computer Problems numbered in blue can be found at bit.ly/2S5cKWu

1. Use the Jacobi Method to solve the sparse system within six correct decimal places (forward error in the infinity norm) for $n = 100$ and $n = 100000$. The correct solution is $[1, \dots, 1]$. Report the number of steps needed and the backward error. The system is

$$
\begin{bmatrix}
3 & -1 & & & \\
-1 & 3 & -1 & & \\
& \ddots & \ddots & \ddots & \\
& & -1 & 3 & -1 \\
& & & -1 & 3
\end{bmatrix}
\begin{bmatrix}
x_1 \\ \vdots \\ \vdots \\ x_n
\end{bmatrix}
=
\begin{bmatrix}
2 \\ 1 \\ \vdots \\ 1 \\ 2
\end{bmatrix}.
$$

2. Use the Jacobi Method to solve the sparse system within three correct decimal places (forward error in the infinity norm) for $n = 100$. The correct solution is $[1, -1, 1, -1, \ldots, 1, -1]$. Report the number of steps needed and the backward error. The system is

$$\begin{bmatrix} 2 & 1 & & & \\ 1 & 2 & 1 & & \\ & \ddots & \ddots & \ddots & \\ & & 1 & 2 & 1 \\ & & & 1 & 2 \end{bmatrix} \begin{bmatrix} x_1 \\ \\ \vdots \\ \\ x_n \end{bmatrix} = \begin{bmatrix} 1 \\ 0 \\ \vdots \\ 0 \\ -1 \end{bmatrix}.$$

3. Rewrite Program 2.2 to carry out Gauss–Seidel iteration. Solve the problem in Example 2.24 to check your work.

4. Rewrite Program 2.2 to carry out SOR. Use $\omega = 1.1$ to recheck Example 2.24.

5. Carry out the steps of Computer Problem 1 with $n = 100$ for (a) Gauss–Seidel Method and (b) SOR with $\omega = 1.2$.

6. Carry out the steps of Computer Problem 2 for (a) Gauss–Seidel Method and (b) SOR with $\omega = 1.5$.

7. Using your program from Computer Problem 3, decide how large a system of type (2.38) you can solve accurately by the Gauss–Seidel Method in one second of computation. Report the time required and forward error for various values of n.

2.6 METHODS FOR SYMMETRIC POSITIVE-DEFINITE MATRICES

Symmetric matrices hold a favored position in linear systems analysis because of their special structure, and because they have only about half as many independent entries as general matrices. That raises the question whether a factorization like the LU can be realized for half the computational complexity, and using only half the memory locations. For symmetric positive-definite matrices, this goal can be achieved with the Cholesky factorization.

Symmetric positive-definite matrices also allow a quite different approach to solving $Ax = b$, one that does not depend on a matrix factorization. This new approach, called the Conjugate Gradient Method, is especially useful for large, sparse matrices, where it falls into the family of iterative methods.

To begin the section, we define the concept of positive-definiteness for symmetric matrices. Then we show that every symmetric positive-definite matrix A can be factored as $A = R^T R$ for an upper-triangular matrix R, the Cholesky factorization. As a result, the problem $Ax = b$ can be solved using two back substitutions, just as with the LU factorization in the nonsymmetric case. We close the section with the conjugate gradient algorithm and an introduction to preconditioning.

2.6.1 Symmetric positive-definite matrices

DEFINITION 2.12   The $n \times n$ matrix A is **symmetric** if $A^T = A$. The matrix A is **positive-definite** if $x^T Ax > 0$ for all vectors $x \neq 0$.   ❐

▶ **EXAMPLE 2.26** Show that the matrix $A = \begin{bmatrix} 2 & 2 \\ 2 & 5 \end{bmatrix}$ is symmetric positive-definite.

Clearly A is symmetric. To show it is positive-definite, one applies the definition:

$$x^T A x = \begin{bmatrix} x_1 & x_2 \end{bmatrix} \begin{bmatrix} 2 & 2 \\ 2 & 5 \end{bmatrix} \begin{bmatrix} x_1 \\ x_2 \end{bmatrix}$$

$$= 2x_1^2 + 4x_1x_2 + 5x_2^2$$

$$= 2(x_1 + x_2)^2 + 3x_2^2$$

This expression is always non-negative, and cannot be zero unless both $x_2 = 0$ and $x_1 + x_2 = 0$, which together imply $x = 0$. ◀

▶ **EXAMPLE 2.27** Show that the symmetric matrix $A = \begin{bmatrix} 2 & 4 \\ 4 & 5 \end{bmatrix}$ is not positive-definite.

Compute $x^T A x$ by completing the square:

$$x^T A x = \begin{bmatrix} x_1 & x_2 \end{bmatrix} \begin{bmatrix} 2 & 4 \\ 4 & 5 \end{bmatrix} \begin{bmatrix} x_1 \\ x_2 \end{bmatrix}$$

$$= 2x_1^2 + 8x_1x_2 + 5x_2^2$$

$$= 2(x_1^2 + 4x_1x_2) + 5x_2^2$$

$$= 2(x_1 + 2x_2)^2 - 8x_2^2 + 5x_2^2$$

$$= 2(x_1 + 2x_2)^2 - 3x_2^2$$

Setting $x_1 = -2$ and $x_2 = 1$, for example, causes the result to be less than zero, contradicting the definition of positive-definite. ◀

Note that a symmetric positive-definite matrix must be nonsingular, since it is impossible for a nonzero vector x to satisfy $Ax = 0$. There are three additional important facts about this class of matrices.

Property 1 If the $n \times n$ matrix A is symmetric, then A is positive-definite if and only if all of its eigenvalues are positive.

Proof. Theorem A.5 says that, the set of unit eigenvectors is orthonormal and spans R^n. If A is positive-definite and $Av = \lambda v$ for a nonzero vector v, then $0 < v^T A v = v^T (\lambda v) = \lambda \|v\|_2^2$, so $\lambda > 0$. On the other hand, if all eigenvalues of A are positive, then write any nonzero $x = c_1 v_1 + \ldots + c_n v_n$ where the v_i are orthonormal unit vectors and not all c_i are zero. Then $x^T A x = (c_1 v_1 + \ldots + c_n v_n)^T (\lambda_1 c_1 v_1 + \ldots + \lambda_n c_n v_n) = \lambda_1 c_1^2 + \ldots + \lambda_n c_n^2 > 0$, so A is positive-definite. ☐

The eigenvalues of A in Example 2.26 are 6 and 1. The eigenvalues of A in Example 2.27 are approximately 7.77 and -0.77.

Property 2 If A is $n \times n$ symmetric positive-definite and X is an $n \times m$ matrix of full rank with $n \geq m$, then $X^T A X$ is $m \times m$ symmetric positive-definite.

Proof. The matrix is symmetric since $(X^T A X)^T = X^T A X$. To prove positive-definite, consider a nonzero m-vector v. Note that $v^T (X^T A X) v = (Xv)^T A(Xv) \geq 0$, with equality only if $Xv = 0$, due to the positive-definiteness of A. Since X has full rank, its columns are linearly independent, so that $Xv = 0$ implies $v = 0$. ☐

DEFINITION 2.13 A **principal** submatrix of a square matrix A is a square submatrix whose diagonal entries are diagonal entries of A. ☐

Property 3 Any principal submatrix of a symmetric positive-definite matrix is symmetric positive-definite.

Proof. Exercise 12. ☐

For example, if

$$\begin{bmatrix} a_{11} & a_{12} & a_{13} & a_{14} \\ a_{21} & a_{22} & a_{23} & a_{24} \\ a_{31} & a_{32} & a_{33} & a_{34} \\ a_{41} & a_{42} & a_{43} & a_{44} \end{bmatrix}$$

is symmetric positive-definite, then so is

$$\begin{bmatrix} a_{22} & a_{23} \\ a_{32} & a_{33} \end{bmatrix}.$$

2.6.2 Cholesky factorization

To demonstrate the main idea, we start with a 2×2 case. All of the important issues arise there; the extension to the general size is only some extra bookkeeping.

Consider the symmetric positive-definite matrix

$$\begin{bmatrix} a & b \\ b & c \end{bmatrix}.$$

By Property 3 of symmetric positive-definite matrices, we know that $a > 0$. In addition, we know that the determinant $ac - b^2$ of A is positive, since the determinant is the product of the eigenvalues, all positive by Property 1. Writing $A = R^T R$ with an upper triangular R implies the form

$$\begin{bmatrix} a & b \\ b & c \end{bmatrix} = \begin{bmatrix} \sqrt{a} & 0 \\ u & v \end{bmatrix} \begin{bmatrix} \sqrt{a} & u \\ 0 & v \end{bmatrix} = \begin{bmatrix} a & u\sqrt{a} \\ u\sqrt{a} & u^2 + v^2 \end{bmatrix},$$

and we want to check whether this is possible. Comparing left and right sides yields the identities $u = b/\sqrt{a}$ and $v^2 = c - u^2$. Note that $v^2 = c - (b/\sqrt{a})^2 = c - b^2/a > 0$ from our knowledge of the determinant. This verifies that v can be defined as a real number and so the Cholesky factorization

$$A = \begin{bmatrix} a & b \\ b & c \end{bmatrix} = \begin{bmatrix} \sqrt{a} & 0 \\ \frac{b}{\sqrt{a}} & \sqrt{c - b^2/a} \end{bmatrix} \begin{bmatrix} \sqrt{a} & \frac{b}{\sqrt{a}} \\ 0 & \sqrt{c - b^2/a} \end{bmatrix} = R^T R$$

exists for 2×2 symmetric positive-definite matrices. The Cholesky factorization is not unique; clearly we could just as well have chosen v to be the negative square root of $c - b^2/a$.

The next result guarantees that the same idea works for the $n \times n$ case.

THEOREM 2.14 (Cholesky Factorization Theorem) If A is a symmetric positive-definite $n \times n$ matrix, then there exists an upper triangular $n \times n$ matrix R such that $A = R^T R$.   ∎

Proof. We construct R by induction on the size n. The case $n = 2$ was done above. Consider A partitioned as

$$
A = \left[\begin{array}{c|c} a & b^T \\ \hline \\ b & C \\ \\ \end{array}\right]
$$

where b is an $(n-1)$-vector and C is an $(n-1) \times (n-1)$ submatrix. We will use block multiplication (see the Appendix section A.2) to simplify the argument. Set $u = b/\sqrt{a}$ as in the 2×2 case. Setting $A_1 = C - uu^T$ and defining the invertible matrix

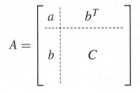

$$
S = \left[\begin{array}{c|c} \sqrt{a} & u^T \\ 0 & \\ \hline \\ \vdots & I \\ 0 & \end{array}\right]
$$

yields

$$
S^T \left[\begin{array}{c|c} 1 & 0 \;\cdots\; 0 \\ 0 & \\ \hline \\ \vdots & A_1 \\ 0 & \end{array}\right] S = \left[\begin{array}{c|c} \sqrt{a} & 0 \;\cdots\; 0 \\ \hline \\ u & I \\ 0 & \end{array}\right] \left[\begin{array}{c|c} 1 & 0 \;\cdots\; 0 \\ 0 & \\ \hline \\ \vdots & A_1 \\ 0 & \end{array}\right] \left[\begin{array}{c|c} \sqrt{a} & u^T \\ 0 & \\ \hline \\ \vdots & I \\ 0 & \end{array}\right]
$$

$$
= \left[\begin{array}{c|c} a & b^T \\ \hline \\ b & uu^T + A_1 \\ \\ \end{array}\right] = A
$$

Notice that A_1 is symmetric positive-definite. This follows from the facts that

$$
\left[\begin{array}{c|c} 1 & 0 \;\cdots\; 0 \\ 0 & \\ \hline \\ \vdots & A_1 \\ 0 & \end{array}\right] = (S^T)^{-1} A S^{-1}
$$

is symmetric positive-definite by Property 2, and therefore so is the $(n-1) \times (n-1)$ principal submatrix A_1 by Property 3. By the induction hypothesis, $A_1 = V^T V$ where V is upper triangular. Finally, define the upper triangular matrix

$$
R = \left[\begin{array}{c|c} \sqrt{a} & u^T \\ 0 & \\ \hline \\ \vdots & V \\ 0 & \end{array}\right]
$$

and check that

$$
R^T R = \left[\begin{array}{c|ccc} \sqrt{a} & 0 & \cdots & 0 \\ \hline u & & V^T & \end{array} \right] \left[\begin{array}{c|c} \sqrt{a} & u^T \\ \hline 0 & \\ \vdots & V \\ 0 & \end{array} \right] = \left[\begin{array}{c|c} a & b^T \\ \hline b & uu^T + V^T V \end{array} \right] = A,
$$

which completes the proof.   ❐

The construction of the proof can be carried out explicitly, in what has become the standard algorithm for the Cholesky factorization. The matrix R is built from the outside in. First we find $r_{11} = \sqrt{a_{11}}$ and set the rest of the top row of R to $u^T = b^T / r_{11}$. Then uu^T is subtracted from the lower principal $(n - 1) \times (n - 1)$ submatrix, and the same steps are repeated on it to fill in the second row of R. These steps are continued until all rows of R are determined. According to the theorem, the new principal submatrix is positive-definite at every stage of the construction, so by Property 3, the top left corner entry is positive, and the square root operation succeeds. This approach can be put directly into the following algorithm. We use the "colon notation" to denote submatrices.

Cholesky factorization

for $k = 1, 2, \ldots, n$
    if $A_{kk} < 0$, **stop, end**
    $R_{kk} = \sqrt{A_{kk}}$
    $u^T = \frac{1}{R_{kk}} A_{k,k+1:n}$
    $R_{k,k+1:n} = u^T$
    $A_{k+1:n,k+1:n} = A_{k+1:n,k+1:n} - uu^T$
end

The resulting R is upper triangular and satisfies $A = R^T R$.

▶ **EXAMPLE 2.28**   Find the Cholesky factorization of $\begin{bmatrix} 4 & -2 & 2 \\ -2 & 2 & -4 \\ 2 & -4 & 11 \end{bmatrix}$.

The top row of R is $R_{11} = \sqrt{a_{11}} = 2$, followed by $R_{1,2:3} = [-2, 2]/R_{11} = [-1, 1]$:

$$
R = \left[\begin{array}{c|cc} 2 & -1 & 1 \\ \hline & & \\ & & \\ \end{array} \right].
$$

Subtracting the outer product $uu^T = \begin{bmatrix} -1 \\ 1 \end{bmatrix} \begin{bmatrix} -1 & 1 \end{bmatrix}$ from the lower principal 2×2 submatrix $A_{2:3,2:3}$ of A leaves

$$
\left[\begin{array}{c|cc} & & \\ \hline & 2 & -4 \\ & -4 & 11 \end{array} \right] - \left[\begin{array}{c|cc} & & \\ \hline & 1 & -1 \\ & -1 & 1 \end{array} \right] = \left[\begin{array}{c|cc} & & \\ \hline & 1 & -3 \\ & -3 & 10 \end{array} \right].
$$

Now we repeat the same steps on the 2×2 submatrix to find $R_{22} = 1$ and $R_{23} = -3/1 = -3$:

$$R = \begin{bmatrix} 2 & -1 & 1 \\ & 1 & -3 \\ & & \end{bmatrix}.$$

The lower 1×1 principal submatrix of A is $10 - (-3)(-3) = 1$, so $R_{33} = \sqrt{1}$. The Cholesky factor of A is

$$R = \begin{bmatrix} 2 & -1 & 1 \\ 0 & 1 & -3 \\ 0 & 0 & 1 \end{bmatrix}.$$
◄

Solving $Ax = b$ for symmetric positive-definite A follows the same idea as the LU factorization. Now that $A = R^T R$ is a product of two triangular matrices, we need to solve the lower triangular system $R^T c = b$ and the upper triangular system $Rx = c$ to determine the solution x.

2.6.3 Conjugate Gradient Method

The introduction of the Conjugate Gradient Method (Hestenes and Steifel, 1952) ushered in a new era for iterative methods to solve sparse matrix problems. Although the method was slow to catch on, once effective preconditioners were developed, huge problems that could not be attacked any other way became feasible. The achievement led shortly to much further progress and a new generation of iterative solvers.

SPOTLIGHT ON

Orthogonality Our first real application of orthogonality in this book uses it in a roundabout way, to solve a problem that has no obvious link to orthogonality. The Conjugate Gradient Method tracks down the solution of a positive-definite $n \times n$ linear system by successively locating and eliminating the n orthogonal components of the error, one by one. The complexity of the algorithm is minimized by using the directions established by pairwise orthogonal residual vectors. We will develop this point of view further in Chapter 4, culminating in the GMRES Method, a nonsymmetric counterpart to conjugate gradients.

The ideas behind conjugate gradients rely on the generalization of the usual idea of inner product. The Euclidean inner product $(v, w) = v^T w$ is symmetric and linear in the inputs v and w, since $(v, w) = (w, v)$ and $(\alpha v + \beta w, u) = \alpha(v, u) + \beta(w, u)$ for scalars α and β. The Euclidean inner product is also positive-definite, in that $(v, v) > 0$ if $v \neq 0$.

DEFINITION 2.15 Let A be a symmetric positive-definite $n \times n$ matrix. For two n-vectors v and w, define the A-**inner product**

$$(v, w)_A = v^T A w.$$

The vectors v and w are A-**conjugate** if $(v, w)_A = 0$. ❑

Note that the new inner product inherits the properties of symmetry, linearity, and positive-definiteness from the matrix A. Because A is symmetric, so is the A-inner product: $(v, w)_A = v^T A w = (v^T A w)^T = w^T A v = (w, v)_A$. The A-inner product is also linear, and positive-definiteness follows from the fact that if A is positive-definite, then

$$(v, v)_A = v^T A v > 0$$

if $v \neq 0$.

Strictly speaking, the Conjugate Gradient Method is a direct method, and arrives at the solution x of the symmetric positive-definite system $Ax = b$ with the following finite loop:

Conjugate Gradient Method

$x_0 =$ initial guess
$d_0 = r_0 = b - Ax_0$
for $k = 0, 1, 2, \ldots, n - 1$
    if $r_k = 0$, **stop, end**
    $\alpha_k = \frac{r_k^T r_k}{d_k^T A d_k}$
    $x_{k+1} = x_k + \alpha_k d_k$
    $r_{k+1} = r_k - \alpha_k A d_k$
    $\beta_k = \frac{r_{k+1}^T r_{k+1}}{r_k^T r_k}$
    $d_{k+1} = r_{k+1} + \beta_k d_k$
end

An informal description of the iteration is next, to be followed by proof of the necessary facts in Theorem 2.16. The conjugate gradient iteration updates three different vectors on each step. The vector x_k is the approximate solution at step k. The vector r_k represents the residual of the approximate solution x_k. This is clear for r_0 by definition, and during the iteration, notice that

$$Ax_{k+1} + r_{k+1} = A(x_k + \alpha_k d_k) + r_k - \alpha_k A d_k$$
$$= Ax_k + r_k,$$

and so by induction $r_k = b - Ax_k$ for all k. Finally, the vector d_k represents the new search direction used to update the approximation x_k to the improved version x_{k+1}.

The method succeeds because each residual is arranged to be orthogonal to all previous residuals. If this can be done, the method runs out of orthogonal directions in which to look, and must reach a zero residual and a correct solution in at most n steps. The key to accomplishing the orthogonality among residuals turns out to be choosing the search directions d_k pairwise conjugate. The concept of conjugacy generalizes orthogonality and gives its name to the algorithm.

Now we explain the choices of α_k and β_k. The directions d_k are chosen from the vector space span of the previous residuals, as seen inductively from the last line of the pseudocode. In order to ensure that the next residual is orthogonal to all past residuals, α_k in chosen precisely so that the new residual r_{k+1} is orthogonal to the direction d_k:

$$x_{k+1} = x_k + \alpha_k d_k$$
$$b - Ax_{k+1} = b - Ax_k - \alpha_k A d_k$$
$$r_{k+1} = r_k - \alpha_k A d_k$$

$$0 = d_k^T r_{k+1} = d_k^T r_k - \alpha_k d_k^T A d_k$$

$$\alpha_k = \frac{d_k^T r_k}{d_k^T A d_k}.$$

This is not exactly how α_k is written in the algorithm, but note that since d_{k-1} is orthogonal to r_k, we have

$$d_k - r_k = \beta_{k-1} d_{k-1}$$
$$r_k^T d_k - r_k^T r_k = 0,$$

which justifies the rewriting $r_k^T d_k = r_k^T r_k$. Secondly, the coefficient β_k is chosen to ensure the pairwise A-conjugacy of the d_k:

$$d_{k+1} = r_{k+1} + \beta_k d_k$$
$$0 = d_k^T A d_{k+1} - d_k^T A r_{k+1} + \beta_k d_k^T A d_k$$
$$\beta_k = -\frac{d_k^T A r_{k+1}}{d_k^T A d_k}.$$

The expression for β_k can be rewritten in the simpler form seen in the algorithm, as shown in (2.47) below.

Theorem 2.16 below verifies that all r_k produced by the conjugate gradient iteration are orthogonal to one another. Since they are n-dimensional vectors, at most n of the r_k can be pairwise orthogonal, so either r_n or a previous r_k must be zero, solving $Ax = b$. Therefore after at most n steps, conjugate gradient arrives at a solution. In theory, the method is a direct, not an iterative, method.

Before turning to the theorem that guarantees the success of the Conjugate Gradient Method, it is instructive to carry out an example in exact arithmetic.

▶ **EXAMPLE 2.29** Solve $\begin{bmatrix} 2 & 2 \\ 2 & 5 \end{bmatrix} \begin{bmatrix} u \\ v \end{bmatrix} = \begin{bmatrix} 6 \\ 3 \end{bmatrix}$ using the Conjugate Gradient Method.

Following the above algorithm we have

$$x_0 = \begin{bmatrix} 0 \\ 0 \end{bmatrix}, r_0 = d_0 = \begin{bmatrix} 6 \\ 3 \end{bmatrix}$$

$$\alpha_0 = \frac{\begin{bmatrix} 6 \\ 3 \end{bmatrix}^T \begin{bmatrix} 6 \\ 3 \end{bmatrix}}{\begin{bmatrix} 6 \\ 3 \end{bmatrix}^T \begin{bmatrix} 2 & 2 \\ 2 & 5 \end{bmatrix} \begin{bmatrix} 6 \\ 3 \end{bmatrix}} = \frac{45}{6 \cdot 18 + 3 \cdot 27} = \frac{5}{21}$$

$$x_1 = \begin{bmatrix} 0 \\ 0 \end{bmatrix} + \frac{5}{21} \begin{bmatrix} 6 \\ 3 \end{bmatrix} = \begin{bmatrix} 10/7 \\ 5/7 \end{bmatrix}$$

$$r_1 = \begin{bmatrix} 6 \\ 3 \end{bmatrix} - \frac{5}{21} \begin{bmatrix} 18 \\ 27 \end{bmatrix} = 12 \begin{bmatrix} 1/7 \\ -2/7 \end{bmatrix}$$

$$\beta_0 = \frac{r_1^T r_1}{r_0^T r_0} = \frac{144 \cdot 5/49}{36 + 9} = \frac{16}{49}$$

$$d_1 = 12 \begin{bmatrix} 1/7 \\ -2/7 \end{bmatrix} + \frac{16}{49} \begin{bmatrix} 6 \\ 3 \end{bmatrix} = \begin{bmatrix} 180/49 \\ -120/49 \end{bmatrix}$$

$$\alpha_1 = \frac{\begin{bmatrix} 12/7 \\ -24/7 \end{bmatrix}^T \begin{bmatrix} 12/7 \\ -24/7 \end{bmatrix}}{\begin{bmatrix} 180/49 \\ -120/49 \end{bmatrix}^T \begin{bmatrix} 2 & 2 \\ 2 & 5 \end{bmatrix} \begin{bmatrix} 180/49 \\ -120/49 \end{bmatrix}} = \frac{7}{10}$$

$$x_2 = \begin{bmatrix} 10/7 \\ 5/7 \end{bmatrix} + \frac{7}{10} \begin{bmatrix} 180/49 \\ -120/49 \end{bmatrix} = \begin{bmatrix} 4 \\ -1 \end{bmatrix}$$

$$r_2 = 12 \begin{bmatrix} 1/7 \\ -2/7 \end{bmatrix} - \frac{7}{10} \begin{bmatrix} 2 & 2 \\ 2 & 5 \end{bmatrix} \begin{bmatrix} 180/49 \\ -120/49 \end{bmatrix} = \begin{bmatrix} 0 \\ 0 \end{bmatrix}$$

Since $r_2 = b - Ax_2 = 0$, the solution is $x_2 = [4, -1]$. ◄

THEOREM 2.16 Let A be a symmetric positive-definite $n \times n$ matrix and let $b \neq 0$ be a vector. In the Conjugate Gradient Method, assume that $r_k \neq 0$ for $k < n$ (if $r_k = 0$ the equation is solved). Then for each $1 \leq k \leq n$,

(a) The following three subspaces of R^n are equal:

$$\langle x_1, \ldots, x_k \rangle = \langle r_0, \ldots, r_{k-1} \rangle = \langle d_0, \ldots, d_{k-1} \rangle,$$

(b) the residuals r_k are pairwise orthogonal: $r_k^T r_j = 0$ for $j < k$,
(c) the directions d_k are pairwise A-conjugate: $d_k^T A d_j = 0$ for $j < k$. ∎

Proof. (a) For $k = 1$, note that $\langle x_1 \rangle = \langle d_0 \rangle = \langle r_0 \rangle$, since $x_0 = 0$. Here we use $\langle \ \rangle$ to denote the span of vectors inside the angle braces. By definition $x_k = x_{k-1} + \alpha_{k-1} d_{k-1}$. This implies by induction that $\langle x_1, \ldots, x_k \rangle = \langle d_0, \ldots, d_{k-1} \rangle$. A similar argument using $d_k = r_k + \beta_{k-1} d_{k-1}$ shows that $\langle r_0, \ldots, r_{k-1} \rangle$ is equal to $\langle d_0, \ldots, d_{k-1} \rangle$.

For (b) and (c), proceed by induction. When $k = 0$ there is nothing to prove. Assume (b) and (c) hold for k, and we will prove (b) and (c) for $k + 1$. Multiply the definition of r_{k+1} by r_j^T on the left:

$$r_j^T r_{k+1} = r_j^T r_k - \frac{r_k^T r_k}{d_k^T A d_k} r_j^T A d_k. \tag{2.46}$$

If $j \leq k - 1$, then $r_j^T r_k = 0$ by the induction hypothesis (b). Since r_j can be expressed as a combination of $d_0, \ldots, d_j$, the term $r_j^T A d_k = 0$ from the induction hypothesis (c), and (b) holds. On the other hand, if $j = k$, then $r_k^T r_{k+1} = 0$ again follows from (2.46) because $d_k^T A d_k = r_k^T A d_k + \beta_{k-1} d_{k-1}^T A d_k = r_k^T A d_k$, using the induction hypothesis (c). This proves (b).

Now that $r_k^T r_{k+1} = 0$, (2.46) with $j = k + 1$ says

$$\frac{r_{k+1}^T r_{k+1}}{r_k^T r_k} = -\frac{r_{k+1}^T A d_k}{d_k^T A d_k}. \tag{2.47}$$

This together with multiplying the definition of d_{k+1} on the left by $d_j^T A$ yields

$$d_j^T A d_{k+1} = d_j^T A r_{k+1} - \frac{r_{k+1}^T A d_k}{d_k^T A d_k} d_j^T A d_k. \tag{2.48}$$

If $j = k$, then $d_k^T A d_{k+1} = 0$ from (2.48), using the symmetry of A. If $j \leq k - 1$, then $A d_j = (r_j - r_{j+1})/\alpha_j$ (from the definition of r_{k+1}) is orthogonal to r_{k+1}, showing the first term on the right-hand side of (2.48) is zero, and the second term is zero by the induction hypothesis, which completes the argument for (c). ☐

In Example 2.29, notice that r_1 is orthogonal to r_0, as guaranteed by Theorem 2.16. This fact is the key to success for the Conjugate Gradient Method: Each new residual r_i is orthogonal to all previous r_i's. If one of the r_i turns out to be zero, then $Ax_i = b$ and x_i is the solution. If not, after n steps through the loop, r_n is orthogonal to a space spanned by the n pairwise orthogonal vectors $r_0, \ldots, r_{n-1}$, which must be all of R^n. So r_n must be the zero vector, and $Ax_n = b$.

The Conjugate Gradient Method is in some ways simpler than Gaussian elimination. For example, writing the code appears to be more foolproof—there are no row operations to worry about, and there is no triple loop as in Gaussian elimination. Both are direct methods, and they both arrive at the theoretically correct solution in a finite number of steps. So two questions remain: Why shouldn't Conjugate Gradient be preferred to Gaussian elimination, and why is Conjugate Gradient often treated as an iterative method?

The answer to both questions begins with an operation count. Moving through the loop requires one matrix-vector product Ad_{n-1} and several additional dot products. The matrix-vector product alone requires n^2 multiplications for each step (along with about the same number of additions), for a total of $2n^3$ operations after n steps. Compared to the count of $2n^3/3$ for Gaussian elimination, this is three times too expensive.

The picture changes if A is sparse. Assume that n is too large for the $2n^3/3$ operations of Gaussian elimination to be feasible. Although Gaussian elimination must be run to completion to give a solution x, Conjugate Gradient gives an approximation x_i on each step.

The backward error, the Euclidean length of the residual, decreases on each step, and so at least by that measure, Ax_i is getting nearer to b on each step. Therefore by monitoring the r_i, a good enough solution x_i may be found to avoid completing all n steps. In this context, Conjugate Gradient becomes indistinguishable from an iterative method.

The method fell out of favor shortly after its discovery because of its susceptibility to accumulation of round-off errors when A is an ill-conditioned matrix. In fact, its performance on ill-conditioned matrices is inferior to Gaussian elimination with partial pivoting. In modern days, this obstruction is relieved by **preconditioning**, which essentially changes the problem to a better-conditioned matrix system, after which Conjugate Gradient is applied. We will investigate the Preconditioned Conjugate Gradient Method in the next section.

The title of the method comes from what the Conjugate Gradient Method is really doing: sliding down the slopes of a quadratic paraboloid in n dimensions. The "gradient" part of the title means it is finding the direction of fastest decline using calculus, and "conjugate" means not quite that its individual steps are orthogonal to one another, but that at least the residuals r_i are. The geometric details of the method and its motivation are interesting. The original article Hestenes and Steifel [1952] gives a complete description.

The MATLAB command `cgs` implements the Conjugate Gradient Method.

► **EXAMPLE 2.30**   Apply the Conjugate Gradient Method to system (2.45) with $n = 100,000$.

After 20 steps of the Conjugate Gradient Method, the difference between the computed solution x and the true solution $(1, \ldots, 1)$ is less than 10^{-9} in the vector infinity norm. The total time of execution was less than one second on a PC.   ◄

2.6.4 Preconditioning

Convergence of iterative methods like the Conjugate Gradient Method can be accelerated by the use of a technique called preconditioning. The convergence rates of iterative methods often depend, directly or indirectly, on the condition number of the coefficient matrix A. The idea of preconditioning is to reduce the effective condition number of the problem.

The preconditioned form of the $n \times n$ linear system $Ax = b$ is

$$M^{-1}Ax = M^{-1}b,$$

where M is an invertible $n \times n$ matrix called the **preconditioner**. All we have done is to left-multiply the equation by a matrix. An effective preconditioner reduces the condition number of the problem by attempting to invert A. Conceptually, it tries to do two things at once: the matrix M should be (1) as close to A as possible and (2) simple to invert. These two goals usually stand in opposition to one another.

The matrix closest to A is A itself. Using $M = A$ would bring the condition number of the problem to 1, but presumably A is not trivial to invert or we would not be using a sophisticated solution method. The easiest matrix to invert is the identity matrix $M = I$, but this does not reduce the condition number. The perfect preconditioner would be a matrix in the middle of the two extremes that combines the best properties of both.

A particularly simple choice is the **Jacobi preconditioner** $M = D$, where D is the diagonal of A. The inverse of D is the diagonal matrix of reciprocals of the entries of D. In a strictly diagonally dominant matrix, for example, the Jacobi preconditioner holds a close resemblance to A while being simple to invert. Note that each diagonal entry of a symmetric positive-definite matrix is strictly positive by Property 3 of Section 2.6.1, so finding reciprocals is not a problem.

When A is a symmetric positive-definite $n \times n$ matrix, we will choose a symmetric positive-definite matrix M for use as a preconditioner. Recall the M-inner product $(v, w)_M = v^T M w$ as defined in Section 2.6.3. The Preconditioned Conjugate Gradient Method is now easy to describe: Replace $Ax = b$ with the preconditioned equation $M^{-1}Ax = M^{-1}b$, and replace the Euclidean inner product with $(v, w)_M$. The reasoning used for the original Conjugate Gradient Method still applies because the matrix $M^{-1}A$ remains symmetric positive-definite in the new inner product.

For example,

$$(M^{-1}Av, w)_M = v^T A M^{-1} M w = v^T A w = v^T M M^{-1} A w = (v, M^{-1}Aw)_M.$$

To convert the algorithm from Section 2.6.3 to the preconditioned version, let $z_k = M^{-1}b - M^{-1}Ax_k = M^{-1}r_k$ be the residual of the preconditioned system. Then

$$\alpha_k = \frac{(z_k, z_k)_M}{(d_k, M^{-1}Ad_k)_M}$$

$$x_{k+1} = x_k + \alpha d_k$$

$$z_{k+1} = z_k - \alpha M^{-1}Ad_k$$

$$\beta_k = \frac{(z_{k+1}, z_{k+1})_M}{(z_k, z_k)_M}$$

$$d_{k+1} = z_{k+1} + \beta_k d_k.$$

Multiplications by M can be reduced by noting that

$$(z_k, z_k)_M = z_k^T M z_k = z_k^T r_k$$

$$(d_k, M^{-1}Ad_k)_M = d_k^T A d_k$$

$$(z_{k+1}, z_{k+1})_M = z_{k+1}^T M z_{k+1} = z_{k+1}^T r_{k+1}.$$

With these simplifications, the pseudocode for the preconditioned version goes as follows.

Preconditioned Conjugate Gradient Method

$x_0 = $ initial guess
$r_0 = b - Ax_0$
$d_0 = z_0 = M^{-1}r_0$
for $k = 0, 1, 2, \ldots, n-1$
    if $r_k = 0$, **stop, end**
    $\alpha_k = r_k^T z_k / d_k^T A d_k$
    $x_{k+1} = x_k + \alpha_k d_k$
    $r_{k+1} = r_k - \alpha_k A d_k$
    $z_{k+1} = M^{-1} r_{k+1}$
    $\beta_k = r_{k+1}^T z_{k+1} / r_k^T z_k$
    $d_{k+1} = z_{k+1} + \beta_k d_k$
end

The approximation to the solution of $Ax = b$ after k steps is x_k. Note that no explicit multiplications by M^{-1} should be carried out. They should be replaced with appropriate back substitutions due to the relative simplicity of M. The Preconditioned Conjugate Gradient Method is implemented in MATLAB with the pcg command.

The Jacobi preconditioner is the simplest of an extensive and growing library of possible choices. We will describe one further family of examples, and direct the reader to the literature for more sophisticated alternatives.

The **symmetric successive over-relaxation (SSOR)** preconditioner is defined by

$$M = (D + \omega L)D^{-1}(D + \omega U)$$

where $A = L + D + U$ is divided into its lower triangular part, diagonal, and upper triangular part. As in the SOR method, ω is a constant between 0 and 2. The special case $\omega = 1$ is called the **Gauss–Seidel preconditioner**.

A preconditioner is of little use if it is difficult to invert. Notice that the SSOR preconditioner is defined as a product $M = (I + \omega L D^{-1})(D + \omega U)$ of a lower triangular and an upper triangular matrix, so that the equation $z = M^{-1}v$ can be solved by two back substitutions:

$$(I + \omega L D^{-1})c = v$$
$$(D + \omega U)z = c$$

For a sparse matrix, the two back substitutions can be done in time proportional to the number of nonzero entries. In other words, multiplication by M^{-1} is not significantly higher in complexity than multiplication by M.

▶ **EXAMPLE 2.31** Let A denote the matrix with diagonal entries $A_{ii} = \sqrt{i}$ for $i = 1, \ldots, n$ and $A_{i,i+10} = A_{i+10,i} = \cos i$ for $i = 1, \ldots, n - 10$, with all other entries zero. Set x to be the vector of n ones, and define $b = Ax$. For $n = 500$, solve $Ax = b$ with the Conjugate Gradient Method in three ways: using no preconditioner, using the Jacobi preconditioner, and using the Gauss–Seidel preconditioner.

The matrix can be defined in MATLAB by
```
A=diag(sqrt(1:n))+ diag(cos(1:(n-10)),10)
                + diag(cos(1:(n-10)),-10).
```

Figure 2.4 shows the three different results. Even with this simply defined matrix, the Conjugate Gradient Method is fairly slow to converge without preconditioning. The Jacobi preconditioner, which is quite easy to apply, makes a significant improvement, while the Gauss–Seidel preconditioner requires only about 10 steps to reach machine accuracy. ◀

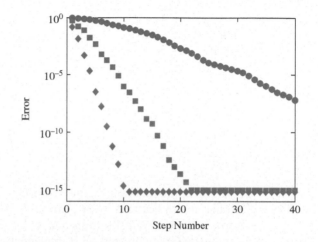

Figure 2.4 Efficiency of Preconditioned Conjugate Gradient Method for the solution of Example 2.31. Error is plotted by step number. Circles: no preconditioner. Squares: Jacobi preconditioner. Diamonds: Gauss–Seidel preconditioner.

▶ **ADDITIONAL EXAMPLES**

1. Find the Cholesky factorization of the symmetric positive-definite matrix $\begin{bmatrix} 4 & -2 \\ -2 & 6 \end{bmatrix}$.

2. Let $n = 100$, and let A be the $n \times n$ matrix with diagonal entries $A(i,i) = i$ and entries $A(i, i+1) = A(i+1, i) = 0.4$ on the super-diagonal and sub-diagonal. Let x_c denote the vector of n ones, and set $b = Ax_c$. Apply the Conjugate Gradient Method (a) with no preconditioner, (b) with the Jacobi preconditioner, and (c) with the Gauss-Seidel preconditioner. Compare errors of the three runs by plotting error versus step number.

Solutions for Additional Examples can be found at `bit.ly/2NMBJuc`

2.6 Exercises

1. Show that the following matrices are symmetric positive-definite by expressing $x^T Ax$ as a sum of squares.

   $(a) \begin{bmatrix} 1 & 0 \\ 0 & 3 \end{bmatrix} (b) \begin{bmatrix} 1 & 3 \\ 3 & 10 \end{bmatrix} (c) \begin{bmatrix} 1 & 0 & 0 \\ 0 & 2 & 0 \\ 0 & 0 & 3 \end{bmatrix}$

2. Show that the following symmetric matrices are not positive-definite by finding a vector $x \neq 0$ such that $x^T Ax < 0$.

   $(a) \begin{bmatrix} 1 & 0 \\ 0 & -3 \end{bmatrix} (b) \begin{bmatrix} 1 & 2 \\ 2 & 2 \end{bmatrix} (c) \begin{bmatrix} 1 & -1 \\ -1 & 0 \end{bmatrix} (d) \begin{bmatrix} 1 & 0 & 0 \\ 0 & -2 & 0 \\ 0 & 0 & 3 \end{bmatrix}$

3. Use the Cholesky factorization procedure to express the matrices in Exercise 1 in the form $A = R^T R$.

4. Show that the Cholesky factorization procedure fails for the matrices in Exercise 2.

5. Find the Cholesky factorization $A = R^T R$ of each matrix.

$$(a) \begin{bmatrix} 1 & 2 \\ 2 & 8 \end{bmatrix} \quad (b) \begin{bmatrix} 4 & -2 \\ -2 & 5/4 \end{bmatrix} \quad (c) \begin{bmatrix} 25 & 5 \\ 5 & 26 \end{bmatrix} \quad (d) \begin{bmatrix} 1 & -2 \\ -2 & 5 \end{bmatrix}$$

6. Find the Cholesky factorization $A = R^T R$ of each matrix.

$$(a) \begin{bmatrix} 4 & -2 & 0 \\ -2 & 2 & -3 \\ 0 & -3 & 10 \end{bmatrix} \quad (b) \begin{bmatrix} 1 & 2 & 0 \\ 2 & 5 & 2 \\ 0 & 2 & 5 \end{bmatrix} \quad (c) \begin{bmatrix} 1 & 1 & 1 \\ 1 & 2 & 2 \\ 1 & 2 & 3 \end{bmatrix} \quad (d) \begin{bmatrix} 1 & -1 & -1 \\ -1 & 2 & 1 \\ -1 & 1 & 2 \end{bmatrix}$$

7. Solve the system of equations by finding the Cholesky factorization of A followed by two back substitutions.

$$(a) \begin{bmatrix} 1 & -1 \\ -1 & 5 \end{bmatrix} \begin{bmatrix} x_1 \\ x_2 \end{bmatrix} = \begin{bmatrix} 3 \\ -7 \end{bmatrix} \quad (b) \begin{bmatrix} 4 & -2 \\ -2 & 10 \end{bmatrix} \begin{bmatrix} x_1 \\ x_2 \end{bmatrix} = \begin{bmatrix} 10 \\ 4 \end{bmatrix}$$

8. Solve the system of equations by finding the Cholesky factorization of A followed by two back substitutions.

$$(a) \begin{bmatrix} 4 & 0 & -2 \\ 0 & 1 & 1 \\ -2 & 1 & 3 \end{bmatrix} \begin{bmatrix} x_1 \\ x_2 \\ x_3 \end{bmatrix} = \begin{bmatrix} 4 \\ 2 \\ 0 \end{bmatrix} \quad (b) \begin{bmatrix} 4 & -2 & 0 \\ -2 & 2 & -1 \\ 0 & -1 & 5 \end{bmatrix} \begin{bmatrix} x_1 \\ x_2 \\ x_3 \end{bmatrix} = \begin{bmatrix} 0 \\ 3 \\ -7 \end{bmatrix}$$

9. Prove that if $d > 4$, the matrix $A = \begin{bmatrix} 1 & 2 \\ 2 & d \end{bmatrix}$ is positive-definite.

10. Find all numbers d such that $A = \begin{bmatrix} 1 & -2 \\ -2 & d \end{bmatrix}$ is positive-definite.

11. Find all numbers d such that $A = \begin{bmatrix} 1 & -1 & 0 \\ -1 & 2 & 1 \\ 0 & 1 & d \end{bmatrix}$ is positive-definite.

12. Prove that a principal submatrix of a symmetric positive-definite matrix is symmetric positive-definite. (Hint: Consider an appropriate X and use Property 2.)

13. Solve the problems by carrying out the Conjugate Gradient Method by hand.

$$(a) \begin{bmatrix} 1 & 2 \\ 2 & 5 \end{bmatrix} \begin{bmatrix} u \\ v \end{bmatrix} = \begin{bmatrix} 1 \\ 1 \end{bmatrix} \quad (b) \begin{bmatrix} 1 & 2 \\ 2 & 5 \end{bmatrix} \begin{bmatrix} u \\ v \end{bmatrix} = \begin{bmatrix} 1 \\ 3 \end{bmatrix}$$

14. Solve the problems by carrying out the Conjugate Gradient Method by hand.

$$(a) \begin{bmatrix} 1 & -1 \\ -1 & 2 \end{bmatrix} \begin{bmatrix} u \\ v \end{bmatrix} = \begin{bmatrix} 0 \\ 1 \end{bmatrix} \quad (b) \begin{bmatrix} 4 & 1 \\ 1 & 4 \end{bmatrix} \begin{bmatrix} u \\ v \end{bmatrix} = \begin{bmatrix} -3 \\ 3 \end{bmatrix}$$

15. Carry out the conjugate gradient iteration in the general scalar case $Ax = b$ where A is a 1×1 matrix. Find α_0, x_1, and confirm that $r_1 = 0$ and $Ax_1 = b$.

2.6 Computer Problems

Solutions for Computer Problems numbered in blue can be found at bit.ly/2Cqvxpy

1. Write a MATLAB version of the Conjugate Gradient Method and use it to solve the systems

$$(a) \begin{bmatrix} 1 & 0 \\ 0 & 2 \end{bmatrix} \begin{bmatrix} u \\ v \end{bmatrix} = \begin{bmatrix} 2 \\ 4 \end{bmatrix} \quad (b) \begin{bmatrix} 1 & 2 \\ 2 & 5 \end{bmatrix} \begin{bmatrix} u \\ v \end{bmatrix} = \begin{bmatrix} 1 \\ 1 \end{bmatrix}$$

2. Use a MATLAB version of conjugate gradient to solve the following problems:

$$\text{(a)} \begin{bmatrix} 1 & -1 & 0 \\ -1 & 2 & 1 \\ 0 & 1 & 2 \end{bmatrix} \begin{bmatrix} u \\ v \\ w \end{bmatrix} = \begin{bmatrix} 0 \\ 2 \\ 3 \end{bmatrix} \quad \text{(b)} \begin{bmatrix} 1 & -1 & 0 \\ -1 & 2 & 1 \\ 0 & 1 & 5 \end{bmatrix} \begin{bmatrix} u \\ v \\ w \end{bmatrix} = \begin{bmatrix} 3 \\ -3 \\ 4 \end{bmatrix}$$

3. Solve the system $Hx = b$ by the Conjugate Gradient Method, where H is the $n \times n$ Hilbert matrix and b is the vector of all ones, for (a) $n = 4$ (b) $n = 8$.

4. Solve the sparse problem of (2.45) by the Conjugate Gradient Method for (a) $n = 6$ (b) $n = 12$.

5. Use the Conjugate Gradient Method to solve Example 2.25 for $n = 100, 1000$, and 10,000. Report the size of the final residual, and the number of steps required.

6. Let A be the $n \times n$ matrix with $n = 1000$ and entries $A(i, i) = i, A(i, i + 1) = A(i + 1, i) = 1/2, A(i, i + 2) = A(i + 2, i) = 1/2$ for all i that fit within the matrix. (a) Print the nonzero structure spy (A). (b) Let x_e be the vector of n ones. Set $b = Ax_e$, and apply the Conjugate Gradient Method, without preconditioner, with the Jacobi preconditioner, and with the Gauss–Seidel preconditioner. Compare errors of the three runs in a plot versus step number.

7. Let $n = 1000$. Start with the $n \times n$ matrix A from Computer Problem 6, and add the nonzero entries $A(i, 2i) = A(2i, i) = 1/2$ for $1 \le i \le n/2$. Carry out steps (a) and (b) as in that problem.

8. Let $n = 500$, and let A be the $n \times n$ matrix with entries $A(i, i) = 2, A(i, i + 2) = A(i + 2, i) = 1/2, A(i, i + 4) = A(i + 4, i) = 1/2$ for all i, and $A(500, i) = A(i, 500) = -0.1$ for $1 \le i \le 495$. Carry out steps (a) and (b) as in Computer Problem 6.

9. Let A be the matrix from Computer Problem 8, but with the diagonal elements replaced by $A(i, i) = \sqrt[3]{i}$. Carry out parts (a) and (b) as in that problem.

10. Let C be the 195×195 matrix block with $C(i, i) = 2, C(i, i + 3) = C(i + 3, i) = 0.1$, $C(i, i + 39) = C(i + 39, i) = 1/2, C(i, i + 42) = C(i + 42, i) = 1/2$ for all i. Define A to be the $n \times n$ matrix with $n = 780$ formed by four diagonally arranged blocks C, and with blocks $\frac{1}{2}C$ on the super- and subdiagonal. Carry out steps (a) and (b) as in Computer Problem 6 to solve $Ax = b$.

2.7 NONLINEAR SYSTEMS OF EQUATIONS

Chapter 1 contains methods for solving one equation in one unknown, usually non-linear. In this Chapter, we have studied solution methods for systems of equations, but required the equations to be linear. The combination of nonlinear and "more than one equation" raises the degree of difficulty considerably. This section describes Newton's Method and variants for the solution of systems of nonlinear equations.

2.7.1 Multivariate Newton's Method

The one-variable Newton's Method

$$x_{k+1} = x_k - \frac{f(x_k)}{f'(x_k)}$$

provides the main outline of the Multivariate Newton's Method. Both are derived from the linear approximation afforded by the Taylor expansion. For example, let

$$f_1(u, v, w) = 0$$
$$f_2(u, v, w) = 0 \qquad\qquad (2.49)$$
$$f_3(u, v, w) = 0$$

be three nonlinear equations in three unknowns u, v, w. Define the vector-valued function $F(u, v, w) = (f_1, f_2, f_3)$, and denote the problem (2.49) by $F(x) = 0$, where $x = (u, v, w)$.

The analogue of the derivative f' in the one-variable case is the **Jacobian matrix** defined by

$$DF(x) = \begin{bmatrix} \dfrac{\partial f_1}{\partial u} & \dfrac{\partial f_1}{\partial v} & \dfrac{\partial f_1}{\partial w} \\[2mm] \dfrac{\partial f_2}{\partial u} & \dfrac{\partial f_2}{\partial v} & \dfrac{\partial f_2}{\partial w} \\[2mm] \dfrac{\partial f_3}{\partial u} & \dfrac{\partial f_3}{\partial v} & \dfrac{\partial f_3}{\partial w} \end{bmatrix}.$$

The Taylor expansion for vector-valued functions around x_0 is

$$F(x) = F(x_0) + DF(x_0) \cdot (x - x_0) + O(x - x_0)^2.$$

For example, the linear expansion of $F(u, v) = (e^{u+v}, \sin u)$ around $x_0 = (0, 0)$ is

$$F(x) = \begin{bmatrix} 1 \\ 0 \end{bmatrix} + \begin{bmatrix} e^0 & e^0 \\ \cos 0 & 0 \end{bmatrix} \begin{bmatrix} u \\ v \end{bmatrix} + O(x^2)$$

$$= \begin{bmatrix} 1 \\ 0 \end{bmatrix} + \begin{bmatrix} u + v \\ u \end{bmatrix} + O(x^2).$$

Newton's Method is based on a linear approximation, ignoring the $O(x^2)$ terms. As in the one-dimensional case, let $x = r$ be the root, and let x_0 be the current guess. Then

$$0 = F(r) \approx F(x_0) + DF(x_0) \cdot (r - x_0),$$

or

$$-DF(x_0)^{-1} F(x_0) \approx r - x_0. \qquad\qquad (2.50)$$

Therefore, a better approximation for the root is derived by solving (2.50) for r.

Multivariate Newton's Method

$$x_0 = \text{initial vector}$$
$$x_{k+1} = x_k - (DF(x_k))^{-1} F(x_k) \quad \text{for } k = 0, 1, 2, \ldots.$$

Since computing inverses is computationally burdensome, we use a trick to avoid it. On each step, instead of following the preceding definition literally, set $x_{k+1} = x_k - s$, where s is the solution of $DF(x_k)s = F(x_k)$. Now, only Gaussian elimination ($n^3/3$ multiplications) is needed to carry out a step, instead of computing an inverse (about three times as many). Therefore, the iteration step for Multivariate Newton's Method is

$$\begin{cases} DF(x_k)s = -F(x_k) \\ x_{k+1} = x_k + s. \end{cases} \qquad\qquad (2.51)$$

▶ **EXAMPLE 2.32**   Use Newton's Method with starting guess $(1, 2)$ to find a solution of the system

$$v - u^3 = 0$$
$$u^2 + v^2 - 1 = 0.$$

Figure 2.5 shows the sets on which $f_1(u, v) = v - u^3$ and $f_2(u, v) = u^2 + v^2 - 1$ are zero and their two intersection points, which are the solutions to the system of equations. The Jacobian matrix is

$$DF(u, v) = \begin{bmatrix} -3u^2 & 1 \\ 2u & 2v \end{bmatrix}.$$

Using starting point $x_0 = (1, 2)$, on the first step we must solve the matrix equation (2.51):

$$\begin{bmatrix} -3 & 1 \\ 2 & 4 \end{bmatrix} \begin{bmatrix} s_1 \\ s_2 \end{bmatrix} = -\begin{bmatrix} 1 \\ 4 \end{bmatrix}.$$

The solution is $s = (0, -1)$, so the first iteration produces $x_1 = x_0 + s = (1, 1)$. The second step requires solving

$$\begin{bmatrix} -3 & 1 \\ 2 & 2 \end{bmatrix} \begin{bmatrix} s_1 \\ s_2 \end{bmatrix} = -\begin{bmatrix} 0 \\ 1 \end{bmatrix}.$$

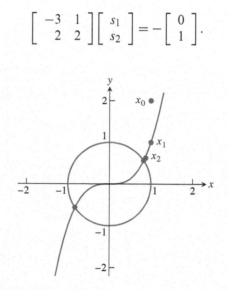

Figure 2.5 Newton's Method for Example 2.32. The two roots are the dots on the circle. Newton's Method produces the dots that are converging to the solution at approximately (0.8260, 0.5636).

The solution is $s = (-1/8, -3/8)$ and $x_2 = x_1 + s = (7/8, 5/8)$. Both iterates are shown in Figure 2.5. Further steps yield the following table:

step	u	v
0	1.00000000000000	2.00000000000000
1	1.00000000000000	1.00000000000000
2	0.87500000000000	0.62500000000000
3	0.82903634826712	0.56434911242604
4	0.82604010817065	0.56361977350284
5	0.82603135773241	0.56362416213163
6	0.82603135765419	0.56362416216126
7	0.82603135765419	0.56362416216126

The familiar doubling of correct decimal places characteristic of quadratic convergence is evident in the output sequence. The symmetry of the equations shows that if (u, v) is a solution, then so is $(-u, -v)$, as is visible in Figure 2.5. The second solution can also be found by applying Newton's Method with a nearby starting guess. ◄

► **EXAMPLE 2.33**   Use Newton's Method to find the solutions of the system

$$f_1(u, v) = 6u^3 + uv - 3v^3 - 4 = 0$$
$$f_2(u, v) = u^2 - 18uv^2 + 16v^3 + 1 = 0.$$

Notice that $(u, v) = (1, 1)$ is one solution. It turns out that there are two others. The Jacobian matrix is

$$DF(u, v) = \begin{bmatrix} 18u^2 + v & u - 9v^2 \\ 2u - 18v^2 & -36uv + 48v^2 \end{bmatrix}.$$

Which solution is found by Newton's Method depends on the starting guess, just as in the one-dimensional case. Using starting point $(u_0, v_0) = (2, 2)$, iterating the preceding formula yields the following table:

step	u	v
0	2.00000000000000	2.00000000000000
1	1.37258064516129	1.34032258064516
2	1.07838681200443	1.05380123264984
3	1.00534968896520	1.00269261871539
4	1.00003367866506	1.00002243772010
5	1.00000000111957	1.00000000057894
6	1.00000000000000	1.00000000000000
7	1.00000000000000	1.00000000000000

Other initial vectors lead to the other two roots, which are approximately $(0.865939, 0.462168)$ and $(0.886809, -0.294007)$. See Computer Problem 2.   ◄

Newton's Method is a good choice if the Jacobian can be calculated. If not, the best alternative is Broyden's Method, the subject of the next section.

2.7.2 Broyden's Method

Newton's Method for solving one equation in one unknown requires knowledge of the derivative. The development of this method in Chapter 1 was followed by the discussion of the Secant Method, for use when the derivative is not available or is too expensive to evaluate.

Now that we have a version of Newton's Method for systems of nonlinear equations $F(x) = 0$, we are faced with the same question: What if the Jacobian matrix DF is not available? Although there is no simple extension of Newton's Method to a Secant Method for systems, Broyden [1965] suggested a method that is generally considered the next best thing.

Suppose A_i is the best approximation available at step i to the Jacobian matrix, and that it has been used to create

$$x_{i+1} = x_i - A_i^{-1} F(x_i). \tag{2.52}$$

To update A_i to A_{i+1} for the next step, we would like to respect the derivative aspect of the Jacobian DF, and satisfy

$$A_{i+1} \delta_{i+1} = \Delta_{i+1}, \tag{2.53}$$

where $\delta_{i+1} = x_{i+1} - x_i$ and $\Delta_{i+1} = F(x_{i+1}) - F(x_i)$. On the other hand, for the orthogonal complement of δ_{i+1}, we have no new information. Therefore, we ask that

$$A_{i+1} w = A_i w \tag{2.54}$$

for every w satisfying $\delta_{i+1}^T w = 0$. One checks that a matrix that satisfies both (2.53) and (2.54) is

$$A_{i+1} = A_i + \frac{(\Delta_{i+1} - A_i \delta_i)\delta_{i+1}^T}{\delta_{i+1}^T \delta_{i+1}}. \tag{2.55}$$

Broyden's Method uses the Newton's Method step (2.52) to advance the current guess, while updating the approximate Jacobian by (2.55). Summarizing, the algorithm starts with an initial guess x_0 and an initial approximate Jacobian A_0, which can be chosen to be the identity matrix if there is no better choice.

Broyden's Method I

$x_0 = $ initial vector
$A_0 = $ initial matrix
for $i = 0, 1, 2, \ldots$
$\quad x_{i+1} = x_i - A_i^{-1} F(x_i)$
$\quad A_{i+1} = A_i + \dfrac{(\Delta_{i+1} - A_i \delta_{i+1})\delta_{i+1}^T}{\delta_{i+1}^T \delta_{i+1}}$
end

$\qquad$ where $\delta_{i+1} = x_{i+1} - x_i$ and $\Delta_{i+1} = F(x_{i+1}) - F(x_i)$.

Note that the Newton-type step is carried out by solving $A_i \delta_{i+1} = F(x_i)$, just as for Newton's Method. Also like Newton's Method, Broyden's Method is not guaranteed to converge to a solution.

A second approach to Broyden's Method avoids the relatively expensive matrix solver step $A_i \delta_{i+1} = F(x_i)$. Since we are at best only approximating the derivative DF during the iteration, we may as well be approximating the inverse of DF instead, which is what is needed in the Newton step.

We redo the derivation of Broyden from the point of view of $B_i = A_i^{-1}$. We would like to have

$$\delta_{i+1} = B_{i+1} \Delta_{i+1}, \tag{2.56}$$

where $\delta_{i+1} = x_{i+1} - x_i$ and $\Delta_{i+1} = F(x_{i+1}) - F(x_i)$, and for every w satisfying $\delta_{i+1}^T w = 0$, still satisfy $A_{i+1} w = A_i w$, or

$$B_{i+1} A_i w = w. \tag{2.57}$$

A matrix that satisfies both (2.56) and (2.57) is

$$B_{i+1} = B_i + \frac{(\delta_{i+1} - B_i \Delta_{i+1})\delta_{i+1}^T B_i}{\delta_{i+1}^T B_i \Delta_{i+1}}. \tag{2.58}$$

The new version of the iteration, which needs no matrix solve, is

$$x_{i+1} = x_i - B_i F(x_i). \tag{2.59}$$

The resulting algorithm is called Broyden's Method II.

Broyden's Method II

$x_0 = $ initial vector
$B_0 = $ initial matrix
for $i = 0, 1, 2, \ldots$
$\quad x_{i+1} = x_i - B_i F(x_i)$
$$\quad B_{i+1} = B_i + \frac{(\delta_{i+1} - B_i \Delta_{i+1}) \delta_{i+1}^T B_i}{\delta_{i+1}^T B_i \Delta_{i+1}}$$
end

$\quad$ where $\delta_i = x_i - x_{i-1}$ and $\Delta_i = F(x_i) - F(x_{i-1})$.

To begin, an initial vector x_0 and an initial guess for B_0 are needed. If it is impossible to compute derivatives, the choice $B_0 = I$ can be used.

A perceived disadvantage of Broyden II is that estimates for the Jacobian, needed for some applications, are not easily available. The matrix B_i is an estimate for the matrix inverse of the Jacobian. Broyden I, on the other hand, keeps track of A_i, which estimates the Jacobian. For this reason, in some circles Broyden I and II are referred to as "Good Broyden" and "Bad Broyden," respectively.

Both versions of Broyden's Method converge superlinearly (to simple roots), slightly slower than the quadratic convergence of Newton's Method. If a formula for the Jacobian is available, it usually speeds convergence to use the inverse of $DF(x_0)$ for the initial matrix B_0.

MATLAB code for Broyden's Method II is as follows:

MATLAB code
shown here can be found
at bit.ly/2P3KEME

```
% Program 2.3 Broyden's Method II
% Input: initial vector x0, max steps k
% Output: solution x
% Example usage: broyden2(f,[1;1],10)
function x=broyden2(f,x0,k)
[n,m]=size(x0);
b=eye(n,n);              % initial b
for i=1:k
  x=x0-b*f(x0);
  del=x-x0;delta=f(x)-f(x0);
  b=b+(del-b*delta)*del'*b/(del'*b*delta);
  x0=x;
end
```

For example, a solution of the system in Example 2.32 is found by defining a function

```
>> f=@(x) [x(2)-x(1)^3;x(1)^2+x(2)^2-1];
```

and calling Broyden's Method II as

```
>> x=broyden2(f,[1;1],10)
```

Broyden's Method, in either implementation, is very useful in cases where the Jacobian is unavailable. A typical instance of this situation is illustrated in the model of pipe buckling in Reality Check 7.

▶ **ADDITIONAL EXAMPLES**

1. Use Multivariate Newton's Method to find the intersection points in R^2 of the circle of radius 2 centered at the origin, and the circle of radius 1 centered at $(1, 1)$.

2. Use the Broyden II method to find the two common intersection points in R^3 of three spheres: the sphere of radius 2 centered at the origin, and the two spheres of radius 1 centered at $(1, 1, 1)$ and $(1, 1, 0)$, respectively.

⌨ **Solutions** for Additional Examples can be found at `bit.ly/2OzesB4`

2.7 Exercises

⌨ **Solutions** for Exercises numbered in blue can be found at `bit.ly/2QXpLQt`

1. Find the Jacobian of the functions (a) $F(u, v) = (u^3, uv^3)$ (b) $F(u, v) = (\sin uv, e^{uv})$ (c) $F(u, v) = (u^2 + v^2 - 1, (u - 1)^2 + v^2 - 1)$ (d) $F(u, v, w) = (u^2 + v - w^2, \sin uvw, uvw^4)$.

2. Use the Taylor expansion to find the linear approximation $L(x)$ to $F(x)$ near x_0.
   (a) $F(u, v) = (1 + e^{u+2v}, \sin(u + v)), x_0 = (0, 0)$
   (b) $F(u, v) = (u + e^{u-v}, 2u + v), x_0 = (1, 1)$

3. Sketch the two curves in the uv-plane, and find all solutions exactly by simple algebra.

   (a) $\begin{cases} u^2 + v^2 = 1 \\ (u - 1)^2 + v^2 = 1 \end{cases}$ (b) $\begin{cases} u^2 + 4v^2 = 4 \\ 4u^2 + v^2 = 4 \end{cases}$ (c) $\begin{cases} u^2 - 4v^2 = 4 \\ (u - 1)^2 + v^2 = 4 \end{cases}$

4. Apply two steps of Newton's Method to the systems in Exercise 3, with starting point $(1, 1)$.

5. Apply two steps of Broyden I to the systems in Exercise 3, with starting point $(1, 1)$, using $A_0 = I$.

6. Apply two steps of Broyden II to the systems in Exercise 3, with starting point $(1, 1)$, using $B_0 = I$.

7. Prove that (2.55) satisfies (2.53) and (2.54).

8. Prove that (2.58) satisfies (2.56) and (2.57).

2.7 Computer Problems

⌨ **Solutions** for Computer Problems numbered in blue can be found at `bit.ly/2CRr008`

1. Implement Newton's Method with appropriate starting points to find all solutions. Check with Exercise 3 to make sure your answers are correct.

   (a) $\begin{cases} u^2 + v^2 = 1 \\ (u - 1)^2 + v^2 = 1 \end{cases}$ (b) $\begin{cases} u^2 + 4v^2 = 4 \\ 4u^2 + v^2 = 4 \end{cases}$ (c) $\begin{cases} u^2 - 4v^2 = 4 \\ (u - 1)^2 + v^2 = 4 \end{cases}$

2. Use Newton's Method to find the three solutions of Example 2.31.

3. Use Newton's Method to find the two solutions of the system $u^3 - v^3 + u = 0$ and $u^2 + v^2 = 1$.

4. Apply Newton's Method to find both solutions of the system of three equations.

$$2u^2 - 4u + v^2 + 3w^2 + 6w + 2 = 0$$
$$u^2 + v^2 - 2v + 2w^2 - 5 = 0$$
$$3u^2 - 12u + v^2 + 3w^2 + 8 = 0$$

5. Use Multivariate Newton's Method to find the two points in common of the three given spheres in three-dimensional space. (a) Each sphere has radius 1, with centers $(1, 1, 0), (1, 0, 1)$, and $(0, 1, 1)$. (Ans. $(1, 1, 1)$ and $(1/3, 1/3, 1/3)$) (b) Each sphere has radius 5, with centers $(1, -2, 0), (-2, 2, -1)$, and $(4, -2, 3)$.

6. Although a generic intersection of three spheres in three-dimensional space is two points, it can be a single point. Apply Multivariate Newton's Method to find the single point of intersection of the spheres with center $(1, 0, 1)$ and radius $\sqrt{8}$, center $(0, 2, 2)$ and radius $\sqrt{2}$, and center $(0, 3, 3)$ and radius $\sqrt{2}$. Does the iteration still converge quadratically? Explain.

7. Apply Broyden I with starting guesses $x_0 = (1, 1)$ and $A_0 = I$ to the systems in Exercise 3. Report the solutions to as much accuracy as possible and the number of steps required.

8. Apply Broyden II with starting guesses $(1, 1)$ and $B_0 = I$ to the systems in Exercise 3. Report the solutions to as much accuracy as possible and the number of steps required.

9. Apply Broyden I to find the sets of two intersection points in Computer Problem 5.

10. Apply Broyden I to find the intersection point in Computer Problem 6. What can you observe about the convergence rate?

11. Apply Broyden II to find the sets of two intersection points in Computer Problem 5.

12. Apply Broyden II to find the intersection point in Computer Problem 6. What can you observe about the convergence rate?

Software and Further Reading

Many excellent texts have appeared on numerical linear algebra, including Stewart [1973] and the comprehensive reference Golub and Van Loan [1996]. Two excellent books with a modern approach to numerical linear algebra are Demmel [1997] and Trefethen and Bau [1997]. Books to consult on iterative methods include Axelsson [1994], Hackbush [1994], Kelley [1995], Saad [1996], Traub [1964], Varga [2000], Young [1971], and Dennis and Schnabel [1983].

LAPACK is a comprehensive, public domain software package containing high-quality routines for matrix algebra computations, including methods for solving $Ax = b$, matrix factorizations, and condition number estimation. It is carefully written to be portable to modern computer architectures, including shared memory vector and parallel processors. See Anderson et al. [1990].

The portability of LAPACK depends on the fact that its algorithms are written in such a way as to maximize use of the Basic Linear Algebra Subprograms (BLAS), a set of primitive matrix/vector computations that can be tuned to optimize performance on particular machines and architectures. BLAS is divided roughly into three parts: Level 1, requiring $O(n)$ operations like dot products; Level 2, operations such as matrix/vector multiplication, that are $O(n^2)$; and Level 3, including full matrix/matrix multiplication, which has complexity $O(n^3)$.

The general dense matrix routine in LAPACK for solving $Ax = b$ in double precision, using the $PA = LU$ factorization, is called DGESV, and there are other versions for sparse and banded matrices. See www.netlib.org/lapack for more details. Implementations of LAPACK routines also form the basis for MATLAB's matrix algebra computations. The Matrix Market (math.nist.gov/MatrixMarket) is a useful repository of test data for numerical linear algebra algorithms.

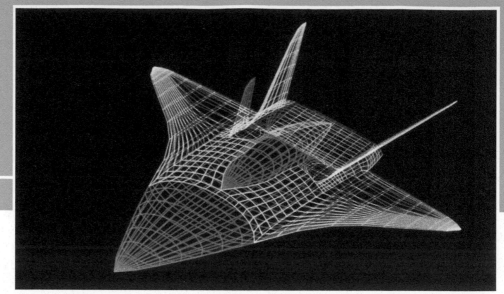

Interpolation

Polynomial interpolation is an ancient practice, but the heavy industrial use of interpolation began with cubic splines in the 20th century. Motivated by practices in the shipbuilding and aircraft industries, engineers Paul de Casteljau and Pierre Bézier at rival European car manufacturers Citroen and Renault, followed by others at General Motors in the United States, spurred the development of what are now called cubic splines and Bézier splines.

Although developed for aerodynamic studies of automobiles, splines have been used for many applications, including computer typesetting. A revolution in printing was caused by two Xerox engineers who formed a company named Adobe and released the PostScript^{TM} language in 1984. It came to the attention of Steve Jobs at Apple Corporation, who was looking for a way to control a newly invented laser printer. Bézier splines were a simple way to adapt the same mathematical curves to fonts with multiple printer resolutions. Later, Adobe used many of the fundamental ideas of PostScript as the basis of a more flexible format called PDF (Portable Document Format), which became a ubiquitous document file type by the early 21st century.

Reality Check Reality Check 3 on page 190 explores how PDF files use Bézier splines to represent printed characters in arbitrary fonts.

Efficient ways of representing data are fundamental to advancing the understanding of scientific problems. At its most fundamental, approximating data by a polynomial is an act of data compression. Suppose that points (x, y) are taken from a given function $y = f(x)$, or perhaps from an experiment where x denotes temperature and y denotes reaction rate. A function on the real numbers represents an infinite amount of information. Finding a polynomial through the set of data means replacing the information with a rule that can be evaluated in a finite number of steps. Although it is unrealistic to expect the polynomial to represent the function exactly at new inputs x, it may be close enough to solve practical problems.

This chapter introduces polynomial interpolation and spline interpolation as convenient tools for finding functions that pass through given data points.

3.1 DATA AND INTERPOLATING FUNCTIONS

A function is said to interpolate a set of data points if it passes through those points. Suppose that a set of (x, y) data points has been collected, such as $(0, 1), (2, 2)$, and $(3, 4)$. There is a parabola that passes through the three points, shown in Figure 3.1. This parabola is called the degree 2 interpolating polynomial passing through the three points.

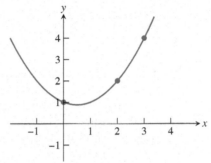

Figure 3.1 Interpolation by parabola. The points (0,1), (2,2), and (3,4) are interpolated by the function $P(x) = \frac{1}{2}x^2 - \frac{1}{2}x + 1$.

DEFINITION 3.1   The function $y = P(x)$ **interpolates** the data points $(x_1, y_1), \ldots, (x_n, y_n)$ if $P(x_i) = y_i$ for each $1 \leq i \leq n$.   ☐

Note that P is required to be a function; that is, each value x corresponds to a single y. This puts a restriction on the set of data points $\{(x_i, y_i)\}$ that can be interpolated—the x_i's must be all distinct in order for a function to pass through them. There is no such restriction on the y_i's.

To begin, we will look for an interpolating polynomial. Does such a polynomial always exist? Assuming that the x-coordinates of the points are distinct, the answer is yes. No matter how many points are given, there is some polynomial $y = P(x)$ that runs through all the points. This and several other facts about interpolating polynomials are proved in this section.

Interpolation is the reverse of evaluation. In polynomial evaluation (such as the nested multiplication of Chapter 0), we are given a polynomial and asked to evaluate a y-value for a given x-value—that is, compute points lying on the curve. Polynomial interpolation asks for the opposite process: Given these points, compute a polynomial that can generate them.

SPOTLIGHT ON

> **Complexity**   Why do we use polynomials? Polynomials are very often used for interpolation because of their straightforward mathematical properties. There is a simple theory about when an interpolating polynomial of a given degree exists for a given set of points. More important, in a real sense, polynomials are the most fundamental of functions for digital computers. Central processing units usually have fast methods in hardware for adding and multiplying floating point numbers, which are the only operations needed to evaluate a polynomial. Complicated functions can be approximated by interpolating polynomials in order to make them computable with these two hardware operations.

3.1.1 Lagrange interpolation

Assume that n data points $(x_1, y_1), \ldots, (x_n, y_n)$ are given, and that we would like to find an interpolating polynomial. There is an explicit formula, called the Lagrange interpolating formula, for writing down a polynomial of degree $d = n - 1$ that interpolates the points. For example, suppose that we are given three points $(x_1, y_1), (x_2, y_2), (x_3, y_3)$. Then the polynomial

$$P_2(x) = y_1 \frac{(x - x_2)(x - x_3)}{(x_1 - x_2)(x_1 - x_3)} + y_2 \frac{(x - x_1)(x - x_3)}{(x_2 - x_1)(x_2 - x_3)} + y_3 \frac{(x - x_1)(x - x_2)}{(x_3 - x_1)(x_3 - x_2)}$$

(3.1)

is the **Lagrange interpolating polynomial** for these points. First notice why the points each lie on the polynomial curve. When x_1 is substituted for x, the terms evaluate to $y_1 + 0 + 0 = y_1$. The second and third numerators are chosen to disappear when x_1 is substituted, and the first denominator is chosen just so to balance the first denominator so that y_1 pops out. It is similar when x_2 and x_3 are substituted. When any other number is substituted for x, we have little control over the result. But then, the job was only to interpolate at the three points—that is the extent of our concern. Second, notice that the polynomial (3.1) is of degree 2 in the variable x.

▶ **EXAMPLE 3.1**   Find an interpolating polynomial for the data points $(0, 1), (2, 2)$, and $(3, 4)$ in Figure 3.1.

Substituting into Lagrange's formula (3.1) yields

$$
\begin{aligned}
P_2(x) &= 1 \frac{(x - 2)(x - 3)}{(0 - 2)(0 - 3)} + 2 \frac{(x - 0)(x - 3)}{(2 - 0)(2 - 3)} + 4 \frac{(x - 0)(x - 2)}{(3 - 0)(3 - 2)} \\
&= \frac{1}{6}(x^2 - 5x + 6) + 2\left(-\frac{1}{2}\right)(x^2 - 3x) + 4\left(\frac{1}{3}\right)(x^2 - 2x) \\
&= \frac{1}{2}x^2 - \frac{1}{2}x + 1.
\end{aligned}
$$

Check that $P_2(0) = 1$, $P_2(2) = 2$, and $P_2(3) = 4$.   ◀

In general, suppose that we are presented with n points $(x_1, y_1), \ldots, (x_n, y_n)$. For each k between 1 and n, define the degree $n - 1$ polynomial

$$L_k(x) = \frac{(x - x_1) \cdots (x - x_{k-1})(x - x_{k+1}) \cdots (x - x_n)}{(x_k - x_1) \cdots (x_k - x_{k-1})(x_k - x_{k+1}) \cdots (x_k - x_n)}.$$

The interesting property of L_k is that $L_k(x_k) = 1$, while $L_k(x_j) = 0$, where x_j is any of the other data points. Then define the degree $n - 1$ polynomial

$$P_{n-1}(x) = y_1 L_1(x) + \cdots + y_n L_n(x).$$

This is a straightforward generalization of the polynomial in (3.1) and works the same way. Substituting x_k for x yields

$$P_{n-1}(x_k) = y_1 L_1(x_k) + \cdots + y_n L_n(x_k) = 0 + \cdots + 0 + y_k L_k(x_k) + 0 + \cdots + 0 = y_k,$$

so it works as designed.

We have constructed a polynomial of degree at most $n - 1$ that passes through any set of n points with distinct x_i's. Interestingly, it is the only one.

THEOREM 3.2   **Main Theorem of Polynomial Interpolation.** Let $(x_1, y_1), \ldots, (x_n, y_n)$ be n points in the plane with distinct x_i. Then there exists one and only one polynomial P of degree $n - 1$ or less that satisfies $P(x_i) = y_i$ for $i = 1, \ldots, n$.   ∎

Proof. The existence is proved by the explicit formula for Lagrange interpolation. To show there is only one, assume for the sake of argument that there are two, say, $P(x)$ and $Q(x)$, that have degree at most $n - 1$ and that both interpolate all n points. That is, we are assuming that $P(x_1) = Q(x_1) = y_1, P(x_2) = Q(x_2) = y_2, \ldots, P(x_n) = Q(x_n) = y_n$. Now define the new polynomial $H(x) = P(x) - Q(x)$. Clearly, the degree of H is also at most $n - 1$, and note that $0 = H(x_1) = H(x_2) = \cdots = H(x_n)$; that is, H has n distinct zeros. According to the Fundamental Theorem of Algebra, a degree d polynomial can have at most d zeros, unless it is the identically zero polynomial. Therefore, H is the identically zero polynomial, and $P(x) \equiv Q(x)$. We conclude that there is a unique $P(x)$ of degree $\leq n - 1$ interpolating the n points (x_i, y_i).   □

▶ **EXAMPLE 3.2**   Find the polynomial of degree 3 or less that interpolates the points $(0, 2), (1, 1), (2, 0)$, and $(3, -1)$.

The Lagrange form is as follows:

$$P(x) = 2\frac{(x - 1)(x - 2)(x - 3)}{(0 - 1)(0 - 2)(0 - 3)} + 1\frac{(x - 0)(x - 2)(x - 3)}{(1 - 0)(1 - 2)(1 - 3)}$$
$$+ 0\frac{(x - 0)(x - 1)(x - 3)}{(2 - 0)(2 - 1)(2 - 3)} - 1\frac{(x - 0)(x - 1)(x - 2)}{(3 - 0)(3 - 1)(3 - 2)}$$
$$= -\frac{1}{3}(x^3 - 6x^2 + 11x - 6) + \frac{1}{2}(x^3 - 5x^2 + 6x) - \frac{1}{6}(x^3 - 3x^2 + 2x)$$
$$= -x + 2.$$

Theorem 3.2 says that there exists exactly one interpolating polynomial of degree 3 or less, but it may or may not be exactly degree 3. In Example 3.2, the data points are collinear, so the interpolating polynomial has degree 1. Theorem 3.2 implies that there are no interpolating polynomials of degree 2 or 3. It may be already intuitively obvious to you that no parabola or cubic curve can pass through four collinear points, but here is the reason.   ◀

3.1.2 Newton's divided differences

The Lagrange interpolation method, as described in the previous section, is a constructive way to write the unique polynomial promised by Theorem 3.2. It is also intuitive; one glance explains why it works. However, it is seldom used for calculation because alternative methods result in more manageable and less computationally complex forms.

Newton's divided differences give a particularly simple way to write the interpolating polynomial. Given n data points, the result will be a polynomial of degree at most $n - 1$, just as Lagrange form does. Theorem 3.2 says that it can be none other than the same as the Lagrange interpolating polynomial, written in a disguised form.

The idea of divided differences is fairly simple, but some notation needs to be mastered first. Assume that the data points come from a function $f(x)$, so that our goal is to interpolate $(x_1, f(x_1)), \ldots, (x_n, f(x_n))$.

List the data points in a table:

$$
\begin{array}{c|c}
x_1 & f(x_1) \\
x_2 & f(x_2) \\
\vdots & \vdots \\
x_n & f(x_n).
\end{array}
$$

Now define the divided differences, which are the real numbers

$$
\begin{aligned}
f[x_k] &= f(x_k) \\
f[x_k\ x_{k+1}] &= \frac{f[x_{k+1}] - f[x_k]}{x_{k+1} - x_k} \\
f[x_k\ x_{k+1}\ x_{k+2}] &= \frac{f[x_{k+1}\ x_{k+2}] - f[x_k\ x_{k+1}]}{x_{k+2} - x_k} \\
f[x_k\ x_{k+1}\ x_{k+2}\ x_{k+3}] &= \frac{f[x_{k+1}\ x_{k+2}\ x_{k+3}] - f[x_k\ x_{k+1}\ x_{k+2}]}{x_{k+3} - x_k},
\end{aligned}
\tag{3.2}
$$

and so on. The **Newton's divided difference formula**

$$
\begin{aligned}
P(x) = f[x_1] &+ f[x_1\ x_2](x - x_1) \\
&+ f[x_1\ x_2\ x_3](x - x_1)(x - x_2) \\
&+ f[x_1\ x_2\ x_3\ x_4](x - x_1)(x - x_2)(x - x_3) \\
&+ \cdots \\
&+ f[x_1 \cdots x_n](x - x_1) \cdots (x - x_{n-1}).
\end{aligned}
\tag{3.3}
$$

is an alternative formula for the unique interpolating polynomial through $(x_1, f(x_1)), \ldots, (x_n, f(x_n))$. The proof that this polynomial interpolates the data is postponed until Section 3.2.2. Notice that the divided difference formula gives the interpolating polynomial as a nested polynomial. It is automatically ready to be evaluated in an efficient way.

Newton's divided differences

Given $x = [x_1, \ldots, x_n], y = [y_1, \ldots, y_n]$

for $j = 1, \ldots, n$
    $f[x_j] = y_j$
end

for $i = 2, \ldots, n$
    for $j = 1, \ldots, n + 1 - i$
        $f[x_j \ldots x_{j+i-1}] = (f[x_{j+1} \ldots x_{j+i-1}] - f[x_j \ldots x_{j+i-2}])/(x_{j+i-1} - x_j)$
    end
end
The interpolating polynomial is

$$
P(x) = \sum_{i=1}^{n} f[x_1 \ldots x_i](x - x_1) \cdots (x - x_{i-1})
$$

The recursive definition of the Newton's divided differences allows arrangement into a convenient table. For three points the table has the form

$$
\begin{array}{c|cccc}
x_1 & f[x_1] \\
 & & f[x_1\ x_2] \\
x_2 & f[x_2] & & f[x_1\ x_2\ x_3] \\
 & & f[x_2\ x_3] \\
x_3 & f[x_3]
\end{array}
$$

The coefficients of the polynomial (3.3) can be read from the top edge of the triangle.

▶ **EXAMPLE 3.3**   Use divided differences to find the interpolating polynomial passing through the points $(0, 1), (2, 2), (3, 4)$.

Applying the definitions of divided differences leads to the following table:

$$
\begin{array}{c|cccc}
0 & 1 \\
 & & \frac{1}{2} \\
2 & 2 & & \frac{1}{2} \\
 & & 2 \\
3 & 4
\end{array}
$$

This table is computed as follows: After writing down the x and y coordinates in separate columns, calculate the next columns, left to right, as divided differences, as in (3.2). For example,

$$
\frac{2-1}{2-0} = \frac{1}{2}
$$

$$
\frac{2-\frac{1}{2}}{3-0} = \frac{1}{2}
$$

$$
\frac{4-2}{3-2} = 2.
$$

After completing the divided difference triangle, the coefficients of the polynomial 1, 1/2, 1/2 can be read from the top edge of the table. The interpolating polynomial can be written as

$$
P(x) = 1 + \frac{1}{2}(x - 0) + \frac{1}{2}(x - 0)(x - 2),
$$

or, in nested form,

$$
P(x) = 1 + (x - 0)\left(\frac{1}{2} + (x - 2)\cdot\frac{1}{2}\right).
$$

The base points for the nested form (see Chapter 0) are $r_1 = 0$ and $r_2 = 2$. Alternatively, we could do more algebra and write the interpolating polynomial as

$$
P(x) = 1 + \frac{1}{2}x + \frac{1}{2}x(x - 2) = \frac{1}{2}x^2 - \frac{1}{2}x + 1,
$$

matching the Lagrange interpolation version shown previously.   ◀

Using the divided difference approach, new data points that arrive after computing the original interpolating polynomial can be easily added.

▶ **EXAMPLE 3.4**   Add the fourth data point $(1, 0)$ to the list in Example 3.3.

We can keep the calculations that were already done and just add a new bottom row to the triangle:

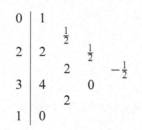

The result is one new term to add to the original polynomial $P_2(x)$. Reading from the top edge of the triangle, we see that the new degree 3 interpolating polynomial is

$$P_3(x) = 1 + \frac{1}{2}(x - 0) + \frac{1}{2}(x - 0)(x - 2) - \frac{1}{2}(x - 0)(x - 2)(x - 3).$$

Note that $P_3(x) = P_2(x) - \frac{1}{2}(x - 0)(x - 2)(x - 3)$, so the previous polynomial can be reused as part of the new one. ◄

It is interesting to compare the extra work necessary to add a new point to the Lagrange formulation versus the divided difference formulation. The Lagrange polynomial must be restarted from the beginning when a new point is added; none of the previous calculation can be used. On the other hand, in divided difference form, we keep the earlier work and add one new term to the polynomial. Therefore, the divided difference approach has a "real-time updating" property that the Lagrange form lacks.

► **EXAMPLE 3.5** Use Newton's divided differences to find the interpolating polynomial passing through $(0, 2), (1, 1), (2, 0), (3, -1)$.

The divided difference triangle is

$$
\begin{array}{c|ccc}
0 & 2 \\
  &   & -1 \\
1 & 1 &   & 0 \\
  &   & -1 &   & 0 \\
2 & 0 &   & 0 \\
  &   & -1 \\
3 & -1 \\
\end{array}
$$

Reading off the coefficients, we find that the interpolating polynomial of degree 3 or less is

$$P(x) = 2 + (-1)(x - 0) = 2 - x,$$

agreeing with Example 3.2, but arrived at with much less work. ◄

3.1.3 How many degree d polynomials pass through n points?

Theorem 3.2, the Main Theorem of Polynomial Interpolation, answers this question if $0 \le d \le n - 1$. Given $n = 3$ points $(0, 1), (2, 2), (3, 4)$, there is one interpolating polynomial of degree 2 or less. Example 3.1 shows that it is degree 2, so there are no degree 0 or 1 interpolating polynomials through the three data points.

How many degree 3 polynomials interpolate the same three points? One way to construct such a polynomial is clear from the previous discussion: Add a fourth point. Extending the Newton's divided difference triangle gives a new top coefficient. In Example 3.4, the point $(1, 0)$ was added. The resulting polynomial,

$$P_3(x) = P_2(x) - \frac{1}{2}(x - 0)(x - 2)(x - 3), \qquad (3.4)$$

passes through the three points in question, in addition to the new point $(1, 0)$. So there is at least one degree 3 polynomial passing through our three original points $(0, 1)$, $(2, 2)$, $(3, 4)$.

Of course, there are many different ways we could have chosen the fourth point. For example, if we keep the same $x_4 = 1$ and simply change y_4 from 0, we *must* get a different degree 3 interpolating polynomial, since a function can only go through one y-value at x_4. Now we know there are infinitely many polynomials that interpolate the three points $(x_1, y_1), (x_2, y_2), (x_3, y_3)$, since for any fixed x_4 there are infinitely many ways y_4 can be chosen, each giving a different polynomial. This line of thinking shows that given n data points (x_i, y_i) with distinct x_i, there are infinitely many degree n polynomials passing through them.

A second look at (3.4) suggests a more direct way to produce interpolating polynomials of degree 3 through three points. Instead of adding a fourth point to generate a new degree 3 coefficient, why not just pencil in an arbitrary degree 3 coefficient? Does the result interpolate the original three points? Yes, because $P_2(x)$ does, and the new term evaluates to zero at x_1, x_2, and x_3. So there is really no need to construct the extra Newton's divided differences for this purpose. Any degree 3 polynomial of the form

$$P_3(x) = P_2(x) + cx(x - 2)(x - 3)$$

with $c \neq 0$ will pass through $(0, 1), (2, 2)$, and $(3, 4)$. This technique will also easily construct (infinitely many) polynomials of degree $\geq n$ for n given data points, as illustrated in the next example.

▶ **EXAMPLE 3.6**   How many polynomials of each degree $0 \leq d \leq 5$ pass through the points $(-1, -5)$, $(0, -1), (2, 1)$, and $(3, 11)$?

The Newton's divided difference triangle is

$$
\begin{array}{ccccc}
-1 & -5 & & & \\
 & & 4 & & \\
0 & -1 & & -1 & \\
 & & 1 & & 1 \\
2 & 1 & & 3 & \\
 & & 10 & & \\
3 & 11 & & & \\
\end{array}
$$

So there are no interpolating polynomials of degree 0, 1, or 2, and the single degree 3 is

$$P_3(x) = -5 + 4(x + 1) - (x + 1)x + (x + 1)x(x - 2).$$

There are infinitely many degree 4 interpolating polynomials

$$P_4(x) = P_3(x) + c_1(x + 1)x(x - 2)(x - 3)$$

for arbitrary $c_1 \neq 0$, and infinitely many degree 5 interpolating polynomials

$$P_5(x) = P_3(x) + c_2(x + 1)x^2(x - 2)(x - 3)$$

for arbitrary $c_2 \neq 0$.   ◀

3.1.4 Code for interpolation

The MATLAB program newtdd.m for computing the coefficients follows:

```
%Program 3.1 Newton Divided Difference Interpolation Method
%Computes coefficients of interpolating polynomial
%Input: x and y are vectors containing the x and y coordinates
%       of the n data points
%Output: coefficients c of interpolating polynomial in nested form
%Use with nest.m to evaluate interpolating polynomial
function c=newtdd(x,y,n)
for j=1:n
  v(j,1)=y(j);            % Fill in y column of Newton triangle
end
for i=2:n                 % For column i,
  for j=1:n+1-i           % fill in column from top to bottom
    v(j,i)=(v(j+1,i-1)-v(j,i-1))/(x(j+i-1)-x(j));
  end
end
for i=1:n
  c(i)=v(1,i);            % Read along top of triangle
end                       % for output coefficients
```

This program can be applied to the data points of Example 3.3 to return the coefficients $1, 1/2, 1/2$ found above. These coefficients can be used in the nested multiplication program to evaluate the interpolating polynomial at various x-values.

For example, the MATLAB code segment

```
x0=[0 2 3];
y0=[1 2 4];
c=newtdd(x0,y0,3);
x=0:.01:4;
y=nest(2,c,x,x0);
plot(x0,y0,'o',x,y)
```

will result in the plot of the polynomial shown in Figure 3.1.

SPOTLIGHT ON

Compression   This is our first encounter with the concept of compression in numerical analysis. At first, interpolation may not seem like compression. After all, we take n points as input and deliver n coefficients (of the interpolating polynomial) as output. What has been compressed?

Think of the data points as coming from somewhere, say as representatives chosen from the multitude of points on a curve $y = f(x)$. The degree $n - 1$ polynomial, characterized by n coefficients, is a "compressed version" of $f(x)$, and may in some cases be used as a fairly simple representative of $f(x)$ for computational purposes.

For example, what happens when the sin key is pushed on a calculator? The calculator has hardware to add and multiply, but how does it compute the sin of a number? Somehow the operation must reduce to the evaluation of a polynomial, which requires exactly those operations. By choosing data points lying on the sine curve, an interpolating polynomial can be calculated and stored in the calculator as a compressed version of the sine function.

This type of compression is "lossy compression," meaning that there will be error involved, since the sine function is not actually a polynomial. How much error is made when a function $f(x)$ is replaced by an interpolating polynomial is the subject of the next section.

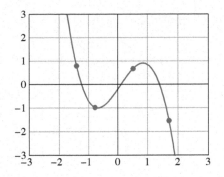

Figure 3.2 Interpolation program 3.2 using mouse input. Screenshot of MATLAB code `clickinterp.m` with four input data points.

Now that we have MATLAB code for finding the coefficients of the interpolating polynomial (`newtdd.m`) and for evaluating the polynomial (`nest.m`), we can put them together to build a polynomial interpolation routine. The program `clickinterp.m` uses MATLAB's graphics capability to plot the interpolation polynomial as it is being created. See Figure 3.2. MATLAB's mouse input command `ginput` is used to facilitate data entry.

MATLAB code shown here can be found at bit.ly/2yKx5aI

```
%Program 3.2 Polynomial Interpolation Program
%Click in MATLAB figure window to locate data point.
%      Continue, to add more points.
%      Press return to terminate program.
function clickinterp
xl=-3;xr=3;yb=-3;yt=3;
plot([xl xr],[0 0],'k',[0 0],[yb yt],'k');grid on;
xlist=[];ylist=[];
k=0;          % initialize counter k
while(0==0)
  [xnew,ynew] = ginput(1);   % get mouse click
  if length (xnew) <1
    break                    % if return pressed, terminate
  end
  k=k+1;          % k counts clicks
  xlist(k)=xnew; ylist(k)=ynew; % add new point to the list
  c=newtdd(xlist,ylist,k); % get interpolation coeffs
  x=xl:.01:xr;    % define x coordinates of curve
  y=nest(k-1,c,x,xlist); % get y coordinates of curve
  plot(xlist,ylist,'o',x,y,[xl xr],[0,0],'k',[0 0],[yb yt],'k');
  axis([xl xr yb yt]);grid on;
end
```

3.1.5 Representing functions by approximating polynomials

A major use of polynomial interpolation is to replace evaluation of a complicated function by evaluation of a polynomial, which involves only elementary computer operations like addition, subtraction, and multiplication. Think of this as a form of compression: Something complex is replaced with something simpler and computable, with perhaps some loss in accuracy that we will have to analyze. We begin with an example from trigonometry.

▶ **EXAMPLE 3.7**   Interpolate the function $f(x) = \sin x$ at 4 equally spaced points on $[0, \pi/2]$.

Let's compress the sine function on the interval $[0, \pi/2]$. Take four data points at equally spaced points and form the divided difference triangle. We list the values to four correct places:

$$
\begin{array}{c|llll}
0 & 0.0000 & & & \\
& & 0.9549 & & \\
\pi/6 & 0.5000 & & -0.2443 & \\
& & 0.6990 & & -0.1139 \\
2\pi/6 & 0.8660 & & -0.4232 & \\
& & 0.2559 & & \\
3\pi/6 & 1.0000 & & &
\end{array}
$$

The degree 3 interpolating polynomial is therefore

$$P_3(x) = 0 + 0.9549x - 0.2443x(x - \pi/6) - 0.1139x(x - \pi/6)(x - \pi/3)$$
$$= 0 + x(0.9549 + (x - \pi/6)(-0.2443 + (x - \pi/3)(-0.1139))). \quad (3.5)$$

This polynomial is graphed together with the sine function in Figure 3.3. At this level of resolution, $P_3(x)$ and $\sin x$ are virtually indistinguishable on the interval $[0, \pi/2]$. We have compressed the infinite amount of information held by the sine curve into a few stored coefficients and the ability to perform the 3 adds and 3 multiplies in (3.5). ◀

How close are we to designing the sin key on a calculator? Certainly we need to handle inputs from the entire real line. But due to the symmetries of the sine function, we have done the hard part. The interval $[0, \pi/2]$ is a so-called **fundamental domain** for sine, meaning that an input from any other interval can be referred back to it. Given an input x from $[\pi/2, \pi]$, say, we can compute $\sin x$ as $\sin(\pi - x)$, since sin is symmetric about $x = \pi/2$. Given an input x from $[\pi, 2\pi]$, $\sin x = -\sin(2\pi - x)$ due to antisymmetry about $x = \pi$. Finally, because sin repeats its behavior on the interval $[0, 2\pi]$ across the entire real line, we can calculate for any input by first reducing modulo 2π. This leads to a straightforward design for the sin key:

⌨ MATLAB code
shown here can be found
at bit.ly/2J3gHqw

```
%Program 3.3 Building a sin calculator key, attempt #1
%Approximates sin curve with degree 3 polynomial
%      (Caution: do not use to build bridges,
%      at least until we have discussed accuracy.)
%Input:  x
%Output: approximation for sin(x)
function y=sin1(x)
%First calculate the interpolating polynomial and
%  store coefficients
b=pi*(0:3)/6;yb=sin(b);     % b holds base points
c=newtdd(b,yb,4);
%For each input x, move x to the fundamental domain and evaluate
%      the interpolating polynomial
s=1;                          % Correct the sign of sin
x1=mod(x,2*pi);
if x1>pi
  x1 = 2*pi-x1;
  s = -1;
end
if x1 > pi/2
  x1 = pi-x1;
end
y = s*nest(3,c,x1,b);
```

Most of the work in Program 3.3 is to place x into the fundamental domain. Then we evaluate the degree 3 polynomial by nested multiplication. Here is some typical output from Program 3.3:

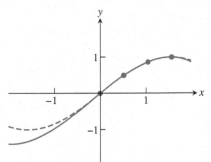

Figure 3.3 Degree 3 interpolation of sin x. The interpolation polynomial (solid curve) is plotted along with $y = \sin x$. Equally spaced interpolation nodes are at $0, \pi/6, 2\pi/6$, and $3\pi/6$. The approximation is very close between 0 and $\pi/2$.

x	$\sin x$	sin1(x)	error
1	0.8415	0.8411	0.0004
2	0.9093	0.9102	0.0009
3	0.1411	0.1428	0.0017
4	−0.7568	−0.7557	0.0011
14	0.9906	0.9928	0.0022
1000	0.8269	0.8263	0.0006

This is not bad for the first try. The error is usually under 1 percent. In order to get enough correct digits to fill the calculator readout, we'll need to know a little more about interpolation error, the topic of the next section.

▶ **ADDITIONAL EXAMPLES**

*1 (a) Find the polynomial of lowest degree that passes through the points $(-2, -9), (-1, -1), (1, -9), (3, -9)$, and $(4, 9)$. (b) Find a degree 6 polynomial that passes through the points in part (a).

2. The National Snow and Ice Center at Boulder, CO estimates the ice extent at the North Pole in units of million square kilometers, using remote sensing from satellites. The January extent from the last few decades is shown in the table.

year	ice extent (M km²)
1980	15.05
1985	14.96
1990	15.07
1995	14.74
2000	14.54
2005	13.81
2010	13.91
2015	13.75

Use newtdd.m and nest.m to plot the degree 7 interpolating polynomial through the 8 data points. Use the polynomial to estimate the ice extent in 2002 and 2012 and compare with the exact values (14.57 and 13.86, respectively).

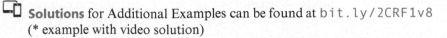

Solutions for Additional Examples can be found at bit.ly/2CRF1v8
(* example with video solution)

3.1 Exercises

Solutions
for Exercises
numbered in blue
can be found at
bit.ly/2J2nGzS

1. Use Lagrange interpolation to find a polynomial that passes through the points.

   (a) $(0, 1), (2, 3), (3, 0)$
   (b) $(-1, 0), (2, 1), (3, 1), (5, 2)$
   (c) $(0, -2), (2, 1), (4, 4)$

2. Use Newton's divided differences to find the interpolating polynomials of the points in Exercise 1, and verify agreement with the Lagrange interpolating polynomial.

3. How many degree d polynomials pass through the four points $(-1, 3), (1, 1), (2, 3), (3, 7)$? Write one down if possible. (a) $d = 2$ (b) $d = 3$ (c) $d = 6$.

4. (a) Find a polynomial $P(x)$ of degree 3 or less whose graph passes through the points $(0, 0), (1, 1), (2, 2), (3, 7)$. (b) Find two other polynomials (of any degree) that pass through these four points. (c) Decide whether there exists a polynomial $P(x)$ of degree 3 or less whose graph passes through the points $(0, 0), (1, 1), (2, 2), (3, 7)$, and $(4, 2)$.

5. (a) Find a polynomial $P(x)$ of degree 3 or less whose graph passes through the four data points $(-2, 8), (0, 4), (1, 2), (3, -2)$. (b) Describe any other polynomials of degree 4 or less which pass through the four points in part (a).

6. Write down a polynomial of degree exactly 5 that interpolates the four points $(1, 1), (2, 3), (3, 3), (4, 4)$.

7. Find $P(0)$, where $P(x)$ is the degree 10 polynomial that is zero at $x = 1, \ldots, 10$ and satisfies $P(12) = 44$.

8. Let $P(x)$ be the degree 9 polynomial that takes the value 112 at $x = 1$, takes the value 2 at $x = 10$, and equals zero for $x = 2, \ldots, 9$. Calculate $P(0)$.

9. Give an example of the following, or explain why no such example exists. (a) A degree 6 polynomial $L(x)$ that is zero at $x = 1, 2, 3, 4, 5, 6$ and equal to 10 at $x = 7$. (b) A degree 6 polynomial $L(x)$ that is zero at $x = 1, 2, 3, 4, 5, 6$, equal to 10 at $x = 7$, and equal to 70 at $x = 8$.

10. Let $P(x)$ be the degree 5 polynomial that takes the value 10 at $x = 1, 2, 3, 4, 5$ and the value 15 at $x = 6$. Find $P(7)$.

11. Let P_1, P_2, P_3, and P_4 be four different points lying on a parabola $y = ax^2 + bx + c$. How many cubic (degree 3) polynomials pass through those four points? Explain your answer.

12. Can a degree 3 polynomial intersect a degree 4 polynomial in exactly five points? Explain.

13. Let $P(x)$ be the degree 10 polynomial through the 11 points $(-5, 5), (-4, 5), (-3, 5), (-2, 5), (-1, 5), (0, 5), (1, 5), (2, 5), (3, 5), (4, 5), (5, 42)$. Calculate $P(6)$.

14. Write down 4 noncollinear points $(1, y_1), (2, y_2), (3, y_3), (4, y_4)$ that do not lie on any polynomial $y = P_3(x)$ of degree exactly three.

15. Write down the degree 25 polynomial that passes through the points $(1, -1), (2, -2), \ldots, (25, -25)$ and has constant term equal to 25.

16. List all degree 42 polynomials that pass through the eleven points $(-5, 5), (-4, 4), \ldots, (4, -4), (5, -5)$ and have constant term equal to 42.

17. The estimated mean atmospheric concentration of carbon dioxide in earth's atmosphere is given in the table that follows, in parts per million by volume. Find the degree 3 interpolating polynomial of the data and use it to estimate the CO_2 concentration in (a) 1950 and (b) 2050. (The actual concentration in 1950 was 310 ppm.)

year	CO_2 (ppm)
1800	280
1850	283
1900	291
2000	370

18. The expected lifetime of an industrial fan when operated at the listed temperature is shown in the table that follows. Estimate the lifetime at 70°C by using (a) the parabola from the last three data points (b) the degree 3 curve using all four points.

temp (°C)	hrs (×1000)
25	95
40	75
50	63
60	54

3.1 Computer Problems

Solutions for Computer Problems numbered in blue can be found at bit.ly/2yobC7Q

1. Apply the following world population figures to estimate the 1980 population, using (a) the straight line through the 1970 and 1990 estimates; (b) the parabola through the 1960, 1970, and 1990 estimates; and (c) the cubic curve through all four data points. Compare with the 1980 estimate of 4452584592.

year	population
1960	3039585530
1970	3707475887
1990	5281653820
2000	6079603571

2. Write a version of Program 3.2 that is a MATLAB function, whose inputs x and y are equal length vectors of data points, and whose output is a plot of the interpolating polynomial. In this way, the points can be entered more accurately than by mouse input. Check your program by replicating Figure 3.2.

3. Write a MATLAB function polyinterp.m that takes as input a set of (x, y) interpolating points and another x_0, and outputs y_0, the value of the interpolating polynomial at x_0. The first line of the file should be function y0 = polyinterp(x,y,x0), where x and y are input vectors of data points. Your function may call newtdd from Program 3.1 and nest from Chapter 0, and may be structured similarly to Program 3.2, but without the graphics. Demonstrate that your function works.

4. Remodel the sin1 calculator key in Program 3.3 to build cos1, a cosine key that follows the same principles. First decide on the fundamental domain for cosine.

5. (a) Use the addition formulas for sin and cos to prove that $\tan(\pi/2 - x) = 1/\tan x$. (b) Show that $[0, \pi/4]$ can be used as a fundamental domain for $\tan x$. (c) Design a tangent key, following the principles of Program 3.3, using degree 3 polynomial interpolation on this fundamental domain. (d) Empirically calculate the maximum error of the tangent key in $[0, \pi/4]$.

3.2 INTERPOLATION ERROR

The accuracy of our sin calculator key depends on the approximation in Figure 3.3. How close is it? We presented a table indicating that, for a few examples, the first two digits are fairly reliable, but after that the digits are not always correct. In this section, we investigate ways to measure this error and determine how to make it smaller.

3.2.1 Interpolation error formula

Assume that we start with a function $y = f(x)$ and take data points from it to build an interpolating polynomial $P(x)$, as we did with $f(x) = \sin x$ in Example 3.7. The **interpolation error** at x is $f(x) - P(x)$, the difference between the original function that provided the data points and the interpolating polynomial, evaluated at x. The interpolation error is the vertical distance between the curves in Figure 3.3. The next theorem gives a formula for the interpolation error that is usually impossible to evaluate exactly, but often can at least lead to an error bound.

THEOREM 3.3   Assume that $P(x)$ is the (degree $n - 1$ or less) interpolating polynomial fitting the n points $(x_1, y_1), \ldots, (x_n, y_n)$. The interpolation error is

$$f(x) - P(x) = \frac{(x - x_1)(x - x_2)\cdots(x - x_n)}{n!} f^{(n)}(c),$$ (3.6)

where c lies between the smallest and largest of the numbers $x, x_1, \ldots, x_n$. ∎

See Section 3.2.2 for a proof of Theorem 3.3. We can use the theorem to assess the accuracy of the `sin` key we built in Example 3.7. Equation (3.6) yields

$$\sin x - P(x) = \frac{(x - 0)\left(x - \frac{\pi}{6}\right)\left(x - \frac{\pi}{3}\right)\left(x - \frac{\pi}{2}\right)}{4!} f''''(c),$$

where $0 < c < \pi/2$. The fourth derivative $f''''(c) = \sin c$ varies from 0 to 1 in this range. At worst, $|\sin c|$ is no more than 1, so we can be assured of an upper bound on interpolation error:

$$|\sin x - P(x)| \le \frac{\left|(x - 0)\left(x - \frac{\pi}{6}\right)\left(x - \frac{\pi}{3}\right)\left(x - \frac{\pi}{2}\right)\right|}{24}|1|.$$

At $x = 1$, the worst-case error is

$$|\sin 1 - P(1)| \le \frac{\left|(1 - 0)\left(1 - \frac{\pi}{6}\right)\left(1 - \frac{\pi}{3}\right)\left(1 - \frac{\pi}{2}\right)\right|}{24}|1| \approx 0.0005348.$$ (3.7)

This is an upper bound for the error, since we used a "worst case" bound for the fourth derivative. Note that the actual error at $x = 1$ was 0.0004, which is within the error bound given by (3.7). We can make some conclusions on the basis of the form of the interpolation error formula. We expect smaller errors when x is closer to the middle of the interval of x_i's than when it is near one of the ends, because there will be more small terms in the product. For example, we compare the preceding error bound to the case $x = 0.2$, which is near the left end of the range of data points. In this case, the error formula is

$$|\sin 0.2 - P(0.2)| \le \frac{\left|(0.2 - 0)\left(0.2 - \frac{\pi}{6}\right)\left(0.2 - \frac{\pi}{3}\right)\left(0.2 - \frac{\pi}{2}\right)\right|}{24}|1| \approx 0.00313,$$

about six times larger. Correspondingly, the actual error is larger, specifically,

$$|\sin 0.2 - P(0.2)| = |0.19867 - 0.20056| = 0.00189.$$

▶ **EXAMPLE 3.8** Find an upper bound for the difference at $x = 0.25$ and $x = 0.75$ between $f(x) = e^x$ and the polynomial that interpolates it at the points $-1, -0.5, 0, 0.5, 1$.

Construction of the interpolating polynomial, shown in Figure 3.4, is not necessary to find the bound. The interpolation error formula (3.6) gives

$$f(x) - P_4(x) = \frac{(x+1)\left(x+\frac{1}{2}\right)x\left(x-\frac{1}{2}\right)(x-1)}{5!}f^{(5)}(c),$$

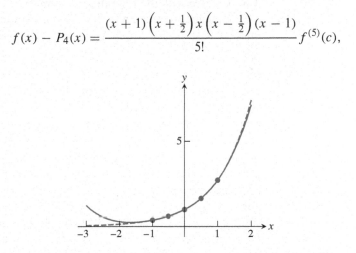

Figure 3.4 Interpolating polynomial for approximating $f(x) = e^x$**.** Equally spaced base points $-1, -0.5, 0, 0.5, 1$. The solid curve is the interpolating polynomial.

where $-1 < c < 1$. The fifth derivative is $f^{(5)}(c) = e^c$. Since e^x is increasing with x, its maximum is at the right-hand end of the interval, so $|f^{(5)}| \le e^1$ on $[-1, 1]$. For $-1 \le x \le 1$, the error formula becomes

$$|e^x - P_4(x)| \le \frac{(x+1)\left(x+\frac{1}{2}\right)x\left(x-\frac{1}{2}\right)(x-1)}{5!}e.$$

At $x = 0.25$, the interpolation error has the upper bound

$$|e^{0.25} - P_4(0.25)| \le \frac{(1.25)(0.75)(0.25)(-0.25)(-0.75)}{120}e$$
$$\approx 0.000995.$$

At $x = 0.75$, the interpolation error is potentially larger:

$$|e^{0.75} - P_4(0.75)| \le \frac{(1.75)(1.25)(0.75)(0.25)(0.25)}{120}e$$
$$\approx 0.002323.$$

Note again that the interpolation error will tend to be smaller close to the center of the interpolation interval. ◀

3.2.2 Proof of Newton form and error formula

In this section, we explain the reasoning behind two important facts used earlier. First we establish the Newton's divided difference form of the interpolating polynomial, and then we prove the interpolation error formula.

Recall what we know so far. If $x_1, \ldots, x_n$ are n distinct points on the real line and $y_1, \ldots, y_n$ are arbitrary, we know by Theorem 3.2 that there is exactly one (degree at most $n-1$) interpolating polynomial $P_{n-1}(x)$ for these points. We also know that the Lagrange interpolating formula gives such a polynomial.

We are missing the proof that the Newton's divided difference formula also gives an interpolating polynomial. Once we prove that it does in Theorem 3.4, we will know it must agree with the Lagrange version.

To begin, we note an interesting fact about divided differences: It does not matter in what order the x_i are listed. Since the definition of divided differences is recursive (see (3.2)), it is not surprising that the proof will be by induction.

FACT 1 $f[x_1 \ldots x_n] = f[\sigma(x_1) \ldots \sigma(x_n)]$ for any permutation σ of the x_i. ☐

Proof. The proof proceeds by induction. The cases $n=1$ and $n=2$ are clear. Assume $n \geq 3$, and that the conclusion has been proved for cases up to $n-1$. Every permutation σ is a composition of transpositions, where two neighboring x_i are exchanged, so the proof reduces to checking transpositions. If the transposition does not affect x_1 or x_n, the result follows immediately from the induction hypothesis applied to the length $n-1$ divided differences. The same is true if the transposition simply exchanges x_1 and x_n. The remaining cases cover when either x_1 or x_n is involved in the transposition. We handle the former case, showing that $f[x_2 x_1 x_3 \cdots x_n] = f[x_1 x_2 x_3 \cdots x_n]$. The latter is done similarly. We can expand $f[x_2 x_1 x_3 \cdots x_n]$ as

$$
\frac{f[x_1 x_3 \cdots x_n] - f[x_2 x_1 x_3 \cdots x_{n-1}]}{x_n - x_2} = \frac{f[x_1 x_3 \cdots x_n] - f[x_1 x_3 \cdots x_{n-1} x_2]}{x_n - x_2}
$$

$$
= \frac{f[x_3 \cdots x_n] - f[x_1 x_3 \cdots x_{n-1}]}{(x_n - x_1)(x_n - x_2)} - \frac{f[x_3 \cdots x_{n-1} x_2] - f[x_1 x_3 \cdots x_{n-1}]}{(x_2 - x_1)(x_n - x_2)}
$$

$$
= \frac{f[x_1 x_3 \cdots x_{n-1}](x_n - x_2) + f[x_3 \cdots x_n](x_2 - x_1) - f[x_3 \cdots x_{n-1} x_2](x_n - x_2 + x_2 - x_1)}{(x_n - x_1)(x_2 - x_1)(x_n - x_2)}
$$

$$
= \frac{\left(f[x_1 x_3 \cdots x_{n-1}] - f[x_3 \cdots x_{n-1} x_2]\right)(x_n - x_2) + \left(f[x_3 \cdots x_n] - f[x_3 \cdots x_{n-1} x_2]\right)(x_2 - x_1)}{(x_n - x_1)(x_2 - x_1)(x_n - x_2)}
$$

$$
= \frac{-f[x_1 x_3 \cdots x_{n-1} x_2] + f[x_2 \cdots x_n]}{x_n - x_1} = f[x_1 x_2 x_3 \cdots x_n],
$$

where we have used the induction hypothesis repeatedly to permute x_i for divided differences of length less than n. ☐

In Example 3.4, we found that $f[0\ 2\ 3\ 1] = -1/2$. Fact 1 implies that $f[0\ 1\ 2\ 3] = -1/2$, $f[1\ 3\ 2\ 0] = -1/2$, and the same with any permutation of $0, 1, 2$ and 3. With this fact, we can verify that the Newton divided differences provide the coefficients for the interpolating polynomial.

THEOREM 3.4 Let $P(x)$ be the interpolating polynomial of $(x_1, f(x_1)), \ldots, (x_n, f(x_n))$ where the x_i are distinct. Then

$$
P(x) = f[x_1] + f[x_1 x_2](x - x_1) + f[x_1 x_2 x_3](x - x_1)(x - x_2) + \ldots
$$

$$
+ f[x_1 x_2 \ldots x_n](x - x_1)(x - x_2) \cdots (x - x_{n-1})
$$

■

Proof. The proof proceeds by induction. The case $n=1$ is clear. The induction hypothesis assumes that the divided difference polynomial interpolates the data for $n-1$ points; we prove it is true for n points. Therefore, we can assume that

$$P_1(x) = f[x_1] + f[x_1x_2](x - x_1) + f[x_1x_2x_3](x - x_1)(x - x_2)$$
$$+ \ldots + f[x_1x_2\cdots x_{n-1}](x - x_1)\cdots(x - x_{n-2}) \qquad (3.8)$$

interpolates the data at $x_1,\ldots,x_{n-1}$, and also that

$$P_2(x) = f[x_1] + f[x_1x_2](x - x_1) + f[x_1x_2x_3](x - x_1)(x - x_2)$$
$$+ \ldots + f[x_1x_2\cdots x_{n-2}x_n](x - x_1)\cdots(x - x_{n-2})$$

interpolates the data at $x_1,\ldots,x_{n-2},x_n$. In particular,

$$P_2(x_n) = C + f[x_1x_2\cdots x_{n-2}x_n](x_n - x_1)(x_n - x_2)\cdots(x_n - x_{n-2}) = f[x_n], \qquad (3.9)$$

where we define

$$C = f[x_1] + f[x_1x_2](x_n - x_1) + f[x_1x_2x_3](x - x_1)(x_n - x_2)$$
$$+ \ldots + f[x_1x_2\cdots x_{n-2}](x_n - x_1)(x_n - x_2)\cdots(x_n - x_{n-3}).$$

The goal of the proof is to substitute $x_1,\ldots,x_n$ into the degree n polynomial

$$P(x) = f[x_1] + f[x_1x_2](x - x_1) + f[x_1x_2x_3](x - x_1)(x - x_2)$$
$$+ \ldots + f[x_1x_2\cdots x_n](x - x_1)(x - x_2)\cdots(x - x_{n-1})$$

and retrieve $f(x_1),\ldots,f(x_n)$, respectively. First note that this follows immediately from the induction hypothesis (3.8) for all but x_n. To check the last case, note that

$$P(x_n) = C + f[x_1\cdots x_{n-1}](x_n - x_1)\cdots(x_n - x_{n-2}) + f[x_1\cdots x_n](x_n - x_1)\cdots(x_n - x_{n-1})$$
$$= C + f[x_1\cdots x_{n-1}](x_n - x_1)\cdots(x_n - x_{n-2})$$
$$+ (f[x_2\cdots x_n] - f[x_1\cdots x_{n-1}])(x_n - x_2)\cdots(x_n - x_{n-1})$$
$$= C + f[x_1\cdots x_{n-1}](x_n - x_2)\cdots(x_n - x_{n-2})[x_n - x_1 - (x_n - x_{n-1})]$$
$$+ f[x_2\cdots x_n](x_n - x_2)\cdots(x_n - x_{n-1})$$
$$= C + (x_n - x_2)\cdots(x_n - x_{n-2}) \cdot$$
$$\{f[x_1\cdots x_{n-1}](x_{n-1} - x_1) + f[x_2\cdots x_n](x_n - x_{n-1})\}$$
$$= C + (x_n - x_2)\cdots(x_n - x_{n-2}) \cdot$$
$$\{f[x_1\cdots x_{n-1}](x_{n-1} - x_1) + f[x_{n-1}x_2x_3\cdots x_{n-2}x_n](x_n - x_{n-1})\}$$

where we have used Fact 1 repeatedly to permute x_i. The term in the braces can be written

$$f[x_2\cdots x_{n-1}] - f[x_1\cdots x_{n-2}] + f[x_2\cdots x_{n-2}x_n] - f[x_{n-1}x_2x_3\cdots x_{n-2}]$$
$$= f[x_2\cdots x_{n-2}x_n] - f[x_1\cdots x_{n-2}] = f[x_1\cdots x_{n-2}x_n](x_n - x_1)$$

where we have used Fact 1 again. Substituting this expression yields

$$P(x_n) = C + (x_n - x_1)(x_n - x_2)\cdots(x_n - x_{n-2})f[x_1\cdots x_{n-2}x_n] = f(x_n),$$

where the induction hypothesis formula (3.9) was used for the last equality. ☐

Next we prove the Interpolation Error Theorem 3.3. Consider adding one more point x to the set of interpolation points. The new interpolation polynomial would be

$$P_n(t) = P_{n-1}(t) + f[x_1\ldots x_n x](t - x_1)\cdots(t - x_n).$$

Evaluated at the extra point x, $P_n(x) = f(x)$, so

$$f(x) = P_{n-1}(x) + f[x_1 \dots x_n x](x - x_1) \cdots (x - x_n). \qquad (3.10)$$

This formula is true for all x. Now define

$$h(t) = f(t) - P_{n-1}(t) - f[x_1 \dots x_n x](t - x_1) \cdots (t - x_n).$$

Note that $h(x) = 0$ by (3.10) and $0 = h(x_1) = \cdots = h(x_n)$ because P_{n-1} interpolates f at these points. Between each neighboring pair of the $n + 1$ points $x, x_1, \dots, x_n$, there must be a new point where $h' = 0$, by Rolle's Theorem (see Chapter 0). There are n of these points. Between each pair of these, there must be a new point where $h'' = 0$; there are $n - 1$ of these. Continuing in this way, there must be one point c for which $h^{(n)}(c) = 0$, where c lies between the smallest and largest of $x, x_1, \dots, x_n$. Note that

$$h^{(n)}(t) = f^{(n)}(t) - n! f[x_1 \dots x_n x],$$

because the nth derivative of the polynomial $P_{n-1}(t)$ is zero. Substituting c gives

$$f[x_1 \dots x_n x] = \frac{f^{(n)}(c)}{n!},$$

which leads to

$$f(x) = P_{n-1}(x) + \frac{f^{(n)}(c)}{n!}(x - x_1) \cdots (x - x_n),$$

using (3.10).

3.2.3 Runge phenomenon

Polynomials can fit any set of data points, as Theorem 3.2 shows. However, there are some shapes that polynomials prefer over others. You can achieve a better understanding of this point by playing with Program 3.2. Try data points that cause the function to be zero at equally spaced points $x = -3, -2.5, -2, -1.5, \dots, 2.5, 3$, except for $x = 0$, where we set a value of 1. The data points are flat along the x-axis, except for a triangular "bump" at $x = 0$, as shown in Figure 3.5.

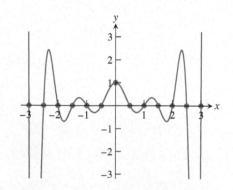

Figure 3.5 Interpolation of triangular bump function. The interpolating polynomial wiggles much more than the input data points.

The polynomial that goes through points situated like this refuses to stay between 0 and 1, unlike the data points. This is an illustration of the so-called **Runge phenomenon**. It is usually used to describe extreme "polynomial wiggle" associated with high-degree polynomial interpolation at evenly spaced points.

▶ **EXAMPLE 3.9**   Interpolate $f(x) = 1/(1 + 12x^2)$ at evenly spaced points in $[-1, 1]$.

This is called the **Runge example**. The function has the same general shape as the triangular bump in Figure 3.5. Figure 3.6 shows the result of the interpolation, behavior that is characteristic of the Runge phenomenon: polynomial wiggle near the ends of the interpolation interval.   ◀

As we have seen, examples with the Runge phenomenon characteristically have large error near the outside of the interval of data points. The cure for this problem is intuitive: Move some of the interpolation points toward the outside of the interval, where the function producing the data can be better fit. We will see how to accomplish this in the next section on Chebyshev interpolation.

▶ **ADDITIONAL EXAMPLES**

1. Let $P_5(x)$ be the degree 5 polynomial that interpolates $f(x) = 2^x$ at the six nodes $0, 0.2, 0.4, 0.6, 0.8, 1.0$ on the interval $0, 1$. Find the best possible upper bounds for the interpolation error $|2^x - P_5(x)|$ at $x = 0.5$ and $x = 0.9$.

2. Use the `newtdd.m` and `nest.m` codes to plot the interpolating polynomial of $f(x) = e^{\sin x}$ at the interpolation points $x_i = 0.4i, i = 0, \ldots, 10$ in the interval $0, 4$. Find the (empirical) maximum interpolation error on the interval.

⌨ **Solutions** for Additional Examples can be found at `bit.ly/2yPMDtx`

3.2 Exercises

⌨ **Solutions** for Exercises numbered in blue can be found at `bit.ly/2QXpMnv`

1. (a) Find the degree 2 interpolating polynomial $P_2(x)$ through the points $(0, 0)$, $(\pi/2, 1)$, and $(\pi, 0)$. (b) Calculate $P_2(\pi/4)$, an approximation for $\sin(\pi/4)$. (c) Use Theorem 3.3 to give an error bound for the approximation in part (b). (d) Using a calculator or MATLAB, compare the actual error to your error bound.

2. (a) Given the data points $(1, 0)$, $(2, \ln 2)$, $(4, \ln 4)$, find the degree 2 interpolating polynomial. (b) Use the result of (a) to approximate $\ln 3$. (c) Use Theorem 3.3 to give an error bound for the approximation in part (b). (d) Compare the actual error to your error bound.

3. Assume that the polynomial $P_9(x)$ interpolates the function $f(x) = e^{-2x}$ at the 10 evenly spaced points $x = 0, 1/9, 2/9, 3/9, \ldots, 8/9, 1$. (a) Find an upper bound for the error $|f(1/2) - P_9(1/2)|$. (b) How many decimal places can you guarantee to be correct if $P_9(1/2)$ is used to approximate e^{-1}?

4. Consider the interpolating polynomial for $f(x) = 1/(x + 5)$ with interpolation nodes $x = 0, 2, 4, 6, 8, 10$. Find an upper bound for the interpolation error at (a) $x = 1$ and (b) $x = 5$.

5. Assume that a function $f(x)$ has been approximated by the degree 5 interpolating polynomial $P(x)$, using the data points $(x_i, f(x_i))$, where $x_1 = 0.1$, $x_2 = 0.2$, $x_3 = 0.3$, $x_4 = 0.4$, $x_5 = 0.5$, $x_6 = 0.6$. Do you expect the interpolation error $|f(x) - P(x)|$ to be smaller for $x = 0.35$ or for $x = 0.55$? Quantify your answer.

6. Assume that the polynomial $P_5(x)$ interpolates a function $f(x)$ at the six data points $(x_i, f(x_i))$ with x-coordinates $x_1 = 0, x_2 = 0.2, x_3 = 0.4, x_4 = 0.6, x_5 = 0.8$, and $x_6 = 1$. Assume that the interpolation error at $x = 0.3$ is $|f(0.3) - P_5(0.3)| = 0.01$. Estimate the new interpolation error $|f(0.3) - P_7(0.3)|$ that would result if two additional interpolation points $(x_6, y_6) = (0.1, f(0.1))$ and $(x_7, y_7) = (0.5, f(0.5))$ are added. What assumptions have you made to produce this estimate?

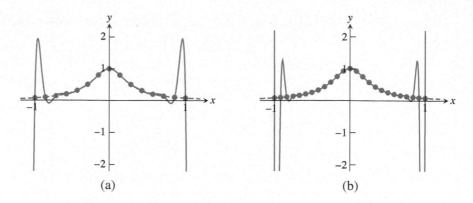

(a)   (b)

Figure 3.6 Runge example. Polynomial interpolation of the Runge function of Example 3.9 at evenly spaced base points causes extreme variation near the ends of the interval, similar to Figure 3.5 (a) 15 base points (b) 25 base points.

3.2 Computer Problems

Solutions for Computer Problems numbered in **blue** can be found at bit.ly/2PH1nCs

1. (a) Use the method of divided differences to find the degree 4 interpolating polynomial $P_4(x)$ for the data $(0.6, 1.433329)$, $(0.7, 1.632316)$, $(0.8, 1.896481)$, $(0.9, 2.247908)$, and $(1.0, 2.718282)$. (b) Calculate $P_4(0.82)$ and $P_4(0.98)$. (c) The preceding data come from the function $f(x) = e^{x^2}$. Use the interpolation error formula to find upper bounds for the error at $x = 0.82$ and $x = 0.98$, and compare the bounds with the actual error. (d) Plot the actual interpolation error $e^{x^2} - P(x)$ on the intervals $[0.5, 1]$ and $[0, 2]$.

2. Plot the interpolation error of the sin1 key from Program 3.3 on the interval $[-2\pi, 2\pi]$.

3. The total world oil production in millions of barrels per day is shown in the table that follows. Determine and plot the degree 9 polynomial through the data. Use it to estimate 2010 oil production. Does the Runge phenomenon occur in this example? In your opinion, is the interpolating polynomial a good model of the data? Explain.

year	bbl/day ($\times 10^6$)
1994	67.052
1995	68.008
1996	69.803
1997	72.024
1998	73.400
1999	72.063
2000	74.669
2001	74.487
2002	74.065
2003	76.777

4. Use the degree 3 polynomial through the first four data points in Computer Problem 3 to estimate the 1998 world oil production. Is the Runge phenomenon present?

3.3 CHEBYSHEV INTERPOLATION

It is common to choose the base points x_i for interpolation to be evenly spaced. In many cases, the data to be interpolated are available only in that form—for example, when the data consist of instrument readings separated by a constant time interval. In other cases—for instance, the sine key—we are free to choose the base points as we see fit. It turns out that the choice of base point spacing can have a significant effect on the interpolation error. Chebyshev interpolation refers to a particular optimal way of spacing the points.

3.3.1 Chebyshev's theorem

The motivation for Chebyshev interpolation is to improve control of the maximum value of the interpolation error

$$\frac{(x - x_1)(x - x_2)\cdots(x - x_n)}{n!} f^{(n)}(c)$$

on the interpolation interval. Let's fix the interval to be $[-1, 1]$ for now.

The numerator

$$(x - x_1)(x - x_2)\cdots(x - x_n) \tag{3.11}$$

of the interpolation error formula is itself a degree n polynomial in x and has some maximum value on $[-1, 1]$. Is it possible to find particular $x_1, \ldots, x_n$ in $[-1, 1]$ that cause the maximum value of (3.11) to be as small as possible? This is called the mini-max problem of interpolation.

For example, Figure 3.7(a) shows a plot of the degree 9 polynomial (3.11) when $x_1, \ldots, x_9$ are evenly spaced. The tendency for this polynomial to be large near the ends of the interval $[-1, 1]$ is a manifestation of the Runge phenomenon. Figure 3.7(b) shows the same polynomial (3.11), but where the points $x_1, \ldots, x_9$ have been chosen in a way that equalizes the size of the polynomial throughout $[-1, 1]$. The points have been chosen according to Theorem 3.8, presented shortly.

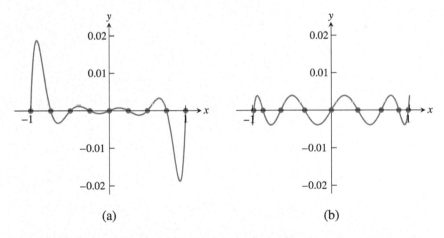

(a)   (b)

Figure 3.7 Part of the interpolation error formula. Plots of $(x-x_1)\cdots(x-x_9)$ for (a) nine evenly spaced base points x_i (b) nine Chebyshev roots x_i.

In fact, this precise positioning, in which the base points x_i are chosen to be $\cos\frac{\pi}{18}, \cos\frac{3\pi}{18}, \ldots, \cos\frac{17\pi}{18}$, makes the maximum absolute value of (3.11) equal to $1/256$, the minimum possible for nine points on the interval $[-1, 1]$. Such positioning, due to Chebyshev, is summarized in the following theorem:

THEOREM 3.5   The choice of real numbers $-1 \leq x_1, \ldots, x_n \leq 1$ that makes the value of

$$\max_{-1 \leq x \leq 1} |(x - x_1)\cdots(x - x_n)|$$

as small as possible is

$$x_i = \cos \frac{(2i-1)\pi}{2n} \quad \text{for } i = 1, \dots, n,$$

and the minimum value is $1/2^{n-1}$. In fact, the minimum is achieved by

$$(x - x_1) \cdots (x - x_n) = \frac{1}{2^{n-1}} T_n(x),$$

where $T_n(x)$ denotes the degree n Chebyshev polynomial. ∎

The proof of this theorem is given later, after we establish a few properties of Chebyshev polynomials. We conclude from the theorem that interpolation error can be minimized if the n interpolation base points in $[-1, 1]$ are chosen to be the roots of the degree n Chebyshev interpolating polynomial $T_n(x)$. These roots are

$$x_i = \cos \frac{\text{odd } \pi}{2n} \tag{3.12}$$

where "odd" stands for the odd numbers from 1 to $2n - 1$. Then we are guaranteed that the absolute value of (3.11) is less than $1/2^{n-1}$ for all x in $[-1, 1]$.

Choosing the Chebyshev roots as the base points for interpolation distributes the interpolation error as evenly as possible across the interval $[-1, 1]$. We will call the interpolating polynomial that uses the Chebyshev roots as base points the **Chebyshev interpolating polynomial**.

▶ **EXAMPLE 3.10**   Find a worst-case error bound for the difference on $[-1, 1]$ between $f(x) = e^x$ and the degree 4 Chebyshev interpolating polynomial.

The interpolation error formula (3.6) gives

$$f(x) - P_4(x) = \frac{(x - x_1)(x - x_2)(x - x_3)(x - x_4)(x - x_5)}{5!} f^{(5)}(c),$$

where

$$x_1 = \cos \frac{\pi}{10}, \quad x_2 = \cos \frac{3\pi}{10}, \quad x_3 = \cos \frac{5\pi}{10}, \quad x_4 = \cos \frac{7\pi}{10}, \quad x_5 = \cos \frac{9\pi}{10}$$

are the Chebyshev roots and where $-1 < c < 1$. According to the Chebyshev Theorem 3.6, for $-1 \leq x \leq 1$,

$$|(x - x_1) \cdots (x - x_5)| \leq \frac{1}{2^4}.$$

In addition, $|f^{(5)}| \leq e^1$ on $[-1, 1]$. The interpolation error is

$$|e^x - P_4(x)| \leq \frac{e}{2^4 5!} \approx 0.00142$$

for all x in the interval $[-1, 1]$.

Compare this result with Example 3.8. The error bound for Chebyshev interpolation for the entire interval is only slightly larger than the bound for a point near the center of the interval, when evenly spaced interpolation is used. Near the ends of the interval, the Chebyshev error is much smaller. ◀

Returning to the Runge Example 3.9, we can eliminate the Runge phenomenon by choosing the interpolation points according to Chebyshev's idea. Figure 3.8 shows that the interpolation error is made small throughout the interval $[-1, 1]$.

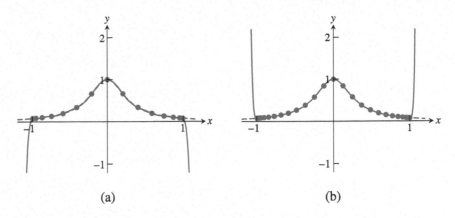

Figure 3.8 Interpolation of Runge example with Chebyshev nodes. The Runge function $f(x) = 1/(1 + 12x^2)$ is graphed along with its Chebyshev interpolation polynomial for (a) 15 points (b) 25 points. The error on $[-1, 1]$ is negligible at this resolution. The polynomial wiggle of Figure 3.6 has vanished, at least between -1 and 1.

3.3.2 Chebyshev polynomials

Define the nth **Chebyshev polynomial** by $T_n(x) = \cos(n \arccos x)$. Despite its appearance, it is a polynomial in the variable x for each n. For example, for $n = 0$ it gives the degree 0 polynomial 1, and for $n = 1$ we get $T_1(x) = \cos(\arccos x) = x$. For $n = 2$, recall the cosine addition formula $\cos(a + b) = \cos a \cos b - \sin a \sin b$. Set $y = \arccos x$, so that $\cos y = x$. Then $T_2(x) = \cos 2y = \cos^2 y - \sin^2 y = 2\cos^2 y - 1 = 2x^2 - 1$, a degree 2 polynomial. In general, note that

$$T_{n+1}(x) = \cos(n + 1)y = \cos(ny + y) = \cos ny \cos y - \sin ny \sin y$$
$$T_{n-1}(x) = \cos(n - 1)y = \cos(ny - y) = \cos ny \cos y - \sin ny \sin(-y). \quad (3.13)$$

Because $\sin(-y) = -\sin y$, we can add the preceding equations to get

$$T_{n+1}(x) + T_{n-1}(x) = 2\cos ny \cos y = 2x T_n(x). \quad (3.14)$$

The resulting relation,

$$T_{n+1}(x) = 2x T_n(x) - T_{n-1}(x), \quad (3.15)$$

is called the **recursion relation** for the Chebyshev polynomials. Several facts follow from (3.15):

FACT 1   The T_n's are polynomials. We showed this explicitly for T_0, T_1, and T_2. Since T_3 is a polynomial combination of T_1 and T_2, T_3 is also a polynomial. The same argument goes for all T_n. The first few Chebyshev polynomials (see Figure 3.9) are

$$T_0(x) = 1$$
$$T_1(x) = x$$
$$T_2(x) = 2x^2 - 1$$
$$T_3(x) = 4x^3 - 3x. \qquad \square$$

FACT 2 $\deg(T_n) = n$, and the leading coefficient is 2^{n-1}. This is clear for $n = 1$ and 2, and the recursion relation extends the fact to all n. $\qquad \square$

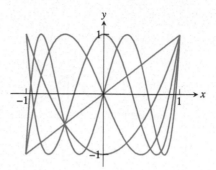

Figure 3.9 Plot of the degree 1 through 5 Chebyshev polynomials. Note that $T_n(1) = 1$ and the maximum absolute value taken on by $T_n(x)$ inside $[-1, 1]$ is 1.

FACT 3 $T_n(1) = 1$ and $T_n(-1) = (-1)^n$. Both are clear for $n = 1$ and 2. In general,

$$T_{n+1}(1) = 2(1)T_n(1) - T_{n-1}(1) = 2(1) - 1 = 1$$

and

$$T_{n+1}(-1) = 2(-1)T_n(-1) - T_{n-1}(-1)$$
$$= -2(-1)^n - (-1)^{n-1}$$
$$= (-1)^{n-1}(2 - 1) = (-1)^{n-1} = (-1)^{n+1}. \qquad \square$$

FACT 4 The maximum absolute value of $T_n(x)$ for $-1 \le x \le 1$ is 1. This follows immediately from the fact that $T_n(x) = \cos y$ for some y. $\qquad \square$

FACT 5 All zeros of $T_n(x)$ are located between -1 and 1. See Figure 3.10. In fact, the zeros are the solution of $0 = \cos(n \arccos x)$. Since $\cos y = 0$ if and only if $y = $ odd integer $\cdot (\pi/2)$, we find that

$$n \arccos x = \text{odd} \cdot \pi/2$$
$$x = \cos \frac{\text{odd} \cdot \pi}{2n}. \qquad \square$$

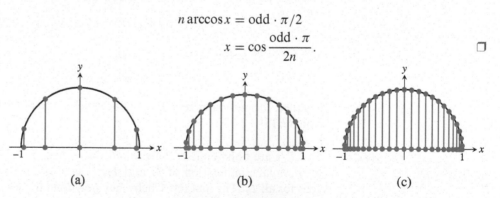

(a) (b) (c)

Figure 3.10 Location of zeros of the Chebyshev polynomial. The roots are the x-coordinates of evenly spaced points around the circle (a) degree 5 (b) degree 15 (c) degree 25.

FACT 6 $T_n(x)$ alternates between -1 and 1 a total of $n + 1$ times. In fact, this happens at $\cos 0, \cos \pi/n, \ldots, \cos(n - 1)\pi/n, \cos \pi$. $\qquad \square$

It follows from Fact 2 that the polynomial $T_n(x)/2^{n-1}$ is monic (has leading coefficient 1). Since, according to Fact 5, all roots of $T_n(x)$ are real, we can write $T_n(x)/2^{n-1}$

in factored form as $(x - x_1) \cdots (x - x_n)$ where the x_i are the Chebyshev nodes as described in Theorem 3.8.

Chebyshev's theorem follows directly from these facts.

Proof of Theorem 3.5. Let $P_n(x)$ be a monic polynomial with an even smaller absolute maximum on $[-1, 1]$; in other words, $|P_n(x)| < 1/2^{n-1}$ for $-1 \le x \le 1$. This assumption leads to a contradiction. Since $T_n(x)$ alternates between -1 and 1 a total of $n + 1$ times (Fact 6), at these $n + 1$ points the difference $P_n - T_n/2^{n-1}$ is alternately positive and negative. Therefore, $P_n - T_n/2^{n-1}$ must cross zero at least n times; that is, it must have at least n roots. This contradicts the fact that, because P_n and $T_n/2^{n-1}$ are monic, their difference is of degree $\le n - 1$.

3.3.3 Change of interval

So far our discussion of Chebyshev interpolation has been restricted to the interval $[-1, 1]$, because Theorem 3.6 is most easily stated for this interval. Next, we will move the whole methodology to a general interval $[a, b]$.

The base points are moved so that they have the same relative positions in $[a, b]$ that they had in $[-1, 1]$. It is best to think of doing this in two steps: (1) *Stretch* the points by the factor $(b - a)/2$ (the ratio of the two interval lengths), and (2) *Translate* the points by $(b + a)/2$ to move the center of mass from 0 to the midpoint of $[a, b]$. In other words, move from the original points

$$\cos \frac{\text{odd } \pi}{2n}$$

to

$$\frac{b - a}{2} \cos \frac{\text{odd } \pi}{2n} + \frac{b + a}{2}.$$

With the new Chebyshev base points $x_1, \ldots, x_n$ in $[a, b]$, the corresponding upper bound on the numerator of the interpolation error formula is changed due to the stretch by $(b - a)/2$ on each factor $x - x_i$. As a result, the minimax value $1/2^{n-1}$ must be replaced by $[(b - a)/2]^n / 2^{n-1}$.

Chebyshev interpolation nodes

On the interval [a,b],

$$x_i = \frac{b + a}{2} + \frac{b - a}{2} \cos \frac{(2i - 1)\pi}{2n}$$

for $i = 1, \ldots, n$. The inequality

$$|(x - x_1) \cdots (x - x_n)| \le \frac{\left(\frac{b-a}{2}\right)^n}{2^{n-1}} \tag{3.16}$$

holds on $[a, b]$.

The next example illustrates the use of Chebyshev interpolation in a general interval.

▶ **EXAMPLE 3.11**   Find the four Chebyshev base points for interpolation on the interval $[0, \pi/2]$, and find an upper bound for the Chebyshev interpolation error for $f(x) = \sin x$ on the interval.

SPOTLIGHT ON

Compression   As shown in this section, Chebyshev interpolation is a good way to turn general functions into a small number of floating point operations, for ease of computation. An upper bound for the error made is easily available, is usually smaller than for evenly spaced interpolation, and can be made as small as desired.

Although we have used the sine function to demonstrate this process, a different approach is taken to construct the actual "sine key" on most calculators and canned software. Special properties of the sine function allow it to be approximated by a simple Taylor expansion, slightly altered to take rounding effects into account. Because sine is an odd function, the even-numbered terms in its Taylor series around zero are missing, making it especially efficient to calculate.

This is a second attempt. We used evenly spaced base points in Example 3.7. The Chebyshev base points are

$$\frac{\frac{\pi}{2} - 0}{2} \cos\left(\frac{\text{odd } \pi}{2(4)}\right) + \frac{\frac{\pi}{2} + 0}{2},$$

or

$$x_1 = \frac{\pi}{4} + \frac{\pi}{4}\cos\frac{\pi}{8}, x_2 = \frac{\pi}{4} + \frac{\pi}{4}\cos\frac{3\pi}{8}, x_3 = \frac{\pi}{4} + \frac{\pi}{4}\cos\frac{5\pi}{8}, x_4 = \frac{\pi}{4} + \frac{\pi}{4}\cos\frac{7\pi}{8}.$$

From (3.16), the worst-case interpolation error for $0 \le x \le \pi/2$ is

$$|\sin x - P_3(x)| = \frac{|(x - x_1)(x - x_2)(x - x_3)(x - x_4)|}{4!}|f''''(c)|$$

$$\le \frac{\left(\frac{\frac{\pi}{2}-0}{2}\right)^4}{4!2^3} 1 \approx 0.00198.$$

The Chebyshev interpolating polynomial for this example is evaluated at several points in the following table:

x	$\sin x$	$P_3(x)$	error
1	0.8415	0.8408	0.0007
2	0.9093	0.9097	0.0004
3	0.1411	0.1420	0.0009
4	−0.7568	−0.7555	0.0013
14	0.9906	0.9917	0.0011
1000	0.8269	0.8261	0.0008

The interpolation errors are well below the worst-case estimate. Figure 3.11 plots the interpolation error as a function of x on the interval $[0, \pi/2]$, compared with the same for evenly spaced interpolation. The Chebyshev error (dashed curve) is a bit smaller and is distributed more evenly throughout the interpolation interval.   ◄

▶ **EXAMPLE 3.12**   Design a sine key that will give output correct to 10 decimal places.

Thanks to our work earlier on setting up a fundamental domain for the sine function, we can continue to concentrate on the interval $[0, \pi/2]$. Repeat the previous calculation, but leave n, the number of base points, as an unknown to be determined. The maximum interpolation error for the polynomial $P_{n-1}(x)$ on the interval $[0, \pi/2]$ is

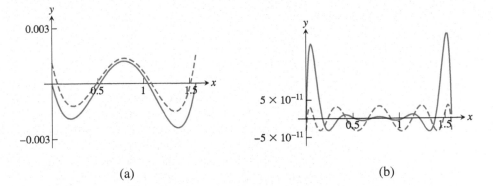

Figure 3.11 Interpolation error for approximating $f(x) = \sin x$. (a) Interpolation error for degree 3 interpolating polynomial with evenly spaced base points (solid curve) and Chebyshev base points (dashed curve). (b) Same as (a), but degree 9.

$$|\sin x - P_{n-1}(x)| = \frac{|(x-x_1)\cdots(x-x_n)|}{n}|f^{(n)}(c)|$$

$$\leq \frac{\left(\frac{\frac{\pi}{2}-0}{2}\right)^n}{n2^{n-1}}1.$$

This equation is not simple to solve for n, but a little trial and error finds that for $n = 9$ the error bound is $\approx 0.1224 \times 10^{-8}$, and for $n = 10$ it is $\approx 0.4807 \times 10^{-10}$. The latter meets our criterion for 10 correct decimal places. Figure 3.11(b) compares the actual error of the Chebyshev interpolation polynomial with the error of the evenly spaced interpolation polynomial.

The 10 Chebyshev base points on $[0, \pi/2]$ are $\pi/4 + (\pi/4)\cos(\text{odd }\pi/20)$. The key can be designed by storing the 10 y-values for sine at the base points and doing a nested multiplication evaluation for each key press. ◄

The following MATLAB code sin2.m carries out the preceding task. The code is a bit awkward as written: We have to do 10 sin evaluations, at the 10 Chebyshev nodes, in order to set up the interpolating polynomial to approximate sin at one point. Of course, in a real implementation, these numbers would be computed once and stored.

MATLAB code
shown here can be found
at bit.ly/2AgCRTk

```
%Program 3.4 Building a sin calculator key, attempt #2
%Approximates sin curve with degree 9 polynomial
%Input:   x
%Output: approximation for sin(x), correct to 10 decimal places
function y=sin2(x)
%First calculate the interpolating polynomial and
%   store coefficients
n=10;
b=pi/4+(pi/4)*cos((1:2:2*n-1)*pi/(2*n));
yb=sin(b);                          % b holds Chebyshev base points
c=newtdd(b,yb,n);
%For each input x, move x to the fundamental domain and evaluate
%   the interpolating polynomial
s=1;                                % Correct the sign of sin
x1=mod(x,2*pi);
if x1>pi
  x1 = 2*pi-x1;
  s = -1;
```

```
end
if x1 > pi/2
  x1 = pi-x1;
end
y = s*nest(n-1,c,x1,b);
```

In this chapter, we have often illustrated polynomial interpolation, either evenly spaced or using Chebyshev nodes, for the purpose of approximating the trigonometric functions. Although polynomial interpolation can be used to approximate sine and cosine to arbitrarily high accuracy, most calculators use a slightly more efficient approach called the CORDIC (Coordinate Rotation Digital Computer) algorithm (Volder [1959]). CORDIC is an elegant iterative method, based on complex arithmetic, that can be applied to several special functions. Polynomial interpolation remains a simple and useful technique for approximating general functions and for representing and compressing data.

► **ADDITIONAL EXAMPLES**

*1 Let $P_5(x)$ be the degree 5 polynomial that interpolates $f(x) = 2^x$ at the six Chebyshev nodes on the interval [0, 1]. Find the best possible upper bound for the interpolation error $|2^x - P_5(x)|$ for x on the entire interval [0, 1].

2. Use the newtdd.m and nest.m codes to plot the interpolating polynomial of $f(x) = e^{\sin x}$ at the 11 Chebyshev interpolation points in the interval [0, 4]. Find the (empirical) maximum interpolation error on the interval.

🖵 **Solutions** for Additional Examples can be found at bit.ly/2RVkpq6
(* example with video solution)

3.3 Exercises

1. List the Chebyshev interpolation nodes $x_1, \ldots, x_n$ in the given interval. (a) $[-1, 1], n = 6$ (b) $[-2, 2], n = 4$ (c) $[4, 12], n = 6$ (d) $[-0.3, 0.7], n = 5$

2. Find the upper bound for $|(x - x_1) \ldots (x - x_n)|$ on the intervals and Chebyshev nodes in Exercise 1.

3. Assume that Chebyshev interpolation is used to find a fifth degree interpolating polynomial $Q_5(x)$ on the interval $[-1, 1]$ for the function $f(x) = e^x$. Use the interpolation error formula to find a worst-case estimate for the error $|e^x - Q_5(x)|$ that is valid for x throughout the interval $[-1, 1]$. How many digits after the decimal point will be correct when $Q_5(x)$ is used to approximate e^x?

4. Answer the same questions as in Exercise 3, but for the interval [0.6, 1.0].

5. Find an upper bound for the error on [0, 2] when the degree 3 Chebyshev interpolating polynomial is used to approximate $f(x) = \sin x$.

6. Assume that you are to use Chebyshev interpolation to find a degree 3 interpolating polynomial $Q_3(x)$ that approximates the function $f(x) = x^{-3}$ on the interval [3, 4]. (a) Write down the (x, y) points that will serve as interpolation nodes for Q_3. (b) Find a worst-case estimate for the error $|x^{-3} - Q_3(x)|$ that is valid for all x in the interval [3, 4]. How many digits after the decimal point will be correct when $Q_3(x)$ is used to approximate x^{-3}?

7. Suppose you are designing the **ln** key for a calculator whose display shows six digits to the right of the decimal point. Find the least degree d for which Chebyshev interpolation on the interval $[1, e]$ will approximate within this accuracy.

8. Let $T_n(x)$ denote the degree n Chebyshev polynomial. Find a formula for $T_n(0)$.

9. Determine the following values: (a) $T_{999}(-1)$ (b) $T_{1000}(-1)$ (c) $T_{999}(0)$ (d) $T_{1000}(0)$ (e) $T_{999}(-1/2)$ (f) $T_{1000}(-1/2)$.

3.3 Computer Problems

Solutions for Computer Problems numbered in blue can be found at bit.ly/2RXJWim

1. Rebuild Program 3.3 to implement the Chebyshev interpolating polynomial with four nodes on the interval $[0, \pi/2]$. (Only one line of code needs to be changed.) Then plot the polynomial and the sine function on the interval $[-2, 2]$.

2. Build a MATLAB program to evaluate the cosine function correct to 10 decimal places using Chebyshev interpolation. Start by interpolating on a fundamental domain $[0, \pi/2]$, and extend your answer to inputs between -10^4 and 10^4. You may want to use some of the MATLAB code written in this chapter.

3. Carry out the steps of Computer Problem 2 for $\ln x$, for inputs x between 10^{-4} and 10^4. Use $[1, e]$ as the fundamental domain. What is the degree of the interpolation polynomial that guarantees 10 correct digits? Your program should begin by finding the integer k such that $e^k \le x < e^{k+1}$. Then xe^{-k} lies in the fundamental domain. Demonstrate the accuracy of your program by comparing it with MATLAB's log command.

4. Let $f(x) = e^{|x|}$. Compare evenly spaced interpolation with Chebyshev interpolation by plotting degree n polynomials of both types on the interval $[-1, 1]$, for $n = 10$ and 20. For evenly spaced interpolation, the left and right interpolation base points should be -1 and 1. By sampling at a 0.01 step size, create the empirical interpolation errors for each type, and plot a comparison. Can the Runge phenomenon be observed in this problem?

5. Carry out the steps of Computer Problem 4 for $f(x) = e^{-x^2}$.

3.4 CUBIC SPLINES

Splines represent an alternative approach to data interpolation. In polynomial interpolation, a single formula, given by a polynomial, is used to meet all data points. The idea of splines is to use several formulas, each a low-degree polynomial, to pass through the data points.

The simplest example of a spline is a linear spline, in which one "connects the dots" with straight-line segments. Assume that we are given a set of data points $(x_1, y_1), \ldots, (x_n, y_n)$ with $x_1 < \cdots < x_n$. A linear spline consists of the $n - 1$ line segments that are drawn between neighboring pairs of points. Figure 3.12(a) shows a linear spline where, between each neighboring pair of points $(x_i, y_i), (x_{i+1}, y_{i+1})$, the linear function $y = a_i + b_i x$ is drawn through the two points. The given data points in the figure are $(1, 2), (2, 1), (4, 4)$, and $(5, 3)$, and the linear spline is given by

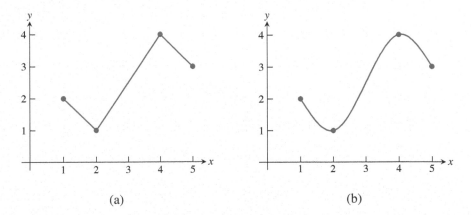

(a)   (b)

Figure 3.12 Splines through four data points. (a) Linear spline through (1,2), (2,1), (4,4), and (5,3) consists of three linear polynomials given by (3.17). (b) Cubic spline through the same points, given by (3.18).

$$S_1(x) = 2 - (x - 1) \text{ on } [1, 2]$$

$$S_2(x) = 1 + \frac{3}{2}(x - 2) \text{ on } [2, 4]$$

$$S_3(x) = 4 - (x - 4) \text{ on } [4, 5]. \tag{3.17}$$

The linear spline successfully interpolates an arbitrary set of n data points. However, linear splines lack smoothness. Cubic splines are meant to address this shortcoming of linear splines. A cubic spline replaces linear functions between the data points by degree 3 (cubic) polynomials.

An example of a cubic spline that interpolates the same points $(1, 2)$, $(2, 1)$, $(4, 4)$, and $(5, 3)$ is shown in Figure 3.12(b). The equations defining the spline are

$$S_1(x) = 2 - \frac{13}{8}(x - 1) + 0(x - 1)^2 + \frac{5}{8}(x - 1)^3 \text{ on } [1, 2]$$

$$S_2(x) = 1 + \frac{1}{4}(x - 2) + \frac{15}{8}(x - 2)^2 - \frac{5}{8}(x - 2)^3 \text{ on } [2, 4]$$

$$S_3(x) = 4 + \frac{1}{4}(x - 4) - \frac{15}{8}(x - 4)^2 + \frac{5}{8}(x - 4)^3 \text{ on } [4, 5]. \tag{3.18}$$

Note in particular the smooth transition from one S_i to the next at the base points, or "knots," $x = 2$ and $x = 4$. This is achieved by arranging for the neighboring pieces S_i and S_{i+1} of the spline to have the same zeroth, first, and second derivatives when evaluated at the knots. Just how to do this is the topic of the next section.

Given n points $(x_1, y_1), \ldots, (x_n, y_n)$, there is obviously one and only one linear spline through the data points. This will not be true for cubic splines. We will find that there are infinitely many through any set of data points. Extra conditions will be added when it is necessary to nail down a particular spline of interest.

3.4.1 Properties of splines

To be a little more precise about the properties of a cubic spline, we make the following definition: Assume that we are given the n data points $(x_1, y_1), \ldots, (x_n, y_n)$, where the x_i are distinct and in increasing order. A **cubic spline** $S(x)$ through the data points $(x_1, y_1), \ldots, (x_n, y_n)$ is a set of cubic polynomials

$$S_1(x) = y_1 + b_1(x - x_1) + c_1(x - x_1)^2 + d_1(x - x_1)^3 \text{ on } [x_1, x_2]$$
$$S_2(x) = y_2 + b_2(x - x_2) + c_2(x - x_2)^2 + d_2(x - x_2)^3 \text{ on } [x_2, x_3] \tag{3.19}$$
$$\vdots$$

$$S_{n-1}(x) = y_{n-1} + b_{n-1}(x - x_{n-1}) + c_{n-1}(x - x_{n-1})^2 + d_{n-1}(x - x_{n-1})^3 \text{on } [x_{n-1}, x_n]$$

with the following properties:

Property 1 $S_i(x_i) = y_i$ and $S_i(x_{i+1}) = y_{i+1}$ for $i = 1, \ldots, n - 1$.

Property 2 $S'_{i-1}(x_i) = S'_i(x_i)$ for $i = 2, \ldots, n - 1$.

Property 3 $S''_{i-1}(x_i) = S''_i(x_i)$ for $i = 2, \ldots, n - 1$.

Property 1 guarantees that the spline $S(x)$ interpolates the data points. Property 2 forces the slopes of neighboring parts of the spline to agree where they meet, and Property 3 does the same for the curvature, represented by the second derivative.

► **EXAMPLE 3.13**   Check that $\{S_1, S_2, S_3\}$ in (3.18) satisfies all cubic spline properties for the data points $(1, 2)$, $(2, 1)$, $(4, 4)$, and $(5, 3)$.

We will check all three properties.

Property 1. There are $n = 4$ data points. We must check

$$S_1(1) = 2 \text{ and } S_1(2) = 1$$
$$S_2(2) = 1 \text{ and } S_2(4) = 4$$
$$S_3(4) = 4 \text{ and } S_3(5) = 3.$$

These follow easily from the defining equations (3.18).

Property 2. The first derivatives of the spline functions are

$$S_1'(x) = -\frac{13}{8} + \frac{15}{8}(x - 1)^2$$
$$S_2'(x) = \frac{1}{4} + \frac{15}{4}(x - 2) - \frac{15}{8}(x - 2)^2$$
$$S_3'(x) = \frac{1}{4} - \frac{15}{4}(x - 4) + \frac{15}{8}(x - 4)^2.$$

We must check $S_1'(2) = S_2'(2)$ and $S_2'(4) = S_3'(4)$. The first is

$$-\frac{13}{8} + \frac{15}{8} = \frac{1}{4},$$

and the second is

$$\frac{1}{4} + \frac{15}{4}(4 - 2) - \frac{15}{8}(4 - 2)^2 = \frac{1}{4},$$

both of which check out.

Property 3. The second derivatives are

$$S_1''(x) = \frac{15}{4}(x - 1)$$
$$S_2''(x) = \frac{15}{4} - \frac{15}{4}(x - 2)$$
$$S_3''(x) = -\frac{15}{4} + \frac{15}{4}(x - 4). \tag{3.20}$$

We must check $S_1''(2) = S_2''(2)$ and $S_2''(4) = S_3''(4)$, both of which are true. Therefore, (3.18) is a cubic spline.   ◄

Constructing a spline from a set of data points means finding the coefficients b_i, c_i, d_i that make Properties 1–3 hold. Before we discuss how to determine the unknown coefficients b_i, c_i, d_i of the spline, let us count the number of conditions imposed by the definition. The first half of Property 1 is already reflected in the form (3.19); it says that the constant term of the cubic S_i must be y_i. The second half of Property 1 consists of $n - 1$ separate equations that must be satisfied by the coefficients, which we consider as unknowns. Each of Properties 2 and 3 add $n - 2$ additional equations, for a total of $n - 1 + 2(n - 2) = 3n - 5$ independent equations to be satisfied.

How many unknown coefficients are there? For each part S_i of the spline, three coefficients b_i, c_i, d_i are needed, for a total of $3(n-1) = 3n - 3$. Therefore, solving for the coefficients is a problem of solving $3n - 5$ linear equations in $3n - 3$ unknowns. Unless there are inconsistent equations in the system (and there are not), the system of equations is underdetermined and so has infinitely many solutions. In other words, there are infinitely many cubic splines passing through the arbitrary set of data points $(x_1, y_1), \ldots, (x_n, y_n)$.

Users of splines normally exploit the shortage of equations by adding two extra to the $3n - 5$ equations to arrive at a system of m equations in m unknowns, where $m = 3n - 3$. Aside from allowing the user to constrain the spline to given specifications, narrowing the field to a single solution simplifies computing and describing the result.

The simplest way of adding two more constraints is to require, in addition to the previous $3n - 5$ constraints, that the spline $S(x)$ have an inflection point at each end of the defining interval $[x_1, x_n]$. The constraints added to Properties 1–3 are

Property 4a   **Natural spline.** $S_1''(x_1) = 0$ and $S_{n-1}''(x_n) = 0$.

A cubic spline that satisfies these two additional conditions is called a **natural cubic spline**. Note that (3.18) is a natural cubic spline, since it is easily verified from (3.20) that $S_1''(1) = 0$ and $S_3''(5) = 0$.

There are several other ways to add two more conditions. Usually, as in the case of the natural spline, they determine extra properties of the left and right ends of the spline, so they are called **end conditions**. We will take up this topic in the next section, but for now we concentrate on natural cubic splines.

Now that we have the right number of equations, $3n - 3$ equations in $3n - 3$ unknowns, we can write a MATLAB function to solve them for the spline coefficients. First we write out the equations in the unknowns b_i, c_i, d_i. Part 2 of Property 1 then implies the $n - 1$ equations:

$$y_2 = S_1(x_2) = y_1 + b_1(x_2 - x_1) + c_1(x_2 - x_1)^2 + d_1(x_2 - x_1)^3$$

$$\vdots$$

$$y_n = S_{n-1}(x_n) = y_{n-1} + b_{n-1}(x_n - x_{n-1}) + c_{n-1}(x_n - x_{n-1})^2$$
$$+ d_{n-1}(x_n - x_{n-1})^3. \tag{3.21}$$

Property 2 generates the $n - 2$ equations,

$$0 = S_1'(x_2) - S_2'(x_2) = b_1 + 2c_1(x_2 - x_1) + 3d_1(x_2 - x_1)^2 - b_2$$

$$\vdots$$

$$0 = S_{n-2}'(x_{n-1}) - S_{n-1}'(x_{n-1}) = b_{n-2} + 2c_{n-2}(x_{n-1} - x_{n-2})$$
$$+ 3d_{n-2}(x_{n-1} - x_{n-2})^2 - b_{n-1}, \tag{3.22}$$

and Property 3 implies the $n - 2$ equations:

$$0 = S_1''(x_2) - S_2''(x_2) = 2c_1 + 6d_1(x_2 - x_1) - 2c_2$$

$$\vdots$$

$$0 = S_{n-2}''(x_{n-1}) - S_{n-1}''(x_{n-1}) = 2c_{n-2} + 6d_{n-2}(x_{n-1} - x_{n-2}) - 2c_{n-1}. \tag{3.23}$$

Instead of solving the equations in this form, the system can be simplified drastically by decoupling the equations. With a little algebra, a much smaller system of equations in the c_i can be solved first, followed by explicit formulas for the b_i and d_i in terms of the known c_i.

It is conceptually simpler if an extra unknown $c_n = S''_{n-1}(x_n)/2$ is introduced. In addition, we introduce the shorthand notation $\delta_i = x_{i+1} - x_i$ and $\Delta_i = y_{i+1} - y_i$. Then (3.23) can be solved for the coefficients

$$d_i = \frac{c_{i+1} - c_i}{3\delta_i} \quad \text{for } i = 1, \ldots, n-1. \tag{3.24}$$

Solving (3.21) for b_i yields

$$\begin{aligned}
b_i &= \frac{\Delta_i}{\delta_i} - c_i\delta_i - d_i\delta_i^2 \\
&= \frac{\Delta_i}{\delta_i} - c_i\delta_i - \frac{\delta_i}{3}(c_{i+1} - c_i) \\
&= \frac{\Delta_i}{\delta_i} - \frac{\delta_i}{3}(2c_i + c_{i+1})
\end{aligned} \tag{3.25}$$

for $i = 1, \ldots, n-1$.

Substituting (3.24) and (3.25) into (3.22) results in the following $n-2$ equations in $c_1, \ldots, c_n$:

$$\delta_1 c_1 + 2(\delta_1 + \delta_2)c_2 + \delta_2 c_3 = 3\left(\frac{\Delta_2}{\delta_2} - \frac{\Delta_1}{\delta_1}\right)$$

$$\vdots$$

$$\delta_{n-2}c_{n-2} + 2(\delta_{n-2} + \delta_{n-1})c_{n-1} + \delta_{n-1}c_n = 3\left(\frac{\Delta_{n-1}}{\delta_{n-1}} - \frac{\Delta_{n-2}}{\delta_{n-2}}\right).$$

Two more equations are given by the natural spline conditions (Property 4a):

$$S''_1(x_1) = 0 \to 2c_1 = 0$$
$$S''_{n-1}(x_n) = 0 \to 2c_n = 0.$$

This gives a total of n equations in n unknowns c_i, which can be written in the matrix form

$$\begin{bmatrix}
1 & 0 & & 0 & & & \\
\delta_1 & 2\delta_1 + 2\delta_2 & \delta_2 & & \ddots & & \\
0 & \delta_2 & 2\delta_2 + 2\delta_3 & \delta_3 & & & \\
& \ddots & \ddots & \ddots & \ddots & & \\
& & & \delta_{n-2} & 2\delta_{n-2} + 2\delta_{n-1} & \delta_{n-1} \\
& & & 0 & 0 & 1
\end{bmatrix}
\begin{bmatrix}
c_1 \\ \\ \vdots \\ \\ \\ c_n
\end{bmatrix}$$

$$= \begin{bmatrix}
0 \\
3\left(\frac{\Delta_2}{\delta_2} - \frac{\Delta_1}{\delta_1}\right) \\
\vdots \\
3\left(\frac{\Delta_{n-1}}{\delta_{n-1}} - \frac{\Delta_{n-2}}{\delta_{n-2}}\right) \\
0
\end{bmatrix}. \tag{3.26}$$

After $c_1, \ldots, c_n$ are obtained from (3.26), $b_1, \ldots, b_{n-1}$ and $d_1, \ldots, d_{n-1}$ are found from (3.24) and (3.25).

Note that (3.26) is always solvable for the c_i. The coefficient matrix is strictly diagonally dominant, so by Theorem 2.10, there is a unique solution for the c_i and therefore also for the b_i and d_i. We have thus proved the following theorem:

THEOREM 3.6 Let $n \geq 2$. For a set of data points $(x_1, y_1), \ldots, (x_n, y_n)$ with distinct x_i, there is a unique natural cubic spline fitting the points. ∎

Natural cubic spline

Given $x = [x_1, \ldots, x_n]$ where $x_1 < \cdots < x_n, y = [y_1, \ldots, y_n]$

for $i = 1, \ldots, n - 1$
$\quad a_i = y_i$
$\quad \delta_i = x_{i+1} - x_i$
$\quad \Delta_i = y_{i+1} - y_i$
end
Solve (3.26) for $c_1, \ldots, c_n$
for $i = 1, \ldots, n - 1$
$$d_i = \frac{c_{i+1} - c_i}{3\delta_i}$$
$$b_i = \frac{\Delta_i}{\delta_i} - \frac{\delta_i}{3}(2c_i + c_{i+1})$$
end
The natural cubic spline is
$S_i(x) = a_i + b_i(x - x_i) + c_i(x - x_i)^2 + d_i(x - x_i)^3$ on $[x_i, x_{i+1}]$ for $i = 1, \ldots, n - 1$.

▶ **EXAMPLE 3.14** Find the natural cubic spline through $(0, 3), (1, -2)$, and $(2, 1)$.

The x-coordinates are $x_1 = 0, x_2 = 1$, and $x_3 = 2$. The y-coordinates are $a_1 = y_1 = 3, a_2 = y_2 = -2$, and $a_3 = y_3 = 1$, and the differences are $\delta_1 = \delta_2 = 1, \Delta_1 = -5$, and $\Delta_2 = 3$. The tridiagonal matrix equation (3.26) is

$$\begin{bmatrix} 1 & 0 & 0 \\ 1 & 4 & 1 \\ 0 & 0 & 1 \end{bmatrix} \begin{bmatrix} c_1 \\ c_2 \\ c_3 \end{bmatrix} = \begin{bmatrix} 0 \\ 24 \\ 0 \end{bmatrix}.$$

The solution is $[c_1, c_2, c_3] = [0, 6, 0]$. Now, (3.24) and (3.25) yield

$$d_1 = \frac{c_2 - c_1}{3\delta_1} = \frac{6}{3} = 2$$

$$d_2 = \frac{c_3 - c_2}{3\delta_2} = \frac{-6}{3} = -2$$

$$b_1 = \frac{\Delta_1}{\delta_1} - \frac{\delta_1}{3}(2c_1 + c_2) = -5 - \frac{1}{3}(6) = -7$$

$$b_2 = \frac{\Delta_2}{\delta_2} - \frac{\delta_2}{3}(2c_2 + c_3) = 3 - \frac{1}{3}(12) = -1.$$

Therefore, the cubic spline is

$$S_1(x) = 3 - 7x + 0x^2 + 2x^3 \text{ on } [0, 1]$$

$$S_2(x) = -2 - 1(x - 1) + 6(x - 1)^2 - 2(x - 1)^3 \text{ on } [1, 2].$$

◀

MATLAB code for this calculation follows. For different (not natural) endpoint conditions, discussed in the next section, the top and bottom rows of (3.26) are replaced by other appropriate rows.

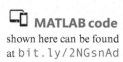

MATLAB code

shown here can be found at bit.ly/2Cpe5SE

```
%Program 3.5 Calculation of spline coefficients
%Calculates coefficients of cubic spline
%Input: x,y vectors of data points
%   plus two optional extra data v1, vn
%Output: matrix of coefficients b1,c1,d1;b2,c2,d2;...
function coeff=splinecoeff(x,y)
n=length(x);v1=0;vn=0;
A=zeros(n,n);            % matrix A is nxn
r=zeros(n,1);
for i=1:n-1              % define the deltas
    dx(i)= x(i+1)-x(i); dy(i)=y(i+1)-y(i);
end
for i=2:n-1             % load the A matrix
    A(i,i-1:i+1)=[dx(i-1) 2*(dx(i-1)+dx(i)) dx(i)];
    r(i)=3*(dy(i)/dx(i)-dy(i-1)/dx(i-1)); % right-hand side
end
% Set endpoint conditions
% Use only one of following 5 pairs:
A(1,1) = 1;            % natural spline conditions
A(n,n) = 1;
%A(1,1)=2;r(1)=v1;      % curvature-adj conditions
%A(n,n)=2;r(n)=vn;
%A(1,1:2)=[2*dx(1) dx(1)];r(1)=3*(dy(1)/dx(1)-v1);  %clamped
%A(n,n-1:n)=[dx(n-1) 2*dx(n-1)];r(n)=3*(vn-dy(n-1)/dx(n-1));
%A(1,1:2)=[1 -1];        % parabol-term conditions, for n>=3
%A(n,n-1:n)=[1 -1];
%A(1,1:3)=[dx(2) -(dx(1)+dx(2)) dx(1)]; % not-a-knot, for n>=4
%A(n,n-2:n)=[dx(n-1) -(dx(n-2)+dx(n-1)) dx(n-2)];
coeff=zeros(n,3);
coeff(:,2)=A\r;         % solve for c coefficients
for i=1:n-1             % solve for b and d
    coeff(i,3)=(coeff(i+1,2)-coeff(i,2))/(3*dx(i));
    coeff(i,1)=dy(i)/dx(i)-dx(i)*(2*coeff(i,2)+coeff(i+1,2))/3;
end
coeff=coeff(1:n-1,1:3);
```

We have taken the liberty of listing other choices for end conditions, although they are commented out for now. The alternative conditions will be discussed in the next section. Another MATLAB function, titled splineplot.m, calls splinecoeff.m to get the coefficients and then plots the cubic spline:

MATLAB code

shown here can be found at bit.ly/2NGsnAd

```
%Program 3.6 Cubic spline plot
%Computes and plots spline from data points
%Input: x,y vectors of data points, number k of plotted points
%   per segment
%Output: x1, y1 spline values at plotted points
function [x1,y1]=splineplot(x,y,k)
n=length(x);
coeff=splinecoeff(x,y);
x1=[]; y1=[];
for i=1:n-1
    xs=linspace(x(i),x(i+1),k+1);
    dx=xs-x(i);
```

```
      ys=coeff(i,3)*dx; % evaluate using nested multiplication
      ys=(ys+coeff(i,2)).*dx;
      ys=(ys+coeff(i,1)).*dx+y(i);
      x1=[x1; xs(1:k)']; y1=[y1;ys(1:k)'];
end
x1=[x1; x(end)];y1=[y1;y(end)];
plot(x,y,'o',x1,y1)
```

Figure 3.13(a) shows a natural cubic spline generated by `splineplot.m`.

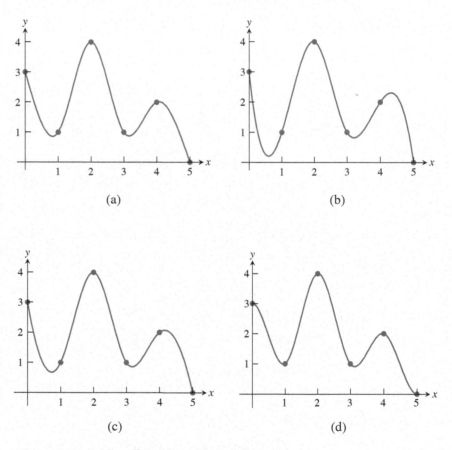

(a)   (b)

(c)   (d)

Figure 3.13 Cubic splines through six points. The plots are generated by `splineplot(x,y,10)` with input vectors `x=[0 1 2 3 4 5]` and `y=[3 1 4 1 2 0]`. (a) Natural cubic spline (notice inflection points at ends) (b) Not-a-knot cubic spline (single cubic equation on [0,2] and on [3,5]) (c) Parabolically terminated spline (d) Clamped cubic spline (clamped at slope 0 at both ends).

3.4.2 Endpoint conditions

The two extra conditions specified in Property 4a are called the "endpoint conditions" for a natural spline. Requiring that these be satisfied along with Properties 1 through 3 narrows the field to exactly one cubic spline, according to Theorem 3.9. It turns out that there are many different versions of Property 4, meaning many other pairs of endpoint conditions, for which an analogous theorem holds. In this section, we present a few of the more popular ones.

Property 4b **Curvature-adjusted cubic spline.** The first alternative to a natural cubic spline requires setting $S_1''(x_1)$ and $S_{n-1}''(x_n)$ to arbitrary values, chosen by the user, instead of zero. This choice corresponds to setting the desired curvatures at the left and right endpoints of the spline. In terms of (3.23), it translates to the two extra conditions

$$2c_1 = v_1$$
$$2c_n = v_n,$$

where v_1, v_n denote the desired values. The equations turn into the two tableau rows

$$\begin{bmatrix} 2 & 0 & 0 & 0 & 0 & \cdots & \cdots & 0 & 0 & | & v_1 \\ 0 & 0 & 0 & 0 & 0 & \cdots & \cdots & 0 & 2 & | & v_n \end{bmatrix}$$

to replace the top and bottom rows of (3.26), which were added for the natural spline. Notice that the new coefficient matrix is again strictly diagonally dominant, so that a generalized form of Theorem 3.9 holds for curvature-adjusted splines. (See Theorem 3.10, presented shortly.) In splinecoeff.m, the two lines

```
A(1,1)=2;r(1)=v1;        % curvature-adj conditions
A(n,n)-2;r(n)=vn;
```

must be substituted in place of the two existing lines for the natural spline.

The next alternative set of end conditions is

Property 4c **Clamped cubic spline.** This alternative is similar to the preceding one, but it is the first derivatives $S_1'(x_1)$ and $S_{n-1}'(x_n)$ that are set to user-specified values v_1 and v_n, respectively. Thus, the slope at the beginning and end of the spline are under the user's control.

Using (3.24) and (3.25), we can write the extra condition $S_1'(x_1) = v_1$ as

$$2\delta_1 c_1 + \delta_1 c_2 = 3\left(\frac{\Delta_1}{\delta_1} - v_1\right)$$

and $S_{n-1}'(x_n) = v_n$ as

$$\delta_{n-1} c_{n-1} + 2\delta_{n-1} c_n = 3\left(v_n - \frac{\Delta_{n-1}}{\delta_{n-1}}\right).$$

The two corresponding tableau rows are

$$\begin{bmatrix} 2\delta_1 & \delta_1 & 0 & 0 & \cdots & \cdots & 0 & 0 & 0 & | & 3(\Delta_1/\delta_1 - v_1) \\ 0 & 0 & 0 & 0 & \cdots & \cdots & 0 & \delta_{n-1} & 2\delta_{n-1} & | & 3(v_n - \Delta_{n-1}/\delta_{n-1}) \end{bmatrix}.$$

Note that strict diagonal dominance holds also for the revised coefficient matrix in (3.26), so Theorem 3.9 also holds with the natural spline replaced with the clamped spline. In splinecoeff.m, the two lines

```
A(1,1:2)=[2*dx(1) dx(1)];r(1)=3*(dy(1)/dx(1)-v1);
A(n,n-1:n)=[dx(n-1) 2*dx(n-1)];r(n)=3*(vn-dy(n-1)/dx(n-1));
```

must be substituted. See Figure 3.13 for a clamped spline with $v_1 = v_n = 0$.

Property 4d **Parabolically terminated cubic spline.** The first and last parts of the spline, S_1 and S_{n-1}, are forced to be at most degree 2, by specifying that $d_1 = 0 = d_{n-1}$. Equivalently, according to (3.24), we can require that $c_1 = c_2$ and $c_{n-1} = c_n$. The equations form the two tableau rows

$$\begin{bmatrix} 1 & -1 & 0 & 0 & 0 & \cdots & \cdots & 0 & 0 & 0 & | & 0 \\ 0 & 0 & 0 & 0 & 0 & \cdots & \cdots & 0 & 1 & -1 & | & 0 \end{bmatrix}$$

to be used as the top and bottom rows of (3.26). Assume that the number n of data points satisfies $n \geq 3$. (See Exercise 19 for the case $n = 2$.) In this case, upon replacing c_1 by c_2 and c_n by c_{n-1}, we find that the matrix equation reduces to a strictly diagonally dominant $n - 2 \times n - 2$ matrix equation in $c_2, \ldots, c_{n-1}$. Therefore, a version of Theorem 3.9 holds for parabolically terminated splines, assuming that $n \geq 3$.

In `splinecoeff.m`, the two lines

```
A(1,1:2)=[1 -1];                    % parabol-term conditions
A(n,n-1:n)=[1 -1];
```

must be substituted.

Property 4e **Not-a-knot cubic spline.** The two added equations are $d_1 = d_2$ and $d_{n-2} = d_{n-1}$, or equivalently, $S_1'''(x_2) = S_2'''(x_2)$ and $S_{n-2}'''(x_{n-1}) = S_{n-1}'''(x_{n-1})$. Since S_1 and S_2 are polynomials of degree 3 or less, requiring their third derivatives to agree at x_2, while their zeroth, first, and second derivatives already agree there, causes S_1 and S_2 to be identical cubic polynomials. (Cubics are defined by four coefficients, and four conditions are specified.) Thus, x_2 is not needed as a base point: The spline is given by the same formula $S_1 = S_2$ on the entire interval $[x_1, x_3]$. The same reasoning shows that $S_{n-2} = S_{n-1}$, so not only x_2, but also x_{n-1}, is "no longer a knot."

Note that $d_1 = d_2$ implies that $(c_2 - c_1)/\delta_1 = (c_3 - c_2)/\delta_2$, or

$$\delta_2 c_1 - (\delta_1 + \delta_2)c_2 + \delta_1 c_3 = 0,$$

and similarly, $d_{n-2} = d_{n-1}$ implies that

$$\delta_{n-1} c_{n-2} - (\delta_{n-2} + \delta_{n-1})c_{n-1} + \delta_{n-2} c_n = 0.$$

It follows that the two tableau rows are

$$\begin{pmatrix} \delta_2 & -(\delta_1 + \delta_2) & \delta_1 & 0 & \cdots & \cdots & 0 & 0 & 0 & 0 & | & 0 \\ 0 & 0 & 0 & 0 & \cdots & \cdots & 0 & \delta_{n-1} & -(\delta_{n-2} + \delta_{n-1}) & \delta_{n-2} & | & 0 \end{pmatrix}.$$

In `splinecoeff.m`, the two lines

```
A(1,1:3)=[dx(2) -(dx(1)+dx(2)) dx(1)]; % not-a-knot conditions
A(n,n-2:n)=[dx(n-1) -(dx(n-2)+dx(n-1)) dx(n-2)];
```

are used. Figure 3.13(b) shows an example of a not-a-knot cubic spline, compared with the natural spline through the same data points in part (a) of the figure.

As mentioned earlier, a theorem analogous to Theorem 3.7 exists for each of the preceding choices of end conditions:

THEOREM 3.7 Assume that $n \geq 2$. Then, for a set of data points $(x_1, y_1), \ldots, (x_n, y_n)$ and for any one of the end conditions given by Properties 4a–4c, there is a unique cubic spline satisfying the end conditions and fitting the points. The same is true assuming that $n \geq 3$ for Property 4d and $n \geq 4$ for Property 4e. ∎

MATLAB's default `spline` command constructs a not-a-knot spline when given four or more points. Let x and y be vectors containing the x_i and y_i data values, respectively. Then the y-coordinate of the not-a-knot spline at another input x0 is calculated by the MATLAB command

```
>> y0 = spline(x,y,x0);
```

If x0 is a vector of x-coordinates, then the output y0 will be a corresponding vector of y-coordinates, suitable for plotting, etc. Alternatively, if the vector input y has exactly two more inputs than x, the clamped cubic spline is calculated, with clamps v_1 and v_n equal to the first and last entries of y.

▶ **ADDITIONAL EXAMPLES**

1. Find c_1 and b_3 in the cubic spline

$$S(x) = \begin{cases} 4 + 12x + c_1 x^2 + x^3 & \text{on } 0,1 \\ 10 + (x-1) - 4(x-1)^2 + (x-1)^3 & \text{on } 1,3 \\ 4 + b_3(x-3) + 2(x-3)^2 + (x-3)^3 & \text{on } 3,5 \end{cases}$$

Is the spline not-a-knot?

2. Use splinecoeff.m and splineplot.m to plot the natural cubic spline through the ice extent data points in Additional Example 3.1.2. Compare the spline estimates for 2002 and 2012 with the exact values.

📖 **Solutions** for Additional Examples can be found at bit.ly/2NOJDDt

3.4 Exercises

1. Decide whether the equations form a cubic spline.

(a) $S(x) = \begin{cases} x^3 + x - 1 & \text{on } [0,1] \\ -(x-1)^3 + 3(x-1)^2 + 3(x-1) + 1 & \text{on } [1,2] \end{cases}$

(b) $S(x) = \begin{cases} 2x^3 + x^2 + 4x + 5 & \text{on } [0,1] \\ (x-1)^3 + 7(x-1)^2 + 12(x-1) + 12 & \text{on } [1,2] \end{cases}$

2. (a) Check the spline conditions for

$$\begin{cases} S_1(x) = 1 + 2x + 3x^2 + 4x^3 & \text{on } [0,1] \\ S_2(x) = 10 + 20(x-1) + 15(x-1)^2 + 4(x-1)^3 & \text{on } [1,2] \end{cases}.$$

(b) Regardless of your answer to (a), decide whether any of the following extra conditions are satisfied for this example: natural, parabolically terminated, not-a-knot.

3. Find c in the following cubic splines. Which of the three end conditions—natural, parabolically terminated, or not-a-knot—if any, are satisfied?

(a) $S(x) = \begin{cases} 4 - \frac{11}{4}x + \frac{3}{4}x^3 & \text{on } [0,1] \\ 2 - \frac{1}{2}(x-1) + c(x-1)^2 - \frac{3}{4}(x-1)^3 & \text{on } [1,2] \end{cases}$

(b) $S(x) = \begin{cases} 3 - 9x + 4x^2 & \text{on } [0,1] \\ -2 - (x-1) + c(x-1)^2 & \text{on } [1,2] \end{cases}$

(c) $S(x) = \begin{cases} -2 - \frac{3}{2}x + \frac{7}{2}x^2 - x^3 & \text{on } [0,1] \\ -1 + c(x-1) + \frac{1}{2}(x-1)^2 - (x-1)^3 & \text{on } [1,2] \\ 1 + \frac{1}{2}(x-2) - \frac{5}{2}(x-2)^2 - (x-2)^3 & \text{on } [2,3] \end{cases}$

4. Find k_1, k_2, k_3 in the following cubic spline. Which of the three end conditions—natural, parabolically terminated, or not-a-knot—if any, are satisfied?

$$S(x) = \begin{cases} 4 + k_1 x + 2x^2 - \frac{1}{6}x^3 & \text{on } [0,1] \\ 1 - \frac{4}{3}(x-1) + k_2(x-1)^2 - \frac{1}{6}(x-1)^3 & \text{on } [1,2]. \\ 1 + k_3(x-2) + (x-2)^2 - \frac{1}{6}(x-2)^3 & \text{on } [2,3] \end{cases}$$

5. How many natural cubic splines on $[0, 2]$ are there for the given data $(0, 0)$, $(1, 1)$, $(2, 2)$? Exhibit one such spline.

6. Find the parabolically terminated cubic spline through the data points $(0,1)$, $(1,1)$, $(2,1)$, $(3,1)$, $(4,1)$. Is this spline also not-a-knot? natural?

7. Solve equations (3.26) to find the natural cubic spline through the three points (a) $(0,0)$, $(1,1)$, $(2,4)$ (b) $(-1,1)$, $(1,1)$, $(2,4)$.

8. Solve equations (3.26) to find the natural cubic spline through the three points (a) $(0,1)$, $(2,3)$, $(3,2)$ (b) $(0,0)$, $(1,1)$, $(2,6)$.

9. Find $S'(0)$ and $S'(3)$ for the cubic spline

$$\begin{cases} S_1(x) = 3 + b_1 x + x^3 & \text{on } [0, 1] \\ S_2(x) = 1 + b_2(x - 1) + 3(x - 1)^2 - 2(x - 1)^3 & \text{on } [1, 3] \end{cases}.$$

10. True or false: Given $n = 3$ data points, the parabolically terminated cubic spline through the points must be not-a-knot.

11. (a) How many parabolically terminated cubic splines on $[0, 2]$ are there for the given data $(0, 2)$, $(1, 0)$, $(2, 2)$? Exhibit one such spline. (b) Answer the same question for not-a-knot.

12. How many not-a-knot cubic splines are there for the given data $(1, 3)$, $(3, 3)$, $(4, 2)$, $(5, 0)$? Exhibit one such spline.

13. (a) Find b_1 and c_3 in the cubic spline

$$S(x) = \begin{cases} -1 + b_1 x - \frac{5}{9} x^2 + \frac{5}{9} x^3 & \text{on } [0, 1] \\ \frac{14}{9}(x - 1) + \frac{10}{9}(x - 1)^2 - \frac{2}{3}(x - 1)^3 & \text{on } [1, 2] \\ 2 + \frac{16}{9}(x - 2) + c_3(x - 2)^2 - \frac{1}{9}(x - 2)^3 & \text{on } [2, 3] \end{cases}$$

(b) Is this spline natural? (c) This spline satisfies "clamped" endpoint conditions. What are the values of the two clamps?

14. Consider the cubic spline

$$\begin{cases} S_1(x) = 6 - 2x + \frac{1}{2} x^3 & \text{on } [0, 2] \\ S_2(x) = 6 + 4(x - 2) + c(x - 2)^2 + d(x - 2)^3 & \text{on } [2, 3] \end{cases}$$

(a) Find c. (b) Does there exist a number d such that the spline is natural? If so, find d.

15. Can a cubic spline be both natural and parabolically terminated? If so, what else can you say about such a spline?

16. Does there exist a (simultaneously) natural, parabolically terminated, not-a-knot cubic spline through each set of data points $(x_1, y_1), \ldots, (x_{100}, y_{100})$ with distinct x_i? If so, give a reason. If not, explain what conditions must hold on the 100 points in order for such a spline to exist.

17. Assume that the leftmost piece of a given natural cubic spline is the constant function $S_1(x) = 1$ on the interval $[-1, 0]$. Find three different possibilities for the neighboring piece $S_2(x)$ of the spline on $[0, 1]$.

18. Assume that a car travels along a straight road from one point to another from a standing start at time $t = 0$ to a standing stop at time $t = 1$. The distance along the road is sampled at certain times between 0 and 1. Which cubic spline (in terms of end conditions) will be most appropriate for describing distance versus time?

19. The case $n = 2$ for parabolically terminated cubic splines is not covered by Theorem 3.8. Discuss existence and uniqueness for the cubic spline in this case.

20. Discuss the existence and uniqueness of a not-a-knot cubic spline when $n = 2$ and $n = 3$.

21. Theorem 3.8 says that there is exactly one not-a-knot spline through any given four points with distinct x_i. (a) How many not-a-knot splines go through any given 3 points with distinct x_i? (b) Find a not-a-knot spline through $(0, 0)$, $(1, 1)$, $(2, 4)$ that is not parabolically terminated.

3.4 Computer Problems

1. Find the equations and plot the natural cubic spline that interpolates the data points (a) $(0, 3)$, $(1, 5)$, $(2, 4)$, $(3, 1)$ (b) $(-1, 3)$, $(0, 5)$, $(3, 1)$, $(4, 1)$, $(5, 1)$.

2. Find and plot the not-a-knot cubic spline that interpolates the data points (a) $(0, 3)$, $(1, 5)$, $(2, 4)$, $(3, 1)$ (b) $(-1, 3)$, $(0, 5)$, $(3, 1)$, $(4, 1)$, $(5, 1)$.

3. Find and plot the cubic spline S satisfying $S(0) = 1$, $S(1) = 3$, $S(2) = 3$, $S(3) = 4$, $S(4) = 2$ and with $S''(0) = S''(4) = 0$.

4. Find and plot the cubic spline S satisfying $S(0) = 1$, $S(1) = 3$, $S(2) = 3$, $S(3) = 4$, $S(4) = 2$ and with $S''(0) = 3$ and $S''(4) = 2$.

5. Find and plot the cubic spline S satisfying $S(0) = 1$, $S(1) = 3$, $S(2) = 3$, $S(3) = 4$, $S(4) = 2$ and with $S'(0) = 0$ and $S'(4) = 1$.

6. Find and plot the cubic spline S satisfying $S(0) = 1$, $S(1) = 3$, $S(2) = 3$, $S(3) = 4$, $S(4) = 2$ and with $S'(0) = -2$ and $S'(4) = 1$.

7. Find the clamped cubic spline that interpolates $f(x) = \cos x$ at five evenly spaced points in $[0, \pi/2]$, including the endpoints. What is the best choice for $S'(0)$ and $S'(\pi/2)$ to minimize interpolation error? Plot the spline and $\cos x$ on $[0, 2]$.

8. Carry out the steps of Computer Problem 7 for the function $f(x) = \sin x$.

9. Find the clamped cubic spline that interpolates $f(x) = \ln x$ at five evenly spaced points in $[1, 3]$, including the endpoints. Empirically find the maximum interpolation error on $[1, 3]$.

10. Find the number of interpolation nodes in Computer Problem 9 required to make the maximum interpolation error at most 0.5×10^{-7}.

11. (a) Consider the natural cubic spline through the world population data points in Computer Problem 3.1.1. Evaluate the year 1980 and compare with the correct population. (b) Using a linear spline, estimate the slopes at 1960 and 2000, and use these slopes to find the clamped cubic spline through the data. Plot the spline and estimate the 1980 population. Which estimates better, natural or clamped?

12. Recall the carbon dioxide data of Exercise 3.1.17. (a) Find and plot the natural cubic spline through the data, and compute the spline estimate for the CO_2 concentration in 1950. (b) Carry out the same analysis for the parabolically terminated spline. (c) How does the not-a-knot spline differ from the solution to Exercise 3.1.17?

13. In a single plot, show the natural, not-a-knot, and parabolically terminated cubic splines through the world oil production data from Computer Problem 3.2.3.

14. Compile a list of 101 consecutive daily close prices of an exchange-traded stock from a financial data website. (a) Plot the interpolating polynomial through every fifth point. That is, let x0=0:5:100 and y0 denote the stock prices on days $0, 5, 10, \ldots, 100$. Plot the degree 20 interpolating polynomial at points x=0:1:100 and compare with the daily price data. What is the maximum interpolation error? Is the Runge phenomenon evident in your plot? (b) Plot the natural cubic spline with interpolating nodes 0:5:100 instead of the interpolating polynomial, along with the daily data. Answer the same two questions. (c) Compare the two approaches of representing the data.

15. Compile a list of 121 hourly temperatures over five consecutive days from a weather data website. Let x0=0:6:120 denote hours, and y0 denote the temperatures at hours $0, 6, 12, \ldots, 120$. Carry out steps (a)–(c) of Computer Problem 14, suitably adapted.

3.5 BÉZIER CURVES

Bézier curves are splines that allow the user to control the slopes at the knots. In return for the extra freedom, the smoothness of the first and second derivatives across the knot, which are automatic features of the cubic splines of the previous section, are no longer guaranteed. Bézier splines are appropriate for cases where corners (discontinuous first derivatives) and abrupt changes in curvature (discontinuous second derivatives) are occasionally needed.

Pierre Bézier developed the idea during his work for the Renault automobile company. The same idea was discovered independently by Paul de Casteljau, working for Citroen, a rival automobile company. It was considered an industrial secret by both companies, and the fact that both had developed the idea came to light only after Bézier published his research. Today the Bézier curve is a cornerstone of computer-aided design and manufacturing.

Each piece of a planar Bézier spline is determined by four points (x_1, y_1), $(x_2, y_2), (x_3, y_3), (x_4, y_4)$. The first and last of the points are endpoints of the spline curve, and the middle two are **control points**, as shown in Figure 3.14. The curve leaves (x_1, y_1) along the tangent direction $(x_2 - x_1, y_2 - y_1)$ and ends at (x_4, y_4) along the tangent direction $(x_4 - x_3, y_4 - y_3)$. The equations that accomplish this are expressed as a parametric curve $(x(t), y(t))$ for $0 \leq t \leq 1$.

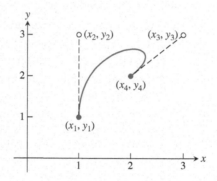

Figure 3.14 Bézier curve of Example 3.15. The points (x_1, y_1) and (x_4, y_4) are spline points, while (x_2, y_2) and (x_3, y_3) are control points.

Bézier curve

Given   endpoints $(x_1, y_1), (x_4, y_4)$
           control points $(x_2, y_2), (x_3, y_3)$
Set

$$b_x = 3(x_2 - x_1)$$
$$c_x = 3(x_3 - x_2) - b_x$$
$$d_x = x_4 - x_1 - b_x - c_x$$
$$b_y = 3(y_2 - y_1)$$
$$c_y = 3(y_3 - y_2) - b_y$$
$$d_y = y_4 - y_1 - b_y - c_y.$$

The Bézier curve is defined for $0 \leq t \leq 1$ by

$$x(t) = x_1 + b_x t + c_x t^2 + d_x t^3$$
$$y(t) = y_1 + b_y t + c_y t^2 + d_y t^3.$$

It is easy to check the claims of the previous paragraph from the equations. In fact, according to Exercise 11,

$$x(0) = x_1$$
$$x'(0) = 3(x_2 - x_1)$$
$$x(1) = x_4$$
$$x'(1) = 3(x_4 - x_3), \tag{3.27}$$

and the analogous facts hold for $y(t)$.

▶ **EXAMPLE 3.15** Find the Bézier curve $(x(t), y(t))$ through the points $(x, y) = (1, 1)$ and $(2, 2)$ with control points $(1, 3)$ and $(3, 3)$.

The four points are $(x_1, y_1) = (1, 1), (x_2, y_2) = (1, 3), (x_3, y_3) = (3, 3)$, and $(x_4, y_4) = (2, 2)$. The Bézier formulas yield $b_x = 0, c_x = 6, d_x = -5$ and $b_y = 6, c_y = -6, d_y = 1$. The Bézier spline

$$x(t) = 1 + 6t^2 - 5t^3$$
$$y(t) = 1 + 6t - 6t^2 + t^3$$

is shown in Figure 3.14 along with the control points. ◀

Bézier curves are building blocks that can be stacked to fit arbitrary function values and slopes. They are an improvement over cubic splines, in the sense that the slopes at the nodes can be specified as the user wants them. However, this freedom comes at the expense of smoothness: The second derivatives from the two different directions generally disagree at the nodes. In some applications, this disagreement is an advantage.

As a special case, when the control points equal the endpoints, the spline is a simple line segment, as shown next.

▶ **EXAMPLE 3.16** Prove that the Bézier spline with $(x_1, y_1) = (x_2, y_2)$ and $(x_3, y_3) = (x_4, y_4)$ is a line segment.

The Bézier formulas show that the equations are

$$x(t) = x_1 + 3(x_4 - x_1)t^2 - 2(x_4 - x_1)t^3 = x_1 + (x_4 - x_1)t^2(3 - 2t)$$
$$y(t) = y_1 + 3(y_4 - y_1)t^2 - 2(y_4 - y_1)t^3 = y_1 + (y_4 - y_1)t^2(3 - 2t)$$

for $0 \le t \le 1$. Every point in the spline has the form

$$(x(t), y(t)) = (x_1 + r(x_4 - x_1), y_1 + r(y_4 - y_1))$$
$$= ((1 - r)x_1 + rx_4, (1 - r)y_1 + ry_4),$$

where $r = t^2(3 - 2t)$. Since $0 \le r \le 1$, each point lies on the line segment connecting (x_1, y_1) and (x_4, y_4). ◀

Bézier curves are simple to program and are often used in drawing software. A freehand curve in the plane can be viewed as a parametric curve $(x(t), y(t))$ and represented by a Bézier spline. The equations are implemented in the following MATLAB freehand drawing program. The user clicks the mouse once to fix a starting point (x_0, y_0) in the plane, and three more clicks to mark the first control point, second control point, and endpoint. A Bézier spline is drawn between the start and end points. Each subsequent triple of mouse clicks extends the curve further, using the previous endpoint as the starting point for the next piece. The MATLAB command ginput is used to read the mouse location. Figure 3.15 shows a screenshot of bezierdraw.m.

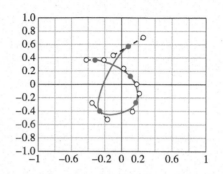

Figure 3.15 Program 3.7 built from Bézier curves. Screenshot of MATLAB code bezierdraw.m, including direction vectors drawn at each control point.

```
%Program 3.7 Freehand Draw Program Using Bezier Splines
%Click in Matlab figure window to locate first point, and click
%      three more times to specify 2 control points and the next
%      spline point. Continue with groups of 3 points to add more
%      to the curve. Press return to terminate program.
function bezierdraw
plot([-1 1],[0,0],'k',[0 0],[-1 1],'k');hold on
t=0:.02:1;
[x,y]=ginput(1);            % get one mouse click
while(0 == 0)
  [xnew,ynew] = ginput(3);  % get three mouse clicks
  if length(xnew) < 3
    break                   % if return pressed, terminate
  end
  x=[x;xnew];y=[y;ynew];    % plot spline points and control pts
  plot([x(1) x(2)],[y(1) y(2)],'r:',x(2),y(2),'rs');
  plot([x(3) x(4)],[y(3) y(4)],'r:',x(3),y(3),'rs');
  plot(x(1),y(1),'bo',x(4),y(4),'bo');
  bx=3*(x(2)-x(1)); by=3*(y(2)-y(1)); % spline equations ...
  cx=3*(x(3)-x(2))-bx;cy=3*(y(3)-y(2))-by;
  dx=x(4)-x(1)-bx-cx;dy=y(4)-y(1)-by-cy;
  xp=x(1)+t.*(bx+t.*(cx+t*dx));     % Horner's method
  yp=y(1)+t.*(by+t.*(cy+t*dy));
  plot(xp,yp,'b').                  % plot spline curve
  x=x(4);y=y(4);                    % promote last to first and repeat
end
hold off
```

Although our discussion has been restricted to two-dimensional Bézier curves, the defining equations are easily extended to three dimensions, in which they are called Bézier space curves. Each piece of the spline requires four (x, y, z) points—two endpoints and two control points—just as in the two-dimensional case. Examples of Bézier space curves are explored in the exercises.

► **ADDITIONAL EXAMPLES**

1. Assume that a one-piece Bezier spline has endpoints $(0, 2)$ and $(0, -1)$, and horizontal tangents at both endpoints. Find all possible points $(x(1/3), y(1/3))$ on such a curve.

2. Assume that a one-piece Bezier spline has endpoints $(0, -1)$ and $(0, 1)$, and control points $(A, -1)$ and $(A, 1)$ for some real number A. Find the maximum x-coordinate on the curve in terms of A.

⌧ **Solutions** for Additional Examples can be found at bit.ly/2NNxn6c

3.5 Exercises

⌧ **Solutions** for Exercises numbered in blue can be found at bit.ly/2NKXYAQ

1. Find the one-piece Bézier curve $(x(t), y(t))$ defined by the given four points.
   (a) $(0,0)$, $(0,2)$, $(2,0)$, $(1,0)$ (b) $(1,1)$, $(0,0)$, $(-2,0)$, $(-2,1)$ (c) $(1,2)$, $(1,3)$, $(2,3)$, $(2,2)$

2. Find the first endpoint, two control points, and last endpoint for the following one-piece Bézier curves.

   (a) $\begin{cases} x(t) = 1 + 6t^2 + 2t^3 \\ y(t) = 1 - t + t^3 \end{cases}$ (b) $\begin{cases} x(t) = 3 + 4t - t^2 + 2t^3 \\ y(t) = 2 - t + t^2 + 3t^3 \end{cases}$

   (c) $\begin{cases} x(t) = 2 + t^2 - t^3 \\ y(t) = 1 - t + 2t^3 \end{cases}$

3. Find the three-piece Bézier curve forming the triangle with vertices $(1, 2)$, $(3, 4)$, and $(5, 1)$.

4. Build a four-piece Bézier spline that forms a square with sides of length 5.

5. Describe the character drawn by the following two-piece Bezier curve:
   $(0,2)$ $(1,2)$ $(1,1)$ $(0,1)$
   $(0,1)$ $(1,1)$ $(1,0)$ $(0,0)$

6. Describe the character drawn by the following three-piece Bezier curve:
   $(0,1)$ $(0,1)$ $(0,0)$ $(0,0)$
   $(0,0)$ $(0,1)$ $(1,1)$ $(1,0)$
   $(1,0)$ $(1,1)$ $(2,1)$ $(2,0)$

7. Find a one-piece Bézier spline that has vertical tangents at its endpoints $(-1, 0)$ and $(1, 0)$ and that passes through $(0, 1)$.

8. Find a one-piece Bézier spline that has a horizontal tangent at endpoint $(0, 1)$ and a vertical tangent at endpoint $(1, 0)$ and that passes through $(1/3, 2/3)$ at $t = 1/3$.

9. Find the one-piece Bézier space curve $(x(t), y(t), z(t))$ defined by the four points.
   (a) $(1, 0, 0)$, $(2, 0, 0)$, $(0, 2, 1)$, $(0, 1, 0)$ (b) $(1, 1, 2)$, $(1, 2, 3)$, $(-1, 0, 0)$, $(1, 1, 1)$
   (c) $(2, 1, 1)$, $(3, 1, 1)$, $(0, 1, 3)$, $(3, 1, 3)$

10. Find the knots and control points for the following Bézier space curves.

    (a) $\begin{cases} x(t) = 1 + 6t^2 + 2t^3 \\ y(t) = 1 - t + t^3 \\ z(t) = 1 + t + 6t^2 \end{cases}$ (b) $\begin{cases} x(t) = 3 + 4t - t^2 + 2t^3 \\ y(t) = 2 - t + t^2 + 3t^3 \\ z(t) = 3 + t + t^2 - t^3 \end{cases}$

    (c) $\begin{cases} x(t) = 2 + t^2 - t^3 \\ y(t) = 1 - t + 2t^3 \\ z(t) = 2t^3 \end{cases}$

11. Prove the facts in (3.27), and explain how they justify the Bézier formulas.

12. Given (x_1, y_1), (x_2, y_2), (x_3, y_3), and (x_4, y_4), show that the equations

$$x(t) = x_1(1 - t)^3 + 3x_2(1 - t)^2 t + 3x_3(1 - t)t^2 + x_4 t^3$$
$$y(t) = y_1(1 - t)^3 + 3y_2(1 - t)^2 t + 3y_3(1 - t)t^2 + y_4 t^3$$

give the Bézier curve with endpoints (x_1, y_1), (x_4, y_4) and control points (x_2, y_2), (x_3, y_3).

3.5 Computer Problems

Solutions for
Computer Problems
numbered in blue can
be found at
bit.ly/2OyapVW

1. Plot the curve in Exercise 7.
2. Plot the curve in Exercise 8.
3. Plot the letter from Bézier curves: (a) W (b) B (c) C (d) D.

Reality Check 3 *Fonts from Bézier Curves*

In this project, we explain how to draw letters and numerals by using two-dimensional Bézier curves. They can be implemented by modifying the MATLAB code in Program 3.7 or by writing a PDF file.

Modern fonts are built directly from Bézier curves, in order to be independent of the printer or imaging device. Bézier curves were a fundamental part of the PostScript language from its start in the 1980s, and the PostScript commands for drawing curves have migrated in slightly altered form to the PDF format. Here is a complete PDF file that illustrates the curve we discussed in Example 3.15.

```
%PDF-1.7
1 0 obj
<<
/Length 2 0 R
>>
stream
100 100 m
100 300 300 300 200 200 c
S
endstream
endobj
2 0 obj
1000
endobj
4 0 obj
<<
/Type /Page
/Parent 5 0 R
/Contents 1 0 R
>>
endobj
5 0 obj
<<
/Kids [4 0 R]
/Count 1
/Type /Pages
/MediaBox [0 0 612 792]
>>
endobj
3 0 obj
```

```
<<
/Pages 5 0 R
/Type /Catalog
>>
endobj
xref
0 6
0000000000 65535 f
0000000100 00000 n
0000000200 00000 n
0000000500 00000 n
0000000300 00000 n
0000000400 00000 n
trailer
<<
/Size 6
/Root 3 0 R
>>
startxref
1000
%%EOF
```

Most of the lines in this template file do various housekeeping chores. For example, the first line identifies the file as a PDF. We will focus on the lines between `stream` and `endstream`, which are the ones that identify the Bézier curve. The move command (m) sets the current plot point to be the (x, y) point specified by the two preceding numbers—in this case, the point $(100, 100)$. The curve command (c) accepts three (x, y) points and constructs the Bézier spline starting at the current plot point, treating the three (x, y) pairs as the two control points and the endpoint, respectively. The stroke command (S) draws the curve.

This text file `sample.pdf` can be downloaded from the textbook website. If it is opened with a PDF viewer, the Bézier curve of Figure 3.14 will be displayed. The coordinates have been multiplied by 100 to match the default conventions of PDF, which are 72 units to the inch. A sheet of letter-sized paper is 612 units wide and 792 high.

At present, characters from hundreds of fonts are drawn on computer screens and printers using Bézier curves. Of course, since PDF files often contain many characters, there are shortcuts for predefined fonts. The Bézier curve information for common fonts is usually stored in the PDF reader rather than the PDF file. We will choose to ignore this fact for now in order to see what we can do on our own.

Let's begin with a typical example. The upper case T character in the Times Roman font is constructed out of the following 16 Bézier curves. Each line consists of the numbers x_1 y_1 x_2 y_2 x_3 y_3 x_4 y_4 that define one piece of the Bézier spline.

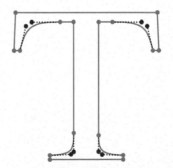

Figure 3.16 Times-Roman T made with Bézier splines. Blue circles are spline endpoints, and black circles are control points.

```
237  620  237  620  237  120  237  120;
237  120  237   35  226   24  143   19;
143   19  143   19  143    0  143    0;
143    0  143    0  435    0  435    0;
435    0  435    0  435   19  435   19;
435   19  353   23  339   36  339  109;
339  109  339  108  339  620  339  620;
339  620  339  620  393  620  393  620;
393  620  507  620  529  602  552  492;
552  492  552  492  576  492  576  492;
576  492  576  492  570  662  570  662;
570  662  570  662    6  662    6  662;
  6  662    6  662    0  492    0  492;
  0  492    0  492   24  492   24  492;
 24  492   48  602   71  620  183  620;
183  620  183  620  237  620  237  620;
```

To create a PDF file that writes the letter T, one needs to add commands within the stream/endstream area of the above template file. First, move to the initial endpoint $(237, 620)$

```
237 620 m
```

after which the first curve is drawn by the command

```
237 620 237 120 237 120 c
```

followed by fifteen more c commands, and the stroke command (S) to finish the letter T, shown in Figure 3.16. Note that the move command is necessary only at the first step; after that the next curve command takes the current plot point as the first point in the next Bézier curve, and needs only three more points to complete the curve command. The next curve command is completed in the same way, and so on. As an alternative to the stroke command S, the f command will fill in the outline if the figure is closed. The command b will both stroke and fill.

The number 5 is drawn by a 21-piece Bézier curve, shown in Figure 3.17.

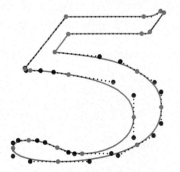

Figure 3.17 Times-Roman 5 made with Bézier splines. Blue circles are spline endpoints, and black circles are control points.

```
149 597 149 597 149 597 345 597;
345 597 361 597 365 599 368 606;
368 606 406 695 368 606 406 695;
406 695 397 702 406 695 397 702;
397 702 382 681 372 676 351 676;
351 676 351 676 351 676 142 676;
142 676  33 439 142 676  33 439;
 33 439  32 438  32 436  32 434;
 32 434  32 428  35 426  44 426;
 44 426  74 426 109 420 149 408;
149 408 269 372 324 310 324 208;
324 208 324 112 264  37 185  37;
185  37 165  37 149  44 119  66;
119  66  86  90  65  99  42  99;
 42  99  14  99   0  87   0  62;
  0  62   0  24  46   0 121   0;
121   0 205   0 282  27 333  78;
333  78 378 123 399 180 399 256;
399 256 399 327 381 372 333 422;
333 422 288 468 232 491 112 512;
112 512 112 512 149 597 149 597;
```

Suggested activities:

1. Use the `bezierdraw.m` program of Section 3.5 to sketch the upper case initial of your first name.

2. Revise the draw program to accept an $n \times 8$ matrix of numbers, each row representing a piece of a Bézier spline. Have the program draw the lower case letter f in the Times-Roman font, using the following 21-piece Bézier curve:

```
289 452 289 452 166 452 166 452;
166 452 166 452 166 568 166 568;
166 568 166 627 185 657 223 657;
223 657 245 657 258 647 276 618;
276 618 292 589 304 580 321 580;
321 580 345 580 363 598 363 621;
363 621 363 657 319 683 259 683;
259 683 196 683 144 656 118 611;
118 611  92 566  84 530  83 450;
 83 450  83 450   1 450   1 450;
  1 450   1 450   1 418   1 418;
  1 418   1 418  83 418  83 418;
 83 418  83 418  83 104  83 104;
 83 104  83  31  72  19   0  15;
  0  15   0  15   0   0   0   0;
  0   0   0   0 260   0 260   0;
260   0 260   0 260  15 260  15;
260  15 178  18 167  29 167 104;
167 104 167 104 167 418 167 418;
167 418 167 418 289 418 289 418;
289 418 289 418 289 452 289 452;
```

3. Using the template above and your favorite text editor, write a PDF file that draws the lower case letter f. The program should begin with an m command to move to the first point, followed by 21 c commands and a stroke or fill command. These commands should lie between the stream and endstream commands. Test your file by opening it in a PDF viewer.

4. Here are some other PDF commands:

```
1.0 0.0 0.0 RG   % set stroke color to red
0.0 1.0 0.0 rg   % set fill color to green
2 w              % set stroke width to 2
b                % both stroke and fill (S is stroke, f is fill,
                                         b both)
```

Colors are represented according to the RGB convention, by three numbers between 0 and 1 embodying the relative contributions of red, green, and blue. Linear transformations may be used to change the size of the Bézier curves, and rotate and skew the results. Such coordinate changes are accomplished with the cm command. Preceding the curve commands with

a b c d e f cm

for real numbers a, b, c, d, e, f will transform the underlying planar coordinate system by

$$x' = ax + by + e$$
$$y' = cx + dy + f.$$

For example, using the cm command with $a = d = 0.5, b = c = e = f = 0$ reduces the size by a factor of 2, and $a = d = -0.5, b = c = 0$, and $e = f = 400$ turns the result upside down and translates by 400 units in the x and y directions. Other choices can perform rotations, reflections, or skews of the original Bézier curves. Coordinate changes are cumulative. In this step, use the coordinate system commands to present a resized, colored, and skewed version of the lower case f or other characters.

5. Although font information was a closely guarded secret for many years, much of it is now freely available on the Web. Search for other fonts, and find Bézier curve data that will draw letters of your choice in PDF or with bezierdraw.m.

6. Design your own letter or numeral. You should begin by drawing the figure on graph paper, respecting any symmetries that might be present. Estimate control points, and be prepared to revise them later as needed.

Software and Further Reading

Interpolation software usually consists of separate codes for determining and evaluating the interpolating polynomial. MATLAB provides the polyfit and polyval commands for this purpose. The MATLAB spline command calculates not-a-knot splines by default, but has options for several other common end conditions. The command interp1 combines several one-dimensional interpolation options. The NAG library contains subroutines e01aef and e01baf for polynomial and spline interpolation.

A classical reference for basic interpolation facts is Davis [1975], and the references Rivlin [1981] and Rivlin [1990] cover function approximation and Chebyshev interpolation. DeBoor [2001] on splines is also a classic; see also Schultz [1973] and Schumaker [1981]. Applications to computer-aided modeling and design are treated in Farin [1990] and Yamaguchi [1988]. The CORDIC Method for approximation of special functions was introduced in Volder [1959]. For more information on PDF files, see the *PDF Reference*, 6th Ed., published by Adobe Systems Inc. [2006].

4

Least Squares

The Global Positioning System (GPS) is a satellite-based location technology that provides accurate positioning at any time, from any point on earth. In just a few years, GPS has gone from a special-purpose navigation technology used by pilots, ship captains, and hikers to everyday use in automobiles, cellphones, and PDAs.

The system consists of 24 satellites following precisely regulated orbits, emitting synchronized signals.

An earth-based receiver picks up the satellite signals, finds its distance from all visible satellites, and uses the data to triangulate its position.

Reality Check Reality Check 4 on page 248 shows the use of equation solvers and least squares calculations to do the location estimation.

T he concept of least squares dates from the pioneering work of Gauss and Legendre in the early 19th century. Its use permeates modern statistics and mathematical modeling. The key techniques of regression and parameter estimation have become fundamental tools in the sciences and engineering.

In this chapter, the normal equations are introduced and applied to a variety of data-fitting problems. Later, a more sophisticated approach, using the QR factorization, is explored, followed by a discussion of nonlinear least squares problems.

4.1 LEAST SQUARES AND THE NORMAL EQUATIONS

The need for least squares methods comes from two different directions, one each from our studies of Chapters 2 and 3. In Chapter 2, we learned how to find the solution of $Ax = b$ when a solution exists. In this chapter, we find out what to do when there is no solution. When the equations are inconsistent, which is likely if the number of equations exceeds the number of unknowns, the answer is to find the next best thing: the least squares approximation.

Chapter 3 addressed finding polynomials that exactly fit data points. However, if the data points are numerous, or the data points are collected only within some margin of error, fitting a high-degree polynomial exactly is rarely the best approach. In such cases, it is more reasonable to fit a simpler model that may only approximate the data points. Both problems, solving inconsistent systems of equations and fitting data approximately, are driving forces behind least squares.

4.1.1 Inconsistent systems of equations

It is not hard to write down a system of equations that has no solutions. Consider the following three equations in two unknowns:

$$
\begin{aligned}
x_1 + x_2 &= 2 \\
x_1 - x_2 &= 1 \\
x_1 + x_2 &= 3.
\end{aligned}
\tag{4.1}
$$

Any solution must satisfy the first and third equations, which cannot both be true. A system of equations with no solution is called **inconsistent**.

What is the meaning of a system with no solutions? Perhaps the coefficients are slightly inaccurate. In many cases, the number of equations is greater than the number of unknown variables, making it unlikely that a solution can satisfy all the equations. In fact, m equations in n unknowns typically have no solution when $m > n$. Even though Gaussian elimination will not give us a solution to an inconsistent system $Ax = b$, we should not completely give up. An alternative in this situation is to find a vector x that comes the closest to being a solution.

If we choose this "closeness" to mean close in Euclidean distance, there is a straightforward algorithm for finding the closest x. This special x will be called the **least squares solution**.

We can get a better picture of the failure of system (4.1) to have a solution by writing it in a different way. The matrix form of the system is $Ax = b$, or

$$
\begin{bmatrix} 1 & 1 \\ 1 & -1 \\ 1 & 1 \end{bmatrix} \begin{bmatrix} x_1 \\ x_2 \end{bmatrix} = \begin{bmatrix} 2 \\ 1 \\ 3 \end{bmatrix}.
\tag{4.2}
$$

The alternative view of matrix/vector multiplication is to write the equivalent equation

$$
x_1 \begin{bmatrix} 1 \\ 1 \\ 1 \end{bmatrix} + x_2 \begin{bmatrix} 1 \\ -1 \\ 1 \end{bmatrix} = \begin{bmatrix} 2 \\ 1 \\ 3 \end{bmatrix}.
\tag{4.3}
$$

In fact, any $m \times n$ system $Ax = b$ can be viewed as a vector equation

$$
x_1 v_1 + x_2 v_2 + \cdots + x_n v_n = b,
\tag{4.4}
$$

which expresses b as a linear combination of the columns v_i of A, with coefficients $x_1, \ldots, x_n$. In our case, we are trying to hit the target vector b as a linear combination of two other three-dimensional vectors. Since the combinations of two three-dimensional vectors form a plane inside R^3, equation (4.3) has a solution only if the vector b lies in that plane. This will always be the situation when we are trying to solve m equations in n unknowns, with $m > n$. Too many equations make the problem overspecified and the equations inconsistent.

Figure 4.1(b) shows a direction for us to go when a solution does not exist. There is no pair x_1, x_2 that solves (4.1), but there is a point in the plane Ax of all possible

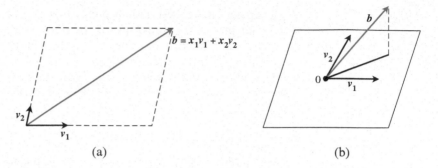

(a)   (b)

Figure 4.1 Geometric solution of a system of three equations in two unknowns. (a) Equation (4.3) requires that the vector b, the right-hand side of the equation, is a linear combination of the columns vectors v_1 and v_2. (b) If b lies outside of the plane defined by v_1 and v_2, there will be no solution. The least squares solution $\overline{x}$ makes the combination vector $A\overline{x}$ the one in the plane Ax that is nearest to b in the sense of Euclidean distance.

candidates that is closest to b. This special vector $A\overline{x}$ is distinguished by the following fact: The residual vector $b - A\overline{x}$ is perpendicular to the plane $\{Ax | x \in R^n\}$. We will exploit this fact to find a formula for $\overline{x}$, the least squares "solution."

First we establish some notation. Recall the concept of the **transpose** A^T of the $m \times n$ matrix A, which is the $n \times m$ matrix whose rows are the columns of A and whose columns are the rows of A, in the same order. The transpose of the sum of two matrices is the sum of the transposes, $(A + B)^T = A^T + B^T$. The transpose of a product of two matrices is the product of the transposes in the reverse order—that is, $(AB)^T = B^T A^T$.

To work with perpendicularity, recall that two vectors are at right angles to one another if their dot product is zero. For two m-dimensional column vectors u and v, we can write the dot product solely in terms of matrix multiplication by

$$u^T v = [u_1, \ldots, u_m] \begin{bmatrix} v_1 \\ \vdots \\ v_m \end{bmatrix}. \tag{4.5}$$

The vectors u and v are perpendicular, or **orthogonal**, if $u^T \cdot v = 0$, using ordinary matrix multiplication.

Now we return to our search for a formula for $\overline{x}$. We have established that

$$(b - A\overline{x}) \perp \{Ax | x \in R^n\}.$$

Expressing the perpendicularity in terms of matrix multiplication, we find that

$$(Ax)^T (b - A\overline{x}) = 0 \text{ for all } x \text{ in } R^n.$$

Using the preceding fact about transposes, we can rewrite this expression as

$$x^T A^T (b - A\overline{x}) = 0 \text{ for all } x \text{ in } R^n,$$

SPOTLIGHT ON

Orthogonality   Least squares is based on orthogonality. The shortest distance from a point to a plane is carried by a line segment orthogonal to the plane. The normal equations are a computational way to locate the line segment, which represents the least squares error.

meaning that the n-dimensional vector $A^T(b - A\overline{x})$ is perpendicular to every vector x in R^n, including itself. There is only one way for that to happen:

$$A^T(b - A\overline{x}) = 0.$$

This gives a system of equations that defines the least squares solution,

$$A^T A\overline{x} = A^T b. \tag{4.6}$$

The system of equations (4.6) is known as the **normal equations**. Its solution $\overline{x}$ is the so-called least squares solution of the system $Ax = b$.

Normal equations for least squares

Given the inconsistent system

$$Ax = b,$$

solve

$$A^T A\overline{x} = A^T b$$

for the least squares solution $\overline{x}$ that minimizes the Euclidean length of the residual $r = b - Ax$.

► **EXAMPLE 4.1**   Use the normal equations to find the least squares solution of the inconsistent system (4.1).

The problem in matrix form $Ax = b$ has

$$A = \begin{bmatrix} 1 & 1 \\ 1 & -1 \\ 1 & 1 \end{bmatrix}, \quad b = \begin{bmatrix} 2 \\ 1 \\ 3 \end{bmatrix}.$$

The components of the normal equations are

$$A^T A = \begin{bmatrix} 1 & 1 & 1 \\ 1 & -1 & 1 \end{bmatrix} \begin{bmatrix} 1 & 1 \\ 1 & -1 \\ 1 & 1 \end{bmatrix} = \begin{bmatrix} 3 & 1 \\ 1 & 3 \end{bmatrix}$$

and

$$A^T b = \begin{bmatrix} 1 & 1 & 1 \\ 1 & -1 & 1 \end{bmatrix} \begin{bmatrix} 2 \\ 1 \\ 3 \end{bmatrix} = \begin{bmatrix} 6 \\ 4 \end{bmatrix}.$$

The normal equations

$$\begin{bmatrix} 3 & 1 \\ 1 & 3 \end{bmatrix} \begin{bmatrix} x_1 \\ x_2 \end{bmatrix} = \begin{bmatrix} 6 \\ 4 \end{bmatrix}$$

can now be solved by Gaussian elimination. The tableau form is

$$\begin{bmatrix} 3 & 1 & | & 6 \\ 1 & 3 & | & 4 \end{bmatrix} \longrightarrow \begin{bmatrix} 3 & 1 & | & 6 \\ 0 & 8/3 & | & 2 \end{bmatrix},$$

which can be solved to get $\overline{x} = (\overline{x}_1, \overline{x}_2) = (7/4, 3/4)$.   ◄

Substituting the least squares solution into the original problem yields

$$\begin{bmatrix} 1 & 1 \\ 1 & -1 \\ 1 & 1 \end{bmatrix} \begin{bmatrix} \frac{7}{4} \\ \frac{3}{4} \end{bmatrix} = \begin{bmatrix} 2.5 \\ 1 \\ 2.5 \end{bmatrix} \neq \begin{bmatrix} 2 \\ 1 \\ 3 \end{bmatrix}.$$

To measure our success at fitting the data, we calculate the residual of the least squares solution $\bar{x}$ as

$$r = b - A\bar{x} = \begin{bmatrix} 2 \\ 1 \\ 3 \end{bmatrix} - \begin{bmatrix} 2.5 \\ 1 \\ 2.5 \end{bmatrix} = \begin{bmatrix} -0.5 \\ 0.0 \\ 0.5 \end{bmatrix}.$$

If the residual is the zero vector, then we have solved the original system $Ax = b$ exactly. If not, the Euclidean length of the residual vector is a backward error measure of how far $\bar{x}$ is from being a solution.

There are at least three ways to express the size of the residual. The Euclidean length of a vector,

$$||r||_2 = \sqrt{r_1^2 + \cdots + r_m^2}, \tag{4.7}$$

is a norm in the sense of Chapter 2, called the **2-norm**. The **squared error**

$$SE = r_1^2 + \cdots + r_m^2,$$

and the **root mean squared error** (the root of the mean of the squared error)

$$RMSE = \sqrt{SE/m} = \sqrt{\left(r_1^2 + \cdots + r_m^2\right)/m}, \tag{4.8}$$

are also used to measure the error of the least squares solution. The three expressions are closely related; namely

$$RMSE = \frac{\sqrt{SE}}{\sqrt{m}} = \frac{||r||_2}{\sqrt{m}},$$

so finding the $\bar{x}$ that minimizes one, minimizes all. For Example 4.1, the $SE = (0.5)^2 + 0^2 + (-0.5)^2 = 0.5$, the 2-norm of the error is $||r||_2 = \sqrt{0.5} \approx 0.707$, and the $RMSE = \sqrt{0.5/3} = 1/\sqrt{6} \approx 0.408$.

▶ **EXAMPLE 4.2** Solve the least squares problem $\begin{bmatrix} 1 & -4 \\ 2 & 3 \\ 2 & 2 \end{bmatrix} \begin{bmatrix} x_1 \\ x_2 \end{bmatrix} = \begin{bmatrix} -3 \\ 15 \\ 9 \end{bmatrix}$.

The normal equations $A^T Ax = A^T b$ are

$$\begin{bmatrix} 9 & 6 \\ 6 & 29 \end{bmatrix} \begin{bmatrix} x_1 \\ x_2 \end{bmatrix} = \begin{bmatrix} 45 \\ 75 \end{bmatrix}.$$

The solution of the normal equations are $\bar{x}_1 = 3.8$ and $\bar{x}_2 = 1.8$. The residual vector is

$$r = b - A\bar{x} = \begin{bmatrix} -3 \\ 15 \\ 9 \end{bmatrix} - \begin{bmatrix} 1 & -4 \\ 2 & 3 \\ 2 & 2 \end{bmatrix} \begin{bmatrix} 3.8 \\ 1.8 \end{bmatrix}$$

$$= \begin{bmatrix} -3 \\ 15 \\ 9 \end{bmatrix} - \begin{bmatrix} -3.4 \\ 13 \\ 11.2 \end{bmatrix} = \begin{bmatrix} 0.4 \\ 2 \\ -2.2 \end{bmatrix},$$

which has Euclidean norm $||e||_2 = \sqrt{(0.4)^2 + 2^2 + (-2.2)^2} = 3$. This problem is solved in an alternative way in Example 4.14. ◀

4.1.2 Fitting models to data

Let $(t_1, y_1), \ldots, (t_m, y_m)$ be a set of points in the plane, which we will often refer to as the "data points." Given a fixed class of models, such as all lines $y = c_1 + c_2 t$, we can seek to locate the specific instance of the model that best fits the data points in the 2-norm. The core of the least squares idea consists of measuring the residual of the fit by the squared errors of the model at the data points and finding the model parameters that minimize this quantity. This criterion is displayed in Figure 4.2.

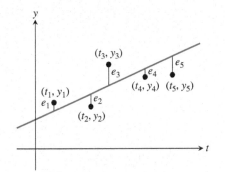

Figure 4.2 Least squares fitting of a line to data. The best line is the one for which the squared error $e_1^2 + e_2^2 + \cdots + e_5^2$ is as small as possible among all lines $y = c_1 + c_2 t$.

▶ **EXAMPLE 4.3** Find the line that best fits the three data points $(t, y) = (1, 2), (-1, 1)$, and $(1, 3)$ in Figure 4.3.

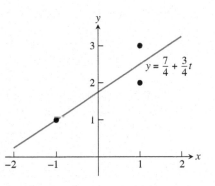

Figure 4.3 Best line in Example 4.3. One each of the data points lies above, on, and below the best line.

The model is $y = c_1 + c_2 t$, and the goal is to find the best c_1 and c_2. Substitution of the data points into the model yields

$$c_1 + c_2(1) = 2$$
$$c_1 + c_2(-1) = 1$$
$$c_1 + c_2(1) = 3,$$

or, in matrix form,

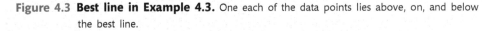

$$\begin{bmatrix} 1 & 1 \\ 1 & -1 \\ 1 & 1 \end{bmatrix} \begin{bmatrix} c_1 \\ c_2 \end{bmatrix} = \begin{bmatrix} 2 \\ 1 \\ 3 \end{bmatrix}.$$

We know this system has no solution (c_1, c_2) for two separate reasons. First, if there is a solution, then the $y = c_1 + c_2 t$ would be a line containing the three data points. However, it is easily seen that the points are not collinear. Second, this is the system of equation (4.2) that we discussed at the beginning of this chapter. We noticed then that the first and third equations are inconsistent, and we found that the best solution in terms of least squares is $(c_1, c_2) = (7/4, 3/4)$. Therefore, the best line is $y = 7/4 + 3/4t$. ◄

We can evaluate the fit by using the statistics defined earlier. The residuals at the data points are

t	y	line	error
1	2	2.5	−0.5
−1	1	1.0	0.0
1	3	2.5	0.5

and the RMSE is $1/\sqrt{6}$, as seen earlier.

The previous example suggests a three-step program for solving least squares data-fitting problems.

Fitting data by least squares

Given a set of m data points $(t_1, y_1), \ldots, (t_m, y_m)$.

STEP 1. Choose a model. Identify a parameterized model, such as $y = c_1 + c_2 t$, which will be used to fit the data.

STEP 2. Force the model to fit the data. Substitute the data points into the model. Each data point creates an equation whose unknowns are the parameters, such as c_1 and c_2 in the line model. This results in a system $Ax = b$, where the unknown x represents the unknown parameters.

STEP 3. Solve the normal equations. The least squares solution for the parameters will be found as the solution to the system of normal equations $A^T A x = A^T b$.

These steps are demonstrated in the following example:

► **EXAMPLE 4.4**   Find the best line and best parabola for the four data points $(-1, 1), (0, 0)$, $(1, 0), (2, -2)$ in Figure 4.4.

In accordance with the preceding program, we will follow three steps:
(1) Choose the model $y = c_1 + c_2 t$ as before. (2) Forcing the model to fit the data yields

SPOTLIGHT ON

Compression   Least squares is a classic example of data compression. The input consists of a set of data points, and the output is a model that, with a relatively few parameters, fits the data as well as possible. Usually, the reason for using least squares is to replace noisy data with a plausible underlying model. The model is then often used for signal prediction or classification purposes.

In Section 4.2, various models are used to fit data, including polynomials, exponentials, and trigonometric functions. The trigonometric approach will be pursued further in Chapters 10 and 11, where elementary Fourier analysis is discussed as an introduction to signal processing.

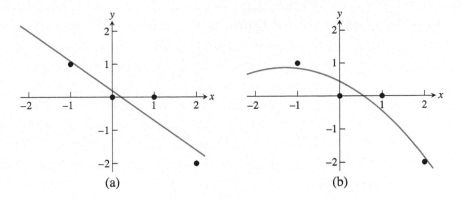

Figure 4.4 Least squares fits to data points in Example 4.4. (a) Best line $y = 0.2 - 0.9t$. RMSE is 0.418. (b) Best parabola $y = 0.45 - 0.65t - 0.25t^2$. RMSE is 0.335.

$$c_1 + c_2(-1) = 1$$
$$c_1 + c_2(0) = 0$$
$$c_1 + c_2(1) = 0$$
$$c_1 + c_2(2) = -2,$$

or, in matrix form,

$$\begin{bmatrix} 1 & -1 \\ 1 & 0 \\ 1 & 1 \\ 1 & 2 \end{bmatrix} \begin{bmatrix} c_1 \\ c_2 \end{bmatrix} = \begin{bmatrix} 1 \\ 0 \\ 0 \\ -2 \end{bmatrix}.$$

(3) The normal equations are

$$\begin{bmatrix} 4 & 2 \\ 2 & 6 \end{bmatrix} \begin{bmatrix} c_1 \\ c_2 \end{bmatrix} = \begin{bmatrix} -1 \\ -5 \end{bmatrix}.$$

Solving for the coefficients c_1 and c_2 results in the best line $y = c_1 + c_2 t = 0.2 - 0.9t$. The residuals are

t	y	line	error
-1	1	1.1	-0.1
0	0	0.2	-0.2
1	0	-0.7	0.7
2	-2	-1.6	-0.4

The error statistics are squared error SE $= (-0.1)^2 + (-0.2)^2 + (0.7)^2 + (-0.4)^2 = 0.7$ and RMSE $= \sqrt{0.7}/\sqrt{4} = 0.418$.

Next, we extend this example by keeping the same four data points, but changing the model. Set $y = c_1 + c_2 t + c_3 t^2$ and substitute the data points to yield

$$c_1 + c_2(-1) + c_3(-1)^2 = 1$$
$$c_1 + c_2(0) + c_3(0)^2 = 0$$
$$c_1 + c_2(1) + c_3(1)^2 = 0$$
$$c_1 + c_2(2) + c_3(2)^2 = -2,$$

Conditioning   Since input data is assumed to be subject to errors in least squares problems, it is especially important to reduce error magnification. We have presented the normal equations as the most straightforward approach to solving the least squares problem, and it is fine for small problems. However, the condition number $\text{cond}(A^T A)$ is approximately the square of the original $\text{cond}(A)$, which will greatly increase the possibility that the problem is ill-conditioned. More sophisticated methods allow computing the least squares solution directly from A without forming $A^T A$. These methods are based on the QR factorization, introduced in Section 4.3, and the singular value decomposition of Chapter 12.

or, in matrix form,

$$\begin{bmatrix} 1 & -1 & 1 \\ 1 & 0 & 0 \\ 1 & 1 & 1 \\ 1 & 2 & 4 \end{bmatrix} \begin{bmatrix} c_1 \\ c_2 \\ c_3 \end{bmatrix} = \begin{bmatrix} 1 \\ 0 \\ 0 \\ -2 \end{bmatrix}.$$

This time, the normal equations are three equations in three unknowns:

$$\begin{bmatrix} 4 & 2 & 6 \\ 2 & 6 & 8 \\ 6 & 8 & 18 \end{bmatrix} \begin{bmatrix} c_1 \\ c_2 \\ c_3 \end{bmatrix} = \begin{bmatrix} -1 \\ -5 \\ -7 \end{bmatrix}.$$

Solving for the coefficients results in the best parabola $y = c_1 + c_2 t + c_3 t^2 = 0.45 - 0.65t - 0.25t^2$. The residual errors are given in the following table:

t	y	parabola	error
-1	1	0.85	0.15
0	0	0.45	-0.45
1	0	-0.45	0.45
2	-2	-1.85	-0.15

The error statistics are squared error $\text{SE} = (0.15)^2 + (-0.45)^2 + (0.45)^2 + (-0.15)^2 = 0.45$ and $\text{RMSE} = \sqrt{.45}/\sqrt{4} \approx 0.335$.   ◀

The MATLAB commands polyfit and polyval are designed not only to interpolate data, but also to fit data with polynomial models. For n input data points, polyfit used with input degree $n - 1$ returns the coefficients of the interpolating polynomial of degree $n - 1$. If the input degree is less than $n - 1$, polyfit will instead find the best least squares polynomial of that degree. For example, the commands

```
>> x0=[-1 0 1 2];
>> y0=[1 0 0 -2];
>> c=polyfit(x0,y0,2);
>> x=-1:.01:2;
>> y=polyval(c,x);
>> plot(x0,y0,'o',x,y)
```

find the coefficients of the least squares degree-two polynomial and plot it along with the given data from Example 4.4.

Example 4.4 shows that least squares modeling need not be restricted to finding best lines. By expanding the definition of the model, we can fit coefficients for any model as long as the coefficients enter the model in a linear way.

4.1.3 Conditioning of least squares

We have seen that the least squares problem reduces to solving the normal equations $A^T A \bar{x} = A^T b$. How accurately can the least squares solution $\bar{x}$ be determined? This is a question about the forward error of the normal equations. We carry out a double precision numerical experiment to test this question, by solving the normal equations in a case where the correct answer is known.

▶ **EXAMPLE 4.5** Let $x_1 = 2.0, x_2 = 2.2, x_3 = 2.4, \ldots, x_{11} = 4.0$ be equally spaced points in $[2, 4]$, and set $y_i = 1 + x_i + x_i^2 + x_i^3 + x_i^4 + x_i^5 + x_i^6 + x_i^7$ for $1 \leq i \leq 11$. Use the normal equations to find the least squares polynomial $P(x) = c_1 + c_2 x + \cdots + c_8 x^7$ fitting the (x_i, y_i).

A degree 7 polynomial is being fit to 11 data points lying on the degree 7 polynomial $P(x) = 1 + x + x^2 + x^3 + x^4 + x^5 + x^6 + x^7$. Obviously, the correct least squares solution is $c_1 = c_2 = \cdots = c_8 = 1$. Substituting the data points into the model $P(x)$ yields the system $Ac = b$:

$$
\begin{bmatrix}
1 & x_1 & x_1^2 & \cdots & x_1^7 \\
1 & x_2 & x_2^2 & \cdots & x_2^7 \\
\vdots & \vdots & \vdots & & \vdots \\
1 & x_{11} & x_{11}^2 & \cdots & x_{11}^7
\end{bmatrix}
\begin{bmatrix}
c_1 \\ c_2 \\ \vdots \\ c_8
\end{bmatrix}
=
\begin{bmatrix}
y_1 \\ y_2 \\ \vdots \\ y_{11}
\end{bmatrix}.
$$

The coefficient matrix A is a **Van der Monde matrix**, a matrix whose jth column consists of the elements of the second column raised to the $(j-1)$st power. We use MATLAB to solve the normal equations:

```
>> x = (2+(0:10)/5)';
>> y = 1+x+x.^2+x.^3+x.^4+x.^5+x.^6+x.^7;
>> A = [x.^0 x x.^2 x.^3 x.^4 x.^5 x.^6 x.^7];
>> c = (A'*A)\(A'*y)

c=
    1.5134
   -0.2644
    2.3211
    0.2408
    1.2592
    0.9474
    1.0059
    0.9997

>> cond(A'*A)

ans=
  1.4359e+019
```

Solving the normal equations in double precision cannot deliver an accurate value for the least squares solution. The condition number of $A^T A$ is too large to deal with in double precision arithmetic, and the normal equations are ill-conditioned, even though the original least squares problem is moderately conditioned. There is clearly room for improvement in the normal equations approach to least squares. In Example 4.15, we revisit this problem after developing an alternative that avoids forming $A^T A$. ◀

▶ **ADDITIONAL EXAMPLES**

1. Find the best least squares parabola for the points $(-2, 17), (0, 6), (1, -9), (2, 3),$ $(4, -2)$. Calculate the RMSE of the fit.

2. Use MATLAB commands to find the best least squares line through the North Pole ice extent data from Additional Example 3.1.2. Use the line to predict the ice extent in the year 2030.

⌨ **Solutions** for Additional Examples can be found at bit.ly/2EI7HJ1

4.1 Exercises

⌨ **Solutions** for Exercises numbered in blue can be found at bit.ly/2yupr55

1. Solve the normal equations to find the least squares solution and 2-norm error for the following inconsistent systems:

(a) $\begin{bmatrix} 1 & 2 \\ 0 & 1 \\ 2 & 1 \end{bmatrix} \begin{bmatrix} x_1 \\ x_2 \end{bmatrix} = \begin{bmatrix} 3 \\ 1 \\ 1 \end{bmatrix}$ (b) $\begin{bmatrix} 1 & 1 \\ 2 & 1 \\ 3 & 1 \end{bmatrix} \begin{bmatrix} x_1 \\ x_2 \end{bmatrix} = \begin{bmatrix} 1 \\ 2 \\ 0 \end{bmatrix}$ (c) $\begin{bmatrix} 1 & 2 \\ 1 & 1 \\ 2 & 1 \\ 2 & 2 \end{bmatrix} \begin{bmatrix} x_1 \\ x_2 \end{bmatrix} = \begin{bmatrix} 3 \\ 3 \\ 3 \\ 2 \end{bmatrix}$

2. Find the least squares solutions and RMSE of the following systems:

(a) $\begin{bmatrix} 1 & 1 & 0 \\ 0 & 1 & 1 \\ 1 & 2 & 1 \\ 1 & 0 & 1 \end{bmatrix} \begin{bmatrix} x_1 \\ x_2 \\ x_3 \end{bmatrix} = \begin{bmatrix} 2 \\ 2 \\ 3 \\ 4 \end{bmatrix}$ (b) $\begin{bmatrix} 1 & 0 & 1 \\ 1 & 0 & 2 \\ 1 & 1 & 1 \\ 2 & 1 & 1 \end{bmatrix} \begin{bmatrix} x_1 \\ x_2 \\ x_3 \end{bmatrix} = \begin{bmatrix} 2 \\ 3 \\ 1 \\ 2 \end{bmatrix}$

3. Find the least squares solution of the inconsistent system

$$\begin{bmatrix} 1 & 0 \\ 1 & 0 \\ 1 & 0 \end{bmatrix} \begin{bmatrix} x_1 \\ x_2 \end{bmatrix} = \begin{bmatrix} 1 \\ 5 \\ 6 \end{bmatrix}.$$

4. Let $m \geq n$, let A be the $m \times n$ identity matrix (the principal submatrix of the $m \times m$ identity matrix), and let $b = [b_1, \ldots, b_m]$ be a vector. Find the least squares solution of $Ax = b$ and the 2-norm error.

5. Prove that the 2-norm is a vector norm. You will need to use the Cauchy–Schwarz inequality $|u \cdot v| \leq ||u||_2 ||v||_2$.

6. Let A be an $n \times n$ nonsingular matrix. (a) Prove that $(A^T)^{-1} = (A^{-1})^T$. (b) Let b be an n-vector; then $Ax = b$ has exactly one solution. Prove that this solution satisfies the normal equations.

7. Find the best line through the set of data points, and find the RMSE:
   (a) $(-3, 3), (-1, 2), (0, 1), (1, -1), (3, -4)$ (b) $(1, 1), (1, 2), (2, 2), (2, 3), (4, 3)$.

8. Find the best line through each set of data points, and find the RMSE:
   (a) $(0, 0), (1, 3), (2, 3), (5, 6)$ (b) $(1, 2), (3, 2), (4, 1), (6, 3)$ (c) $(0, 5), (1, 3), (2, 3), (3, 1)$.

9. Find the best parabola through each data point set in Exercise 8, and compare the RMSE with the best-line fit.

10. Find the best degree 3 polynomial through each set in Exercise 8. Also, find the degree 3 interpolating polynomial, and compare.

11. Assume that the height of a model rocket is measured at four times, and the measured times and heights are $(t, h) = (1, 135), (2, 265), (3, 385), (4, 485)$, in seconds and meters. Fit the model $h = a + bt - 4.905t^2$ to estimate the eventual maximum height of the object and when it will return to earth.

12. Given data points $(x, y, z) = (0, 0, 3), (0, 1, 2), (1, 0, 3), (1, 1, 5), (1, 2, 6)$, find the plane in three dimensions (model $z = c_0 + c_1 x + c_2 y$) that best fits the data.

4.1 Computer Problems

Solutions for Computer Problems numbered in blue can be found at bit.ly/2EznQAb

1. Form the normal equations, and compute the least squares solution and 2-norm error for the following inconsistent systems:

   (a) $\begin{bmatrix} 3 & -1 & 2 \\ 4 & 1 & 0 \\ -3 & 2 & 1 \\ 1 & 1 & 5 \\ -2 & 0 & 3 \end{bmatrix} \begin{bmatrix} x_1 \\ x_2 \\ x_3 \end{bmatrix} = \begin{bmatrix} 10 \\ 10 \\ -5 \\ 15 \\ 0 \end{bmatrix}$ (b) $\begin{bmatrix} 4 & 2 & 3 & 0 \\ -2 & 3 & -1 & 1 \\ 1 & 3 & -4 & 2 \\ 1 & 0 & 1 & -1 \\ 3 & 1 & 3 & -2 \end{bmatrix} \begin{bmatrix} x_1 \\ x_2 \\ x_3 \\ x_4 \end{bmatrix} = \begin{bmatrix} 10 \\ 0 \\ 2 \\ 0 \\ 5 \end{bmatrix}$

2. Consider the world oil production data of Computer Problem 3.2.3. Find the best least squares (a) line, (b) parabola, and (c) cubic curve through the 10 data points and the RMSE of the fits. Use each to estimate the 2010 production level. Which fit best represents the data in terms of RMSE?

3. Consider the world population data of Computer Problem 3.1.1. Find the best least squares (a) line, (b) parabola through the data points, and the RMSE of the fit. In each case, estimate the 1980 population. Which fit gives the best estimate?

4. Consider the carbon dioxide concentration data of Exercise 3.1.13. Find the best least squares (a) line, (b) parabola, and (c) cubic curve through the data points and the RMSE of the fit. In each case, estimate the 1950 CO_2 concentration.

5. A company test-markets a new soft drink in 22 cities of approximately equal size. The selling price (in dollars) and the number sold per week in the cities are listed as follows:

city	price	sales/week	city	price	sales/week
1	0.59	3980	12	0.49	6000
2	0.80	2200	13	1.09	1190
3	0.95	1850	14	0.95	1960
4	0.45	6100	15	0.79	2760
5	0.79	2100	16	0.65	4330
6	0.99	1700	17	0.45	6960
7	0.90	2000	18	0.60	4160
8	0.65	4200	19	0.89	1990
9	0.79	2440	20	0.79	2860
10	0.69	3300	21	0.99	1920
11	0.79	2300	22	0.85	2160

   (a) First, the company wants to find the "demand curve": how many it will sell at each potential price. Let P denote price and S denote sales per week. Find the line $S = c_1 + c_2 P$ that best fits the data from the table in the sense of least squares. Find the normal equations and the coefficients c_1 and c_2 of the least squares line. Plot the least squares line along with the data, and calculate the root mean square error.

   (b) After studying the results of the test marketing, the company will set a single selling price P throughout the country. Given a manufacturing cost of \$0.23 per unit, the total profit (per city, per week) is $S(P - 0.23)$ dollars. Use the results of the preceding least squares approximation to find the selling price for which the company's profit will be maximized.

6. What is the "slope" of the parabola $y = x^2$ on 0, 1? Find the best least squares line that fits the parabola at n evenly spaced points in the interval for (a) $n = 10$ and (b) $n = 20$. Plot the parabola and the lines. What do you expect the result to be as $n \to \infty$? (c) Find the minimum of the function $F(c_1, c_2) = \int_0^1 (x^2 - c_1 - c_2 x)^2 \, dx$, and explain its relation to the problem.

7. Find the least squares (a) line (b) parabola through the 13 data points of Figure 3.5 and the RMSE of each fit.

8. Let A be the $10 \times n$ matrix formed by the first n columns of the 10×10 Hilbert matrix. Let c be the n-vector $[1, \ldots, 1]$, and set $b = Ac$. Use the normal equations to solve the least squares problem $Ax = b$ for (a) $n = 6$ (b) $n = 8$, and compare with the correct least squares solution $\bar{x} = c$. How many correct decimal places can be computed? Use condition number to explain the results. (This least squares problem is revisited in Computer Problem 4.3.7.)

9. Let $x_1, \ldots, x_{11}$ be 11 evenly spaced points in $[2, 4]$ and $y_i = 1 + x_i + x_i^2 + \cdots + x_i^d$. Use the normal equations to compute the best degree d polynomial, where (a) $d = 5$ (b) $d = 6$ (c) $d = 8$. Compare with Example 4.5. How many correct decimal places of the coefficients can be computed? Use condition number to explain the results. (This least squares problem is revisited in Computer Problem 4.3.8.)

10. The following data, collected by U.S. Bureau of Economic Analysis, lists the year-over-year percent change in mean disposable personal income in the United States during 15 election years. Also, the proportion of the U.S. electorate that voted for the incumbent party's presidential candidate is listed. The first line of the table says that income increased by 1.49% from 1951 to 1952, and that 44.6% of the electorate voted for Adlai Stevenson, the incumbent Democratic party's candidate for president. Find the best least squares linear model for incumbent party vote as a function of income change. Plot this line along with the 15 data points. How many percentage points of vote can the incumbent party expect for each additional percent of change in personal income?

year	% income change	% incumbent vote
1952	1.49	44.6
1956	3.03	57.8
1960	0.57	49.9
1964	5.74	61.3
1968	3.51	49.6
1972	3.73	61.8
1976	2.98	49.0
1980	−0.18	44.7
1984	6.23	59.2
1988	3.38	53.9
1992	2.15	46.5
1996	2.10	54.7
2000	3.93	50.3
2004	2.47	51.2
2008	−0.41	45.7

4.2 A SURVEY OF MODELS

The previous linear and polynomial models illustrate the use of least squares to fit data. The art of data modeling includes a wide variety of models, some derived from physical principles underlying the source of the data and others based on empirical factors.

4.2.1 Periodic data

Periodic data calls for periodic models. Outside air temperatures, for example, obey cycles on numerous timescales, including daily and yearly cycles governed by the rotation of the earth and the revolution of the earth around the sun. As a first example, hourly temperature data are fit to sines and cosines.

► **EXAMPLE 4.6**   Fit the recorded temperatures in Washington, D.C., on January 1, 2001, as listed in the following table, to a periodic model:

time of day	t	temp (C)
12 mid.	0	−2.2
3 am	$\frac{1}{8}$	−2.8
6 am	$\frac{1}{4}$	−6.1
9 am	$\frac{3}{8}$	−3.9
12 noon	$\frac{1}{2}$	0.0
3 pm	$\frac{5}{8}$	1.1
6 pm	$\frac{3}{4}$	−0.6
9 pm	$\frac{7}{8}$	−1.1

We choose the model $y = c_1 + c_2 \cos 2\pi t + c_3 \sin 2\pi t$ to match the fact that temperature is roughly periodic with a period of 24 hours, at least in the absence of longer-term temperature movements. The model uses this information by fixing the period to be exactly one day, where we are using days for the t units. The variable t is listed in these units in the table.

Substituting the data into the model results in the following overdetermined system of linear equations:

$$c_1 + c_2 \cos 2\pi(0) + c_3 \sin 2\pi(0) = -2.2$$

$$c_1 + c_2 \cos 2\pi \left(\frac{1}{8}\right) + c_3 \sin 2\pi \left(\frac{1}{8}\right) = -2.8$$

$$c_1 + c_2 \cos 2\pi \left(\frac{1}{4}\right) + c_3 \sin 2\pi \left(\frac{1}{4}\right) = -6.1$$

$$c_1 + c_2 \cos 2\pi \left(\frac{3}{8}\right) + c_3 \sin 2\pi \left(\frac{3}{8}\right) = -3.9$$

$$c_1 + c_2 \cos 2\pi \left(\frac{1}{2}\right) + c_3 \sin 2\pi \left(\frac{1}{2}\right) = 0.0$$

$$c_1 + c_2 \cos 2\pi \left(\frac{5}{8}\right) + c_3 \sin 2\pi \left(\frac{5}{8}\right) = 1.1$$

$$c_1 + c_2 \cos 2\pi \left(\frac{3}{4}\right) + c_3 \sin 2\pi \left(\frac{3}{4}\right) = -0.6$$

$$c_1 + c_2 \cos 2\pi \left(\frac{7}{8}\right) + c_3 \sin 2\pi \left(\frac{7}{8}\right) = -1.1$$

SPOTLIGHT ON   **Orthogonality**   The least squares problem can be simplified considerably by special choices of basis functions. The choices in Examples 4.6 and 4.7, for instance, yield normal equations already in diagonal form. This property of orthogonal basis functions is explored in detail in Chapter 10. Model (4.9) is a Fourier expansion.

The corresponding inconsistent matrix equation is $Ax = b$, where

$$A = \begin{bmatrix} 1 & \cos 0 & \sin 0 \\ 1 & \cos \frac{\pi}{4} & \sin \frac{\pi}{4} \\ 1 & \cos \frac{\pi}{2} & \sin \frac{\pi}{2} \\ 1 & \cos \frac{3\pi}{4} & \sin \frac{3\pi}{4} \\ 1 & \cos \pi & \sin \pi \\ 1 & \cos \frac{5\pi}{4} & \sin \frac{5\pi}{4} \\ 1 & \cos \frac{3\pi}{2} & \sin \frac{3\pi}{2} \\ 1 & \cos \frac{7\pi}{4} & \sin \frac{7\pi}{4} \end{bmatrix} = \begin{bmatrix} 1 & 1 & 0 \\ 1 & \sqrt{2}/2 & \sqrt{2}/2 \\ 1 & 0 & 1 \\ 1 & -\sqrt{2}/2 & \sqrt{2}/2 \\ 1 & -1 & 0 \\ 1 & -\sqrt{2}/2 & -\sqrt{2}/2 \\ 1 & 0 & -1 \\ 1 & \sqrt{2}/2 & -\sqrt{2}/2 \end{bmatrix} \quad \text{and} \quad b = \begin{bmatrix} -2.2 \\ -2.8 \\ -6.1 \\ -3.9 \\ 0.0 \\ 1.1 \\ -0.6 \\ -1.1 \end{bmatrix}.$$

The normal equations $A^T A c = A^T b$ are

$$\begin{bmatrix} 8 & 0 & 0 \\ 0 & 4 & 0 \\ 0 & 0 & 4 \end{bmatrix} \begin{bmatrix} c_1 \\ c_2 \\ c_3 \end{bmatrix} = \begin{bmatrix} -15.6 \\ -2.9778 \\ -10.2376 \end{bmatrix},$$

which are easily solved as $c_1 = -1.95$, $c_2 = -0.7445$, and $c_3 = -2.5594$. The best version of the model, in the sense of least squares, is $y = -1.9500 - 0.7445 \cos 2\pi t - 2.5594 \sin 2\pi t$, with RMSE ≈ 1.063. Figure 4.5(a) compares the least squares fit model with the actual hourly recorded temperatures. ◀

▶ **EXAMPLE 4.7** Fit the temperature data to the improved model

$$y = c_1 + c_2 \cos 2\pi t + c_3 \sin 2\pi t + c_4 \cos 4\pi t. \tag{4.9}$$

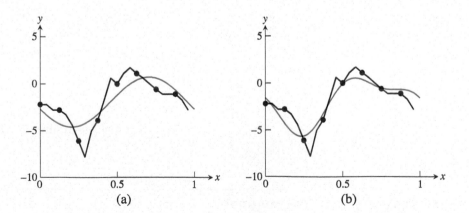

(a) (b)

Figure 4.5 Least squares fits to periodic data in Example 4.6. (a) Sinusoid model $y = -1.95 - 0.7445 \cos 2\pi t - 2.5594 \sin 2\pi t$ shown in bold, along with recorded temperature trace on Jan. 1, 2001. (b) Improved sinusoid $y = -1.95 - 0.7445 \cos 2\pi t - 2.5594 \sin 2\pi t + 1.125 \cos 4\pi t$ fits the data more closely.

The system of equations is now

$$c_1 + c_2 \cos 2\pi (0) + c_3 \sin 2\pi (0) + c_4 \cos 4\pi (0) = -2.2$$

$$c_1 + c_2 \cos 2\pi \left(\frac{1}{8}\right) + c_3 \sin 2\pi \left(\frac{1}{8}\right) + c_4 \cos 4\pi \left(\frac{1}{8}\right) = -2.8$$

$$c_1 + c_2 \cos 2\pi \left(\frac{1}{4}\right) + c_3 \sin 2\pi \left(\frac{1}{4}\right) + c_4 \cos 4\pi \left(\frac{1}{4}\right) = -6.1$$

$$c_1 + c_2 \cos 2\pi \left(\frac{3}{8}\right) + c_3 \sin 2\pi \left(\frac{3}{8}\right) + c_4 \cos 4\pi \left(\frac{3}{8}\right) = -3.9$$

$$c_1 + c_2 \cos 2\pi \left(\frac{1}{2}\right) + c_3 \sin 2\pi \left(\frac{1}{2}\right) + c_4 \cos 4\pi \left(\frac{1}{2}\right) = 0.0$$

$$c_1 + c_2 \cos 2\pi \left(\frac{5}{8}\right) + c_3 \sin 2\pi \left(\frac{5}{8}\right) + c_4 \cos 4\pi \left(\frac{5}{8}\right) = 1.1$$

$$c_1 + c_2 \cos 2\pi \left(\frac{3}{4}\right) + c_3 \sin 2\pi \left(\frac{3}{4}\right) + c_4 \cos 4\pi \left(\frac{3}{4}\right) = -0.6$$

$$c_1 + c_2 \cos 2\pi \left(\frac{7}{8}\right) + c_3 \sin 2\pi \left(\frac{7}{8}\right) + c_4 \cos 4\pi \left(\frac{7}{8}\right) = -1.1,$$

leading to the following normal equations:

$$\begin{bmatrix} 8 & 0 & 0 & 0 \\ 0 & 4 & 0 & 0 \\ 0 & 0 & 4 & 0 \\ 0 & 0 & 0 & 4 \end{bmatrix} \begin{bmatrix} c_1 \\ c_2 \\ c_3 \\ c_4 \end{bmatrix} = \begin{bmatrix} -15.6 \\ -2.9778 \\ -10.2376 \\ 4.5 \end{bmatrix}.$$

The solutions are $c_1 = -1.95$, $c_2 = -0.7445, c_3 = -2.5594$, and $c_4 = 1.125$, with RMSE ≈ 0.705. Figure 4.5(b) shows that the extended model $y = -1.95 - 0.7445 \cos 2\pi t - 2.5594 \sin 2\pi t + 1.125 \cos 4\pi t$ substantially improves the fit.   ◄

4.2.2 Data linearization

Exponential growth of a population is implied when its rate of change is proportional to its size. Under perfect conditions, when the growth environment is unchanging and when the population is well below the carrying capacity of the environment, the model is a good representation.

The **exponential model**

$$y = c_1 e^{c_2 t} \tag{4.10}$$

cannot be directly fit by least squares because c_2 does not appear linearly in the model equation. Once the data points are substituted into the model, the difficulty is clear: The set of equations to solve for the coefficients are nonlinear and cannot be expressed as a linear system $Ax = b$. Therefore, our derivation of the normal equations is irrelevant.

There are two ways to deal with the problem of nonlinear coefficients. The more difficult way is to directly minimize the least square error, that is, solve the nonlinear least squares problem. We return to this problem in Section 4.5. The simpler way is to change the problem. Instead of solving the original least squares problem, we can solve a different problem, which is related to the original, by "linearizing" the model.

In the case of the exponential model (4.10), the model is linearized by applying the natural logarithm:

$$\ln y = \ln(c_1 e^{c_2 t}) = \ln c_1 + c_2 t. \qquad (4.11)$$

Note that for an exponential model, the graph of $\ln y$ is a linear plot in t. At first glance, it appears that we have only traded one problem for another. The c_2 coefficient is now linear in the model, but c_1 no longer is. However, by renaming $k = \ln c_1$, we can write

$$\ln y = k + c_2 t. \qquad (4.12)$$

Now both coefficients k and c_2 are linear in the model. After solving the normal equations for the best k and c_2, we can find the corresponding $c_1 = e^k$ if we wish.

It should be noted that our way out of the difficulty of nonlinear coefficients was to change the problem. The original least squares problem we posed was to fit the data to (4.10)—that is, to find c_1, c_2 that minimize

$$(c_1 e^{c_2 t_1} - y_1)^2 + \cdots + (c_1 e^{c_2 t_m} - y_m)^2, \qquad (4.13)$$

the sum of squares of the residuals of the equations $c_1 e^{c_2 t_i} = y_i$ for $i = 1, \ldots, m$. For now, we solve the revised problem minimizing least squares error in "log space"—that is, by finding c_1, c_2 that minimizes

$$(\ln c_1 + c_2 t_1 - \ln y_1)^2 + \cdots + (\ln c_1 + c_2 t_m - \ln y_m)^2, \qquad (4.14)$$

the sum of squares of the residuals of the equations $\ln c_1 + c_2 t_i = \ln y_i$ for $i = 1, \ldots, m$. These are two different minimizations and have different solutions, meaning that they generally result in different values of the coefficients c_1, c_2.

Which method is correct for this problem, the nonlinear least squares of (4.13) or the model-linearized version (4.14)? The former is least squares, as we have defined it. The latter is not. However, depending on the context of the data, either may be the more natural choice. To answer the question, the user needs to decide which errors are most important to minimize, the errors in the original sense or the errors in "log space." In fact, the log model is linear, and it may be argued that only after log-transforming the data to a linear relation is it natural to evaluate the fitness of the model.

▶ **EXAMPLE 4.8** Use model linearization to find the best least squares exponential fit $y = c_1 e^{c_2 t}$ to the following world automobile supply data:

year	cars ($\times 10^6$)
1950	53.05
1955	73.04
1960	98.31
1965	139.78
1970	193.48
1975	260.20
1980	320.39

The data describe the number of automobiles operating throughout the world in the given year. Define the time variable t in terms of years since 1950. Solving the linear least squares problem yields $k_1 \approx 3.9896, c_2 \approx 0.06152$. Since $c_1 \approx e^{3.9896} \approx 54.03$, the model

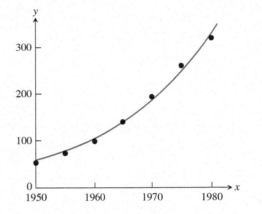

Figure 4.6 Exponential fit of world automobile supply data, using linearization. The best least squares fit is $y = 54.03e^{0.06152t}$. Compare with Figure 4.14.

is $y = 54.03e^{0.06152t}$. The RMSE of the log-linearized model in log space is ≈ 0.0357, while RMSE of the original exponential model is ≈ 9.56. The best model and data are plotted in Figure 4.6. ◀

▶ **EXAMPLE 4.9** The number of transistors on Intel central processing units since the early 1970s is given in the table that follows. Fit the model $y = c_1 e^{c_2 t}$ to the data.

CPU	year	transistors
4004	1971	2,250
8008	1972	2,500
8080	1974	5,000
8086	1978	29,000
286	1982	120,000
386	1985	275,000
486	1989	1,180,000
Pentium	1993	3,100,000
Pentium II	1997	7,500,000
Pentium III	1999	24,000,000
Pentium 4	2000	42,000,000
Itanium	2002	220,000,000
Itanium 2	2003	410,000,000

Parameters will be fit by using model linearization (4.11). Linearizing the model gives

$$\ln y = k + c_2 t.$$

We will let $t = 0$ correspond to the year 1970. Substituting the data into the linearized model yields

$$k + c_2(1) = \ln 2250$$
$$k + c_2(2) = \ln 2500$$
$$k + c_2(4) = \ln 5000$$
$$k + c_2(8) = \ln 29000, \tag{4.15}$$

and so forth. The matrix equation is $Ax = b$, where $x = (k, c_2)$,

$$A = \begin{bmatrix} 1 & 1 \\ 1 & 2 \\ 1 & 4 \\ 1 & 8 \\ \vdots & \vdots \\ 1 & 33 \end{bmatrix}, \text{ and } b = \begin{bmatrix} \ln 2250 \\ \ln 2500 \\ \ln 5000 \\ \ln 29000 \\ \vdots \\ \ln 410000000 \end{bmatrix}. \tag{4.16}$$

The normal equations $A^T A x = A^T b$ are

$$\begin{bmatrix} 13 & 235 \\ 235 & 5927 \end{bmatrix} \begin{bmatrix} k \\ c_2 \end{bmatrix} = \begin{bmatrix} 176.90 \\ 3793.23 \end{bmatrix},$$

which has solution $k \approx 7.197$ and $c_2 \approx 0.3546$, leading to $c_1 = e^k \approx 1335.3$. The exponential curve $y = 1335.3 e^{0.3546 t}$ is shown in Figure 4.7 along with the data. The doubling time for the law is $\ln 2 / c_2 \approx 1.95$ years. Gordon C. Moore, cofounder of Intel, predicted in 1965 that over the ensuing decade, computing power would double every 2 years. Astoundingly, that exponential rate has continued for 40 years. There is some evidence in Figure 4.7 that this rate has accelerated since 2000.

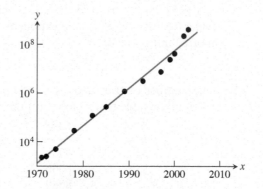

Figure 4.7 Semilog plot of Moore's law. Number of transistors on CPU chip versus year.

◀

Another important example with nonlinear coefficients is the **power law** model $y = c_1 t^{c_2}$. This model also can be simplified with linearization by taking logs of both sides:

$$\ln y = \ln c_1 + c_2 \ln t$$
$$= k + c_2 \ln t. \tag{4.17}$$

Substitution of data into the model will give

$$k + c_2 \ln t_1 = \ln y_1 \tag{4.18}$$

$$\vdots$$

$$k + c_2 \ln t_n = \ln y_n, \tag{4.19}$$

resulting in the matrix form

$$A = \begin{bmatrix} 1 & \ln t_1 \\ \vdots & \vdots \\ 1 & \ln t_n \end{bmatrix} \text{ and } b = \begin{bmatrix} \ln y_1 \\ \vdots \\ \ln y_n \end{bmatrix}. \tag{4.20}$$

The normal equations allow determination of k and c_2, and $c_1 = e^k$.

▶ **EXAMPLE 4.10**   Use linearization to fit the given height–weight data with a power law model.

The mean height and weight of boys ages 2–11 were collected in the U.S. National Health and Nutrition Examination Survey by the Centers for Disease Control (CDC) in 2002, resulting in the following table:

age (yrs.)	height (m)	weight (kg)
2	0.9120	13.7
3	0.9860	15.9
4	1.0600	18.5
5	1.1300	21.3
6	1.1900	23.5
7	1.2600	27.2
8	1.3200	32.7
9	1.3800	36.0
10	1.4100	38.6
11	1.4900	43.7

Following the preceding strategy, the resulting power law for weight versus height is $W = 16.3H^{2.42}$. The relationship is graphed in Figure 4.8. Since weight is a proxy for volume, the coefficient $c_2 \approx 2.42$ can be viewed as the "effective dimension" of the human body.

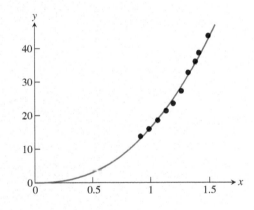

Figure 4.8 Power law of weight versus height for 2–11-year-olds. The best fit formula is $W = 16.3H^{2.42}$.

◀

The time course of drug concentration y in the bloodstream is well described by

$$y = c_1 t e^{c_2 t}, \tag{4.21}$$

where t denotes time after the drug was administered. The characteristics of the model are a quick rise as the drug enters the bloodstream, followed by slow exponential decay. The **half-life** of the drug is the time from the peak concentration to the time it drops to half that level. The model can be linearized by applying the natural logarithm to both sides, producing

$$\ln y = \ln c_1 + \ln t + c_2 t$$
$$k + c_2 t = \ln y - \ln t,$$

where we have set $k = \ln c_1$. This leads to the matrix equation $Ax = b$, where

$$A = \begin{bmatrix} 1 & t_1 \\ \vdots & \vdots \\ 1 & t_m \end{bmatrix} \quad \text{and} \quad b = \begin{bmatrix} \ln y_1 - \ln t_1 \\ \vdots \\ \ln y_m - \ln t_m \end{bmatrix}. \tag{4.22}$$

The normal equations are solved for k and c_2, and $c_1 = e^k$.

▶ **EXAMPLE 4.11** Fit the model (4.21) with the measured level of the drug norfluoxetine in a patient's bloodstream, given in the following table:

hour	concentration (ng/ml)
1	8.0
2	12.3
3	15.5
4	16.8
5	17.1
6	15.8
7	15.2
8	14.0

Solving the normal equations yields $k \approx 2.28$ and $c_2 \approx -0.215$, and $c_1 \approx e^{2.28} \approx 9.77$. The best version of the model is $y = 9.77te^{-0.215t}$, plotted in Figure 4.9. From the model, the timing of the peak concentration and the half-life can be estimated. (See Computer Problem 5.)

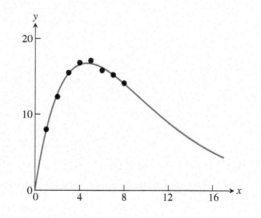

Figure 4.9 Plot of drug concentration in blood. Model (4.21) shows exponential decay after initial peak.

◀

It is important to realize that model linearization changes the least squares problem. The solution obtained will minimize the RMSE with respect to the linearized problem, not necessarily the original problem, which in general will have a different set of optimal parameters. If they enter the model nonlinearly, they cannot be computed from the normal equations, and we need nonlinear techniques to solve the original least squares problem. This is done in the Gauss–Newton Method in Section 4.5, where we revisit the automobile supply data and compare fitting the exponential model in linearized and nonlinearized forms.

► **ADDITIONAL EXAMPLES**

*1 Fit the data points $(-1, 4), (0, 2), (2, 1), (3, 0.5)$ with an exponential model, using data linearization. Calculate the RMSE of the fit.

2. The National Snow and Ice Data Center estimates of the ice extent at the North Pole (in 10^6 km^2), during the years of 2015–16, are shown in the table.

month	2015	2016
Jan.	13.75	13.64
Feb.	14.51	14.32
Mar.	14.49	14.53
Apr.	13.98	13.83
May	12.69	12.08
Jun.	11.05	10.60
Jul.	8.83	8.13
Aug.	5.66	5.60
Sept.	4.68	4.72
Oct.	7.79	6.45
Nov.	10.11	9.08
Dec.	12.33	12.09

Write MATLAB code to fit the model $y = c_1 + c_2 t + c_3 \sin 2\pi t + c_4 \cos 2\pi t$ to the data, where y denotes the ice extent and t is time in years beginning Jan. 2015. Calculate the RMSE. What does c_2 denote? Add a further term $c_5 \cos 4\pi t$ to the model, and discuss changes in RMSE and c_2.

Solutions for Additional Examples can be found at bit.ly/2AhmTZc
(* example with video solution)

4.2 Exercises

Solutions for Exercises numbered in blue can be found at bit.ly/2PCQZLP

1. Fit data to the periodic model $y = F_3(t) = c_1 + c_2 \cos 2\pi t + c_3 \sin 2\pi t$. Find the 2-norm error and the RMSE.

(a)

t	y
0	1
1/4	3
1/2	2
3/4	0

(b)

t	y
0	1
1/4	3
1/2	2
3/4	1

(c)

t	y
0	3
1/2	1
1	3
3/2	2

2. Fit the data to the periodic models $F_3(t) = c_1 + c_2 \cos 2\pi t + c_3 \sin 2\pi t$ and $F_4(t) = c_1 + c_2 \cos 2\pi t + c_3 \sin 2\pi t + c_4 \cos 4\pi t$. Find the 2-norm errors $||e||_2$ and compare the fits of F_3 and F_4.

(a)

t	y
0	0
1/6	2
1/3	0
1/2	−1
2/3	1
5/6	1

(b)

t	y
0	4
1/6	2
1/3	0
1/2	−5
2/3	−1
5/6	3

3. Fit data to the exponential model by using linearization. Find the 2-norm of the difference between the data points y_i and the best model $c_1 e^{c_2 t_i}$.

(a)

t	y
−2	1
0	2
1	2
2	5

(b)

t	y
0	1
1	1
1	2
2	4

4. Fit data to the exponential model by using linearization. Find the 2-norm of the difference between the data points y_i and the best model $c_1 e^{c_2 t_i}$.

(a)

t	y
−2	4
−1	2
1	1
2	1/2

(b)

t	y
0	10
1	5
2	2
3	1

5. Fit data to the power law model by using linearization. Find the RMSE of the fit.

(a)

t	y
1	6
2	2
3	1
4	1

(b)

t	y
1	2
1	4
2	5
3	6
5	10

6. Fit data to the drug concentration model (1.21). Find the RMSE of the fit.

(a)

t	y
1	3
2	4
3	5
4	5

(b)

t	y
1	2
2	4
3	3
4	2

4.2 Computer Problems

1. Fit the monthly data for Japan 2003 oil consumption, shown in the following table, with the periodic model (4.9), and calculate the RMSE:

month	oil use (10^6 bbl/day)
Jan.	6.224
Feb.	6.665
Mar.	6.241
Apr.	5.302
May	5.073
Jun.	5.127
Jul.	4.994
Aug.	5.012
Sept.	5.108
Oct.	5.377
Nov.	5.510
Dec.	6.372

2. The temperature data in Example 4.6 was taken from the Weather Underground website www.wunderground.com. Find a similar selection of hourly temperature data from a location and date of your choice, and fit it with the two sinusoidal models of the example.

3. Consider the world population data of Computer Problem 3.1.1. Find the best exponential fit of the data points by using linearization. Estimate the 1980 population, and find the estimation error.

4. Consider the carbon dioxide concentration data of Exercise 3.1.17. Find the best exponential fit of the difference between the CO_2 level and the background (279 ppm) by using linearization. Estimate the 1950 CO_2 concentration, and find the estimation error.

5. (a) Find the time at which the maximum concentration is reached in model (1.21). (b) Use an equation solver to estimate the half-life from the model in Example 4.11.

6. The bloodstream concentration of a drug, measured hourly after administration, is given in the accompanying table. Fit the model (4.21). Find the estimated maximum and the half-life. Suppose that the therapeutic range for the drug is 4–15 ng/ml. Use the equation solver of your choice to estimate the time the drug concentration stays within therapeutic levels.

hour	concentration (ng/ml)
1	6.2
2	9.5
3	12.3
4	13.9
5	14.6
6	13.5
7	13.3
8	12.7
9	12.4
10	11.9

7. The file windmill.txt, available from the textbook website, is a list of 60 numbers which represent the monthly megawatt-hours generated from Jan. 2005 to Dec. 2009 by a wind turbine owned by the Minnkota Power Cooperative near Valley City, ND. The data is currently available at http://www.minnkota.com. For reference, a typical home uses around 1 MWh per month.

   (a) Find a rough model of power output as a yearly periodic function. Fit the data to equation (4.9),

   $$f(t) = c_1 + c_2 \cos 2\pi t + c_3 \sin 2\pi t + c_4 \cos 4\pi t$$

   where the units of t are years, that is $0 \le t \le 5$, and write down the resulting function.

   (b) Plot the data and the model function for years $0 \le t \le 5$. What features of the data are captured by the model?

8. The file scrippsy.txt, available from the textbook website, is a list of 50 numbers which represent the concentration of atmospheric carbon dioxide, in parts per million by volume (ppv), recorded at Mauna Loa, Hawaii, each May 15 of the years 1961 to 2010. The data is part of a data collection effort initiated by Charles Keeling of the Scripps Oceanographic Institute (Keeling et al. [2001]). Subtract the background level 279 ppm as in Computer Problem 4, and fit the data to an exponential model. Plot the data along with the best fit exponential function, and report the RMSE.

9. The file scrippsm.txt, available from the textbook website, is a list of 180 numbers which represent the concentration of atmospheric carbon dioxide, in parts per million by volume (ppv), recorded monthly at Mauna Loa from Jan. 1996 to Dec. 2010, taken from the same Scripps study as Computer Problem 8.

(a) Carry out a least squares fit of the CO_2 data using the model

$$f(t) = c_1 + c_2 t + c_3 \cos 2\pi t + c_4 \sin 2\pi t$$

where t is measured in months. Report the best fit coefficients c_i and the RMSE of the fit. Plot the continuous curve from Jan. 1989 to the end of this year, including the 180 data points in the plot.

(b) Use your model to predict the CO_2 concentration in May 2004, Sept. 2004, May 2005, and Sept. 2005. These months tend to contain the yearly maxima and minima of the CO_2 cycle. The actual recorded values are 380.63, 374.06, 382.45, and 376.73 ppv, respectively. Report the model error at these four points.

(c) Add the extra term $c_5 \cos 4\pi t$ and redo parts (a) and (b). Compare the new RMSE and four model errors.

(d) Repeat part (c) using the extra term $c_5 t^2$. Which term leads to more improvement in the model, part (c) or (d)?

(e) Add both terms from (c) and (d) and redo parts (a) and (b). Prepare a table summarizing your results from all parts of the problem, and try to provide an explanation for the results.

See the website `http://scrippsco2.ucsd.edu` for much more data and analysis of the Scripps carbon dioxide study.

4.3 QR FACTORIZATION

In Chapter 2, the LU factorization was used to solve matrix equations. The factorization is useful because it encodes the steps of Gaussian elimination. In this section, we develop the QR factorization as a way to solve least squares calculations that is superior to the normal equations.

After introducing the factorization by way of Gram–Schmidt orthogonalization, we return to Example 4.5, for which the normal equations turned out to be inadequate. Later in this section, Householder reflections are introduced as a more efficient method of computing Q and R.

4.3.1 Gram–Schmidt orthogonalization and least squares

The Gram–Schmidt method orthogonalizes a set of vectors. Given an input set of m-dimensional vectors, the goal is to find an orthogonal coordinate system for the subspace spanned by the set. More precisely, given n linearly independent input vectors, it computes n mutually perpendicular unit vectors spanning the same subspace as the input vectors. The unit length is with respect to the Euclidean or 2-norm (4.7), which is used throughout Chapter 4.

Let $A_1, \ldots, A_n$ be linearly independent vectors from R^m. Thus $n \leq m$. The Gram–Schmidt method begins by dividing A_1 by its length to make it a unit vector. Define

$$y_1 = A_1 \quad \text{and} \quad q_1 = \frac{y_1}{\|y_1\|_2}. \tag{4.23}$$

To find the second unit vector, subtract away the projection of A_2 in the direction of q_1, and normalize the result:

$$y_2 = A_2 - q_1(q_1^T A_2) \quad \text{and} \quad q_2 = \frac{y_2}{\|y_2\|_2}. \tag{4.24}$$

Then $q_1^T y_2 = q_1^T(A_2 - q_1(q_1^T A_2)) = q_1^T A_2 - q_1^T A_2 = 0$, so q_1 and q_2 are pairwise orthogonal, as shown in Figure 4.10.

At the jth step, define

$$y_j = A_j - q_1(q_1^T A_j) - q_2(q_2^T A_j) - \ldots - q_{j-1}(q_{j-1}^T A_j) \quad \text{and} \quad q_j = \frac{y_j}{\|y_j\|_2}. \quad (4.25)$$

It is clear that q_j is orthogonal to each of the previously produced q_i for $i = 1, \ldots, j-1$, since (4.25) implies

$$q_i^T y_j = q_i^T A_j - q_i^T q_1 q_1^T A_j - \ldots - q_i^T q_{j-1} q_{j-1}^T A_j$$
$$= q_i^T A_j - q_i^T q_i q_i^T A_j = 0,$$

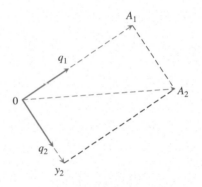

Figure 4.10 Gram–Schmidt orthogonalization. The input vectors are A_1 and A_2, and the output is the orthonormal set consisting of q_1 and q_2. The second orthogonal vector q_2 is formed by subtracting the projection of A_2 in the direction of q_1 from A_2, followed by normalizing.

where by induction hypothesis, the q_i are pairwise orthogonal for $i < j$. Geometrically, (4.25) corresponds to subtracting from A_j the projections of A_j onto the previously determined orthogonal vectors $q_i, i = 1, \ldots, j-1$. What remains is orthogonal to the q_i and, after dividing by its length to become a unit vector, is used as q_j. Therefore, the set $\{q_1, \ldots, q_n\}$ consists of mutually orthogonal vectors spanning the same subspace of R^m as $\{A_1, \ldots, A_n\}$.

The result of Gram–Schmidt orthogonalization can be put into matrix form by introducing new notation for the dot products in the above calculation. Define $r_{jj} = \|y_j\|_2$ and $r_{ij} = q_i^T A_j$. Then (4.23) and (4.24) can be written

$$A_1 = r_{11} q_1$$
$$A_2 = r_{12} q_1 + r_{22} q_2,$$

and the general case (4.25) translates to

$$A_j = r_{1j} q_1 + \cdots + r_{j-1,j} q_{j-1} + r_{jj} q_j.$$

Therefore, the result of Gram–Schmidt orthogonalization can be written in matrix form as

$$(A_1|\cdots|A_n) = (q_1|\cdots|q_n) \begin{bmatrix} r_{11} & r_{12} & \cdots & r_{1n} \\ & r_{22} & \cdots & r_{2n} \\ & & \ddots & \vdots \\ & & & r_{nn} \end{bmatrix}, \quad (4.26)$$

or $A = QR$, where we consider A to be the matrix consisting of the columns A_j. We call this the **reduced QR factorization**; the full version is just ahead. The assumption that the vectors A_j are linearly independent guarantees that the main diagonal coefficients r_{jj} are nonzero. Conversely, if A_j lies in the span of $A_1, \ldots, A_{j-1}$, then the projections onto the latter vectors make up the entire vector, and $r_{jj} = \|y_j\|_2 = 0$.

▶ **EXAMPLE 4.12**   Find the reduced QR factorization by applying Gram–Schmidt orthogonalization to the columns of $A = \begin{bmatrix} 1 & -4 \\ 2 & 3 \\ 2 & 2 \end{bmatrix}$.

Set $y_1 = A_1 = \begin{bmatrix} 1 \\ 2 \\ 2 \end{bmatrix}$. Then $r_{11} = \|y_1\|_2 = \sqrt{1^2 + 2^2 + 2^2} = 3$, and the first unit vector is

$$q_1 = \frac{y_1}{\|y_1\|_2} = \begin{bmatrix} \frac{1}{3} \\ \frac{2}{3} \\ \frac{2}{3} \end{bmatrix}.$$

To find the second unit vector, set

$$y_2 = A_2 - q_1 q_1^T A_2 = \begin{bmatrix} -4 \\ 3 \\ 2 \end{bmatrix} - \begin{bmatrix} \frac{1}{3} \\ \frac{2}{3} \\ \frac{2}{3} \end{bmatrix} 2 = \begin{bmatrix} -\frac{14}{3} \\ \frac{5}{3} \\ \frac{2}{3} \end{bmatrix}$$

and

$$q_2 = \frac{y_2}{\|y_2\|_2} = \frac{1}{5} \begin{bmatrix} -\frac{14}{3} \\ \frac{5}{3} \\ \frac{2}{3} \end{bmatrix} = \begin{bmatrix} -\frac{14}{15} \\ \frac{1}{3} \\ \frac{2}{15} \end{bmatrix}.$$

Since $r_{12} = q_1^T A_2 = 2$ and $r_{22} = \|y_2\|_2 = 5$, the result written in matrix form (4.26) is

$$A = \begin{bmatrix} 1 & -4 \\ 2 & 3 \\ 2 & 2 \end{bmatrix} = \begin{bmatrix} 1/3 & -14/15 \\ 2/3 & 1/3 \\ 2/3 & 2/15 \end{bmatrix} \begin{bmatrix} 3 & 2 \\ 0 & 5 \end{bmatrix} = QR. \qquad ◀$$

We use the term "classical" for this version of Gram–Schmidt, since we will provide an upgraded, or "modified," version in the next section.

Classical Gram-Schmidt orthogonalization

Let $A_j, j = 1, \ldots, n$ be linearly independent vectors.
for $j = 1, 2, \ldots, n$
    $y = A_j$
    for $i = 1, 2, \ldots, j - 1$
        $r_{ij} = q_i^T A_j$
        $y = y - r_{ij} q_i$
    end
    $r_{jj} = \|y\|_2$
    $q_j = y/r_{jj}$
end

This algorithm is expressed in the MATLAB code `clgs.m` of Program 4.1.

MATLAB code

shown here can be found at bit.ly/2yQjTBg

```
% Program 4.1  Classical Gram-Schmidt orthogonalization
% Input: mxn matrix A with linearly independent columns
% Output: orthogonal matrix Q, upper triangular R
% Example usage: [Q,R]=clgs(A)
function [Q,R]=clgs(A)
[m,n]=size(A);
for j=1:n
    y=A(:,j);
    for i=1:j-1
        R(i,j)=Q(:,i)'*A(:,j);
        y=y-R(i,j)*Q(:,i);
    end
    R(j,j)=norm(y);
    Q(:,j)=y/R(j,j);
end
```

When the method is successful, it is customary to fill out the matrix of orthogonal unit vectors to a complete basis of R^m, to achieve the "full" QR factorization. This can be done, for example, by adding $m - n$ extra vectors to the A_j, so that the m vectors span R^m, and carrying out the Gram–Schmidt method. In terms of the basis of R^m formed by $q_1, \ldots, q_m$, the original vectors can be expressed as

$$(A_1|\cdots|A_n) = (q_1|\cdots|q_m) \begin{bmatrix} r_{11} & r_{12} & \cdots & r_{1n} \\ & r_{22} & \cdots & r_{2n} \\ & & \ddots & \vdots \\ & & & r_{nn} \\ 0 & \cdots & \cdots & 0 \\ \vdots & & & \vdots \\ 0 & \cdots & \cdots & 0 \end{bmatrix}. \tag{4.27}$$

This matrix equation is the **full QR factorization** of the matrix $A = (A_1|\cdots|A_n)$, formed by the original input vectors. Note the matrix sizes in the full QR factorization: A is $m \times n$, Q is a square $m \times m$ matrix, and the upper triangular matrix R is $m \times n$, the same size as A. The matrix Q in the full QR factorization has a special place in numerical analysis and is given a special definition.

DEFINITION 4.1 A square matrix Q is **orthogonal** if $Q^T = Q^{-1}$. ◻

Note that a square matrix is orthogonal if and only if its columns are pairwise orthogonal unit vectors (Exercise 9). Therefore, a full QR factorization is the equation $A = QR$, where Q is an orthogonal square matrix and R is an upper triangular matrix the same size as A.

The key property of an orthogonal matrix is that it preserves the Euclidean norm of a vector.

LEMMA 4.2 If Q is an orthogonal $m \times m$ matrix and x is an m-dimensional vector, then $\|Qx\|_2 = \|x\|_2$. ∎

Proof. $\|Qx\|_2^2 = (Qx)^T Qx = x^T Q^T Qx = x^T x = \|x\|_2^2$. ◻

The product of two orthogonal $m \times m$ matrices is again orthogonal (Exercise 10). The QR factorization of an $m \times m$ matrix by the Gram–Schmidt method requires approximately m^3 multiplication/divisions, three times more than the LU factorization, plus about the same number of additions (Exercise 11).

▶ **EXAMPLE 4.13** Find the full QR factorization of $A = \begin{bmatrix} 1 & -4 \\ 2 & 3 \\ 2 & 2 \end{bmatrix}$.

SPOTLIGHT ON

> **Orthogonality**   In Chapter 2, we found that the LU factorization is an efficient means of encoding the information of Gaussian elimination. In the same way, the QR factorization records the orthogonalization of a matrix, namely, the construction of an orthogonal set that spans the space of column vectors of A. Doing calculations with orthogonal matrices is preferable because (1) they are easy to invert by definition, and (2) by Lemma 4.2, they do not magnify errors.

In Example 4.12, we found the orthogonal unit vectors $q_1 = \begin{bmatrix} \frac{1}{3} \\ \frac{2}{3} \\ \frac{2}{3} \end{bmatrix}$ and

$q_2 = \begin{bmatrix} -\frac{14}{15} \\ \frac{1}{3} \\ \frac{2}{15} \end{bmatrix}$. Adding a third vector $A_3 = \begin{bmatrix} 1 \\ 0 \\ 0 \end{bmatrix}$ leads to

$$y_3 = A_3 - q_1 q_1^T A_3 - q_2 q_2^T A_3$$

$$= \begin{bmatrix} 1 \\ 0 \\ 0 \end{bmatrix} - \begin{bmatrix} \frac{1}{3} \\ \frac{2}{3} \\ \frac{2}{3} \end{bmatrix} \frac{1}{3} - \begin{bmatrix} -\frac{14}{15} \\ \frac{1}{3} \\ -\frac{2}{15} \end{bmatrix} \left(-\frac{14}{15} \right) = \frac{2}{225} \begin{bmatrix} 2 \\ 10 \\ -11 \end{bmatrix}$$

and $q_3 = y_3/\|y_3\| = \begin{bmatrix} \frac{2}{15} \\ \frac{10}{15} \\ -\frac{11}{15} \end{bmatrix}$. Putting the parts together, we obtain the full QR factorization

$$A = \begin{bmatrix} 1 & -4 \\ 2 & 3 \\ 2 & 2 \end{bmatrix} = \begin{bmatrix} 1/3 & -14/15 & 2/15 \\ 2/3 & 1/3 & 2/3 \\ 2/3 & 2/15 & -11/15 \end{bmatrix} \begin{bmatrix} 3 & 2 \\ 0 & 5 \\ 0 & 0 \end{bmatrix} = QR.$$

Note that the choice of A_3 was arbitrary. Any third column vector linearly independent of the first two columns could be used. Compare this result with the reduced QR factorization in Example 4.12. ◀

The MATLAB command qr carries out the QR factorization on an $m \times n$ matrix. It does not use Gram–Schmidt orthogonalization, but uses more efficient and stable methods that will be introduced in a later subsection. The command

```
>> [Q,R]=qr(A,0)
```

returns the reduced QR factorization, and

```
>> [Q,R]=qr(A)
```

returns the full QR factorization.

There are three major applications of the QR factorization. We will describe two of them here; the third is the QR algorithm for eigenvalue calculations, introduced in Chapter 12.

First, the QR factorization can be used to solve a system of n equations in n unknowns $Ax = b$. Just factor $A = QR$, and the equation $Ax = b$ becomes $QRx = b$ and $Rx = Q^T b$. Assuming that A is nonsingular, the diagonal entries of the upper triangular matrix R are nonzero, so that R is nonsingular. A triangular back substitution yields the solution x. As mentioned before, this approach is about three times more expensive in terms of complexity when compared with the LU approach.

The second application is to least squares. Let A be an $m \times n$ matrix with $m \geq n$. To minimize $||Ax - b||_2$, rewrite as $||QRx - b||_2 = ||Rx - Q^T b||_2$ by Lemma 4.2. The vector inside the Euclidean norm is

$$
\begin{bmatrix} e_1 \\ \vdots \\ e_n \\ \hdashline e_{n+1} \\ \vdots \\ e_m \end{bmatrix} = \begin{bmatrix} r_{11} & r_{12} & \cdots & r_{1n} \\ & r_{22} & \cdots & r_{2n} \\ & & \ddots & \vdots \\ & & & r_{nn} \\ \hdashline 0 & \cdots & \cdots & 0 \\ \vdots & & & \vdots \\ 0 & \cdots & \cdots & 0 \end{bmatrix} \begin{bmatrix} x_1 \\ \vdots \\ x_n \end{bmatrix} - \begin{bmatrix} d_1 \\ \vdots \\ d_n \\ \hdashline d_{n+1} \\ \vdots \\ d_m \end{bmatrix} \tag{4.28}
$$

where $d = Q^T b$. Assume that $r_{ii} \neq 0$. Then the upper part $(e_1, \ldots, e_n)$ of the error vector e can be made zero by back substitution. The choice of the x_i makes no difference for the lower part of the error vector; clearly, $(e_{n+1}, \ldots, e_m) = (-d_{n+1}, \ldots, -d_m)$. Therefore, the least squares solution is minimized by using the x from back-solving the upper part, and the least squares error is $||e||_2^2 = d_{n+1}^2 + \cdots + d_m^2$.

Least squares by QR factorization

Given the $m \times n$ inconsistent system

$$Ax = b,$$

find the full QR factorization $A = QR$ and set

$$\hat{R} = \text{upper } n \times n \text{ submatrix of } R$$
$$\hat{d} = \text{upper } n \text{ entries of } d = Q^T b$$

Solve $\hat{R}\bar{x} = \hat{d}$ for least squares solution $\bar{x}$.

▶ **EXAMPLE 4.14**   Use the full QR factorization to solve the least squares problem $\begin{bmatrix} 1 & -4 \\ 2 & 3 \\ 2 & 2 \end{bmatrix} \begin{bmatrix} x_1 \\ x_2 \end{bmatrix}$

$$= \begin{bmatrix} -3 \\ 15 \\ 9 \end{bmatrix}.$$

We need to solve $Rx = Q^T b$, or

$$\begin{bmatrix} 3 & 2 \\ 0 & 5 \\ \hline 0 & 0 \end{bmatrix} \begin{bmatrix} x_1 \\ x_2 \end{bmatrix} = \frac{1}{15} \begin{bmatrix} 5 & 10 & 10 \\ -14 & 5 & 2 \\ 2 & 10 & -11 \end{bmatrix} \begin{bmatrix} -3 \\ 15 \\ 9 \end{bmatrix} = \begin{bmatrix} 15 \\ 9 \\ \hline 3 \end{bmatrix}.$$

The least squares error will be $||e||_2 = ||(0, 0, 3)||_2 = 3$. Equating the upper parts yields

$$\begin{bmatrix} 3 & 2 \\ 0 & 5 \end{bmatrix} \begin{bmatrix} x_1 \\ x_2 \end{bmatrix} = \begin{bmatrix} 15 \\ 9 \end{bmatrix},$$

whose solution is $\overline{x}_1 = 3.8, \overline{x}_2 = 1.8$. This least squares problem was solved by the normal equations in Example 4.2. ◀

Finally, we return to the problem in Example 4.5 that led to an ill-conditioned system of normal equations.

SPOTLIGHT ON

> **Conditioning**   In Chapter 2, we found that the best way to handle ill-conditioned problems is to avoid them. Example 4.15 is a classic case of that advice. While the normal equations of Example 4.5 are ill-conditioned, the QR approach solves least squares without constructing $A^T A$.

▶ **EXAMPLE 4.15**   Use the full QR factorization to solve the least squares problem of Example 4.5.

The normal equations were notably unsuccessful in solving this least squares problem of 11 equations in 8 variables. We use the MATLAB qr command to carry out an alternative approach:

```
>> x=(2+(0:10)/5)';
>> y=1+x+x.^2+x.^3+x.^4+x.^5+x.^6+x.^7;
>> A=[x.^0 x x.^2 x.^3 x.^4 x.^5 x.^6 x.^7];
>> [Q,R]=qr(A);
>> b=Q'*y;
>> c=R(1:8,1:8)\b(1:8)

c=
   0.99999991014308
   1.00000021004107
   0.99999979186557
   1.00000011342980
   0.99999996325039
   1.00000000708455
   0.99999999924685
   1.00000000003409
```

Six decimal places of the correct solution $c = [1, \ldots, 1]$ are found by using QR factorization. This approach finds the least squares solution without forming the normal equations, which have a condition number of about 10^{19}. ◀

4.3.2 Modified Gram-Schmidt orthogonalization

A slight modification to Gram–Schmidt turns out to enhance its accuracy in machine calculations. The new algorithm called modified Gram–Schmidt is mathematically equivalent to the original, or "classical" Gram–Schmidt algorithm.

Modified Gram-Schmidt orthogonalization

Let A_j, $j = 1, \ldots, n$ be linearly independent vectors.

for $j = 1, 2, \ldots, n$
    $y = A_j$
    for $i = 1, 2, \ldots, j - 1$
        $r_{ij} = q_i^T y$
        $y = y - r_{ij} q_i$
    end
    $r_{jj} = \|y\|_2$
    $q_j = y/r_{jj}$
end

The only difference from classical Gram–Schmidt is that A_j is replaced by y in the innermost loop. Geometrically speaking, when projecting away the part of vector A_j in the direction of q_2, for example, one should subtract away the projection of the remainder y of A_j with the q_1 part already removed, instead of the projection of A_j itself on q_2. Modified Gram–Schmidt is the version that will be used in the GMRES algorithm in Section 4.4.

▶ **EXAMPLE 4.16** Compare the results of classical Gram–Schmidt and modified Gram–Schmidt, computed in double precision, on the matrix of almost-parallel vectors

$$\begin{bmatrix} 1 & 1 & 1 \\ \delta & 0 & 0 \\ 0 & \delta & 0 \\ 0 & 0 & \delta \end{bmatrix}$$

where $\delta = 10^{-10}$.

First, we apply classical Gram–Schmidt.

$$y_1 = A_1 = \begin{bmatrix} 1 \\ \delta \\ 0 \\ 0 \end{bmatrix} \quad \text{and} \quad q_1 = \frac{1}{\sqrt{1 + \delta^2}} \begin{bmatrix} 1 \\ \delta \\ 0 \\ 0 \end{bmatrix} = \begin{bmatrix} 1 \\ \delta \\ 0 \\ 0 \end{bmatrix}.$$

Note that $\delta^2 = 10^{-20}$ is a perfectly acceptable double precision number, but $1 + \delta^2 = 1$ after rounding. Then

$$y_2 = \begin{bmatrix} 1 \\ 0 \\ \delta \\ 0 \end{bmatrix} - \begin{bmatrix} 1 \\ \delta \\ 0 \\ 0 \end{bmatrix} q_1^T A_2 = \begin{bmatrix} 1 \\ 0 \\ \delta \\ 0 \end{bmatrix} - \begin{bmatrix} 1 \\ \delta \\ 0 \\ 0 \end{bmatrix} = \begin{bmatrix} 0 \\ -\delta \\ \delta \\ 0 \end{bmatrix} \quad \text{and} \quad q_2 = \begin{bmatrix} 0 \\ -\frac{1}{\sqrt{2}} \\ \frac{1}{\sqrt{2}} \\ 0 \end{bmatrix}$$

after dividing by $||y_2||_2 = \sqrt{\delta^2 + \delta^2} = \sqrt{2}\delta$. Completing classical Gram–Schmidt,

$$y_3 = \begin{bmatrix} 1 \\ 0 \\ 0 \\ \delta \end{bmatrix} - \begin{bmatrix} 1 \\ \delta \\ 0 \\ 0 \end{bmatrix} q_1^T A_3 - \begin{bmatrix} 0 \\ -\frac{1}{\sqrt{2}} \\ \frac{1}{\sqrt{2}} \\ 0 \end{bmatrix} q_2^T A_3 = \begin{bmatrix} 1 \\ 0 \\ 0 \\ \delta \end{bmatrix} - \begin{bmatrix} 1 \\ \delta \\ 0 \\ 0 \end{bmatrix} = \begin{bmatrix} 0 \\ -\delta \\ 0 \\ \delta \end{bmatrix} \text{ and } q_3 = \begin{bmatrix} 0 \\ -\frac{1}{\sqrt{2}} \\ 0 \\ \frac{1}{\sqrt{2}} \end{bmatrix}.$$

Unfortunately, due to the double precision rounding done in the first step, q_2 and q_3 turn out to be not orthogonal:

$$q_2^T q_3 = \begin{bmatrix} 0 \\ -\frac{1}{\sqrt{2}} \\ \frac{1}{\sqrt{2}} \\ 0 \end{bmatrix}^T \begin{bmatrix} 0 \\ -\frac{1}{\sqrt{2}} \\ 0 \\ \frac{1}{\sqrt{2}} \end{bmatrix} = \frac{1}{2}.$$

On the other hand, modified Gram–Schmidt does much better. While q_1 and q_2 are calculated the same way, q_3 is found as

$$y_3^1 = \begin{bmatrix} 1 \\ 0 \\ 0 \\ \delta \end{bmatrix} - \begin{bmatrix} 1 \\ \delta \\ 0 \\ 0 \end{bmatrix} q_1^T A_3 = \begin{bmatrix} 0 \\ -\delta \\ 0 \\ \delta \end{bmatrix},$$

$$y_3 = y_3^1 - \begin{bmatrix} 0 \\ -\frac{1}{\sqrt{2}} \\ \frac{1}{\sqrt{2}} \\ 0 \end{bmatrix} q_2^T y_3^1 = \begin{bmatrix} 0 \\ -\delta \\ 0 \\ \delta \end{bmatrix} - \begin{bmatrix} 0 \\ -\frac{1}{\sqrt{2}} \\ \frac{1}{\sqrt{2}} \\ 0 \end{bmatrix} \frac{\delta}{\sqrt{2}}$$

$$= \begin{bmatrix} 0 \\ -\frac{\delta}{2} \\ -\frac{\delta}{2} \\ \delta \end{bmatrix} \text{ and } q_3 = \begin{bmatrix} 0 \\ -\frac{1}{\sqrt{6}} \\ -\frac{1}{\sqrt{6}} \\ \frac{2}{\sqrt{6}} \end{bmatrix}.$$

Now $q_2^T q_3 = 0$ as desired. Note that for both classical and modified Gram–Schmidt, $q_1^T q_2$ is on the order of δ, so even modified Gram–Schmidt leaves room for improvement. Orthogonalization by Householder reflectors, described in the next section, is widely considered to be more computationally stable. ◀

4.3.3 Householder reflectors

Although the modified Gram–Schmidt orthogonalization method is an improved way to calculate the QR factorization of a matrix, it is not the best way. An alternative method using Householder reflectors requires fewer operations and is more stable, in the sense of amplification of rounding errors. In this section, we will define the reflectors and show how they are used to factorize a matrix.

A Householder reflector is an orthogonal matrix that reflects all m-vectors through an $m - 1$ dimensional plane. This means that the length of each vector is unchanged when multiplied by the matrix, making Householder reflectors ideal for moving vectors. Given a vector x that we would like to relocate to a vector w of equal length, the recipe for Householder reflectors gives a matrix H such that $Hx = w$.

The origin of the recipe is clear in Figure 4.11. Draw the $m - 1$ dimensional plane bisecting x and w, and perpendicular to the vector connecting them. Then reflect all vectors through the plane.

LEMMA 4.3 Assume that x and w are vectors of the same Euclidean length, $||x||_2 = ||w||_2$. Then $w - x$ and $w + x$ are perpendicular. ∎

> **Proof.** $(w - x)^T (w + x) = w^T w - x^T w + w^T x - x^T x = ||w||^2 - ||x||^2 = 0.$ □

Define the vector $v = w - x$, and consider the projection matrix

$$P = \frac{vv^T}{v^T v}. \tag{4.29}$$

A **projection matrix** is a matrix that satisfies $P^2 = P$. Exercise 13 asks the reader to verify that P in (4.29) is a symmetric projection matrix and that $Pv = v$. Geometrically, for any vector u, Pu is the projection of u onto v. Figure 4.11 hints that if we subtract twice the projection Px from x, we should get w. To verify this, set $H = I - 2P$. Then

$$
\begin{aligned}
Hx &= x - 2Px \\
&= w - v - \frac{2vv^T x}{v^T v} \\
&= w - v - \frac{vv^T x}{v^T v} - \frac{vv^T (w - v)}{v^T v} \\
&= w - \frac{vv^T (w + x)}{v^T v} \\
&= w,
\end{aligned} \tag{4.30}
$$

the latter equality following from Lemma 4.3, since $w + x$ is orthogonal to $v = w - x$.

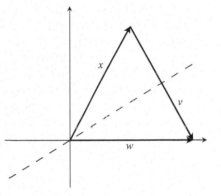

Figure 4.11 Householder reflector. Given equal length vectors x and w, reflection through the bisector of the angle between them (dotted line) exchanges them.

The matrix H is called a **Householder reflector**. Note that H is a symmetric (Exercise 14) and orthogonal matrix, since

$$
\begin{aligned}
H^T H = HH &= (I - 2P)(I - 2P) \\
&= I - 4P + 4P^2 \\
&= I.
\end{aligned}
$$

These facts are summarized in the following theorem:

THEOREM 4.4 **Householder reflectors.** Let x and w be vectors with $||x||_2 = ||w||_2$ and define $v = w - x$. Then $H = I - 2vv^T/v^T v$ is a symmetric orthogonal matrix and $Hx = w$. ∎

► **EXAMPLE 4.17** Let $x = [3, 4]$ and $w = [5, 0]$. Find a Householder reflector H that satisfies $Hx = w$.

Set

$$v = w - x = \begin{bmatrix} 5 \\ 0 \end{bmatrix} - \begin{bmatrix} 3 \\ 4 \end{bmatrix} = \begin{bmatrix} 2 \\ -4 \end{bmatrix},$$

and define the projection matrix

$$P = \frac{vv^T}{v^T v} = \frac{1}{20} \begin{bmatrix} 4 & -8 \\ -8 & 16 \end{bmatrix} = \begin{bmatrix} 0.2 & -0.4 \\ -0.4 & 0.8 \end{bmatrix}.$$

Then

$$H = I - 2P = \begin{bmatrix} 1 & 0 \\ 0 & 1 \end{bmatrix} - \begin{bmatrix} 0.4 & -0.8 \\ -0.8 & 1.6 \end{bmatrix} = \begin{bmatrix} 0.6 & 0.8 \\ 0.8 & -0.6 \end{bmatrix}.$$

Check that H moves x to w and vice versa:

$$Hx = \begin{bmatrix} 0.6 & 0.8 \\ 0.8 & -0.6 \end{bmatrix} \begin{bmatrix} 3 \\ 4 \end{bmatrix} = \begin{bmatrix} 5 \\ 0 \end{bmatrix} = w$$

and

$$Hw = \begin{bmatrix} 0.6 & 0.8 \\ 0.8 & -0.6 \end{bmatrix} \begin{bmatrix} 5 \\ 0 \end{bmatrix} = \begin{bmatrix} 3 \\ 4 \end{bmatrix} = x. \qquad \blacktriangleleft$$

As a first application of Householder reflectors, we will develop a new way to do the QR factorization. In Chapter 12, we apply Householder to the eigenvalue problem, to put matrices into upper Hessenberg form. In both applications, we will use reflectors for a single purpose: to move a column vector x to a coordinate axis as a way of putting zeros into a matrix.

We start with a matrix A that we want to write in the form $A = QR$. Let x_1 be the first column of A. Let $w = \pm(||x_1||_2, 0, \ldots, 0)$ be a vector along the first coordinate axis of identical Euclidean length. (Either sign works in theory. For numerical stability, the sign is often chosen to be the opposite of the sign of the first component of x to avoid the possibility of subtracting nearly equal numbers when forming v.) Create the Householder reflector H_1 such that $H_1 x = w$. In the 4×3 case, multiplying H_1 by A results in

$$H_1 A = H_1 \begin{bmatrix} \times & \times & \times \\ \times & \times & \times \\ \times & \times & \times \\ \times & \times & \times \end{bmatrix} = \begin{bmatrix} \times & \times & \times \\ 0 & \times & \times \\ 0 & \times & \times \\ 0 & \times & \times \end{bmatrix}.$$

We have introduced some zeros into A. We want to continue in this way until A becomes upper triangular; then we will have R of the QR factorization. Find the Householder reflector $\hat{H}_2$ that moves the $(m - 1)$-vector x_2 consisting of the lower $m - 1$ entries in column 2 of $H_1 A$ to $\pm(||x_2||_2, 0, \ldots, 0)$. Since $\hat{H}_2$ is an $(m - 1) \times (m - 1)$-matrix, define H_2 to be the $m \times m$ matrix formed by putting $\hat{H}_2$ into the lower part of the identity matrix. Then

$$\left(\begin{array}{c|ccc} 1 & 0 & 0 & 0 \\ \hline 0 & & & \\ 0 & & \hat{H}_2 & \\ 0 & & & \end{array} \right) \left(\begin{array}{c|cc} \times & \times & \times \\ \hline 0 & \times & \times \\ 0 & \times & \times \\ 0 & \times & \times \end{array} \right) = \left(\begin{array}{c|cc} \times & \times & \times \\ \hline 0 & \times & \times \\ 0 & 0 & \times \\ 0 & 0 & \times \end{array} \right)$$

The result $H_2 H_1 A$ is one step from upper triangularity. One more step gives

$$
\begin{pmatrix}
1 & 0 & 0 & 0 \\
0 & 1 & 0 & 0 \\
\hline
0 & 0 & & \\
0 & 0 & & \hat{H}_3
\end{pmatrix}
\begin{pmatrix}
\times & \times & \times \\
0 & \times & \times \\
\hline
0 & 0 & \times \\
0 & 0 & \times
\end{pmatrix}
=
\begin{pmatrix}
\times & \times & \times \\
0 & \times & \times \\
\hline
0 & 0 & \times \\
0 & 0 & 0
\end{pmatrix}
$$

and the result

$$H_3 H_2 H_1 A = R,$$

an upper triangular matrix. Multiplying on the left by the inverses of the Householder reflectors allows us to rewrite the result as

$$A = H_1 H_2 H_3 R = QR,$$

where $Q = H_1 H_2 H_3$. Note that $H_i^{-1} = H_i$ since H_i is symmetric orthogonal. A MATLAB implementation of the Householder method to calculate the QR factorization follows.

MATLAB code shown here can be found at bit.ly/2AhFIeA

```
% Program 4.2  Orthogonalization by Householder reflectors
% Input: mxn matrix A with linearly independent columns
% Output: orthogonal matrix Q, upper triangular R
% Example usage: [Q,R]=houseqr(A)
function [Q,R]=houseqr(A)
[m,n]=size(A);
Q=eye(m,m);
for i=1:min(n,m-1)
  x=A(i:m,i);
  w=[-sign(x(1))*norm(x);zeros(m-i,1)];
  v=w-x;
  H=eye(m,m);
  H(i:m,i:m)=eye(m-i+1,m-i+1)-2*v*v'/(v'*v);
  Q=Q*H;
  A-H*A;
end
R=A;
```

► **EXAMPLE 4.18** Use Householder reflectors to find the QR factorization of

$$
A = \begin{bmatrix} 3 & 1 \\ 4 & 3 \end{bmatrix}.
$$

We need to find a Householder reflector that moves the first column $[3, 4]$ onto the x-axis. We found such a reflector H_1 in Example 4.17, and

$$
H_1 A = \begin{bmatrix} 0.6 & 0.8 \\ 0.8 & -0.6 \end{bmatrix} \begin{bmatrix} 3 & 1 \\ 4 & 3 \end{bmatrix} = \begin{bmatrix} 5 & 3 \\ 0 & -1 \end{bmatrix}.
$$

Multiplying both sides on the left by $H_1^{-1} = H_1$ yields

$$
A = \begin{bmatrix} 3 & 1 \\ 4 & 3 \end{bmatrix} = \begin{bmatrix} 0.6 & 0.8 \\ 0.8 & -0.6 \end{bmatrix} \begin{bmatrix} 5 & 3 \\ 0 & -1 \end{bmatrix} = QR,
$$

where $Q = H_1^T = H_1$.   ◄

► **EXAMPLE 4.19** Use Householder reflectors to find the QR factorization of $A = \begin{bmatrix} 1 & -4 \\ 2 & 3 \\ 2 & 2 \end{bmatrix}$.

We need to find a Householder reflector that moves the first column $x = [1, 2, 2]$ to the vector $w = [||x||_2, 0, 0]$. Set $v = w - x = [3, 0, 0] - [1, 2, 2] = [2, -2, -2]$. Referring to Theorem 4.4, we have

$$H_1 = \begin{bmatrix} 1 & 0 & 0 \\ 0 & 1 & 0 \\ 0 & 0 & 1 \end{bmatrix} - \frac{2}{12} \begin{bmatrix} 4 & -4 & -4 \\ -4 & 4 & 4 \\ -4 & 4 & 4 \end{bmatrix} = \begin{bmatrix} \frac{1}{3} & \frac{2}{3} & \frac{2}{3} \\ \frac{2}{3} & \frac{1}{3} & -\frac{2}{3} \\ \frac{2}{3} & -\frac{2}{3} & \frac{1}{3} \end{bmatrix}$$

and

$$H_1 A = \begin{bmatrix} \frac{1}{3} & \frac{2}{3} & \frac{2}{3} \\ \frac{2}{3} & \frac{1}{3} & -\frac{2}{3} \\ \frac{2}{3} & -\frac{2}{3} & \frac{1}{3} \end{bmatrix} \begin{bmatrix} 1 & -4 \\ 2 & 3 \\ 2 & 2 \end{bmatrix} = \begin{bmatrix} 3 & 2 \\ 0 & -3 \\ 0 & -4 \end{bmatrix}.$$

The remaining step is to move the vector $\hat{x} = [-3, -4]$ to $\hat{w} = [5, 0]$. Calculating $\hat{H}_2$ from Theorem 4.4 yields

$$\begin{bmatrix} -0.6 & -0.8 \\ -0.8 & 0.6 \end{bmatrix} \begin{bmatrix} -3 \\ -4 \end{bmatrix} = \begin{bmatrix} 5 \\ 0 \end{bmatrix},$$

leading to

$$H_2 H_1 A = \begin{bmatrix} 1 & 0 & 0 \\ 0 & -0.6 & -0.8 \\ 0 & -0.8 & 0.6 \end{bmatrix} \begin{bmatrix} \frac{1}{3} & \frac{2}{3} & \frac{2}{3} \\ \frac{2}{3} & \frac{1}{3} & -\frac{2}{3} \\ \frac{2}{3} & -\frac{2}{3} & \frac{1}{3} \end{bmatrix} \begin{bmatrix} 1 & -4 \\ 2 & 3 \\ 2 & 2 \end{bmatrix} = \begin{bmatrix} 3 & 2 \\ 0 & 5 \\ 0 & 0 \end{bmatrix} = R.$$

Multiplying both sides on the left by $H_1^{-1} H_2^{-1} = H_1 H_2$ yields the QR factorization:

$$\begin{bmatrix} 1 & -4 \\ 2 & 3 \\ 2 & 2 \end{bmatrix} = H_1 H_2 R = \begin{bmatrix} \frac{1}{3} & \frac{2}{3} & \frac{2}{3} \\ \frac{2}{3} & \frac{1}{3} & -\frac{2}{3} \\ \frac{2}{3} & -\frac{2}{3} & \frac{1}{3} \end{bmatrix} \begin{bmatrix} 1 & 0 & 0 \\ 0 & -0.6 & -0.8 \\ 0 & -0.8 & 0.6 \end{bmatrix} \begin{bmatrix} 3 & 2 \\ 0 & 5 \\ 0 & 0 \end{bmatrix}$$

$$= \begin{bmatrix} 1/3 & -14/15 & -2/15 \\ 2/3 & 1/3 & -2/3 \\ 2/3 & 2/15 & 11/15 \end{bmatrix} \begin{bmatrix} 3 & 2 \\ 0 & 5 \\ 0 & 0 \end{bmatrix} = QR.$$

Compare this result with the factorization from Gram–Schmidt orthogonalization in Example 4.13. ◄

The QR factorization is not unique for a given $m \times n$ matrix A. For example, define $D = \text{diag}(d_1, \ldots, d_m)$, where each d_i is either $+1$ or -1. Then $A = QR = QDDR$, and we check that QD is orthogonal and DR is upper triangular.

Exercise 12 asks for an operation count of QR factorization by Householder reflections, which comes out to $(2/3)m^3$ multiplications and the same number of additions—lower complexity than Gram–Schmidt orthogonalization. Moreover, the Householder method is known to deliver better orthogonality in the unit vectors and has lower memory requirements. For these reasons, it is the method of choice for factoring typical matrices into QR.

▶ **ADDITIONAL EXAMPLES**

*1 Find the QR factorization of the matrix $\begin{bmatrix} 4 & -1 & 1 \\ -2 & 2 & 7 \\ 4 & 2 & -2 \end{bmatrix}$ by classical Gram–Schmidt.

2. Use MATLAB to apply each of the three methods, classical Gram–Schmidt, modified Gram–Schmidt, and Householder, to orthogonalize the 4×3 matrix

$$A = \begin{bmatrix} 1 & 1 & 1 \\ \delta & \delta/2 & \delta/3 \\ \delta/2 & \delta/3 & \delta/4 \\ \delta/3 & \delta/4 & \delta/5 \end{bmatrix}$$

where $\delta = 10^{-10}$. Compare accuracy of the results by computing $Q^T Q$ for each method.

⌨ **Solutions** for Additional Examples can be found at bit.ly/2Ow82mh (* example with video solution)

4.3 Exercises

⌨ **Solutions** for Exercises numbered in blue can be found at bit.ly/2P0yAM7

1. Apply classical Gram–Schmidt orthogonalization to find the full QR factorization of the following matrices:

   (a) $\begin{bmatrix} 4 & 0 \\ 3 & 1 \end{bmatrix}$ (b) $\begin{bmatrix} 1 & 2 \\ 1 & 1 \end{bmatrix}$ (c) $\begin{bmatrix} 2 & 1 \\ 1 & -1 \\ 2 & 1 \end{bmatrix}$ (d) $\begin{bmatrix} 4 & 8 & 1 \\ 0 & 2 & -2 \\ 3 & 6 & 7 \end{bmatrix}$

2. Apply classical Gram–Schmidt orthogonalization to find the full QR factorization of the following matrices:

   (a) $\begin{bmatrix} 2 & 3 \\ -2 & -6 \\ 1 & 0 \end{bmatrix}$ (b) $\begin{bmatrix} -4 & -4 \\ -2 & 7 \\ 4 & -5 \end{bmatrix}$

3. Apply modified Gram–Schmidt orthogonalization to find the full QR factorization of the matrices in Exercise 1.

4. Apply modified Gram–Schmidt orthogonalization to find the full QR factorization of the matrices in Exercise 2.

5. Apply Householder reflectors to find the full QR factorization of the matrices in Exercise 1.

6. Apply Householder reflectors to find the full QR factorization of the matrices in Exercise 2.

7. Use the QR factorization from Exercise 2, 4, or 6 to solve the least squares problem.

   (a) $\begin{bmatrix} 2 & 3 \\ -2 & -6 \\ 1 & 0 \end{bmatrix} \begin{bmatrix} x_1 \\ x_2 \end{bmatrix} = \begin{bmatrix} 3 \\ -3 \\ 6 \end{bmatrix}$ (b) $\begin{bmatrix} -4 & -4 \\ -2 & 7 \\ 4 & -5 \end{bmatrix} \begin{bmatrix} x_1 \\ x_2 \end{bmatrix} = \begin{bmatrix} 3 \\ 9 \\ 0 \end{bmatrix}$

8. Find the QR factorization and use it to solve the least squares problem.

   (a) $\begin{bmatrix} 1 & 4 \\ -1 & 1 \\ 1 & 1 \\ 1 & 0 \end{bmatrix} \begin{bmatrix} x_1 \\ x_2 \end{bmatrix} = \begin{bmatrix} 3 \\ 1 \\ 1 \\ -3 \end{bmatrix}$ (b) $\begin{bmatrix} 2 & 4 \\ 0 & -1 \\ 2 & -1 \\ 1 & 3 \end{bmatrix} \begin{bmatrix} x_1 \\ x_2 \end{bmatrix} = \begin{bmatrix} -1 \\ 3 \\ 2 \\ 1 \end{bmatrix}$

9. Prove that a square matrix is orthogonal if and only if its columns are pairwise orthogonal unit vectors.

10. Prove that the product of two orthogonal $m \times m$ matrices is again orthogonal.

11. Show that the Gram–Schmidt orthogonalization of an $m \times m$ matrix requires approximately m^3 multiplications and m^3 additions.

12. Show that the Householder reflector method for the QR factorization requires approximately $(2/3)m^3$ multiplications and $(2/3)m^3$ additions.

13. Let P be the matrix defined in (1.29). Show (a) $P^2 = P$ (b) P is symmetric (c) $Pv = v$.

14. Prove that Householder reflectors are symmetric matrices.

15. Verify that classical and modified Gram–Schmidt are mathematically identical (in exact arithmetic).

4.3 Computer Problems

Solutions for Computer Problems numbered in blue can be found at `bit.ly/2QXa0c8`

1. Write a MATLAB program that implements the modified Gram–Schmidt method to find the reduced QR factorization. Only one line from the classical version needs to be changed. Check your work by comparing factorizations of the matrices in Exercise 1 with the MATLAB `qr(A,0)` command or equivalent. The factorization is unique up to signs of the entries of Q and R.

2. Apply the classical Gram–Schmidt, modified Gram–Schmidt, and Householder versions of orthogonalization to the matrix in Example 4.16. Reproduce the theoretical results of the example. Calculate $Q^T Q$ to check orthogonality, and rank the three approaches in terms of accuracy.

3. (a) Consider the $(n + 1) \times n$ matrix composed of $\delta I_{n \times n}$, a scalar multiple of the identity matrix, with a row of ones placed across the top. This is the general analogue of Example 4.16. Repeat Computer Problem 2 for the matrix with $\delta = 10^{-10}$, $n = 5, 10$, and 50. (b) Repeat (a), replacing $\delta I_{n \times n}$ with δH, where H is the $n \times n$ Hilbert matrix. Compare the results with (a).

4. Write a MATLAB program that implements (a) classical and (b) modified Gram–Schmidt to find the full QR factorization. Check your work by comparing factorizations of the matrices in Exercise 1 with the MATLAB `qr(A)` command or equivalent.

5. Use the MATLAB QR factorization to find the least squares solutions and 2-norm error of the following inconsistent systems:

(a) $\begin{bmatrix} 1 & 1 \\ 2 & 1 \\ 1 & 2 \\ 0 & 3 \end{bmatrix} \begin{bmatrix} x_1 \\ x_2 \end{bmatrix} = \begin{bmatrix} 3 \\ 5 \\ 5 \\ 5 \end{bmatrix}$ (b) $\begin{bmatrix} 1 & 2 & 2 \\ 2 & -1 & 2 \\ 3 & 1 & 1 \\ 1 & 1 & -1 \end{bmatrix} \begin{bmatrix} x_1 \\ x_2 \\ x_3 \end{bmatrix} = \begin{bmatrix} 10 \\ 5 \\ 10 \\ 3 \end{bmatrix}$

6. Use the MATLAB QR factorization to find the least squares solutions and 2-norm error of the following inconsistent systems:

(a) $\begin{bmatrix} 3 & -1 & 2 \\ 4 & 1 & 0 \\ -3 & 2 & 1 \\ 1 & 1 & 5 \\ -2 & 0 & 3 \end{bmatrix} \begin{bmatrix} x_1 \\ x_2 \\ x_3 \end{bmatrix} = \begin{bmatrix} 10 \\ 10 \\ -5 \\ 15 \\ 0 \end{bmatrix}$ (b) $\begin{bmatrix} 4 & 2 & 3 & 0 \\ -2 & 3 & -1 & 1 \\ 1 & 3 & -4 & 2 \\ 1 & 0 & 1 & -1 \\ 3 & 1 & 3 & -2 \end{bmatrix} \begin{bmatrix} x_1 \\ x_2 \\ x_3 \\ x_4 \end{bmatrix} = \begin{bmatrix} 10 \\ 0 \\ 2 \\ 0 \\ 5 \end{bmatrix}$

7. Let A be the $10 \times n$ matrix formed by the first n columns of the 10×10 Hilbert matrix. Let c be the n-vector $1, \ldots, 1$, and set $b = Ac$. Use the QR factorization to solve the least squares problem $Ax = b$ for (a) $n = 6$ (b) $n = 8$, and compare with the correct least squares solution $\bar{x} = c$. How many correct decimal places can be computed? See Computer Problem 4.1.8, where the normal equations are used.

8. Let $x_1, \ldots, x_{11}$ be 11 evenly spaced points in $[2, 4]$ and $y_i = 1 + x_i + x_i^2 + \cdots + x_i^d$. Use the QR factorization to compute the best degree d polynomial, where (a) $d = 5$ (b) $d = 6$ (c) $d = 8$. Compare with Example 4.5 and Computer Problem 4.1.9. How many correct decimal places of the coefficients can be computed?

4.4 Generalized Minimum Residual (GMRES) Method

In Chapter 2, we saw that the Conjugate Gradient Method can be viewed as an iterative method specially designed to solve the matrix system $Ax = b$ for a symmetric square matrix A. If A is not symmetric, the conjugate gradient theory fails. However, there are several alternatives that work for the nonsymmetric problem. One of the most popular is the Generalized Minimum Residual Method, or GMRES for short. This method is a good choice for the solution of large, sparse, nonsymmetric linear systems $Ax = b$.

At first sight, it might seem strange to be discussing a method for solving linear systems in the chapter on least squares. Why should orthogonality matter to a problem that has no apparent connection with it? The answer lies in the fact, as we found in Chapter 2, that matrices with almost-parallel column vectors tend to be ill-conditioned, which in turn causes great magnification of error in solving $Ax = b$.

In fact, orthogonalization is built into GMRES in two separate ways. First, the backward error of the system is minimized at each iteration step using a least squares formulation. Second and more subtle, the basis of the search space is reorthogonalized at each step in order to avoid inaccuracy from ill-conditioning. GMRES is an interesting example of a method that exploits ideas of orthogonality in places where they are not obviously present.

4.4.1 Krylov methods

GMRES is a member of the family of **Krylov methods**. These methods rely on accurate computation of the **Krylov space**, which is the vector space spanned by $\{r, Ar, \ldots, A^k r\}$, where $r = b - Ax_0$ is the residual vector of the initial guess. Since the vectors $A^k r$ tend toward a common direction for large k, a basis for the Krylov space must be calculated carefully. Finding an accurate basis for the Krylov space requires the use of orthogonalization methods like Gram–Schmidt or Householder reflections.

The idea behind GMRES is to look for improvements to the initial guess x_0 in a particular vector space, the Krylov space spanned by the residual r and its products under the nonsingular matrix A. At step k of the method, we enlarge the Krylov space by adding $A^k r$, reorthogonalize the basis, and then use least squares to find the best improvement to add to x_0.

Generalized Minimum Residual Method (GMRES)

$x_0 = $ initial guess
$r = b - Ax_0$
$q_1 = r/||r||_2$
for $k = 1, 2, \ldots, m$
    $y = Aq_k$
    for $j = 1, 2, \ldots, k$
        $h_{jk} = q_j^T y$
        $y = y - h_{jk} q_j$

> **end**
> $h_{k+1,k} = ||y||_2$ (If $h_{k+1,k} = 0$, skip next line and terminate at bottom.)
> $q_{k+1} = y/h_{k+1,k}$
> Minimize $||Hc_k - [||r||_2 \ 0 \ 0 \ldots \ 0]^T||_2$ for c_k
> $x_k = Q_k c_k + x_0$
> **end**

The iterates x_k are approximate solutions to the system $Ax = b$. In the kth step of the pseudocode, the matrix H is a $(k+1) \times k$ matrix. The minimization step that yields c is a least squares problem of $k + 1$ equations in k unknowns that can be solved using techniques of this chapter. The matrix Q_k in the code is $n \times k$, consisting of the k orthonormal columns $q_1, \ldots, q_k$. If $h_{k+1,k} = 0$, then step k is the final step and the minimization will arrive at the exact solution of $Ax = b$.

SPOTLIGHT ON

> ## Orthogonality
> GMRES is our first example of a Krylov method, which depends on accurate calculation of the Krylov space. We found in Chapter 2 that nearly parallel column vectors of a matrix cause ill-conditioning. The defining vectors $A^k r$ of the Krylov space tend to become more parallel as k grows, so the use of the orthogonalization techniques of Section 4.3 is essential to build stable, efficient algorithms like GMRES.

To approximate the space, the most direct approach is not the best. In Chapter 12, we will exploit the fact that the vectors $A^k r$ asymptotically tend toward the same direction to compute eigenvalues. In order to generate an efficient basis for the Krylov space $\{r, Ar, \ldots, A^k r\}$, we rely on the power of Gram–Schmidt orthogonalization as the simplest approach.

The application of modified Gram–Schmidt to $\{r, Ar, \ldots, A^k r\}$, beginning with $q_1 = r/||r||_2$, is carried out in the inner loop of the pseudocode. It results in the matrix equality $AQ_k = Q_{k+1} H_k$, or

$$
\left[A \middle| \begin{array}{ccc} q_1 & \cdots & q_k \end{array} \right]
=
\left[\begin{array}{cccc} q_1 & \cdots & q_k & q_{k+1} \end{array} \right]
\begin{bmatrix}
h_{11} & h_{12} & \cdots & h_{1k} \\
h_{21} & h_{22} & \cdots & h_{2k} \\
 & h_{32} & \cdots & h_{3k} \\
 & & \ddots & \vdots \\
 & & & h_{k+1,k}
\end{bmatrix}.
$$

Here A is $n \times n$, Q_k is $n \times k$, and H_k is $(k+1) \times k$. In most cases, k will be much smaller than n.

The columns of Q_k span the k-dimensional Krylov space that will be searched for additions x_{add} to the original approximation x_0. Vectors in this space are written as $x_{\text{add}} = Q_k c$. To minimize the residual

$$b - A(x_0 + x_{\text{add}}) = r - Ax_{\text{add}},$$

of the original problem $Ax = b$ means finding c that minimizes

$$||Ax_{\text{add}} - r||_2 = ||AQ_k c - r||_2 = ||Q_{k+1} H_k c - r||_2 = ||H_k c - Q_{k+1}^T r||_2,$$

where the last equality follows from the norm-preserving property of orthonormal columns. Note that $Q_{k+1}^T r = [||r||_2 \ 0 \ 0 \ldots 0]^T$, since $q_1 = r/||r||_2$ as noted above, and all but the first column of Q_{k+1} is orthogonal to r. The least squares problem is now

$$
\begin{bmatrix}
h_{11} & h_{12} & \cdots & h_{1k} \\
h_{21} & h_{22} & \cdots & h_{2k} \\
 & h_{32} & \cdots & h_{3k} \\
 & & \ddots & \vdots \\
 & & & h_{k+1,k}
\end{bmatrix}
\begin{bmatrix}
c_1 \\ c_2 \\ \\ \vdots \\ \\ c_k
\end{bmatrix}
=
\begin{bmatrix}
\|r\|_2 \\ 0 \\ \\ \vdots \\ \\ 0
\end{bmatrix}.
$$

Using the least squares solution c gives the kth step approximate solution $x_k = x_0 + x_{\text{add}} = x_0 + Q_k c$ to the original problem $Ax = b$.

It is important to note the respective sizes of the subproblems in GMRES. The part of the algorithm with the highest computational complexity is the least squares computation, which minimizes the error of $k + 1$ equations in k unknowns. The size k will be small compared to the total problem size n in most applications. In the special case $h_{k+1,k} = 0$, the least squares problem becomes square, and the approximate solution x_k is exact.

A convenient feature of GMRES is that the backward error $\|b - Ax_k\|_2$ decreases monotonically with k. The reason is clear from the fact that the least squares problem in step k minimizes $\|r - Ax_{\text{add}}\|_2$ for x_{add} in the k-dimensional Krylov space. As GMRES proceeds, the Krylov space is enlarged, so the next approximation cannot do worse.

Concerning the above GMRES pseudocode, several other implementation details are worth mentioning. First, note that the least squares minimization step is only warranted when an approximate solution x_k is needed. Therefore it may be done only intermittently, in order to monitor progress toward the solution, or at the extreme, the least squares computation can be taken out of the loop and done only at the end, since $x_{\text{add}} = Q_k c$ does not depend on previous least squares calculations. This corresponds to moving the final **end** statement above the previous two lines. Second, the Gram–Schmidt orthogonalization step carried out in the inner loop can be substituted with Householder orthogonalization at slightly increased computational complexity, if conditioning is a significant issue.

The typical use of GMRES is for a large and sparse $n \times n$ matrix A. In theory, the algorithm terminates after n steps at the correct solution x as long as A is nonsingular. In most cases, however, the goal is to run the method for k steps, where k is much smaller than n. Note that the matrix Q_k is $n \times k$ and not guaranteed to be sparse. Thus memory considerations may also limit the number k of GMRES steps.

These conditions lead to a variation of the algorithm known as **Restarted GMRES**. If not enough progress is made toward the solution after k iterations, and if the $n \times k$ matrix Q_k is becoming too large to handle, the idea is simple: Discard Q_k and start GMRES from the beginning, *using the current best guess x_k as the new x_0*.

The GMRES algorithm, with options including restart, is included in MATLAB with the command `gmres`.

4.4.2 Preconditioned GMRES

The concept behind preconditioning GMRES is very similar to the conjugate gradient case. Begin with a nonsymmetric linear system $Ax = b$. We again try to solve $M^{-1}Ax = M^{-1}b$, where M is one of the preconditioners discussed in Section 2.6.4.

Very few changes need to be made to the GMRES pseudocode of the previous section. In the preconditioned version, the starting residual is now $r = M^{-1}(b - Ax_0)$. The Krylov space iteration step is changed to $w = M^{-1}Aq_k$. Note that neither of these steps require the explicit formation of M^{-1}. They should be carried out by back substitution, assuming that M is in a simple or factored form. With these changes, the resulting algorithm is as follows.

Preconditioned GMRES

$x_0 = $ initial guess
$r = M^{-1}(b - Ax_0)$
$q_1 = r/||r||_2$
for $k = 1, 2, \ldots, m$
$\quad w = M^{-1}Aq_k$
$\quad$ **for** $j = 1, 2, \ldots, k$
$\quad\quad h_{jk} = w^T q_j$
$\quad\quad w = w - h_{jk}q_j$
$\quad$ **end**
$\quad h_{k+1,k} = ||w||_2$
$\quad q_{k+1} = w/h_{k+1,k}$
$\quad$ Minimize $||Hc_k - [||r||_2\, 0\, 0 \ldots 0]^T||_2$ for c_k
$\quad x_k = Qc_k + x_0$
end

▶ **EXAMPLE 4.20**   Let A denote the matrix with diagonal entries $A_{ii} = \sqrt{i}$ for $i = 1, \ldots, n$ and $A_{i,i+10} = \cos i$, $A_{i+10,i} = \sin i$ for $i = 1, \ldots, n - 10$, with all other entries zero. Set x to be the vector of n ones, and define $b = Ax$. For $n = 500$, solve $Ax = b$ with GMRES in three ways: using no preconditioner, using the Jacobi preconditioner, and using the Gauss–Seidel preconditioner.

The matrix can be defined in MATLAB by

```
A=diag(sqrt(1:n))+diag(cos(1:(n-10)),10)
                +diag(sin(1:(n-10)),-10).
```

Figure 4.12 shows the three different results. GMRES is slow to converge without preconditioning. The Jacobi preconditioner makes a significant improvement, and GMRES with the Gauss–Seidel preconditioner requires only about 10 steps to reach machine accuracy.

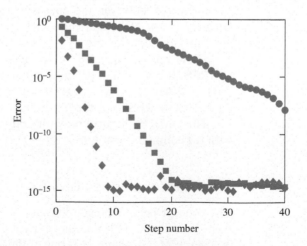

Figure 4.12 Efficiency of preconditioned GMRES Method for the solution of Example 4.20. Error is plotted by step number. Circles: no preconditioner. Squares: Jacobi preconditioner. Diamonds: Gauss–Seidel preconditioner.

◀

► **ADDITIONAL EXAMPLES**

1. Use the GMRES algorithm to solve the system $\begin{bmatrix} 0 & 1 & 0 \\ 1 & 1 & 1 \\ -1 & 1 & 1 \end{bmatrix} \begin{bmatrix} x_1 \\ x_2 \\ x_3 \end{bmatrix} = \begin{bmatrix} 0 \\ 1 \\ 0 \end{bmatrix}$.

2. Let A be the $n \times n$ matrix with $n = 100$ and entries ($A(i,i) = i$, $A(i,i+1) = 3$, $A(i,i+2) = -2$, $A(i+1,i) = -7$, $A(i+2,i) = 5$) for all i that fit within the matrix. Let x_c be the vector of n ones, and set $b = Ax_c$. Apply GMRES without preconditioner, with the Jacobi preconditioner, and with the Gauss–Seidel preconditioner to solve $Ax = b$. Compare errors of the three runs.

⌐□ **Solutions** for Additional Examples can be found at `bit.ly/2yo91L9`

4.4 Exercises

⌐□ **Solutions** for Exercises numbered in blue can be found at `bit.ly/2NJ9DQp`

1. Solve $Ax = b$ for the following A and $b = [1,0,0]^T$, using GMRES with $x_0 = [0,0,0]^T$. Report all approximations x_k up to and including the correct solution.

$$(a) \begin{bmatrix} 1 & 1 & 0 \\ 0 & 1 & 0 \\ 1 & 1 & 1 \end{bmatrix} \quad (b) \begin{bmatrix} 1 & 1 & 0 \\ -1 & 1 & 2 \\ 0 & 0 & 1 \end{bmatrix} \quad (c) \begin{bmatrix} 0 & 0 & 1 \\ 1 & 0 & 0 \\ 0 & 1 & 0 \end{bmatrix}$$

2. Repeat Exercise 1 with $b = [0,0,1]^T$.

3. Let $A = \begin{bmatrix} 1 & 0 & a_{13} \\ 0 & 1 & a_{23} \\ 0 & 0 & 1 \end{bmatrix}$. Prove that for any x_0 and b, GMRES converges to the exact solution after two steps.

4. Generalize Exercise 3 by showing that for $A = \begin{bmatrix} I & C \\ 0 & I \end{bmatrix}$ and any x_0 and b, GMRES converges to the exact solution after two steps. Here C is an $m_1 \times m_2$ submatrix, 0 denotes the $m_2 \times m_1$ matrix of zeros, and I denotes the appropriate-sized identity matrix.

4.4 Computer Problems

⌐□ **Solutions** for Computer Problems numbered in blue can be found at `bit.ly/2R0XKHA`

1. Let A be the $n \times n$ matrix with $n = 1000$ and entries $A(i,i) = i$, $A(i,i+1) = A(i+1,i) = 1/2$, $A(i,i+2) = A(i+2,i) = 1/2$ for all i that fit within the matrix. (a) Print the nonzero structure spy(A). (b) Let x_e be the vector of n ones. Set $b = Ax_e$, and apply the GMRES Method, without preconditioner, with the Jacobi preconditioner, and with the Gauss–Seidel preconditioner. Compare errors of the three runs in a plot versus step number.

2. Let $n = 1000$. Start with the $n \times n$ matrix A from Computer Problem 1, and add the nonzero entries $A(i,2i) = A(2i,i) = 1/2$ for $1 \le i \le n/2$. Carry out steps (a) and (b) as in that problem.

3. Let $n = 500$, and let A be the $n \times n$ matrix with entries $A(i,i) = 2$, $A(i,i+2) = A(i+2,i) = 1/2$, $A(i,i+4) = A(i+4,i) = 1/2$ for all i, and $A(500,i) = A(i,500) = -0.1$ for $1 \le i \le 495$. Carry out steps (a) and (b) as in Computer Problem 1.

4. Let A be the matrix from Computer Problem 3, but with the diagonal elements replaced by $A(i,i) = \sqrt[3]{i}$. Carry out parts (a) and (b) as in that problem.

5. Let C be the 195×195 matrix block with $C(i,i) = 2$, $C(i,i+3) = C(i+3,i) = 0.1$, $C(i,i+39) = C(i+39,i) = 1/2$, $C(i,i+42) = C(i+42,i) = 1/2$ for all i. Define A to be the $n \times n$ matrix with $n = 780$ formed by four diagonally arranged blocks C, and with blocks $\frac{1}{2}C$ on the super- and subdiagonal. Carry out steps (a) and (b) as in Computer Problem 1 to solve $Ax = b$.

4.5 NONLINEAR LEAST SQUARES

The least squares solution of a linear system of equations $Ax = b$ minimizes the Euclidean norm of the residual $||Ax - b||_2$. We have learned two methods to find the solution $\bar{x}$, one based on the normal equations and another on the QR factorization.

Neither method can be applied if the equations are nonlinear. In this section, we develop the Gauss–Newton Method for solving nonlinear least squares problems. In addition to illustrating the use of the method to solve circle intersection problems, we apply Gauss–Newton to fitting models with nonlinear coefficients to data.

4.5.1 Gauss–Newton Method

Consider the system of m equations in n unknowns

$$r_1(x_1, \ldots, x_n) = 0$$
$$\vdots$$
$$r_m(x_1, \ldots, x_n) = 0. \tag{4.31}$$

The sum of the squares of the errors is represented by the function

$$E(x_1, \ldots, x_n) = \frac{1}{2}(r_1^2 + \cdots + r_m^2) = \frac{1}{2}r^T r,$$

where $r = [r_1, \ldots, r_m]^T$. The constant $1/2$ has been included in the definition to simplify later formulas. To minimize E, we set the gradient $F(x) = \nabla E(x)$ to zero:

$$0 = F(x) = \nabla E(x) = \nabla\left(\frac{1}{2}r(x)^T r(x)\right) = r(x)^T Dr(x). \tag{4.32}$$

Observe that we have used the dot product rule for the gradient (see Appendix A).

We begin by recalling Multivariate Newton's Method, and apply it to the function viewed as a column vector $F(x)^T = (r^T Dr)^T = (Dr)^T r$. The matrix/vector product rule (see Appendix A) can be applied to yield

$$DF(x)^T = D((Dr)^T r) = (Dr)^T \cdot Dr + \sum_{i=1}^{m} r_i Dc_i,$$

where c_i is the ith column of Dr. Note that $Dc_i = H_{r_i}$, the matrix of second partial derivatives, or **Hessian**, of r_i:

$$H_{r_i} = \begin{bmatrix} \frac{\partial^2 r_i}{\partial x_1 \partial x_1} & \cdots & \frac{\partial^2 r_i}{\partial x_1 \partial x_n} \\ \vdots & & \vdots \\ \frac{\partial^2 r_i}{\partial x_n \partial x_1} & \cdots & \frac{\partial^2 r_i}{\partial x_n \partial x_n} \end{bmatrix}.$$

The application of Newton's Method can be simplified by dropping some of the terms. Without the above m-term summation, we have the following.

Gauss–Newton Method

To minimize

$$r_1(x)^2 + \cdots + r_m(x)^2.$$

Set $x^0 = $ initial vector,
for $k = 0, 1, 2, \ldots$

$$A = Dr(x^k) \tag{4.33}$$
$$A^T A v^k = -A^T r(x^k)$$
$$x^{k+1} = x^k + v^k \tag{4.34}$$

end

Notice that each step of the Gauss–Newton Method is reminiscent of the normal equations, where the coefficient matrix has been replaced by Dr. The Gauss–Newton Method solves for a root of the gradient of the squared error. Although the gradient must be zero at the minimum, the converse is not true, so it is possible for the method to converge to a maximum or a neutral point. Caution must be used in interpreting the algorithm's result.

The following three examples illustrate use of the Gauss–Newton Method, as well as Multivariate Newton's Method of Chapter 2. Two intersecting circles intersect in one or two points, unless the circles coincide. Three circles in the plane, however, typically have no points of common intersection. In such a case, we can ask for the point in the plane that comes closest to being an intersection point in the sense of least squares. For three circles, this is a question of three nonlinear equations in the two unknowns x, y.

Example 4.21 shows how the Gauss–Newton Method solves this nonlinear least squares problem. Example 4.22 defines the best point in a different way: Find the unique point of intersection of the 3 circles, allowing their radii to be changed by a common amount K. This is a question of three equations in three unknowns x, y, K, not a least squares problem, and is solved using Multivariate Newton's Method.

Finally, Example 4.23 adds a fourth circle. The solution of four equations in the three unknowns x, y, K is again a least squares problem that requires Gauss–Newton. This last formulation is relevant to calculations in GPS, as shown in Reality Check 4.

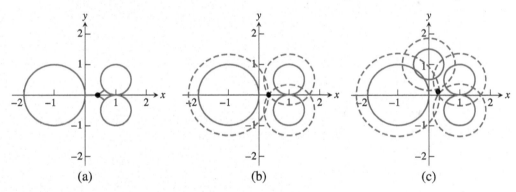

Figure 4.13 Near-intersection points of three circles. (a) The least squares near-intersection point, found by the Gauss-Newton Method. (b) Expanding the radii by a common amount gives a different type of near-intersection point by Multivariate Newton's Method. (c) The four circles of Example 4.23 with least squares solution point found by the Gauss-Newton Method.

▶ **EXAMPLE 4.21**   Consider the three circles in the plane with centers $(x_1, y_1) = (-1, 0), (x_2, y_2) = (1, 1/2), (x_3, y_3) = (1, -1/2)$ and radii $R_1 = 1, R_2 = 1/2, R_3 = 1/2$, respectively. Use the Gauss–Newton Method to find the point for which the sum of the squared distances to the three circles is minimized.

The circles are shown in Figure 4.13(a). The point (x, y) in question minimizes the sum of the squares of the residual errors:

$$r_1(x, y) = \sqrt{(x - x_1)^2 + (y - y_1)^2} - R_1$$

$$r_2(x, y) = \sqrt{(x - x_2)^2 + (y - y_2)^2} - R_2$$

$$r_3(x, y) = \sqrt{(x - x_3)^2 + (y - y_3)^2} - R_3.$$

This follows from the fact that the distance from a point (x, y) to a circle with center (x_1, y_1) and radius R_1 is $|\sqrt{(x - x_1)^2 + (y - y_1)^2} - R_1|$ (see Exercise 3). The Jacobian of $r(x, y)$ is

$$Dr(x, y) = \begin{bmatrix} \frac{x-x_1}{S_1} & \frac{y-y_1}{S_1} \\ \frac{x-x_2}{S_2} & \frac{y-y_2}{S_2} \\ \frac{x-x_3}{S_3} & \frac{y-y_3}{S_3} \end{bmatrix},$$

where $S_i = \sqrt{(x - x_i)^2 + (y - y_i)^2}$ for $i = 1, 2, 3$. The Gauss–Newton iteration with initial vector $(x^0, y^0) = (0, 0)$ converges to $(\overline{x}, \overline{y}) = (0.412891, 0)$ within six correct decimal places after seven steps. ◀

A related problem for three circles gives a different type of answer. Instead of looking for points that most resemble intersection points, we can expand (or contract) the circles' radii by a common amount until they have a common intersection. This is equivalent to solving the system

$$r_1(x, y, K) = \sqrt{(x - x_1)^2 + (y - y_1)^2} - (R_1 + K) = 0$$

$$r_2(x, y, K) = \sqrt{(x - x_2)^2 + (y - y_2)^2} - (R_2 + K) = 0$$

$$r_3(x, y, K) = \sqrt{(x - x_3)^2 + (y - y_3)^2} - (R_3 + K) = 0. \tag{4.35}$$

The point (x, y) identified in this way is in general different from the least squares solution of Example 4.21.

▶ **EXAMPLE 4.22**   Solve the system (4.35) for (x, y, K), using the circles from Example 4.21.

The system consists of three nonlinear equations in three unknowns, calling for Multivariate Newton's Method. The Jacobian is

$$Dr(x, y, K) = \begin{bmatrix} \frac{x-x_1}{S_1} & \frac{y-y_1}{S_1} & -1 \\ \frac{x-x_2}{S_2} & \frac{y-y_2}{S_2} & -1 \\ \frac{x-x_3}{S_3} & \frac{y-y_3}{S_3} & -1 \end{bmatrix}.$$

Newton's Method yields the solution $(x, y, K) = (1/3, 0, 1/3)$ in three steps. The intersection point $(1/3, 0)$ and the three circles with radii expanded by $K = 1/3$ appear in Figure 4.13(b). ◀

Examples 4.21 and 4.22 show two different viewpoints on the meaning of the "near-intersection point" of a group of circles. Example 4.23 combines the two different approaches.

▶ **EXAMPLE 4.23**   Consider the four circles with centers $(-1, 0), (1, 1/2), (1, -1/2), (0, 1)$ and radii $1, 1/2, 1/2, 1/2$, respectively. Find the point (x, y) and constant K for which the sum of the squared distances from the point to the four circles with radii increased by K (thus $1 + K, 1/2 + K, 1/2 + K, 1/2 + K$, respectively) is minimized.

This is a straightforward combination of the previous two examples. There are four equations in the three unknowns x, y, K. The least squares residual is similar to (4.35), but with four terms, and the Jacobian is

$$Dr(x, y, K) = \begin{bmatrix} \frac{x-x_1}{S_1} & \frac{y-y_1}{S_1} & -1 \\ \frac{x-x_2}{S_2} & \frac{y-y_2}{S_2} & -1 \\ \frac{x-x_3}{S_3} & \frac{y-y_3}{S_3} & -1 \\ \frac{x-x_4}{S_4} & \frac{y-y_4}{S_4} & -1 \end{bmatrix}.$$

The Gauss–Newton Method provides the solution $(\overline{x}, \overline{y}) = (0.311385, 0.112268)$ with $\overline{K} = 0.367164$, pictured in Figure 4.13(c). ◄

The analogue of Example 4.23 for spheres in three dimensions forms the mathematical foundation of the Global Positioning System (GPS). See Reality Check 4.

4.5.2 Models with nonlinear parameters

An important application of the Gauss–Newton Method is to fit models that are nonlinear in the coefficients. Let $(t_1, y_1), \ldots, (t_m, y_m)$ be data points and $y = f_c(x)$ the function to be fit, where $c = [c_1, \ldots, c_p]$ is a set of parameters to be chosen to minimize the sum of the squares of the residuals

$$r_1(c) = f_c(t_1) - y_1$$
$$\vdots$$
$$r_m(c) = f_c(t_m) - y_m.$$

This particular case of (4.31) is seen commonly enough to warrant special treatment here.

If the parameters $c_1, \ldots, c_p$ enter the model in a linear way, then this is a set of linear equations in the c_i, and the normal equations, or QR-factorization solution, gives the optimal choice of parameters c. If the parameters c_i are nonlinear in the model, the same treatment results in a system of equations that is nonlinear in the c_i. For example, fitting the model $y = c_1 t^{c_2}$ to the data points (t_i, y_i) yields the nonlinear equations

$$y_1 = c_1 t_1^{c_2}$$
$$y_2 = c_1 t_2^{c_2}$$
$$\vdots$$
$$y_m = c_1 t_m^{c_2}.$$

Because c_2 enters the model nonlinearly, the system of equations cannot be put in matrix form.

In Section 4.2, we handled this difficulty by changing the problem: We "linearized the model" by taking log of both sides of the model and minimized the error in these log-transformed coordinates by least squares. In cases where the log-transformed coordinates are really the proper coordinates in which to be minimizing error, this is appropriate.

To solve the original least squares problem, however, we turn to the Gauss–Newton Method. It is used to minimize the error function E as a function of the vector of parameters c. The matrix Dr is the matrix of partial derivatives of the errors r_i with respect to the parameters c_j, which are

$$(Dr)_{ij} = \frac{\partial r_i}{\partial c_j} = f_{c_j}(t_i).$$

With this information, the Gauss–Newton Method (4.33) can be implemented.

► **EXAMPLE 4.24** Use the Gauss–Newton Method to fit the world automobile supply data of Example 4.8 with a (nonlinearized) exponential model.

Finding the best least squares fit of the data to an exponential model means finding c_1, c_2 that minimize the RMSE for errors $r_i = c_1 e^{c_2 t_i} - y_i$, $i = 1, \ldots, m$. Using model linearization in the previous section, we minimized the RMSE for the errors of the log model $\ln y_i - (\ln c_1 + c_2 t_i)$. The values of c_i that minimize the RMSE in the two different senses are different in general.

To compute the best least squares fit by the Gauss–Newton Method, define

$$
r = \begin{bmatrix} c_1 e^{c_2 t_1} - y_1 \\ \vdots \\ c_1 e^{c_2 t_m} - y_m \end{bmatrix},
$$

and take derivatives with respect to the parameters c_1 and c_2 to get

$$
Dr = - \begin{bmatrix} e^{c_2 t_1} & c_1 t_1 e^{c_2 t_1} \\ \vdots & \vdots \\ e^{c_2 t_m} & c_1 t_m e^{c_2 t_m} \end{bmatrix}.
$$

SPOTLIGHT ON

Convergence   Nonlinearity in least squares problems causes extra challenges. The normal equations and QR approach find the single solution as long as the coefficient matrix A has full rank. On the other hand, Gauss–Newton iteration applied to a nonlinear problem may converge to one of several different relative minima of the least squares error. Using a reasonable approximation for the initial vector, if available, aids convergence to the absolute minimum.

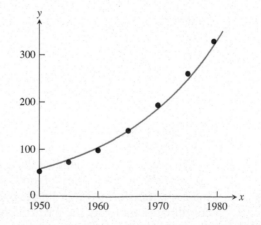

Figure 4.14 Exponential fit of world automobile supply data, without using linearization. The best least squares fit is $y = 58.51 e^{0.05772t}$.

This model is fit with the world automobile supply data, where t is measured in years since 1970, and cars in millions. Five steps of the Gauss–Newton Method (4.33) from initial guess $(c_1, c_2) = (50, 0.1)$ yields $(c_1, c_2) \approx (58.51, 0.05772)$ with four digits of precision. The best least squares exponential model for the data is

$$
y = 58.51 e^{0.05772t}. \tag{4.36}
$$

The RMSE is 7.68, meaning an average modeling error, in the least squares sense, of 7.68 million cars (see Figure 4.14).

The best model (4.36) can be compared with the best linearized exponential model

$$y = 54.03e^{0.06152t}$$

calculated in Example 4.8. This was obtained from the normal equations applied to the linearized model $\ln y = \ln c_1 + c_2 t$. The RMSE of the errors r_i of the linearized model is 9.56, greater than the RMSE of (4.36), as necessary. However, the linearized model minimizes the RMSE of the errors $\ln y_i - (\ln c_1 + c_2 t_i)$, giving a value of 0.0357, lower than the corresponding value 0.0568 for model (4.36), also as required. Each of the models is the optimal fit in its data space.

The moral is that there are computational algorithms for solving either problem. Minimizing the r_i is the standard least squares problem, but the user must decide on the basis of the data context whether it is more appropriate to minimize errors or log errors. ◄

4.5.3 The Levenberg–Marquardt Method

Least squares minimization is especially challenging when the coefficient matrix turns out to be ill-conditioned. In Example 4.5, large errors were encountered in the least squares solution of $Ax = b$ when using the normal equations, since $A^T A$ had large condition number.

The problem is often worse for nonlinear least squares minimization. Many plausible model definitions yield poorly conditioned Dr matrices. The Levenberg–Marquardt Method uses a "regularization term" to partially remedy the conditioning problem. It can be thought of as a mixture of Gauss–Newton and the steepest descent method, which will be introduced for general optimization problems in Chapter 13.

The algorithm is a simple modification of the Gauss–Newton Method.

Levenberg-Marquardt Method

To minimize

$$r_1(x)^2 + \cdots + r_m(x)^2.$$

Set x^0 = initial vector, λ = constant
for $k = 0, 1, 2, \ldots$

$$A = Dr(x^k)$$
$$(A^T A + \lambda \operatorname{diag}(A^T A))v^k = -A^T r(x^k)$$
$$x^{k+1} = x^k + v^k$$

end

The $\lambda = 0$ case is identical to Gauss–Newton. Increasing the regularization parameter λ accentuates the effect of the diagonal of the matrix $A^T A$, which improves the condition number and generally allows the method to converge from a broader set of initial guesses x_0 than Gauss–Newton.

► **EXAMPLE 4.25**   Use Levenberg–Marquardt to fit the model $y = c_1 e^{-c_2(t-c_3)^2}$ to the data points $(t_i, y_i) = \{(1, 3), (2, 5), (2, 7), (3, 5), (4, 1)\}$.

We must find the c_1, c_2, c_3 that minimize the RMSE for error vector

$$r = \begin{bmatrix} c_1 e^{-c_2(t_1-c_3)^2} - y_1 \\ \vdots \\ c_1 e^{-c_2(t_5-c_3)^2} - y_5 \end{bmatrix}.$$

The derivative of r evaluated at the five data points is the 5×3 matrix

$$Dr = \begin{bmatrix} e^{-c_2(t_1-c_3)^2} & -c_1(t_1-c_3)^2 e^{-c_2(t_1-c_3)^2} & 2c_1c_2(t_1-c_3)e^{-c_2(t_1-c_3)^2} \\ \vdots & \vdots & \vdots \\ e^{-c_2(t_5-c_3)^2} & -c_1(t_5-c_3)^2 e^{-c_2(t_5-c_3)^2} & 2c_1c_2(t_5-c_3)e^{-c_2(t_5-c_3)^2} \end{bmatrix}.$$

Levenberg–Marquardt with initial guess $(c_1, c_2, c_3) = (1, 1, 1)$ and λ fixed at 50 converges to the best least squares model

$$y = 6.301 e^{-0.5088(t-2.249)^2}.$$

The best model is plotted along with the data points in Figure 4.15. The corresponding Gauss–Newton Method diverges to infinity from this initial guess.   ◄

The method originated by a suggestion in Levenberg [1944] to add λI to $A^T A$ in Gauss–Newton to improve its conditioning. Several years later, D. Marquardt, a statistician at DuPont, improved on Levenberg's suggestion by replacing the identity matrix with the diagonal of $A^T A$ (Marquardt [1963]).

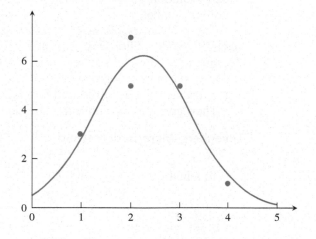

Figure 4.15 Model fit of Example 4.25. The Levenberg–Marquardt Method is used to find the best least squares model $y = 6.301 e^{-0.5088}(t - 2.249)^2$, plotted as the solid curve.

Although we have treated λ as a constant for simplicity, the method is often applied adaptively with a varying λ. A common strategy is to continue to decrease λ by a factor of 10 on each iteration step as long as the residual sum of squared errors is decreased by the step, and if the sum increases, to reject the step and increase λ by a factor of 10.

▶ **ADDITIONAL EXAMPLES**

1. Find the first iterate of the Gauss–Newton Method applied to find the least-squares near-intersection point of the three circles of radius 1 with centers $(-2, 0), (1, 1), (1, -1)$. Use initial guess $(0, 0)$.

2. Implement the Levenberg–Marquardt Method in MATLAB with $\lambda = 0.1$ to fit an exponential model to the data points $(-1, 4), (0, 2), (2, 1), (3, 0.5)$. Calculate the RMSE of the fit.

📖 **Solutions** for Additional Examples can be found at bit.ly/2q20SG9

4.5 Exercises

📖 **Solutions** for Exercises numbered in blue can be found at bit.ly/2P6BUVT

1. The Gauss–Newton Method can be applied to find the point $\overline{x}, \overline{y}$ for which the sum of the squared distances to the three circles is minimized. Using initial vector $(x_0, y_0) = (0, 0)$, carry out the first step to find (x_1, y_1) (a) centers $(0, 1), (1, 1), (0, -1)$ and all radii 1 (b) centers $(-1, 0), (1, 1), (1, -1)$ and all radii 1. (Computer Problem 1 asks for $(\overline{x}, \overline{y})$.)

2. Carry out the first step of Multivariate Newton's Method applied to the system (4.35) for the three circles in Exercise 1. Use $(x_0, y_0, K_0) = (0, 0, 0)$. (Computer Problem 1 asks for the solution (x, y, K).)

3. Prove that the distance from a point (x, y) to a circle $(x - x_1)^2 + (y - y_1)^2 = R_1^2$ is $|\sqrt{(x - x_1)^2 + (y - y_1)^2} - R_1|$.

4. Prove that the Gauss–Newton Method applied to the linear system $Ax = b$ converges in one step to the solution of the normal equations.

5. Find the matrix Dr needed for the application of Gauss–Newton iteration to the model-fitting problem with three data points $(t_1, y_1), (t_2, y_2), (t_3, y_3)$ (a) power law $y = c_1 t^{c_2}$ (b) $y = c_1 t e^{c_2 t}$.

6. Find the matrix Dr needed for the application of Gauss–Newton iteration to the model-fitting problem with three data points $(t_1, y_1), (t_2, y_2), (t_3, y_3)$ (a) translated exponential $y = c_3 + c_1 e^{c_2 t}$ (b) translated power law $y = c_3 + c_1 t^{c_2}$

7. Prove that the number of real solutions (x, y, K) of (4.35) is either infinity or at most two.

4.5 Computer Problems

📖 **Solutions** for Computer Problems numbered in blue can be found at bit.ly/2yQiFFM

1. Apply the Gauss–Newton Method to find the point $(\overline{x}, \overline{y})$ for which the sum of the squared distances to the three circles is minimized. Use initial vector $(x_0, y_0) = (0, 0)$. (a) Centers $(0, 1), (1, 1), (0, -1)$ and all radii 1. (b) Centers $(-1, 0), (1, 1), (1, -1)$ and all radii 1.

2. Apply Multivariate Newton's Method to the system (4.35) for the three circles in Computer Problem 1. Use initial vector $(x_0, y_0, K_0) = (0, 0, 0)$.

3. Find the point (x, y) and distance K that minimizes the sum of squares distance to the circles with radii increased by K, as in Example 4.23 (a) circles with centers $(-1, 0), (1, 0), (0, 1), (0, -2)$ and all radii 1 (b) circles with centers $(-2, 0), (3, 0), (0, 2), (0, -2)$ and all radii 1.

4. Carry out the steps of Computer Problem 3 with the following circles and plot the results (a) centers $(-2, 0), (2, 0), (0, 2), (0, -2)$, and $(2, 2)$ with radii $1, 1, 1, 1$, and 2, respectively (b) centers $(1, 1), (1, -1), (-1, 1), (-1, -1), (2, 0)$ and all radii 1.

5. Use the Gauss–Newton Method to fit a power law to the height–weight data of Example 4.10 without linearization. Compute the RMSE.

6. Use the Gauss–Newton Method to fit the blood concentration model (4.21) to the data of Example 4.11 without linearization.

7. Use the Levenberg–Marquardt Method with $\lambda = 1$ to fit a power law to the height–weight data of Example 4.10 without linearization. Compute the RMSE.

8. Use the Levenberg–Marquardt Method with $\lambda = 1$ to fit the blood concentration model (4.21) to the data of Example 4.11 without linearization.

9. Apply Levenberg–Marquardt to fit the model $y = c_1 e^{-c_2(t-c_3)^2}$ to the following data points, with an appropriate initial guess. State the initial guess, the regularization parameter λ used, and the RMSE. Plot the best least squares curve and the data points.
   (a) $(t_i, y_i) = \{(-1, 1), (0, 5), (1, 10), (3, 8), (6, 1)\}$
   (b) $(t_i, y_i) = \{(1, 1), (2, 3), (4, 7), (5, 12), (6, 13), (8, 5), (9, 2), (11, 1)\}$

10. Further investigate Example 4.25 by determining the initial guesses from the grid $0 \le c_1 \le 10$ with a grid spacing of 1, and $0 \le c_2 \le 1$ with a grid spacing of 0.1, $c_3 = 1$, for which the Levenberg–Marquardt Method converges to the correct least squares solution. Use the MATLAB mesh command to plot your answers, 1 for a convergent initial guess and 0 otherwise. Make plots for $\lambda = 50, \lambda = 1$, and the Gauss–Newton case $\lambda = 0$. Comment on the differences you find.

11. Apply Levenberg–Marquardt to fit the model $y = c_1 e^{-c_2 t} \cos(c_3 t + c_4)$ to the following data points, with an appropriate initial guess. State the initial guess, the regularization parameter λ used, and the RMSE. Plot the best least squares curve and the data points. This problem has multiple solutions with the same RMSE, since c_4 is only determined modulo 2π.
   (a) $(t_i, y_i) = \{(0, 3), (2, -5), (3, -2), (5, 2), (6, 1), (8, -1), (10, 0)\}$
   (b) $(t_i, y_i) = \{(1, 2), (3, 6), (4, 4), (5, 2), (6, -1), (8, -3)\}$

Reality Check **4** GPS, Conditioning, and Nonlinear Least Squares

The Global Positioning System (GPS) consists of 24 satellites carrying atomic clocks, orbiting the earth at an altitude of 20,200 km. Four satellites in each of six planes, slanted at $55°$ with respect to the poles, make two revolutions per day. At any time, from any point on earth, five to eight satellites are in the direct line of sight. Each satellite has a simple mission: to transmit carefully synchronized signals from predetermined positions in space, to be picked up by GPS receivers on earth. The receivers use the information, with some mathematics (described shortly), to determine accurate (x, y, z) coordinates of the receiver.

At a given instant, the receiver collects the synchronized signal from the ith satellite and determines its transmission time t_i, the difference between the times the signal was transmitted and received. The nominal speed of the signal is the speed of light, $c \approx 299792.458$ km/sec. Multiplying transmission time by c gives the distance of the satellite from the receiver, putting the receiver on the surface of a sphere centered at the satellite position and with radius ct_i. If three satellites are available, then three spheres are known, whose intersection consists of two points, as shown in Figure 4.16. One intersection point is the location of the receiver. The other is normally far from the earth's surface and can be safely disregarded. In theory, the problem is reduced to computing this intersection, the common solution of three sphere equations.

However, there is a major problem with this analysis. First, although the transmissions from the satellites are timed nearly to the nanosecond by onboard atomic clocks, the clock in the typical low-cost receiver on earth has relatively poor accuracy. If we solve the three equations with slightly inaccurate timing, the calculated position could be wrong by several kilometers. Fortunately, there is a way to fix this problem. The price to pay is one extra satellite. Define d to be the difference between the synchronized time on the (now four) satellite clocks and the earth-bound receiver clock. Denote the location of satellite i by (A_i, B_i, C_i). Then the true intersection point (x, y, z) satisfies

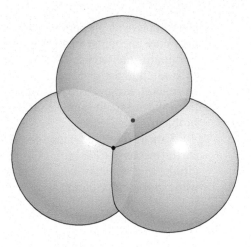

Figure 4.16 Three intersecting spheres. Generically, only two points lie on all three spheres.

$$r_1(x, y, z, d) = \sqrt{(x - A_1)^2 + (y - B_1)^2 + (z - C_1)^2} - c(t_1 - d) = 0$$

$$r_2(x, y, z, d) = \sqrt{(x - A_2)^2 + (y - B_2)^2 + (z - C_2)^2} - c(t_2 - d) = 0$$

$$r_3(x, y, z, d) = \sqrt{(x - A_3)^2 + (y - B_3)^2 + (z - C_3)^2} - c(t_3 - d) = 0$$

$$r_4(x, y, z, d) = \sqrt{(x - A_4)^2 + (y - B_4)^2 + (z - C_4)^2} - c(t_4 - d) = 0 \quad (4.37)$$

to be solved for the unknowns x, y, z, d. Solving the system reveals not only the receiver location, but also the correct time from the satellite clocks, due to knowing d. Therefore, the inaccuracy in the GPS receiver clock can be fixed by using one extra satellite.

Geometrically speaking, four spheres may not have a common intersection point, but they will if the radii are expanded or contracted by the right common amount. The system (4.37) representing the intersection of four spheres is the three-dimensional analogue of (4.35), representing the intersection point of three circles in the plane.

The system (4.37) can be seen to have two solutions (x, y, z, d). The equations can be equivalently written

$$(x - A_1)^2 + (y - B_1)^2 + (z - C_1)^2 = [c(t_1 - d)]^2$$

$$(x - A_2)^2 + (y - B_2)^2 + (z - C_2)^2 = [c(t_2 - d)]^2$$

$$(x - A_3)^2 + (y - B_3)^2 + (z - C_3)^2 = [c(t_3 - d)]^2$$

$$(x - A_4)^2 + (y - B_4)^2 + (z - C_4)^2 = [c(t_4 - d)]^2. \quad (4.38)$$

Note that by subtracting the last three equations from the first, three *linear* equations are obtained. Each linear equation can be used to eliminate a variable x, y, z, and by substituting into any of the original equations, a quadratic equation in the single variable d results. Therefore, system (4.37) has at most two real solutions, and they can be found by the quadratic formula.

Two further problems emerge when GPS is deployed. First is the conditioning of the system of equations (4.37). We will find that solving for (x, y, z, d) is ill-conditioned when the satellites are bunched closely in the sky.

The second difficulty is that the transmission speed of the signals is not precisely c. The signals pass through 100 km of ionosphere and 10 km of troposphere, whose electromagnetic properties may affect the transmission speed. Furthermore, the signals may encounter obstacles on earth before reaching the receiver, an effect called multipath interference. To the extent that these obstacles have an equal impact on each satellite path, introducing the time correction d on the right side of (4.37) helps. In general, however, this assumption is not viable and will lead us to add information from more satellites and consider applying Gauss–Newton to solve a least squares problem.

Consider a three-dimensional coordinate system whose origin is the center of the earth (radius $\approx$ 6370 km). GPS receivers convert these coordinates into latitude, longitude, and elevation data for readout and more sophisticated mapping applications using global information system (GIS), a process we will not consider here.

Suggested activities:

1. Solve the system (4.37) by using Multivariate Newton's Method. Find the receiver position (x, y, z) near earth and time correction d for known, simultaneous satellite positions $(15600, 7540, 20140)$, $(18760, 2750, 18610)$, $(17610, 14630, 13480)$, $(19170, 610, 18390)$ in km, and measured time intervals $0.07074, 0.07220, 0.07690, 0.07242$ in seconds, respectively. Set the initial vector to be $(x_0, y_0, z_0, d_0) = (0, 0, 6370, 0)$. As a check, the answers are approximately $(x, y, z) = (-41.77271, -16.78919, 6370.0596)$, and $d = -3.201566 \times 10^{-3}$ seconds.

2. Write a MATLAB program to solve the problem in Step 1 via the quadratic formula. Hint: Subtracting the last three equations of (4.38) from the first yields three linear equations in the four unknowns $u = [x, y, z, d]^T$, or in matrix form, $Au = b$ where A is a 3×4 matrix. The MATLAB rref command applied to the augmented matrix $[A \mid b]$ returns

$$r = \begin{bmatrix} 1 & 0 & 0 & r_{14} & | & r_{15} \\ 0 & 1 & 0 & r_{24} & | & r_{25} \\ 0 & 0 & 1 & r_{34} & | & r_{35} \end{bmatrix},$$

giving expressions

$$x = -r_{14}d + r_{15}$$
$$y = -r_{24}d + r_{25}$$
$$z = -r_{34}d + r_{35}.$$

Substitute these expressions into the first equation of (4.38) to get a quadratic equation in one variable d.

3. If the MATLAB Symbolic Toolbox is available (or a symbolic package such as Maple or Mathematica), an alternative to Step 2 is possible. Define symbolic variables by using the syms command and solve the simultaneous equations with the Symbolic Toolbox command solve. Use subs to evaluate the symbolic result as a floating point number.

4. Now set up a test of the conditioning of the GPS problem. Define satellite positions (A_i, B_i, C_i) from spherical coordinates (ρ, ϕ_i, θ_i) as

$$A_i = \rho \cos \phi_i \cos \theta_i$$
$$B_i = \rho \cos \phi_i \sin \theta_i$$
$$C_i = \rho \sin \phi_i,$$

where $\rho = 26570$ km is fixed, while $0 \le \phi_i \le \pi/2$ and $0 \le \theta_i \le 2\pi$ for $i = 1, \ldots, 4$ are chosen arbitrarily. The ϕ coordinate is restricted so that the four satellites are in the upper hemisphere. Set $x = 0, y = 0, z = 6370, d = 0.0001$, and calculate the corresponding satellite ranges $R_i = \sqrt{A_i^2 + B_i^2 + (C_i - 6370)^2}$ and travel times $t_i = d + R_i/c$.

We will define an error magnification factor specially tailored to the situation. The atomic clocks aboard the satellites are correct up to about 10 nanoseconds, or 10^{-8} second. Therefore, it is important to study the effect of changes in the transmission time of this magnitude. Let the backward, or input error be the input change in meters. At the speed of light, $\Delta t_i = 10^{-8}$ second corresponds to $10^{-8} c \approx 3$ meters. Let the forward, or output error be the change in position $\|(\Delta x, \Delta y, \Delta z)\|_\infty$, caused by such a change in t_i, also in meters. Then we can define the dimensionless

$$\text{error magnification factor} = \frac{\|(\Delta x, \Delta y, \Delta z)\|_\infty}{c\|(\Delta t_1, \ldots, \Delta t_m)\|_\infty},$$

and the condition number of the problem to be the maximum error magnification factor for all small Δt_i (say, 10^{-8} or less).

Change each t_i defined in the foregoing by $\Delta t_i = +10^{-8}$ or -10^{-8}, not all the same. Denote the new solution of the equations (4.37) by $(\overline{x}, \overline{y}, \overline{z}, \overline{d})$, and compute the difference in position $\|(\Delta x, \Delta y, \Delta z)\|_\infty$ and the error magnification factor. Try different variations of the Δt_i's. What is the maximum position error found, in meters? Estimate the condition number of the problem, on the basis of the error magnification factors you have computed.

5. Now repeat Step 4 with a more tightly grouped set of satellites. Choose all ϕ_i within 5 percent of one another and all θ_i within 5 percent of one another. Solve with and without the same input error as in Step 4. Find the maximum position error and error magnification factor. Compare the conditioning of the GPS problem when the satellites are tightly or loosely bunched.

6. Decide whether the GPS error and condition number can be reduced by adding satellites. Return to the unbunched satellite configuration of Step 4, and add four more. (At all times and at every position on earth, 5 to 12 GPS satellites are visible.) Design a Gauss–Newton iteration to solve the least squares system of eight equations in four variables (x, y, z, d). What is a good initial vector? Find the maximum GPS position error, and estimate the condition number. Summarize your results from four unbunched, four bunched, and eight unbunched satellites. What configuration is best, and what is the maximum GPS error, in meters, that you should expect solely on the basis of satellite signals?

Software and Further Reading

Least squares approximation dates from the early 19th century. Like polynomial interpolation, it can be viewed as a form of lossy data compression, finding a simple representation for a complicated or noisy data set. Lines, polynomials, exponential functions, and power laws are commonly implemented models. Periodic data call for trigonometric representations, which, taken to the extreme, lead to trigonometric interpolation and trigonometric least squares fits, pursued in Chapter 10.

Any function that is linear in its coefficients can be used to fit data by applying the three-step method of Section 4.2, resulting in solution of the normal equations. For ill-conditioned problems, the normal equations are not recommended, due to the fact that the condition number is roughly squared in this approach. The matrix factorization preferred in this case is the QR factorization and, in some cases, the singular value decomposition, introduced in Chapter 12. Golub and Van Loan [2012] is an excellent reference for the QR and other matrix factorizations. Lawson and Hanson [1995] is a good source for the fundamentals of least squares. The statistical aspects of least squares fitting the linear and multiple regression are covered in the more specialized texts Draper and Smith [2001], Fox [1997], and Ryan [1997].

MATLAB's backslash command applied to $Ax = b$ carries out Gaussian elimination if the system is consistent, and solves the least squares problem by QR factorization if inconsistent. MATLAB's qr command is based on the LAPACK routine DGEQRF. The NAG library routine E02ADF carries out least squares approximation to polynomials, as does MATLAB's polyfit. Statistical packages such as S^+, SAS, SPSS, and Minitab carry out a variety of regression analyses.

Nonlinear least squares refers to fitting coefficients that are nonlinear in the model. The Gauss–Newton Method and its variants like Levenberg–Marquardt are the preferred tools for this calculation, although convergence is not guaranteed, and even when convergence occurs, no unique optimum is implied. See Strang and Borre [1997] for an introduction to the mathematics of GPS, and Hoffman-Wellenhof et al. [2001] for general information on the topic.

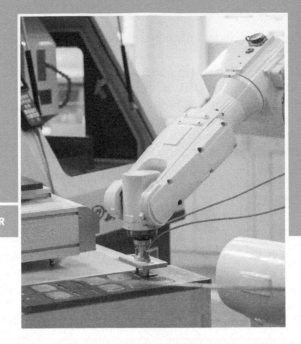

Numerical Differentiation and Integration

Computer-aided manufacturing depends on precise control of motion along a prescribed path. For example, lathes or milling machines under numerical control rely on parametric curves, often given by cubic or Bézier splines from computer-aided design software, to describe the path of cutting or shaping tools. Computer-generated animation in filmmaking, computer games, and virtual reality applications face similar problems.

Reality Check Reality Check 5 on page 289 considers the problem of controlling the velocity along an arbitrary parametric path. For the path parameter to traverse the curve at a desired rate, the curve is reparametrized with respect to arc length. Adaptive Quadrature applied to the arc length integral provides an efficient way to achieve the control.

The main problem of computational calculus is to compute derivatives and integrals of functions. There are two directions that we can take for such problems, numerical computing and symbolic computing. We will discuss both in this chapter, but go into the most detail on numerical computing issues. Both derivatives and integrals have clear mathematical definitions, but the type of answer wanted by a user often depends on the way in which the function is specified.

The derivatives of functions like $f(x) = \sin x$ are the topic of introductory calculus. If the function is known in terms of elementary functions, say, $f(x) = \sin^3(x^{\tan x} \cosh x)$, its third derivative may be found more quickly by symbolic computing methods, where the calculus rules are carried out by computer. The same is true for antiderivatives in cases where the answer can be expressed in terms of elementary functions.

In practice, there are two other common ways for a function to be known. A function may be specified as a tabulated list, for example, a list $\{(t_1, T_1), \ldots, (t_n, T_n)\}$ of time/temperature pairs measured from an experiment, perhaps at evenly spaced times. In this case, finding the derivative or antiderivative from the rules of freshman calculus is impossible. Finally, a function may be specified as the output of an experiment or computer simulation whose input is specified by the user. In the latter two cases, symbolic computing methods cannot be applied, and numerical differentiation and integration are required to solve the problem.

5.1 NUMERICAL DIFFERENTIATION

To begin, we develop finite difference formulas for approximating derivatives. In some cases, that is the goal of the calculation. In Chapters 7 and 8, these formulas will be used to discretize ordinary and partial differential equations.

5.1.1 Finite difference formulas

By definition, the derivative of $f(x)$ at a value x is

$$f'(x) = \lim_{h \to 0} \frac{f(x+h) - f(x)}{h}, \tag{5.1}$$

provided that the limit exists. This leads to a useful formula for approximating the derivative at x. Taylor's Theorem says that if f is twice continuously differentiable, then

$$f(x+h) = f(x) + hf'(x) + \frac{h^2}{2} f''(c), \tag{5.2}$$

where c is between x and $x + h$. Equation (5.2) implies the following formula:

Two-point forward-difference formula

$$f'(x) = \frac{f(x+h) - f(x)}{h} - \frac{h}{2} f''(c), \tag{5.3}$$

where c is between x and $x + h$.

In a finite calculation, we cannot take the limit in (5.1), but (5.3) implies that the quotient will closely approximate the derivative if h is small. We use (5.3) by computing the approximation

$$f'(x) \approx \frac{f(x+h) - f(x)}{h} \tag{5.4}$$

and treating the last term in (5.3) as error. Because the error made by the approximation is proportional to the increment h, we can make the error small by making h small. The two-point-forward-difference formula is a first-order method for approximating the first derivative. In general, if the error is $O(h^n)$, we call the formula an **order** n approximation.

A subtle point about calling the formula "first order" is that c depends on h. The idea of first order is that the error should be proportional to h as $h \to 0$. As $h \to 0$, c is a moving target, and as a result, the proportionality constant changes. But as long as f'' is continuous, the proportionality constant $f''(c)$ tends toward $f''(x)$ as $h \to 0$, making it legitimate to call the formula first order.

SPOTLIGHT ON

Convergence   What good is the error formula $-hf''(c)/2$ of the two-point forward-difference method? We are trying to approximate $f'(x)$, so $f''(x)$ is likely to be out of our reach. There are two answers. First, when verifying code and software, a good check is to run it on a completely solved example, where the correct answers are known and even the errors can be compared with what is expected. In such a case we may know $f''(x)$ as well as $f'(x)$. Second, even when we can't evaluate the entire formula, it is often helpful to know how the error scales with h. The fact that the formula is first order means that cutting h in half should cut the error approximately in half, even if we have no way of computing the proportionality constant $f''(c)/2$.

▶ **EXAMPLE 5.1**   Use the two-point forward-difference formula with $h = 0.1$ to approximate the derivative of $f(x) = 1/x$ at $x = 2$.

The two-point forward-difference formula (5.4) evaluates to

$$f'(x) \approx \frac{f(x+h) - f(x)}{h} = \frac{\frac{1}{2.1} - \frac{1}{2}}{0.1} \approx -0.2381.$$

The difference between this approximation and the correct derivative $f'(x) = -x^{-2}$ at $x = 2$ is the error

$$-0.2381 - (-0.2500) = 0.0119.$$

Compare this to the error predicted by the formula, which is $hf''(c)/2$ for some c between 2 and 2.1. Since $f''(x) = 2x^{-3}$, the error must be between

$$(0.1)2^{-3} \approx 0.0125 \quad \text{and} \quad (0.1)(2.1)^{-3} \approx 0.0108,$$

which is consistent with our result. However, this information is usually not available.
◀

A second-order formula can be developed by a more advanced strategy. According to Taylor's Theorem, if f is three times continuously differentiable, then

$$f(x+h) = f(x) + hf'(x) + \frac{h^2}{2}f''(x) + \frac{h^3}{6}f'''(c_1)$$

and

$$f(x-h) = f(x) - hf'(x) + \frac{h^2}{2}f''(x) - \frac{h^3}{6}f'''(c_2),$$

where $x - h < c_2 < x < c_1 < x + h$. Subtracting the two equations gives the following three-point formula with an explicit error term:

$$f'(x) = \frac{f(x+h) - f(x-h)}{2h} - \frac{h^2}{12}f'''(c_1) - \frac{h^2}{12}f'''(c_2). \tag{5.5}$$

In order to be more precise about the error term for the new formula, we will use the following theorem:

THEOREM 5.1   **Generalized Intermediate Value Theorem.** Let f be a continuous function on the interval $[a, b]$. Let $x_1, \ldots, x_n$ be points in $[a, b]$, and $a_1, \ldots, a_n > 0$. Then there exists a number c between a and b such that

$$(a_1 + \cdots + a_n)f(c) = a_1 f(x_1) + \cdots + a_n f(x_n). \tag{5.6}$$

∎

Proof. Let $f(x_i)$ equal the minimum and $f(x_j)$ the maximum of the n function values. Then

$$a_1 f(x_i) + \cdots + a_n f(x_i) \leq a_1 f(x_1) + \cdots + a_n f(x_n) \leq a_1 f(x_j) + \cdots + a_n f(x_j)$$

implies that

$$f(x_i) \leq \frac{a_1 f(x_1) + \cdots + a_n f(x_n)}{a_1 + \cdots + a_n} \leq f(x_j).$$

By the Intermediate Value Theorem, there is a number c between x_i and x_j such that

$$f(c) = \frac{a_1 f(x_1) + \cdots + a_n f(x_n)}{a_1 + \cdots + a_n},$$

and (5.6) is satisfied. ❑

Theorem 5.1 says that we can combine the last two terms of (5.5), yielding a second-order formula:

Three-point centered-difference formula

$$f'(x) = \frac{f(x+h) - f(x-h)}{2h} - \frac{h^2}{6} f'''(c), \qquad (5.7)$$

where $x - h < c < x + h$.

▶ **EXAMPLE 5.2** Use the three-point centered-difference formula with $h = 0.1$ to approximate the derivative of $f(x) = 1/x$ at $x = 2$.

The three-point centered-difference formula evaluates to

$$f'(x) \approx \frac{f(x+h) - f(x-h)}{2h} = \frac{\frac{1}{2.1} - \frac{1}{1.9}}{0.2} \approx -0.2506.$$

The error is 0.0006, an improvement on the two-point forward-difference formula in Example 5.1. ◀

Approximation formulas for higher derivatives can be obtained in the same way. For example, the Taylor expansions

$$f(x+h) = f(x) + hf'(x) + \frac{h^2}{2} f''(x) + \frac{h^3}{6} f'''(x) + \frac{h^4}{24} f^{(iv)}(c_1)$$

and

$$f(x-h) = f(x) - hf'(x) + \frac{h^2}{2} f''(x) - \frac{h^3}{6} f'''(x) + \frac{h^4}{24} f^{(iv)}(c_2),$$

SPOTLIGHT ON

Convergence   The two- and three-point approximations converge to the derivative as $h \to 0$, although at different rates. The formulas break the cardinal rule of floating point computing by subtracting nearly equal numbers, but it can't be helped, as finding derivatives is an inherently unstable process. For very small values of h, roundoff error will affect the calculation, as shown in Example 5.3.

where $x - h < c_2 < x < c_1 < x + h$ can be added together to eliminate the first derivative terms to get

$$f(x+h) + f(x-h) - 2f(x) = h^2 f''(x) + \frac{h^4}{24}f^{(iv)}(c_1) + \frac{h^4}{24}f^{(iv)}(c_2).$$

Using Theorem 5.1 to combine the error terms and dividing by h^2 yields the following formula:

Three-point centered-difference formula for second derivative

$$f''(x) = \frac{f(x-h) - 2f(x) + f(x+h)}{h^2} - \frac{h^2}{12}f^{(iv)}(c) \qquad (5.8)$$

for some c between $x - h$ and $x + h$.

5.1.2 Rounding error

So far, all of this chapter's formulas break the rule from Chapter 0 that advises against subtracting nearly equal numbers. This is the greatest difficulty with numerical differentiation, but it is essentially impossible to avoid. To understand the problem better, consider the following example:

▶ **EXAMPLE 5.3**   Approximate the derivative of $f(x) = e^x$ at $x = 0$.

The two-point formula (5.4) gives

$$f'(x) \approx \frac{e^{x+h} - e^x}{h}, \qquad (5.9)$$

and the three-point formula (5.7) yields

$$f'(x) \approx \frac{e^{x+h} - e^{x-h}}{2h}. \qquad (5.10)$$

The results of these formulas for $x = 0$ and a wide range of increment size h, along with errors compared with the correct value $e^0 = 1$, are given in the following table:

h	formula (5.9)	error	formula (5.10)	error
10^{-1}	1.05170918075648	−0.05170918075648	1.00166750019844	−0.00166750019844
10^{-2}	1.00501670841679	−0.00501670841679	1.00001666674999	−0.00001666674999
10^{-3}	1.00050016670838	−0.00050016670838	1.00000016666668	−0.00000016666668
10^{-4}	1.00005000166714	−0.00005000166714	1.00000000166689	−0.00000000166689
10^{-5}	1.00000500000696	−0.00000500000696	1.00000000001210	−0.00000000001210
10^{-6}	1.00000049996218	−0.00000049996218	0.99999999997324	0.00000000002676
10^{-7}	1.00000004943368	−0.00000004943368	0.99999999947364	0.00000000052636
10^{-8}	0.99999999392253	0.00000000607747	0.99999999392253	0.00000000607747
10^{-9}	1.00000008274037	−0.00000008274037	1.00000002722922	−0.00000002722922

At first, the error decreases as h decreases, following closely the expected errors $O(h)$ and $O(h^2)$, respectively, for the two-point forward-difference formula (5.4) and the three-point centered-difference formula (5.7). However, notice the deterioration of the approximations as h is decreased still further.

The reason that the approximations lose accuracy for very small h is loss of significance. Both formulas subtract nearly equal numbers, lose significant digits, and then, to make matters worse, magnify the effect by dividing by a small number. ◄

To get a better idea of the degree to which numerical differentiation formulas are susceptible to loss of significance, we analyze the three-point centered-difference formula in detail. Denote the floating point version of the input $f(x + h)$ by $\hat{f}(x + h)$, which will differ from the correct value $f(x + h)$ by a number on the order of machine epsilon in relative terms. We will assume the function values are on the order of 1 for the present discussion, so that relative and absolute errors are about equal.

Since $\hat{f}(x + h) = f(x + h) + \epsilon_1$ and $\hat{f}(x - h) = f(x - h) + \epsilon_2$, where $|\epsilon_1|, |\epsilon_2| \approx \epsilon_{\text{mach}}$, the difference between the correct $f'(x)$ and the machine version of the three-point centered-difference formula (5.7) is

$$
\begin{aligned}
f'(x)_{\text{correct}} - f'(x)_{\text{machine}} &= f'(x) - \frac{\hat{f}(x + h) - \hat{f}(x - h)}{2h} \\
&= f'(x) - \frac{f(x + h) + \epsilon_1 - (f(x - h) + \epsilon_2)}{2h} \\
&= \left(f'(x) - \frac{f(x + h) - f(x - h)}{2h} \right) + \frac{\epsilon_2 - \epsilon_1}{2h} \\
&= \left(f'(x)_{\text{correct}} - f'(x)_{\text{formula}} \right) + \text{error}_{\text{rounding}}.
\end{aligned}
$$

We can view the total error as a sum of the truncation error, the difference between the correct derivative and the correct approximating formula, and the rounding error, which accounts for the loss of significance of the computer-implemented formula. The rounding error has absolute value

$$
\left| \frac{\epsilon_2 - \epsilon_1}{2h} \right| \leq \frac{2\epsilon_{\text{mach}}}{2h} = \frac{\epsilon_{\text{mach}}}{h},
$$

where ϵ_{mach} represents machine epsilon. Therefore, the absolute value of the error of the machine approximation of $f'(x)$ is bounded above by

$$
E(h) \equiv \frac{h^2}{6} f'''(c) + \frac{\epsilon_{\text{mach}}}{h}, \tag{5.11}
$$

where $x - h < c < x + h$. Previously we had considered only the first term of the error, the mathematical error. The preceding table forces us to consider the loss of significance term as well.

It is instructive to plot the function $E(h)$, shown in Figure 5.1. The minimum of $E(h)$ occurs at the solution of

$$
0 = E'(h) = -\frac{\epsilon_{\text{mach}}}{h^2} + \frac{M}{3} h, \tag{5.12}
$$

where we have approximated $|f'''(c)| \approx |f'''(x)|$ by M. Solving (5.12) yields

$$
h = (3\epsilon_{\text{mach}}/M)^{1/3}
$$

for the increment size h that gives smallest overall error, including the effects of computer rounding. In double precision, this is approximately $\epsilon_{\text{mach}}^{1/3} \approx 10^{-5}$, consistent with the table.

The main message is that the three-point centered-difference formula will improve in accuracy as h is decreased until h becomes about the size of the cube root of machine epsilon. As h drops below this size, the error may begin increasing again.

Similar results on rounding analysis can be derived for other formulas. Exercise 18 asks the reader to analyze rounding effects for the two-point forward-difference formula.

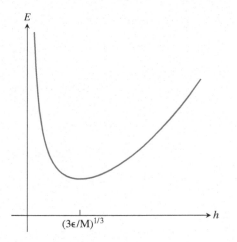

Figure 5.1 The effect of rounding error on numerical differentiation. For sufficiently small h, the error is dominated by rounding error.

5.1.3 Extrapolation

Assume that we are presented with an order n formula $F(h)$ for approximating a given quantity Q. The order means that

$$Q \approx F(h) + Kh^n,$$

where K is roughly constant over the range of h in which we are interested. A relevant example is

$$f'(x) = \frac{f(x+h) - f(x-h)}{2h} - \frac{f'''(c_h)}{6}h^2, \tag{5.13}$$

where we have emphasized the fact that the unknown point c_h lies between x and $x + h$, but depends on h. Even though c_h is not constant, if the function f is reasonably smooth and h is not too large, the values of the error coefficient $f'''(c_h)/6$ should not vary far from $f'''(x)/6$.

In a case like this, a little bit of algebra can be used to leverage an order n formula into one of higher order. Because we know the order of the formula $F(h)$ is n, if we apply the formula again with $h/2$ instead of h, our error should be reduced from a constant times h^n to a constant times $(h/2)^n$, or reduced by a factor of 2^n. In other words, we expect

$$Q - F(h/2) \approx \frac{1}{2^n}(Q - F(h)). \tag{5.14}$$

We are relying on the assumption that K is roughly constant. Notice that (5.14) is readily solved for the quantity Q in question to give the following formula:

Extrapolation for order n formula

$$Q \approx \frac{2^n F(h/2) - F(h)}{2^n - 1}. \tag{5.15}$$

This is the **extrapolation** formula for $F(h)$. Extrapolation, sometimes called **Richardson extrapolation**, typically gives a higher-order approximation of Q than $F(h)$. To understand why, assume that the nth-order formula $F_n(h)$ can be written

$$Q = F_n(h) + Kh^n + O(h^{n+1}).$$

Then cutting h in half yields

$$Q = F_n(h/2) + K\frac{h^n}{2^n} + O(h^{n+1}),$$

and the extrapolated version, which we call $F_{n+1}(h)$, will satisfy

$$
\begin{aligned}
F_{n+1}(h) &= \frac{2^n F_n(h/2) - F_n(h)}{2^n - 1} \\
&= \frac{2^n(Q - Kh^n/2^n - O(h^{n+1})) - (Q - Kh^n - O(h^{n+1}))}{2^n - 1} \\
&= Q + \frac{-Kh^n + Kh^n + O(h^{n+1})}{2^n - 1} = Q + O(h^{n+1}).
\end{aligned}
$$

Therefore, $F_{n+1}(h)$ is (at least) an order $n + 1$ formula for approximating the quantity Q.

▶ **EXAMPLE 5.4**   Apply extrapolation to formula (5.13).

We start with the second-order centered-difference formula $F_2(h)$ for the derivative $f'(x)$. The extrapolation formula (5.15) gives a new formula for $f'(x)$ as

$$
\begin{aligned}
F_4(x) &= \frac{2^2 F_2(h/2) - F_2(h)}{2^2 - 1} \\
&= \left[4\frac{f(x + h/2) - f(x - h/2)}{h} - \frac{f(x + h) - f(x - h)}{2h} \right] \Big/ 3 \\
&= \frac{f(x - h) - 8f(x - h/2) + 8f(x + h/2) - f(x + h)}{6h}.
\end{aligned}
\tag{5.16}
$$

This is a five-point centered-difference formula. The previous argument guarantees that this formula is of order at least three, but it turns out to have order four, because the order three error terms cancel out. In fact, since $F_4(h) = F_4(-h)$ by inspection, the error must be the same for h as for $-h$. Therefore, the error terms can be even powers of h only.   ◀

▶ **EXAMPLE 5.5**   Apply extrapolation to the second derivative formula (5.8).

Again, the method is second order, so the extrapolation formula (5.15) is used with $n = 2$. The extrapolated formula is

$$
\begin{aligned}
F_4(x) &= \frac{2^2 F_2(h/2) - F_2(h)}{2^2 - 1} \\
&= \left[4\frac{f(x + h/2) - 2f(x) + f(x - h/2)}{h^2/4} \right. \\
&\quad \left. - \frac{f(x + h) - 2f(x) + f(x - h)}{h^2} \right] \Big/ 3 \\
&= \frac{-f(x - h) + 16f(x - h/2) - 30f(x) + 16f(x + h/2) - f(x + h)}{3h^2}.
\end{aligned}
$$

The new method for approximating second derivatives is fourth order, for the same reason as the previous example.   ◀

5.1.4 Symbolic differentiation and integration

The MATLAB Symbolic Toolbox contains commands for obtaining the symbolic derivative of symbolically written functions. The following commands are illustrative:

```
>> syms x;
>> f=sin(3*x);
>> f1=diff(f)

f1=

3*cos(3*x)

>>
```

The third derivative is also easily found:

```
>>f3=diff(f,3)

f3=

-27*cos(3*x)
```

Integration uses the MATLAB symbolic command `int`:

```
>>syms x
>>f=sin(x)

f=

sin(x)

>>int(f)

ans=

-cos(x)

>>int(f,0,pi)

ans=

2
```

With more complicated functions, the MATLAB command `pretty`, to view the resulting answer, and `simple`, to simplify it, are helpful, as in the following code:

```
>>syms x

>>f=sin(x)^7

f=

sin(x)^7

>>int(f)

ans=
```

```
-1/7*sin(x)^6*cos(x)-6/35*sin(x)^4*cos(x)-8/35*sin(x)^2*cos(x)
    -16/35*cos(x)

>>pretty(simple(int(f)))
                      3          5            7
    -cos(x) + cos(x)  - 3/5 cos(x)  + 1/7 cos(x)
```

Of course, for some integrands, there is no expression for the indefinite integral in terms of elementary functions. Try the function $f(x) = e^{\sin x}$ to see MATLAB give up. In a case like this, there is no alternative but the numerical methods of the next section.

▶ **ADDITIONAL EXAMPLES**

1. Use the three point centered-difference formula to approximate the derivative $f'(\pi/2)$ where $f(x) = e^{\cos x}$, where (a) $h = 0.1$ (b) $h = 0.01$.

2. Develop a first-order formula for estimating $f''(x)$ that uses the data $f(x - 2h)$, $f(x)$, and $f(x + h)$ only.

⊡ **Solutions** for Additional Examples can be found at bit.ly/2RYcZm2

5.1 Exercises

⊡ **Solutions** for Exercises numbered in blue can be found at bit.ly/2PJK9EE

1. Use the two-point forward-difference formula to approximate $f'(1)$, and find the approximation error, where $f(x) = \ln x$, for (a) $h = 0.1$ (b) $h = 0.01$ (c) $h = 0.001$.

2. Use the three-point centered-difference formula to approximate $f'(0)$, where $f(x) = e^x$, for (a) $h = 0.1$ (b) $h = 0.01$ (c) $h = 0.001$.

3. Use the two-point forward-difference formula to approximate $f'(\pi/3)$, where $f(x) = \sin x$, and find the approximation error. Also, find the bounds implied by the error term and show that the approximation error lies between them (a) $h = 0.1$ (b) $h = 0.01$ (c) $h = 0.001$.

4. Carry out the steps of Exercise 3, using the three-point centered-difference formula.

5. Use the three-point centered-difference formula for the second derivative to approximate $f''(1)$, where $f(x) = x^{-1}$, for (a) $h = 0.1$ (b) $h = 0.01$ (c) $h = 0.001$. Find the approximation error.

6. Use the three-point centered-difference formula for the second derivative to approximate $f''(0)$, where $f(x) = \cos x$, for (a) $h = 0.1$ (b) $h = 0.01$ (c) $h = 0.001$. Find the approximation error.

7. Develop a formula for a two-point backward-difference formula for approximating $f'(x)$, including error term.

8. Prove the second-order formula for the first derivative

$$f'(x) = \frac{-f(x + 2h) + 4f(x + h) - 3f(x)}{2h} + O(h^2).$$

9. Develop a second-order formula for the first derivative $f'(x)$ in terms of $f(x)$, $f(x - h)$, and $f(x - 2h)$.

10. Find the error term and order for the approximation formula

$$f'(x) = \frac{4f(x + h) - 3f(x) - f(x - 2h)}{6h}.$$

11. Find a second-order formula for approximating $f'(x)$ by applying extrapolation to the two-point forward-difference formula.

12. (a) Compute the two-point forward-difference formula approximation to $f'(x)$ for $f(x) = 1/x$, where x and h are arbitrary. (b) Subtract the correct answer to get the error explicitly, and show that it is approximately proportional to h. (c) Repeat parts (a) and (b), using the three-point centered-difference formula instead. Now the error should be proportional to h^2.

13. Develop a second-order method for approximating $f'(x)$ that uses the data $f(x - h), f(x),$ and $f(x + 3h)$ only.

14. (a) Extrapolate the formula developed in Exercise 13. (b) Demonstrate the order of the new formula by approximating $f'(\pi/3)$, where $f(x) = \sin x$, with $h = 0.1$ and $h = 0.01$.

15. Develop a first-order method for approximating $f''(x)$ that uses the data $f(x - h), f(x),$ and $f(x + 3h)$ only.

16. (a) Apply extrapolation to the formula developed in Exercise 15 to get a second-order formula for $f''(x)$. (b) Demonstrate the order of the new formula by approximating $f''(0)$, where $f(x) = \cos x$, with $h = 0.1$ and $h = 0.01$.

17. Develop a second-order method for approximating $f'(x)$ that uses the data $f(x - 2h), f(x),$ and $f(x + 3h)$ only.

18. Find $E(h)$, an upper bound for the error of the machine approximation of the two-point forward-difference formula for the first derivative. Follow the reasoning preceding (5.11). Find the h corresponding to the minimum of $E(h)$.

19. Prove the second-order formula for the third derivative

$$f'''(x) = \frac{-f(x - 2h) + 2f(x - h) - 2f(x + h) + f(x + 2h)}{2h^3} + O(h^2).$$

20. Prove the second-order formula for the third derivative

$$f'''(x) = \frac{f(x - 3h) - 6f(x - 2h) + 12f(x - h) - 10f(x) + 3f(x + h)}{2h^3} + O(h^2).$$

21. Prove the second-order formula for the fourth derivative

$$f^{(iv)}(x) = \frac{f(x - 2h) - 4f(x - h) + 6f(x) - 4f(x + h) + f(x + 2h)}{h^4} + O(h^2).$$

This formula is used in Reality Check 2.

22. This exercise justifies the beam equations (2.33) and (2.34) in Reality Check 2. Let $f(x)$ be a six-times continuously differentiable function.

(a) Prove that if $f(x) = f'(x) = 0$, then

$$f^{(iv)}(x + h) - \frac{16f(x + h) - 9f(x + 2h) + \frac{8}{3}f(x + 3h) - \frac{1}{4}f(x + 4h)}{h^4} = O(h^2).$$

(Hint: First show that if $f(x) = f'(x) = 0$, then $f(x - h) - 10f(x + h) + 5f(x + 2h) - \frac{5}{3}f(x + 3h) + \frac{1}{4}f(x + 4h) = O(h^6)$. Then apply Exercise 21.)

(b) Prove that if $f''(x) = f'''(x) = 0$, then

$$f^{(iv)}(x + h) - \frac{-28f(x) + 72f(x + h) - 60f(x + 2h) + 16f(x + 3h)}{17h^4} = O(h^2).$$

(Hint: First show that if $f''(x) = f'''(x) = 0$, then $17f(x - h) - 40f(x) + 30f(x + h) - 8f(x + 2h) + f(x + 3h) = O(h^6)$. Then apply Exercise 21.)

(c) Prove that if $f''(x) = f'''(x) = 0$, then

$$f^{(iv)}(x) - \frac{72f(x) - 156f(x + h) + 96f(x + 2h) - 12f(x + 3h)}{17h^4} = O(h^2).$$

(Hint: First show that if $f''(x) = f'''(x) = 0$, then
$17f(x - 2h) - 130f(x) + 208f(x + h) - 111f(x + 2h) + 16f(x + 3h) = O(h^6)$. Then apply part (b) together with Exercise 21.)

23. Use Taylor expansions to prove that (5.16) is a fourth-order formula.

24. The error term in the two-point forward-difference formula for $f'(x)$ can be written in other ways. Prove the alternative result

$$f'(x) = \frac{f(x + h) - f(x)}{h} - \frac{h}{2}f''(x) - \frac{h^2}{6}f'''(c),$$

where c is between x and $x + h$. We will use this error form in the derivation of the Crank–Nicolson Method in Chapter 8.

25. Investigate the reason for the name extrapolation. Assume that $F(h)$ is an nth order formula for approximating a quantity Q, and consider the points $(Kh^2, F(h))$ and $(K(h/2)^2, F(h/2))$ in the xy-plane, where error is plotted on the x-axis and the formula output on the y-axis. Find the line through the two points (the best functional approximation for the relationship between error and F). The y-intercept of this line is the value of the formula when you extrapolate the error to zero. Show that this extrapolated value is given by formula (5.15).

5.1 Computer Problems

1. Make a table of the error of the three-point centered-difference formula for $f'(0)$, where $f(x) = \sin x - \cos x$, with $h = 10^{-1}, \ldots, 10^{-12}$, as in the table in Section 5.1.2. Draw a plot of the results. Does the minimum error correspond to the theoretical expectation?

2. Make a table and plot of the error of the three-point centered-difference formula for $f'(1)$, as in Computer Problem 1, where $f(x) = (1 + x)^{-1}$.

3. Make a table and plot of the error of the two-point forward-difference formula for $f'(0)$, as in Computer Problem 1, where $f(x) = \sin x - \cos x$. Compare your answers with the theory developed in Exercise 18.

4. Make a table and plot as in Problem 3, but approximate $f'(1)$, where $f(x) = x^{-1}$. Compare your answers with the theory developed in Exercise 18.

5. Make a plot as in Problem 1 to approximate $f''(0)$ for $f(x) = \cos x$, using the three-point centered-difference formula. Where does the minimum error appear to occur, in terms of machine epsilon?

5.2 NEWTON–COTES FORMULAS FOR NUMERICAL INTEGRATION

The numerical calculation of definite integrals relies on many of the same tools we have already seen. In Chapters 3 and 4, methods were developed for finding function approximation to a set of data points, using interpolation and least squares modeling. We will discuss methods for **numerical integration**, or **quadrature**, based on both of these ideas.

For example, given a function f defined on an interval $[a, b]$, we can draw an interpolating polynomial through some of the points of $f(x)$. Since it is simple to evaluate the definite integral of a polynomial, this calculation can be used to approximate the integral of $f(x)$. This is the Newton–Cotes approach to approximating integrals. Alternatively, we could find a low-degree polynomial that approximates the function well in the sense of least squares and use the integral as the approximation, in a method called Gaussian Quadrature. Both of these approaches will be described in this chapter.

To develop the Newton–Cotes formulas, we need the values of three simple definite integrals, pictured in Figure 5.2.

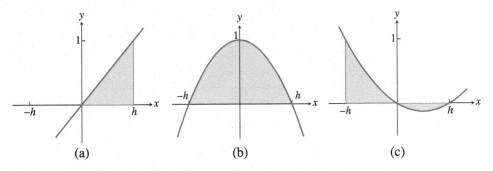

(a)   (b)   (c)

Figure 5.2 Three simple integrals (5.17), (5.18), and (5.19). Net positive area is (a) $h/2$, (b) $4h/3$, and (c) $h/3$.

Figure 5.2(a) shows the region under the line interpolating the data points $(0, 0)$ and $(h, 1)$. The region is a triangle of height 1 and base h, so the area is

$$\int_0^h \frac{x}{h}\, dx = h/2. \tag{5.17}$$

Figure 5.2(b) shows the region under the parabola $P(x)$ interpolating the data points $(-h, 0), (0, 1)$, and $(h, 0)$, which has area

$$\int_{-h}^h P(x)\, dx = x - \frac{x^3}{3h^2} = \frac{4}{3}h. \tag{5.18}$$

Figure 5.2(c) shows the region between the x-axis and the parabola interpolating the data points $(-h, 1), (0, 0)$, and $(h, 0)$, with net positive area

$$\int_{-h}^h P(x)\, dx = \frac{1}{3}h. \tag{5.19}$$

5.2.1 Trapezoid Rule

We begin with the simplest application of interpolation-based numerical integration. Let $f(x)$ be a function with a continuous second derivative, defined on the interval $[x_0, x_1]$, as shown in Figure 5.3(a). Denote the corresponding function values by $y_0 = f(x_0)$ and $y_1 = f(x_1)$. Consider the degree 1 interpolating polynomial $P_1(x)$ through (x_0, y_0) and (x_1, y_1). Using the Lagrange formulation, we find that the interpolating polynomial with error term is

$$f(x) = y_0 \frac{x - x_1}{x_0 - x_1} + y_1 \frac{x - x_0}{x_1 - x_0} + \frac{(x - x_0)(x - x_1)}{2!} f''(c_x) = P(x) + E(x).$$

It can be proved that the "unknown point" c_x depends continuously on x.

Integrating both sides on the interval of interest $[x_0, x_1]$ yields

$$\int_{x_0}^{x_1} f(x)\, dx = \int_{x_0}^{x_1} P(x)\, dx + \int_{x_0}^{x_1} E(x)\, dx.$$

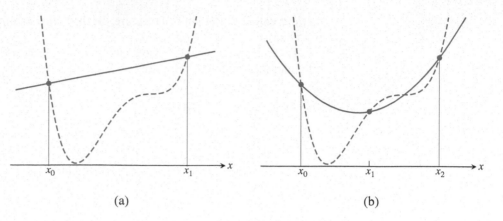

Figure 5.3 Newton-Cotes formulas are based on interpolation. (a) Trapezoid Rule replaces the function with the line interpolating $(x_0, f(x_0))$ and $(x_1, f(x_1))$. (b) Simpson's Rule uses the parabola interpolating the function at three points $(x_0, f(x_0))$, $(x_1, f(x_1))$, and $(x_2, f(x_2))$.

Computing the first integral gives

$$\int_{x_0}^{x_1} P(x)\, dx = y_0 \int_{x_0}^{x_1} \frac{x - x_1}{x_0 - x_1}\, dx + y_1 \int_{x_0}^{x_1} \frac{x - x_0}{x_1 - x_0}\, dx$$

$$= y_0 \frac{h}{2} + y_1 \frac{h}{2} = h \frac{y_0 + y_1}{2}, \qquad (5.20)$$

where we have defined $h = x_1 - x_0$ to be the interval length and computed the integrals by using the fact (5.17). For example, substituting $w = -x + x_1$ into the first integral gives

$$\int_{x_0}^{x_1} \frac{x - x_1}{x_0 - x_1}\, dx = \int_h^0 \frac{-w}{-h}(-dw) = \int_0^h \frac{w}{h}\, dw = \frac{h}{2},$$

and the second integral, after substituting $w = x - x_0$, is

$$\int_{x_0}^{x_1} \frac{x - x_0}{x_1 - x_0}\, dx = \int_0^h \frac{w}{h}\, dw = \frac{h}{2}.$$

Formula (5.20) calculates the area of a trapezoid, which gives the rule its name.

The error term is

$$\int_{x_0}^{x_1} E(x)\, dx = \frac{1}{2!} \int_{x_0}^{x_1} (x - x_0)(x - x_1) f''(c(x))\, dx$$

$$= \frac{f''(c)}{2} \int_{x_0}^{x_1} (x - x_0)(x - x_1)\, dx$$

$$= \frac{f''(c)}{2} \int_0^h u(u - h)\, du$$

$$= -\frac{h^3}{12} f''(c),$$

where we have used Theorem 0.9, the Mean Value Theorem for Integrals. We have shown:

Trapezoid Rule

$$\int_{x_0}^{x_1} f(x)\,dx = \frac{h}{2}(y_0 + y_1) - \frac{h^3}{12}f''(c), \tag{5.21}$$

where $h = x_1 - x_0$ and c is between x_0 and x_1.

5.2.2 Simpson's Rule

Figure 5.3(b) illustrates **Simpson's Rule**, which is similar to the Trapezoid Rule, except that the degree 1 interpolant is replaced by a parabola. As before, we can write the integrand $f(x)$ as the sum of the interpolating parabola and the interpolation error:

$$
\begin{aligned}
f(x) &= y_0 \frac{(x - x_1)(x - x_2)}{(x_0 - x_1)(x_0 - x_2)} + y_1 \frac{(x - x_0)(x - x_2)}{(x_1 - x_0)(x_1 - x_2)} \\
&\quad + y_2 \frac{(x - x_0)(x - x_1)}{(x_2 - x_0)(x_2 - x_1)} + \frac{(x - x_0)(x - x_1)(x - x_2)}{3!} f'''(c_x) \\
&= P(x) + E(x).
\end{aligned}
$$

Integrating gives

$$\int_{x_0}^{x_2} f(x)\,dx = \int_{x_0}^{x_2} P(x)\,dx + \int_{x_0}^{x_2} E(x)\,dx,$$

where

$$
\begin{aligned}
\int_{x_0}^{x_2} P(x)\,dx &= y_0 \int_{x_0}^{x_2} \frac{(x - x_1)(x - x_2)\,dx}{(x_0 - x_1)(x_0 - x_2)} + y_1 \int_{x_0}^{x_2} \frac{(x - x_0)(x - x_2)\,dx}{(x_1 - x_0)(x_1 - x_2)} \\
&\quad + y_2 \int_{x_0}^{x_2} \frac{(x - x_0)(x - x_1)\,dx}{(x_2 - x_0)(x_2 - x_1)} \\
&= y_0 \frac{h}{3} + y_1 \frac{4h}{3} + y_2 \frac{h}{3}.
\end{aligned}
$$

We have set $h = x_2 - x_1 = x_1 - x_0$ and used (5.18) for the middle integral and (5.19) for the first and third. The error term can be computed (proof omitted) as

$$\int_{x_0}^{x_2} E(x)\,dx = -\frac{h^5}{90} f^{(iv)}(c)$$

for some c in the interval $[x_0, x_2]$, provided that $f^{(iv)}$ exists and is continuous. Concluding the derivation yields Simpson's Rule:

Simpson's Rule

$$\int_{x_0}^{x_2} f(x)\,dx = \frac{h}{3}(y_0 + 4y_1 + y_2) - \frac{h^5}{90} f^{(iv)}(c), \tag{5.22}$$

where $h = x_2 - x_1 = x_1 - x_0$ and c is between x_0 and x_2.

▶ **EXAMPLE 5.6** Apply the Trapezoid Rule and Simpson's Rule to approximate

$$\int_1^2 \ln x \, dx,$$

and find an upper bound for the error in your approximations.

The Trapezoid Rule estimates that

$$\int_1^2 \ln x \, dx \approx \frac{h}{2}(y_0 + y_1) = \frac{1}{2}(\ln 1 + \ln 2) = \frac{\ln 2}{2} \approx 0.3466.$$

The error for the Trapezoid Rule is $-h^3 f''(c)/12$, where $1 < c < 2$. Since $f''(x) = -1/x^2$, the magnitude of the error is at most

$$\frac{1^3}{12c^2} \leq \frac{1}{12} \approx 0.0834.$$

In other words, the Trapezoid Rule says that

$$\int_1^2 \ln x \, dx = 0.3466 \pm 0.0834.$$

The integral can be computed exactly by using integration by parts:

$$\int_1^2 \ln x \, dx = x \ln x \big|_1^2 - \int_1^2 dx$$
$$= 2\ln 2 - 1\ln 1 - 1 \approx 0.386294. \qquad (5.23)$$

The Trapezoid Rule approximation and error bound are consistent with this result. Simpson's Rule yields the estimate

$$\int_1^2 \ln x \, dx \approx \frac{h}{3}(y_0 + 4y_1 + y_2) = \frac{0.5}{3}\left(\ln 1 + 4\ln \frac{3}{2} + \ln 2\right) \approx 0.3858.$$

The error for Simpson's Rule is $-h^5 f^{(iv)}(c)/90$, where $1 < c < 2$. Since $f^{(iv)}(x) = -6/x^4$, the error is at most

$$\frac{6(0.5)^5}{90c^4} \leq \frac{6(0.5)^5}{90} = \frac{1}{480} \approx 0.0021.$$

Thus, Simpson's Rule says that

$$\int_1^2 \ln x \, dx = 0.3858 \pm 0.0021,$$

which is again consistent with the correct value and more accurate than the Trapezoid Rule approximation. ◀

One way of comparing numerical integration rules like the Trapezoid Rule or Simpson's Rule is by comparing error terms. This information is conveyed simply through the following definition:

DEFINITION 5.2 The **degree of precision** of a numerical integration method is the greatest integer k for which all degree k or less polynomials are integrated exactly by the method. ☐

For example, the error term of the Trapezoid Rule, $-h^3 f''(c)/12$, shows that if $f(x)$ is a polynomial of degree 1 or less, the error will be zero, and the polynomial will be integrated exactly. So the degree of precision of the Trapezoid Rule is 1. This is intuitively obvious from geometry, since the area under a linear function is approximated exactly by a trapezoid.

It is less obvious that the degree of precision of Simpson's Rule is three, but that is what the error term in (5.22) shows. The geometric basis of this surprising result is the fact that a parabola intersecting a cubic curve at three equally spaced points has the same integral as the cubic curve over that interval (Exercise 17).

▶ **EXAMPLE 5.7** Find the degree of precision of the degree 3 Newton–Cotes formula, called the **Simpson's 3/8 Rule**

$$\int_{x_0}^{x_3} f(x)dx \approx \frac{3h}{8}(y_0 + 3y_1 + 3y_2 + y_3).$$

It suffices to test monomials in succession. We will leave the details to the reader. For example, when $f(x) = x^2$, we check the identity

$$\frac{3h}{8}(x^2 + 3(x+h)^2 + 3(x+2h)^2 + (x+3h)^2) = \frac{(x+3h)^3 - x^3}{3},$$

the latter being the correct integral of x^2 on $[x, x+3h]$. Equality holds for $1, x, x^2, x^3$, but fails for x^4. Therefore, the degree of precision of the rule is 3. ◀

The Trapezoid Rule and Simpson's Rule are examples of "closed" Newton–Cotes formulas, because they include evaluations of the integrand at the interval endpoints. The open Newton–Cotes formulas are useful for circumstances where that is not possible, for example, when approximating an improper integral. We discuss open formulas in Section 5.2.4.

5.2.3 Composite Newton–Cotes formulas

The Trapezoid and Simpson's Rules are limited to operating on a single interval. Of course, since definite integrals are additive over subintervals, we can evaluate an integral by dividing the interval up into several subintervals, applying the rule separately on each one, and then totaling up. This strategy is called **composite numerical integration**.

The composite Trapezoid Rule is simply the sum of Trapezoid Rule approximations on adjacent subintervals, or **panels**. To approximate

$$\int_a^b f(x)\, dx,$$

consider an evenly spaced grid

$$a = x_0 < x_1 < x_2 < \cdots < x_{m-2} < x_{m-1} < x_m = b$$

along the horizontal axis, where $h = x_{i+1} - x_i$ for each i as shown in Figure 5.4. On each subinterval, we make the approximation with error term

$$\int_{x_i}^{x_{i+1}} f(x)\, dx = \frac{h}{2}(f(x_i) + f(x_{i+1})) - \frac{h^3}{12} f''(c_i),$$

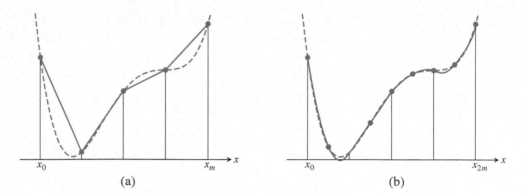

Figure 5.4 Newton–Cotes composite formulas. (a) Composite Trapezoid Rule sums the Trapezoid Rule formula (solid care) on m adjacent subintervals. (b) Composite Simpson's Rule does the same for Simpson's Rule.

assuming that f'' is continuous. Adding up over all subintervals (note the overlapping on the interior subintervals) yields

$$\int_a^b f(x)\,dx = \frac{h}{2}\left[f(a) + f(b) + 2\sum_{i=1}^{m-1} f(x_i) \right] - \sum_{i=0}^{m-1} \frac{h^3}{12} f''(c_i).$$

The error term can be written

$$\frac{h^3}{12}\sum_{i=0}^{m-1} f''(c_i) = \frac{h^3}{12} m f''(c),$$

according to Theorem 5.1, for some $a < c < b$. Since $mh = (b - a)$, the error term is $(b-a)h^2 f''(c)/12$. To summarize, if f'' is continuous on $[a, b]$, then the following holds:

Composite Trapezoid Rule

$$\int_a^b f(x)\,dx = \frac{h}{2}\left(y_0 + y_m + 2\sum_{i=1}^{m-1} y_i \right) - \frac{(b-a)h^2}{12} f''(c), \qquad (5.24)$$

where $h = (b - a)/m$ and c is between a and b.

The composite Simpson's Rule follows the same strategy. Consider an evenly spaced grid

$$a = x_0 < x_1 < x_2 < \cdots < x_{2m-2} < x_{2m-1} < x_{2m} = b$$

along the horizontal axis, where $h = x_{i+1} - x_i$ for each i. On each length $2h$ panel $[x_{2i}, x_{2i+2}]$, for $i = 0, \ldots, m - 1$, a Simpson's Method is carried out. In other words, the integrand $f(x)$ is approximated on each subinterval by the interpolating parabola fit at x_{2i}, x_{2i+1}, and x_{2i+2}, which is integrated and added to the sum. The approximation with error term on the subinterval is

$$\int_{x_{2i}}^{x_{2i+2}} f(x)\,dx = \frac{h}{3}[f(x_{2i}) + 4f(x_{2i+1}) + f(x_{2i+2})] - \frac{h^5}{90} f^{(iv)}(c_i).$$

This time, the overlapping is over even-numbered x_j only. Adding up over all subintervals yields

$$\int_a^b f(x)\, dx = \frac{h}{3}\left[f(a) + f(b) + 4\sum_{i=1}^{m} f(x_{2i-1}) + 2\sum_{i=1}^{m-1} f(x_{2i}) \right] - \sum_{i=0}^{m-1}\frac{h^5}{90} f^{(iv)}(c_i).$$

The error term can be written

$$\frac{h^5}{90}\sum_{i=0}^{m-1} f^{(iv)}(c_i) = \frac{h^5}{90}mf^{(iv)}(c),$$

according to Theorem 5.1, for some $a < c < b$. Since $m \cdot 2h = (b - a)$, the error term is $(b - a)h^4 f^{(iv)}(c)/180$. Assuming that $f^{(iv)}$ is continuous on $[a, b]$, the following holds:

Composite Simpson's Rule

$$\int_a^b f(x)\, dx = \frac{h}{3}\left[y_0 + y_{2m} + 4\sum_{i=1}^{m} y_{2i-1} + 2\sum_{i=1}^{m-1} y_{2i} \right] - \frac{(b-a)h^4}{180} f^{(iv)}(c), \quad (5.25)$$

where c is between a and b.

▶ **EXAMPLE 5.8** Carry out four-panel approximations of

$$\int_1^2 \ln x\, dx,$$

using the composite Trapezoid Rule and composite Simpson's Rule.

For the composite Trapezoid Rule on $[1, 2]$, four panels means that $h = 1/4$. The approximation is

$$\int_1^2 \ln x\, dx \approx \frac{1/4}{2}\left[y_0 + y_4 + 2\sum_{i=1}^{3} y_i \right]$$

$$= \frac{1}{8}[\ln 1 + \ln 2 + 2(\ln 5/4 + \ln 6/4 + \ln 7/4)]$$

$$\approx 0.3837.$$

The error is at most

$$\frac{(b-a)h^2}{12}|f''(c)| = \frac{1/16}{12}\frac{1}{c^2} \leq \frac{1}{(16)(12)(1^2)} = \frac{1}{192} \approx 0.0052.$$

A four-panel Simpson's Rule sets $h = 1/8$. The approximation is

$$\int_1^2 \ln x\, dx \approx \frac{1/8}{3}\left[y_0 + y_8 + 4\sum_{i=1}^{4} y_{2i-1} + 2\sum_{i=1}^{3} y_{2i} \right]$$

$$= \frac{1}{24}[\ln 1 + \ln 2 + 4(\ln 9/8 + \ln 11/8 + \ln 13/8 + \ln 15/8)$$

$$+ 2(\ln 5/4 + \ln 6/4 + \ln 7/4)]$$

$$\approx 0.386292.$$

This agrees within five decimal places with the correct value 0.386294 from (5.23). Indeed, the error cannot be more than

$$\frac{(b-a)h^4}{180}|f^{(iv)}(c)| = \frac{(1/8)^4}{180}\frac{6}{c^4} \le \frac{6}{8^4 \cdot 180 \cdot 1^4} \approx 0.000008.$$ ◀

▶ **EXAMPLE 5.9** Find the number of panels m necessary for the composite Simpson's Rule to approximate

$$\int_0^\pi \sin^2 x \, dx$$

within six correct decimal places.

We require the error to satisfy

$$\frac{(\pi-0)h^4}{180}|f^{(iv)}(c)| < 0.5 \times 10^{-6}.$$

Since the fourth derivative of $\sin^2 x$ is $-8\cos 2x$, we need

$$\frac{\pi h^4}{180}8 < 0.5 \times 10^{-6},$$

or $h < 0.0435$. Therefore, $m = \texttt{ceil}(\pi/(2h)) = 37$ panels will be sufficient. ◀

5.2.4 Open Newton–Cotes Methods

The so-called closed Newton–Cotes Methods like Trapezoid and Simpson's Rules require input values from the ends of the integration interval. Some integrands that have a removable singularity at an interval endpoint may be more easily handled with an open Newton–Cotes Method, which does not use values from the endpoints. The following rule is applicable to functions f whose second derivative f'' is continuous on $[a, b]$:

Midpoint Rule

$$\int_{x_0}^{x_1} f(x) \, dx = hf(w) + \frac{h^3}{24}f''(c), \tag{5.26}$$

where $h = (x_1 - x_0)$, w is the midpoint $x_0 + h/2$, and c is between x_0 and x_1.

The Midpoint Rule is also useful for cutting the number of function evaluations needed. Compared with the Trapezoid Rule, the closed Newton–Cotes Method of the same order, it requires one function evaluation rather than two. Moreover, the error term is half the size of the Trapezoid Rule error term.

The proof of (5.26) follows the same lines as the derivation of the Trapezoid Rule. Set $h = x_1 - x_0$. The degree 1 Taylor expansion of $f(x)$ about the midpoint $w = x_0 + h/2$ of the interval is

$$f(x) = f(w) + (x - w)f'(w) + \frac{1}{2}(x - w)^2 f''(c_x),$$

where c_x depends on x and lies between x_0 and x_1. Integrating both sides yields

$$\int_{x_0}^{x_1} f(x)\,dx = (x_1 - x_0)f(w) + f'(w)\int_{x_0}^{x_1}(x-w)\,dx + \frac{1}{2}\int_{x_0}^{x_1} f''(c_x)(x-w)^2\,dx$$

$$= hf(w) + 0 + \frac{f''(c)}{2}\int_{x_0}^{x_1}(x-w)^2\,dx$$

$$= hf(w) + \frac{h^3}{24}f''(c),$$

where $x_0 < c < x_1$. Again, we have used the Mean Value Theorem for Integrals to pull the second derivative outside of the integral. This completes the derivation of (5.26).

The proof of the composite version is left to the reader (Exercise 12).

Composite Midpoint Rule

$$\int_a^b f(x)\,dx = h\sum_{i=1}^m f(w_i) + \frac{(b-a)h^2}{24}f''(c), \tag{5.27}$$

where $h = (b-a)/m$ and c is between a and b. The w_i are the midpoints of the m equal subintervals of a, b.

▶ **EXAMPLE 5.10** Approximate $\int_0^1 \sin x/x\,dx$ by using the Composite Midpoint Rule with $m = 10$ panels.

First note that we cannot apply a closed method directly to the problem, without special handling at $x = 0$. The midpoint method can be applied directly. The midpoints are $0.05, 0.15, \ldots, 0.95$, so the Composite Midpoint Rule delivers

$$\int_0^1 f(x)\,dx \approx 0.1\sum_1^{10} f(m_i) = 0.94620858.$$

The correct answer to eight places is 0.94608307. ◀

Another useful open Newton–Cotes Rule is

$$\int_{x_0}^{x_4} f(x)\,dx = \frac{4h}{3}[2f(x_1) - f(x_2) + 2f(x_3)] + \frac{14h^5}{45}f^{(iv)}(c), \tag{5.28}$$

where $h = (x_4 - x_0)/4$, $x_1 = x_0 + h$, $x_2 = x_0 + 2h$, $x_3 = x_0 + 3h$, and where $x_0 < c < x_4$. The rule has degree of precision three. Exercise 11 asks you to extend it to a composite rule.

▶ **ADDITIONAL EXAMPLES**

1. Show by direct calculation that the open Newton–Cotes formula $\int_0^b f(x)\,dx \approx \frac{b}{3}[2f(b/4) - f(b/2) + 2f(3b/4)]$ has degree of precision 3.

*2 Use the Composite Trapezoid Method error formula to find the number of panels required to estimate $\int_0^1 e^{-x^2}\,dx$ to 6 correct decimal places, and carry out the estimate.

⌐❏ **Solutions** for Additional Examples can be found at bit.ly/2Cunk41
(* example with video solution)

5.2 Exercises

Solutions
for Exercises
numbered in blue
can be found at
bit.ly/2RVYxef

1. Apply the composite Trapezoid Rule with $m = 1, 2$, and 4 panels to approximate the integral. Compute the error by comparing with the exact value from calculus.

$$\text{(a)} \quad \int_0^1 x^2 \, dx \quad \text{(b)} \quad \int_0^{\pi/2} \cos x \, dx \quad \text{(c)} \quad \int_0^1 e^x \, dx$$

2. Apply the Composite Midpoint Rule with $m = 1, 2$, and 4 panels to approximate the integrals in Exercise 1, and report the errors.

3. Apply the composite Simpson's Rule with $m = 1, 2$, and 4 panels to the integrals in Exercise 1, and report the errors.

4. Apply the composite Simpson's Rule with $m = 1, 2$, and 4 panels to the integrals, and report the errors.

$$\text{(a)} \quad \int_0^1 x e^x \, dx \quad \text{(b)} \quad \int_0^1 \frac{dx}{1 + x^2} \, dx \quad \text{(c)} \quad \int_0^\pi x \cos x \, dx$$

5. Apply the Composite Midpoint Rule with $m = 1, 2$, and 4 panels to approximate the integrals. Compute the error by comparing with the exact value from calculus.

$$\text{(a)} \quad \int_0^1 \frac{dx}{\sqrt{x}} \quad \text{(b)} \quad \int_0^1 x^{-1/3} \, dx \quad \text{(c)} \quad \int_0^2 \frac{dx}{\sqrt{2 - x}}$$

6. Apply the Composite Midpoint Rule with $m = 1, 2$, and 4 panels to approximate the integrals.

$$\text{(a)} \quad \int_0^{\pi/2} \frac{1 - \cos x}{x^2} \, dx \quad \text{(b)} \quad \int_0^1 \frac{e^x - 1}{x} \, dx \quad \text{(c)} \quad \int_0^{\pi/2} \frac{\cos x}{\frac{\pi}{2} - x} \, dx$$

7. Apply the open Newton–Cotes Rule (5.28) to approximate the integrals of Exercise 5, and report the errors.

8. Apply the open Newton–Cotes Rule (5.28) to approximate the integrals of Exercise 6.

9. Apply Simpson's Rule approximation to $\int_0^1 x^4 \, dx$, and show that the approximation error matches the error term from (5.22).

10. Integrate Newton's divided-difference interpolating polynomial to prove the formula (a) (5.18) (b) (5.19).

11. Find the degree of precision of the following approximation for $\int_{-1}^1 f(x) \, dx$:
    (a) $f(1) + f(-1)$ (b) $2/3 f(-1) + f(0) + f(1)$ (c) $f(-1/\sqrt{3}) + f(1/\sqrt{3})$.

12. Find c_1, c_2, and c_3 such that the rule

$$\int_0^1 f(x) \, dx \approx c_1 f(0) + c_2 f(0.5) + c_3 f(1)$$

has degree of precision greater than one. (Hint: Substitute $f(x) = 1, x$, and x^2.) Do you recognize the method that results?

13. Develop a composite version of the rule (5.28), with error term.

14. Prove the Composite Midpoint Rule (5.27).

15. Find the degree of precision of the degree four Newton–Cotes Rule (often called Boole's Rule)

$$\int_{x_0}^{x_4} f(x) \, dx \approx \frac{2h}{45} (7 y_0 + 32 y_1 + 12 y_2 + 32 y_3 + 7 y_4).$$

16. Use the fact that the error term of Boole's Rule is proportional to $f^{(6)}(c)$ to find the exact error term, by the following strategy: Compute Boole's approximation for $\int_0^{4h} x^6 \, dx$, find the approximation error, and write it in terms of h and $f^{(6)}(c)$.

17. Let $P_3(x)$ be a degree 3 polynomial, and let $P_2(x)$ be its interpolating polynomial at the three points $x = -h, 0$, and h. Prove directly that $\int_{-h}^{h} P_3(x) \, dx = \int_{-h}^{h} P_2(x) \, dx$. What does this fact say about Simpson's Rule?

5.2 Computer Problems

Solutions for Computer Problems numbered in blue can be found at bit.ly/2CS2DQv

1. Use the composite Trapezoid Rule with $m = 16$ and 32 panels to approximate the definite integral. Compare with the correct integral and report the two errors.

   (a) $\displaystyle\int_0^4 \frac{x \, dx}{\sqrt{x^2 + 9}}$ (b) $\displaystyle\int_0^1 \frac{x^3 \, dx}{x^2 + 1}$ (c) $\displaystyle\int_0^1 x e^x \, dx$ (d) $\displaystyle\int_1^3 x^2 \ln x \, dx$

   (e) $\displaystyle\int_0^\pi x^2 \sin x \, dx$ (f) $\displaystyle\int_2^3 \frac{x^3 \, dx}{\sqrt{x^4 - 1}}$ (g) $\displaystyle\int_0^{2\sqrt{3}} \frac{dx}{\sqrt{x^2 + 4}} \, dx$ (h) $\displaystyle\int_0^1 \frac{x \, dx}{\sqrt{x^4 + 1}}$

2. Apply the composite Simpson's Rule to the integrals in Computer Problem 1. Use $m = 16$ and 32, and report errors.

3. Use the composite Trapezoid Rule with $m = 16$ and 32 panels to approximate the definite integral.

   (a) $\displaystyle\int_0^1 e^{x^2} \, dx$ (b) $\displaystyle\int_0^{\sqrt{\pi}} \sin x^2 \, dx$ (c) $\displaystyle\int_0^\pi e^{\cos x} \, dx$ (d) $\displaystyle\int_0^1 \ln(x^2 + 1) \, dx$

   (e) $\displaystyle\int_0^1 \frac{x \, dx}{2e^x - e^{-x}}$ (f) $\displaystyle\int_0^\pi \cos e^x \, dx$ (g) $\displaystyle\int_0^1 x^x \, dx$ (h) $\displaystyle\int_0^{\pi/2} \ln(\cos x + \sin x) \, dx$

4. Apply the composite Simpson's Rule to the integrals of Computer Problem 3, using $m = 16$ and 32.

5. Apply the Composite Midpoint Rule to the improper integrals of Exercise 5, using $m = 10, 100$, and 1000. Compute the error by comparing with the exact value.

6. Apply the Composite Midpoint Rule to the improper integrals of Exercise 6, using $m = 16$ and 32.

7. Apply the Composite Midpoint Rule to the improper integrals

   (a) $\displaystyle\int_0^{\frac{\pi}{2}} \frac{x}{\sin x} \, dx$ (b) $\displaystyle\int_0^{\frac{\pi}{2}} \frac{e^x - 1}{\sin x} \, dx$ (c) $\displaystyle\int_0^1 \frac{\arctan x}{x} \, dx$,

   using $m = 16$ and 32.

8. The arc length of the curve defined by $y = f(x)$ from $x = a$ to $x = b$ is given by the integral $\int_a^b \sqrt{1 + f'(x)^2} \, dx$. Use the composite Simpson's Rule with $m = 32$ panels to approximate the lengths of the curves

   (a) $y = x^3$ on 0, 1   (b) $y = \tan x$ on $0, \pi/4$ (c) $y = \arctan x$ on 0, 1.

9. For the integrals in Computer Problem 1, calculate the approximation error of the composite Trapezoid Rule for $h = b - a, h/2, h/4, \ldots, h/2^8$, and plot. Make a log–log plot, using, for example, MATLAB's loglog command. What is the slope of the plot, and does it agree with theory?

10. Carry out Computer Problem 9, but use the composite Simpson's Rule instead of the composite Trapezoid Rule.

5.3 ROMBERG INTEGRATION

In this section, we begin discussing efficient methods for calculating definite integrals that can be extended by adding data until the required accuracy is attained. Romberg Integration is the result of applying extrapolation to the composite Trapezoid Rule. Recall from Section 5.1 that, given a rule $N(h)$ for approximating a quantity M, depending on a step size h, the rule can be extrapolated if the order of the rule is known. Equation (5.24) shows that the composite Trapezoid Rule is a second-order rule in h. Therefore, extrapolation can be applied to achieve a new rule of (at least) third order.

Examining the error of the Trapezoid Rule (5.24) more carefully, it can be shown that, for an infinitely differentiable function f,

$$\int_a^b f(x)\,dx = \frac{h}{2}\left(y_0 + y_m + 2\sum_{i=1}^{m-1} y_i\right) + c_2 h^2 + c_4 h^4 + c_6 h^6 + \cdots, \qquad (5.29)$$

where the c_i depend only on higher derivatives of f at a and b, and not on h. For example, $c_2 = (f'(a) - f'(b))/12$. The absence of odd powers in the error gives an extra bonus when extrapolation is done. Since there are no odd-power terms, extrapolation with the second-order formula given by the composite Trapezoid Rule yields a fourth-order formula; extrapolation with the resulting fourth-order formula gives a sixth-order formula, and so on.

Extrapolation involves combining the formula evaluated once at h and once at $h/2$, half the step size. Foreshadowing where we are headed, define the following series of step sizes:

$$h_1 = b - a$$

$$h_2 = \frac{1}{2}(b - a)$$

$$\vdots$$

$$h_j = \frac{1}{2^{j-1}}(b - a). \qquad (5.30)$$

The quantity being approximated is $M = \int_a^b f(x)\,dx$. Define the approximating formulas R_{j1} to be the composite Trapezoid Rule, using h_j. Thus, $R_{j+1,1}$ is exactly R_{j1} with step size cut in half, as needed to apply extrapolation. Second, notice the overlapping of the formulas. Some of the same function evaluations $f(x)$ are needed in both R_{j1} and $R_{j+1,1}$. For example, we have

$$R_{11} = \frac{h_1}{2}(f(a) + f(b))$$

$$R_{21} = \frac{h_2}{2}\left(f(a) + f(b) + 2f\left(\frac{a+b}{2}\right)\right)$$

$$= \frac{1}{2}R_{11} + h_2 f\left(\frac{a+b}{2}\right).$$

We prove (see Exercise 5) that for $j = 2, 3, \dots$.

$$R_{j1} = \frac{1}{2}R_{j-1,1} + h_j \sum_{i=1}^{2^{j-2}} f(a + (2i - 1)h_j). \qquad (5.31)$$

Equation (5.31) gives an efficient way to calculate the composite Trapezoid Rule incrementally. The second feature of Romberg Integration is extrapolation. Form the tableau

$$
\begin{matrix}
R_{11} \\
R_{21} & R_{22} \\
R_{31} & R_{32} & R_{33} \\
R_{41} & R_{42} & R_{43} & R_{44} \\
\vdots & & & & \ddots
\end{matrix}
\tag{5.32}
$$

where we define the second column R_{i2} as the extrapolations of the first column:

$$
R_{22} = \frac{2^2 R_{21} - R_{11}}{3}
$$

$$
R_{32} = \frac{2^2 R_{31} - R_{21}}{3}
$$

$$
R_{42} = \frac{2^2 R_{41} - R_{31}}{3}.
\tag{5.33}
$$

The third column consists of fourth-order approximations of M, so they can be extrapolated as

$$
R_{33} = \frac{4^2 R_{32} - R_{22}}{4^2 - 1}
$$

$$
R_{43} = \frac{4^2 R_{42} - R_{32}}{4^2 - 1}
$$

$$
R_{53} = \frac{4^2 R_{52} - R_{42}}{4^2 - 1},
\tag{5.34}
$$

and so forth. The general jkth entry is given by the formula (see Exercise 6)

$$
R_{jk} = \frac{4^{k-1} R_{j,k-1} - R_{j-1,k-1}}{4^{k-1} - 1}.
\tag{5.35}
$$

The tableau is a lower triangular matrix that extends infinitely down and across. The best approximation for the definite integral M is R_{jj}, the bottom rightmost entry computed so far, which is a $2j$th-order approximation. The Romberg Integration calculation is just a matter of writing formulas (5.31) and (5.35) in a loop.

Romberg Integration

$$
R_{11} = (b - a)\frac{f(a) + f(b)}{2}
$$

for $j = 2, 3, \ldots$

$$
h_j = \frac{b - a}{2^{j-1}}
$$

$$
R_{j1} = \frac{1}{2} R_{j-1,1} + h_j \sum_{i=1}^{2^{j-2}} f(a + (2i - 1)h_j)
$$

    for $k = 2, \ldots, j$

$$
R_{jk} = \frac{4^{k-1} R_{j,k-1} - R_{j-1,k-1}}{4^{k-1} - 1}
$$

    end

end

The MATLAB code is a straightforward implementation of the preceding algorithm.

MATLAB code

shown here can be found at bit.ly/2NLs1Ik

```
%Program 5.1 Romberg integration
% Computes approximation to definite integral
% Inputs: Matlab function specifying integrand f,
%   a,b integration interval, n=number of rows
% Output: Romberg tableau r
function r=romberg(f,a,b,n)
h=(b-a)./(2.^(0:n-1));
r(1,1)=(b-a)*(f(a)+f(b))/2;
for j=2:n
  subtotal = 0;
  for i=1:2^(j-2)
    subtotal = subtotal + f(a+(2*i-1)*h(j));
  end
  r(j,1) = r(j-1,1)/2+h(j)*subtotal;
  for k=2:j
    r(j,k)=(4^(k-1)*r(j,k-1)-r(j-1,k-1))/(4^(k-1)-1);
  end
end
```

► **EXAMPLE 5.11**   Apply Romberg Integration to approximate $\int_1^2 \ln x \, dx$.

We use the MATLAB built-in function log. Its function handle is designated by @log. Running the foregoing code results in

```
>> romberg(@log,1,2,4)

ans =

    0.34657359027997   0                  0                0
    0.37601934919407   0.38583460216543   0                0
    0.38369950940944   0.38625956281457   0.38628789352451 0
    0.38564390995210   0.38629204346631   0.38629420884310 0.38629430908625
```

Note the agreement of R_{43} and R_{44} in their first six decimal places. This is a sign of convergence of the Romberg Method to the correct value of the definite integral. Compare with the exact value $2\ln 2 - 1 \approx 0.38629436$.   ◄

Comparing the results of Example 5.11 with those of Example 5.8 shows a match between the last entry in the second column of Romberg and the composite Simpson's Rule results. This is not a coincidence. In fact, just as the first column of Romberg is defined to be successive composite trapezoidal rule entries, the second column is composite Simpson's entries. In other words, the extrapolation of the composite Trapezoid Rule is the composite Simpson's Rule. See Exercise 3.

A common stopping criterion for Romberg Integration is to compute new rows until two successive diagonal entries R_{jj} differ by less than a preset error tolerance.

► **ADDITIONAL EXAMPLES**

*1 Apply Romberg Integration to hand-calculate R_{33} for the integral $\int_0^{\pi/2} \sin x \, dx$.

2. Use romberg.m to compute R_{55} for the integral $\int_0^{\pi/4} \tan x \, dx$. How many decimal places of the approximation are correct?

Solutions for Additional Examples can be found at bit.ly/20x48cV
(* example with video solution)

5.3 Exercises

1. Apply Romberg Integration to find R_{33} for the integrals.

   (a) $\int_0^1 x^2 \, dx$ (b) $\int_0^{\pi/2} \cos x \, dx$ (c) $\int_0^1 e^x \, dx$

2. Apply Romberg Integration to find R_{33} for the integrals.

   (a) $\int_0^1 xe^x \, dx$ (b) $\int_0^1 \frac{dx}{1+x^2} \, dx$ (c) $\int_0^\pi x \cos x \, dx$

3. Show that the extrapolation of the composite Trapezoid Rules in R_{11} and R_{21} yields the composite Simpson's Rule (with step size h_2) in R_{22}.

4. Show that R_{33} of Romberg Integration can be expressed as Boole's Rule (with step size h_3), defined in Exercise 5.2.13.

5. Prove formula (5.31).

6. Prove formula (5.35).

5.3 Computer Problems

1. Use Romberg Integration approximation R_{55} to approximate the definite integral. Compare with the correct integral, and report the error.

   (a) $\int_0^4 \frac{x \, dx}{\sqrt{x^2 + 9}}$ (b) $\int_0^1 \frac{x^3 \, dx}{x^2 + 1}$ (c) $\int_0^1 xe^x \, dx$ (d) $\int_1^3 x^2 \ln x \, dx$

   (e) $\int_0^\pi x^2 \sin x \, dx$ (f) $\int_2^3 \frac{x^3 \, dx}{\sqrt{x^4 - 1}}$ (g) $\int_0^{2\sqrt{3}} \frac{dx}{\sqrt{x^2 + 4}} \, dx$ (h) $\int_0^1 \frac{x \, dx}{\sqrt{x^4 + 1}} \, dx$

2. Use Romberg Integration to approximate the definite integral. As a stopping criterion, continue until two successive diagonal entries differ by less than 0.5×10^{-8}.

   (a) $\int_0^1 e^{x^2} \, dx$ (b) $\int_0^{\sqrt{\pi}} \sin x^2 \, dx$ (c) $\int_0^\pi e^{\cos x} \, dx$ (d) $\int_0^1 \ln(x^2 + 1) \, dx$

   (e) $\int_0^1 \frac{x \, dx}{2e^x - e^{-x}}$ (f) $\int_0^\pi \cos e^x \, dx$ (g) $\int_0^1 x^x \, dx$ (h) $\int_0^{\pi/2} \ln(\cos x + \sin x) \, dx$

3. (a) Test the order of the second column of Romberg. If they are fourth-order approximations, how should a log–log plot of the error versus h look? Carry this out for the integral in Example 5.11. (b) Test the order of the third column of Romberg.

5.4 ADAPTIVE QUADRATURE

The approximate integration methods we have learned so far use equal step sizes. Smaller step sizes improve accuracy, in general. A wildly varying function will require more steps, and therefore more computing time, because of the smaller steps needed to keep track of the variations.

Although we have error formulas for the composite methods, using them to directly calculate the value of h that meets a given error tolerance is often difficult. The formulas involve higher derivatives, which may be complicated and hard to estimate over the interval in question. The higher derivative may not even be available if the function is known only through a list of values.

A second problem with applying the composite formulas with equal step sizes is that functions often vary wildly over some of their domain and vary more slowly through other parts. (See Figure 5.5.) A step size that is sufficient to meet the error tolerance in the former section may be overkill in the latter section.

Fortunately, there is a way to solve both problems. By using the information from the integration error formulas, a criterion can be developed for deciding during the calculation what step size is appropriate for a particular subinterval. The idea behind this method, called **Adaptive Quadrature**, is closely related to the extrapolation ideas we have studied in this chapter.

According to (5.21), the Trapezoid Rule $S_{[a,b]}$ on the interval $[a,b]$ satisfies the formula

$$\int_a^b f(x)\, dx = S_{[a,b]} - h^3 \frac{f''(c_0)}{12} \tag{5.36}$$

for some $a < c_0 < b$, where $h = b - a$. Setting c to be the midpoint of $[a,b]$, we could apply the Trapezoid Rule to both half-intervals and, by the same formula, get

$$\int_a^b f(x)\, dx = S_{[a,c]} - \frac{h^3}{8}\frac{f''(c_1)}{12} + S_{[c,b]} - \frac{h^3}{8}\frac{f''(c_2)}{12}$$

$$= S_{[a,c]} + S_{[c,b]} - \frac{h^3}{4}\frac{f''(c_3)}{12}, \tag{5.37}$$

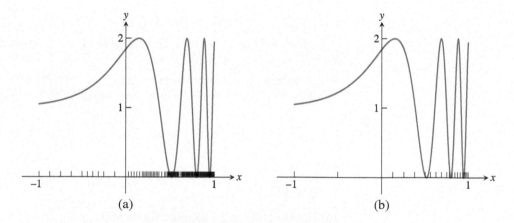

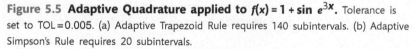

Figure 5.5 Adaptive Quadrature applied to $f(x) = 1 + \sin e^{3x}$. Tolerance is set to TOL = 0.005. (a) Adaptive Trapezoid Rule requires 140 subintervals. (b) Adaptive Simpson's Rule requires 20 subintervals.

where c_1 and c_2 lie in $[a,c]$ and $[c,b]$, respectively. We have applied Theorem 5.1 to consolidate the error terms. Subtracting (5.37) from (5.36) yields

$$S_{[a,b]} - (S_{[a,c]} + S_{[c,b]}) = -\frac{h^3}{4}\frac{f''(c_3)}{12} + h^3 \frac{f''(c_0)}{12}$$

$$\approx \frac{3}{4} h^3 \frac{f''(c_3)}{12}, \tag{5.38}$$

where the approximation $f''(c_3) \approx f''(c_0)$ has been made.

By subtracting the exact integral out of the equation, we have written the error (approximately) in terms of things we can compute. For example, note that $S_{a,b} - (S_{a,c} + S_{c,b})$ is approximately three times the size of the integration error of the formula $S_{a,c} + S_{c,b}$ on a, b, from (5.37). Therefore, we can check whether the former expression is less than 3*TOL for some error tolerance as an approximate way of checking whether the latter approximates the unknown exact integral within TOL.

If the criterion is not met, we can subdivide again. Now that there is a criterion for accepting an approximation over a given subinterval, we can continue breaking intervals in half and applying the criterion to the halves recursively. For each half, the required error tolerance goes down by a factor of 2, while the error (for the Trapezoid Rule) should drop by a factor of $2^3 = 8$, so a sufficient number of halvings should allow the original tolerance to be met with an adaptive composite approach.

Adaptive Quadrature

To approximate $\int_a^b f(x)\,dx$ within tolerance TOL:

$$c = \frac{a+b}{2}$$
$$S_{a,b} = (b-a)\frac{f(a)+f(b)}{2}$$
if $|S_{a,b} - S_{a,c} - S_{c,b}| < 3 \cdot \text{TOL} \cdot \left(\frac{b-a}{b_{\text{orig}} - a_{\text{orig}}}\right)$

    accept $S_{a,c} + S_{c,b}$ as approximation over a, b

else

    repeat above recursively for a, c and c, b

end

The MATLAB programming strategy works as follows: A list is established of subintervals yet to be processed. The list originally consists of one interval, a, b. In general, choose the last subinterval on the list and apply the criterion. If met, the approximation of the integral over that subinterval is added to a running sum, and the interval is crossed off the list. If unmet, the subinterval is replaced on the list by two subintervals, lengthening the list by one, and we move to the end of the list and repeat. The following MATLAB code carries out this strategy:

MATLAB code
shown here can be found
at bit.ly/20tu0q5

```
%Program 5.2 Adaptive Quadrature
% Computes approximation to definite integral
% Inputs: Matlab function f, interval [a0,b0],
%   error tolerance tol0
% Output: approximate definite integral
function int=adapquad(f,a0,b0,tol0)
int=0; n=1; a(1)=a0; b(1)=b0; tol(1)=tol0; app(1)=trap(f,a,b);
while n>0                   % n is current position at end of the list
    c=(a(n)+b(n))/2; oldapp=app(n);
    app(n)=trap(f,a(n),c);app(n+1)=trap(f,c,b(n));
    if abs(oldapp-(app(n)+app(n+1)))<3*tol(n)
        int=int+app(n)+app(n+1);        % success
        n=n-1;                          % done with interval
    else                                % divide into two intervals
        b(n+1)=b(n); b(n)=c;            % set up new intervals
        a(n+1)=c;
    tol(n)=tol(n)/2; tol(n+1)=tol(n);
    n=n+1;                              % go to end of list, repeat
```

```
      end
   end

   function s=trap(f,a,b)
   s=(f(a)+f(b))*(b-a)/2;
```

▶ **EXAMPLE 5.12** Use Adaptive Quadrature to approximate the integral

$$\int_{-1}^{1} (1 + \sin e^{3x})\, dx.$$

Figure 5.5(a) shows the result of the Adaptive Quadrature algorithm for $f(x)$, with an error tolerance of 0.005. Although 140 intervals are required, only 11 of them lie in the "calm" region $[-1, 0]$. The approximate definite integral is 2.502 ± 0.005. In a second run, we change the error tolerance to 0.5×10^{-4} and get 2.5008, reliable to four decimal places, computed over 1316 subintervals. ◄

Of course, the Trapezoid Rule can be replaced by more sophisticated rules. For example, let $S_{[a,b]}$ denote Simpson's Rule (5.22) on the interval $[a, b]$:

$$\int_{a}^{b} f(x)\, dx = S_{[a,b]} - \frac{h^5}{90} f^{(iv)}(c_0). \tag{5.39}$$

Applying Simpson's Rule to two halves of $[a, b]$ yields

$$\int_{a}^{b} f(x)\, dx = S_{[a,c]} - \frac{h^5}{32}\frac{f^{(iv)}(c_1)}{90} + S_{[c,b]} - \frac{h^5}{32}\frac{f^{(iv)}(c_2)}{90}$$

$$= S_{[a,c]} + S_{[c,b]} - \frac{h^5}{16}\frac{f^{(iv)}(c_3)}{90}, \tag{5.40}$$

where we have applied Theorem 5.1 to consolidate the error terms. Subtracting (5.40) from (5.39) yields

$$S_{[a,b]} - (S_{[a,c]} + S_{[c,b]}) = h^5 \frac{f^{(iv)}(c_0)}{90} - \frac{h^5}{16}\frac{f^{(iv)}(c_3)}{90}$$

$$\approx \frac{15}{16} h^5 \frac{f^{(iv)}(c_3)}{90}, \tag{5.41}$$

where we make the approximation $f^{(iv)}(c_3) \approx f^{(iv)}(c_0)$.

Since $S_{[a,b]} - (S_{[a,c]} + S_{[c,b]})$ is now 15 times the error of the approximation $S_{[a,c]} + S_{[c,b]}$ for the integral, we can make our new criterion

$$|S_{[a,b]} - (S_{[a,c]} + S_{[c,b]})| < 15 * \text{TOL} \tag{5.42}$$

and proceed as before. It is traditional to replace the 15 by 10 in the criterion to make the algorithm more conservative. Figure 5.5(b) shows an application of Adaptive Simpson's Quadrature to the same integral. The approximate integral is 2.500 when a tolerance of 0.005 is used, using 20 subintervals, a considerable savings over adaptive Trapezoid Rule Quadrature. Decreasing the tolerance to 0.5×10^{-4} yields 2.5008, using just 58 subintervals.

▶ **ADDITIONAL EXAMPLES**

1. Carry out Adaptive Quadrature by hand, using Simpson's Rule with tolerance $\text{TOL} = 0.01$ to approximate the integral $\int_{1}^{3} \frac{1}{x}\, dx.$

2. Use the integral $\int_0^1 \sqrt{1-x^2}\,dx$ and adapquad.m to calculate π to 8 correct decimal places.

⬛ **Solutions** for Additional Examples can be found at bit.ly/2P7PwjF

5.4 Exercises

1. Apply Adaptive Quadrature by hand, using the Trapezoid Rule with tolerance TOL = 0.05 to approximate the integrals. Find the approximation error.

   (a) $\int_0^1 x^2\,dx$ (b) $\int_0^{\pi/2} \cos x\,dx$ (c) $\int_0^1 e^x\,dx$

2. Apply Adaptive Quadrature by hand, using Simpson's Rule with tolerance TOL = 0.01 to approximate the integrals. Find the approximation error.

   (a) $\int_0^1 xe^x\,dx$ (b) $\int_0^1 \frac{dx}{1+x^2}\,dx$ (c) $\int_0^\pi x\cos x\,dx$

3. Develop an Adaptive Quadrature method for the Midpoint Rule (5.26). Begin by finding a criterion for meeting the tolerance on subintervals.

4. Develop an Adaptive Quadrature method for rule (5.28).

5.4 Computer Problems

1. Use Adaptive Trapezoid Quadrature to approximate the definite integral within 0.5×10^{-8}. Report the answer with eight correct decimal places and the number of subintervals required.

   (a) $\int_0^4 \frac{x\,dx}{\sqrt{x^2+9}}$ (b) $\int_0^1 \frac{x^3\,dx}{x^2+1}$ (c) $\int_0^1 xe^x\,dx$ (d) $\int_1^3 x^2\ln x\,dx$

   (e) $\int_0^\pi x^2\sin x\,dx$ (f) $\int_2^3 \frac{x^3\,dx}{\sqrt{x^4-1}}$ (g) $\int_0^{2\sqrt{3}} \frac{dx}{\sqrt{x^2+4}}\,dx$ (h) $\int_0^1 \frac{x\,dx}{\sqrt{x^4+1}}\,dx$

2. Modify the MATLAB code for Adaptive Trapezoid Rule Quadrature to use Simpson's Rule instead, applying the criterion (5.42) with the 15 replaced by 10. Approximate the integral in Example 5.12 within 0.005, and compare with Figure 5.5(b). How many subintervals were required?

3. Carry out the steps of Computer Problem 1 for adaptive Simpson's Rule, developed in Computer Problem 2.

4. Carry out the steps of Computer Problem 1 for the adaptive Midpoint Rule, developed in Exercise 3.

5. Carry out the steps of Computer Problem 1 for the adaptive open Newton–Cotes Rule developed in Exercise 4. Use criterion (5.42) with the 15 replaced by 10.

6. Use Adaptive Trapezoid Quadrature to approximate the definite integral within 0.5×10^{-8}.

   (a) $\int_0^1 e^{x^2}\,dx$ (b) $\int_0^{\sqrt{\pi}} \sin x^2\,dx$ (c) $\int_0^\pi e^{\cos x}\,dx$ (d) $\int_0^1 \ln(x^2+1)\,dx$

(e) $\int_0^1 \dfrac{x\,dx}{2e^x - e^{-x}}$ (f) $\int_0^\pi \cos e^x\,dx$ (g) $\int_0^1 x^x\,dx$ (h) $\int_0^{\pi/2} \ln(\cos x + \sin x)\,dx$

7. Carry out the steps of Problem 6, using Adaptive Simpson's Quadrature.

8. The probability within σ standard deviations of the mean of the normal distribution is

$$\frac{1}{\sqrt{2\pi}} \int_{-\sigma}^{\sigma} e^{-x^2/2}\,dx.$$

Use Adaptive Simpson's Quadrature to find, within eight correct decimal places, the probability within (a) 1 (b) 2 (c) 3 standard deviations.

9. Write a MATLAB function called myerf.m that uses Adaptive Simpson's Rule to calculate the value of

$$\text{erf}(x) = \frac{2}{\sqrt{\pi}} \int_0^x e^{-s^2}\,ds$$

within eight correct decimal places for arbitrary input x. Test your program for $x = 1$ and $x = 3$ by comparing with MATLAB's function erf.

5.5 GAUSSIAN QUADRATURE

The degree of precision of a quadrature method is the degree for which all polynomial functions are integrated by the method with no error. Newton–Cotes Methods of degree n have degree of precision n (for n odd) and $n + 1$ (for n even). The Trapezoid Rule (Newton–Cotes for $n = 1$) has degree of precision one. Simpson's Rule ($n = 2$) is correct up to and including third degree polynomials.

To achieve this degree of precision, the Newton–Cotes formulas use $n + 1$ function evaluations, done at evenly spaced points. The question we ask is reminiscent of our discussion in Chapter 3 about Chebyshev polynomials. Are the Newton–Cotes formulas optimal for their degree of precision, or can more powerful formulas be developed? In particular, if the requirement that evaluation points be evenly spaced is relaxed, are there better methods?

At least from the point of view of degree of precision, there are more powerful and sophisticated methods. We pick out the most famous one to discuss in this section. Gaussian Quadrature has degree of precision $2n + 1$ when $n + 1$ points are used, double that of Newton–Cotes. The evaluation points are not evenly spaced. Explaining how Gaussian Quadrature works involves a short digression into orthogonal functions, which is not only interesting in its own right, but the tip of an iceberg of numerical methods inspired by the benefits of orthogonality.

DEFINITION 5.3   The set of nonzero functions $\{p_0, \ldots, p_n\}$ on the interval $[a, b]$ is **orthogonal** on $[a, b]$ if

$$\int_a^b p_j(x) p_k(x)\,dx = \begin{cases} 0 & j \neq k \\ \neq 0 & j = k. \end{cases}$$

   ❏

THEOREM 5.4   If $\{p_0, p_1, \ldots, p_n\}$ is an orthogonal set of polynomials on the interval $[a, b]$, where deg $p_i = i$, then $\{p_0, p_1, \ldots, p_n\}$ is a basis for the vector space of degree at most n polynomials on $[a, b]$.   ■

Proof. We must show that the polynomials span the vector space and are linearly independent. An easy induction argument shows that any set of polynomials $\{p_0, p_1, \ldots, p_n\}$, where deg $p_i = i$, spans the space of polynomials of degree at most n. To show linear independence, we will assume that there is a linear dependency $\sum_{i=0}^{n} c_i p_i(x) = 0$ and show that all c_i must be zero, using the orthogonality assumption. For any $0 \leq k \leq n$, since p_k is orthogonal to every polynomial but itself, we get

$$0 = \int_a^b p_k \sum_{i=0}^{n} c_i p_i(x) \, dx = \sum_{i=0}^{n} c_i \int_a^b p_k p_i \, dx = c_k \int_a^b p_k^2 \, dx. \qquad (5.43)$$

Therefore, $c_k = 0$. □

THEOREM 5.5 If $\{p_0, \ldots, p_n\}$ is an orthogonal set of polynomials on $[a, b]$ and if deg $p_i = i$, then p_i has i distinct roots in the interval (a, b). ∎

Proof. Let $x_1, \ldots x_r$ be all distinct roots of $p_i(x)$ in (a, b) with odd multiplicity (e.g., double roots if any, are not in the set). The sign of $p_i(x)$ changes only at the x_j's. Define the degree r polynomial $q(x) = (x - x_1) \cdots (x - x_r)$. Then $p_i(x)q(x)$ is a polynomial that never changes sign on (a, b) except for becoming zero at a finite set of points, and therefore

$$\int_a^b p_i(x)q(x) \, dx \neq 0.$$

If $r < i$, then $q(x)$ is a linear combination of terms in $\{p_k(x) : k < i\}$ by Theorem 5.4, contradicting the integral above by the orthogonality hypothesis. Therefore $r = i$, and so all roots of $p_i(x)$ lie in (a, b) and are distinct. □

▶ **EXAMPLE 5.13** Find a set of three orthogonal polynomials on the interval $[-1, 1]$.

Guessing $p_0(x) = 1$ and $p_1(x) = x$ is a good start, because

$$\int_{-1}^{1} 1 \cdot x \, dx = 0.$$

Trying $p_2(x) = x^2$ doesn't quite work, since it lacks orthogonality with $p_0(x)$:

$$\int_{-1}^{1} p_0(x)x^2 \, dx = 2/3 \neq 0.$$

SPOTLIGHT ON

Orthogonality In Chapter 4, we found that orthogonality of finite-dimensional vectors was helpful in formulating and solving least squares problems. For quadrature, we need orthogonality in infinite-dimensional spaces like the vector space of polynomials in one variable. One basis is the monomial basis $\{1, x, x^2, \ldots\}$. However, a more useful basis is one that is also an orthogonal set. For orthogonality on the interval $[-1, 1]$, the right choice is the Legendre polynomials.

Adjusting to $p_2(x) = x^2 + c$, we find that

$$\int_{-1}^{1} p_0(x)(x^2 + c) \, dx = 2/3 + 2c = 0,$$

as long as $c = -1/3$. Check that p_1 and p_2 are orthogonal. (See Exercise 7.) Therefore, the set $\{1, x, x^2 - 1/3\}$ is an orthogonal set on $[-1, 1]$. ◄

The three polynomials in Example 5.13 belong to a set discovered by Legendre.

► **EXAMPLE 5.14**   Show that the set of **Legendre polynomials**

$$p_i(x) = \frac{1}{2^i i!} \frac{d^i}{dx^i} [(x^2 - 1)^i]$$

for $0 \le i \le n$ is orthogonal on $[-1, 1]$.

Notice first that $p_i(x)$ is a degree i polynomial (as the ith derivative of a degree $2i$ polynomial). Second, notice that the ith derivative of $(x^2 - 1)^j$ is divisible by $(x^2 - 1)$ if $i < j$.

We want to show that if $i < j$, then the integral

$$\int_{-1}^{1} [(x^2 - 1)^i]^{(i)} [(x^2 - 1)^j]^{(j)} \, dx$$

is zero. Integrating by parts with $u = [(x^2 - 1)^i]^{(i)}$ and $dv = [(x^2 - 1)^j]^{(j)} \, dx$ yields

$$uv - \int_{-1}^{1} v \, du = [(x^2 - 1)^i]^{(i)} [(x^2 - 1)^j]^{(j-1)} |_{-1}^{1}$$

$$- \int_{-1}^{1} [(x^2 - 1)^i]^{(i+1)} [(x^2 - 1)^j]^{(j-1)} \, dx$$

$$= - \int_{-1}^{1} [(x^2 - 1)^i]^{(i+1)} [(x^2 - 1)^j]^{(j-1)} \, dx,$$

since $[(x^2 - 1)^j]^{(j-1)}$ is divisible by $(x^2 - 1)$.

After $i + 1$ repeated integration by parts, we are left with

$$(-1)^{i+1} \int_{-1}^{1} [(x^2 - 1)^i]^{(2i+1)} [(x^2 - 1)^j]^{(j-i-1)} \, dx = 0,$$

because the $(2i + 1)$st derivative of $(x^2 - 1)^i$ is zero. ◄

By Theorem 5.5, the nth Legendre polynomial has n roots $x_1, \ldots, x_n$ in $[-1, 1]$. Gaussian Quadrature of a function is simply a linear combination of function evaluations at the Legendre roots. We achieve this by approximating the integral of the desired function by the integral of the interpolating polynomial, whose nodes are the Legendre roots.

Fix an n, and let $Q(x)$ be the interpolating polynomial for the integrand $f(x)$ at the nodes $x_1, \ldots, x_n$. Using the Lagrange formulation, we can write

$$Q(x) = \sum_{i=1}^{n} L_i(x) f(x_i), \text{ where } L_i(x) = \frac{(x - x_1) \cdots \overline{(x - x_i)} \cdots (x - x_n)}{(x_i - x_1) \cdots \overline{(x_i - x_i)} \cdots (x_i - x_n)},$$

and the overbar denotes the term is omitted.

Integrating both sides yields the following approximation for the integral:

n	roots x_i		coefficients c_i	
2	$-\sqrt{1/3} =$	-0.57735026918963	1	$= 1.00000000000000$
	$\sqrt{1/3} =$	0.57735026918963	1	$= 1.00000000000000$
3	$-\sqrt{3/5} =$	-0.77459666924148	$5/9$	$= 0.55555555555555$
	$0 =$	0.00000000000000	$8/9$	$= 0.88888888888888$
	$\sqrt{3/5} =$	0.77459666924148	$5/9$	$= 0.55555555555555$
4	$-\sqrt{\frac{15+2\sqrt{30}}{35}} =$	-0.86113631159405	$\frac{90-5\sqrt{30}}{180} =$	0.34785484513745
	$-\sqrt{\frac{15-2\sqrt{30}}{35}} =$	-0.33998104358486	$\frac{90+5\sqrt{30}}{180} =$	0.65214515486255
	$\sqrt{\frac{15-2\sqrt{30}}{35}} =$	0.33998104358486	$\frac{90+5\sqrt{30}}{180} =$	0.65214515486255
	$\sqrt{\frac{15+2\sqrt{30}}{35}} =$	0.86113631159405	$\frac{90-5\sqrt{30}}{180} =$	0.34785484513745

Table 5.1   Gaussian Quadrature coefficients. The roots x_i of the **n**th Legendre polynomials, and the coefficients c_i in (5.44).

Gaussian Quadrature

$$\int_{-1}^{1} f(x)\, dx \approx \sum_{i=1}^{n} c_i f(x_i), \tag{5.44}$$

where

$$c_i = \int_{-1}^{1} L_i(x)\, dx, \quad i = 1, \ldots, n.$$

The c_i are tabulated to great accuracy. Values are given in Table 5.1 up to $n = 4$.

▶ **EXAMPLE 5.15**   Approximate

$$\int_{-1}^{1} e^{-\frac{x^2}{2}}\, dx,$$

using Gaussian Quadrature.

The correct answer to 14 digits is 1.71124878378430. For the integrand $f(x) = e^{-x^2/2}$, the $n = 2$ Gaussian Quadrature approximation is

$$\int_{-1}^{1} e^{-\frac{x^2}{2}}\, dx \approx c_1 f(x_1) + c_2 f(x_2)$$

$$= 1 \cdot f(-\sqrt{1/3}) + 1 \cdot f(\sqrt{1/3}) \approx 1.69296344978123.$$

The $n = 3$ approximation is

$$\frac{5}{9} f(-\sqrt{3/5}) + \frac{8}{9} f(0) + \frac{5}{9} f(\sqrt{3/5}) \approx 1.71202024520191,$$

and the $n = 4$ approximation is

$$c_1 f(x_1) + c_2 f(x_2) + c_3 f(x_3) + c_4 f(x_4) \approx 1.71122450459949.$$

This approximation, using four function evaluations, is much closer than the Romberg approximation R_{33}, which uses five evenly spaced function evaluations on $[-1, 1]$:

```
1.21306131942527    0                  0
1.60653065971263    1.73768710647509   0
1.68576223244091    1.71217275668367   1.71047180003091
```
◄

The secret of the accuracy of Gaussian Quadrature is revealed by the next theorem.

THEOREM 5.6   The Gaussian Quadrature Method, using the degree n Legendre polynomial on $[-1, 1]$, has degree of precision $2n - 1$. ■

Proof. Let $P(x)$ be a polynomial of degree at most $2n - 1$. We must show it is integrated exactly by Gaussian Quadrature.

Using long division of polynomials, we can express

$$P(x) = S(x)p_n(x) + R(x), \tag{5.45}$$

where the $S(x)$ and $R(x)$ are polynomials of degree less than n. Note that Gaussian Quadrature will be exact on the polynomial $R(x)$, since it is just integration of the interpolating polynomial of degree $n - 1$, which is identical to $R(x)$.

At the roots x_i of the nth Legendre polynomial, $P(x_i) = R(x_i)$, since $p_n(x_i) = 0$ for all i. This implies that their Gaussian Quadrature approximations will be the same. But their integrals are also identical: Integrating (5.45) gives

$$\int_{-1}^{1} P(x)\, dx = \int_{-1}^{1} S(x)p_n(x)\, dx + \int_{-1}^{1} R(x)\, dx = 0 + \int_{-1}^{1} R(x)\, dx,$$

since by Theorem 5.4, $S(x)$ can be written as a linear combination of polynomials of degree less than n, which are orthogonal to $p_n(x)$. Since Gaussian Quadrature is exact on $R(x)$, it must also be for $P(x)$. ❑

To approximate integrals on a general interval $[a, b]$, the problem needs to be translated back to $[-1, 1]$. Using the substitution $t = (2x - a - b)/(b - a)$, we find it easy to check that

$$\int_{a}^{b} f(x)\, dx = \int_{-1}^{1} f\left(\frac{(b - a)t + b + a}{2}\right) \frac{b - a}{2}\, dt. \tag{5.46}$$

We demonstrate with an example.

► **EXAMPLE 5.16**   Approximate the integral

$$\int_{1}^{2} \ln x\, dx,$$

using Gaussian Quadrature.

From (5.46),

$$\int_{1}^{2} \ln x\, dx = \int_{-1}^{1} \ln\left(\frac{t + 3}{2}\right) \frac{1}{2}\, dt.$$

Now we can set $f(t) = \ln((t + 3)/2)/2$ and use the standard roots and coefficients. The result for $n = 4$ is 0.38629449693871, compared with the correct value $2\ln 2 - 1 \approx$

0.38629436111989. Again, this is more accurate than the Romberg Integration using four points in Example 5.11. ◄

▶ **ADDITIONAL EXAMPLES**

1. Translate the integral $\int_1^2 x \ln x \, dx$ to the interval $-1, 1$ and approximate with degree 4 Gaussian Quadrature.

2. Use the integral $\int_0^1 \frac{1}{1+x^2} \, dx$ and Gaussian Quadrature of degree 4 to approximate π. How many decimal places are correct?

⊡ **Solutions** for Additional Examples can be found at bit.ly/2CWAKXe

5.5 Exercises

⊡ **Solutions** for Exercises numbered in blue can be found at bit.ly/2PCrZEv

1. Approximate the integrals, using $n = 2$ Gaussian Quadrature. Compare with the correct value, and give the approximation error.

   (a) $\int_{-1}^1 (x^3 + 2x) \, dx$ (b) $\int_{-1}^1 x^4 \, dx$ (c) $\int_{-1}^1 e^x \, dx$ (d) $\int_{-1}^1 \cos \pi x \, dx$

2. Approximate the integrals in Exercise 1, using $n = 3$ Gaussian Quadrature, and give the error.

3. Approximate the integrals in Exercise 1, using $n = 4$ Gaussian Quadrature, and give the error.

4. Change variables, using the substitution (5.46) to rewrite as an integral over $-1, 1$.

   (a) $\int_0^4 \frac{x \, dx}{\sqrt{x^2 + 9}}$ (b) $\int_0^1 \frac{x^3 \, dx}{x^2 + 1}$ (c) $\int_0^1 xe^x \, dx$ (d) $\int_1^3 x^2 \ln x \, dx$

5. Approximate the integrals in Exercise 4, using $n = 3$ Gaussian Quadrature.

6. Approximate the integrals, using $n = 4$ Gaussian Quadrature.

   (a) $\int_0^1 (x^3 + 2x) \, dx$ (b) $\int_1^4 \ln x \, dx$ (c) $\int_{-1}^2 x^5 \, dx$ (d) $\int_{-3}^3 e^{-\frac{x^2}{2}} \, dx$

7. Show that the Legendre polynomials $p_1(x) = x$ and $p_2(x) = x^2 - 1/3$ are orthogonal on $-1, 1$.

8. Find the Legendre polynomials up to degree 3 and compare with Example 5.13.

9. Verify the coefficients c_i and x_i in Table 5.1 for degree $n = 3$.

10. Verify the coefficients c_i and x_i in Table 5.1 for degree $n = 4$.

Reality Check **1** *Motion Control in Computer-Aided Modeling*

Computer-aided modeling and manufacturing requires precise control of spatial position along a prescribed motion path. We will illustrate the use of Adaptive Quadrature to solve a fundamental piece of the problem: equipartition, or the division of an arbitrary path into equal-length subpaths.

In numerical machining problems, it is preferable to maintain constant speed along the path. During each second, progress should be made along an equal length of the machine–material interface. In other motion planning applications, including computer animation, more complicated progress curves may be required: A hand reaching for a doorknob might begin and end with low velocity and have higher velocity in between. Robotics and virtual reality applications require the construction of parametrized curves and surfaces to be navigated. Building a table of small equal increments in path distance is often a necessary first step.

Assume that a parametric path $P = \{x(t), y(t) | 0 \le t \le 1\}$ is given. Figure 5.6 shows the example path

$$P = \begin{cases} x(t) = 0.5 + 0.3t + 3.9t^2 - 4.7t^3 \\ y(t) = 1.5 + 0.3t + 0.9t^2 - 2.7t^3 \end{cases},$$

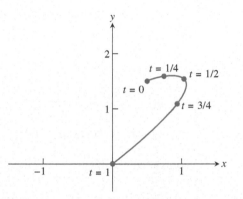

Figure 5.6 Parametrized curve given by Bézier spline. Typically, equal intervals of the parameter t do not divide the path into segments of equal length.

which is the Bézier curve defined by the four points $(0.5, 1.5), (0.6, 1.6), (2, 2), (0, 0)$. (See Section 3.5.) Points defined by evenly spaced parameter values $t = 0, 1/4, 1/2, 3/4, 1$ are shown. Note that even spacing in parameter does not imply even spacing in arc length. Your goal is to apply quadrature methods to divide this path into n equal lengths.

Recall from calculus that the arc length of the path from t_1 to t_2 is

$$\int_{t_1}^{t_2} \sqrt{x'(t)^2 + y'(t)^2} \, dt.$$

Only rarely does the integral yield a closed-form expression, and normally an Adaptive Quadrature technique is used to control the parametrization of the path.

Suggested activities:

1. Write a MATLAB function that uses Adaptive Quadrature to compute the arc length from $t = 0$ to $t = T$ for a given $T \le 1$.

2. Write a program that, for any input s between 0 and 1, finds the parameter $t^*(s)$ that is s of the way along the curve. In other words, the arc length from $t = 0$ to $t = t^*(s)$ divided by the arc length from $t = 0$ to $t = 1$ should be equal to s. Use the Bisection Method to locate the point $t^*(s)$ to three correct decimal places. What function is being set to zero? What bracketing interval should be used to start the Bisection Method?

3. Equipartition the path of Figure 5.6 into n subpaths of equal length, for $n = 4$ and $n = 20$. Plot analogues of Figure 5.6, showing the equipartitions. If your computations

are too slow, consider speeding up the Adaptive Quadrature with Simpson's Rule, as suggested in Computer Problem 5.4.2.

4. Replace the Bisection Method in Step 2 with Newton's Method, and repeat Steps 2 and 3. What is the derivative needed? What is a good choice for the initial guess? Is computation time decreased by this replacement?

5. Appendix A demonstrates animation commands available in MATLAB. For example, the commands

```
set(gca,'XLim',[-2 2],'YLim',[-2 2],'Drawmode','fast',...
    'Visible','on');
cla
axis square
ball=line('color','r','Marker','o','MarkerSize',10,...
    'LineWidth',2, 'erase','xor','xdata',[],'ydata',[]);
```

define an object "ball" that is assigned position (x, y) by the following commands:

```
set(ball,'xdata',x,'ydata',y); drawnow;pause(0.01)
```

Putting this line in a loop that changes x and y causes the ball to move along the path in the MATLAB figure window.

Use MATLAB's animation commands to demonstrate traveling along the path, first at the original parameter $0 \le t \le 1$ speed and then at the (constant) speed given by $t^*(s)$ for $0 \le s \le 1$.

6. Experiment with equipartitioning a path of your choice. Build a design, initial, etc. of your choice out of Bézier curves, partition it into equal arc length segments, and animate as in Step 5.

7. Write a program that traverses the path P according to an arbitrary **progress curve** $C(s), 0 \le s \le 1$, with $C(0) = 0$ and $C(1) = 1$. The object is to move along the curve C in such a way that the proportion $C(s)$ of the path's total arc length is traversed between 0 and s. For example, constant speed along the path would be represented by $C(s) = s$. Try progress curves $C(s) = s^{1/3}$, $C(s) = s^2$, $C(s) = \sin s\pi/2$, or $C(s) = 1/2 + (1/2)\sin(2s - 1)\pi/2$, for example.

Consult Wang et al. [2003] and Guenter and Parent [1990] for more details and applications of reparametrization of curves in the plane and space.

Software and Further Reading

The closed and open Newton–Cotes Methods are basic tools for approximating definite integrals. Romberg Integration is an accelerated version. Most commercial software implementations involve Adaptive Quadrature in some form. Classic texts on numerical differentiation and integration include Davis and Rabinowitz [1984], Stroud and Secrest [1966], Krommer and Ueberhuber [1998], Engels [1980], and Evans [1993].

Many effective quadrature techniques are implemented by Fortran subroutines in the public–domain software package Quadpack (Piessens et al. [1983]), available in Netlib (www.netlib.org/quadpack). The Gauss–Kronrod Method is an adaptive method based on Gaussian Quadrature. Quadpack provides nonadaptive and adaptive methods QNG and QAG, respectively, the latter based on Gauss–Kronrod.

MATLAB's `quad` command is an implementation of adaptive composite Simpson's Quadrature, and `dblquad` handles double integrals. MATLAB's Symbolic Toolbox has commands `diff` and `int` for symbolic differentiation and integration, respectively.

Integration of functions of several variables can be done by extending the one-dimensional methods in a straightforward way, as long as the integration region is simple; for example, see Davis and Rabinowitz [1984] and Haber [1970]. For some complicated regions, Monte Carlo integration is indicated. Monte Carlo is easier to implement, but converges more slowly in general. These issues are discussed further in Chapter 9.

CHAPTER

6

Ordinary Differential Equations

By November 7, 1940, the Tacoma Narrows Bridge, the third longest suspension bridge in the world, had become famous for its pronounced vertical oscillations during high winds. Around 11 A.M. on that day, it fell into Puget Sound.

But the motion which preceded the collapse was primarily torsional, twisting from side to side. This motion, which had been seldom seen prior to that day, continued for 45 minutes before the collapse. The twisting motion eventually became large enough to snap a support cable, and the bridge disintegrated rapidly.

The debate among architects and engineers about the reason for the collapse has continued unabated since that time. High winds caused vertical oscillation for aerodynamic reasons, with the bridge acting like an airplane wing, but the bridge's integrity was not in danger from strictly vertical movements. The mystery is how the torsional oscillation arose.

Reality Check Reality Check 6 on page 337 proposes a differential equations model that explores possible mechanisms for the torsional oscillation.

A differential equation is an equation involving derivatives. In the form

$$y'(t) = f(t, y(t)),$$

a first-order differential equation expresses the rate of change of a quantity y in terms of the present time and the current value of the quantity. Differential equations are used to model, understand, and predict systems that change with time.

A wide majority of interesting equations have no closed-form solution, which leaves approximations as the only recourse. This chapter covers the approximate solution of ordinary differential equations (ODE) by computational methods. After introductory ideas on differential equations, Euler's Method is described and analyzed in detail. Although too simple to be heavily used in applications, Euler's Method is crucial, since most of the important issues in the subject can be easily understood in its very simple context.

More sophisticated methods follow, and interesting examples of systems of differential equations are explored. Variable step-size protocols are important for efficient solution, and special methods are necessary for stiff problems. The chapter ends with an introduction to implicit and multistep methods.

6.1 INITIAL VALUE PROBLEMS

Many physical laws that have been successful in modeling nature are expressed in the form of differential equations. Sir Isaac Newton wrote his laws of motion in that form: $F = ma$ is an equation connecting the composite force acting on an object and the object's acceleration, which is the second derivative of the position. In fact, Newton's postulation of his laws, together with development of the infrastructure needed to write them down (calculus), constituted one of the most important revolutions in the history of science.

A simple model known as the **logistic equation** models the rate of change of a population as

$$y' = cy(1 - y), \tag{6.1}$$

where y' denotes the derivative with respect to time t. If we think of y as representing the population as a proportion of the carrying capacity of the animal's habitat, then we expect y to grow to near that capacity and then level off. The differential equation (6.1) shows the rate of change y' as being proportional to the product of the current population y and the "remaining capacity" $1 - y$. Therefore, the rate of change is small both when the population is small (y near 0) and also when the population nears capacity (y near 1).

The ordinary differential equation (6.1) is typical in that it has infinitely many solutions $y(t)$. By specifying an initial condition, we can identify which of the infinite family we are interested in. (We will get more precise about existence and uniqueness in the next section.) An **initial value problem** for a first-order ordinary differential equation is the equation together with an initial condition on a specific interval $a \le t \le b$:

$$\begin{cases} y' = f(t, y) \\ y(a) = y_a \\ t \text{ in } [a, b] \end{cases} \tag{6.2}$$

It will be helpful to think of a differential equation as a field of slopes, as in Figure 6.1(a). Equation (6.1) can be viewed as specifying a slope for any current values of (t, y). If we use an arrow to plot the slope at each point in the plane, we get the **slope field**, or **direction field**, of the differential equation. The equation is **autonomous** if the right-hand side $f(t, y)$ is independent of t. This is apparent in Figure 6.1.

When an initial condition is specified on a slope field, one out of the infinite family of solutions can be identified. In Figure 6.1(b), two different solutions are plotted starting at two different initial values, $y(0) = 0.2$ and $y(0) = 1.4$, respectively.

Equation (6.1) has a solution that can be written in terms of elementary functions. We check, by differentiating and substituting, that as long as the initial condition $y_0 \neq 1$,

$$y(t) = 1 - \frac{1}{1 + \frac{y_0}{1-y_0} e^{ct}} \tag{6.3}$$

is the solution of the initial value problem

$$\begin{cases} y' = cy(1-y) \\ y(0) = y_0 \\ t \text{ in } [0, T] \end{cases} \tag{6.4}$$

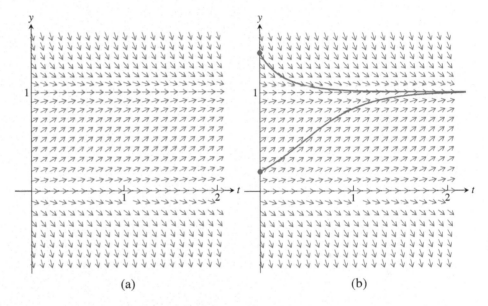

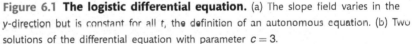

(a)   (b)

Figure 6.1 The logistic differential equation. (a) The slope field varies in the y-direction but is constant for all t, the definition of an autonomous equation. (b) Two solutions of the differential equation with parameter $c = 3$.

The solution follows the arrows in Figure 6.1(b). If $y_0 = 1$, the solution is $y(t) = 1$, which is checked the same way.

6.1.1 Euler's Method

The logistic equation had an explicit, fairly simple solution. A much more common scenario is a differential equation with no explicit solution formula. The geometry of Figure 6.1 suggests an alternative approach: to computationally "solve" the differential equation by following arrows. Start at the initial condition (t_0, y_0), and follow the direction specified there. After moving a short distance, re-evaluate the slope at the new point (t_1, y_1), move farther according to the new slope, and repeat the process. There will be some error associated with the process, since, in between evaluations of the slope, we will not be moving along a completely accurate slope. But if the slopes change slowly, we may get a fairly good approximation to the solution of the initial value problem.

▶ **EXAMPLE 6.1** Draw the slope field of the initial value problem

$$\begin{cases} y' = ty + t^3 \\ y(0) = y_0 \\ t \text{ in } [0, 1] \end{cases} . \tag{6.5}$$

Figure 6.2(a) shows the slope field. For each point (t, y) in the plane, an arrow is plotted with slope equal to $ty + y^3$. This initial value problem is nonautonomous because t appears explicitly in the right-hand side of the equation. It is also clear from the slope field, which varies according to both t and y. The exact solution $y(t) = 3e^{t^2/2} - t^2 - 2$ is shown for initial condition $y(0) = 1$. See Example 6.6 for derivation of the explicit solution. ◀

Figure 6.2(b) shows an implementation of the method of computationally following the slope field, which is known as Euler's Method. We begin with a grid of $n + 1$ points

$$t_0 < t_1 < t_2 < \cdots < t_n$$

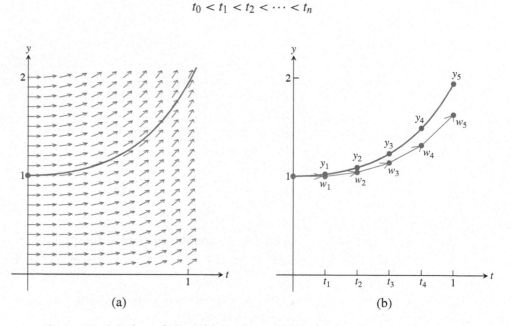

(a)   (b)

Figure 6.2 Solution of the initial value problem (6.5). (a) Slope field for a nonautonomous equation varies with t. Solution satisfying $y(0) = 1$ is shown. (b) Application of Euler's Method to the equation, with step size $h = 0.2$.

along the t-axis with equal step size h. In Figure 6.2(b), the t values were selected to be

$$t_0 = 0.0 \quad t_1 = 0.2 \quad t_2 = 0.4 \quad t_3 = 0.6 \quad t_4 = 0.8 \quad t_5 = 1.0 \tag{6.6}$$

with step size $h = 0.2$.

Begin with $w_0 = y_0$. Following the slope field at each t_i yields the approximation

$$w_{i+1} = w_i + hf(t_i, w_i)$$

at t_{i+1}, since $f(t_i, w_i)$ represents the slope of the solution. Note that the change in y is the horizontal distance h multiplied by the slope. As shown in Figure 6.2(b), each w_i is an approximation to the solution at t_i.

The formula for this method can be expressed as follows:

Euler's Method

$$w_0 = y_0$$
$$w_{i+1} = w_i + hf(t_i, w_i). \tag{6.7}$$

▶ **EXAMPLE 6.2** Apply Euler's Method to initial value problem (6.5), with initial condition $y_0 = 1$.

The right-hand side of the differential equation is $f(t, y) = ty + t^3$. Therefore, Euler's Method will be the iteration

$$w_0 = 1$$
$$w_{i+1} = w_i + h(t_i w_i + t_i^3). \tag{6.8}$$

Using the grid (6.6) with step size $h = 0.2$, we calculate the approximate solution iteratively from (6.8). The values w_i given by Euler's Method and plotted in Figure 6.2(b) are compared with the true values y_i in the following table:

step	t_i	w_i	y_i	e_i
0	0.0	1.0000	1.0000	0.0000
1	0.2	1.0000	1.0206	0.0206
2	0.4	1.0416	1.0899	0.0483
3	0.6	1.1377	1.2317	0.0939
4	0.8	1.3175	1.4914	0.1739
5	1.0	1.6306	1.9462	0.3155

The table also shows the error $e_i = |y_i - w_i|$ at each step. The error tends to grow, from zero at the initial condition to its largest value at the end of the interval, although the maximum error will not always be found at the end.

Applying Euler's Method with step size $h = 0.1$ causes the error to decrease, as is apparent from Figure 6.3(a). Again using (6.8), we calculate the following values:

step	t_i	w_i	y_i	e_i
0	0.0	1.0000	1.0000	0.0000
1	0.1	1.0000	1.0050	0.0050
2	0.2	1.0101	1.0206	0.0105
3	0.3	1.0311	1.0481	0.0170
4	0.4	1.0647	1.0899	0.0251
5	0.5	1.1137	1.1494	0.0357
6	0.6	1.1819	1.2317	0.0497
7	0.7	1.2744	1.3429	0.0684
8	0.8	1.3979	1.4914	0.0934
9	0.9	1.5610	1.6879	0.1269
10	1.0	1.7744	1.9462	0.1718

Compare the error e_{10} for the $h = 0.1$ calculation with the error e_5 for the $h = 0.2$ calculation. Note that cutting the step size h in half results in cutting the error at $t = 1.0$ approximately in half. ◀

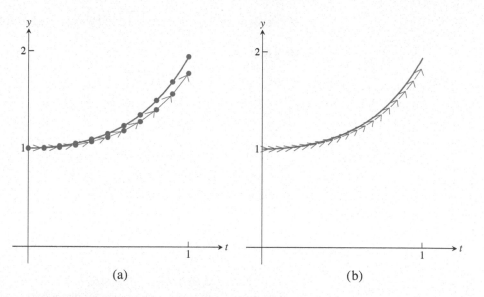

Figure 6.3 Euler's Method applied to IVP (6.5). The arrows show the Euler steps, exactly as in Figure 6.2, except for the step size. (a) Ten steps of size $h = 0.1$ (b) Twenty steps of size $h = 0.05$.

Euler's Method is implemented in the MATLAB code that follows, which has been written in somewhat modular form to highlight the three individual components. The plotting program calls a subprogram to execute each Euler step, which in turn calls the function containing the right-hand side f of the differential equation. In this form, it will be easy later to trade both the right-hand side for another differential equation and the Euler Method for another more sophisticated method. Here is the code:

MATLAB code
shown here can be found
at bit.ly/2P9BFt4

```
%Program 6.1 Euler's Method for Solving Initial Value Problems
%Use with ydot.m to evaluate rhs of differential equation
% Input: interval inter, initial value y0, number of steps n
% Output: time steps t, solution y
% Example usage: euler1([0 1],1,10);
function [t,y]=euler1(inter,y0,n)
t(1)=inter(1); y(1)=y0;
h=(inter(2)-inter(1))/n;
for i=1:n
  t(i+1)=t(i)+h;
  y(i+1)=eulerstep(t(i),y(i),h);
end
plot(t,y)

function y=eulerstep(t,y,h)
%one step of Euler's Method
%Input: current time t, current value y,  stepsize h
%Output: approximate solution value at time t+h
y=y+h*ydot(t,y);

function z=ydot(t,y)
%right-hand side of differential equation
z=t*y+t^3;
```

Comparing the Euler's Method approximation for (6.5) with the exact solution at $t = 1$ gives us the following table, extending our previous results for $n = 5$ and 10:

steps n	step size h	error at $t = 1$
5	0.20000	0.3155
10	0.10000	0.1718
20	0.05000	0.0899
40	0.02500	0.0460
80	0.01250	0.0233
160	0.00625	0.0117
320	0.00312	0.0059
640	0.00156	0.0029

Two facts are evident from the table and Figures 6.3 and 6.4. First, the error is nonzero. Since Euler's Method takes noninfinitesimal steps, the slope changes along the step and the approximation does not lie exactly on the solution curve. Second, the error decreases as the step size is decreased, as can also be seen in Figure 6.3. It appears from the table that the error is proportional to h; this fact will be confirmed in the next section.

▶ **EXAMPLE 6.3**   Find the Euler's Method formula for the following initial value problem:

$$\begin{cases} y' = cy \\ y(0) = y_0 \, . \\ t \text{ in } [0, 1] \end{cases} \qquad (6.9)$$

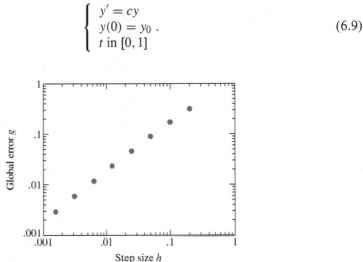

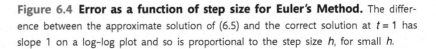

Figure 6.4 Error as a function of step size for Euler's Method. The difference between the approximate solution of (6.5) and the correct solution at $t = 1$ has slope 1 on a log–log plot and so is proportional to the step size h, for small h.

For $f(t, y) = cy$ where c is a constant, Euler's Method gives

$$w_0 = y_0$$
$$w_{i+1} = w_i + hcw_i = (1 + hc)w_i \text{ for } i = 1, 2, 3 \ldots. \qquad ◀$$

The exact solution of the equation $y' = cy$ can be found by using the method of **separation of variables**. Assuming that $y \neq 0$, divide both sides by y, separate variables, and integrate, as follows:

$$\frac{dy}{y} = c \, dt$$
$$\ln |y| = ct + k.$$
$$|y| = e^{ct+k} = e^k e^{ct}$$

The initial condition $y(0) = y_0$ implies $y = y_0 e^{ct}$.

In this simple case, we can show that Euler's Method converges to the correct solution as the number of steps $n \to \infty$. Note that

$$w_i = (1 + hc)w_{i-1} = (1 + hc)^2 w_{i-2} = \cdots = (1 + hc)^i w_0.$$

For a fixed t, set the step size $h = t/n$ for an integer n. Then the approximate value at t is

$$w_n = (1 + hc)^n y_0$$
$$= \left(1 + \frac{ct}{n}\right)^n y_0.$$

The classical formula says that

$$\lim_{n \to \infty} \left(1 + \frac{ct}{n}\right)^n = e^{ct},$$

which shows that, as $n \to \infty$, Euler's Method will converge to the correct value.

6.1.2 Existence, uniqueness, and continuity for solutions

This section provides some theoretical background for computational initial value problem methods. Before we start out to compute a solution to a problem, it is helpful to know that (1) the solution exists and (2) there is only one solution, so that the algorithm is not confused about which one to calculate. Under the right circumstances, initial value problems have exactly one solution.

DEFINITION 6.1 A function $f(t, y)$ is **Lipschitz continuous** in the variable y on the rectangle $S = [a, b] \times [\alpha, \beta]$ if there exists a constant L (called the **Lipschitz constant**) satisfying

$$|f(t, y_1) - f(t, y_2)| \le L|y_1 - y_2|$$

for each $(t, y_1), (t, y_2)$ in S. ❏

A function that is Lipschitz continuous in y is continuous in y, but not necessarily differentiable.

▶ **EXAMPLE 6.4** Find the Lipschitz constant for the right-hand side $f(t, y) = ty + t^3$ of (6.5).

The function $f(t, y) = ty + t^3$ is Lipschitz continuous in the variable y on the set $0 \le t \le 1, -\infty < y < \infty$. Check that

$$|f(t, y_1) - f(t, y_2)| = |ty_1 - ty_2| \le |t||y_1 - y_2| \le |y_1 - y_2| \tag{6.10}$$

on the set. The Lipschitz constant is $L = 1$. ◀

Although Definition 6.1 specifies the set S to be a rectangle, more generally S can be a **convex** set, one that contains the line segment connecting any two points in the set. If the function f is continuously differentiable in the variable y, the maximum absolute value of the partial derivative $\partial f/\partial y$ is a Lipschitz constant. According to the Mean Value Theorem, for each fixed t, there is a c between y_1 and y_2 such that

$$\frac{f(t, y_1) - f(t, y_2)}{y_1 - y_2} = \frac{\partial f}{\partial y}(t, c).$$

Therefore, L can be taken to be the maximum of

$$\left| \frac{\partial f}{\partial y}(t, c) \right|$$

on the set.

The Lipschitz continuity hypothesis guarantees the existence and uniqueness of solutions of initial value problems. We refer to Birkhoff and Rota [1989] for a proof of the following theorem:

THEOREM 6.2   Assume that $f(t, y)$ is Lipschitz continuous in the variable y on the set $[a, b] \times [\alpha, \beta]$ and that $\alpha < y_a < \beta$. Then there exists c between a and b such that the initial value problem

$$\begin{cases} y' = f(t, y) \\ y(a) = y_a \\ t \text{ in } [a, c] \end{cases} \qquad (6.11)$$

has exactly one solution $y(t)$. Moreover, if f is Lipschitz on $[a, b] \times (-\infty, \infty)$, then there exists exactly one solution on $[a, b]$. ∎

A careful reading of Theorem 6.2 is important, especially if the goal is to calculate the solution numerically. The fact that the initial value problem satisfies a Lipschitz condition on $[a, b] \times [\alpha, \beta]$ containing the initial condition does not guarantee a solution for t in *the entire interval* $[a, b]$. The simple reason is that the solution may wander outside the y range $[\alpha, \beta]$ for which the Lipschitz constant is valid. The best that can be said is that the solution exists on some shorter interval $[a, c]$. This point is illustrated by the following example:

▶ **EXAMPLE 6.5**   On which intervals $[0, c]$ does the initial value problem have a unique solution?

$$\begin{cases} y' = y^2 \\ y(0) = 1 \\ t \text{ in } [0, 2]. \end{cases} \qquad (6.12)$$

The partial derivative of f with respect to y is $2y$. The Lipschitz constant $\max |2y| = 20$ is valid on the set $0 \le t \le 2, -10 \le y \le 10$. Theorem 6.2 guarantees a solution starting at $t = 0$ and existing on some interval $[a, c]$ for $c > 0$, but a solution is not guaranteed on the entire interval $[0, 2]$.

In fact, the unique solution of the differential equation (6.12) is $y(t) = 1/(1 - t)$, which can be found by separation of variables. This solution goes to infinity as t approaches 1. In other words, the solution exists on the interval $0 \le t \le c$ for any $0 < c < 1$, but not for any larger c. This example explains the role of c in Theorem 6.2: The Lipschitz constant 20 is valid for $|y| \le 10$, but the solution y exceeds 10 before t reaches 2. ◀

Theorem 6.3 is the basic fact about stability (error amplification) for ordinary differential equations. If a Lipschitz constant exists for the right-hand side of the differential equation, then the solution at a later time is a Lipschitz function of the initial value, with a new Lipschitz constant that is exponential in the original one. This is a version of the Gronwall inequality.

THEOREM 6.3 Assume that $f(t, y)$ is Lipschitz in the variable y on the set $S = [a, b] \times [\alpha, \beta]$. If $Y(t)$ and $Z(t)$ are solutions in S of the differential equation

$$y' = f(t, y)$$

with initial conditions $Y(a)$ and $Z(a)$ respectively, then

$$|Y(t) - Z(t)| \leq e^{L(t-a)}|Y(a) - Z(a)|. \qquad (6.13)$$

■

Proof. If $Y(a) = Z(a)$, then $Y(t) = Z(t)$ by uniqueness of solutions, and (6.13) is trivially satisfied. We may assume that $Y(a) \neq Z(a)$, in which case $Y(t) \neq Z(t)$ for all t in the interval, to avoid contradicting uniqueness.

Define $u(t) = Y(t) - Z(t)$. Since $u(t)$ is either strictly positive or strictly negative, and because (6.13) depends only on $|u|$, we may assume that $u > 0$. Then $u(a) = Y(a) - Z(a)$,

SPOTLIGHT ON

> **Conditioning**   Error magnification was discussed in Chapters 1 and 2 as a way to quantify the effects of small input changes on the solution. The analogue of that question for initial value problems is given a precise answer by Theorem 6.3. When initial condition (input data) $Y(a)$ is changed to $Z(a)$, the greatest possible change in output t time units later, $Y(t) - Z(t)$, is exponential in t and linear in the initial condition difference. The latter implies that we can talk of a "condition number" equal to $e^{L(t-a)}$ for a fixed time t.

and the derivative is $u'(t) = Y'(t) - Z'(t) = f(t, Y(t)) - f(t, Z(t))$. The Lipschitz condition implies that

$$u' = |f(t, Y) - f(t, Z)| \leq L|Y(t) - Z(t)| = L|u(t)| = Lu(t),$$

and therefore $(\ln u)' = u'/u \leq L$. By the Mean Value Theorem,

$$\frac{\ln u(t) - \ln u(a)}{t - a} \leq L,$$

which simplifies to

$$\ln \frac{u(t)}{u(a)} \leq L(t - a)$$

$$u(t) \leq u(a)e^{L(t-a)}.$$

This is the desired result.   □

Returning to Example 6.4, Theorem 6.3 implies that solutions $Y(t)$ and $Z(t)$, starting at different initial values, must not grow apart any faster than a multiplicative factor of e^t for $0 \leq t \leq 1$. In fact, the solution at initial value Y_0 is $Y(t) = (2 + Y_0)e^{t^2/2} - t^2 - 2$, and so the difference between two solutions is

$$|Y(t) - Z(t)| \leq |(2 + Y_0)e^{t^2/2} - t^2 - 2 - ((2 + Z_0)e^{t^2/2} - t^2 - 2)|$$

$$\leq |Y_0 - Z_0|e^{t^2/2}, \qquad (6.14)$$

which is less than $|Y_0 - Z_0|e^t$ for $0 \leq t \leq 1$, as prescribed by Theorem 6.3.

6.1.3 First-order linear equations

A special class of ordinary differential equations that can be readily solved provides a handy set of illustrative examples. They are the first-order equations whose right-hand sides are linear in the y variable. Consider the initial value problem

$$\begin{cases} y' = g(t)y + h(t) \\ y(a) = y_a \\ t \text{ in } [a, b] \end{cases}. \tag{6.15}$$

First note that if $g(t)$ is continuous on $[a, b]$, a unique solution exists by Theorem 6.2, using $L = \max_{[a,b]} g(t)$ as the Lipschitz constant. The solution is found by a trick, multiplying the equation through by an "integrating factor."

The integrating factor is $e^{-\int g(t)\,dt}$. Multiplying both sides by it yields

$$(y' - g(t)y)e^{-\int g(t)\,dt} = e^{-\int g(t)\,dt}h(t)$$

$$\left(ye^{-\int g(t)\,dt}\right)' = e^{-\int g(t)\,dt}h(t)$$

$$ye^{-\int g(t)\,dt} = \int e^{-\int g(t)\,dt}h(t)\,dt,$$

which can be solved as

$$y(t) = e^{\int g(t)\,dt}\int e^{-\int g(t)\,dt}h(t)\,dt. \tag{6.16}$$

If the integrating factor can be expressed simply, this method allows an explicit solution of the first-order linear equation (6.15).

▶ **EXAMPLE 6.6** Solve the first-order linear differential equation

$$\begin{cases} y' = ty + t^3 \\ y(0) = y_0 \end{cases}. \tag{6.17}$$

The integrating factor is

$$e^{-\int g(t)\,dt} = e^{-\frac{t^2}{2}}.$$

According to (6.16), the solution is

$$y(t) = e^{\frac{t^2}{2}}\int e^{-\frac{t^2}{2}}t^3\,dt$$

$$= e^{\frac{t^2}{2}}\int e^{-u}(2u)\,du$$

$$= 2e^{\frac{t^2}{2}}\left[-\frac{t^2}{2}e^{-\frac{t^2}{2}} - e^{-\frac{t^2}{2}} + C\right]$$

$$= -t^2 - 2 + 2Ce^{\frac{t^2}{2}},$$

where the substitution $u = t^2/2$ was made. Solving for the integration constant C yields $y_0 = -2 + 2C$, so $C = (2 + y_0)/2$. Therefore,

$$y(t) = (2 + y_0)e^{\frac{t^2}{2}} - t^2 - 2.$$

◀

▶ **ADDITIONAL**
EXAMPLES

1. Find the solution of the first-order linear initial value problem $\begin{cases} y' = 3t^2y + 4t^2 \\ y(0) = 5. \end{cases}$

2. Plot the Euler's Method approximate solution for the initial value problem in Additional Example 1 for step sizes $h = 0.1, 0.05$, and 0.01, along with the exact solution.

⬛ **Solutions** for Additional Examples can be found at `bit.ly/2Owq1ZO`

6.1 Exercises

⬛ **Solutions**
for Exercises
numbered in blue
can be found at
`bit.ly/2CQ5nNY`

1. Show that the function $y(t) = t \sin t$ is a solution of the differential equations
   (a) $y + t^2 \cos t = ty'$ (b) $y'' = 2\cos t - y$ (c) $t(y'' + y) = 2y' - 2\sin t$.

2. Show that the function $y(t) = e^{\sin t}$ is a solution of the initial value problems
   (a) $y' = y\cos t, y(0) = 1$ (b) $y'' = (\cos t)y' - (\sin t)y, y(0) = 1, y'(0) = 1$
   (c) $y'' = y(1 - \ln y - (\ln y)^2), y(\pi) = 1, y'(\pi) = -1$.

3. Use separation of variables to find solutions of the IVP given by $y(0) = 1$ and the following differential equations:

$$\text{(a)} \quad y' = t \quad \text{(b)} \quad y' = t^2y \quad \text{(c)} \quad y' = 2(t + 1)y$$

$$\text{(d)} \quad y' = 5t^4y \quad \text{(e)} \quad y' = 1/y^2 \quad \text{(f)} \quad y' = t^3/y^2$$

4. Find the solutions of the IVP given by $y(0) = 0$ and the following first-order linear differential equations:

$$\text{(a)} \quad y' = t + y \quad \text{(b)} \quad y' = t - y \quad \text{(c)} \quad y' = 4t - 2y$$

5. Apply Euler's Method with step size $h = 1/4$ to the IVPs in Exercise 3 on the interval $[0,1]$. List the $w_i, i = 0, \ldots, 4$, and find the error at $t = 1$ by comparing with the correct solution.

6. Apply Euler's Method with step size $h = 1/4$ to the IVPs in Exercise 3 on the interval $[0,1]$. Find the error at $t = 1$ by comparing with the correct solution.

7. (a) Show that $y = \tan(t + c)$ is a solution of the differential equation $y' = 1 + y^2$ for each c. (b) For each real number y_0, find c in the interval $(-\pi/2, \pi/2)$ such that the initial value problem $y' = 1 + y^2, y(0) = y_0$ has a solution $y = \tan(t + c)$.

8. (a) Show that $y = \tanh(t + c)$ is a solution of the differential equation $y' = 1 - y^2$ for each c. (b) For each real number y_0 in the interval $(-1, 1)$, find c such that the initial value problem $y' = 1 - y^2, y(0) = y_0$ has a solution $y = \tanh(t + c)$.

9. For which of these initial value problems on $[0,1]$ does Theorem 6.2 guarantee a unique solution? Find the Lipschitz constants if they exist (a) $y' = t$ (b) $y' = y$ (c) $y' = -y$ (d) $y' = -y^3$.

10. Sketch the slope field of the differential equations in Exercise 9, and draw rough approximations to the solutions, starting at the initial conditions $y(0) = 1, y(0) = 0$, and $y(0) = -1$.

11. Find the solutions of the initial value problems in Exercise 9. For each equation, use the Lipschitz constants from Exercise 9, and verify, if possible, the inequality of Theorem 6.3 for the pair of solutions with initial conditions $y(0) = 0$ and $y(0) = 1$.

12. (a) Show that if $a \neq 0$, the solution of the initial value problem $y' = ay + b, y(0) = y_0$ is $y(t) = (b/a)(e^{at} - 1) + y_0e^{at}$. (b) Verify the inequality of Theorem 6.3 for solutions $y(t), z(t)$ with initial values y_0 and z_0, respectively.

13. Use separation of variables to solve the initial value problem $y' = y^2, y(0) = 1$.

14. Find the solution of the initial value problem $y' = ty^2$ with $y(0) = 1$. What is the largest interval $[0,b]$ for which the solution exists?

15. Consider the initial value problem $y' = \sin y$, $y(a) = y_a$ on $a \leq t \leq b$.
    (a) On what subinterval of $[a,b]$ does Theorem 6.2 guarantee a unique solution?
    (b) Show that $y(t) = 2\arctan(e^{t-a}\tan(y_a/2)) + 2\pi[(y_a + \pi)/2\pi]$ is the solution of the initial value problem, where $[\;]$ denotes the greatest integer function.

16. Consider the initial value problem $y' = \sinh y$, $y(a) = y_a$ on $a \leq t \leq b$.
    (a) On what subinterval of a, b does Theorem 6.2 guarantee a unique solution?
    (b) Show that $y(t) = 2\,\text{arctanh}(e^{t-a}\tanh(y_a/2))$ is a solution of the initial value problem.
    (c) On what interval a, c does the solution exist?

17. (a) Show that $y = \sec(t + c) + \tan(t + c)$ is a solution of $2y' = 1 + y^2$ for each c. (b) Show that the solution with $c = 0$ satisfies the initial value problem $2y' = 1 + y^2$, $y(0) = 1$. (c) What initial value problem is satisfied by the solution with $c = \pi/6$?

6.1 Computer Problems

Solutions for Computer Problems numbered in blue can be found at
bit.ly/2S1dj3m

1. Apply Euler's Method with step size $h = 0.1$ on $[0,1]$ to the initial value problems in Exercise 3. Print a table of the t values, Euler approximations, and error (difference from exact solution) at each step.

2. Plot the Euler's Method approximate solutions for the IVPs in Exercise 3 on $[0,1]$ for step sizes $h = 0.1, 0.05$, and 0.025, along with the exact solution.

3. Plot the Euler's Method approximate solutions for the IVPs in Exercise 4 on $[0,1]$ for step sizes $h = 0.1, 0.05$, and 0.025, along with the exact solution.

4. For the IVPs in Exercise 3, make a log–log plot of the error of Euler's Method at $t = 1$ as a function of $h = 0.1 \times 2^{-k}$ for $0 \leq k \leq 5$. Use the MATLAB loglog command as in Figure 6.4.

5. For the IVPs in Exercise 4, make a log–log plot of the error of Euler's Method at $t = 1$ as a function of $h = 0.1 \times 2^{-k}$ for $0 \leq k \leq 5$.

6. For the initial value problems in Exercise 4, make a log–log plot of the error of Euler's Method at $t = 2$ as a function of $h - 0.1 \times 2^{-k}$ for $0 \leq k \leq 5$.

7. Plot the Euler's Method approximate solution on $[0,1]$ for the differential equation $y' = 1 + y^2$ and initial condition (a) $y_0 = 0$ (b) $y_0 = 1/2$, along with the exact solution (see Exercise 7). Use step sizes $h = 0.1$ and 0.05.

8. Plot the Euler's Method approximate solution on $[0,1]$ for the differential equation $y' = 1 - y^2$ and initial condition (a) $y_0 = 0$ (b) $y_0 = -1/2$, along with the exact solution (see Exercise 8). Use step sizes $h = 0.1$ and 0.05.

9. Calculate the Euler's Method approximate solution on $[0,4]$ for the differential equation $y' = \sin y$ and initial condition (a) $y_0 = 0$ (b) $y_0 = 100$, using step sizes $h = 0.1 \times 2^{-k}$ for $0 \leq k \leq 5$. Plot the $k = 0$ and $k = 5$ approximate solutions along with the exact solution (see Exercise 15), and make a log–log plot of the error at $t = 4$ as a function of h.

10. Calculate the Euler's Method approximate solution of the differential equation $y' = \sinh y$ and initial condition (a) $y_0 = 1/4$ on the interval $[0,2]$ (b) $y_0 = 2$ on the interval $[0,1/4]$, using step sizes $h = 0.1 \times 2^{-k}$ for $0 \leq k \leq 5$. Plot the $k = 0$ and $k = 5$ approximate solutions along with the exact solution (see Exercise 16), and make a log–log plot of the error at the end of the time interval as a function of h.

11. Plot the Euler's Method approximate solution on $[0,1]$ for the initial value problem $2y' = 1 + y^2$, $y(0) = y_0$, along with the exact solution (see Exercise 6.1.17) for initial values (a) $y_0 = 1$, and (b) $y_0 = \sqrt{3}$. Use step sizes $h = 0.1$ and 0.05.

6.2 ANALYSIS OF IVP SOLVERS

Figure 6.4 shows consistently decreasing error in the Euler's Method approximation as a function of decreasing step size for Example 6.1. Is this generally true? Can we make the error as small as we want, just by decreasing the step size? A careful investigation of error in Euler's Method will illustrate the issues for IVP solvers in general.

6.2.1 Local and global truncation error

Figure 6.5 shows a schematic picture for one step of a solver like Euler's Method when solving an initial value problem of the form

$$\begin{cases} y' = f(t, y) \\ y(a) = y_a \\ t \text{ in } [a, b] \end{cases}. \tag{6.18}$$

At step i, the accumulated error from the previous steps is carried along and perhaps amplified, while new error from the Euler approximation is added. To be precise, let us define the **global truncation error**

$$g_i = |w_i - y_i|$$

to be the difference between the ODE solver (Euler's Method, for example) approximation and the correct solution of the initial value problem. Also, we will define the **local truncation error**, or one-step error, to be

$$e_{i+1} = |w_{i+1} - z(t_{i+1})|, \tag{6.19}$$

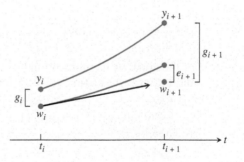

Figure 6.5 One step of an ODE solver. The Euler Method follows a line segment with the slope of the vector field at the current point to the next point (t_{i+1}, w_{i+1}). The upper curve represents the true solution to the differential equation. The global truncation error g_{i+1} is the sum of the local truncation error e_{i+1} and the accumulated, amplified error from previous steps.

the difference between the value of the solver on that interval and the correct solution of the "one-step initial value problem"

$$\begin{cases} y' = f(t, y) \\ y(t_i) = w_i \\ t \text{ in } [t_i, t_{i+1}] \end{cases}. \tag{6.20}$$

(We give the solution the name z because y is already being used for the solution to the same initial value problem starting at the exact initial condition $y(t_i) = y_i$.) The local truncation error is the error occurring just from a single step, taking the previous solution approximation w_i as the starting point. The global truncation error is the accumulated error from the first i steps. The local and global truncation errors are illustrated in Figure 6.5. At each step, the new global error is the combination of the amplified global error from the previous step and the new local error. Because of the amplification, the global error is not simply the sum of the local truncation errors.

▶ **EXAMPLE 6.7**   Find the local truncation error for Euler's Method.

According to the definition, this is the new error made on a single step of Euler's Method. Assume that the previous step w_i is correct, solve the initial value problem (6.20) exactly, and compare the exact solution $y(t_{i+1})$ with the Euler Method approximation.

Assuming that y'' is continuous, the exact solution at $t_{i+1} = t_i + h$ is

$$y(t_i + h) = y(t_i) + hy'(t_i) + \frac{h^2}{2}y''(c),$$

according to Taylor's Theorem, for some (unknown) c satisfying $t_i < c < t_{i+1}$. Since $y(t_i) = w_i$ and $y'(t_i) = f(t_i, w_i)$, this can be written as

$$y(t_{i+1}) = w_i + hf(t_i, w_i) + \frac{h^2}{2}y''(c).$$

Meanwhile, Euler's Method says that

$$w_{i+1} = w_i + hf(t_i, w_i).$$

Subtracting the two expressions yields the local truncation error

$$e_{i+1} = |w_{i+1} - y(t_{i+1})| = \frac{h^2}{2}|y''(c)|$$

for some c in the interval. If M is an upper bound for y'' on $[a, b]$, then the local truncation error satisfies $e_i \leq Mh^2/2$.   ◀

Now let's investigate how the local errors accumulate to form global errors. At the initial condition $y(a) = y_a$, the global error is $g_0 = |w_0 - y_0| = |y_a - y_a| = 0$. After one step, there is no accumulated error from previous steps, and the global error is equal to the first local error, $g_1 = e_1 = |w_1 - y_1|$. After two steps, break down g_2 into the local truncation error plus the accumulated error from the earlier step, as in Figure 6.5. Define $z(t)$ to be the solution of the initial value problem

$$\begin{cases} y' = f(t, y) \\ y(t_1) = w_1 \\ t \text{ in } [t_1, t_2] \end{cases} \tag{6.21}$$

Thus, $z(t_2)$ is the exact value of the solution starting at initial condition (t_1, w_1). Note that if we used the initial condition (t_1, y_1), we would get y_2, which is on the actual solution curve, unlike $z(t_2)$. Then $e_2 = |w_2 - z(t_2)|$ is the local truncation error of step $i = 2$. The other difference $|z(t_2) - y_2|$ is covered by Theorem 6.3, since it is the difference between two solutions of the same equation with different initial conditions w_1 and y_1. Therefore,

$$\begin{aligned} g_2 = |w_2 - y_2| &= |w_2 - z(t_2) + z(t_2) - y_2| \\ &\leq |w_2 - z(t_2)| + |z(t_2) - y_2| \\ &\leq e_2 + e^{Lh}g_1 \\ &= e_2 + e^{Lh}e_1. \end{aligned}$$

The argument is the same for step $i = 3$, which yields

$$g_3 = |w_3 - y_3| \leq e_3 + e^{Lh} g_2 \leq e_3 + e^{Lh} e_2 + e^{2Lh} e_1. \tag{6.22}$$

Likewise, the global truncation error at step i satisfies

$$g_i = |w_i - y_i| \leq e_i + e^{Lh} e_{i-1} + e^{2Lh} e_{i-2} + \cdots + e^{(i-1)Lh} e_1. \tag{6.23}$$

In Example 6.7, we found that Euler's Method has local truncation error proportional to h^2. More generally, assume that the local truncation error satisfies

$$e_i \leq Ch^{k+1}$$

for some integer k and a constant $C > 0$. Then

$$g_i \leq Ch^{k+1}\left(1 + e^{Lh} + \cdots + e^{(i-1)Lh}\right)$$

$$= Ch^{k+1} \frac{e^{iLh} - 1}{e^{Lh} - 1}$$

$$\leq Ch^{k+1} \frac{e^{L(t_i-a)} - 1}{Lh}$$

$$= \frac{Ch^k}{L}(e^{L(t_i-a)} - 1). \tag{6.24}$$

Note how the local truncation error is related to the global truncation error. The local truncation error is proportional to h^{k+1} for some k. Roughly speaking, the global truncation error "adds up" the local truncation errors over a number of steps

---

SPOTLIGHT ON

Convergence   Theorem 6.4 is the main theorem on convergence of one-step differential equation solvers. The dependence of global error on h shows that we can expect error to decrease as h is decreased, so that (at least in exact arithmetic) error can be made as small as desired. This brings us to the other important point: the exponential dependence of global error on b. As time increases, the global error bound may grow extremely large. For large t_i, the step size h required to keep global error small may be so tiny as to be impractical.

---

proportional to h^{-1}, the reciprocal of the step size. Thus, the global error turns out to be proportional to h^k. This is the major finding of the preceding calculation, and we state it in the following theorem:

THEOREM 6.4   Assume that $f(t, y)$ has a Lipschitz constant L for the variable y and that the value y_i of the solution of the initial value problem (6.2) at t_i is approximated by w_i from a one-step ODE solver with local truncation error $e_i \leq Ch^{k+1}$, for some constant C and $k \geq 0$. Then, for each $a < t_i < b$, the solver has global truncation error

$$g_i = |w_i - y_i| \leq \frac{Ch^k}{L}(e^{L(t_i-a)} - 1). \tag{6.25}$$

■

If an ODE solver satisfies (6.25) as $h \to 0$, we say that the solver has **order** k. Example 6.7 shows that the local truncation error of Euler's Method is of size bounded by $Mh^2/2$, so the order of Euler's Method is 1. Restating the theorem in the Euler's Method case gives the following corollary:

COROLLARY 6.5 (Euler's Method convergence) Assume that $f(t,y)$ has a Lipschitz constant L for the variable y and that the solution y_i of the initial value problem (6.2) at t_i is approximated by w_i, using Euler's Method. Let M be an upper bound for $|y''(t)|$ on $[a,b]$. Then

$$|w_i - y_i| \le \frac{Mh}{2L}(e^{L(t_i-a)} - 1). \tag{6.26}$$

■

▶ **EXAMPLE 6.8** Find an error bound for Euler's Method applied to Example 6.1.

The Lipschitz constant on $[0,1]$ is $L=1$. Now that the solution $y(t) = 3e^{t^2/2} - t^2 - 2$ is known, the second derivative is determined to be $y''(t) = (t^2 + 2)e^{t^2/2} - 2$, whose absolute value is bounded above on $[0,1]$ by $M = y''(1) = 3\sqrt{e} - 2$. Corollary 6.5 implies that the global truncation error at $t=1$ must be smaller than

$$\frac{Mh}{2L}e^L(1-0) = \frac{(3\sqrt{e}-2)}{2}eh \approx 4.004h. \tag{6.27}$$

This upper bound is confirmed by the actual global truncation errors, shown in Figure 6.4, which are roughly 2 times h for small h. ◀

So far, Euler's Method seems to be foolproof. It is intuitive in construction, and the errors it makes get smaller when the step size decreases, according to Corollary 6.5. However, for more difficult IVPs, Euler's Method is rarely used. There exist more sophisticated methods whose order, or power of h in (6.25), is greater than one. This leads to vastly reduced global error, as we shall see. We close this section with an innocent-looking example in which such a reduction in error is needed.

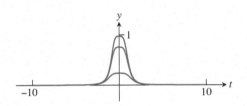

Figure 6.6 Approximation of Example 6.9 by Euler's Method. From bottom to top, approximate solutions with step sizes $h = 10^{-3}, 10^{-4}$, and 10^{-5}. The correct solution has $y(0) = 1$. Extremely small steps are needed to get a reasonable approximation.

▶ **EXAMPLE 6.9** Apply Euler's Method to the initial value problem

$$\begin{cases} y' = -4t^3y^2 \\ y(-10) = 1/10001 \\ t \text{ in } [-10,0]. \end{cases} \tag{6.28}$$

It is easy to check by substitution that the exact solution is $y(t) = 1/(t^4 + 1)$. The solution is very well behaved on the interval of interest. We will assess the ability of Euler's Method to approximate the solution at $t=0$.

Figure 6.6 shows Euler's Method approximations to the solution, with step sizes $h = 10^{-3}, 10^{-4}$, and 10^{-5}, from bottom to top. The value of the correct solution at $t = 0$ is $y(0) = 1$. Even the best approximation, which uses one million steps to reach $t = 0$ from the initial condition, is noticeably incorrect. ◀

This example shows that more accurate methods are needed to achieve accuracy in a reasonable amount of computation. The remainder of the chapter is devoted to developing more sophisticated methods that require fewer steps to get the same or better accuracy.

6.2.2 The Explicit Trapezoid Method

A small adjustment in the Euler's Method formula makes a great improvement in accuracy. Consider the following geometrically motivated method:

Explicit Trapezoid Method

$$w_0 = y_0$$
$$w_{i+1} = w_i + \frac{h}{2}(f(t_i, w_i) + f(t_i + h, w_i + hf(t_i, w_i))). \qquad (6.29)$$

For Euler's Method, the slope $y'(t_i)$ governing the discrete step is taken from the slope field at the left-hand end of the interval $[t_i, t_{i+1}]$. For the Trapezoid Method, as illustrated in Figure 6.7, this slope is replaced by the average between the contribution $y'(t_i)$ from the left-hand endpoint and the slope $f(t_i + h, w_i + hf(t_i, w_i))$ from the right-hand point that Euler's Method would have given. The Euler's Method "prediction" is used as the w-value to evaluate the slope function f at $t_{i+1} = t_i + h$. In a sense, the Euler's Method prediction is corrected by the Trapezoid Method, which is more accurate, as we will show.

The Trapezoid Method is called explicit because the new approximation w_{i+1} can be determined by an explicit formula in terms of previous w_i, t_i, and h. Euler's Method is also an explicit method.

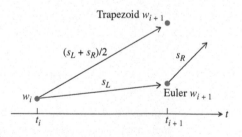

Figure 6.7 Schematic view of single step of the Explicit Trapezoid Method. The slopes $s_L = f(t_i, w_i)$ and $s_R = f(t_i + h, w_i + hf(t_i, w_i))$ are averaged to define the slope used to advance the solution to t_{i+1}.

The reason for the name "Trapezoid Method" is that in the special case where $f(t, y)$ is independent of y, the method

$$w_{i+1} = w_i + \frac{h}{2}[f(t_i) + f(t_i + h)]$$

can be viewed as adding a Trapezoid Rule approximation of the integral $\int_{t_i}^{t_i+h} f(t)\, dt$ to the current w_i. Since

$$\int_{t_i}^{t_i+h} f(t)\, dt = \int_{t_i}^{t_i+h} y'(t)\, dt = y(t_i + h) - y(t_i),$$

this corresponds to solving the differential equation $y' = f(t)$ by integrating both sides with the use of the Trapezoid Rule (5.21). The Explicit Trapezoid Method is also called the improved Euler Method and the Heun Method in the literature, but we will use the more descriptive and more easily remembered title.

▶ **EXAMPLE 6.10**

Apply the Explicit Trapezoid Method to the initial value problem (6.5) with initial condition $y(0) = 1$.

Formula (6.29) for $f(t, y) = ty + t^3$ is

$$w_0 = y_0 = 1$$
$$w_{i+1} = w_i + \frac{h}{2}(f(t_i, w_i) + f(t_i + h, w_i + hf(t_i, w_i)))$$
$$= w_i + \frac{h}{2}(t_i y_i + t_i^3 + (t_i + h)(w_i + h(t_i y_i + t_i^3)) + (t_i + h)^3).$$

Using step size $h = 0.1$, the iteration yields the following table:

step	t_i	w_i	y_i	e_i
0	0.0	1.0000	1.0000	0.0000
1	0.1	1.0051	1.0050	0.0001
2	0.2	1.0207	1.0206	0.0001
3	0.3	1.0483	1.0481	0.0002
4	0.4	1.0902	1.0899	0.0003
5	0.5	1.1499	1.1494	0.0005
6	0.6	1.2323	1.2317	0.0006
7	0.7	1.3437	1.3429	0.0008
8	0.8	1.4924	1.4914	0.0010
9	0.9	1.6890	1.6879	0.0011
10	1.0	1.9471	1.9462	0.0010

◀

The comparison of Example 6.10 with the results of Euler's Method on the same problem in Example 6.2 is striking. In order to quantify the improvement that the Trapezoid Method brings toward solving initial value problems, we need to calculate its local truncation error (6.19).

The local truncation error is the error made on a single step. Starting at an assumed correct solution point (t_i, y_i), the correct extension of the solution at t_{i+1} can be given by the Taylor expansion

$$y_{i+1} = y(t_i + h) = y_i + hy'(t_i) + \frac{h^2}{2}y''(t_i) + \frac{h^3}{6}y'''(c), \tag{6.30}$$

for some number c between t_i and t_{i+1}, assuming that y''' is continuous. In order to compare these terms with the Trapezoid Method, we will write them a little differently. From the differential equation $y'(t) = f(t, y)$, differentiate both sides with respect to t, using the chain rule:

$$y''(t) = \frac{\partial f}{\partial t}(t, y) + \frac{\partial f}{\partial y}(t, y)y'(t)$$
$$= \frac{\partial f}{\partial t}(t, y) + \frac{\partial f}{\partial y}(t, y)f(t, y).$$

The new version of (6.30) is

$$y_{i+1} = y_i + hf(t_i, y_i) + \frac{h^2}{2}\left(\frac{\partial f}{\partial t}(t_i, y_i) + \frac{\partial f}{\partial y}(t_i, y_i)f(t_i, y_i)\right) + \frac{h^3}{6}y'''(c). \tag{6.31}$$

We want to compare this expression with the Explicit Trapezoid Method, using the two-dimensional Taylor theorem to expand the term

$$f(t_i + h, y_i + hf(t_i, y_i)) = f(t_i, y_i) + h\frac{\partial f}{\partial t}(t_i, y_i) + hf(t_i, y_i)\frac{\partial f}{\partial y}(t_i, y_i) + O(h^2).$$

The Trapezoid Method can be written

$$
\begin{aligned}
w_{i+1} &= y_i + \frac{h}{2}\left(f(t_i, y_i) + f(t_i + h, y_i + hf(t_i, y_i))\right) \\
&= y_i + \frac{h}{2}f(t_i, y_i) + \frac{h}{2}\left(f(t_i, y_i) + h\left(\frac{\partial f}{\partial t}(t_i, y_i)\right.\right. \\
&\quad + \left.\left. f(t_i, y_i)\frac{\partial f}{\partial y}(t_i, y_i)\right) + O(h^2)\right) \\
&= y_i + hf(t_i, y_i) + \frac{h^2}{2}\left(\frac{\partial f}{\partial t}(t_i, y_i) + f(t_i, y_i)\frac{\partial f}{\partial y}(t_i, y_i)\right) + O(h^3). \quad (6.32)
\end{aligned}
$$

SPOTLIGHT ON

Complexity   Is a second-order method more efficient or less efficient than a first-order method? On each step, the error is smaller, but the computational work is greater, since ordinarily two function evaluations (of $f(t, y)$) are required instead of one. A rough comparison goes like this: Suppose that an approximation has been run with step size h, and we want to double the amount of computation to improve the approximation. For the same number of function evaluations, we can (a) halve the step size of the first-order method, multiplying the global error by $1/2$, or (b) keep the same step size, but use a second-order method, replacing the h in Theorem 6.4 by h^2, essentially multiplying the global error by h. For small h, (b) wins.

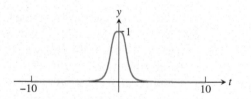

Figure 6.8 Approximation of Example 6.9 by the Trapezoid Method. Step size is $h = 10^{-3}$. Note the significant improvement in accuracy compared with Euler's Method in Figure 6.6.

Subtracting (6.32) from (6.31) gives the local truncation error as

$$y_{i+1} - w_{i+1} = O(h^3).$$

Theorem 6.4 shows that the global error of the Trapezoid Method is proportional to h^2, meaning that the method is of order two, compared with order one for Euler's Method. For small h this is a significant difference, as shown by returning to Example 6.9.

▶ **EXAMPLE 6.11**   Apply the Trapezoid Method to Example 6.9:

$$
\begin{cases}
y' = -4t^3 y^2 \\
y(-10) = 1/10001. \\
t \text{ in } [-10, 0]
\end{cases}
$$

Revisiting Example 6.9 with a more powerful method yields a great improvement in approximating the solution, for example, at $x = 0$. The correct value $y(0) = 1$ is attained within .0015 with a step size of $h = 10^{-3}$ with the Trapezoid Method, as shown in Figure 6.8. This is already better than Euler with a step size of $h = 10^{-5}$. Using the Trapezoid Method with $h = 10^{-5}$ yields an error on the order of 10^{-7} for this relatively difficult initial value problem. ◀

6.2.3 Taylor Methods

So far, we have learned two methods for approximating solutions of ordinary differential equations. The Euler Method has order one, and the apparently superior Trapezoid Method has order two. In this section, we show that methods of all orders exist. For each positive integer k, there is a Taylor Method of order k, which we will describe next.

The basic idea is a straightforward exploitation of the Taylor expansion. Assume that the solution $y(t)$ is $(k + 1)$ times continuously differentiable. Given the current point $(t, y(t))$ on the solution curve, the goal is to express $y(t + h)$ in terms of $y(t)$ for some step size h, using information about the differential equation. The Taylor expansion of $y(t)$ about t is

$$y(t + h) = y(t) + hy'(t) + \frac{1}{2}h^2 y''(t) + \cdots + \frac{1}{k!}h^k y^{(k)}(t)$$
$$+ \frac{1}{(k + 1)!}h^{k+1} y^{(k+1)}(c), \tag{6.33}$$

where c lies between t and $t + h$. The last term is the Taylor remainder term. This equation motivates the following method:

Taylor Method of order k

$$w_0 = y_0$$
$$w_{i+1} = w_i + hf(t_i, w_i) + \frac{h^2}{2}f'(t_i, w_i) + \cdots + \frac{h^k}{k!}f^{(k-1)}(t_i, w_i). \tag{6.34}$$

The prime notation refers to the total derivative of $f(t, y(t))$ with respect to t. For example,

$$f'(t, y) = f_t(t, y) + f_y(t, y)y'(t)$$
$$= f_t(t, y) + f_y(t, y)f(t, y).$$

We use the notation f_t to denote the partial derivative of f with respect to t, and similarly for f_y. To find the local truncation error of the Taylor Method, set $w_i = y_i$ in (6.34) and compare with the Taylor expansion (6.33) to get

$$y_{i+1} - w_{i+1} = \frac{h^{k+1}}{(k + 1)!}y^{(k+1)}(c).$$

We conclude that the Taylor Method of order k has local truncation error h^{k+1} and has order k, according to Theorem 6.4.

The first-order Taylor Method is

$$w_{i+1} = w_i + hf(t_i, w_i),$$

which is identified as Euler's Method. The second-order Taylor Method is

$$w_{i+1} = w_i + hf(t_i, w_i) + \frac{1}{2}h^2(f_t(t_i, w_i) + f_y(t_i, w_i)f(t_i, w_i)).$$

▶ **EXAMPLE 6.12** Determine the second-order Taylor Method for the first-order linear equation

$$\begin{cases} y' = ty + t^3 \\ y(0) = y_0. \end{cases} \tag{6.35}$$

Since $f(t, y) = ty + t^3$, it follows that

$$f'(t, y) = f_t + f_y f$$
$$= y + 3t^2 + t(ty + t^3),$$

and the method gives

$$w_{i+1} = w_i + h(t_i w_i + t_i^3) + \frac{1}{2}h^2(w_i + 3t_i^2 + t_i(t_i w_i + t_i^3)). \qquad \blacktriangleleft$$

Although second-order Taylor Method is a second-order method, notice that manual labor on the user's part was required to determine the partial derivatives. Compare this with the other second-order method we have learned, where (6.29) requires only calls to a routine that computes values of $f(t, y)$ itself.

Conceptually, the lesson represented by Taylor Methods is that ODE methods of arbitrary order exist, as shown in (6.34). However, they suffer from the problem that extra work is needed to compute the partial derivatives of f that show up in the formula. Since formulas of the same orders can be developed that do not require these partial derivatives, the Taylor Methods are used only for specialized purposes.

▶ **ADDITIONAL EXAMPLES**

*1 Calculate the Trapezoid Method approximation on the interval [0,1] to the initial value problem $y' = ty^2, y(0) = -1$ for step size $h = 1/4$. Find the error at $t = 1$ by comparing with the exact solution $y(t) = -2/(t^2 + 2)$.

2. Plot the Trapezoid Method approximation to the solution of the initial value problem

$$\begin{cases} y' = 3t^2 y + 4t^2 \\ y(0) = 5 \end{cases}$$

on the interval [0,1] with step size $h = 0.1$, along with the exact solution $y(t) = -\frac{4}{3} + \frac{19}{3}e^{t^3}$.

📱 **Solutions** for Additional Examples can be found at bit.ly/2EyK9WR
(* example with video solution)

6.2 Exercises

📱 **Solutions** for Exercises numbered in blue can be found at bit.ly/2CNYBZ9

1. Using initial condition $y(0) = 1$ and step size $h = 1/4$, calculate the Trapezoid Method approximation $w_0, \ldots, w_4$ on the interval [0,1]. Find the error at $t = 1$ by comparing with the correct solution found in Exercise 6.1.3.

(a) $y' = t$ (b) $y' = t^2 y$ (c) $y' = 2(t + 1)y$
(d) $y' = 5t^4 y$ (e) $y' = 1/y^2$ (f) $y' = t^3/y^2$

2. Using initial condition $y(0) = 0$ and step size $h = 1/4$, calculate the Trapezoid Method approximation on the interval $0, 1$. Find the error at $t = 1$ by comparing with the correct solution found in Exercise 6.1.4.

$$\text{(a)} \quad y' = t + y \quad \text{(b)} \quad y' = t - y \quad \text{(c)} \quad y' = 4t - 2y$$

3. Find the formula for the second-order Taylor Method for the following differential equations: (a) $y' = ty$ (b) $y' = ty^2 + y^3$ (c) $y' = y \sin y$ (d) $y' = e^{yt^2}$

4. Apply the second-order Taylor Method to the initial value problems in Exercise 1. Using step size $h = 1/4$, calculate the second-order Taylor Method approximation on the interval $0, 1$. Compare with the correct solution found in Exercise 6.1.3, and find the error at $t = 1$.

5. (a) Prove (6.22) (b) Prove (6.23).

6.2 Computer Problems

Solutions for Computer Problems numbered in blue can be found at bit.ly/2J5xZmX

1. Apply the Explicit Trapezoid Method on a grid of step size $h = 0.1$ in $0, 1$ to the initial value problems in Exercise 1. Print a table of the t values, approximations, and global truncation error at each step.

2. Plot the approximate solutions for the IVPs in Exercise 1 on $0, 1$ for step sizes $h = 0.1, 0.05$, and 0.025, along with the true solution.

3. For the IVPs in Exercise 1, plot the global truncation error of the Explicit Trapezoid Method at $t = 1$ as a function of $h = 0.1 \times 2^{-k}$ for $0 \le k \le 5$. Use a log–log plot as in Figure 6.4.

4. For the IVPs in Exercise 1, plot the global truncation error of the second-order Taylor Method at $t = 1$ as a function of $h = 0.1 \times 2^{-k}$ for $0 \le k \le 5$.

5. Plot the Trapezoid Method approximate solution on $0, 1$ for the differential equation $y' = 1 + y^2$ and initial condition (a) $y_0 = 0$ (b) $y_0 = 1/2$, along with the exact solution (see Exercise 6.1.7). Use step sizes $h = 0.1$ and 0.05.

6. Plot the Trapezoid Method approximate solution on $0, 1$ for the differential equation $y' = 1 - y^2$ and initial condition (a) $y_0 = 0$ (b) $y_0 = -1/2$, along with the exact solution (see Exercise 6.1.8). Use step sizes $h = 0.1$ and 0.05.

7. Calculate the Trapezoid Method approximate solution on $0, 4$ for the differential equation $y' = \sin y$ and initial condition (a) $y_0 = 0$ (b) $y_0 = 100$, using step sizes $h = 0.1 \times 2^{-k}$ for $0 \le k \le 5$. Plot the $k = 0$ and $k = 5$ approximate solutions along with the exact solution (see Exercise 6.1.15), and make a log–log plot of the error at $t = 4$ as a function of h.

8. Calculate the Trapezoid Method approximate solution of the differential equation $y' = \sinh y$ and initial condition (a) $y_0 = 1/4$ on the interval $0, 2$ (b) $y_0 = 2$ on the interval $0, 1/4$, using step sizes $h = 0.1 \times 2^{-k}$ for $0 \le k \le 5$. Plot the $k = 0$ and $k = 5$ approximate solutions along with the exact solution (see Exercise 6.1.16), and make a log–log plot of the error at the end of the time interval as a function of h.

9. Calculate the Trapezoid Method approximate solution of the differential equation $2y' = 1 + y^2$ and initial condition (a) $y_0 = 1$ and (b) $y_0 = \sqrt{3}$ on the interval $0, 1$, using step sizes $h = 0.1 \times 2^{-k}$ for $0 \le k \le 5$. Plot the $k = 0$ and $k = 5$ approximate solution along with the exact solution (see Exercise 6.1.17), and make a log–log plot of the error at the end of the time interval as a function of h.

6.3 SYSTEMS OF ORDINARY DIFFERENTIAL EQUATIONS

Approximation of systems of differential equations can be done as a simple extension of the methodology for a single differential equation. Treating systems of equations greatly extends our ability to model interesting dynamical behavior.

The ability to solve systems of ordinary differential equations lies at the core of the art and science of computer simulation. In this section, we introduce two physical systems whose simulation has motivated a great deal of development of ODE solvers: the pendulum and orbital mechanics. The study of these examples will provide the reader some practical experience in the capabilities and limitations of the solvers.

The **order** of a differential equation refers to the highest order derivative appearing in the equation. A first-order system has the form

$$y_1' = f_1(t, y_1, \ldots, y_n)$$
$$y_2' = f_2(t, y_1, \ldots, y_n)$$
$$\vdots$$
$$y_n' = f_n(t, y_1, \ldots, y_n).$$

In an initial value problem, each variable needs its own initial condition.

► **EXAMPLE 6.13**   Apply Euler's Method to the first-order system of two equations:

$$y_1' = y_2^2 - 2y_1$$
$$y_2' = y_1 - y_2 - ty_2^2$$
$$y_1(0) = 0$$
$$y_2(0) = 1. \tag{6.36}$$

Check that the solution of the system (6.36) is the vector-valued function

$$y_1(t) = te^{-2t}$$
$$y_2(t) = e^{-t}.$$

For the moment, forget that we know the solution, and apply Euler's Method. The scalar Euler's Method formula is applied to each component in turn as follows:

$$w_{i+1,1} = w_{i,1} + h(w_{i,2}^2 - 2w_{i,1})$$
$$w_{i+1,2} = w_{i,2} + h(w_{i,1} - w_{i,2} - t_i w_{i,2}^2).$$

Figure 6.9 shows the Euler Method approximations of y_1 and y_2, along with the correct solution. The MATLAB code that carries this out is essentially the same as Program 6.1, with a few adjustments to treat y as a vector:

⎙ **MATLAB code**

shown here can be found
at bit.ly/2PyPuOF

```
% Program 6.2  Vector version of Euler Method
% Input: interval inter, initial vector y0, number of steps n
% Output: time steps t, solution y
% Example usage: euler2([0 1],[0 1],10);
function [t,y]=euler2(inter,y0,n)
t(1)=inter(1); y(1,:)=y0;
h=(inter(2)-inter(1))/n;
for i=1:n
  t(i+1)=t(i)+h;
```

```
   y(i+1,:)=eulerstep(t(i),y(i,:),h);
end
plot(t,y(:,1),t,y(:,2));

function y=eulerstep(t,y,h)
%one step of the Euler Method
%Input: current time t, current vector y, step size h
%Output: the approximate solution vector at time t+h
y=y+h*ydot(t,y);

function z=ydot(t,y)
%right-hand side of differential equation
z(1)=y(2)^2-2*y(1);
z(2)=y(1)-y(2)-t*y(2)^2;
```

◀

6.3.1 Higher order equations

A single differential equation of higher order can be converted to a system. Let

$$y^{(n)} = f(t, y, y', y'', \ldots, y^{(n-1)})$$

be an nth-order ordinary differential equation. Define new variables

$$
\begin{aligned}
y_1 &= y \\
y_2 &= y' \\
y_3 &= y'' \\
&\vdots \\
y_n &= y^{(n-1)},
\end{aligned}
$$

and notice that the original differential equation can be written

$$y_n' = f(t, y_1, y_2, \ldots, y_n).$$

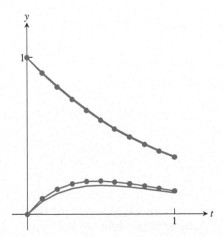

Figure 6.9 Equation (6.36) approximated by Euler Method. Step size $h = 0.1$. The upper curve is $y_1(t)$, along with its approximate solution $w_{i,1}$ (circles), while the lower curve is $y_2(t)$ and $w_{i,2}$.

Taken together, the equations

$$y_1' = y_2$$
$$y_2' = y_3$$
$$y_3' = y_4$$
$$\vdots$$
$$y_{n-1}' = y_n,$$
$$y_n' = f(t, y_1, \ldots, y_n)$$

convert the nth-order differential equation into a system of first-order equations, which can be solved by using methods like the Euler or Trapezoid Methods.

▶ **EXAMPLE 6.14**   Convert the third-order differential equation

$$y''' = a(y'')^2 - y' + yy'' + \sin t \tag{6.37}$$

to a system.

Set $y_1 = y$ and define the new variables

$$y_2 = y'$$
$$y_3 = y''.$$

Then, in terms of first derivatives, (6.37) is equivalent to

$$y_1' = y_2$$
$$y_2' = y_3$$
$$y_3' = ay_3^2 - y_2 + y_1 y_3 + \sin t. \tag{6.38}$$

The solution $y(t)$ of the third-order equation (6.37) can be found by solving the system (6.38) for $y_1(t), y_2(t), y_3(t)$.   ◀

Because of the possibility of converting higher-order equations to systems, we will restrict our attention to systems of first-order equations. Note also that a system of several higher-order equations can be converted to a system of first-order equations in the same way.

6.3.2 Computer simulation: the pendulum

Figure 6.10 shows a pendulum swinging under the influence of gravity. Assume that the pendulum is hanging from a rigid rod that is free to swing through 360 degrees. Denote by y the angle of the pendulum with respect to the vertical, so that $y = 0$ corresponds to straight down. Therefore, y and $y + 2\pi$ are considered the same angle.

Newton's second law of motion $F = ma$ can be used to find the pendulum equation. The motion of the pendulum bob is constrained to be along a circle of radius l, where l is the length of the pendulum rod. If y is measured in radians, then the component of acceleration tangent to the circle is ly'', because the component of position tangent to the circle is ly. The component of force along the direction of motion is $mg \sin y$. It is a restoring force, meaning that it is directed in the opposite direction from the displacement of the variable y. The differential equation governing the frictionless pendulum is therefore

$$mly'' = F = -mg \sin y. \tag{6.39}$$

This is a second-order differential equation for the angle y of the pendulum. The initial conditions are given by the initial angle $y(0)$ and angular velocity $y'(0)$.

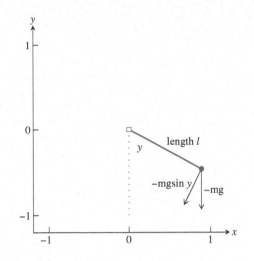

Figure 6.10 The pendulum. Component of force in the tangential direction is $F = -mg \sin y$, where y is the angle the pendulum bob makes with the vertical.

By setting $y_1 = y$ and introducing the new variable $y_2 = y'$, the second-order equation is converted to a first-order system:

$$y_1' = y_2$$
$$y_2' = -\frac{g}{l} \sin y_1. \tag{6.40}$$

The system is autonomous because there is no t dependence in the right-hand side. If the pendulum is started from a position straight out to the right, the initial conditions are $y_1(0) = \pi/2$ and $y_2(0) = 0$. In MKS units, the gravitational acceleration at the earth's surface is about 9.81m/sec^2. Using these parameters, we can test the suitability of Euler's Method as a solver for this system.

Figure 6.11 shows Euler's Method approximations to the pendulum equations with two different step sizes. The pendulum rod is assigned to be $l = 1$ meter in length. The smaller curve represents the angle y as a function of time, and the larger amplitude curve is the instantaneous angular velocity. Note that the zeros of the angle, representing the vertical position of the pendulum, correspond to the largest angular velocity, positive or negative. The pendulum is traveling fastest as it swings through the lowest point. When the pendulum is extended to the far right, the peak of the smaller curve, the velocity is zero as it turns from positive to negative.

The inadequacy of Euler's Method is apparent in Figure 6.11. The step size $h = 0.01$ is clearly too large to achieve even qualitative correctness. An undamped pendulum started with zero velocity should swing back and forth forever, returning to its starting position with a regular periodicity. The amplitude of the angle in Figure 6.11(a) is growing, which violates the conservation of energy. Using 10 times more steps, as in Figure 6.11(b), improves the situation at least visually, but a total of 10^4 steps are needed, an extreme number for the routine dynamical behavior shown by the pendulum.

A second-order ODE solver like the Trapezoid Method improves accuracy at a much lower cost. We will rewrite the MATLAB code to use the Trapezoid Method and take the opportunity to illustrate the ability of MATLAB to do simple animations.

The code pend.m that follows contains the same differential equation information, but eulerstep is replaced by trapstep. In addition, the variables rod and bob are introduced to represent the rod and pendulum bob, respectively. The MATLAB set command assigns attributes to variables. The drawnow command plots

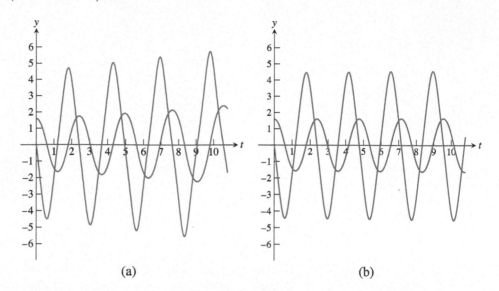

(a)   (b)

Figure 6.11 Euler Method applied to the pendulum equation (6.40). The curve of smaller amplitude is the angle y_1 in radians; the curve of larger amplitude is the angular velocity y_2. (a) Step size $h = 0.01$ is too large; energy is growing. (b) Step size $h = 0.001$ shows more accurate trajectories.

the rod and bob variables. Note that the erase mode of both variables is set to xor, meaning that when the plotted variable is redrawn somewhere else, the previous position is erased. Figure 6.10 is a screen shot of the animation. Here is the code:

MATLAB code
shown here can be found
at bit.ly/2RUk1rX

```
% Program 6.3 Animation program for pendulum
% Inputs: time interval inter,
%   initial values ic = [y(1,1) y(1,2)], number of steps n
% Calls a one-step method such as trapstep.m
% Example usage: pend([0 10],[pi/2 0],200)
function pend(inter,ic,n)
h=(inter(2)-inter(1))/n;  % plot n points in total
y(1,:)=ic;                     % enter initial conds in y
t(1)=inter(1);
set(gca,'xlim',[-1.2 1.2],'ylim',[-1.2 1.2], ...
  'XTick',[-1 0 1],'YTick',[-1 0 1])
bob=animatedline('color','r','Marker','.','markersize',40);
rod=animatedline('color','b','LineStyle','-','LineWidth',3);
axis square               % make aspect ratio 1 - 1
for k=1:n
  t(k+1)=t(k)+h;
  y(k+1,:)=trapstep(t(k),y(k,:),h);
  xbob = sin(y(k+1,1)); ybob = -cos(y(k+1,1));
  xrod = [0 xbob]; yrod = [0 ybob];
  clearpoints(bob);addpoints(bob,xbob,ybob);
  clearpoints(rod);addpoints(rod,xrod,yrod);
  drawnow; pause(h)
end

function y = trapstep(t,x,h)
%one step of the Trapezoid Method
z1=ydot(t,x);
```

```
g=x+h*z1;
z2=ydot(t+h,g);
y=x+h*(z1+z2)/2;

function z=ydot(t,y)
g=9.81;length=1;
z(1) = y(2);
z(2) = -(g/length)*sin(y(1));
```

Using the Trapezoid Method in the pendulum equation allows fairly accurate solutions to be found with larger step size. This section ends with several interesting variations on the basic pendulum simulation, which the reader is encouraged to experiment with in the Computer Problems.

▶ **EXAMPLE 6.15**   The damped pendulum.

The force of damping, such as air resistance or friction, is often modeled as being proportional and in the opposite direction to velocity. The pendulum equation becomes

$$y_1' = y_2$$
$$y_2' = -\frac{g}{l}\sin y_1 - dy_2, \tag{6.41}$$

where $d > 0$ is the damping coefficient. Unlike the undamped pendulum, this one will lose energy through damping and in time approach the limiting equilibrium solution $y_1 = y_2 = 0$, from any initial condition. Computer Problem 3 asks you to run a damped version of pend.m.   ◀

▶ **EXAMPLE 6.16**   The forced damped pendulum.

Adding a time-dependent term to (6.41) represents outside forcing on the damped pendulum. Consider adding the sinusoidal term $A\sin t$ to the right-hand side of y_2', yielding

$$y_1' = y_2$$
$$y_2' = -\frac{g}{l}\sin y_1 - dy_2 + A\sin t. \tag{6.42}$$

This can be considered as a model of a pendulum that is affected by an oscillating magnetic field, for example.

A host of new dynamical behaviors becomes possible when forcing is added. For a two-dimensional autonomous system of differential equations, the Poincaré–Bendixson Theorem (from the theory of differential equations) says that trajectories can tend toward only regular motion, such as stable equilibria like the down position of the pendulum, or stable periodic cycles like the pendulum swinging back and forth forever. The forcing makes the system nonautonomous (it can be rewritten as a three-dimensional autonomous system, but not as a two-dimensional one), so that a third type of trajectories is allowed, namely, chaotic trajectories.

Setting the damping coefficient to $d = 1$ and the forcing coefficient to $A = 10$ results in interesting periodic behavior, explored in Computer Problem 4. Moving the parameter to $A = 15$ introduces chaotic trajectories.   ◀

▶ **EXAMPLE 6.17**   The double pendulum.

The double pendulum is composed of a simple pendulum, with another simple pendulum hanging from its bob. If y_1 and y_3 are the angles of the two bobs with respect to the vertical, the system of differential equations is

$$y_1' = y_2$$
$$y_2' = \frac{-3g \sin y_1 - g \sin(y_1 - 2y_3) - 2\sin(y_1 - y_3)(y_4^2 + y_2^2 \cos(y_1 - y_3))}{3 - \cos(2y_1 - 2y_3)} - dy_2$$
$$y_3' = y_4$$
$$y_4' = \frac{2\sin(y_1 - y_3)[2y_2^2 + 2g \cos y_1 + y_4^2 \cos(y_1 - y_3)]}{3 - \cos(2y_1 - 2y_3)},$$

where $g = 9.81$ and the length of both rods has been set to 1. The parameter d represents friction at the pivot. For $d = 0$, the double pendulum exhibits sustained nonperiodicity for many initial conditions and is mesmerizing to observe. See Computer Problem 8.   ◀

6.3.3 Computer simulation: orbital mechanics

As a second example, we simulate the motion of an orbiting satellite. Newton's second law of motion says that the acceleration a of the satellite is related to the force F applied to the satellite as $F = ma$, where m is the mass. The law of gravitation expresses the force on a body of mass m_1 due to a body of mass m_2 by an inverse-square law

$$F = \frac{gm_1 m_2}{r^2},$$

where r is the distance separating the masses. In the **one-body problem**, one of the masses is considered negligible compared with the other, as in the case of a small satellite orbiting a large planet. This simplification allows us to neglect the force of the satellite on the planet, so that the planet may be regarded as fixed.

Place the large mass at the origin, and denote the position of the satellite by (x, y). The distance between the masses is $r = \sqrt{x^2 + y^2}$, and the force on the satellite is central—that is, in the direction of the large mass. The direction vector, a unit vector in this direction, is

$$\left(-\frac{x}{\sqrt{x^2 + y^2}}, -\frac{y}{\sqrt{x^2 + y^2}} \right).$$

Therefore, the force on the satellite in terms of components is

$$(F_x, F_y) = \left(\frac{gm_1 m_2}{x^2 + y^2} \frac{-x}{\sqrt{x^2 + y^2}}, \frac{gm_1 m_2}{x^2 + y^2} \frac{-y}{\sqrt{x^2 + y^2}} \right). \tag{6.43}$$

Inserting these forces into Newton's law of motion yields the two second-order equations

$$m_1 x'' = -\frac{gm_1 m_2 x}{(x^2 + y^2)^{3/2}}$$
$$m_1 y'' = -\frac{gm_1 m_2 y}{(x^2 + y^2)^{3/2}}.$$

Introducing the variables $v_x = x'$ and $v_y = y'$ allows the two second-order equations to be reduced to a system of four first-order equations:

$$x' = v_x$$
$$v_x' = -\frac{gm_2 x}{(x^2 + y^2)^{3/2}}$$
$$y' = v_y$$
$$v_y' = -\frac{gm_2 y}{(x^2 + y^2)^{3/2}}. \qquad (6.44)$$

The following MATLAB program orbit.m calls eulerstep.m and sequentially plots the satellite orbit.

MATLAB code

shown here can be found at bit.ly/2QXpM6Z

```
%Program 6.4 Plotting program for one-body problem
% Inputs: time interval inter, initial conditions
% ic = [x0 vx0 y0 vy0], x position, x velocity, y pos, y vel,
% number of steps n, steps per point plotted p
% Calls a one-step method such as trapstep.m
% Example usage: orbit([0 100],[0 1 2 0],10000,5)
function z=orbit(inter,ic,n,p)
h=(inter(2)-inter(1))/n;          % plot n points
x0=ic(1);vx0=ic(2);y0=ic(3);vy0=ic(4);   % grab initial conds
y(1,:)=[x0 vx0 y0 vy0];t(1)=inter(1);     % build y vector
set(gca,'XLim',[-5 5],'YLim',[-5 5],...
    'XTick',[-5 0 5],'YTick',[-5 0 5]);
sun=animatedline('color','y','Marker','.','markersize',50);
addpoints(sun,0,0)
head=animatedline('color','r','Marker','.','markersize',35);
tail=animatedline('color','b','LineStyle','-');
%[px,py]=ginput(1);                % include these three lines
%[px1,py1]=ginput(1);              % to enable mouse support
%y(1,:)=[px px1-px py py1-py];     % 2 clicks set direction
for k=1:n/p
  for i=1:p
    t(i+1)=t(i)+h;
    y(i+1,:)=eulerstep(t(i),y(i,:),h);
  end
  y(1,:)=y(p+1,:);t(1)=t(p+1);
  clearpoints(head);addpoints(head,y(1,1),y(1,3))
  addpoints(tail,y(1,1),y(1,3))
  drawnow;
end

function y=eulerstep(t,x,h)
%one step of the Euler method
y=x+h*ydot(t,x);

function y = trapstep(t,x,h)
%one step of the Trapezoid Method
z1=ydot(t,x);
g=x+h*z1;
z2=ydot(t+h,g);
y=x+h*(z1+z2)/2;

function z = ydot(t,x)
m2=3;g=1;mg2=m2*g;px2=0;py2=0;
```

```
px1=x(1);py1=x(3);vx1=x(2);vy1=x(4);
dist=sqrt((px2-px1)^2+(py2-py1)^2);
z=zeros(1,4);
z(1)=vx1;
z(2)=(mg2*(px2-px1))/(dist^3);
z(3)=vy1;
z(4)=(mg2*(py2-py1))/(dist^3);
```

Running the MATLAB script `orbit.m` immediately shows the limitations of Euler's Method for approximating interesting problems. Figure 6.12(a) shows the outcome of running `orbit([0 100],[0 1 2 0],10000,5)`. In other words, we follow the orbit over the time interval $[a, b] = [0, 100]$, the initial position is $(x_0, y_0) = (0, 2)$, the initial velocity is $(v_x, v_y) = (1, 0)$, and the Euler step size is $h = 100/10000 = 0.01$.

Solutions to the one-body problem must be conic sections—either ellipses, parabolas, or hyperbolas. The spiral seen in Figure 6.12(a) is a numerical artifact, meaning a misrepresentation caused by errors of computation. In this case, it is the truncation error of Euler's Method that leads to the failure of the orbit to close up into an ellipse. If the step size is cut by a factor of 10 to $h = 0.001$, the result is improved, as shown in Figure 6.12(b). It is clear that even with the greatly decreased step size, the accumulated error is noticeable.

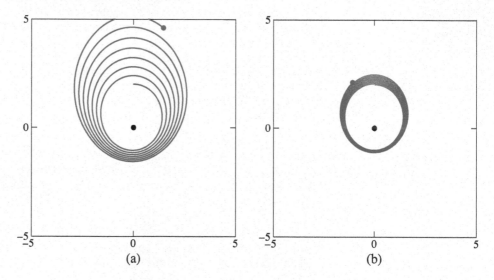

Figure 6.12 Euler's Method applied to one-body problem. (a) $h = 0.01$ and (b) $h = 0.001$.

Corollary 6.5 says that the Euler Method, in principle, can approximate a solution with as much accuracy as desired, if the step size h is sufficiently small. However, results like those represented by Figures 6.6 and 6.12 show that the method is seriously limited in practice.

Figure 6.13 shows the clear improvement in the one-body problem resulting from the replacement of the Euler step with the Trapezoid step. The plot was made by replacing the function `eulerstep` by `trapstep` in the foregoing code.

The one-body problem is fictional, in the sense that it ignores the force of the satellite on the (much larger) planet. When the latter is included as well, the motion of the two objects is called the two-body problem.

The case of three objects interacting gravitationally, called the **three-body problem**, holds an important position in the history of science. Even when all motion is

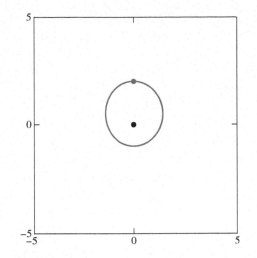

Figure 6.13 One-body problem approximated by the Trapezoid Method.
Step size $h = 0.01$. The orbit appears to close, at least to the resolution visible in the plot.

confined to a plane (the **restricted** three-body problem) the long-term trajectories may be essentially unpredictable. In 1889, King Oscar II of Sweden and Norway held a competition for work proving the stability of the solar system. The prize was awarded to Henri Poincaré, who showed that it would be impossible to prove any such thing, due to phenomena seen even for three interacting bodies.

The unpredictability stems from **sensitive dependence on initial conditions**, a term which denotes the fact that small uncertainties in the initial positions and velocities can lead to large deviations at a later time. In our terms, this is the statement that the solution of the system of differential equations is ill-conditioned with respect to the input of initial conditions.

The restricted three-body problem is a system of 12 equations, 4 for each body, that are also derived from Newton's second law. For example, the equations of the first body are

$$x_1' = v_{1x}$$
$$v_{1x}' = \frac{gm_2(x_2 - x_1)}{((x_2 - x_1)^2 + (y_2 - y_1)^2)^{3/2}} + \frac{gm_3(x_3 - x_1)}{((x_3 - x_1)^2 + (y_3 - y_1)^2)^{3/2}}$$
$$y_1' = v_{1y}$$
$$v_{1y}' = \frac{gm_2(y_2 - y_1)}{((x_2 - x_1)^2 + (y_2 - y_1)^2)^{3/2}} + \frac{gm_3(y_3 - y_1)}{((x_3 - x_1)^2 + (y_3 - y_1)^2)^{3/2}}. \quad (6.45)$$

The second and third bodies, at (x_2, y_2) and (x_3, y_3), respectively, satisfy similar equations.

Computer Problems 9 and 10 ask the reader to computationally solve the two- and three-body problems. The latter problem illustrates severe sensitive dependence on initial conditions.

▶ **ADDITIONAL**
EXAMPLES

1. Show that $y(t) = e^{-2t} + 4e^t$ is a solution of the initial value problem

$$\begin{cases} y'' + y' - 2y = 0 \\ y(0) = 5 \\ y'(0) = 2, \end{cases}$$

and convert the differential equation to an equivalent first-order system.

2. Apply Euler's Method with step sizes $h = 0.1$ and 0.05 to approximate the solution of the first-order system in Additional Example 1 on the interval $[0, 1]$. Plot both approximate solutions $y(t)$ along with the exact solution.

⌨ **Solutions** for Additional Examples can be found at `bit.ly/2q1lyjp`

6.3 Exercises

⌨ **Solutions**
for Exercises
numbered in blue
can be found at
`bit.ly/2EsIVfI`

1. Apply the Euler's Method with step size $h = 1/4$ to the initial value problem on $[0, 1]$.

(a)
$$\begin{cases} y_1' = y_1 + y_2 \\ y_2' = -y_1 + y_2 \\ y_1(0) = 1 \\ y_2(0) = 0 \end{cases}$$
(b)
$$\begin{cases} y_1' = -y_1 - y_2 \\ y_2' = y_1 - y_2 \\ y_1(0) = 1 \\ y_2(0) = 0 \end{cases}$$

(c)
$$\begin{cases} y_1' = -y_2 \\ y_2' = y_1 \\ y_1(0) = 1 \\ y_2(0) = 0 \end{cases}$$
(d)
$$\begin{cases} y_1' = y_1 + 3y_2 \\ y_2' = 2y_1 + 2y_2 \\ y_1(0) = 5 \\ y_2(0) = 0 \end{cases}$$

Find the global truncation errors of y_1 and y_2 at $t = 1$ by comparing with the correct solutions (a) $y_1(t) = e^t \cos t$, $y_2(t) = -e^t \sin t$ (b) $y_1(t) = e^{-t} \cos t$, $y_2(t) = e^{-t} \sin t$ (c) $y_1(t) = \cos t$, $y_2(t) = \sin t$ (d) $y_1(t) = 3e^{-t} + 2e^{4t}$, $y_2(t) = -2e^{-t} + 2e^{4t}$.

2. Apply the Trapezoid Method with $h = 1/4$ to the initial value problems in Exercise 1. Find the global truncation error at $t = 1$ by comparing with the correct solutions.

3. Convert the higher-order ordinary differential equation to a first-order system of equations. (a) $y'' - ty = 0$ (Airy's equation) (b) $y'' - 2ty' + 2y = 0$ (Hermite's equation) (c) $y'' - ty' - y = 0$

4. Apply the Trapezoid Method with $h = 1/4$ to the initial value problems in Exercise 3, using $y(0) = y'(0) = 1$.

5. (a) Show that $y(t) = (e^t + e^{-t} - t^2)/2 - 1$ is the solution of the initial value problem $y''' - y' = t$, with $y(0) = y'(0) = y''(0) = 0$. (b) Convert the differential equation to a system of three first-order equations. (c) Use Euler's Method with step size $h = 1/4$ to approximate the solution on $[0, 1]$. (d) Find the global truncation error at $t = 1$.

6.3 Computer Problems

⌨ **Solutions** for
Computer Problems
numbered in blue can
be found at
`bit.ly/2S2TbOh`

1. Apply Euler's Method with step sizes $h = 0.1$ and 0.01 to the initial value problems in Exercise 1. Plot the approximate solutions and the correct solution on $[0, 1]$, and find the global truncation error at $t = 1$. Is the reduction in error for $h = 0.01$ consistent with the order of Euler's Method?

2. Carry out Computer Problem 1 for the Trapezoid Method.

3. Adapt `pend.m` to model the damped pendulum. Run the resulting code with $d = 0.1$. Except for the initial condition $y_1(0) = \pi$, $y_2(0) = 0$, all trajectories move toward the straight-down position as time progresses. Check the exceptional initial condition: Does the simulation agree with theory? with a physical pendulum?

4. Adapt `pend.m` to build a forced, damped version of the pendulum. Run the Trapezoid Method in the following: (a) Set damping $d = 1$ and the forcing parameter $A = 10$. Set the step size $h = 0.005$ and the initial condition of your choice. After moving through

some transient behavior, the pendulum will settle into a periodic (repeating) trajectory. Describe this trajectory qualitatively. Try different initial conditions. Do all solutions end up at the same "attracting" periodic trajectory? (b) Now increase the step size to $h = 0.1$, and repeat the experiment. Try initial condition $[\pi/2, 0]$ and others. Describe what happens, and give a reasonable explanation for the anomalous behavior at this step size.

5. Run the forced damped pendulum as in Computer Problem 4, but set $A = 12$. Use the Trapezoid Method with $h = 0.005$. There are now two periodic attractors that are mirror images of one another. Describe the two attracting trajectories, and find two initial conditions $(y_1, y_2) = (a, 0)$ and $(b, 0)$, where $|a - b| \le 0.1$, that are attracted to different periodic trajectories. Set $A = 15$ to view chaotic motion of the forced damped pendulum.

6. Adapt pend.m to build a damped pendulum with oscillating pivot. The goal is to investigate the phenomenon of parametric resonance, by which the inverted pendulum becomes stable! The equation is

$$y'' + dy' + \left(\frac{g}{l} + A\cos 2\pi t\right)\sin y = 0,$$

where A is the forcing strength. Set $d = 0.1$ and the length of the pendulum to be 2.5 meters. In the absence of forcing $A = 0$, the downward pendulum $y = 0$ is a stable equilibrium, and the inverted pendulum $y = \pi$ is an unstable equilibrium. Find as accurately as possible the range of parameter A for which the inverted pendulum becomes stable. (Of course, $A = 0$ is too small; it turns out that $A = 30$ is too large.) Use the initial condition $y = 3.1$ for your test, and call the inverted position "stable" if the pendulum does not pass through the downward position.

7. Use the parameter settings of Computer Problem 6 to demonstrate the other effect of parametric resonance: The stable equilibrium can become unstable with an oscillating pivot. Find the smallest (positive) value of the forcing strength A for which this happens. Classify the downward position as unstable if the pendulum eventually travels to the inverted position.

8. Adapt pend.m to build the double pendulum. A new pair of rod and bob must be defined for the second pendulum. Note that the pivot end of the second rod is equal to the formerly free end of the first rod: The (x, y) position of the free end of the second rod can be calculated by using simple trigonometry.

9. Adapt orbit.m to solve the two-body problem. Set the masses to $m_1 = 0.03$, $m_2 = 0.3$, and plot the trajectories with initial conditions $(x_1, y_1) = (2, 2)$, $(x_1', y_1') = (0.2, -0.2)$ and $(x_2, y_2) = (0, 0)$, $(x_2', y_2') = (-0.02, 0.02)$.

10. Use the two-body problem code developed in Computer Problem 9 to investigate the following example trajectories. Set the initial positions and velocities of the two bodies to be $(x_1, y_1) = (0, 1)$, $(x_1', y_1') = (0.1, -0.1)$ and $(x_2, y_2) = (-2, -1)$, $(x_2', y_2') = (-0.01, 0.01)$. The masses are given by (a) $m_1 = 0.03, m_2 = 0.3$ (b) $m_1 = 0.05, m_2 = 0.5$ (c) $m_1 = 0.08$, $m_2 = 0.8$. In each case, apply the Trapezoid Method with step sizes $h = 0.01$ and 0.001 and compare the results on the time interval $[0, 500]$. Do you believe the Trapezoid Method is giving reliable estimates of the trajectories? Are there particular points in the trajectories that cause problems? These examples are considered further in Computer Problem 6.4.14.

11. Answer the same questions as in Computer Problem 10 for the two-body problem with initial positions and velocities given by $(x_1, y_1) = (0, 1)$, $(x_1', y_1') = (0.2, -0.2)$ and $(x_2, y_2) = (-2, -1)$, $(x_2', y_2') = (-0.2, 0.2)$. The masses are given by (a) $m_1 = m_2 = 2$ (b) $m_1 = m_2 = 1$ (c) $m_1 = m_2 = 0.5$.

12. Adapt orbit.m to solve the three-body problem. Set the masses to $m_2 = 0.3$, $m_1 = m_3 = 0.03$. (a) Plot the trajectories with initial conditions $(x_1, y_1) = (2, 2)$, $(x_1', y_1') = (0.2, -0.2)$, $(x_2, y_2) = (0, 0)$, $(x_2', y_2') = (0, 0)$ and $(x_3, y_3) = (-2, -2)$,

$(x_3', y_3') = (-0.2, 0.2)$. (b) Change the initial condition of x_1' to 0.20001, and compare the resulting trajectories. This is a striking visual example of sensitive dependence.

13. Add a third body to the code developed in Computer Problem 10 to investigate the three-body problem. The third body has initial position and velocity $(x_3, y_3) = (4, 3)$, $(x_3', y_3') = (-0.2, 0)$, and mass $m_3 = 10^{-4}$. For each case (a)–(c), does the original trajectory from Computer Problem 10 change appreciably? Describe the trajectory of the third body m_3. Do you predict that it will stay near m_1 and m_2 indefinitely?

14. Investigate the three-body problem of a sun and two planets. The initial conditions and velocities are $(x_1, y_1) = (0, 2)$, $(x_1', y_1') = (0.6, 0.05)$, $(x_2, y_2) = (0, 0)$, $(x_2', y_2') = (-0.03, 0)$, and $(x_3, y_3) = (4, 3)$, $(x_3', y_3') = (0, -0.5)$. The masses are $m_1 = 0.05, m_2 = 1$, and $m_3 = 0.005$. Apply the Trapezoid Method with step sizes $h = 0.01$ and 0.001 and compare the results on the time interval $[0, 500]$. Do you believe the Trapezoid Method is giving reliable estimates of the trajectories? Are there particular parts of the trajectories that cause problems?

15. Investigate the three-body problem of a sun, planet, and comet. The initial conditions and velocities are $(x_1, y_1) = (0, 2)$, $(x_1', y_1') = (0.6, 0)$, $(x_2, y_2) = (0, 0)$, $(x_2', y_2') = (-0.03, 0)$, and $(x_3, y_3) = (4, 3)$, $(x_3', y_3') = (-0.2, 0)$. The masses are $m_1 = 0.05, m_2 = 1$, and $m_3 = 10^{-5}$. Answer the same questions raised in Computer Problem 14.

16. A remarkable three-body figure-eight orbit was discovered by C. Moore in 1993. In this configuration, three bodies of equal mass chase one another along a single figure-eight loop. Set the masses to $m_1 = m_2 = m_3 = 1$ and gravity $g = 1$. (a) Adapt orbit.m to plot the trajectory with initial conditions $(x_1, y_1) = (-0.970, 0.243)$, $(x_1', y_1') = (-0.466, -0.433)$, $(x_2, y_2) = (-x_1, -y_1)$, $(x_2', y_2') = (x_1', y_1')$ and $(x_3, y_3) = (0, 0)$, $(x_3', y_3') = (-2x_1', -2y_1')$. (b) Are the trajectories sensitive to small changes in initial conditions? Investigate the effect of changing x_3' by 10^{-k} for $1 \leq k \leq 5$. For each k, decide whether the figure-eight pattern persists, or a catastrophic change eventually occurs.

6.4 RUNGE–KUTTA METHODS AND APPLICATIONS

The Runge–Kutta Methods are a family of ODE solvers that include the Euler and Trapezoid Methods, and also more sophisticated methods of higher order. In this section, we introduce a variety of one-step methods and apply them to simulate trajectories of some key applications.

6.4.1 The Runge–Kutta family

We have seen that the Euler Method has order one and the Trapezoid Method has order two. In addition to the Trapezoid Method, there are other second-order methods of the Runge–Kutta type. One important example is the Midpoint Method.

Midpoint Method

$$w_0 = y_0$$
$$w_{i+1} = w_i + hf\left(t_i + \frac{h}{2}, w_i + \frac{h}{2}f(t_i, w_i)\right). \tag{6.46}$$

To verify the order of the Midpoint Method, we must compute its local truncation error. When we did this for the Trapezoid Method, we found the expression (6.31) useful:

$$y_{i+1} = y_i + hf(t_i, y_i) + \frac{h^2}{2}\left(\frac{\partial f}{\partial t}(t_i, y_i) + \frac{\partial f}{\partial y}(t_i, y_i)f(t_i, y_i)\right) + \frac{h^3}{6}y'''(c). \quad (6.47)$$

To compute the local truncation error at step i, we assume that $w_i = y_i$ and calculate $y_{i+1} - w_{i+1}$. Repeating the use of the Taylor series expansion as for the Trapezoid Method, we can write

$$w_{i+1} = y_i + hf\left(t_i + \frac{h}{2}, y_i + \frac{h}{2}f(t_i, y_i)\right)$$
$$= y_i + h\left(f(t_i, y_i) + \frac{h}{2}\frac{\partial f}{\partial t}(t_i, y_i) + \frac{h}{2}f(t_i, y_i)\frac{\partial f}{\partial y}(t_i, y_i) + O(h^2)\right). (6.48)$$

Comparing (6.47) and (6.48) yields

$$y_{i+1} - w_{i+1} = O(h^3),$$

so the Midpoint Method is of order two by Theorem 6.4.

Each function evaluation of the right-hand side of the differential equation is called a **stage** of the method. The Trapezoid and Midpoint Methods are members of the family of two-stage, second-order Runge–Kutta Methods, having form

$$w_{i+1} = w_i + h\left(1 - \frac{1}{2\alpha}\right)f(t_i, w_i) + \frac{h}{2\alpha}f(t_i + \alpha h, w_i + \alpha h f(t_i, w_i)) \quad (6.49)$$

for some $\alpha \neq 0$. Setting $\alpha = 1$ corresponds to the Explicit Trapezoid Method and $\alpha = 1/2$ to the Midpoint Method. Exercise 5 asks you to verify the order of methods in this family.

Figure 6.14 illustrates the intuition behind the Trapezoid and Midpoint Methods. The Trapezoid Method uses an Euler step to the right endpoint of the interval, evaluates the slope there, and then averages with the slope from the left endpoint. The Midpoint Method uses an Euler step to move to the midpoint of the interval, evaluates the slope there as $f(t_i + h/2, w_i + (h/2)f(t_i, w_i))$, and uses that slope to move from w_i to the new approximation w_{i+1}. These methods use different approaches to solving the same problem: acquiring a slope that represents the entire interval better than the Euler Method, which uses only the slope estimate from the left end of the interval.

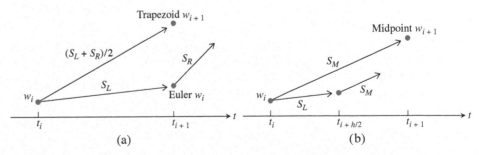

Figure 6.14 Schematic view of two members of the RK2 family. (a) The Trapezoid Method uses an average from the left and right endpoints to traverse the interval. (b) The Midpoint Method uses a slope from the interval midpoint.

SPOTLIGHT ON

Convergence   The convergence properties of a fourth-order method, like RK4, are far superior to those of the order 1 and 2 methods we have discussed so far. Convergence here means how fast the (global) error of the ODE approximation at some fixed time t goes to zero as the step size h goes to zero. Fourth order means that for every halving of the step size, the error drops by approximately a factor of $2^4 = 16$, as is clear from Figure 6.15.

There are Runge–Kutta Methods of all orders. A particularly ubiquitous example is the method of fourth order.

Runge-Kutta Method of order four (RK4)

$$w_{i+1} = w_i + \frac{h}{6}(s_1 + 2s_2 + 2s_3 + s_4) \qquad (6.50)$$

where

$$s_1 = f(t_i, w_i)$$
$$s_2 = f\left(t_i + \frac{h}{2}, w_i + \frac{h}{2}s_1\right)$$
$$s_3 = f\left(t_i + \frac{h}{2}, w_i + \frac{h}{2}s_2\right)$$
$$s_4 = f(t_i + h, w_i + hs_3).$$

The popularity of this method stems from its simplicity and ease of programming. It is a one-step method, so that it requires only an initial condition to get started; yet, as a fourth-order method, it is considerably more accurate than either the Euler or Trapezoid Methods.

The quantity $h(s_1 + 2s_2 + 2s_3 + s_4)/6$ in the fourth-order Runge–Kutta Method takes the place of slope in the Euler Method. This quantity can be considered as an improved guess for the slope of the solution in the interval $[t_i, t_i + h]$. Note that s_1 is the slope at the left end of the interval, s_2 is the slope used in the Midpoint Method, s_3 is an improved slope at the midpoint, and s_4 is an approximate slope at the right-hand endpoint $t_i + h$. The algebra needed to prove that this method is order four is similar to our derivation of the Trapezoid and Midpoint Methods, but is a bit lengthy, and can be found, for example, in Henrici [1962]. We return one more time to differential equation (6.5) for purposes of comparison.

▶ **EXAMPLE 6.18**   Apply Runge–Kutta of order four to the initial value problem

$$\begin{cases} y' = ty + t^3 \\ y(0) = 1 \end{cases}. \qquad (6.51)$$

Computing the global truncation error at $t = 1$ for a variety of step sizes gives the following table:

steps n	step size h	error at $t = 1$
5	0.20000	2.3788×10^{-5}
10	0.10000	1.4655×10^{-6}
20	0.05000	9.0354×10^{-8}
40	0.02500	5.5983×10^{-9}
80	0.01250	3.4820×10^{-10}
160	0.00625	2.1710×10^{-11}
320	0.00312	1.3491×10^{-12}
640	0.00156	7.2609×10^{-14}

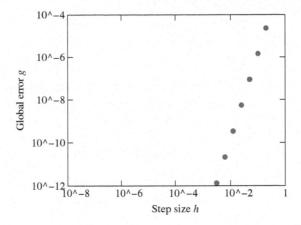

Figure 6.15 Error as a function of step size for Runge-Kutta of order 4.
The difference between the approximate solution of (6.5) and the correct solution at $t = 1$ has slope 4 on a log-log plot, so is proportional to h^4, for small h.

Compare with the corresponding table for Euler's Method on page 299. The difference is remarkable and easily makes up for the extra complexity of RK4, which requires four function calls per step, compared with only one for Euler. Figure 6.15 displays the same information in a way that exhibits the fact that the global truncation error is proportional to h^4, as expected for a fourth-order method. ◄

6.4.2 Computer simulation: the Hodgkin–Huxley neuron

Computers were in their early development stages in the middle of the 20th century. Some of the first applications were to help solve hitherto intractable systems of differential equations.

A.L. Hodgkin and A.F. Huxley gave birth to the field of computational neuroscience by developing a realistic firing model for nerve cells, or neurons. They were able to approximate solutions of the differential equations model even with the rudimentary computers that existed at the time. For this work, they won the Nobel Prize in Biology in 1963.

The model is a system of four coupled differential equations, one of which models the voltage difference between the interior and exterior of the cell. The three other equations model activation levels of ion channels, which do the work of exchanging sodium and potassium ions between the inside and outside. The **Hodgkin–Huxley equations** are

$$Cv' = -g_1 m^3 h(v - E_1) - g_2 n^4 (v - E_2) - g_3 (v - E_3) + I_{\text{in}}$$
$$m' = (1 - m)\alpha_m (v - E_0) - m\beta_m (v - E_0)$$
$$n' = (1 - n)\alpha_n (v - E_0) - n\beta_n (v - E_0)$$
$$h' = (1 - h)\alpha_h (v - E_0) - h\beta_h (v - E_0), \tag{6.52}$$

where

$$\alpha_m(v) = \frac{2.5 - 0.1v}{e^{2.5 - 0.1v} - 1}, \quad \beta_m(v) = 4e^{-v/18},$$

$$\alpha_n(v) = \frac{0.1 - 0.01v}{e^{1 - 0.1v} - 1}, \quad \beta_n(v) = \frac{1}{8}e^{-v/80},$$

and

$$\alpha_h(v) = 0.07e^{-v/20}, \quad \beta_h(v) = \frac{1}{e^{3-0.1v} + 1}.$$

The coefficient C denotes the capacitance of the cell, and I_{in} denotes the input current from other cells. Typical coefficient values are capacitance $C = 1$ microfarads, conductances $g_1 = 120$, $g_2 = 36$, $g_3 = 0.3$ siemens, and voltages $E_0 = -65$, $E_1 = 50$, $E_2 = -77$, $E_3 = -54.4$ millivolts.

The v' equation is an equation of current per unit area, in units of milliamperes/cm^2, while the three other activations m, n, and h are unitless. The coefficient C is the capacitance of the neuron membrane, g_1, g_2, g_3 are conductances, and E_1, E_2, and E_3 are the "reversal potentials," which are the voltage levels that form the boundary between currents flowing inward and outward.

Hodgkin and Huxley carefully chose the form of the equations to match experimental data, which was acquired from the giant axon of the squid. They also fit parameters to the model. Although the particulars of the squid axon differ from mammal neurons, the model has held up as a realistic depiction of neural dynamics. More generally, it is useful as an example of excitable media that translates continuous input into an all-or-nothing response. The MATLAB code implementing the model is as follows:

MATLAB code shown here can be found at bit.ly/2pYDpqX

```
% Program 6.5 Hodgkin-Huxley equations
% Inputs: time interval inter,
% ic=initial voltage v and 3 gating variables, steps n
% Output: solution y
% Calls a one-step method such as rk4step.m
% Example usage: hh([0,100],[-65,0,0.3,0.6],2000);
function y=hh(inter,ic,n)
global pa pb pulse
inp=input('pulse start, end, muamps in [ ], e.g. [50 51 7]: ');
pa=inp(1);pb=inp(2);pulse=inp(3);
a=inter(1); b=inter(2); h=(b-a)/n; % plot n points in total
y(1,:)=ic;                         % enter initial conds in y
t(1)=a;
for i=1:n
  t(i+1)=t(i)+h;
  y(i+1,:)=rk4step(t(i),y(i,:),h);
end
subplot(3,1,1);
plot([a pa pa pb pb b],[0 0 pulse pulse 0 0]);
grid;axis([0 100 0 2*pulse])
ylabel('input pulse')
subplot(3,1,2);
plot(t,y(:,1));grid;axis([0 100 -100 100])
ylabel('voltage (mV)')
subplot(3,1,3);
plot(t,y(:,2),t,y(:,3),t,y(:,4));grid;axis([0 100 0 1])
ylabel('gating variables')
legend('m','n','h')
xlabel('time (msec)')

function y=rk4step(t,w,h)
%one step of the Runge-Kutta order 4 method
s1=ydot(t,w);
s2=ydot(t+h/2,w+h*s1/2);
s3=ydot(t+h/2,w+h*s2/2);
```

```
s4=ydot(t+h,w+h*s3);
y=w+h*(s1+2*s2+2*s3+s4)/6;

function z=ydot(t,w)
global pa pb pulse
c=1;g1=120;g2=36;g3=0.3;T=(pa+pb)/2;len=pb-pa;
e0=-65;e1=50;e2=-77;e3=-54.4;
in=pulse*(1-sign(abs(t-T)-len/2))/2;
% square pulse input on interval [pa,pb] of pulse muamps
v=w(1);m=w(2);n=w(3);h=w(4);
z=zeros(1,4);
z(1)=(in-g1*m*m*m*h*(v-e1)-g2*n*n*n*n*(v-e2)-g3*(v-e3))/c;
v=v-e0;
z(2)=(1-m)*(2.5-0.1*v)/(exp(2.5-0.1*v)-1)-m*4*exp(-v/18);
z(3)=(1-n)*(0.1-0.01*v)/(exp(1-0.1*v)-1)-n*0.125*exp(-v/80);
z(4)=(1-h)*0.07*exp(-v/20)-h/(exp(3-0.1*v)+1);
```

Without input, the Hodgkin–Huxley neuron stays quiescent, at a voltage of approximately E_0. Setting I_{in} to be a square current pulse of length 1 msec and strength 7 μA is sufficient to cause a spike, a large depolarizing deflection of the voltage. This is illustrated in Figure 6.16. Run the program to check that 6.9 μA is not sufficient to cause a full spike. Hence, the all-or-nothing response. It is this property of greatly magnifying the effect of small differences in input that may explain the neuron's success at information processing. Figure 6.16(b) shows that if the input current is sustained, the neuron will fire a periodic volley of spikes. Computer Problem 10 is an investigation of the thresholding capabilities of this virtual neuron.

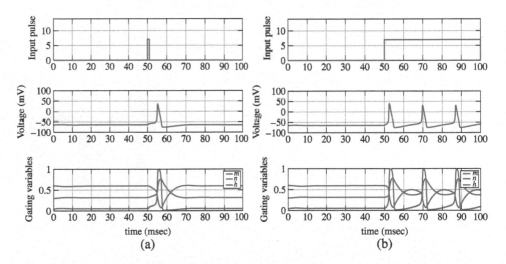

Figure 6.16 Screen shots of Hodgkin-Huxley program. (a) Square wave input of size $I_{in} = 7$ μA at time 50 msecs, 1 msec duration, causes the model neuron to fire once. (b) Sustained square wave, with $I_{in} = 7$ μA, causes the model neuron to fire periodically.

6.4.3 Computer simulation: the Lorenz equations

In the late 1950s, MIT meteorologist E. Lorenz acquired one of the first commercially available computers. It was the size of a refrigerator and operated at the speed of 60 multiplications per second. This unprecedented cache of personal computing power allowed him to develop and meaningfully evaluate weather models consisting of several differential equations that, like the Hodgkin–Huxley equations, could not be analytically solved.

The **Lorenz equations** are a simplification of a miniature atmosphere model that he designed to study Rayleigh–Bénard convection, the movement of heat in a fluid, such as air, from a lower warm medium (such as the ground) to a higher cool medium (like the upper atmosphere). In this model of a two-dimensional atmosphere, a circulation of air develops that can be described by the following system of three equations:

$$
\begin{aligned}
x' &= -sx + sy \\
y' &= -xz + rx - y \\
z' &= xy - bz.
\end{aligned}
$$

(6.53)

The variable x denotes the clockwise circulation velocity, y measures the temperature difference between the ascending and descending columns of air, and z measures the deviation from a strictly linear temperature profile in the vertical direction. The Prandtl number s, the Rayleigh number r, and b are parameters of the system. The most common setting for the parameters is $s = 10, r = 28$, and $b = 8/3$. These settings were used for the trajectory shown in Figure 6.17, computed by order four Runge–Kutta, using the following code to describe the differential equation.

```
function z=ydot(t,y)
%Lorenz equations
s=10; r=28; b=8/3;
z(1)=-s*y(1)+s*y(2);
z(2)=-y(1)*y(3)+r*y(1)-y(2);
z(3)=y(1)*y(2)-b*y(3);
```

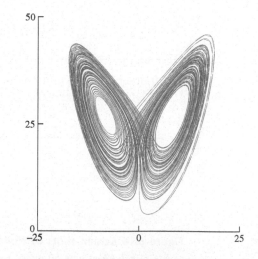

Figure 6.17 One trajectory of the Lorenz equations (6.53), projected to the xz-plane. Parameters are set to $s = 10, r = 28$, and $b = 8/3$.

The Lorenz equations are an important example because the trajectories show great complexity, despite the fact that the equations are deterministic and fairly simple (almost linear). The explanation for the complexity is similar to that of the double pendulum or three-body problem: sensitive dependence on initial conditions. Computer Problems 12 and 13 explore the sensitive dependence of this so-called chaotic attractor.

▶ **ADDITIONAL EXAMPLES**

1. Consider the initial value problem

$$\begin{cases} y' = ty \\ y(0) = 1. \end{cases}$$

Find the approximate solution on 0, 1 by the Runge–Kutta Order 4 Method with step size $h = 1/4$. Report the global truncation error at $t = 1$.

2. Show that $y(t) = e^{\sin t}$ is a solution of the initial value problem

$$\begin{cases} y'' + y \sin t - y' \cos t = 0 \\ y(0) = 1 \\ y'(0) = 1. \end{cases}$$

Convert the differential equation to an equivalent first-order system, and plot the Runge–Kutta Order 4 approximate solution on the interval 0, 4 for step size $h = 1/2$. Compute the global truncation error $|y(4) - w(4)|$ at $t = 4$ for the step sizes $h = 1/2, 1/4, 1/8$, and $1/16$.

⌸ Solutions for Additional Examples can be found at `bit.ly/2yoJ8Lu`

6.4 Exercises

⌸ **Solutions** for Exercises numbered in blue can be found at `bit.ly/2yKx5Yg`

1. Apply the Midpoint Method for the IVPs

   (a) $y' = t$ (b) $y' = t^2 y$ (c) $y' = 2(t + 1)y$

   (d) $y' = 5t^4 y$ (e) $y' = 1/y^2$ (f) $y' = t^3/y^2$

   with initial condition $y(0) = 1$. Using step size $h = 1/4$, calculate the Midpoint Method approximation on the interval 0, 1. Compare with the correct solution found in Exercise 6.1.3, and find the global truncation error at $t = 1$.

2. Carry out the steps of Exercise 1 for the IVPs

   (a) $y' = t + y$ (b) $y' = t - y$ (c) $y' = 4t - 2y$

   with initial condition $y(0) = 0$. The exact solutions were found in Exercise 6.1.4.

3. Apply fourth-order Runge–Kutta Method to the IVPs in Exercise 1. Using step size $h = 1/4$, calculate the approximation on the interval 0, 1. Compare with the correct solution found in Exercise 6.1.3, and find the global truncation error at $t = 1$.

4. Carry out the steps of Exercise 3 for the IVPs in Exercise 2.

5. Prove that for any $\alpha \neq 0$, the method (6.49) is second order.

6. Consider the initial value problem $y' = \lambda y$. The solution is $y(t) = y_0 e^{\lambda t}$. (a) Calculate w_1 for RK4 in terms of w_0 for this differential equation. (b) Calculate the local truncation error by setting $w_0 = y_0 = 1$ and determining $y_1 - w_1$. Show that the local truncation error is of size $O(h^5)$, as expected for a fourth-order method.

7. Assume that the right-hand side $f(t, y) = f(t)$ does not depend on y. Show that $s_2 = s_3$ in fourth-order Runge–Kutta and that RK4 is equivalent to Simpson's Rule for the integral $\int_{t_i}^{t_i+h} f(s) \, ds$.

6.4 Computer Problems

⌸ **Solutions** for Computer Problems numbered in blue can be found at `bit.ly/2P80tSj`

1. Apply the Midpoint Method on a grid of step size $h = 0.1$ in 0, 1 for the initial value problems in Exercise 1. Print a table of the t values, approximations, and global truncation error at each step.

2. Apply the fourth-order Runge–Kutta Method solution on a grid of step size $h = 0.1$ in $[0, 1]$ for the initial value problems in Exercise 1. Print a table of the t values, approximations, and global truncation error at each step.

3. Carry out the steps of Computer Problem 2, but plot the approximate solutions on $[0, 1]$ for step sizes $h = 0.1, 0.05$, and 0.025, along with the true solution.

4. Carry out the steps of Computer Problem 2 for the equations of Exercise 2.

5. Plot the fourth-order Runge–Kutta Method approximate solution on $[0, 1]$ for the differential equation $y' = 1 + y^2$ and initial condition (a) $y_0 = 0$ (b) $y_0 = 1/2$, along with the exact solution (see Exercise 6.1.7). Use step sizes $h = 0.1$ and 0.05.

6. Plot the fourth-order Runge–Kutta Method approximate solution on $[0, 1]$ for the differential equation $y' = 1 - y^2$ and initial condition (a) $y_0 = 0$ (b) $y_0 = -1/2$, along with the exact solution (see Exercise 6.1.8). Use step sizes $h = 0.1$ and 0.05.

7. Calculate the fourth-order Runge–Kutta Method approximate solution on $[0, 4]$ for the differential equation $y' = \sin y$ and initial condition (a) $y_0 = 0$ (b) $y_0 = 100$, using step sizes $h = 0.1 \times 2^{-k}$ for $0 \le k \le 5$. Plot the $k = 0$ and $k = 5$ approximate solutions along with the exact solution (see Exercise 6.1.15), and make a log–log plot of the error as a function of h.

8. Calculate the fourth-order Runge–Kutta Method approximate solution of the differential equation $y' = \sinh y$ and initial condition (a) $y_0 = 1/4$ on the interval $[0, 2]$ (b) $y_0 = 2$ on the interval $[0, 1/4]$, using step sizes $h = 0.1 \times 2^{-k}$ for $0 \le k \le 5$. Plot the $k = 0$ and $k = 5$ approximate solutions along with the exact solution (see Exercise 6.1.16), and make a log–log plot of the error as a function of h.

9. For the IVPs in Exercise 1, plot the global error of the RK4 method at $t = 1$ as a function of h, as in Figure 6.4.

10. Consider the Hodgkin–Huxley equations (6.52) with default parameters. (a) Find as accurately as possible the minimum threshold, in microamps, for generating a spike with a 1 msec pulse. (b) Does the answer change if the pulse is 5 msec long? (c) Experiment with the shape of the pulse. Does a triangular pulse of identical enclosed area cause the same effect as a square pulse? (d) Discuss the existence of a threshold for constant sustained input.

11. Adapt the `orbit.m` MATLAB program to animate a solution to the Lorenz equations by the order four Runge–Kutta Method with step size $h = 0.001$. Draw the trajectory with initial condition $(x_0, y_0, z_0) = (5, 5, 5)$.

12. Assess the conditioning of the Lorenz equations by following two trajectories from two nearby initial conditions. Consider the initial conditions $(x, y, z) = (5, 5, 5)$ and another initial condition at a distance $\Delta = 10^{-5}$ from the first. Compute both trajectories by fourth-order Runge–Kutta with step size $h = 0.001$, and calculate the error magnification factor after $t = 10$ and $t = 20$ time units.

13. Follow two trajectories of the Lorenz equations with nearby initial conditions, as in Computer Problem 12. For each, construct the binary symbol sequence consisting of 0 if the trajectory traverses the "negative x" loop in Figure 6.17 and 1 if it traverses the positive loop. For how many time units do the symbol sequences of the two trajectories agree?

14. Repeat Computer Problem 6.3.10, but replace the Trapezoid Method with Runge–Kutta Order 4. Compare results using the two methods.

15. Repeat Computer Problem 6.3.11, but replace the Trapezoid Method with Runge–Kutta Order 4. Compare results using the two methods.

16. Repeat Computer Problem 6.3.14, but replace the Trapezoid Method with Runge–Kutta Order 4. Compare results using the two methods.

17. Repeat Computer Problem 6.3.15, but replace the Trapezoid Method with Runge–Kutta Order 4. Compare results using the two methods.

18. More complicated versions of the periodic three-body orbits of Computer Problem 6.3.16 have been recently published by X. Li and S. Liao. Set the masses to $m_1 = m_2 = m_3 = 1$ and gravity $g = 1$. Adapt `orbit.m` to use Runge–Kutta Order 4 with the following initial conditions. (a) $(x_1, y_1) = (-1, 0), (x_2, y_2) = (1, 0), (x_1', y_1') = (x_2', y_2') = (v_x, v_y), (x_3, y_3) = (0, 0), (x_3', y_3') = (-2v_x, -2v_y)$ where $v_x = 0.6150407229, v_y = 0.5226158545$. How small must the step size h be to plot four complete periods of the orbit, and what is the period? (The orbit is sensitive to small changes; accumulated error will cause the bodies to diverge eventually.) (b) Same as (a), but set $v_x = 0.5379557207, v_y = 0.3414578545$.

Reality Check 6 *The Tacoma Narrows Bridge*

A mathematical model that attempts to capture the Tacoma Narrows Bridge incident was proposed by McKenna and Tuama [2001]. The goal is to explain how torsional, or twisting, oscillations can be magnified by forcing that is strictly vertical.

Consider a roadway of width $2l$ hanging between two suspended cables, as in Figure 6.18(a). We will consider a two-dimensional slice of the bridge, ignoring the dimension of the bridge's length for this model, since we are only interested in the side-to-side motion. At rest, the roadway hangs at a certain equilibrium height due to gravity; let y denote the current distance the center of the roadway hangs below this equilibrium.

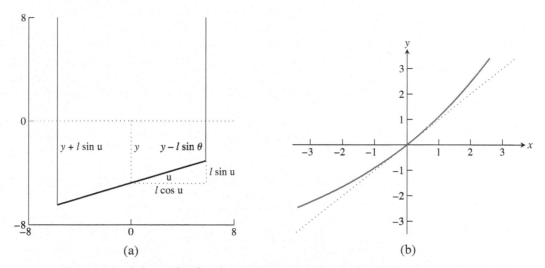

(a) (b)

Figure 6.18 Schematics for the McKenna–Tuama model of the Tacoma Narrows Bridge. (a) Denote the distance from the roadway center of mass to its equilibrium position by y, and the angle of the roadway with the horizontal by θ. (b) Exponential Hooke's law curve $f(y) = (K/a)(e^{ay} - 1)$.

Hooke's law postulates a linear response, meaning that the restoring force the cables apply will be proportional to the deviation. Let θ be the angle the roadway makes with the horizontal. There are two suspension cables, stretched $y - l\sin\theta$ and $y + l\sin\theta$ from equilibrium, respectively. Assume that a viscous damping term is given that is proportional to the velocity. Using Newton's law $F = ma$ and denoting Hooke's constant by K, the equations of motion for y and θ are as follows:

$$y'' = -dy' - \left[\frac{K}{m}(y - l\sin\theta) + \frac{K}{m}(y + l\sin\theta)\right]$$

$$\theta'' = -d\theta' + \frac{3\cos\theta}{l}\left[\frac{K}{m}(y - l\sin\theta) - \frac{K}{m}(y + l\sin\theta)\right].$$

However, Hooke's law is designed for springs, where the restoring force is more or less equal whether the springs are compressed or stretched. McKenna and Tuama hypothesize that cables pull back with more force when stretched than they push back when compressed. (Think of a string as an extreme example.) They replace the linear Hooke's law restoring force $f(y) = Ky$ with a nonlinear force $f(y) = (K/a)(e^{ay} - 1)$, as shown in Figure 6.18(b). Both functions have the same slope K at $y = 0$; but for the nonlinear force, a positive y (stretched cable) causes a stronger restoring force than the corresponding negative y (slackened cable). Making this replacement in the preceding equations yields

$$y'' = -dy' - \frac{K}{ma}\left[e^{a(y-l\sin\theta)} - 1 + e^{a(y+l\sin\theta)} - 1\right]$$

$$\theta'' = -d\theta' + \frac{3\cos\theta}{l}\frac{K}{ma}\left[e^{a(y-l\sin\theta)} - e^{a(y+l\sin\theta)}\right]. \tag{6.54}$$

As the equations stand, the state $y = y' = \theta = \theta' = 0$ is an equilibrium. Now turn on the wind. Add the forcing term $0.2W\sin\omega t$ to the right-hand side of the y equation, where W is the wind speed in km/hr. This adds a strictly vertical oscillation to the bridge.

Useful estimates for the physical constants can be made. The mass of a one-foot length of roadway was about 2500 kg, and the spring constant K has been estimated at 1000 Newtons. The roadway was about 12 meters wide. For this simulation, the damping coefficient was set at $d = 0.01$, and the Hooke's nonlinearity coefficient $a = 0.2$. An observer counted 38 vertical oscillations of the bridge in one minute shortly before the collapse—set $\omega = 2\pi(38/60)$. These coefficients are only guesses, but they suffice to show ranges of motion that tend to match photographic evidence of the bridge's final oscillations. MATLAB code that runs the model (6.54) is as follows:

⌨ MATLAB code
shown here can be found
at bit.ly/2J5kx28

```
% Program 6.6 Animation program for bridge using IVP solver
% Inputs: inter = time interval inter,
%    ic = [y(1,1) y(1,2) y(1,3) y(1,4)],
%    number of steps n, steps per point plotted p
% Calls a one-step method such as trapstep.m
% Example usage: tacoma([0 1000],[1 0 0.001 0],25000,5);
function tacoma(inter,ic,n,p)
clf                                    % clear figure window
h=(inter(2)-inter(1))/n;
y(1,:)=ic;                             % enter initial conds in y
t(1)=inter(1);len=6;
set(gca,'XLim',[-8 8],'YLim',[-8 8], ...
    'XTick',[-8 0 8],'YTick',[-8 0 8]);
cla;                                   % clear screen
axis square                            % make aspect ratio 1 - 1
road=animatedline('color','b','LineStyle','-','LineWidth',1);
lcable=animatedline('color','r','LineStyle','-','LineWidth',1);
rcable=animatedline('color','r','LineStyle','-','LineWidth',1);
for k=1:n
  for i=1:p
    t(i+1) = t(i)+h;
    y(i+1,:) = trapstep(t(i),y(i,:),h);
```

```
    end
    y(1,:) = y(p+1,:);t(1)=t(p+1);
    z1(k)=y(1,1);z3(k)=y(1,3);
    c=len*cos(y(1,3));s=len*sin(y(1,3));
    clearpoints(road);addpoints(road,[-c c],[-s-y(1,1) s-y(1,1)])
    clearpoints(lcable);addpoints(lcable,[-c -c],[-s-y(1,1) 8])
    clearpoints(rcable);addpoints(rcable,[c c],[s-y(1,1) 8])
    drawnow; pause(h)
end

function y = trapstep(t,x,h)
%one step of the Trapezoid Method
z1=ydot(t,x);
g=x+h*z1;
z2=ydot(t+h,g);
y=x+h*(z1+z2)/2;

function ydot=ydot(t,y)
len=6; a=0.2; W=80; omega=2*pi*38/60;
a1=exp(a*(y(1)-len*sin(y(3))));
a2=exp(a*(y(1)+len*sin(y(3))));
ydot(1) = y(2);
ydot(2) = -0.01*y(2)-0.4*(a1+a2-2)/a+0.2*W*sin(omega*t);
ydot(3) = y(4);
ydot(4) = -0.01*y(4)+1.2*cos(y(3))*(a1-a2)/(len*a);
```

Run `tacoma.m` with the default parameter values to see the phenomenon postulated earlier. If the angle θ of the roadway is set to any small nonzero value, vertical forcing causes θ to eventually grow to a macroscopic value, leading to significant torsion of the roadway. The interesting point is that there is no torsional forcing applied to the equation; the unstable "torsional mode" is excited completely by vertical forcing.

Suggested activities:

1. Run `tacoma.m` with wind speed $W = 80$ km/hr and initial conditions $y = y' = \theta' = 0$, $\theta = 0.001$. The bridge is stable in the torsional dimension if small disturbances in θ die out; unstable if they grow far beyond original size. Which occurs for this value of W?

2. Replace the Trapezoid Method by fourth-order Runge–Kutta to improve accuracy. Also, add new figure windows to plot $y(t)$ and $\theta(t)$.

3. The system is torsionally stable for $W = 50$ km/hr. Find the magnification factor for a small initial angle. That is, set $\theta(0) = 10^{-3}$ and find the ratio of the maximum angle $\theta(t), 0 \leq t < \infty$, to $\theta(0)$. Is the magnification factor approximately consistent for initial angles $\theta(0) = 10^{-4}, 10^{-5},\ldots$?

4. Find the minimum wind speed W for which a small disturbance $\theta(0) = 10^{-3}$ has a magnification factor of 100 or more. Can a consistent magnification factor be defined for this W?

5. Design and implement a method for computing the minimum wind speed in Step 4, to within 0.5×10^{-3} km/hr. You may want to use an equation solver from Chapter 1.

6. Try some larger values of W. Do all extremely small initial angles eventually grow to catastrophic size?

7. What is the effect of increasing the damping coefficient? Double the current value and find the change in the critical wind speed W. Can you suggest possible changes in design that might have made the bridge less susceptible to torsion?

This project is an example of experimental mathematics. The equations are too difficult to derive closed-form solutions, and even too difficult to prove qualitative results about. Equipped with reliable ODE solvers, we can generate numerical trajectories for various parameter settings to illustrate the types of phenomena available to this model. Used in this way, differential equation models can predict behavior and shed light on mechanisms in scientific and engineering problems. ◗

6.5 VARIABLE STEP-SIZE METHODS

Up to this point, the step size h has been treated as a constant in the implementation of the ODE solver. However, there is no reason that h cannot be changed during the solution process. A good reason to want to change the step size is for a solution that moves between periods of slow change and periods of fast change. To make the fixed step size small enough to track the fast changes accurately may mean that the rest of the solution is solved intolerably slowly.

In this section, we discuss strategies for controlling the step size of ODE solvers. The most common approach uses two solvers of different orders, called embedded pairs.

6.5.1 Embedded Runge–Kutta pairs

The key idea of a variable step-size method is to monitor the error produced by the current step. The user sets an error tolerance that must be met by the current step. Then the method is designed to (1) reject the step and cut the step size if the error tolerance is exceeded, or (2) if the error tolerance is met, to accept the step and then choose a step size h that should be appropriate for the next step. The key need is for some way to approximate the error made on each step. First let's assume that we have found such a way and explain how to change the step size.

The simplest way to vary step size is to double or halve the step size, depending on the current error. Compare the error estimate e_i, or relative error estimate $e_i/|w_i|$, with the error tolerance. (Here, as in the rest of this section, we will assume the ODE system being solved consists of one equation. It is fairly easy to generalize the ideas of this section to higher dimensions.) If the tolerance is not met, the step is repeated with new step size equal to $h_i/2$. If the tolerance is met too well—say, if the error is less than 1/10 the tolerance—after accepting the step, the step size is doubled for the next step.

In this way, the step size will be adjusted automatically to a size that maintains the (relative) local truncation error near the user-requested level. Whether the absolute or relative error is used depends on the context; a good general-purpose technique is to use the hybrid $e_i/\max(|w_i|, \theta)$ to compare with the error tolerance, where the constant $\theta > 0$ protects against very small values of w_i.

A more sophisticated way to choose the appropriate step size follows from knowledge of the order of the ODE solver. Assume that the solver has order p, so that the local truncation error $e_i = O(h^{p+1})$. Let T be the relative error tolerance allowed by the user for each step. That means the goal is to ensure $e_i/|w_i| < T$.

If the goal $e_i/|w_i| < T$ is met, then the step is accepted and a new step size for the next step is needed. Assuming that

$$e_i \approx ch_i^{p+1} \tag{6.55}$$

for some constant c, the step size h that best meets the tolerance satisfies

$$T|w_i| = ch^{p+1}. \tag{6.56}$$

Solving the equations (6.55) and (6.56) for h and c yields

$$h_* = 0.8 \left(\frac{T|w_i|}{e_i} \right)^{\frac{1}{p+1}} h_i, \tag{6.57}$$

where we have added a safety factor of 0.8 to be conservative. Thus, the next step size will be set to $h_{i+1} = h_*$.

On the other hand, if the goal $e_i/|w_i| < T$ is not met by the relative error, then h_i is set to h_* for a second try. This should suffice, because of the safety factor. However, if the second try also fails to meet the goal, then the step size is simply cut in half. This continues until the goal is achieved. As stated for general purposes, the relative error should be replaced by $e_i/\max(|w_i|, \theta)$.

Both the simple and sophisticated methods described depend heavily on some way to estimate the error of the current step of the ODE solver $e_i = |w_{i+1} - y_{i+1}|$. An important constraint is to gain the estimate without requiring a large amount of extra computation.

The most widely used way for obtaining such an error estimate is to run a higher order ODE solver in parallel with the ODE solver of interest. The higher order method's estimate for w_{i+1}—call it z_{i+1}—will be significantly more accurate than the original w_{i+1}, so that the difference

$$e_{i+1} \approx |z_{i+1} - w_{i+1}| \tag{6.58}$$

is used as an error estimate for the current step from t_i to t_{i+1}.

Following this idea, several "pairs" of Runge–Kutta Methods, one of order p and another of order $p + 1$, have been developed that share much of the needed computations. In this way, the extra cost of step-size control is kept low. Such a pair is often called an **embedded Runge–Kutta pair**.

▶ **EXAMPLE 6.19**   RK2/3, An example of a Runge–Kutta order 2/order 3 embedded pair.

The Explicit Trapezoid Method can be paired with a third-order RK method to make an embedded pair suitable for step-size control. Set

$$w_{i+1} = w_i + h\frac{s_1 + s_2}{2}$$

$$z_{i+1} = w_i + h\frac{s_1 + 4s_3 + s_2}{6},$$

where

$$s_1 = f(t_i, w_i)$$
$$s_2 = f(t_i + h, w_i + hs_1)$$
$$s_3 = f\left(t_i + \frac{1}{2}h, w_i + \frac{1}{2}h\frac{s_1 + s_2}{2}\right).$$

In the preceding equations, w_{i+1} is the trapezoid step, and z_{i+1} represents a third-order method, which requires the three Runge–Kutta stages shown. The third-order method is just an application of Simpson's Rule for numerical integration to the context of differential equations. From the two ODE solvers, an estimate for the error can be found by subtracting the two approximations:

$$e_{i+1} \approx |w_{i+1} - z_{i+1}| = \left| h \frac{s_1 - 2s_3 + s_2}{3} \right|. \tag{6.59}$$

Using this estimate for the local truncation error allows the implementation of either of the step-size control protocols previously described. Note that the local truncation error estimate for the Trapezoid Method is achieved at the cost of only one extra evaluation of f, used to compute S_3. ◀

Although the step-size protocol has been worked out for w_{i+1}, it makes even better sense to use the higher order approximation z_{i+1} to advance the step, since it is available. This is called **local extrapolation**.

▶ **EXAMPLE 6.20**   The Bogacki–Shampine order 2/order 3 embedded pair.

MATLAB uses a different embedded pair in its ode23 command. Let

$$
\begin{aligned}
s_1 &= f(t_i, w_i) \\
s_2 &= f\left(t_i + \frac{1}{2}h, w_i + \frac{1}{2}hs_1\right) \\
s_3 &= f\left(t_i + \frac{3}{4}h, w_i + \frac{3}{4}hs_2\right) \\
z_{i+1} &= w_i + \frac{h}{9}(2s_1 + 3s_2 + 4s_3) \\
s_4 &= f(t + h, z_{i+1}) \\
w_{i+1} &= w_i + \frac{h}{24}(7s_1 + 6s_2 + 8s_3 + 3s_4).
\end{aligned}
\tag{6.60}
$$

It can be checked that z_{i+1} is an order 3 approximation, and w_{i+1}, despite having four stages, is order 2. The error estimate needed for step-size control is

$$e_{i+1} = |z_{i+1} - w_{i+1}| = \frac{h}{72}|-5s_1 + 6s_2 + 8s_3 - 9s_4|. \tag{6.61}$$

Note that s_4 becomes s_1 on the next step if it is accepted, so that there are no wasted stages—at least three stages are needed, anyway, for a third-order Runge–Kutta Method. This design of the second-order method is called FSAL, for First Same As Last. ◀

6.5.2 Order 4/5 methods

▶ **EXAMPLE 6.21**   The Runge–Kutta–Fehlberg order 4/order 5 embedded pair.

$$
\begin{aligned}
s_1 &= f(t_i, w_i) \\
s_2 &= f\left(t_i + \frac{1}{4}h, w_i + \frac{1}{4}hs_1\right)
\end{aligned}
$$

$$s_3 = f\left(t_i + \frac{3}{8}h, w_i + \frac{3}{32}hs_1 + \frac{9}{32}hs_2\right)$$

$$s_4 = f\left(t_i + \frac{12}{13}h, w_i + \frac{1932}{2197}hs_1 - \frac{7200}{2197}hs_2 + \frac{7296}{2197}hs_3\right)$$

$$s_5 = f\left(t_i + h, w_i + \frac{439}{216}hs_1 - 8hs_2 + \frac{3680}{513}hs_3 - \frac{845}{4104}hs_4\right)$$

$$s_6 = f\left(t_i + \frac{1}{2}h, w_i - \frac{8}{27}hs_1 + 2hs_2 - \frac{3544}{2565}hs_3 + \frac{1859}{4104}hs_4 - \frac{11}{40}hs_5\right)$$

$$w_{i+1} = w_i + h\left(\frac{25}{216}s_1 + \frac{1408}{2565}s_3 + \frac{2197}{4104}s_4 - \frac{1}{5}s_5\right)$$

$$z_{i+1} = w_i + h\left(\frac{16}{135}s_1 + \frac{6656}{12825}s_3 + \frac{28561}{56430}s_4 - \frac{9}{50}s_5 + \frac{2}{55}s_6\right). \qquad (6.62)$$

It can be checked that z_{i+1} is an order 5 approximation, and that w_{i+1} is order 4. The error estimate needed for step-size control is

$$e_{i+1} = |z_{i+1} - w_{i+1}| = h\left|\frac{1}{360}s_1 - \frac{128}{4275}s_3 - \frac{2197}{75240}s_4 + \frac{1}{50}s_5 + \frac{2}{55}s_6\right|. \qquad (6.63)$$

◀

The Runge–Kutta–Fehlberg Method (RKF45) is currently the best-known variable step-size one-step method. Implementation is simple, given the preceding formulas. The user must set a relative error tolerance T and an initial step size h. After computing w_1, z_1, and e_1, the relative error test

$$\frac{e_i}{|w_i|} < T \qquad (6.64)$$

is checked for $i = 1$. If successful, the new w_1 is replaced with the locally extrapolated version z_1, and the program moves on to the next step. On the other hand, if the relative error test (6.64) fails, the step is tried again with step size h given by (6.57) with $p = 4$, the order of the method producing w_i. (A repeated failure, which is unlikely, is treated by cutting step size in half until success is reached.) In any case, the step size h_1 for the next step should be calculated from (6.57).

▶ **EXAMPLE 6.22**   The Dormand–Prince order 4/order 5 embedded pair.

$$s_1 = f(t_i, w_i)$$

$$s_2 = f\left(t_i + \frac{1}{5}h, w_i + \frac{1}{5}hs_1\right)$$

$$s_3 = f\left(t_i + \frac{3}{10}h, w_i + \frac{3}{40}hs_1 + \frac{9}{40}hs_2\right)$$

$$s_4 = f\left(t_i + \frac{4}{5}h, w_i + \frac{44}{45}hs_1 - \frac{56}{15}hs_2 + \frac{32}{9}hs_3\right)$$

$$s_5 = f\left(t_i + \frac{8}{9}h, w_i + h\left(\frac{19372}{6561}s_1 - \frac{25360}{2187}s_2 + \frac{64448}{6561}s_3 - \frac{212}{729}s_4\right)\right)$$

$$s_6 = f\left(t_i + h, w_i + h\left(\frac{9017}{3168}s_1 - \frac{355}{33}s_2 + \frac{46732}{5247}s_3 + \frac{49}{176}s_4 - \frac{5103}{18656}s_5\right)\right)$$

$$z_{i+1} = w_i + h\left(\frac{35}{384}s_1 + \frac{500}{1113}s_3 + \frac{125}{192}s_4 - \frac{2187}{6784}s_5 + \frac{11}{84}s_6\right)$$

$$s_7 = f(t_i + h, z_{i+1})$$

$$w_{i+1} = w_i + h\left(\frac{5179}{57600}s_1 + \frac{7571}{16695}s_3 + \frac{393}{640}s_4 - \frac{92097}{339200}s_5 + \frac{187}{2100}s_6 + \frac{1}{40}s_7\right).$$

$$(6.65)$$

It can be checked that z_{i+1} is an order 5 approximation, and that w_{i+1} is order 4. The error estimate needed for step-size control is

$$e_{i+1} = |z_{i+1} - w_{i+1}|$$

$$= h\left|\frac{71}{57600}s_1 - \frac{71}{16695}s_3 + \frac{71}{1920}s_4 - \frac{17253}{339200}s_5 + \frac{22}{525}s_6 - \frac{1}{40}s_7\right|. \quad (6.66)$$

Again, local extrapolation is used, meaning that the step is advanced with z_{i+1} instead of w_{i+1}. Note that, in fact, w_{i+1} need not be computed—only e_{i+1} is necessary for error control. This is a FSAL method, like the Bogacki–Shampine Method, since s_7 becomes s_1 on the next step, if it is accepted. There are no wasted stages; it can be shown that at least six stages are needed for a fifth-order Runge–Kutta Method. ◄

The MATLAB command ode45 uses the Dormand–Prince embedded pair along with step-size control, roughly as just described. The user can set the relative tolerance T as desired. The right-hand side of the differential equation must be specified as a MATLAB function. For example, the commands

```
>> opts=odeset('RelTol',1e-4,'Refine',1,'MaxStep',1);
>> [t,y]=ode45(@(t,y) t*y+t^3,[0 1],1,opts);
```

will solve the initial value problem of Example 6.1 with initial condition $y_0 = 1$ and relative error tolerance $T = 0.0001$. If the parameter RelTol is not set, the default of 0.001 is used. Note that the function f input to ode45 must be a function of two variables, in this case t and y, even if one of them is absent in the definition of the function.

The output from ode45, using the foregoing parameter settings for this problem, is

step	t_i	w_i	y_i	e_i
0	0.00000000	1.00000000	1.00000000	0.00000000
1	0.54021287	1.17946818	1.17946345	0.00000473
2	1.00000000	1.94617812	1.94616381	0.00001431

If a relative tolerance of 10^{-6} is used, the following output results:

step	t_i	w_i	y_i	e_i
0	0.00000000	1.00000000	1.00000000	0.00000000
1	0.21506262	1.02393440	1.02393440	0.00000000
2	0.43012524	1.10574441	1.10574440	0.00000001
3	0.68607729	1.32535658	1.32535653	0.00000005
4	0.91192246	1.71515156	1.71515144	0.00000012
5	1.00000000	1.94616394	1.94616381	0.00000013

The approximate solutions more than meet the relative error tolerance because of local extrapolation, meaning that the z_{i+1} is being used instead of w_{i+1}, even though the step size is designed to be sufficient for w_{i+1}. This is the best we can do; if we had an error estimate for z_{i+1}, we could use it to tune the step size even better, but we don't have one. Note also that the solutions stop exactly at the end of the interval [0, 1], since ode45 detects the end of the interval and truncates the step as necessary.

In order to see ode45 do its step-size selection, we had to turn off some basic default settings, using the odeset command. The Refine parameter normally increases the number of solution values reported beyond what is computed by the method, to make a more beautiful graph, if and when the output is used for that purpose. The default value is 4, which causes four times the necessary number of points to be provided as output. The MaxStep parameter puts an upper limit on the step size h, and defaults to one-tenth the interval length. Using the default values for both of these parameters would mean that a step size of $h = 0.1$ would be used, and after refining by a factor of 4, the solution would be shown with a step size of 0.025. In fact, running the command without an output variable specified, as in the code

```
>> opts=odeset('RelTol',1e-6);
>> ode45(@(t,y) t*y+t^3,[0 1],1,opts);
```

will cause MATLAB to automatically plot the solution on a grid of constant step size 0.025, as shown in Figure 6.19.

An alternative way to define the right-hand side function f is to create a function file, for example f.m, and use the @ character to designate its function handle:

```
function y=f(t,y)
y=t*y+t^3;
```

The command

```
>> [t,y]=ode45(@f,[0 1],1,opts);
```

causes ode45 to run as before. This alternative will be convenient when the number of independent variables in the differential equation increases.

While it is tempting to crown variable step size Runge–Kutta Methods as the champion ODE solvers, there are a few types of equations that they do not handle very well. Here is a particularly simple but vexing example:

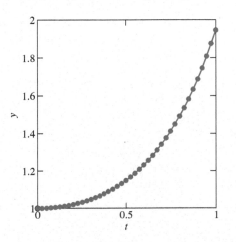

Figure 6.19 MATLAB's ode45 command. Solution of the initial value problem of Example 6.1 is computed, correct to within 10^{-6}.

▶ **EXAMPLE 6.23**   Use `ode45` to solve the initial value problem within a relative tolerance of 10^{-4}:

$$\begin{cases} y' = 10(1 - y) \\ y(0) = 1/2 \\ t \text{ in } [0, 100]. \end{cases} \qquad (6.67)$$

This can be accomplished with the following three lines of MATLAB code:

```
>> opts=odeset('RelTol',1e-4);
>> [t,y]=ode45(@(t,y) 10*(1-y),[0 100],.5,opts);
>> length(t)

ans=          1241
>>
```

We have printed the number of steps because it seems excessive. The solution to the initial value problem is easy to determine: $y(t) = 1 - e^{-10t}/2$. For $t > 1$, the solution has already reached its equilibrium 1 within 4 decimal places, and it never moves any farther away from 1. Yet `ode45` moves at a snail's pace, using an average step size of less than 0.1. Why such a conservative step size selection for a tame solution?

Part of the answer becomes clear by viewing the output from `ode45` in Figure 6.20. Although the solution is very close to 1, the solver continually overshoots in trying to approximate closely. The differential equation is "stiff," a term we will formally define in the next section. For stiff equations, a different strategy in numerical solution greatly increases solving efficiency. For example, note the difference in steps needed when one of MATLAB's stiff solvers are used:

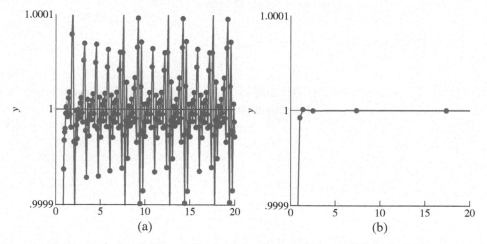

Figure 6.20 Numerical solution of the initial value problem of Example 6.23.
(a) Using `ode45` requires over 10 steps per unit time to stay within relative tolerance 10^{-4}. (b) With `ode23s`, far fewer steps are needed.

```
>> opts=odeset('RelTol',1e-4);
>> [t,y]=ode23s(@(t,y) 10*(1-y),[0 100],.5,opts);
>> length(t)

ans=
        39
```

Figure 6.20(b) plots the solution points from the solver `ode23s`. Relatively few points are needed to keep the numerical solution within the tolerance. We will investigate how to build methods that handle this type of difficulty in the next section.   ◀

▶ **ADDITIONAL EXAMPLES**

*1 Consider the differential equation $y' = \lambda y$. Find the one-step results w_1 and z_1 for the RK23 Method in terms of w_0 and z_0, respectively. Calculate the local truncation errors $|y_1 - w_1|$ and $|y_1 - z_1|$, and show that they are $O(h^3)$ and $O(h^4)$, respectively.

2. The initial value problem

$$\begin{cases} y' = (1 + y^2)/2 \\ y(0) = \sqrt{3} \end{cases}$$

has solution $y(t) = \sec(t + \pi/6) + \tan(t + \pi/6)$ on the interval $0, 1$. Compare the variable step-size MATLAB solvers `ode23` and `ode45`, with `RelTol = 1e-8`, in terms of accuracy at $t = 1$ and the number of steps needed.

⌐⊡ **Solutions** for Additional Examples can be found at `bit.ly/2yOWtft`
(* example with video solution)

6.5 Computer Problems

⌐⊡ **Solutions** for Computer Problems numbered in blue can be found at `bit.ly/2CrkMUa`

1. Write a MATLAB implementation of RK23 (Example 6.19), and apply to approximating the solutions of the IVPs in Exercise 6.1.3 with a relative tolerance of 10^{-8} on $0, 1$. Ask the program to stop exactly at the endpoint $t = 1$. Report the maximum step size used and the number of steps.

2. Compare the results of Computer Problem 1 with the application of MATLAB's `ode23` to the same problem.

3. Carry out the steps of Computer Problem 1 for the Runge–Kutta–Fehlberg Method RKF45.

4. Compare the results of Computer Problem 3 with the application of MATLAB's `ode45` to the same problem.

5. Apply a MATLAB implementation of RKF45 to approximating the solutions of the systems in Exercise 6.3.1 with a relative tolerance of 10^{-6} on $0, 1$. Report the maximum step size used and the number of steps.

6.6 IMPLICIT METHODS AND STIFF EQUATIONS

The differential equations solvers we have presented so far are **explicit**, meaning that there is an explicit formula for the new approximation w_{i+1} in terms of known data, such as h, t_i, and w_i. It turns out that some differential equations are poorly served by explicit methods, and our first goal is to explain why. In Example 6.23, a sophisticated variable step-size solver seems to spend most of its energy overshooting the correct solution in one direction or another.

The stiffness phenomenon can be more easily understood in a simpler context. Accordingly, we begin with Euler's Method.

▶ **EXAMPLE 6.24** Apply Euler's Method to Example 6.23.

Euler's Method for the right-hand side $f(t, y) = 10(1 - y)$ with step size h is

$$\begin{aligned} w_{i+1} &= w_i + hf(t_i, w_i) \\ &= w_i + h(10)(1 - w_i) \\ &= w_i(1 - 10h) + 10h. \end{aligned} \tag{6.68}$$

Since the solution is $y(t) = 1 - e^{-10t}/2$, the approximate solution must approach 1 in the long run. Here we get some help from Chapter 1. Notice that (6.68) can be viewed as a fixed-point iteration with $g(x) = x(1 - 10h) + 10h$. This iteration will converge to the fixed point at $x = 1$ as long as $|g'(1)| = |1 - 10h| < 1$. Solving this inequality yields $0 < h < 0.2$. For any larger h, the fixed point 1 will repel nearby guesses, and the solution will have no hope of being accurate. ◀

Figure 6.21 shows this effect for Example 6.24. The solution is very tame: an attracting equilibrium at $y = 1$. An Euler step of size $h = 0.3$ has difficulty finding the equilibrium because the slope of the nearby solution changes greatly between the beginning and the end of the h interval. This causes overshoot in the numerical solution.

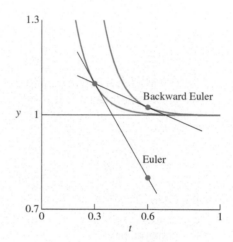

Figure 6.21 Comparison of Euler and Backward Euler steps. The differential equation in Example 6.23 is stiff. The equilibrium solution $y = 1$ is surrounded by other solutions with large curvature (fast-changing slope). The Euler step overshoots, while the Backward Euler step is more consistent with the system dynamics.

Differential equations with this property—that attracting solutions are surrounded with fast-changing nearby solutions—are called **stiff**. This is often a sign of multiple timescales in the system. Quantitatively, it corresponds to the linear part of the right-hand side f of the differential equation, in the variable y, being large and negative. (For a system of equations, this corresponds to an eigenvalue of the linear part being large and negative.) This definition is a bit relative, but that is the nature of stiffness—the more negative, the smaller the step size must be to avoid overshoot. For Example 6.24, stiffness is measured by evaluating $\partial f / \partial y = -10$ at the equilibrium solution $y = 1$.

One way to solve the problem depicted in Figure 6.21 is to somehow bring in information from the right side of the interval $[t_i, t_i + h]$, instead of relying solely on information from the left side. That is the motivation behind the following variation on Euler's Method:

Backward Euler Method

$$w_0 = y_0$$
$$w_{i+1} = w_i + hf(t_{i+1}, w_{i+1}). \tag{6.69}$$

Note the difference: While Euler's Method uses the left-end slope to step across the interval, Backward Euler would like to somehow cross the interval so that the slope is correct at the right end.

A price must be paid for this improvement. Backward Euler is our first example of an **implicit** method, meaning that the method does not directly give a formula for the new approximation w_{i+1}. Instead, we must work a little to get it. For the example $y' = 10(1 - y)$, the Backward Euler Method gives

$$w_{i+1} = w_i + 10h(1 - w_{i+1}),$$

which, after a little algebra, can be expressed as

$$w_{i+1} = \frac{w_i + 10h}{1 + 10h}.$$

Setting $h = 0.3$, for example, the Backward Euler Method gives $w_{i+1} = (w_i + 3)/4$. We can again evaluate the behavior as a fixed point iteration $w \to g(w) = (w + 3)/4$. There is a fixed point at 1, and $g'(1) = 1/4 < 1$, verifying convergence to the true equilibrium solution $y = 1$. Unlike the Euler Method with $h = 0.3$, at least the correct qualitative behavior is followed by the numerical solution. In fact, note that the Backward Euler Method solution converges to $y = 1$ no matter how large the step size h (Exercise 3).

Because of the better behavior of implicit methods like Backward Euler in the presence of stiff equations, it is worthwhile performing extra work to evaluate the next step, even though it is not explicitly available. Example 6.24 was not challenging to solve for w_{i+1}, due to the fact that the differential equation is linear, and it was possible to change the original implicit formula to an explicit one for evaluation. In general, however, this is not possible, and we need to use more indirect means.

If the implicit method leaves a nonlinear equation to solve, we must refer to Chapter 1. Both Fixed-Point Iteration and Newton's Method are often used to solve for w_{i+1}. This means that there is an equation-solving loop within the loop advancing the differential equation. The next example shows how this can be done.

► **EXAMPLE 6.25** Apply the Backward Euler Method to the initial value problem

$$\begin{cases} y' = y + 8y^2 - 9y^3 \\ y(0) = 1/2 \\ t \text{ in } [0, 3]. \end{cases}$$

This equation, like the previous example, has an equilibrium solution $y = 1$. The partial derivative $\partial f/\partial y = 1 + 16y - 27y^2$ evaluates to -10 at $y = 1$, identifying this equation as moderately stiff. There will be an upper bound, similar to that of the previous example, for h, such that Euler's Method is successful. Thus, we are motivated to try the Backward Euler Method

$$\begin{aligned} w_{i+1} &= w_i + hf(t_{i+1}, w_{i+1}) \\ &= w_i + h(w_{i+1} + 8w_{i+1}^2 - 9w_{i+1}^3). \end{aligned}$$

This is a nonlinear equation in w_{i+1}, which we need to solve in order to advance the numerical solution. Renaming $z = w_{i+1}$, we must solve the equation $z = w_i + h(z + 8z^2 - 9z^3)$, or

$$9hz^3 - 8hz^2 + (1 - h)z - w_i = 0 \tag{6.70}$$

for the unknown z. We will demonstrate with Newton's Method.

       To start Newton's Method, an initial guess is needed. Two choices that come to mind are the previous approximation w_i and the Euler's Method approximation for w_{i+1}. Although the latter is accessible since Euler is explicit, it may not be the best choice for stiff problems, as shown in Figure 6.21. In this case, we will use w_i as the starting guess.

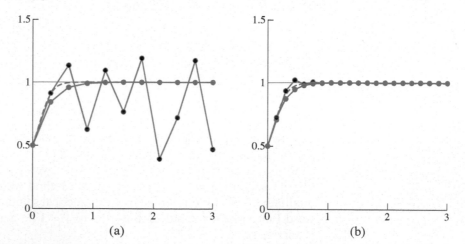

Figure 6.22 Numerical solution of the initial value problem of Example 6.25.
True solution is the dashed curve. The black circles denote the Euler Method approximation; the blue circles denote Backward Euler. (a) $h = 0.3$ (b) $h = 0.15$.

Assembling Newton's Method for (6.70) yields

$$z_{new} = z - \frac{9hz^3 - 8hz^2 + (1-h)z - w_i}{27hz^2 - 16hz + 1 - h}. \tag{6.71}$$

After evaluating (6.71), replace z with z_{new} and repeat. For each Backward Euler step, Newton's Method is run until $z_{new} - z$ is smaller than a preset tolerance (smaller than the errors that are being made in approximating the differential equation solution).

       Figure 6.22 shows the results for two different step sizes. In addition, numerical solutions from Euler's Method are shown. Clearly, $h = 0.3$ is too large for Euler on this stiff problem. On the other hand, when h is cut to 0.15, both methods perform at about the same level. ◀

       So-called stiff solvers like Backward Euler allow sufficient error control with comparatively large step size, increasing efficiency. MATLAB's ode23s is a higher order version with a built-in variable step-size strategy.

▶ **ADDITIONAL EXAMPLES**

1. Using initial condition $y(0) = 1$ and step size $h = 1/4$, calculate the Backward Euler approximation to $y' = ty$ on the interval $[0, 1]$. Find the error at $t = 1$ by comparing with the exact solution.

2. Consider the IVP

$$\begin{cases} y' = 10y^5 - 10y^{10} \\ y(0) = 1/2 \end{cases}$$

on the interval $[0, 10]$. Compare the variable step-size MATLAB solvers ode23s and ode45, with RelTol = 1e-6. Do both methods approach the stable equilibrium

$y = 1$ at $t = 1$? Report the number of steps needed for each method. What happens for RelTol = 1e-8?

📱 **Solutions** for Additional Examples can be found at bit.ly/2q236qD

6.6 Exercises

1. Using initial condition $y(0) = 0$ and step size $h = 1/4$, calculate the Backward Euler approximation on the interval $0, 1$. Find the error at $t = 1$ by comparing with the correct solution found in Exercise 6.1.4.

   (a) $y' = t + y$ (b) $y' = t - y$ (c) $y' = 4t - 2y$

2. Find all equilibrium solutions and the value of the Jacobian at the equilibria. Is the equation stiff? (a) $y' = y - y^2$ (b) $y' = 10y - 10y^2$ (c) $y' = -10\sin y$

3. Show that for every step size h, the Backward Euler approximate solution converges to the equilibrium solution $y = 1$ as $t_i \to \infty$ for Example 6.24.

4. Consider the linear differential equation $y' = ay + b$ for $a < 0$. (a) Find the equilibrium. (b) Write down the Backward Euler Method for the equation. (c) View Backward Euler as a Fixed-Point Iteration to prove that the method's approximate solution will converge to the equilibrium as $t \to \infty$.

6.6 Computer Problems

1. Apply Backward Euler, using Newton's Method as a solver, for the initial value problems. Which of the equilibrium solutions are approached by the approximate solution? Apply Euler's Method. For what approximate range of h can Euler be used successfully to converge to the equilibrium? Plot approximate solutions given by Backward Euler, and by Euler with an excessive step size.

   (a) $\begin{cases} y' = y^2 - y^3 \\ y(0) = 1/2 \\ t \text{ in } 0, 20 \end{cases}$ (b) $\begin{cases} y' = 6y - 6y^2 \\ y(0) = 1/2 \\ t \text{ in } 0, 20 \end{cases}$

2. Carry out the steps in Computer Problem 1 for the following initial value problems:

   (a) $\begin{cases} y' = 6y - 3y^2 \\ y(0) = 1/2 \\ t \text{ in } 0, 20 \end{cases}$ (b) $\begin{cases} y' = 10y^3 - 10y^4 \\ y(0) = 1/2 \\ t \text{ in } 0, 20 \end{cases}$

6.7 MULTISTEP METHODS

The Runge–Kutta family that we have studied consists of one-step methods, meaning that the newest step w_{i+1} is produced on the basis of the differential equation and the value of the previous step w_i. This is in the spirit of initial value problems, for which Theorem 6.2 guarantees a unique solution starting at an arbitrary w_0.

The multistep methods suggest a different approach: using the knowledge of more than one of the previous w_i to help produce the next step. This will lead to ODE solvers that have order as high as the one-step methods, but much of the necessary computation will be replaced with interpolation of already computed values on the solution path.

6.7.1 Generating multistep methods

As a first example, consider the following two-step method:

Adams–Bashforth Two-Step Method

$$w_{i+1} = w_i + h\left[\frac{3}{2}f(t_i, w_i) - \frac{1}{2}f(t_{i-1}, w_{i-1})\right].$$ (6.72)

While the second-order Midpoint Method,

$$w_{i+1} = w_i + hf\left(t_i + \frac{h}{2}, w_i + \frac{h}{2}f(t_i, w_i)\right),$$

needs two function evaluations of the ODE right-hand side f per step, the Adams–Bashforth Two-Step Method requires only one new evaluation per step (one is stored from the previous step). We will see subsequently that (6.72) is also a second-order method. Therefore, multistep methods can achieve the same order with less computational effort—usually just one function evaluation per step.

Since multistep methods use more than one previous w value, they need help getting started. The start-up phase for an s-step method typically consists of a one-step method that uses w_0 to produce $s - 1$ values $w_1, w_2, \ldots, w_{s-1}$, before the multistep method can be used. The Adams–Bashforth Two-Step Method (6.72) needs w_1, along with the given initial condition w_0, in order to begin. The following MATLAB code uses the Trapezoid Method to provide the start-up value w_1.

MATLAB code
shown here can be found
at bit.ly/2AfGxon

```
% Program 6.7 Multistep method
% Inputs: time interval inter,
%   ic=[y0]  initial condition, number of steps n,
%   s=number of (multi)steps, e.g. 2 for 2-step method
% Output: time steps t, solution y
% Calls a multistep method such as ab2step.m
% Example usage: [t,y]=exmultistep([0,1],1,20,2)
function [t,y]=exmultistep(inter,ic,n,s)
h=(inter(2)-inter(1))/n;
% Start-up phase
y(1,:)=ic;t(1)=inter(1);
for i=1:s-1                    % start-up phase, using one-step method
  t(i+1)=t(i)+h;
  y(i+1,:)=trapstep(t(i),y(i,:),h);
  f(i,:)=ydot(t(i),y(i,:));
end
for i=s:n                      % multistep method loop
  t(i+1)=t(i)+h;
  f(i,:)=ydot(t(i),y(i,:));
  y(i+1,:)=ab2step(t(i),i,y,f,h);
end
plot(t,y)

function y=trapstep(t,x,h)
%one step of the Trapezoid Method from section 6.2
z1=ydot(t,x);
g=x+h*z1;
z2=ydot(t+h,g);
y=x+h*(z1+z2)/2;
```

```
function z=ab2step(t,i,y,f,h)
%one step of the Adams-Bashforth 2-step method
z=y(i,:)+h*(3*f(i,:)/2-f(i-1,:)/2);

function z=unstable2step(t,i,y,f,h)
%one step of an unstable 2-step method
z=-y(i,:)+2*y(i-1,:)+h*(5*f(i,:)/2+f(i-1,:)/2);

function z=weaklystable2step(t,i,y,f,h)
%one step of a  weakly-stable 2-step method
z=y(i-1,:)+h*2*f(i,:);

function z=ydot(t,y)  % IVP from section 6.1
z=t*y+t^3;
```

Figure 6.23(a) shows the result of applying the Adams–Bashforth Two-Step Method to the initial value problem (6.5) from earlier in the chapter, using step size $h = 0.05$ and applying the Trapezoid Method for start-up. Part (b) of the figure shows the use of a different two-step method. Its instability will be the subject of our discussion of stability analysis in the next sections.

A general s-step method has the form

$$w_{i+1} = a_1 w_i + a_2 w_{i-1} + \cdots + a_s w_{i-s+1} + h[b_0 f_{i+1} + b_1 f_i$$
$$+ b_2 f_{i-1} + \cdots + b_s f_{i-s+1}]. \qquad (6.73)$$

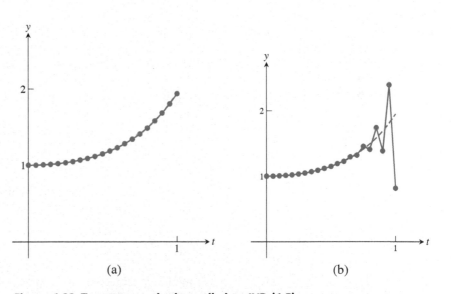

(a)   (b)

Figure 6.23 Two-step methods applied to IVP (6.5). Dashed curve shows the correct solution. Step size $h = 0.05$. (a) Adams–Bashforth Two-Step Method plotted as circles. (b) Unstable method (6.81) in circles.

The step size is h, and we use the notational convenience

$$f_i \equiv f(t_i, w_i).$$

If $b_0 = 0$, the method is explicit. If $b_0 \neq 0$, the method is implicit. We will discuss how to use implicit methods shortly.

First, we want to show how multistep methods are derived and how to decide which ones will work best. The main issues that arise with multistep methods can be introduced in the relatively simple case of two-step methods, so we begin there, and follow later with the general case. A general two-step method (setting $s = 2$ in (6.73)) has the form

$$w_{i+1} = a_1 w_i + a_2 w_{i-1} + h[b_0 f_{i+1} + b_1 f_i + b_2 f_{i-1}]. \tag{6.74}$$

To develop a multistep method, we need to refer to Taylor's Theorem, since we try to match as many terms of the solution's Taylor expansion as possible with the terms of the method. What remains will be the local truncation error.

We assume that all previous w_i are correct—that is, $w_i = y_i$ and $w_{i-1} = y_{i-1}$ in (6.74). The differential equation says that $y_i' = f_i$, so all terms can be expanded in a Taylor expansion as follows:

$$
\begin{aligned}
w_{i+1} &= a_1 w_i + a_2 w_{i-1} + h[b_0 f_{i+1} + b_1 f_i + b_2 f_{i-1}] \\
&= a_1[y_i] \\
&\quad + a_2[y_i \quad - \quad hy_i' \quad + \quad \tfrac{h^2}{2}y_i'' \quad - \quad \tfrac{h^3}{6}y_i''' \quad + \quad \tfrac{h^4}{24}y_i'''' \quad - \quad \cdots] \\
&\quad + b_0[\qquad\quad hy_i' \quad + \quad h^2 y_i'' \quad + \quad \tfrac{h^3}{2}y_i''' \quad + \quad \tfrac{h^4}{6}y_i'''' \quad + \quad \cdots] \\
&\quad + b_1[\qquad\quad hy_i'] \\
&\quad + b_2[\qquad\quad hy_i' \quad - \quad h^2 y_i'' \quad + \quad \tfrac{h^3}{2}y_i''' \quad - \quad \tfrac{h^4}{6}y_i'''' \quad + \quad \cdots].
\end{aligned}
$$

Adding up yields

$$w_{i+1} = (a_1 + a_2)y_i + (b_0 + b_1 + b_2 - a_2)hy_i' + (a_2 - 2b_2 + 2b_0)\frac{h^2}{2}y_i''$$

$$+ (-a_2 + 3b_0 + 3b_2)\frac{h^3}{6}y_i''' + (a_2 + 4b_0 - 4b_2)\frac{h^4}{24}y_i'''' + \cdots. \tag{6.75}$$

By choosing the a_i and b_i appropriately, the local truncation error $y_{i+1} - w_{i+1}$, where

$$y_{i+1} = y_i + hy_i' + \frac{h^2}{2}y_i'' + \frac{h^3}{6}y_i''' + \cdots, \tag{6.76}$$

can be made as small as possible, assuming that the derivatives involved actually exist. Next, we will investigate the possibilities.

6.7.2 Explicit multistep methods

To look for explicit methods, set $b_0 = 0$. A second-order method can be developed by matching terms in (6.75) and (6.76) up to and including the h^2 term, making the local truncation error of size $O(h^3)$. Comparing terms yields the system

$$
\begin{aligned}
a_1 + a_2 &= 1 \\
-a_2 + b_1 + b_2 &= 1 \\
a_2 - 2b_2 &= 1.
\end{aligned} \tag{6.77}
$$

There are three equations in four unknowns a_1, a_2, b_1, b_2, so it will be possible to find infinitely many different explicit order-two methods. (One of the solutions corresponds to an order-three method. See Exercise 3.) Note that the equations can be written in terms of a_1 as follows:

$$a_2 = 1 - a_1$$
$$b_1 = 2 - \frac{1}{2}a_1$$
$$b_2 = -\frac{1}{2}a_1. \tag{6.78}$$

The local truncation error will be

$$y_{i+1} - w_{i+1} = \frac{1}{6}h^3 y_i''' - \frac{3b_2 - a_2}{6}h^3 y_i''' + O(h^4)$$
$$= \frac{1 - 3b_2 + a_2}{6}h^3 y_i''' + O(h^4)$$
$$= \frac{4 + a_1}{12}h^3 y_i''' + O(h^4). \tag{6.79}$$

We are free to set a_1 arbitrarily—any choice leads to a second-order method, as we have just shown. Setting $a_1 = 1$ yields the second-order Adams–Bashforth Method (6.72). Note that $a_2 = 0$ by the first equation, and $b_2 = -1/2$ and $b_1 = 3/2$. According to (6.79), the local truncation error is $5/12h^3 y'''(t_i) + O(h^4)$.

Alternatively, we could set $a_1 = 1/2$ to get another two-step second-order method with $a_2 = 1/2, b_1 = 7/4$, and $b_2 = -1/4$:

$$w_{i+1} = \frac{1}{2}w_i + \frac{1}{2}w_{i-1} + h\left[\frac{7}{4}f_i - \frac{1}{4}f_{i-1}\right]. \tag{6.80}$$

This method has local truncation error $3/8h^3 y'''(t_i) + O(h^4)$.

SPOTLIGHT ON

> **Complexity**   The advantage of multistep methods to one-step methods is clear. After the first few steps, only one new evaluation of the right-hand side function need to be made. For one-step methods, it is typical for several function evaluations to be needed. Fourth-order Runge–Kutta, for example, needs four evaluations per step, while the fourth-order Adams–Bashforth Method needs only one after the start-up phase.

A third choice, $a_1 = -1$, gives the second-order two-step method

$$w_{i+1} = -w_i + 2w_{i-1} + h\left[\frac{5}{2}f_i + \frac{1}{2}f_{i-1}\right] \tag{6.81}$$

that was used in Figure 6.23(b). The failure of (6.81) brings out an important stability condition that must be met by multistep solvers. Consider the even simpler IVP

$$\begin{cases} y' = 0 \\ y(0) = 0 \\ t \text{ in } [0,1] \end{cases}. \tag{6.82}$$

Applying method (6.81) to this example yields

$$w_{i+1} = -w_i + 2w_{i-1} + h[0]. \tag{6.83}$$

One solution $\{w_i\}$ to (6.83) is $w_i \equiv 0$. However, there are others. Substituting the form $w_i = c\lambda^i$ into (6.83) yields

$$c\lambda^{i+1} + c\lambda^i - 2c\lambda^{i-1} = 0$$
$$c\lambda^{i-1}(\lambda^2 + \lambda - 2) = 0. \tag{6.84}$$

The solutions of the "characteristic polynomial" $\lambda^2 + \lambda - 2 = 0$ of this recurrence relation are 1 and -2. The latter is a problem—it means that solutions of form $(-2)^i c$ are solutions of the method for constant c. This allows small rounding and truncation errors to quickly grow to observable size and swamp the computation, as seen in Figure 6.23. To avoid this possibility, it is important that the roots of the characteristic polynomial of the method are bounded by 1 in absolute value. This leads to the following definition:

DEFINITION 6.6 The multistep method (6.73) is **stable** if the roots of the polynomial $P(x) = x^s - a_1 x^{s-1} - \ldots - a_s$ are bounded by 1 in absolute value, and any roots of absolute value 1 are simple roots. A stable method for which 1 is the only root of absolute value 1 is called **strongly stable**; otherwise it is **weakly stable**. ❑

The Adams–Bashforth Method (6.72) has roots 0 and 1, making it strongly stable, while (6.81) has roots -2 and 1, making it unstable.

The characteristic polynomial of the general two-step formula, using the fact that $a_1 = 1 - a_2$ from (6.78), is

$$P(x) = x^2 - a_1 x - a_2$$
$$= x^2 - a_1 x - 1 + a_1$$
$$= (x - 1)(x - a_1 + 1),$$

whose roots are 1 and $a_1 - 1$. Returning to (6.78), we can find a weakly stable second-order method by setting $a_1 = 0$. Then the roots are 1 and -1, leading to the following weakly stable second-order two-step method:

$$w_{i+1} = w_{i-1} + 2hf_i. \tag{6.85}$$

▶ **EXAMPLE 6.26** Apply strongly stable method (6.72), weakly stable method (6.85), and unstable method (6.81) to the initial value problem

$$\begin{cases} y' = -3y \\ y(0) = 1 \\ t \text{ in } [0, 2] \end{cases}. \tag{6.86}$$

The solution is the curve $y = e^{-3t}$. We will use Program 6.7 to follow the solutions, where ydot.m has been changed to

```
function z=ydot(t,y)
z=-3*y;
```

and ab2step is replaced by one of the three calls ab2step, weaklystable2step, or unstable2step.

Figure 6.24 shows the three solution approximations for step size $h = 0.1$. The weakly stable and unstable methods seem to follow closely for a while and then move quickly away from the correct solution. Reducing the step size does not eliminate the problem, although it may delay the onset of instability. ◀

With two more definitions, we can state the fundamental theorem of multistep solvers.

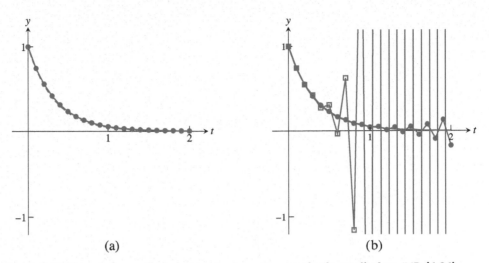

Figure 6.24 Comparison of second-order, two-step methods applied to IVP (6.86).
(a) Adams–Bashforth Method. (b) Weakly stable method (in circles) and unstable method (in squares).

DEFINITION 6.7   A multistep method is **consistent** if it has order at least 1. A solver is **convergent** if the approximate solutions converge to the exact solution for each t, as $h \to 0$. ☐

THEOREM 6.8   (**Dahlquist**) Assume that the starting values are correct. Then a multistep method (6.73) is convergent if and only if it is stable and consistent. ∎

For a proof of Dahlquist's theorem, see Hairer and Wanner [1996]. Theorem 6.8 tells us that avoiding a catastrophe like Figure 6.24(b) for a second-order two-step method is as simple as checking the method's stability.

One root of the characteristic polynomial must be at 1 (see Exercise 6). The Adams–Bashforth Methods are the ones whose other roots are all at 0. For this reason, the Adams–Bashforth Two-Step Method is considered the most stable of the two-step methods.

The derivation of higher order methods, using more steps, is precisely analogous to our previous derivation of two-step methods. Assume that all previous w_i are correct – that is, $w_i = y_i, \ldots, w_{i-s+1} = y_{i-s+1}$ in (6.73). The differential equation says that $y_i' = f_i$, so all terms can be represented by Taylor expansions as follows:

$$w_{i+1} = a_1 w_i + a_2 w_{i-1} + \ldots + a_s w_{i-s+1} + h[b_0 f_{i+1} + b_1 f_i + b_2 f_{i-1} + \ldots + b_s f_{i-s+1}]$$

$$= a_1 y_i$$

$$+ a_2[y_i + \frac{(-h)}{1!} y_i' + \frac{(-h)^2}{2!} y_i'' + \ldots + \frac{(-h)^s}{s!} y_i^{(s)} + O(h^{s+1})]$$

$$+ a_3[y_i + \frac{(-2h)}{1!} y_i' + \frac{(-2h)^2}{2!} y_i'' + \ldots + \frac{(-2h)^s}{s!} y_i^{(s)} + O(h^{s+1})] + \ldots$$

$$+ a_s[y_i + \frac{(-(s-1)h)}{1!} y_i' + \frac{(-(s-1)h)^2}{2!} y_i'' + \ldots + \frac{(-(s-1)h)^s}{s!} y_i^{(s)} + O(h^{s+1})]$$

$$+ b_0 h[y_i' + \frac{h}{1!} y_i'' + \frac{h^2}{2!} y_i''' + \ldots + \frac{h^{s-1}}{(s-1)!} y_i^{(s)} + O(h^s)]$$

$$+ b_1 h y_i'$$

$$+ b_2 h[y_i' + \frac{(-h)}{1!} y_i'' + \frac{(-h)^2}{2!} y_i''' + \ldots + \frac{(-h)^{s-1}}{(s-1)!} y_i^{(s)} + O(h^s)] + \ldots$$

$$+ b_s h[y_i' + \frac{(-(s-1)h)}{1!} y_i'' + \frac{(-(s-1)h)^2}{2!} y_i''' + \ldots + \frac{(-(s-1)h)^{s-1}}{(s-1)!} y_i^{(s)} + O(h^s)].$$

For example, in the case of a two-step explicit method, we set $s = 2$ and $b_0 = 0$ to reduce to

$$w_{i+1} = a_1 y(t_i)$$

$$+ a_2 [y(t_i) + \frac{(-h)}{1!} y'(t_i) + \frac{(-h)^2}{2!} y''(t_i) + O(h^3)]$$

$$+ b_1 h[y'(t_i)]$$

$$+ b_2 h[y'(t_i) + \frac{(-h)}{1!} y''(t_i) + O(h^2)]$$

The Adams–Bashforth family of explicit methods sets $a_1 = 1, b_0 = a_2 = \ldots = a_s = 0$, to achieve a characteristic polynomial $P(x) = x^s - x^{s-1} = x^{s-1}(x-1)$. Therefore all methods in the family are strongly stable by Defn. 6.6.

We can derive the formula of the s-step Adams–Bashforth Method from the expansion above. Using the assumptions of the last paragraph, we must match

$$w_{i+1} = a_1 w_i + a_2 w_{i-1} + \ldots + a_s w_{i-s+1} + h[b_0 f_{i+1} + b_1 f_i + b_2 f_{i-1} + \ldots + b_s f_{i-s+1}]$$

$$= y_i$$

$$+ b_1 h[y_i']$$

$$+ b_2 h[y_i' + \frac{(-h)}{1!} y_i'' + \frac{(-h)^2}{2!} y_i''' + \ldots + \frac{(-h)^{s-1}}{(s-1)!} y_i^{(s)} + O(h^s)] + \ldots$$

$$+ b_s h[y_i' + \frac{(-(s-1)h)}{1!} y_i'' + \frac{(-(s-1)h)^2}{2!} y_i''' + \ldots + \frac{(-(s-1)h)^{s-1}}{(s-1)!} y_i^{(s)} + O(h^s)].$$

with

$$y_{i+1} = y_i + \frac{h}{1!} y_i' + \frac{h^2}{2!} y_i'' + \ldots + \frac{h^s}{s!} y_i^{(s)} + O(h^{s+1}).$$

The y_i' terms imply $b_1 + b_2 + \ldots + b_s = 1$. For $1 \leq j \leq s - 1$, the $y_i^{(j+1)}$ terms yield

$$\frac{h b_2 (-h)^j}{j!} + \frac{h b_3 (-2h)^j}{j!} + \ldots + \frac{h b_s (-(s-1)h)^j}{j!} = \frac{h^{j+1}}{(j+1)!},$$

or

$$(-1)^j b_2 + (-2)^j b_3 + \ldots + (-(s-1))^j b_s = \frac{1}{j+1}.$$

Matching terms results in s linear equations in $b_1, \ldots, b_s$, leaving a remainder of terms proportional to h^{s+1}, which become the truncation error. Therefore the s-step Adams–Bashforth Method is an order-s solver, and we can determine $b_1, \ldots, b_s$ by solving the matrix formulation

$$\begin{bmatrix} -1 & -2 & -3 & \cdots & -(s-1) \\ (-1)^2 & (-2)^2 & (-3)^2 & \cdots & (-(s-1))^2 \\ \vdots & \vdots & \vdots & \vdots & \vdots \\ (-1)^{(s-1)} & (-2)^{(s-1)} & (-3)^{(s-1)} & \cdots & (-(s-1))^{(s-1)} \end{bmatrix} \begin{bmatrix} b_2 \\ b_3 \\ \vdots \\ b_s \end{bmatrix} = \begin{bmatrix} 1/2 \\ 1/3 \\ \vdots \\ 1/s \end{bmatrix}$$

and then setting $b_1 = 1 - b_2 - \ldots - b_s$.

For example, the Adams–Bashforth Two-Step Method coefficients arise from solving the 1×1 system

$$[-1][b_2] = \frac{1}{2},$$

which yields $b_1 = 3/2, b_2 = -1/2$. This agrees with our earlier calculation. The Adams–Bashforth Three-Step Method requires solving

$$\begin{bmatrix} -1 & -2 \\ 1 & 4 \end{bmatrix} \begin{bmatrix} b_2 \\ b_3 \end{bmatrix} = \begin{bmatrix} 1/2 \\ 1/3 \end{bmatrix},$$

which yields $b_1 = 23/12, b_2 = -4/3$, and $b_3 = 5/12$.

Exercises 13 and 14 ask for verification that the following methods are strongly stable:

Adams–Bashforth Three-Step Method (third order)

$$w_{i+1} = w_i + \frac{h}{12}[23f_i - 16f_{i-1} + 5f_{i-2}]. \tag{6.87}$$

Adams–Bashforth Four-Step Method (fourth order)

$$w_{i+1} = w_i + \frac{h}{24}[55f_i - 59f_{i-1} + 37f_{i-2} - 9f_{i-3}]. \tag{6.88}$$

6.7.3 Implicit multistep methods

When the coefficient b_0 in (6.73) is nonzero, the method is implicit. The simplest second-order implicit method (see Exercise 5) is the Implicit Trapezoid Method:

Implicit Trapezoid Method (second order)

$$w_{i+1} = w_i + \frac{h}{2}[f_{i+1} + f_i]. \tag{6.89}$$

If the f_{i+1} term is replaced by evaluating f at the "prediction" for w_{i+1} made by Euler's Method, then this becomes the Explicit Trapezoid Method. The Implicit Trapezoid Method is also called the Adams–Moulton One-Step Method, by analogy with what follows. An example of a two-step implicit method is the Adams–Moulton Two-Step Method:

Adams–Moulton Two-Step Method (third order)

$$w_{i+1} = w_i + \frac{h}{12}[5f_{i+1} + 8f_i - f_{i-1}]. \tag{6.90}$$

There are significant differences between the implicit and explicit methods. First, it is possible to get a stable third-order implicit method by using only two previous steps, unlike the explicit case. Second, the corresponding local truncation error formula is smaller for implicit methods. On the other hand, the implicit method has the inherent difficulty that extra processing is necessary to evaluate the implicit part.

For these reasons, implicit methods are often used as the corrector in a "predictor–corrector" pair. Implicit and explicit methods of the same order are used together.

Each step is the combination of a prediction by the explicit method and a correction by the implicit method, where the implicit method uses the predicted w_{i+1} to calculate f_{i+1}. Predictor–corrector methods take approximately twice the computational effort, since an evaluation of the differential equation right-hand side f is done on both the prediction and the correction parts of the step. However, the added accuracy and stability often make the price worth paying.

A simple predictor–corrector method pairs the Adams–Bashforth Two-Step Explicit Method as predictor with the Adams–Moulton One-Step Implicit Method as corrector. Both are second-order methods. The MATLAB code looks similar to the Adams–Bashforth code used earlier, but with a corrector step added:

⌷ MATLAB code
shown here can be found
at bit.ly/2CnyI1l

```
% Program 6.8 Adams-Bashforth-Moulton second-order p-c
% Inputs: time interval inter,
%   ic=[y0]   initial condition
%   number of steps n, number of (multi)steps s for explicit method
% Output: time steps t, solution y
% Calls multistep methods such as ab2step.m and am1step.m
% Example usage: [t,y]=predcorr([0 1],1,20,2)
function [t,y]=predcorr(inter,ic,n,s)
h=(inter(2)-inter(1))/n;
% Start-up phase
y(1,:)=ic;t(1)=inter(1);
for i=1:s-1                    % start-up phase, using one-step method
  t(i+1)=t(i)+h;
  y(i+1,:)=trapstep(t(i),y(i,:),h);
  f(i,:)=ydot(t(i),y(i,:));
end
for i=s:n                      % multistep method loop
  t(i+1)=t(i)+h;
  f(i,:)=ydot(t(i),y(i,:));
  y(i+1,:)=ab2step(t(i),i,y,f,h);  % predict
  f(i+1,:)=ydot(t(i+1),y(i+1,:));
  y(i+1,:)=am1step(t(i),i,y,f,h);  % correct
end
plot(t,y)

function y=trapstep(t,x,h)
%one step of the Trapezoid Method from section 6.2
z1=ydot(t,x);
g=x+h*z1;
z2=ydot(t+h,g);
y=x+h*(z1+z2)/2;

function z=ab2step(t,i,y,f,h)
%one step of the Adams-Bashforth 2-step method
z=y(i,:)+h*(3*f(i,:)-f(i-1,:))/2;

function z=am1step(t,i,y,f,h)
%one step of the Adams-Moulton 1-step method
z=y(i,:)+h*(f(i+1,:)+f(i,:))/2;

function z=ydot(t,y)  % IVP
z=t*y+t^3;
```

The Adams–Moulton Two-Step Method is derived just as the explicit methods were established. Redo the set of equations (6.77), but without requiring that $b_0 = 0$.

Since there is an extra parameter now (b_0), we are able to match up (6.75) and (6.76) through the degree 3 terms with only a two-step method, putting the local truncation error in the h^4 term. The analogue to (6.77) is

$$a_1 + a_2 = 1$$
$$-a_2 + b_0 + b_1 + b_2 = 1$$
$$a_2 + 2b_0 - 2b_2 = 1$$
$$-a_2 + 3b_0 + 3b_2 = 1. \qquad (6.91)$$

Satisfying these equations results in a third-order two-step implicit method.

The equations can be written in terms of a_1 as follows:

$$a_2 = 1 - a_1$$
$$b_0 = \frac{1}{3} + \frac{1}{12}a_1$$
$$b_1 = \frac{4}{3} - \frac{2}{3}a_1$$
$$b_2 = \frac{1}{3} - \frac{5}{12}a_1. \qquad (6.92)$$

The local truncation error is

$$
\begin{aligned}
y_{i+1} - w_{i+1} &= \frac{1}{24}h^4 y_i'''' - \frac{4b_0 - 4b_2 + a_2}{24}h^4 y_i'''' + O(h^5)\\
&= \frac{1 - a_2 - 4b_0 + 4b_2}{24}h^4 y_i'''' + O(h^5)\\
&= -\frac{a_1}{24}h^4 y_i'''' + O(h^5).
\end{aligned}
$$

The order of the method will be three, as long as $a_1 \neq 0$. Since a_1 is a free parameter, there are infinitely many third-order two-step implicit methods. The Adams–Moulton Two-Step Method uses the choice $a_1 = 1$. Exercise 8 asks for a verification that this method is strongly stable. Exercise 9 explores other choices of a_1.

Note one more special choice, $a_1 = 0$. From the local truncation formula, we see that this two-step method will be fourth order.

Milne–Simpson Method

$$w_{i+1} = w_{i-1} + \frac{h}{3}[f_{i+1} + 4f_i + f_{i-1}]. \qquad (6.93)$$

Exercise 10 asks you to check that it is only weakly stable. For this reason, it is susceptible to error magnification.

The suggestive terminology of the Implicit Trapezoid Method (6.89) and Milne–Simpson Method (6.93) should remind the reader of the numerical integration formulas from Chapter 5. In fact, although we have not emphasized this approach, many of the multistep formulas we have presented can be alternatively derived by integrating approximating interpolants, in a close analogy to numerical integration schemes.

The basic idea behind this approach is that the differential equation $y' = f(t, y)$ can be integrated on the interval $[t_i, t_{i+1}]$ to give

$$y(t_{i+1}) - y(t_i) = \int_{t_i}^{t_{i+1}} f(t, y)\, dt. \qquad (6.94)$$

Applying a numerical integration scheme to approximate the integral in (6.94) results in a multistep ODE method. For example, using the Trapezoid Rule for numerical integration from Chapter 5 yields

$$y(t_{i+1}) - y(t_i) = \frac{h}{2}(f_{i+1} + f_i) + O(h^2),$$

which is the second-order Trapezoid Method for ODEs. If we approximate the integral by Simpson's Rule, the result is

$$y(t_{i+1}) - y(t_i) = \frac{h}{3}(f_{i+1} + 4f_i + f_{i-1}) + O(h^4),$$

the fourth-order Milne–Simpson Method (6.93). Essentially, we are approximating the right-hand side of the ODE by a polynomial and integrating, just as is done in numerical integration. This approach can be extended to recover a number of the multistep methods we have already presented, by changing the degree of interpolation and the location of the interpolation points. Although this approach is a more geometric way of deriving some the multistep methods, it gives no particular insight into the stability of the resulting ODE solver.

The derivation of the Adams–Moulton formulas follow from the same derivation as for the Adams–Bashforth equations, setting $a_1 = 1$, and allowing b_0 to float instead of setting it to zero. In this case the matrix formulation is

$$\begin{bmatrix} 1 & -1 & -2 & \cdots & -(s-1) \\ 1 & (-1)^2 & (-2)^2 & \cdots & (-(s-1))^2 \\ \vdots & \vdots & \vdots & \vdots & \vdots \\ 1 & (-1)^s & (-2)^s & \cdots & (-(s-1))^s \end{bmatrix} \begin{bmatrix} b_0 \\ b_2 \\ \vdots \\ b_s \end{bmatrix} = \begin{bmatrix} 1/2 \\ 1/3 \\ \vdots \\ 1/(s+1) \end{bmatrix}$$

and then setting $b_1 = 1 - b_0 - b_2 - \ldots - b_s$. Note that the unmatched terms are proportional to h^{s+2}, so the Adams–Moulton s-Step Method is an order $s + 1$ solver.

For example, the Adams–Moulton Two-Step Method receives coefficients from

$$\begin{bmatrix} 1 & -1 \\ 1 & (-1)^2 \end{bmatrix} \begin{bmatrix} b_0 \\ b_2 \end{bmatrix} = \begin{bmatrix} 1/2 \\ 1/3 \end{bmatrix},$$

which yields $b_0 = 5/12, b_1 = 2/3$, and $b_2 = -1/12$. Coefficients for the Adams–Moulton Three-Step and Four-Step Methods are found the same way.

Adams–Moulton Three-Step Method (fourth order)

$$w_{i+1} = w_i + \frac{h}{24}[9f_{i+1} + 19f_i - 5f_{i-1} + f_{i-2}]. \tag{6.95}$$

Adams–Moulton Four-Step Method (fifth order)

$$w_{i+1} = w_i + \frac{h}{720}[251f_{i+1} + 646f_i - 264f_{i-1} + 106f_{i-2} - 19f_{i-3}]. \tag{6.96}$$

These methods are heavily used in predictor–corrector methods, along with an Adams–Bashforth predictor of the same order. Computer Problems 9 and 10 ask for MATLAB code to implement this idea.

▶ **ADDITIONAL EXAMPLES**

1. Use the matrix formulation for implicit methods to derive the Adams–Moulton Four-Step Method.

2. The initial value problem

$$\begin{cases} y' = (1 + y^2)/2 \\ y(0) = \sqrt{3} \end{cases}$$

has solution $y(t) = \sec(t + \pi/6) + \tan(t + \pi/6)$ on the interval 0, 1. Use the Adams–Bashforth Four-Step Method to construct the solution. How many steps are needed so that $y(1)$ is approximated within 4 correct decimal places?

Solutions for Additional Examples can be found at bit.ly/2AilDF3

6.7 Exercises

Solutions for Exercises numbered in blue can be found at bit.ly/2OtdQNF

1. Apply the Adams–Bashforth Two-Step Method to the IVPs

   (a) $y' = t$ (b) $y' = t^2 y$ (c) $y' = 2(t + 1)y$
   (d) $y' = 5t^4 y$ (e) $y' = 1/y^2$ (f) $y' = t^3/y^2$

   with initial condition $y(0) = 1$. Use step size $h = 1/4$ on the interval 0, 1. Use the Explicit Trapezoid Method to create w_1. Using the correct solution in Exercise 6.1.3, find the global truncation error at $t = 1$.

2. Carry out the steps of Exercise 1 on the IVPs

   (a) $y' = t + y$ (b) $y' = t - y$ (c) $y' = 4t - 2y$

   with initial condition $y(0) = 0$. Use the correct solution from Exercise 6.1.4 to find the global truncation error at $t = 1$.

3. Find a two-step, third-order explicit method. Is the method stable?

4. Find a second-order, two-step explicit method whose characteristic polynomial has a double root at 1.

5. Show that the Implicit Trapezoid Method (6.89) is a second-order method.

6. Explain why the characteristic polynomial of an explicit or implicit s-step method, for $s \geq 2$, must have a root at 1.

7. (a) For which a_1 does there exist a strongly stable second-order, two-step explicit method? (b) Answer the same question for weakly stable such method.

8. Show that the coefficients of the Adams–Moulton Two-Step Implicit Method satisfy (6.92) and that the method is strongly stable.

9. Find the order and stability type for the following two-step implicit methods:

   (a) $w_{i+1} = 3w_i - 2w_{i-1} + \frac{h}{12} 13 f_{i+1} - 20 f_i - 5 f_{i-1}$
   (b) $w_{i+1} = \frac{4}{3} w_i - \frac{1}{3} w_{i-1} + \frac{2}{3} h f_{i+1}$
   (c) $w_{i+1} = \frac{4}{3} w_i - \frac{1}{3} w_{i-1} + \frac{h}{9} 4 f_{i+1} + 4 f_i - 2 f_{i-1}$
   (d) $w_{i+1} = 3w_i - 2w_{i-1} + \frac{h}{12} 7 f_{i+1} - 8 f_i - 11 f_{i-1}$
   (e) $w_{i+1} = 2w_i - w_{i-1} + \frac{h}{2} f_{i+1} - f_{i-1}$

10. Derive the Milne–Simpson Method (6.93) from (6.92), and show that it is fourth order and weakly stable.

11. Find a second-order, two-step implicit method that is weakly stable.

12. The Milne–Simpson Method is a weakly stable fourth-order, two-step implicit method. Are there any weakly stable third-order, two-step implicit methods?

13. Find a weakly stable third-order, three-step explicit method, and verify these properties.

14. (a) Use the matrix formulation to find the conditions (analogous to (6.77)) on a_i, b_i required for a fourth-order, four-step explicit method. (b) Use the conditions to derive the Adams–Bashforth Four-Step Method.

15. (a) Use the matrix formulation to find the conditions on a_i, b_i required for a fourth-order, three-step implicit method. (b) Use the conditions to derive the Adams–Moulton Three-Step Method. (c) Show that the Adams–Moulton Three-Step Method is strongly stable.

16. (a) Use the matrix formulation to find the conditions on a_i, b_i required for a fifth-order, four-step implicit method. (b) Show that the Adams–Moulton Four-Step Method satisfies the conditions. (c) Show that the Adams–Moulton Four-Step Method is strongly stable.

6.7 Computer Problems

Solutions for Computer Problems numbered in blue can be found at bit.ly/2CsRZ1m

1. Adapt the exmultistep.m program to apply the Adams–Bashforth Two-Step Method to the IVPs in Exercise 1. Using step size $h = 0.1$, calculate the approximation on the interval $[0, 1]$. Print a table of the t values, approximations, and global truncation error at each step.

2. Adapt the exmultistep.m program to apply the Adams–Bashforth Two-Step Method to the IVPs in Exercise 2. Using step size $h = 0.1$, calculate the approximation on the interval $[0, 1]$. Print a table of the t values, approximations, and global truncation error at each step.

3. Carry out the steps of Computer Problem 2, using the unstable two-step method (6.81).

4. Carry out the steps of Computer Problem 2, using the Adams–Bashforth Three-Step Method. Use order-four Runge–Kutta to compute w_1 and w_2.

5. Plot the Adams–Bashforth Three-Step Method approximate solution on $[0, 1]$ for the differential equation $y' = 1 + y^2$ and initial condition (a) $y_0 = 0$ (b) $y_0 = 1/2$, along with the exact solution (see Exercise 6.1.7). Use step sizes $h = 0.1$ and 0.05.

6. Plot the Adams–Bashforth Three-Step Method approximate solution on $[0, 1]$ for the differential equation $y' = 1 - y^2$ and initial condition (a) $y_0 = 0$ (b) $y_0 = -1/2$, along with the exact solution (see Exercise 6.1.8). Use step sizes $h = 0.1$ and 0.05.

7. Calculate the Adams–Bashforth Three-Step Method approximate solution on $[0, 4]$ for the differential equation $y' = \sin y$ and initial condition (a) $y_0 = 0$ (b) $y_0 = 100$, using step sizes $h = 0.1 \times 2^{-k}$ for $0 \le k \le 5$. Plot the $k = 0$ and $k = 5$ approximate solutions along with the exact solution (see Exercise 6.1.15), and make a log–log plot of the error as a function of h.

8. Calculate the Adams–Bashforth Three-Step Method approximate solution of the differential equation $y' = \sinh y$ and initial condition (a) $y_0 = 1/4$ on the interval $[0, 2]$ (b) $y_0 = 2$ on the interval $[0, 1/4]$, using step sizes $h = 0.1 \times 2^{-k}$ for $0 \le k \le 5$. Plot the $k = 0$ and $k = 5$ approximate solutions along with the exact solution (see Exercise 6.1.16), and make a log–log plot of the error as a function of h.

9. Change Program 6.8 into a third-order predictor–corrector method, using the Adams–Bashforth Three-Step Method and the Adams–Moulton Two-Step Method with step size 0.05. Plot the approximation and the correct solution of IVP (6.5) on the interval $[0, 5]$.

10. Change Program 6.8 into a fourth-order predictor–corrector method, using the Adams–Bashforth Four-Step Method and the Adams–Moulton Three-Step Method with step size 0.05. Plot the approximation and the correct solution of IVP (6.5) on the interval $[0, 5]$.

Software and Further Reading

Traditional sources for fundamentals on ordinary differential equations are Blanchard et al. [2011], Boyce and DiPrima [2012], Braun [1993], Edwards and Penny [2014], and Kostelich and Armbruster [1997]. Many books teach the basics of ODEs along with ample computational and graphical help; we mention *ODE Architect* [1999] as a good example. The MATLAB codes in Polking [2003] are an excellent way to learn and visualize ODE concepts.

To supplement our tour through one-step and multistep numerical methods for solving ordinary differential equations, there are many intermediate and advanced texts. Henrici [1962] and Gear [1971] are classics. A contemporary MATLAB approach is taken by Shampine et al. [2003]. Other recommended texts are Iserles [1996], Shampine [1994], Ascher and Petzold [1998], Lambert [1991], Dormand [1996], Butcher [1987], and the comprehensive two-volume set Hairer et al. [2011] and Hairer and Wanner [2004].

There is a great deal of sophisticated software available for solving ODEs. Details on the solvers used by MATLAB can be found in Shampine and Reichelt [1997] and Ashino et al. [2000]. Variable-step-size explicit methods of the Runge–Kutta type are usually successful for nonstiff or mildly stiff problems. In addition to Runge–Kutta–Fehlberg and Dormand–Prince, the variant Runge–Kutta–Verner, an order 5/6 method, is often used. For stiff problems, backward-difference methods and extrapolation methods are called for.

The NAG library provides a driver routine D02BJF that runs standard Runge–Kutta steps. The multistep driver is D02CJF, which includes Adams–style programs with error control. For stiff problems, the D02EJF routine is recommended, where the user has an option to specify the Jacobian for faster computation.

The Netlib repository contains a Fortran routine RKF45 for the Runge–Kutta–Fehlberg Method and DVERK for the Runge–Kutta–Verner Method. The Netlib package ODE contains several multistep routines. The routine VODE handles stiff problems.

The collection ODEPACK is a public-domain set of Fortran code implementing ODE solvers, developed at Lawrence Livermore National Laboratory (LLNL). The basic solver LSODE and its variants are suitable for stiff and nonstiff problems. The routines are freely available at the LLNL website `http://www.llnl.gov/CASC/odepack`.

Boundary Value Problems

Underground and undersea pipelines must be desig-ned to withstand pressure from the outside environ-ment. The deeper the pipe, the more expensive a failure due to collapse will be. The oil pipelines con-necting North Sea platforms to the coast lie at a 70-meter depth. The increasing importance of natural gas, and the danger and expense of transportation by ship, may lead to the construction of intercontinental gas pipelines. Mid-Atlantic depths exceed 5 kilometers, where the hydrostatic pressure of 7000 psi will require innovation in pipe materials and construction to avoid buckling.

The theory of pipe buckling is central to a wide ar-ray of applications, from architectural supports to coro-nary stents. Numerical models of buckling are valuable when direct experimentation is expensive and difficult.

Reality Check Reality Check 7 on page 374 represents a cross-sectional slice of a pipe as a circular ring and examines when and how buckling occurs.

Chapter 6 described methods for calculating the solution to an initial value prob-lem (IVP), a differential equation together with initial data, specified at the left end of the solution interval. The methods we proposed were all "marching" techniques—the approximate solution began at the left end and progressed forward in the independent variable t. An equally important set of problems arises when a dif-ferential equation is presented along with boundary data, specified at both ends of the solution interval.

Chapter 7 describes methods for approximating solutions of a boundary value problem (BVP). The methods are of three types. First, shooting methods are pre-sented, a combination of the IVP solvers from Chapter 6 and equation solvers from Chapter 1. Then, finite difference methods are explored, which convert the differential equation and boundary conditions into a system of linear or nonlinear equations to be solved. The final section is focused on collocation methods and the Finite Element Method, which solve the problem by expressing the solution in terms of elementary basis functions.

7.1 SHOOTING METHOD

The first method converts the boundary value problem into an initial value problem by determining the missing initial values that are consistent with the boundary values. Methods that we have already developed in Chapters 1 and 6 can be combined to carry this out.

7.1.1 Solutions of boundary value problems

A general second-order boundary value problem asks for a solution of

$$\begin{cases} y'' = f(t, y, y') \\ y(a) = y_a \\ y(b) = y_b \end{cases} \tag{7.1}$$

on the interval $a \le t \le b$, as shown in Figure 7.1. In Chapter 6, we learned that a differential equation under typical smoothness conditions has infinitely many solutions, and that extra data is needed to pin down a particular solution. In (7.1), the equation is second order, and two extra constraints are needed. They are given as boundary conditions for the solution $y(t)$ at a and b.

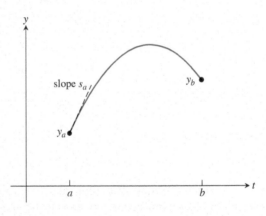

Figure 7.1 Comparison of IVP and BVP. In an initial value problem, the initial value $y_a = y(a)$ and initial slope $s_a = y'(a)$ are specified as part of the problem. In a boundary value problem, boundary values y_a and y_b are specified instead; s_a is unknown.

To aid your intuition, consider a projectile, which satisfies the second-order differential equation $y''(t) = -g$ as it moves, where y is the projectile height and g is the acceleration of gravity. Specifying the initial position and velocity uniquely determines the projectile's motion, as an initial value problem. On the other hand, a time interval $[a, b]$ and the positions $y(a)$ and $y(b)$ could be specified. The latter problem, a boundary value problem, also has a unique solution in this instance.

▶ **EXAMPLE 7.1** Find the maximum height of a projectile that is thrown from the top of a 30-meter tall building and reaches the ground 4 seconds later.

The differential equation is derived from Newton's second law $F = ma$, where the force of gravity is $F = -mg$ and $g = 9.81$ m/sec^2. Let $y(t)$ be the height at time t. The trajectory can be expressed as the solution of the IVP

$$\begin{cases} y'' = -g \\ y(0) = 30 \\ y'(0) = v_0 \end{cases}$$

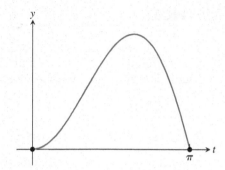

Figure 7.2 Solution of BVP (7.2). Plot of solution $y(t) = t \sin t$ along with boundary values $y(0) = 0$ and $y(\pi) = 0$.

or the BVP

$$\begin{cases} y'' = -g \\ y(0) = 30 \\ y(4) = 0. \end{cases}$$

Since we don't know the initial velocity v_0, we must solve the boundary value problem. Integrating twice gives

$$y(t) = -\frac{1}{2}gt^2 + v_0 t + y_0.$$

Use of the boundary conditions yields

$$30 = y(0) = y_0$$

$$0 = y(4) = -\frac{16}{2}g + 4v_0 + 30,$$

which implies that $v_0 \approx 12.12$ m/sec. The solution trajectory is $y(t) = -\frac{1}{2}gt^2 + 12.12t + 30$. Now it is easy to use calculus to find the maximum of the trajectory, which is about 37.5 m. ◄

▶ **EXAMPLE 7.2** Show that $y(t) = t \sin t$ is a solution of the boundary value problem

$$\begin{cases} y'' = -y + 2\cos t \\ y(0) = 0 \\ y(\pi) = 0. \end{cases} \tag{7.2}$$

The function $y(t) = t \sin t$ is shown in Figure 7.2. This function solves the differential equation because

$$y''(t) = -t \sin t + 2\cos t.$$

Checking the boundary conditions gives $y(0) = 0 \sin 0 = 0$ and $y(\pi) = \pi \sin \pi = 0$. ◄

The existence and uniqueness theory of boundary value problems is more complicated than the corresponding theory for initial value problems. Seemingly reasonable BVPs may have no solutions or infinitely many solutions, a situation that is rare for IVPs. The next two examples show the possibilities for a very simple differential equation.

▶ **EXAMPLE 7.3**   Show that the boundary value problem

$$\begin{cases} y'' = -y \\ y(0) = 0 \\ y(\pi) = 1 \end{cases}$$

has no solutions.

     The differential equation has a two-dimensional family of solutions, generated by the linearly independent solutions $\cos t$ and $\sin t$. All solutions of the equation must have the form $y(t) = a \cos t + b \sin t$. Substituting the first boundary condition, $0 = y(0) = a$ implies that $a = 0$ and $y(t) = b \sin t$. The second boundary condition $1 = y(\pi) = b \sin \pi = 0$ gives a contradiction. There is no solution, and existence fails.   ◀

▶ **EXAMPLE 7.4**   Show that the boundary value problem

$$\begin{cases} y'' = -y \\ y(0) = 0 \\ y(\pi) = 0 \end{cases}$$

has infinitely many solutions.

     Check that $y(t) = k \sin t$ is a solution of the differential equation and satisfies the boundary conditions, for every real number k. In particular, there is no uniqueness of solutions for this example.   ◀

     The next theorem is an example of an existence and uniqueness result, for a common form of boundary value problem.

THEOREM 7.1   For the boundary value problem

$$\begin{cases} y'' = f(t, y, y') \\ y(a) = y_a \\ y(b) = y_b \end{cases}$$

assume that for some $M > 0$, $f(t, y, z)$ satisfies (i) the partial derivatives f_t, f_y, f_z are continuous, (ii) $f_y(t, y, z) > 0$, and (iii) $|f_z(t, y, z)| \le M$ for t in $[a, b]$ and all y, z. Then the BVP has a unique solution $y(t)$.   ■

▶ **EXAMPLE 7.5**   Prove that the boundary value problem

$$\begin{cases} y'' = 4y \\ y(0) = 1 \\ y(1) = 3 \end{cases} \tag{7.3}$$

has a unique solution

     Theorem 7.1 applies with $f(t, y, z) = 4y$. Hypotheses (i)–(iii) are clearly satisfied with $f_y = 4 > 0$, so a unique solution must exist.

     This example is simple enough to solve exactly, yet interesting enough to serve as an example for our BVP solution methods to follow. We can guess two solutions to the differential equation, $y = e^{2t}$ and $y = e^{-2t}$. Since the solutions are not multiples of one another, they are linearly independent; therefore, from elementary differential equations theory, all solutions of the differential equation are linear combinations $c_1 e^{2t} + c_2 e^{-2t}$. The two constants c_1 and c_2 are evaluated by enforcing the two boundary conditions

$$1 = y(0) = c_1 + c_2$$

and

$$3 = y(1) = c_1 e^2 + c_2 e^{-2}.$$

Solving for the constants yields the solution:

$$y(t) = \frac{3 - e^{-2}}{e^2 - e^{-2}} e^{2t} + \frac{e^2 - 3}{e^2 - e^{-2}} e^{-2t}. \tag{7.4}$$

◀

7.1.2 Shooting Method implementation

The Shooting Method solves the BVP (7.1) by finding the IVP that has the same solution. A sequence of IVPs is produced, converging to the correct one. The sequence begins with an initial guess for the slope s_a, provided to go along with the initial value y_a. The IVP that results from this initial slope is solved and compared with the boundary value y_b. By trial and error, the initial slope is improved until the boundary value is matched. To put a more formal structure on this method, define the following function:

$$F(s) = \begin{cases} \text{difference between } y_b \text{ and} \\ y(b), \text{ where } y(t) \text{ is the} \\ \text{solution of the IVP with} \\ y(a) = y_a \text{ and } y'(a) = s. \end{cases}$$

With this definition, the boundary value problem is reduced to solving the equation

$$F(s) = 0, \tag{7.5}$$

as shown in Figure 7.3.

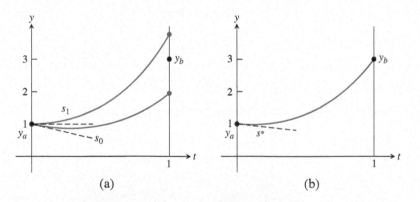

Figure 7.3 The Shooting Method. (a) To solve the BVP, the IVP with initial conditions $y(a) = y_a, y'(a) = s_0$ is solved with initial guess s_0. The value of $F(s_0)$ is $y(b) - y_b$. Then a new s_1 is chosen, and the process is repeated with the goal of solving $F(s) = 0$ for s. (b) The MATLAB command ode45 is used with root s^* to plot the solution of the BVP (7.7).

An equation-solving method from Chapter 1 may now be used to solve the equation. The Bisection Method or a more sophisticated method like Brent's Method may be chosen. Two values of s, called s_0 and s_1, should be found for which $F(s_0)F(s_1) < 0$.

Then s_0 and s_1 bracket a root of (7.5), and a root s^* can be located within the required tolerance by the chosen equation solver. Finally, the solution to the BVP (7.1) can be traced (by an IVP solver from Chapter 6, for example) as the solution to the initial value problem

$$\begin{cases} y'' = f(t, y, y') \\ y(a) = y_a \\ y'(a) = s^*. \end{cases} \tag{7.6}$$

We show a MATLAB implementation of the Shooting Method in the next example.

▶ **EXAMPLE 7.6**   Apply the Shooting Method to the boundary value problem

$$\begin{cases} y'' = 4y \\ y(0) = 1 \\ y(1) = 3. \end{cases} \tag{7.7}$$

Write the differential equation as a first-order system in order to use MATLAB's ode45 IVP solver:

$$y' = v$$
$$v' = 4y. \tag{7.8}$$

Write a function file F.m representing the function in (7.5):

```
function z=F(s)
a=0;b=1;yb=3;
ydot=@(t,y) [y(2);4*y(1)];
[t,y]=ode45(ydot,[a,b],[1,s]);
z=y(end,1)-yb; % end means last entry of solution y
```

Compute $F(-1) \approx -1.05$ and $F(0) \approx 0.76$, as can be viewed in Figure 7.3(a). Therefore, there is a root of F between -1 and 0. Run an equation solver such as bisect.m from Chapter 1 or the MATLAB command fzero with starting interval $[-1, 0]$ to find s within desired precision. For example,

```
>> sstar=fzero(@F,[-1,0])
```

returns approximately -0.4203. (Recall that fzero requires as input the function handle from the function F, which is @F.) Then the solution can be plotted as the solution of an initial value problem (see Figure 7.3(b)). The exact solution of (7.7) is given in (7.4) and $s^* = y'(0) \approx -0.4203$.   ◀

For systems of ordinary differential equations, boundary value problems arise in many forms. To conclude this section, we explore one possible form and refer the reader to the exercises and Reality Check 7 for further examples.

▶ **EXAMPLE 7.7**   Apply the Shooting Method to the boundary value problem

$$\begin{cases} y_1' = (4 - 2y_2)/t^3 \\ y_2' = -e^{y_1} \\ y_1(1) = 0 \\ y_2(2) = 0 \\ t \text{ in } [1, 2]. \end{cases} \tag{7.9}$$

If the initial condition $y_2(1)$ were present, this would be an initial value problem. We will apply the Shooting Method to determine the unknown $y_2(1)$, using MATLAB

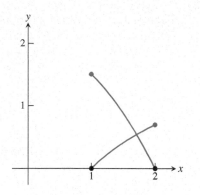

Figure 7.4 Solution of Example 7.7 from the Shooting Method. The curves $y_1(t)$ and $y_2(t)$ are shown. The black circles denote the given boundary data.

routine ode45 as in Example 7.6 to solve the initial value problems. Define the function $F(s)$ to be the end condition $y_2(2)$, where the IVP is solved with initial conditions $y_1(1) = 0$ and $y_2(1) = s$. The objective is to solve $F(s) = 0$.

The solution is bracketed by noting that $F(0) \approx -3.97$ and $F(2) \approx 0.87$. An application of fzero(@F,[0 2]) finds $s^* = 1.5$. Using ode45 with initial values $y_1(1) = 0$ and $y_2(1) = 1.5$ results in the solution depicted in Figure 7.4. The exact solutions are $y_1(t) = \ln t, y_2(t) = 2 - t^2/2$. ◀

▶ **ADDITIONAL EXAMPLES**

1. (a) Use Theorem 7.1 to prove that the boundary value problem has a solution $y(t)$, and that the solution is unique.

$$\begin{cases} y'' = -e^{6-2y} \\ y(1) = 3 \\ y(e) = 4. \end{cases}$$

   (b) Show that the solution is $y(t) = \ln t + 3$.

*2. Implement the shooting method to plot the solution of the boundary value problem

$$\begin{cases} y'' = y^3 + y \\ y(0) = 1 \\ y(2) = -2. \end{cases}$$

📱 **Solutions** for Additional Examples can be found at bit.ly/2ypW3N3 (* example with video solution)

7.1 Exercises

1. Use Theorem 7.1 to prove that the boundary value problems have unique solutions $y(t)$.

(a) $\begin{cases} y'' = y + \frac{2}{3}e^t \\ y(0) = 0 \\ y(1) = \frac{1}{3}e \end{cases}$ (b) $\begin{cases} y'' = (2 + 4t^2)y \\ y(0) = 1 \\ y(1) = e \end{cases}$

(c) $\begin{cases} y'' = 4y + (1 - 4t)e^{-2t} \\ y(0) = 0 \\ y(1) = \frac{1}{2}e^{-2} \end{cases}$ (d) $\begin{cases} y'' = y + 1/t + 2(y - e^t)^3 \\ y(1) = e - 1 \\ y(2) = e^2 - \frac{1}{2} \end{cases}$

2. Show that the solutions to the BVPs in Exercise 1 are (a) $y = \frac{1}{3}te^t$ (b) $y = e^{t^2}$ (c) $y = \frac{1}{2}t^2e^{-2t}$ (d) $y = e^t - 1/t$.

3. Consider the BVP $\begin{cases} y'' = cy \\ y(a) = y_a \\ y(b) = y_b \end{cases}$ where $c > 0, a < b$.

   (a) Show that there exists a unique solution $y(t)$ on a, b. (b) Find a solution of form $y(t) = c_1e^{d_1t} + c_2e^{d_2t}$ for some c_1, c_2, d_1, d_2.

4. Consider the BVP $\begin{cases} y'' = -cy \\ y(0) = 0 \\ y(b) = 0 \end{cases}$ where $c > 0$.

   For each c, find $b > 0$ such that the BVP has at least two different solutions $y_1(t)$ and $y_2(t)$. Exhibit the solutions.

5. (a) For any real number d, prove that the BVP

$$\begin{cases} y'' = -e^{2d-2y} \\ y(1) = d \\ y(e) = d + 1 \end{cases}$$

   has a unique solution on $1, e$. (b) Show that $y(t) = \ln t + d$ is the solution.

6. Express, as the solution of a second-order boundary value problem, the height of a projectile that is thrown from the top of a 60-meter tall building and takes 5 seconds to reach the ground. Then solve the boundary value problem and find the maximum height reached by the projectile.

7. Show that the solutions to the linear BVPs

   (a) $\begin{cases} y'' = -y + 2\cos t \\ y(0) = 0 \\ y(\frac{\pi}{2}) = \frac{\pi}{2} \end{cases}$
   (b) $\begin{cases} y'' = 2 - 4y \\ y(0) = 0 \\ y(\frac{\pi}{2}) = 1 \end{cases}$

   are (a) $y = t\sin t$, (b) $y = \sin^2 t$, respectively.

8. Show that solutions to the BVPs

   (a) $\begin{cases} y'' = \frac{3}{2}y^2 \\ y(1) = 4 \\ y(2) = 1 \end{cases}$
   (b) $\begin{cases} y'' = 2yy' \\ y(0) = 0 \\ y(\frac{\pi}{4}) = 1 \end{cases}$
   (c) $\begin{cases} y'' = -e^{-2y} \\ y(1) = 0 \\ y(e) = 1 \end{cases}$
   (d) $\begin{cases} y'' = 6y^{\frac{1}{3}} \\ y(1) = 1 \\ y(2) = 8 \end{cases}$

   are (a) $y = 4t^{-2}$, (b) $y = \tan t$, (c) $y = \ln t$, (d) $y = t^3$, respectively.

9. Consider the boundary value problem

$$\begin{cases} y'' = -4y \\ y(a) = y_a \\ y(b) = y_b \end{cases}$$

   (a) Find two linearly independent solutions to the differential equation. (b) Assume that $a = 0$ and $b = \pi$. What conditions on y_a, y_b must be satisfied in order for a solution to exist? (c) Same question as (b), for $b = \pi/2$. (d) Same question as (b), for $b = \pi/4$.

7.1 Computer Problems

1. Apply the Shooting Method to the linear BVPs. Begin by finding an interval s_0, s_1 that brackets a solution. Use the MATLAB command `fzero` or the Bisection Method to find the solution. Plot the approximate solution on the specified interval.

   (a) $\begin{cases} y'' = y + \frac{2}{3}e^t \\ y(0) = 0 \\ y(1) = \frac{1}{3}e \end{cases}$
   (b) $\begin{cases} y'' = (2 + 4t^2)y \\ y(0) = 1 \\ y(1) = e \end{cases}$

2. Carry out the steps of Computer Problem 1 for the BVPs.

(a) $\begin{cases} 9y'' + \pi^2 y = 0 \\ y(0) = -1 \\ y\left(\frac{3}{2}\right) = 3 \end{cases}$ (b) $\begin{cases} y'' = 3y - 2y' \\ y(0) = e^3 \\ y(1) = 1 \end{cases}$

3. Apply the Shooting Method to the nonlinear BVPs. Find a bracketing interval $[s_0, s_1]$ and apply an equation solver to find and plot the solution.

(a) $\begin{cases} y'' = 18y^2 \\ y(1) = \frac{1}{3} \\ y(2) = \frac{1}{12} \end{cases}$ (b) $\begin{cases} y'' = 2e^{-2y}(1 - t^2) \\ y(0) = 0 \\ y(1) = \ln 2 \end{cases}$

4. Carry out the steps of Computer Problem 3 for the nonlinear BVPs.

(a) $\begin{cases} y'' = e^y \\ y(0) = 1 \\ y(1) = 3 \end{cases}$ (b) $\begin{cases} y'' = \sin y' \\ y(0) = 1 \\ y(1) = -1 \end{cases}$

5. Apply the Shooting Method to the nonlinear systems of boundary value problems. Follow the method of Example 7.7.

(a) $\begin{cases} y_1' = 1/y_2 \\ y_2' = t + \tan y_1 \\ y_1(0) = 0 \\ y_2(1) = 2 \end{cases}$ (b) $\begin{cases} y_1' = y_1 - 3y_1 y_2 \\ y_2' = -6(ty_2 + \ln y_1) \\ y_1(0) = 1 \\ y_2(1) = -\frac{2}{3} \end{cases}$

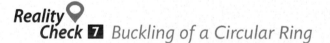

Reality Check **7** *Buckling of a Circular Ring*

Boundary value problems are natural models for structure calculations. A system of seven differential equations serves as a model for a circular ring with compressibility c, under hydrostatic pressure p coming from all directions. The model will be nondimensionalized for simplicity, and we will assume that the ring has radius 1 with horizontal and vertical symmetry in the absence of external pressure. Although simplified, the model is useful for the study of the phenomenon of **buckling**, or collapse of the circular ring shape. This example and many other structural boundary value problems can be found in Huddleston [2000].

The model accounts for only the **upper left quarter of the ring**—the rest can be filled in by the symmetry assumption. The independent variable s represents arc length along the original centerline of the ring, which goes from $s = 0$ to $s = \pi/2$. The dependent variables at the point specified by arc length s are as follows:

$y_1(s) =$ angle of centerline with respect to horizontal

$y_2(s) = x$-coordinate

$y_3(s) = y$-coordinate

$y_4(s) =$ arc length along deformed centerline

$y_5(s) =$ internal axial force

$y_6(s) =$ internal normal force

$y_7(s) =$ bending moment.

Figure 7.5(a) shows the ring and the first four variables. The boundary value problem (see, for example, Huddleston [2000]) is

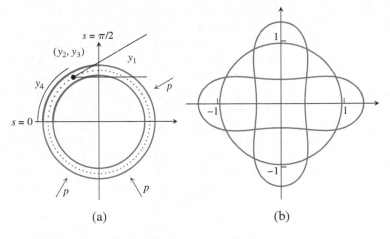

$s = \pi/2$
(y_2, y_3)
y_1
y_4
p
$s = 0$
p p

(a)   (b)

Figure 7.5 Schematics for Buckling Ring. (a) The s variable represents arc length along the dotted centerline of the top left quarter of the ring. (b) Three different solutions for the BVP with parameters $c = 0.01$, $p = 3.8$. The two buckled solutions are stable.

$$y_1' = -1 - cy_5 + (c+1)y_7 \qquad\qquad y_1(0) = \tfrac{\pi}{2} \qquad y_1(\tfrac{\pi}{2}) = 0$$
$$y_2' = (1 + c(y_5 - y_7))\cos y_1 \qquad\qquad\qquad\qquad\quad y_2(\tfrac{\pi}{2}) = 0$$
$$y_3' = (1 + c(y_5 - y_7))\sin y_1 \qquad\qquad y_3(0) = 0$$
$$y_4' = 1 + c(y_5 - y_7) \qquad\qquad\qquad\quad y_4(0) = 0$$
$$y_5' = -y_6(-1 - cy_5 + (c+1)y_7)$$
$$y_6' = y_7 y_5 - (1 + c(y_5 - y_7))(y_5 + p) \qquad y_6(0) = 0 \qquad y_6(\tfrac{\pi}{2}) = 0$$
$$y_7' = (1 + c(y_5 - y_7))y_6.$$

Under no pressure ($p = 0$), note that $y_1 = \pi/2 - s$, $(y_2, y_3) = (-\cos s, \sin s)$, $y_4 = s$, $y_5 = y_6 = y_7 = 0$ is a solution. This solution is a perfect quarter-circle, which corresponds to a perfectly circular ring with the symmetries.

In fact, the following circular solution to the boundary value problem exists for any choice of parameters c and p:

$$y_1(s) = \frac{\pi}{2} - s$$
$$y_2(s) = \frac{c+1}{cp + c + 1}(-\cos s)$$
$$y_3(s) = \frac{c+1}{cp + c + 1}\sin s$$
$$y_4(s) = \frac{c+1}{cp + c + 1}s$$
$$y_5(s) = -\frac{c+1}{cp + c + 1}p$$
$$y_6(s) = 0$$
$$y_7(s) = -\frac{cp}{cp + c + 1}. \tag{7.10}$$

As pressure increases from zero, the radius of the circle decreases. As the pressure parameter p is increased further, there is a **bifurcation**, or change of possible states, of the ring. The circular shape of the ring remains mathematically possible, but unstable, meaning that small perturbations cause the ring to move to another possible configuration (solution of the BVP) that is stable.

For applied pressure p below the bifurcation point, or **critical pressure** p_c, only solution (7.10) exists. For $p > p_c$, three different solutions of the BVP exist, shown in Figure 7.5(b). Beyond critical pressure, the role of the circular ring as an unstable state is similar to that of the inverted pendulum (Computer Problem 6.3.6) or the bridge without torsion in Reality Check 6.

The critical pressure depends on the compressibility of the ring. The smaller the parameter c, the less compressible the ring is, and the lower the critical pressure at which it changes shape instead of compressing in original shape. Your job is to use the Shooting Method paired with Broyden's Method to find the critical pressure p_c and the resulting buckled shapes obtained by the ring.

Suggested activities:

1. Verify that (7.10) is a solution of the BVP for each compressibility c and pressure p.

2. Set compressibility to the moderate value $c = 0.01$. Solve the BVP by the Shooting Method for pressures $p = 0$ and 3. The function F in the Shooting Method should use the three missing initial values $(y_2(0), y_5(0), y_7(0))$ as input and the three final values $(y_1(\pi/2), y_2(\pi/2), y_6(\pi/2))$ as output. The multivariate solver Broyden II from Chapter 2 can be used to solve for the roots of F. Compare with the correct solution (7.10). Note that, for both values of p, various initial conditions for Broyden's Method all result in the same solution trajectory. How much does the radius decrease when p increases from 0 to 3?

3. Plot the solutions in Step 2. The curve $(y_2(s), y_3(s))$ represents the upper left quarter of the ring. Use the horizontal and vertical symmetry to plot the entire ring.

4. Change pressure to $p = 3.5$, and resolve the BVP. Note that the solution obtained depends on the initial condition used for Broyden's Method. Plot each different solution found.

5. Find the critical pressure p_c for the compressibility $c = 0.01$, accurate to two decimal places. For $p > p_c$, there are three different solutions. For $p < p_c$, there is only one solution (7.10).

6. Carry out Step 5 for the reduced compressibility $c = 0.001$. The ring now is more brittle. Is the change in p_c for the reduced compressibility case consistent with your intuition?

7. Carry out Step 5 for increased compressibility $c = 0.05$.

7.2 FINITE DIFFERENCE METHODS

The fundamental idea behind finite difference methods is to replace derivatives in the differential equation by discrete approximations, and evaluate on a grid to develop a system of equations. The approach of discretizing the differential equation will also be used in Chapter 8 on PDEs.

7.2.1 Linear boundary value problems

Let $y(t)$ be a function with at least four continuous derivatives. In Chapter 5, we developed discrete approximations for the first derivative

$$y'(t) = \frac{y(t+h) - y(t-h)}{2h} - \frac{h^2}{6}y'''(c) \qquad (7.11)$$

and for the second derivative

$$y''(t) = \frac{y(t+h) - 2y(t) + y(t-h)}{h^2} + \frac{h^2}{12}f''''(c). \qquad (7.12)$$

Both are accurate up to an error proportional to h^2.

The Finite Difference Method consists of replacing the derivatives in the differential equation with the discrete versions, and solving the resulting simpler, algebraic equations for approximations w_i to the correct values y_i, $y_i = y(t_i)$, as shown in Figure 7.6. Here we assume that $t_0 < t_1 < \ldots < t_{n+1}$ is an evenly spaced partition on the t-axis with spacing $h = t_{i+1} - t_i$. The boundary conditions are substituted in the system of equations where they are needed.

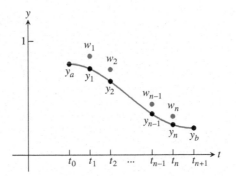

Figure 7.6 The Finite Difference Method for BVPs. Approximations $w_i, i = 1, \ldots, n$ for the correct values y_i at discrete points t_i are calculated by solving a linear system of equations.

After the substitutions, there are two possible situations. If the original boundary value problem was linear, then the resulting system of equations is linear and can be solved by Gaussian elimination or iterative methods. If the original problem was nonlinear, then the algebraic system is a system of nonlinear equations, requiring more sophisticated approaches. We begin with a linear example.

▶ **EXAMPLE 7.8** Solve the BVP (7.7)

$$\begin{cases} y'' = 4y \\ y(0) = 1 \\ y(1) = 3, \end{cases}$$

using finite differences.

Consider the discrete form of the differential equation $y'' = 4y$, using the centered-difference form for the second derivative. The finite difference version at t_i is

$$\frac{w_{i+1} - 2w_i + w_{i-1}}{h^2} - 4w_i = 0$$

or equivalently

$$w_{i-1} + (-4h^2 - 2)w_i + w_{i+1} = 0.$$

For $n = 3$, the interval size is $h = 1/(n+1) = 1/4$ and there are three equations. Inserting the boundary conditions $w_0 = 1$ and $w_4 = 3$, we are left with the following system to solve for w_1, w_2, w_3:

$$1 + (-4h^2 - 2)w_1 + w_2 = 0$$
$$w_1 + (-4h^2 - 2)w_2 + w_3 = 0$$
$$w_2 + (-4h^2 - 2)w_3 + 3 = 0.$$

Substituting for h yields the tridiagonal matrix equation

$$\begin{bmatrix} -\frac{9}{4} & 1 & 0 \\ 1 & -\frac{9}{4} & 1 \\ 0 & 1 & -\frac{9}{4} \end{bmatrix} \begin{bmatrix} w_1 \\ w_2 \\ w_3 \end{bmatrix} = \begin{bmatrix} -1 \\ 0 \\ -3 \end{bmatrix}.$$

Solving this system by Gaussian elimination gives the approximate solution values 1.0249, 1.3061, 1.9138 at three points. The following table shows the approximate values w_i of the solution at t_i compared with the correct solution values y_i (note that the boundary values, w_0 and w_4, are known ahead of time and are not computed):

i	t_i	w_i	y_i
0	0.00	1.0000	1.0000
1	0.25	1.0249	1.0181
2	0.50	1.3061	1.2961
3	0.75	1.9138	1.9049
4	1.00	3.0000	3.0000

The differences are on the order of 10^{-2}. To get even smaller errors, we need to use larger n. In general, $h = (b - a)/(n + 1) = 1/(n + 1)$, and the tridiagonal matrix equation is

$$\begin{bmatrix} -4h^2 - 2 & 1 & 0 & \cdots & 0 & 0 & 0 \\ 1 & -4h^2 - 2 & \ddots & & 0 & 0 & 0 \\ 0 & 1 & \ddots & \ddots & 0 & 0 & 0 \\ \vdots & & \ddots & \ddots & \ddots & & \vdots \\ 0 & 0 & 0 & \ddots & \ddots & 1 & 0 \\ 0 & 0 & 0 & & \ddots & -4h^2 - 2 & 1 \\ 0 & 0 & 0 & \cdots & 0 & 1 & -4h^2 - 2 \end{bmatrix} \begin{bmatrix} w_1 \\ w_2 \\ w_3 \\ \vdots \\ w_{n-1} \\ w_n \end{bmatrix} = \begin{bmatrix} -1 \\ 0 \\ 0 \\ \vdots \\ 0 \\ 0 \\ -3 \end{bmatrix}.$$

As we add more subintervals, we expect the approximations w_i to be closer to the corresponding y_i. ◀

The potential sources of error in the Finite Difference Method are the truncation error made by the centered-difference formulas and the error made in solving the system of equations. For step sizes h greater than the square root of machine epsilon, the former error dominates. This error is $O(h^2)$, so we expect the error to decrease as $O(n^{-2})$ as the number of subintervals $n + 1$ gets large.

We test this expectation for the problem (7.7). Figure 7.7 shows the magnitude of the error E of the solution at $t = 3/4$, for various numbers of subintervals $n + 1$. On a log–log plot, the error as a function of number of subintervals is essentially a straight line with slope -2, meaning that $\log E \approx a + b \log n$, where $b = -2$; in other words, the error $E \approx Kn^{-2}$, as was expected.

7.2.2 Nonlinear boundary value problems

When the Finite Difference Method is applied to a nonlinear differential equation, the result is a system of nonlinear algebraic equations to solve. In Chapter 2, we used

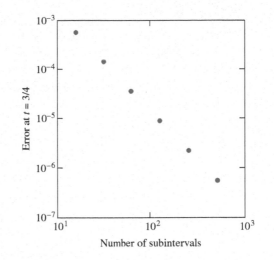

Figure 7.7 Convergence of the Finite Difference Method. The error $|w_i - y_i|$ at $t_i = 3/4$ in Example 7.8 is graphed versus the number of subintervals n. The slope is -2, confirming that the error is $O(n^{-2}) = O(h^2)$.

Multivariate Newton's Method to solve such systems. We demonstrate the use of Newton's Method to approximate the following nonlinear boundary value problem:

▶ **EXAMPLE 7.9**　Solve the nonlinear BVP

$$\begin{cases} y'' = y - y^2 \\ y(0) = 1 \\ y(1) = 4 \end{cases} \tag{7.13}$$

by finite differences.

　　We use the same uniform partition as in Example 7.8. The discretized form of the differential equation at t_i is

$$\frac{w_{i+1} - 2w_i + w_{i\ 1}}{h^2} - w_i + w_i^2 = 0$$

or

$$w_{i-1} - (2 + h^2)w_i + h^2 w_i^2 + w_{i+1} = 0$$

for $2 \le i \le n - 1$, together with the first and last equations

$$y_a - (2 + h^2)w_1 + h^2 w_1^2 + w_2 = 0$$
$$w_{n-1} - (2 + h^2)w_n + h^2 w_n^2 + y_b = 0$$

which carry the boundary condition information.

SPOTLIGHT ON

Convergence　　Figure 7.7 illustrates the second-order convergence of the Finite Difference Method. This follows from the use of the second-order formulas (7.11) and (7.12). Knowledge of the order allows us to apply extrapolation, as introduced in Chapter 5. For any fixed t and step size h, the approximation $w_h(t)$ from the Finite Difference Method is second order in h and can be extrapolated with a simple formula. Computer Problems 7 and 8 explore this opportunity to speed convergence.

Solving the discretized version of the boundary value problem means solving $F(w) = 0$, which we carry out by Newton's Method. Multivariate Newton's Method is the iteration $w^{k+1} = w^k - DF(w^k)^{-1}F(w^k)$. As usual, it is best to carry out the iteration by solving for $\Delta w = w^{k+1} - w^k$ in the equation $DF(w^k)\Delta w = -F(w^k)$.

The function $F(w)$ is given by

$$
F \begin{bmatrix} w_1 \\ w_2 \\ \vdots \\ w_{n-1} \\ w_n \end{bmatrix} = \begin{bmatrix} y_a - (2 + h^2)w_1 + h^2 w_1^2 + w_2 \\ w_1 - (2 + h^2)w_2 + h^2 w_2^2 + w_3 \\ \vdots \\ w_{n-2} - (2 + h^2)w_{n-1} + h^2 w_{n-1}^2 + w_n \\ w_{n-1} - (2 + h^2)w_n + h^2 w_n^2 + y_b \end{bmatrix},
$$

where $y_a = 1$ and $y_b = 4$. The Jacobian $DF(w)$ of F is

$$
\begin{bmatrix} 2h^2 w_1 - (2 + h^2) & 1 & 0 & \cdots & & 0 \\ 1 & 2h^2 w_2 - (2 + h^2) & \ddots & & \ddots & \vdots \\ 0 & 1 & \ddots & 1 & & 0 \\ \vdots & & \ddots & 2h^2 w_{n-1} - (2 + h^2) & 1 \\ 0 & \cdots & 0 & 1 & 2h^2 w_n - (2 + h^2) \end{bmatrix}.
$$

The ith row of the Jacobian is determined by taking the partial derivative of the ith equation (the ith component of F) with respect to each w_j.

Figure 7.8(a) shows the result of using Multivariate Newton's Method to solve $F(w) = 0$, for $n = 40$. The MATLAB code is given in Program 7.1. Twenty steps of Newton's Method are sufficient to reach convergence within machine precision.

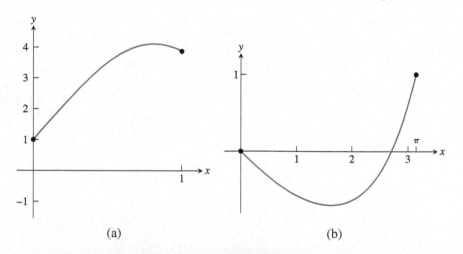

(a)   (b)

Figure 7.8 Solutions of Nonlinear BVPs by the Finite Difference Method.
(a) Solution of Example 7.9 with $n = 40$, after convergence of Newton's Method.
(b) Same for Example 7.10.

```
% Program 7.1 Nonlinear Finite Difference Method for BVP
% Uses Multivariate Newton's Method to solve nonlinear equation
% Inputs: interval inter, boundary values bv, number of steps n
% Output: solution w
% Example usage: w=nlbvpfd([0 1],[1 4],40)
function w=nlbvpfd(inter,bv,n);
```

```
a=inter(1); b=inter(2); ya=bv(1); yb=bv(2);
h=(b-a)/(n+1);                    % h is step size
w=zeros(n,1);                     % initialize solution array w
for i=1:20                       % loop of Newton step
  w=w-jac(w,inter,bv,n)\f(w,inter,bv,n);
end
plot([a a+(1:n)*h b],[ya w' yb]);  % plot w with boundary data

function y=f(w,inter,bv,n)
y=zeros(n,1);h=(inter(2)-inter(1))/(n+1);
y(1)=bv(1)-(2+h^2)*w(1)+h^2*w(1)^2+w(2);
y(n)=w(n-1)-(2+h^2)*w(n)+h^2*w(n)^2+bv(2);
for i=2:n-1
    y(i)=w(i-1)-(2+h^2)*w(i)+h^2*w(i)^2+w(i+1);
end

function a=jac(w,inter,bv,n)
a=zeros(n,n);h=(inter(2)-inter(1))/(n+1);
for i=1:n
  a(i,i)=2*h^2*w(i)-2-h^2;
end
for i=1:n-1
  a(i,i+1)=1;
  a(i+1,i)=1;
end
```

◄

▶ **EXAMPLE 7.10**   Use finite differences to solve the nonlinear boundary value problem

$$\begin{cases} y'' = y' + \cos y \\ y(0) = 0 \\ y(\pi) = 1. \end{cases} \tag{7.14}$$

The discretized form of the differential equation at t_i is

$$\frac{w_{i+1} - 2w_i + w_{i-1}}{h^2} - \frac{w_{i+1} - w_{i-1}}{2h} - \cos(w_i) = 0,$$

or

$$(1 + h/2)w_{i-1} - 2w_i + (1 - h/2)w_{i+1} - h^2 \cos w_i = 0,$$

for $2 \leq i \leq n - 1$, together with the first and last equations,

$$(1 + h/2)y_a - 2w_1 + (1 - h/2)w_2 - h^2 \cos w_1 = 0$$
$$(1 + h/2)w_{n-1} - 2w_n + (1 - h/2)y_b - h^2 \cos w_n = 0,$$

where $y_a = 0$ and $y_b = 1$. The left-hand sides of the n equations form a vector-valued function

$$F(w) = \begin{bmatrix} (1 + h/2)y_a - 2w_1 + (1 - h/2)w_2 - h^2 \cos w_1 \\ \vdots \\ (1 + h/2)w_{i-1} - 2w_i + (1 - h/2)w_{i+1} - h^2 \cos w_i \\ \vdots \\ (1 + h/2)w_{n-1} - 2w_n + (1 - h/2)y_b - h^2 \cos w_n \end{bmatrix}.$$

The Jacobian $DF(w)$ of F is

$$\begin{bmatrix} -2+h^2\sin w_1 & 1-h/2 & 0 & \cdots & & 0 \\ 1+h/2 & -2+h^2\sin w_2 & \ddots & & \ddots & \vdots \\ 0 & 1+h/2 & \ddots & & 1-h/2 & 0 \\ \vdots & & \ddots & \ddots & -2+h^2\sin w_{n-1} & 1-h/2 \\ 0 & & \cdots & 0 & 1+h/2 & -2+h^2\sin w_n \end{bmatrix}.$$

The following code can be inserted into Program 7.1, along with appropriate changes to the boundary condition information, to handle the nonlinear boundary value problem:

```
function y=f(w,inter,bv,n)
  y=zeros(n,1);h=(inter(2)-inter(1))/(n+1);
  y(1)=-2*w(1)+(1+h/2)*bv(1)+(1-h/2)*w(2)-h*h*cos(w(1));
  y(n)=(1+h/2)*w(n-1)-2*w(n)-h*h*cos(w(n))+(1-h/2)*bv(2);
  for j=2:n-1
    y(j)=-2*w(j)+(1+h/2)*w(j-1)+(1-h/2)*w(j+1)-h*h*cos(w(j));
  end

function a=jac(w,inter,bv,n)
  a=zeros(n,n);h=(inter(2)-inter(1))/(n+1);
  for j=1:n
    a(j,j)=-2+h*h*sin(w(j));
  end
  for j=1:n-1
    a(j,j+1)=1-h/2;
    a(j+1,j)=1+h/2;
  end
```

Figure 7.8(b) shows the resulting solution curve $y(t)$. ◄

▶ **ADDITIONAL EXAMPLES**

1. Use finite differences to approximate the solution $y(t)$ to the linear BVP

$$\begin{cases} y'' = (4t^2+6)y \\ y(0) = 0 \\ y(1) = e \end{cases}$$

for $n=9$. Plot the approximate solution together with the exact solution $y = te^{t^2}$. Plot the approximation errors on the interval in a separate semilog plot for $n = 9, 19$, and 39.

2. Use finite differences to approximate the solution $y(t)$ to the nonlinear BVP

$$\begin{cases} y'' = \frac{3}{2}t^2 y^3 \\ y(1) = 2 \\ y(2) = 1/2. \end{cases}$$

Plot the solution for $n=9$ together with the exact solution $y = 2/t^2$. Plot the approximation errors on the interval in a separate semilog plot for $n = 9, 19$, and 39.

⌨ **Solutions** for Additional Examples can be found at bit.ly/2PcjLpE

7.2 Computer Problems

1. Use finite differences to approximate solutions to the linear BVPs for $n = 9, 19$, and 39.

    (a) $\begin{cases} y'' = y + \frac{2}{3}e^t \\ y(0) = 0 \\ y(1) = \frac{1}{3}e \end{cases}$ (b) $\begin{cases} y'' = (2 + 4t^2)y \\ y(0) = 1 \\ y(1) = e \end{cases}$

    Plot the approximate solutions together with the exact solutions (a) $y(t) = te^t/3$ and (b) $y(t) = e^{t^2}$, and display the errors as a function of t in a separate semilog plot.

2. Use finite differences to approximate solutions to the linear BVPs for $n = 9, 19$, and 39.

    (a) $\begin{cases} 9y'' + \pi^2 y = 0 \\ y(0) = -1 \\ y(\frac{3}{2}) = 3 \end{cases}$ (b) $\begin{cases} y'' = 3y - 2y' \\ y(0) = e^3 \\ y(1) = 1 \end{cases}$

    Plot the approximate solutions together with the exact solutions (a) $y(t) = 3\sin\frac{\pi t}{3} - \cos\frac{\pi t}{3}$ and (b) $y(t) = e^{3-3t}$, and display the errors as a function of t in a separate semilog plot.

3. Use finite differences to approximate solutions to the nonlinear boundary value problems for $n = 9, 19$, and 39.

    (a) $\begin{cases} y'' = 18y^2 \\ y(1) = \frac{1}{3} \\ y(2) = \frac{1}{12} \end{cases}$ (b) $\begin{cases} y'' = 2e^{-2y}(1 - t^2) \\ y(0) = 0 \\ y(1) = \ln 2 \end{cases}$

    Plot the approximate solutions together with the exact solutions (a) $y(t) = 1/(3t^2)$ and (b) $y(t) = \ln(t^2 + 1)$, and display the errors as a function of t in a separate semilog plot.

4. Use finite differences to plot solutions to the nonlinear BVPs for $n = 9, 19$, and 39.

    (a) $\begin{cases} y'' = e^y \\ y(0) = 1 \\ y(1) = 3 \end{cases}$ (b) $\begin{cases} y'' = \sin y' \\ y(0) = 1 \\ y(1) = -1 \end{cases}$

5. (a) Find the solution of the BVP $y'' = y$, $y(0) = 0$, $y(1) = 1$ analytically. (b) Implement the finite difference version of the equation, and plot the approximate solution for $n = 15$. (c) Compare the approximation with the exact solution by making a log–log plot of the error at $t = 1/2$ versus n for $n = 2^p - 1$, $p = 2, \ldots, 7$.

6. Solve the nonlinear BVP $4y'' = ty^4$, $y(1) = 2$, $y(2) = 1$ by finite differences. Plot the approximate solution for $n = 15$. Compare your approximation with the exact solution $y(t) = 2/t$ to make a log–log plot of the error at $t = 3/2$ for $n = 2^p - 1$, $p = 2, \ldots, 7$.

7. Extrapolate the approximate solutions in Computer Problem 5. Apply Richardson extrapolation (Section 5.1) to the formula $N(h) = w_h(1/2)$, the finite difference approximation with step size h. How close can extrapolation get to the exact value $y(1/2)$ by using only the approximate values from $h = 1/4, 1/8$, and $1/16$?

8. Extrapolate the approximate solutions in Computer Problem 6. Use the formula $N(h) = w_h(3/2)$, the finite difference approximation with step size h. How close can extrapolation get to the exact value $y(3/2)$ by using only the approximate values from $h = 1/4, 1/8$, and $1/16$?

9. Solve the nonlinear boundary value problem $y'' = \sin y$, $y(0) = 1$, $y(\pi) = 0$ by finite differences. Plot approximations for $n = 9, 19$, and 39.

10. Use finite differences to solve the equation

$$\begin{cases} y'' = 10y(1-y) \\ y(0) = 0 \\ y(1) = 1 \end{cases}.$$

Plot approximations for $n = 9, 19$, and 39.

11. Solve

$$\begin{cases} y'' = cy(1-y) \\ y(0) = 0 \\ y(1/2) = 1/4 \\ y(1) = 1 \end{cases}$$

for $c > 0$, within three correct decimal places. (Hint: Consider the BVP formed by fixing two of the three boundary conditions. Let $G(c)$ be the discrepancy at the third boundary condition, and use the Bisection Method to solve $G(c) = 0$.)

7.3 COLLOCATION AND THE FINITE ELEMENT METHOD

Like the Finite Difference Method, the idea behind Collocation and the Finite Element Method is to reduce the boundary value problem to a set of solvable algebraic equations. However, instead of discretizing the differential equation by replacing derivatives with finite differences, the solution is given a functional form whose parameters are fit by the method.

Choose a set of basis functions $\phi_1(t), \ldots, \phi_n(t)$, which may be polynomials, trigonometric functions, splines, or other simple functions. Then consider the possible solution

$$y(t) = c_1\phi_1(t) + \cdots + c_n\phi_n(t). \tag{7.15}$$

Finding an approximate solution reduces to determining values for the c_i. We will consider two different ways to find the coefficients.

The collocation approach is to substitute (7.15) into the boundary value problem and evaluate at a grid of points. This method is straightforward, reducing the problem to solving a system of equations in c_i, linear if the original problem was linear. Each point gives an equation, and solving them for c_i is a type of interpolation.

A second approach, the Finite Element Method, proceeds by treating the fitting as a least squares problem instead of interpolation. The Galerkin projection is employed to minimize the difference between (7.15) and the exact solution in the sense of squared error. The Finite Element Method is revisited in Chapter 8 to solve boundary value problems in partial differential equations.

7.3.1 Collocation

Consider the BVP

$$\begin{cases} y'' = f(t, y, y') \\ y(a) = y_a \\ y(b) = y_b. \end{cases} \tag{7.16}$$

Choose n points, beginning and ending with the boundary points a and b, say,

$$a = t_1 < t_2 < \cdots < t_n = b. \tag{7.17}$$

The Collocation Method works by substituting the candidate solution (7.15) into the differential equation (7.16) and evaluating the differential equation at the points (7.17) to get n equations in the n unknowns $c_1, \ldots, c_n$.

To start as simply as possible, we choose the basis functions $\phi_j(t) = t^{j-1}$ for $1 \leq j \leq n$. The solution will be of form

$$y(t) = \sum_{j=1}^{n} c_j \phi_j(t) = \sum_{j=1}^{n} c_j t^{j-1}. \tag{7.18}$$

We will write n equations in the n unknowns $c_1, \ldots, c_n$. The first and last are the boundary conditions:

$$i = 1: \sum_{j=1}^{n} c_j a^{j-1} = y(a)$$

$$i = n: \sum_{j=1}^{n} c_j b^{j-1} = y(b).$$

The remaining $n - 2$ equations come from the differential equation evaluated at t_i for $2 \leq i \leq n - 1$. The differential equation $y'' = f(t, y, y')$ applied to $y(t) = \sum_{j=1}^{n} c_j t^{j-1}$ is

$$\sum_{j=1}^{n} (j-1)(j-2) c_j t^{j-3} = f\left(t, \sum_{j=1}^{n} c_j t^{j-1}, \sum_{j=1}^{n} c_j (j-1) t^{j-2}\right). \tag{7.19}$$

Evaluating at t_i for each i yields n equations to solve for the c_i. If the differential equation is linear, then the equations in the c_i will be linear and can be readily solved. We illustrate the approach with the following example.

▶ **EXAMPLE 7.11** Solve the boundary value problem

$$\begin{cases} y'' = 4y \\ y(0) = 1 \\ y(1) = 3 \end{cases}$$

by the Collocation Method.

The first and last equations are the boundary conditions

$$c_1 = \sum_{j=1}^{n} c_j \phi_j(0) = y(0) = 1$$

$$c_1 + \cdots + c_n = \sum_{j=1}^{n} c_j \phi_j(1) = y(1) = 3.$$

The other $n - 2$ equations come from (7.19), which has the form

$$\sum_{j=1}^{n} (j-1)(j-2) c_j t^{j-3} - 4 \sum_{j=1}^{n} c_j t^{j-1} = 0.$$

Evaluating at t_i for each i yields

$$\sum_{j=1}^{n} [(j-1)(j-2) t_i^{j-3} - 4 t_i^{j-1}] c_j = 0.$$

The n equations form a linear system $Ac = g$, where the coefficient matrix A is defined by

$$A_{ij} = \begin{cases} 1 & 0 & 0 & \dots & 0 & \text{row } i = 1 \\ (j-1)(j-2)t_i^{j-3} - 4t_i^{j-1} & & & & & \text{rows } i = 2 \text{ through } n-1 \\ 1 & 1 & 1 & \dots & 1 & \text{row } i = n \end{cases}$$

and $g = (1, 0, 0, \dots, 0, 3)^T$. It is common to use the evenly spaced grid points

$$t_i = a + \frac{i-1}{n-1}(b-a) = \frac{i-1}{n-1}.$$

After solving for the c_j, we obtain the approximate solution $y(t) = \sum c_j t^{j-1}$.
For $n = 2$ the system $Ac = g$ is

$$\begin{bmatrix} 1 & 0 \\ 1 & 1 \end{bmatrix} \begin{bmatrix} c_1 \\ c_2 \end{bmatrix} = \begin{bmatrix} 1 \\ 3 \end{bmatrix},$$

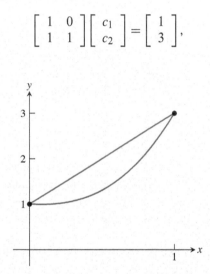

Figure 7.9 Solutions of the linear BVP of Example 7.11 by the Collocation Method. Solutions with $n = 2$ (upper curve) and $n = 4$ (lower) are shown.

and the solution is $c = [1, 2]^T$. The approximate solution (7.18) is the straight line $y(t) = c_1 + c_2 t = 1 + 2t$. The computation for $n = 4$ yields the approximate solution $y(t) \approx 1 - 0.1886t + 1.0273t^2 + 1.1613t^3$. The solutions for $n = 2$ and $n = 4$ are plotted in Figure 7.9. Already for $n = 4$ the approximation is very close to the exact solution (7.4) shown in Figure 7.3(b). More precision can be achieved by increasing n. ◄

The equations to be solved for c_i in Example 7.11 are linear because the differential equation is linear. Nonlinear boundary value problems can be solved by collocation in a similar way. Newton's Method is used to solve the resulting nonlinear system of equations, exactly as in the finite difference approach.

Although we have illustrated the use of collocation with monomial basis functions for simplicity, there are many better choices. Polynomial bases are generally not recommended. Since collocation is essentially doing interpolation of the solution, the use of polynomial basis functions makes the method susceptible to the Runge phenomenon (Chapter 3). The fact that the monomial basis elements t^j are not orthogonal to one another as functions makes the coefficient matrix of the linear equations ill-conditioned when n is large. Using the roots of Chebyshev polynomials as evaluation points, rather than evenly spaced points, improves the conditioning.

The choice of trigonometric functions as basis functions in collocation leads to Fourier analysis and **spectral methods**, which are heavily used for both boundary value problems and partial differential equations. This is a "global" approach, where the basis functions are nonzero over a large range of t, but have good orthogonality properties. We will study discrete Fourier approximations in Chapter 10.

7.3.2 Finite Elements and the Galerkin Method

The choice of splines as basis functions leads to the **Finite Element Method**. In this approach, each basis function is nonzero only over a short range of t. Finite Element Methods are heavily used for BVPs and PDEs in higher dimensions, especially when irregular boundaries make parametrization by standard basis functions inconvenient.

In collocation, we assumed a functional form $y(t) = \sum c_i \phi_i(t)$ and solved for the coefficients c_i by forcing the solution to satisfy the boundary conditions and exactly satisfy the differential equation at discrete points. On the other hand, the Galerkin approach minimizes the squared error of the differential equation along the solution. This leads to a different system of equations for the c_i.

The finite element approach to the BVP

$$\begin{cases} y'' = f(t, y, y') \\ y(a) = y_a \\ y(b) = y_b. \end{cases}$$

is to choose the approximate solution y so that the **residual** $r = y'' - f$, the difference in the two sides of the differential equation, is as small as possible. In analogy with the least squares methods of Chapter 4, this is accomplished by choosing y to make the residual orthogonal to the vector space of potential solutions.

For an interval $[a, b]$, define the vector space of square integrable functions

$$L^2[a, b] = \left\{ \text{functions } y(t) \text{ on } [a, b] \,\middle|\, \int_a^b y(t)^2 \, dt \text{ exists and is finite} \right\}.$$

The L^2 function space has an **inner product**

$$\langle y_1, y_2 \rangle = \int_a^b y_1(t) y_2(t) \, dt$$

that has the usual properties:

1. $\langle y_1, y_1 \rangle \geq 0$;

2. $\langle \alpha y_1 + \beta y_2, z \rangle = \alpha \langle y_1, z \rangle + \beta \langle y_2, z \rangle$ for scalars α, β;

3. $\langle y_1, y_2 \rangle = \langle y_2, y_1 \rangle$.

Two functions y_1 and y_2 are **orthogonal** in $L^2[a, b]$ if $\langle y_1, y_2 \rangle = 0$. Since $L^2[a, b]$ is an infinite-dimensional vector space, we cannot make the residual $r = y'' - f$ orthogonal to all of $L^2[a, b]$ by a finite computation. However, we can choose a basis that spans as much of L^2 as possible with the available computational resources. Let the set of $n + 2$ basis functions be denoted by $\phi_0(t), \ldots, \phi_{n+1}(t)$. We will specify these later.

The Galerkin Method consists of two main ideas. The first is to minimize r by forcing it to be orthogonal to the basis functions, in the sense of the L^2 inner product. This means forcing $\int_a^b (y'' - f)\phi_i \, dt = 0$, or

$$\int_a^b y''(t)\phi_i(t) \, dt = \int_a^b f(t, y, y')\phi_i(t) \, dt \qquad (7.20)$$

for each $0 \leq i \leq n+1$. The form (7.20) is called the **weak form** of the boundary value problem.

The second idea of Galerkin is to use integration by parts to eliminate the second derivatives. Note that

$$\int_a^b y''(t)\phi_i(t)\,dt = \phi_i(t)y'(t)|_a^b - \int_a^b y'(t)\phi_i'(t)\,dt$$

$$= \phi_i(b)y'(b) - \phi_i(a)y'(a) - \int_a^b y'(t)\phi_i'(t)\,dt. \qquad (7.21)$$

Using (7.20) and (7.21) together gives a set of equations

$$\int_a^b f(t, y, y')\phi_i(t)\,dt = \phi_i(b)y'(b) - \phi_i(a)y'(a) - \int_a^b y'(t)\phi_i'(t)\,dt \qquad (7.22)$$

for each i that can be solved for the c_i in the functional form

$$y(t) = \sum_{i=0}^{n+1} c_i\phi_i(t). \qquad (7.23)$$

The two ideas of Galerkin make it convenient to use extremely simple functions as the finite elements $\phi_i(t)$. We will introduce piecewise-linear B-splines only and direct the reader to the literature for more elaborate choices.

Start with a grid $t_0 < t_1 < \cdots < t_n < t_{n+1}$ of points on the t axis. For $i = 1, \ldots, n$ define

$$\phi_i(t) = \begin{cases} \dfrac{t - t_{i-1}}{t_i - t_{i-1}} & \text{for } t_{i-1} < t \leq t_i \\ \dfrac{t_{i+1} - t}{t_{i+1} - t_i} & \text{for } t_i < t < t_{i+1}. \\ 0 & \text{otherwise} \end{cases}$$

Also define

$$\phi_0(t) = \begin{cases} \dfrac{t_1 - t}{t_1 - t_0} & \text{for } t_0 \leq t < t_1 \\ 0 & \text{otherwise} \end{cases} \quad \text{and} \quad \phi_{n+1}(t) = \begin{cases} \dfrac{t - t_n}{t_{n+1} - t_n} & \text{for } t_n < t \leq t_{n+1} \\ 0 & \text{otherwise} \end{cases}.$$

The piecewise-linear "tent" functions ϕ_i, shown in Figure 7.10, satisfy the following interesting property:

$$\phi_i(t_j) = \begin{cases} 1 & \text{if } i = j \\ 0 & \text{if } i \neq j. \end{cases} \qquad (7.24)$$

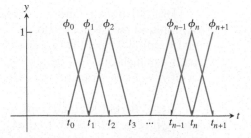

Figure 7.10 Piecewise-linear B-splines used as finite elements. Each $\phi_i(t)$, for $1 \leq i \leq n$, has support on the interval from t_{i-1} to t_{i+1}.

For a set of data points (t_i, c_i), define the **piecewise-linear B-spline**

$$S(t) = \sum_{i=0}^{n+1} c_i \phi_i(t).$$

It follows immediately from (7.24) that $S(t_j) = \sum_{i=0}^{n+1} c_i \phi_i(t_j) = c_j$. Therefore, $S(t)$ is a piecewise-linear function that interpolates the data points (t_i, c_i). In other words, the y-coordinates are the coefficients! This will simplify the interpretation of the solution (7.23). The c_i are not only the coefficients, but also the solution values at the grid points t_i.

SPOTLIGHT ON

Orthogonality We saw in Chapter 4 that the distance from a point to a plane is minimized by drawing the perpendicular segment from the point to the plane. The plane represents candidates to approximate the point; the distance between them is approximation error. This simple fact about orthogonality permeates numerical analysis. It is the core of least squares approximation and is fundamental to the Galerkin approach to boundary value problems and partial differential equations, as well as Gaussian quadrature (Chapter 5), compression (see Chapters 10 and 11), and the solutions of eigenvalue problems (Chapter 12).

Now we show how the c_i are calculated to solve the BVP (7.16). The first and last of the c_i are found by collocation:

$$y(a) = \sum_{i=0}^{n+1} c_i \phi_i(a) = c_0 \phi_0(a) = c_0$$

$$y(b) = \sum_{i=0}^{n+1} c_i \phi_i(b) = c_{n+1} \phi_{n+1}(b) = c_{n+1}.$$

For $i = 1, \ldots, n$, use the finite element equations (7.22):

$$\int_a^b f(t, y, y') \phi_i(t) \, dt + \int_a^b y'(t) \phi_i'(t) \, dt = 0,$$

or substituting the functional form $y(t) = \sum c_i \phi_i(t)$,

$$\int_a^b \phi_i(t) f\left(t, \sum c_j \phi_j(t), \sum c_j \phi_j'(t)\right) dt + \int_a^b \phi_i'(t) \sum c_j \phi_j'(t) \, dt = 0. \qquad (7.25)$$

Note that the boundary terms of (7.22) are zero for $i = 1, \ldots, n$.

Assume that the grid is evenly spaced with step size h. We will need the following integrals, for $i = 1, \ldots, n$:

$$\int_a^b \phi_i(t) \phi_{i+1}(t) \, dt = \int_0^h \frac{t}{h} \left(1 - \frac{t}{h}\right) dt = \int_0^h \left(\frac{t}{h} - \frac{t^2}{h^2}\right) dt$$

$$= \frac{t^2}{2h} - \frac{t^3}{3h^2} \bigg|_0^h = \frac{h}{6} \qquad (7.26)$$

$$\int_a^b (\phi_i(t))^2 \, dt = 2 \int_0^h \left(\frac{t}{h}\right)^2 dt = \frac{2}{3} h \qquad (7.27)$$

$$\int_a^b \phi_i'(t)\phi_{i+1}'(t)\,dt = \int_0^h \frac{1}{h}\left(-\frac{1}{h}\right)\,dt = -\frac{1}{h} \tag{7.28}$$

$$\int_a^b (\phi_i'(t))^2\,dt = 2\int_0^h \left(\frac{1}{h}\right)^2\,dt = \frac{2}{h}. \tag{7.29}$$

The formulas (7.26)–(7.29) are used to simplify (7.25) once the details of the differential equation $y'' = f(t, y, y')$ are substituted. As long as the differential equation is linear, the resulting equations for the c_i will be linear.

▶ **EXAMPLE 7.12**   Apply the Finite Element Method to the BVP

$$\begin{cases} y'' = 4y \\ y(0) = 1 \\ y(1) = 3. \end{cases}$$

Substituting the differential equation into (7.25) yields for each i, the equation

$$0 = \int_0^1 \left(4\phi_i(t)\sum_{j=0}^{n+1} c_j\phi_j(t) + \sum_{j=0}^{n+1} c_j\phi_j'(t)\phi_i'(t)\right)\,dt$$

$$= \sum_{j=0}^{n+1} c_j\left[4\int_0^1 \phi_i(t)\phi_j(t)\,dt + \int_0^1 \phi_j'(t)\phi_i'(t)\,dt\right].$$

Using the B-spline relations (7.26)–(7.29) for $i = 1, \ldots, n$, and the relations $c_0 = f(a), c_{n+1} = f(b)$, we find that the equations are

$$\left[\frac{2}{3}h - \frac{1}{h}\right]c_0 + \left[\frac{8}{3}h + \frac{2}{h}\right]c_1 + \left[\frac{2}{3}h - \frac{1}{h}\right]c_2 = 0$$

$$\left[\frac{2}{3}h - \frac{1}{h}\right]c_1 + \left[\frac{8}{3}h + \frac{2}{h}\right]c_2 + \left[\frac{2}{3}h - \frac{1}{h}\right]c_3 = 0$$

$$\vdots$$

$$\left[\frac{2}{3}h - \frac{1}{h}\right]c_{n-1} + \left[\frac{8}{3}h + \frac{2}{h}\right]c_n + \left[\frac{2}{3}h - \frac{1}{h}\right]c_{n+1} = 0. \tag{7.30}$$

Note that we have $c_0 = y_a = 1$ and $c_{n+1} = y_b = 3$, so the matrix form of the equations is symmetric tridiagonal

$$\begin{bmatrix} \alpha & \beta & 0 & \cdots & 0 \\ \beta & \alpha & \ddots & \ddots & \vdots \\ 0 & \beta & \ddots & \beta & 0 \\ \vdots & \ddots & \ddots & \alpha & \beta \\ 0 & \cdots & 0 & \beta & \alpha \end{bmatrix} \begin{bmatrix} c_1 \\ c_2 \\ \vdots \\ c_{n-1} \\ c_n \end{bmatrix} = \begin{bmatrix} -y_a\beta \\ 0 \\ \vdots \\ 0 \\ -y_b\beta \end{bmatrix}$$

where

$$\alpha = \frac{8}{3}h + \frac{2}{h} \quad \text{and} \quad \beta = \frac{2}{3}h - \frac{1}{h}.$$

Recalling the MATLAB command `spdiags` used in Chapter 2, we can write a sparse implementation that is very compact:

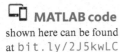

MATLAB code shown here can be found at bit.ly/2J5kwLC

```
% Program 7.2 Finite element solution of linear BVP
% Inputs: interval inter, boundary values bv, number of steps n
% Output: solution values c
% Example usage: c=bvpfem ([0 1],[1 3],9);
function c=bvpfem(inter,bv,n)
a=inter(1); b=inter(2); ya=bv(1); yb=bv(2);
h=(b-a)/(n+1);
alpha=(8/3)*h+2/h; beta=(2/3)*h-1/h;
e=ones(n,1);
M=spdiags([beta*e alpha*e beta*e],-1:1,n,n);
d=zeros(n,1);
d(1)=-ya*beta;
d(n)=-yb*beta;
c=M\d;
```

For $n = 3$, the MATLAB code gives the following c_i:

i	t_i	$w_i = c_i$	y_i
0	0.00	1.0000	1.0000
1	0.25	1.0109	1.0181
2	0.50	1.2855	1.2961
3	0.75	1.8955	1.9049
4	1.00	3.0000	3.0000

The approximate solution w_i at t_i has the value c_i, which is compared with the exact solution y_i. The errors are around 10^{-2}, the same size as the errors for the Finite Difference Method. In fact, Figure 7.11 shows that running the Finite Element Method with larger values of n gives a convergence curve almost identical to that of the Finite Difference Method in Figure 7.7, showing $O(n^{-2})$ convergence.

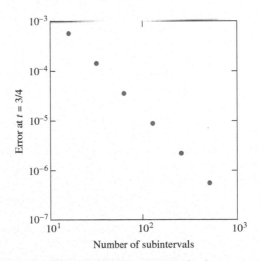

Figure 7.11 Convergence of the Finite Element Method. The error $|w_i - y_i|$ for Example 7.12 at $t_i = 3/4$ is graphed versus the number of subintervals n. According to the slope, the error is $O(n^{-2}) = O(h^2)$.

▶ **ADDITIONAL EXAMPLES**

*1. Use the Collocation Method with $n = 8$ to approximate the solution of the linear BVP

$$\begin{cases} y'' = (4t^2 + 6)y \\ y(0) = 0 \\ y(1) = e \end{cases}$$

Plot the approximate solution together with the exact solution $y = te^{t^2}$. Display the approximation errors on the interval in a separate semilog plot for $n = 8, 16$, and 32.

2. Use the Finite Element Method to plot the approximate solution $y(t)$ of the linear BVP in Additional Example 1 for $n = 9$. Display the approximation errors on the interval in a separate semilog plot for $n = 9, 19$, and 39.

⊡ **Solutions** for Additional Examples can be found at bit.ly/2R36gGg (* example with video solution)

7.3 Computer Problems

⊡ **Solutions** for Computer Problems numbered in blue can be found at bit.ly/2CU55FQ

1. Use the Collocation Method with $n = 8$ and 16 to approximate solutions to the linear boundary value problems

(a) $\begin{cases} y'' = y + \frac{2}{3}e^t \\ y(0) = 0 \\ y(1) = \frac{1}{3}e \end{cases}$ (b) $\begin{cases} y'' = (2 + 4t^2)y \\ y(0) = 1 \\ y(1) = e \end{cases}$

Plot the approximate solutions together with the exact solutions (a) $y(t) = te^t/3$ and (b) $y(t) = e^{t^2}$, and display the errors as a function of t in a separate semilog plot.

2. Use the Collocation Method with $n = 8$ and 16 to approximate solutions to the linear boundary value problems

(a) $\begin{cases} 9y'' + \pi^2 y = 0 \\ y(0) = -1 \\ y(\frac{3}{2}) = 3 \end{cases}$ (b) $\begin{cases} y'' = 3y - 2y' \\ y(0) = e^3 \\ y(1) = 1 \end{cases}$

Plot the approximate solutions together with the exact solutions (a) $y(t) = 3\sin \pi t/3 - \cos \pi t/3$ and (b) $y(t) = e^{3-3t}$, and display the errors as a function of t in a separate semilog plot.

3. Carry out the steps of Computer Problem 1, using the Finite Element Method.

4. Carry out the steps of Computer Problem 2, using the Finite Element Method.

Software and Further Reading

Boundary value problems are discussed in most texts on ordinary differential equations. Ascher et al. [1995] is a comprehensive survey of techniques for ODE boundary value problems, including multiple-shooting methods that are not covered in this chapter. Other good references on shooting methods and finite difference methods for BVPs include Keller [1968], Bailey et al. [1968], and Roberts and Shipman [1972].

The NAG program D02HAF implements a shooting method for the two-point BVP, using the Runge–Kutta–Merson Method and Newton iteration. The routine

D02GAF implements a finite difference technique with Newton iteration to solve the resulting equations. The Jacobian matrix is calculated by numerical differentiation. Finally, D02JAF solves a linear BVP for a single nth-order ODE by collocation.

The Netlib library contains two user-callable Fortran subroutines: MUSL, for linear problems, and MUSN, for nonlinear problems. Each is based upon shooting methods.

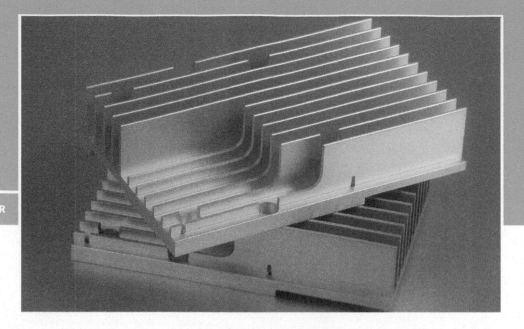

Partial Differential Equations

The 8086 central processing units manufactured by Intel Corp. in the 1970s ran at 5 MHz and required less than 5 watts of power. Today, at speeds increased by a factor of several hundred, chips dissipate over 50 watts. To avoid damage to the processor from excessively high temperatures, it is essential to distribute the heat by using a sink and fan. Cooling considerations are a constant obstacle to extending Moore's Law to faster processing speeds.

The time course of heat dissipation is well modeled by a parabolic PDE. When the heat reaches an equilibrium, an elliptic equation models the steady-state distribution.

Reality Check Reality Check 8 on page 424 shows how to model a simple heat sink configuration, using an elliptic partial differential equation with thermal convection boundary conditions.

A partial differential equation is a differential equation with more than one independent variable. While the topic is vast, we will limit our discussion to equations with two independent variables having the form

$$Au_{xx} + Bu_{xy} + Cu_{yy} + F(u_x, u_y, u, x, y) = 0, \tag{8.1}$$

where the partial derivatives are denoted by subscripts x and y for the independent variables, and u denotes the solution. When one of the variables represents time, as in the heat equation, we prefer to call the independent variables x and t.

Depending on the leading order terms of (8.1), solutions have quite different properties. Second-order PDEs with two independent variables are classified as follows:

(1) Parabolic if $B^2 - 4AC = 0$
(2) Hyperbolic if $B^2 - 4AC > 0$
(3) Elliptic if $B^2 - 4AC < 0$

The practical difference is that parabolic and hyperbolic equations are defined on an open region. Boundary conditions for one variable—in most cases the time variable—are specified at one end of the region, and the system solution is solved moving away from that boundary. Elliptic equations, on the other hand, are customarily specified with boundary conditions on the entire boundary of a closed region. We will study some examples of each type and illustrate the numerical methods available to approximate solutions.

8.1 PARABOLIC EQUATIONS

The **heat equation**

$$u_t = Du_{xx} \tag{8.2}$$

represents temperature x measured along a one-dimensional homogeneous rod. The constant $D > 0$ is called the **diffusion coefficient**, representing the thermal diffusivity of the material making up the rod. The heat equation models the spread of heat from regions of higher concentration to regions of lower concentration. The independent variables are x and t.

We use the variable t instead of y in (8.2) because it represents time. From the foregoing classification, we have $B^2 - 4AC = 0$, so the heat equation is parabolic. The so-called heat equation is an example of a **diffusion equation**, which models the diffusion of a substance. In materials science, the same equation is known as Fick's second law and describes diffusion of a substance within a medium.

Similar to the case of ODEs, the PDE (8.2) has infinitely many solutions, and extra conditions are needed to pin down a particular solution. Chapters 6 and 7 treated the solution of ODEs, where initial conditions or boundary conditions were used, respectively. In order to properly pose a PDE, various combinations of initial and boundary conditions can be used.

For the heat equation, a straightforward analysis suggests which conditions should be required. To specify the situation uniquely, we need to know the initial temperature distribution along the rod and what is happening at the ends of the rod as time progresses. The properly posed heat equation on a finite interval has the form

$$\begin{cases} u_t = Du_{xx} \text{ for all } a \le x \le b, t \ge 0 \\ u(x,0) = f(x) \text{ for all } a \le x \le b \\ u(a,t) = l(t) \text{ for all } t \ge 0 \\ u(b,t) = r(t) \text{ for all } t \ge 0 \end{cases}, \tag{8.3}$$

where the rod extends along the interval $a \le x \le b$. The diffusion coefficient D governs the rate of heat transfer. The function $f(x)$ on $[a, b]$ gives the initial temperature distribution along the rod, and $l(t), r(t)$ for $t \ge 0$ give the temperature at the ends. Here, we have used a combination of initial conditions $f(x)$ and boundary conditions $l(t)$ and $r(t)$ to specify a unique solution of the PDE.

8.1.1 Forward Difference Method

The use of Finite Difference Methods to approximate the solution of a partial differential equation follows the direction established in the previous two chapters. The idea is to lay down a grid in the independent variables and discretize the PDE. The continuous problem is changed into a discrete problem of a finite number of equations. If the PDE is linear, the discrete equations are linear and can be solved by the methods of Chapter 2.

To discretize the heat equation on the time interval $[0, T]$, we consider a grid, or mesh, of points as shown in Figure 8.1. The closed circles represent values of the

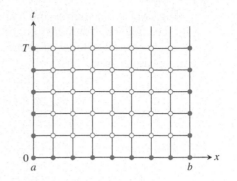

Figure 8.1 Mesh for the Finite Difference Method. The filled circles represent known initial and boundary conditions. The open circles represent unknown values that must be determined.

solution $u(x,t)$ already known from the initial and boundary conditions, and the open circles are mesh points that will be filled in by the method. We will denote the exact solution by $u(x_i, t_j)$ and its approximation at (x_i, t_j) by w_{ij}. Let M and N be the total number of steps in the x and t directions, and let $h = (b - a)/M$ and $k = T/N$ be the step sizes in the x and t directions.

The discretization formulas from Chapter 5 can be used to approximate derivatives in the x and t directions. For example, applying the centered-difference formula for the second derivative to the x variable yields

$$u_{xx}(x,t) \approx \frac{1}{h^2}(u(x+h,t) - 2u(x,t) + u(x-h,t)), \tag{8.4}$$

with error $h^2 u_{xxxx}(c_1,t)/12$; and the forward-difference formula for the first derivative used for the time variable gives

$$u_t(x,t) \approx \frac{1}{k}(u(x,t+k) - u(x,t)), \tag{8.5}$$

with error $k u_{tt}(x,c_2)/2$, where $x - h < c_1 < x + h$ and $t < c_2 < t + h$. Substituting into the heat equation at the point (x_i, t_j) yields

$$\frac{D}{h^2}(w_{i+1,j} - 2w_{ij} + w_{i-1,j}) \approx \frac{1}{k}(w_{i,j+1} - w_{ij}), \tag{8.6}$$

with the local truncation errors given by $O(k) + O(h^2)$. Just as in our study of ordinary differential equations, the local truncation errors will give a good picture of the total errors, as long as the method is stable. We will investigate the stability of the Finite Difference Method after presenting the implementation details.

Note that initial and boundary conditions give known quantities w_{i0} for $i = 0, \ldots, M$, and w_{0j} and w_{Mj} for $j = 0, \ldots, N$, which correspond to the bottom and sides of the rectangle in Figure 8.1. The discrete version (8.6) can be solved by stepping forward in time. Rearrange (8.6) as

$$w_{i,j+1} = w_{ij} + \frac{Dk}{h^2}(w_{i+1,j} - 2w_{ij} + w_{i-1,j})$$
$$= \sigma w_{i+1,j} + (1 - 2\sigma)w_{ij} + \sigma w_{i-1,j}, \tag{8.7}$$

where we have defined $\sigma = Dk/h^2$. Figure 8.2 shows the set of mesh points involved in (8.7), often called the **stencil** of the method.

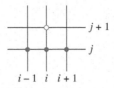

Figure 8.2 Stencil for Forward Difference Method. The open circle represents $w_{i,j+1}$, which can be determined from the values $w_{i-1,j}, w_{ij}$, and $w_{i+1,j}$ at the closed circles by (8.7).

The Forward Difference Method (8.7) is **explicit**, since there is a way to determine new values (in the sense of time) directly from the previously known values. A method that is not explicit is called **implicit**. The stencil of the method shows that this method is explicit. In matrix terms, we can get the values $w_{i,j+1}$ at time t_{j+1} by computing a matrix multiplication $w_{j+1} = Aw_j + s_j$, or

$$
\begin{bmatrix} w_{1,j+1} \\ \vdots \\ w_{m,j+1} \end{bmatrix} = \begin{bmatrix} 1-2\sigma & \sigma & 0 & \cdots & 0 \\ \sigma & 1-2\sigma & \sigma & \ddots & \vdots \\ 0 & \sigma & 1-2\sigma & \ddots & 0 \\ \vdots & \ddots & \ddots & \ddots & \sigma \\ 0 & \cdots & 0 & \sigma & 1-2\sigma \end{bmatrix} \begin{bmatrix} w_{1j} \\ \vdots \\ w_{mj} \end{bmatrix} + \sigma \begin{bmatrix} w_{0,j} \\ 0 \\ \vdots \\ 0 \\ w_{m+1,j} \end{bmatrix}.
$$

$$(8.8)$$

Here, the matrix A is $m \times m$, where $m = M - 1$. The vector s_j on the right represents the side conditions imposed by the problem, in this case the temperature at the ends of the rod.

The solution reduces to iterating a matrix formula, which allows us to fill in the empty circles in Figure 8.1 row by row. Iterating the matrix formula $w_{j+1} = Aw_j + s_j$ is similar to the iterative methods for linear systems described in Chapter 2. There we learned that convergence of the iteration depends on the eigenvalues of the matrix. In our present situation, we are interested in the eigenvalues for the analysis of error magnification.

Consider the heat equation for $D = 1$, with initial condition $f(x) = \sin^2 2\pi x$ and boundary conditions $u(0, t) = u(1, t) = 0$ for all time t. MATLAB code to carry out the calculation in (8.8) is given in Program 8.1.

⌷ **MATLAB code**

shown here can be found at bit.ly/2pYDqev

```
% Program 8.1 Forward difference method for heat equation
% input: space interval [xl,xr], time interval [yb,yt],
%         number of space steps M, number of time steps N
% output: solution w
% Example usage: w=heatfd(0,1,0,1,10,250)
function w=heatfd(xl,xr,yb,yt,M,N)
f=@(x) sin(2*pi*x).^2;
l=@(t) 0*t;
r=@(t) 0*t;
D=1;                                 % diffusion coefficient
h=(xr-xl)/M; k=(yt-yb)/N; m=M-1; n=N;
sigma=D*k/(h*h);
a=diag(1-2*sigma*ones(m,1))+diag(sigma*ones(m-1,1),1);
a=a+diag(sigma*ones(m-1,1),-1);     % define matrix a
lside=l(yb+(0:n)*k); rside=r(yb+(0:n)*k);
w(:,1)=f(xl+(1:m)*h)';              % initial conditions
for j=1:n
  w(:,j+1)=a*w(:,j)+sigma*[lside(j);zeros(m-2,1);rside(j)];
end
```

```
w=[lside;w;rside];                      % attach boundary conds
x=(0:m+1)*h;t=(0:n)*k;
mesh(x,t,w')                            % 3-D plot of solution w
view(60,30);axis([xl xr yb yt -1 1])
```

The initial temperature peaks should diffuse away with time, yielding a graph like the one shown in Figure 8.3(a). In that graph, formulas (8.8) are used with step sizes $h = 0.1$ along the rod and $k = 0.004$ in time. The explicit Forward Difference Method (8.7) gives an approximate solution in Figure 8.3(a), showing the smooth flow of the heat to a near equilibrium after less than one time unit. This corresponds to the temperature of the rod $u \to 0$ as $t \to \infty$.

In Figure 8.3(b), a slightly larger time step $k > .005$ is used. At first, the heat bumps start to die down as expected; but after more time steps, small errors in the approximation become magnified by the Forward Difference Method, causing the

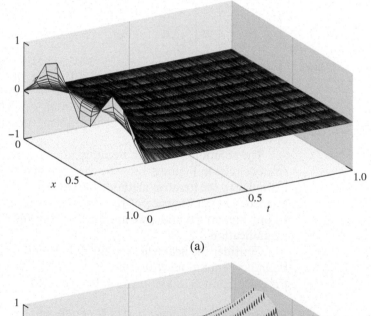

(a)

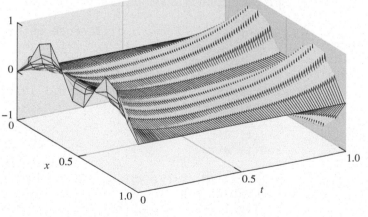

(b)

Figure 8.3 Heat Equation (8.2) approximation by Forward Finite Difference Method of Program 8.1. The diffusion parameter is $D = 1$, with initial condition $f(x) = \sin^2 2\pi x$. Space step size is $h = 0.1$. The Forward Difference Method is (a) stable for time step $k = 0.0040$, (b) unstable for $k > 0.005$.

solution to move away from the correct equilibrium of zero. This is an artifact of the solution process, a sign that the method is unstable. If the simulation were allowed to proceed further, these errors would grow without bound. Therefore, we are constrained to keep the time step k rather small to ensure convergence.

8.1.2 Stability analysis of Forward Difference Method

The strange behavior shown by the preceding heat equation simulation has led us to the core of the problem. In solving partial differential equations by the Forward Difference Method, controlling the error magnification for practical step sizes turns out to be a crucial aspect of efficient solution.

Just as in the ODE case studied earlier, there are two types of error involved. The discretization itself contributes truncation errors due to the derivative approximations. We know the size of these errors from the Taylor error formula, as in (8.4) and (8.5). In addition, there is magnification of the errors due to the method itself. To investigate this magnification, we need to look more closely at what the Finite Difference Method is doing. **Von Neumann stability analysis** measures the error magnification, or amplification. For a stable method, step sizes must be chosen so that the amplification factor is no larger than 1.

Let y_j be the exact solution that satisfies $y_{j+1} = Ay_j + s_j$ in equation (8.8), and let w_j be the computed approximation, satisfying $w_{j+1} = Aw_j + s_j$. The difference $e_j = w_j - y_j$ satisfies

$$
\begin{aligned}
e_j = w_j - y_j &= Aw_{j-1} + s_{j-1} - (Ay_{j-1} + s_{j-1}) \\
&= A(w_{j-1} - y_{j-1}) \\
&= Ae_{j-1}.
\end{aligned}
\tag{8.9}
$$

Theorem A.7 from Appendix A says that, to ensure that the errors e_j are not amplified, we must require the spectral radius $\rho(A) < 1$. This requirement puts limits on the step sizes h and k of the Finite Difference Method. To determine these limits, we need information on the eigenvalues of symmetric tridiagonal matrices.

Consider the following fundamental example:

$$
T = \begin{bmatrix}
1 & -1 & 0 & \cdots & 0 \\
-1 & 1 & -1 & \ddots & \vdots \\
0 & -1 & 1 & \ddots & 0 \\
\vdots & \ddots & \ddots & \ddots & -1 \\
0 & \cdots & 0 & -1 & 1
\end{bmatrix}.
\tag{8.10}
$$

THEOREM 8.1   The eigenvectors of the matrix T in (8.10) are the vectors v_j in (8.12) for $j = 1, \ldots, m$ with corresponding eigenvalues $\lambda_j = 1 - 2\cos\pi j/(m+1)$. ∎

    Proof. First, recall the sine addition formula from trigonometry. For any integer i and real number x, we can add the two equations

$$
\sin(i-1)x = \sin ix \cos x - \cos ix \sin x
$$
$$
\sin(i+1)x = \sin ix \cos x + \cos ix \sin x
$$

to get

$$
\sin(i-1)x + \sin(i+1)x = 2\sin ix \cos x,
$$

which can be rewritten as

$$-\sin(i-1)x + \sin ix - \sin(i+1)x = (1 - 2\cos x)\sin ix. \qquad (8.11)$$

Equation (8.11) can be viewed as a fact about matrix multiplication by T. Fix an integer j, and define the vector

$$v_j = \left[\sin\frac{j\pi}{m+1}, \sin\frac{2\pi j}{m+1}, \ldots, \sin\frac{m\pi j}{m+1}\right]. \qquad (8.12)$$

Note the pattern: The entries are of form $\sin ix$ as in (8.11), where $x = \pi j/(m+1)$. Now (8.11) implies that

$$Tv_j = \left(1 - 2\cos\frac{\pi j}{m+1}\right)v_j \qquad (8.13)$$

for $j = 1, \ldots, m$, which exhibits the m eigenvectors and eigenvalues. □

For j starting at $m+1$, the vectors v_j repeat, so there are exactly m eigenvectors, as expected. (See Exercise 6.) The eigenvalues of T all lie between -1 and 3.

Theorem 8.1 can be exploited to find the eigenvalues of any symmetric tridiagonal matrix whose main diagonal and superdiagonal are constant. For example, the matrix A in (8.8) can be expressed as $A = -\sigma T + (1-\sigma)I$. According to Theorem 8.1, the eigenvalues of A are $-\sigma(1 - 2\cos\pi j/(m+1)) + 1 - \sigma = 2\sigma(\cos\pi j/(m+1) - 1) + 1$ for $j = 1, \ldots, m$. Here we have used the fact that the eigenvalues of a matrix that is shifted by adding a multiple of the identity matrix are shifted by the same multiple.

Now we can apply the criterion of Theorem A.7. Since $-2 < \cos x - 1 < 0$ for the given arguments $x = \pi j/(m+1)$, where $1 \leq j \leq m$, the eigenvalues of A can range from $-4\sigma + 1$ to 1. Assuming that the diffusion coefficient $D > 0$, we need to restrict $\sigma < 1/2$ to ensure that the absolute values of all eigenvalues of A are less than 1—that is, that $\rho(A) < 1$.

We can state the result of the Von Neumann stability analysis as follows:

THEOREM 8.2 Let h be the space step and k be the time step for the Forward Difference Method applied to the heat equation (8.2) with $D > 0$. If $\frac{Dk}{h^2} < \frac{1}{2}$, the Forward Difference Method is stable. ∎

Our analysis confirms what we observed in Figure 8.3. By definition, $\sigma = Dk/h^2 = (1)(0.004)/(0.1)^2 = 0.4 < 1/2$ in Figure 8.3(a), while k is slightly larger than 0.005 in Figure 8.3(b), leading to $\sigma > (1)(0.005)/(0.1)^2 = 1/2$ and noticeable error magnification. The explicit Forward Difference Method is called **conditionally stable**, because its stability depends on the choice of step sizes.

8.1.3 Backward Difference Method

As an alternative, the finite difference approach can be redone with better error magnification properties by using an implicit method. As before, we replace u_{xx} in the heat equation with the centered-difference formula, but this time we use the backward-difference formula

$$u_t = \frac{1}{k}(u(x,t) - u(x,t-k)) + \frac{k}{2}u_{tt}(x,c_0),$$

where $t - k < c_0 < t$, to approximate u_t. Our motivation follows from Chapter 6, where we improved on the stability characteristics of the (explicit) Euler Method by using the (implicit) backward Euler Method, which uses a backward difference.

Substituting the difference formulas into the heat equation at the point (x_i, t_j) gives

$$\frac{1}{k}(w_{ij} - w_{i,j-1}) = \frac{D}{h^2}(w_{i+1,j} - 2w_{ij} + w_{i-1,j}), \tag{8.14}$$

with local truncation error of $O(k) + O(h^2)$, the same error that the Forward Difference Method gives. Equation (8.14) can be rearranged as

$$-\sigma w_{i+1,j} + (1 + 2\sigma)w_{ij} - \sigma w_{i-1,j} = w_{i,j-1},$$

with $\sigma = Dk/h^2$, and written as the $m \times m$ matrix equation

$$\begin{bmatrix} 1+2\sigma & -\sigma & 0 & \cdots & 0 \\ -\sigma & 1+2\sigma & -\sigma & \ddots & \vdots \\ 0 & -\sigma & 1+2\sigma & \ddots & 0 \\ \vdots & \ddots & \ddots & \ddots & -\sigma \\ 0 & \cdots & 0 & -\sigma & 1+2\sigma \end{bmatrix} \begin{bmatrix} w_{1j} \\ \vdots \\ \vdots \\ w_{mj} \end{bmatrix} = \begin{bmatrix} w_{1,j-1} \\ \vdots \\ \vdots \\ w_{m,j-1} \end{bmatrix} + \sigma \begin{bmatrix} w_{0j} \\ 0 \\ \vdots \\ 0 \\ w_{m+1,j} \end{bmatrix}.$$

$$\tag{8.15}$$

With small changes, Program 8.1 can be adapted to follow the **Backward Difference Method**.

MATLAB code shown here can be found at bit.ly/2J1ABlz

```
% Program 8.2 Backward difference method for heat equation
% input: space interval [xl,xr], time interval [yb,yt],
%         number of space steps M, number of time steps N
% output: solution w
% Example usage: w=heatbd(0,1,0,1,10,10)
function w=heatbd(xl,xr,yb,yt,M,N)
f=@(x) sin(2*pi*x).^2;
l=@(t) 0*t;
r=@(t) 0*t;
D=1;                                     % diffusion coefficient
h=(xr-xl)/M; k=(yt-yb)/N; m=M-1; n=N;
sigma=D*k/(h*h);
a=diag(1+2*sigma*ones(m,1))+diag(-sigma*ones(m-1,1),1);
a=a+diag(-sigma*ones(m-1,1),-1);       % define matrix a
lside=l(yb+(0:n)*k); rside=r(yb+(0:n)*k);
w(:,1)=f(xl+(1:m)*h)';                  % initial conditions
for j=1:n
  w(:,j+1)=a\(w(:,j)+sigma*[lside(j);zeros(m-2,1);rside(j)]);
end
w=[lside;w;rside];                      % attach boundary conds
x=(0:m+1)*h;t=(0:n)*k;
mesh(x,t,w')                            % 3-D plot of solution w
view(60,30);axis([xl xr yb yt -1 2])
```

▶ **EXAMPLE 8.1**   Apply the Backward Difference Method to the heat equation

$$\begin{cases} u_t = u_{xx} \text{ for all } 0 \leq x \leq 1, t \geq 0 \\ u(x,0) = \sin^2 2\pi x \text{ for all } 0 \leq x \leq 1 \\ u(0,t) = 0 \text{ for all } t \geq 0 \\ u(1,t) = 0 \text{ for all } t \geq 0 \end{cases}.$$

Using step sizes $h = k = 0.1$, we arrive at the approximate solution shown in Figure 8.4. Compare this with the performance of the Forward Difference Method in Figure 8.3, where $h = 0.1$ and k must be much smaller to avoid instability. ◄

What is the reason for the improved performance of the implicit method? The stability analysis for the Backward Difference Method proceeds similarly to the explicit case. The Backward Difference Method (8.15) can be viewed as the matrix iteration

$$w_j = A^{-1} w_{j-1} + b,$$

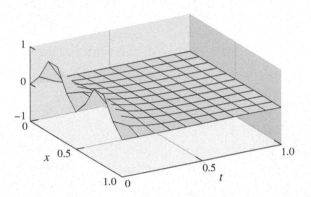

Figure 8.4 Approximate solution of Example 8.1 by the Backward Difference Method. Diffusion coefficient is $D = 1$, and step sizes are $h = 0.1, k = 0.1$.

where

$$A = \begin{bmatrix} 1 + 2\sigma & -\sigma & 0 & \cdots & 0 \\ -\sigma & 1 + 2\sigma & -\sigma & \ddots & \vdots \\ 0 & -\sigma & 1 + 2\sigma & \ddots & 0 \\ \vdots & \ddots & \ddots & \ddots & -\sigma \\ 0 & \cdots & 0 & -\sigma & 1 + 2\sigma \end{bmatrix}. \tag{8.16}$$

As in the Von Neumann stability analysis of the Forward Difference Method, the relevant quantities are the eigenvalues of A^{-1}. Since $A = \sigma T + (1 + \sigma)I$, Lemma 8.1 implies that the eigenvalues of A are

$$\sigma \left(1 - 2\cos \frac{\pi j}{m + 1} \right) + 1 + \sigma = 1 + 2\sigma - 2\sigma \cos \frac{\pi j}{m + 1},$$

and the eigenvalues of A^{-1} are the reciprocals. To ensure that the spectral radius of A^{-1} is less than 1, we need

$$|1 + 2\sigma (1 - \cos x)| > 1, \tag{8.17}$$

which is true for all σ, since $1 - \cos x > 0$ and $\sigma = Dk/h^2 > 0$. Therefore, the implicit method is stable for all σ, and thus for all choices of step sizes h and k, which is the definition of **unconditionally stable**. The step size then can be made much larger, limited only by local truncation error considerations.

THEOREM 8.3   Let h be the space step and k be the time step for the Backward Difference Method applied to the heat equation (8.2) with $D > 0$. For any h, k, the Backward Difference Method is stable. ∎

▶ **EXAMPLE 8.2**   Apply the Backward Difference Method to solve the heat equation

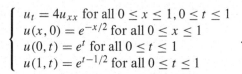

$$\begin{cases} u_t = 4u_{xx} \text{ for all } 0 \le x \le 1, 0 \le t \le 1 \\ u(x,0) = e^{-x/2} \text{ for all } 0 \le x \le 1 \\ u(0,t) = e^t \text{ for all } 0 \le t \le 1 \\ u(1,t) = e^{t-1/2} \text{ for all } 0 \le t \le 1 \end{cases}$$

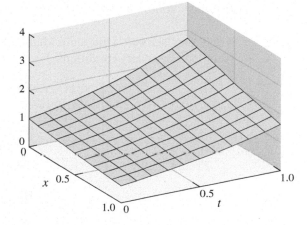

Figure 8.5 Approximate solution of Example 8.2 by Backward Difference Method. Step sizes are $h = 0.1, k = 0.1$.

Check that the correct solution is $u(x,t) = e^{t-x/2}$. Setting $h = k = 0.1$ and $D = 4$ implies that $\sigma = Dk/h^2 = 40$. The matrix A is 9×9, and at each of 10 time steps, (8.15) is solved by using Gaussian elimination. The solution is shown in Figure 8.5.   ◀

Since the Backward Difference Method is stable for any step size, we can discuss the size of the truncation errors that are made by discretizing in space and time. The errors from the time discretization are of order $O(k)$, and the errors from the space discretization are of order $O(h^2)$. This means that, for small step sizes $h \approx k$, the error from the time step will dominate, since $O(h^2)$ will be negligible compared with $O(k)$. In other words, the error from the Backward Difference Method can be roughly described as $O(k) + O(h^2) \approx O(k)$.

To demonstrate this conclusion, we used the implicit Finite Difference Method to produce solutions of Example 8.2 for fixed $h = 0.1$ and a series of decreasing k. The accompanying table shows that the error measured at $(x,t) = (0.5, 1)$ decreases linearly with k; that is, when k is cut in half, so is the error. If the size of h were decreased, the amount of computation would increase, but the errors for a given k would look virtually the same.

h	k	$u(0.5, 1)$	$w(0.5, 1)$	error
0.10	0.10	2.11700	2.12015	0.00315
0.10	0.05	2.11700	2.11861	0.00161
0.10	0.01	2.11700	2.11733	0.00033

The boundary conditions we have been applying to the heat equation are called **Dirichlet** boundary conditions. They specify the values of the solution $u(x,t)$ on the boundary of the solution domain. In the last example, Dirichlet conditions $u(0,t) = e^t$

and $u(1,t) = e^{t-1/2}$ set the required temperature values at the boundaries of the domain 0, 1. Considering the heat equation as a model of heat conduction, this corresponds to holding the temperature at the boundary at a prescribed level.

An alternative type of boundary condition corresponds to an insulated boundary. Here the temperature is not specified, but the assumption is that heat may not conduct across the boundary. In general, a **Neumann** boundary condition specifies the value of a derivative at the boundary. For example, on the domain a, b, requiring $u_x(a,t) = u_x(b,t) = 0$ for all t corresponds to an insulated, or no-flux, boundary. In general, boundary conditions set to zero are called **homogeneous** boundary conditions.

▶ **EXAMPLE 8.3**   Apply the Backward Difference Method to solve the heat equation with homogeneous Neumann boundary conditions

$$\begin{cases} u_t = u_{xx} \text{ for all } 0 \le x \le 1, 0 \le t \le 1 \\ u(x,0) = \sin^2 2\pi x \text{ for all } 0 \le x \le 1 \\ u_x(0,t) = 0 \text{ for all } 0 \le t \le 1 \\ u_x(1,t) = 0 \text{ for all } 0 \le t \le 1. \end{cases} \tag{8.18}$$

From Chapter 5, we recall the second-order formula for the first derivative

$$f'(x) = \frac{-3f(x) + 4f(x+h) - f(x+2h)}{2h} + O(h^2). \tag{8.19}$$

This formula is useful for situations where function values from both sides of x are not available. We are in just this position with Neumann boundary conditions. Therefore, we will use the second-order approximations

$$u_x(0,t) \approx \frac{-3u(0,t) + 4u(0+h,t) - u(0+2h,t)}{2h}$$

$$u_x(1,t) \approx \frac{-u(1-2h,t) + 4u(1-h,t) - 3u(1,t)}{-2h}$$

for the Neumann conditions. Setting these derivative approximations to zero translates to the formulas

$$-3w_0 + 4w_1 - w_2 = 0$$

$$-w_{M-2} + 4w_{M-1} - 3w_M = 0$$

to be added to the nonboundary parts of the equations. For bookkeeping purposes, note that as we move from Dirichlet boundary conditions to Neumann, the new feature is that we need to solve for the two boundary points w_0 and w_M. That means that while for Dirichlet, the matrix size in the Backward Difference Method is $m \times m$ where $m = M - 1$ when we move to Neumann boundary conditions, $m = M + 1$, and the matrix is slightly larger. These details are visible in the following Program 8.3. The first and last equations are replaced by the Neumann conditions.

MATLAB code
shown here can be found
at bit.ly/2pXqVzP

```
% Program 8.3   Backward difference method for heat equation
%               with Neumann boundary conditions
% input: space interval [xl,xr], time interval [yb,yt],
%        number of space steps M, number of time steps N
% output: solution w
% Example usage: w=heatbdn(0,1,0,1,20,20)
function w=heatbdn(xl,xr,yb,yt,M,N)
f=@(x) sin(2*pi*x).^2;
D=1;                                    % diffusion coefficient
```

```
h=(xr-xl)/M; k=(yt-yb)/N; m=M+1; n=N;
sigma=D*k/(h*h);
a=diag(1+2*sigma*ones(m,1))+diag(-sigma*ones(m-1,1),1);
a=a+diag(-sigma*ones(m-1,1),-1);  % define matrix a
a(1,:)=[-3 4 -1 zeros(1,m-3)];    % Neumann conditions
a(m,:)=[zeros(1,m-3) -1 4 -3];
w(:,1)=f(xl+(0:M)*h)';            % initial conditions
for j=1:n
  b=w(:,j);b(1)=0;b(m)=0;
  w(:,j+1)=a\b;
end
x=(0:M)*h;t=(0:n)*k;
mesh(x,t,w')                      % 3-D plot of solution w
view(60,30);axis([xl xr yb yt -1 1])
```

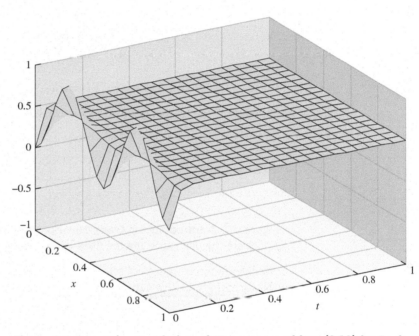

Figure 8.6 Approximate solution of Neumann problem (8.18) by Backward Difference Method. Step sizes are $h = k = 0.05$.

Figure 8.6 shows the results of Program 8.3. With Neumann conditions, the boundary values are no longer fixed at zero, and the solution floats to meet the value of the initial data that is being averaged by diffusion, which is $1/2$. ◀

8.1.4 Crank-Nicolson Method

So far, our methods for the heat equation consist of an explicit method that is sometimes stable and an implicit method that is always stable. Both have errors of size $O(k + h^2)$ when stable. The time step size k needs to be fairly small to obtain good accuracy.

The **Crank–Nicolson Method** is a combination of the explicit and implicit methods, is unconditionally stable, and has error $O(h^2) + O(k^2)$. The formulas are slightly more complicated, but worth the trouble because of the increased accuracy and guaranteed stability.

Crank–Nicolson uses the backward-difference formula for the time derivative, and a evenly weighted combination of forward-difference and backward-difference approximations for the remainder of the equation. In the heat equation (8.2), for example, replace u_t with the backward difference formula

$$\frac{1}{k}(w_{ij} - w_{i,j-1})$$

and u_{xx} with the mixed difference

$$\frac{1}{2}\left(\frac{w_{i+1,j} - 2w_{ij} + w_{i-1,j}}{h^2}\right) + \frac{1}{2}\left(\frac{w_{i+1,j-1} - 2w_{i,j-1} + w_{i-1,j-1}}{h^2}\right).$$

Again setting $\sigma = Dk/h^2$, we can rearrange the heat equation approximation in the form

$$2w_{ij} - 2w_{i,j-1} = \sigma[w_{i+1,j} - 2w_{ij} + w_{i-1,j} + w_{i+1,j-1} - 2w_{i,j-1} + w_{i-1,j-1}],$$

or

$$-\sigma w_{i-1,j} + (2 + 2\sigma)w_{ij} - \sigma w_{i+1,j} = \sigma w_{i-1,j-1} + (2 - 2\sigma)w_{i,j-1} + \sigma w_{i+1,j-1},$$

which leads to the template shown in Figure 8.7.

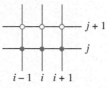

Figure 8.7 Mesh points for Crank-Nicolson Method. At each time step, the open circles are the unknowns and the filled circles are known from the previous step.

Set $w_j = [w_{1j}, \ldots, w_{mj}]^T$. In matrix form, the Crank–Nicolson Method is

$$Aw_j = Bw_{j-1} + \sigma(s_{j-1} + s_j),$$

where

$$A = \begin{bmatrix} 2+2\sigma & -\sigma & 0 & \cdots & 0 \\ -\sigma & 2+2\sigma & -\sigma & \ddots & \vdots \\ 0 & -\sigma & 2+2\sigma & \ddots & 0 \\ \vdots & \ddots & \ddots & \ddots & -\sigma \\ 0 & \cdots & 0 & -\sigma & 2+2\sigma \end{bmatrix},$$

$$B = \begin{bmatrix} 2-2\sigma & \sigma & 0 & \cdots & 0 \\ \sigma & 2-2\sigma & \sigma & \ddots & \vdots \\ 0 & \sigma & 2-2\sigma & \ddots & 0 \\ \vdots & \ddots & \ddots & \ddots & \sigma \\ 0 & \cdots & 0 & \sigma & 2-2\sigma \end{bmatrix},$$

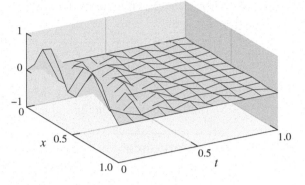

Figure 8.8 Approximate solution of Heat Equation (8.2) computed by Crank–Nicolson Method. Step sizes $h = 0.1, k = 0.1$.

and $s_j = {w_{0j}, 0, \ldots, 0, w_{m+1,j}}^T$. Applying Crank–Nicolson to the heat equation gives the result shown in Figure 8.8, for step sizes $h = 0.1$ and $k = 0.1$. MATLAB code for the method is given in Program 8.4.

MATLAB code shown here can be found at bit.ly/2Af6CUL

```
% Program 8.4  Crank-Nicolson method
%          with Dirichlet boundary conditions
% input: space interval [xl,xr], time interval [yb,yt],
%          number of space steps M, number of time steps N
% output: solution w
% Example usage: w=crank(0,1,0,1,10,10)
function w=crank(xl,xr,yb,yt,M,N)
f=@(x) sin(2*pi*x).^2;
l=@(t) 0*t;
r=@(t) 0*t;
D=1;                                % diffusion coefficient
h=(xr-xl)/M;k=(yt-yb)/N;            % step sizes
sigma=D*k/(h*h); m=M-1; n=N;
a=diag(2+2*sigma*ones(m,1))+diag(-sigma*ones(m-1,1),1);
a=a+diag(-sigma*ones(m-1,1),-1);    % define tridiagonal matrix a
b=diag(2-2*sigma*ones(m,1))+diag(sigma*ones(m-1,1),1);
b=b+diag(sigma*ones(m-1,1),-1);     % define tridiagonal matrix b
lside=l(yb+(0:n)*k); rside=r(yb+(0:n)*k);
w(:,1)=f(xl+(1:m)*h)';              % initial conditions
for j=1:n
    sides=[lside(j)+lside(j+1);zeros(m-2,1);rside(j)+rside(j+1)];
    w(:,j+1)=a\(b*w(:,j)+sigma*sides);
end
w=[lside;w;rside];
x=xl+(0:M)*h;t=yb+(0:N)*k;
mesh(x,t,w');
view (60,30); axis([xl xr yb yt -1 1])
```

To investigate the stability of Crank–Nicolson, we must find the spectral radius of the matrix $A^{-1}B$, for A and B given in the previous paragraph. Once again, the matrix in question can be rewritten in terms of T. Note that $A = \sigma T + (2 + \sigma)I$ and $B = -\sigma T + (2 - \sigma)I$. Multiplying $A^{-1}B$ to the jth eigenvector v_j of T yields

$$A^{-1}Bv_j = (\sigma T + (2 + \sigma)I)^{-1}(-\sigma \lambda_j v_j + (2 - \sigma)v_j)$$

$$= \frac{1}{\sigma \lambda_j + 2 + \sigma}(-\sigma \lambda_j + 2 - \sigma)v_j,$$

where λ_j is the eigenvalue of T associated with v_j. The eigenvalues of $A^{-1}B$ are

$$\frac{-\sigma\lambda_j + 2 - \sigma}{\sigma\lambda_j + 2 + \sigma} = \frac{4 - (\sigma(\lambda_j + 1) + 2)}{\sigma(\lambda_j + 1) + 2} = \frac{4}{L} - 1, \qquad (8.20)$$

where $L = \sigma(\lambda_j + 1) + 2 > 2$, since $\lambda_j > -1$. The eigenvalues (8.20) are therefore between -1 and 1. The Crank–Nicolson Method, like the implicit Finite Difference Method, is unconditionally stable.

SPOTLIGHT ON

> **Convergence**   Crank–Nicolson is a convenient Finite Difference Method for the heat equation due to its unconditional stability (Theorem 8.4) and second-order convergence, shown in (8.23). It is not straightforward to derive such a method, due to the first partial derivative u_t in the equation. For the wave equation and Poisson equation discussed later in the chapter, only second-order derivatives appear, and it is much easier to find stable second-order methods.

THEOREM 8.4   The Crank–Nicolson Method applied to the heat equation (8.2) with $D > 0$ is stable for any step sizes $h, k > 0$.   ∎

To finish this section, we derive the truncation error for the Crank–Nicolson Method, which is $O(h^2) + O(k^2)$. In addition to its unconditional stability, this makes the method in general superior to the Forward and Backward Difference Methods for the heat equation $u_t = Du_{xx}$.

The next four equations are needed for the derivation. We assume the existence of higher derivatives of the solution u as needed. From Exercise 5.1.24, we have the backward-difference formula

$$u_t(x,t) = \frac{u(x,t) - u(x,t-k)}{k} + \frac{k}{2}u_{tt}(x,t) - \frac{k^2}{6}u_{ttt}(x,t_1), \qquad (8.21)$$

where $t - k < t_1 < t$, assuming that the partial derivatives exist. Expanding u_{xx} in a Taylor series in the variable t yields

$$u_{xx}(x,t-k) = u_{xx}(x,t) - ku_{xxt}(x,t) + \frac{k^2}{2}u_{xxtt}(x,t_2),$$

where $t - k < t_2 < t$, or

$$u_{xx}(x,t) = u_{xx}(x,t-k) + ku_{xxt}(x,t) - \frac{k^2}{2}u_{xxtt}(x,t_2). \qquad (8.22)$$

The centered-difference formula for second derivatives gives both

$$u_{xx}(x,t) = \frac{u(x+h,t) - 2u(x,t) + u(x-h,t)}{h^2} + \frac{h^2}{12}u_{xxxx}(x_1,t) \qquad (8.23)$$

and

$$u_{xx}(x,t-k) = \frac{u(x+h,t-k) - 2u(x,t-k) + u(x-h,t-k)}{h^2}$$
$$+ \frac{h^2}{12}u_{xxxx}(x_2,t-k), \qquad (8.24)$$

where x_1 and x_2 lie between x and $x + h$.

Substitute from the preceding four equations into the heat equation

$$u_t = D\left(\frac{1}{2}u_{xx} + \frac{1}{2}u_{xx}\right),$$

where we have split the right side into two. The strategy is to replace the left side by using (8.21), the first half of the right side with (8.23), and the second half of the right side with (8.22) in combination with (8.24). This results in

$$\frac{u(x,t) - u(x,t-k)}{k} + \frac{k}{2}u_{tt}(x,t) - \frac{k^2}{6}u_{ttt}(x,t_1)$$

$$= \frac{1}{2}D\left[\frac{u(x+h,t) - 2u(x,t) + u(x-h,t)}{h^2} + \frac{h^2}{12}u_{xxxx}(x_1,t)\right]$$

$$+ \frac{1}{2}D\left[ku_{xxt}(x,t) - \frac{k^2}{2}u_{xxtt}(x,t_2)\right.$$

$$\left. + \frac{u(x+h,t-k) - 2u(x,t-k) + u(x-h,t-k)}{h^2} + \frac{h^2}{12}u_{xxxx}(x_2,t-k)\right].$$

Therefore, the error associated with equating the difference quotients is the remainder

$$-\frac{k}{2}u_{tt}(x,t) + \frac{k^2}{6}u_{ttt}(x,t_1) + \frac{Dh^2}{24}[u_{xxxx}(x_1,t) + u_{xxxx}(x_2,t-k)]$$

$$+ \frac{Dk}{2}u_{xxt}(x,t) - \frac{Dk^2}{4}u_{xxtt}(x,t_2).$$

This expression can be simplified by using the fact $u_t = Du_{xx}$. For example, note that $Du_{xxt} = (Du_{xx})_t = u_{tt}$, which causes the first and fourth terms in the expression for the error to cancel. The truncation error is

$$\frac{k^2}{6}u_{ttt}(x,t_1) - \frac{Dk^2}{4}u_{xxtt}(x,t_2) + \frac{Dh^2}{24}[u_{xxxx}(x_1,t) + u_{xxxx}(x_2,t-k)]$$

$$= \frac{k^2}{6}u_{ttt}(x,t_1) - \frac{k^2}{4}u_{ttt}(x,t_2) + \frac{h^2}{24D}[u_{tt}(x_1,t) + u_{tt}(x_2,t-k)].$$

A Taylor expansion in the variable t yields

$$u_{tt}(x_2,t-k) = u_{tt}(x_2,t) - ku_{ttt}(x_2,t_4),$$

making the truncation error equal to $O(h^2) + O(k^2)+$ higher-order terms. We conclude that the Crank–Nicolson is a second-order, unconditionally stable method for the heat equation.

To illustrate the fast convergence of Crank–Nicolson, we return to the equation of Example 8.2. See also Computer Problems 5 and 6 to explore the convergence rate.

▶ **EXAMPLE 8.4** Apply the Crank–Nicolson Method to the heat equation

$$\begin{cases} u_t = 4u_{xx} \text{ for all } 0 \leq x \leq 1, 0 \leq t \leq 1 \\ u(x,0) = e^{-x/2} \text{ for all } 0 \leq x \leq 1 \\ u(0,t) = e^t \text{ for all } 0 \leq t \leq 1 \\ u(1,t) = e^{t-1/2} \text{ for all } 0 \leq t \leq 1 \end{cases}. \tag{8.25}$$

The next table demonstrates the $O(h^2) + O(k^2)$ error convergence predicted by the preceding calculation. The correct solution $u(x,t) = e^{t-x/2}$ evaluated at $(x,t) = (0.5,1)$ is $u = e^{3/4}$. Note that the error is reduced by a factor of 4 when the step sizes h and k are halved. Compare errors with the table in Example 8.2.

h	k	$u(0.5, 1)$	$w(0.5, 1)$	error
0.10	0.10	2.11700002	2.11706765	0.00006763
0.05	0.05	2.11700002	2.11701689	0.00001687
0.01	0.01	2.11700002	2.11700069	0.00000067

◄

To summarize, we have introduced three numerical methods for parabolic equations using the heat equation as our primary example. The Forward Difference Method is the most straightforward, the Backward Difference Method is unconditionally stable and just as accurate, and Crank–Nicolson is unconditionally stable and second-order accurate in both space and time. Although the heat equation is representative, there is a vast array of parabolic equations for which these methods are applicable.

One important application area for diffusive equations concerns the spatio-temporal evolution of biological populations. Consider a population (of bacteria, prairie dogs, etc.) living on a patch of substrate or terrain. To start simply, the patch will be a line segment $[0, L]$. We will use a partial differential equation to model $u(x, t)$, the population density for each point $0 \leq x \leq L$. Populations tend to act like heat in the sense that they spread out, or diffuse, from high density areas to lower density areas when possible. They also may grow or die, as in the following representative example.

► **EXAMPLE 8.5**   Consider the diffusion equation with proportional growth

$$\begin{cases} u_t = Du_{xx} + Cu \\ u(x, 0) = \sin^2 \frac{\pi}{L}x \text{ for all } 0 \leq x \leq L \\ u(0, t) = 0 \text{ for all } t \geq 0 \\ u(L, t) = 0 \text{ for all } t \geq 0. \end{cases} \qquad (8.26)$$

The population density at time t and position x is denoted $u(x, t)$. Our use of Dirichlet boundary conditions represents the assumption that the population cannot live outside the patch $0 \leq x \leq L$.   ◄

This is perhaps the simplest possible example of a **reaction-diffusion** equation. The diffusion term Du_{xx} causes the population to spread along the x-direction, while the reaction term Cu contributes population growth of rate C. Because of the Dirichlet boundary conditions, the population is wiped out as it reaches the boundary. In reaction-diffusion equations, there is a competition between the smoothing tendency of the diffusion and the growth contribution of the reaction. Whether the population survives or proceeds toward extinction depends on the competition between the diffusion parameter D, the growth rate C, and the patch size L.

We apply Crank–Nicolson to the problem. The left-hand side of the equation is replaced with

$$\frac{1}{k}(w_{ij} - w_{i,j-1})$$

and the right-hand side with the mixed forward/backward difference

$$\frac{1}{2}\left(D\frac{w_{i+1,j} - 2w_{ij} + w_{i-1,j}}{h^2} + Cw_{ij}\right)$$
$$+\frac{1}{2}\left(D\frac{w_{i+1,j-1} - 2w_{i,j-1} + w_{i-1,j-1}}{h^2} + Cw_{i,j-1}\right).$$

Setting $\sigma = Dk/h^2$, we can rearrange to

$$-\sigma w_{i-1,j} + (2 + 2\sigma - kC)w_{ij} - \sigma w_{i+1,j} = \sigma w_{i-1,j-1} + (2 - 2\sigma + kC)w_{i,j-1}$$
$$+\sigma w_{i+1,j-1}.$$

Comparing with the Crank–Nicolson equations for the heat equation above, we need only to subtract kC from the diagonal entries of matrix A and add kC to the diagonal entries of matrix B. This leads to changes in two lines of Program 8.4.

Figure 8.9 shows the results of Crank–Nicolson applied to (8.26) with diffusion coefficient $D = 1$, on the patch $[0, 1]$. For the choice $C = 9.5$, the original population density tends to zero in the long run. For $C = 10$, the population flourishes. Although it is beyond the scope of our discussion here, it can be shown that the model population survives as long as

$$C > \pi^2 D/L^2. \qquad (8.27)$$

In our case, that translates to $C > \pi^2$, which is between 9.5 and 10, explaining the results we see in Figure 8.9. In modeling of biological populations, the information is often used in reverse: Given known population growth rate and diffusion rate, an ecologist studying species survival might want to know the smallest patch that can support the population.

Computer Problems 7 and 8 ask the reader to investigate this reaction-diffusion system further. Nonlinear reaction-diffusion equations are a focus of Section 8.4.

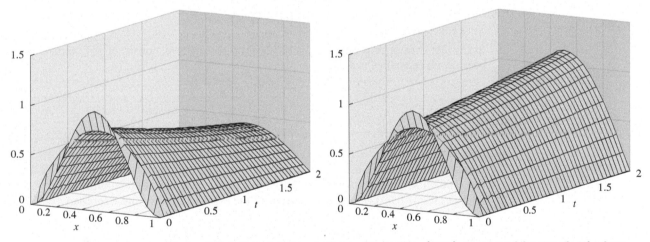

Figure 8.9 Approximate solutions of equation (8.26) computed by Crank–Nicolson Method. The parameters are $D = 1, L = 1$, and the step sizes used are $h = k = 0.05$. (a) $C = 9.5$ (b) $C = 10$.

▶ **ADDITIONAL EXAMPLES**

1. Prove that $u(x, t) = \frac{1}{\sqrt{t}} e^{-kx^2/t}$ satisfies the heat equation $u_t = \frac{1}{4k} u_{xx}$ on $(0, \infty)$.

2. Apply Crank–Nicolson to approximate the solution of the heat equation with boundary conditions

$$\begin{cases} u_t = \frac{1}{9\pi^2} u_{xx} \\ u(x, 0) = \sin 3\pi x \\ u(0, t) = u(1, t) = 0 \end{cases}$$

on $0 \leq x \leq 1$ and $0 \leq t \leq 4$. Plot the solution for step sizes $h = k = 0.05$. Compare the approximate solution to the exact solution $u(x, t) = e^{-t} \sin 3\pi x$ by plotting the error at $(x, t) = (1/2, 2)$ for $h = k = 0.05 \times 2^{-i}$ for $i = 0, \ldots, 6$.

Solutions for Additional Examples can be found at `bit.ly/2EA72cl`

8.1 Exercises

1. Prove that the functions (a) $u(x, t) = e^{2t+x} + e^{2t-x}$, (b) $u(x, t) = e^{2t+x}$ are solutions of the heat equation $u_t = 2u_{xx}$ with the specified initial boundary conditions:

   (a) $\begin{cases} u(x, 0) = 2\cosh x \text{ for } 0 \leq x \leq 1 \\ u(0, t) = 2e^{2t} \text{ for } 0 \leq t \leq 1 \\ u(1, t) = (e^2 + 1)e^{2t-1} \text{ for } 0 \leq t \leq 1 \end{cases}$ (b) $\begin{cases} u(x, 0) = e^x \text{ for } 0 \leq x \leq 1 \\ u(0, t) = e^{2t} \text{ for } 0 \leq t \leq 1 \\ u(1, t) = e^{2t+1} \text{ for } 0 \leq t \leq 1 \end{cases}$

2. Prove that the functions (a) $u(x, t) = e^{-\pi t} \sin \pi x$, (b) $u(x, t) = e^{-\pi t} \cos \pi x$ are solutions of the heat equation $\pi u_t = u_{xx}$ with the specified initial boundary conditions:

   (a) $\begin{cases} u(x, 0) = \sin \pi x \text{ for } 0 \leq x \leq 1 \\ u(0, t) = 0 \text{ for } 0 \leq t \leq 1 \\ u(1, t) = 0 \text{ for } 0 \leq t \leq 1 \end{cases}$ (b) $\begin{cases} u(x, 0) = \cos \pi x \text{ for all } 0 \leq x \leq 1 \\ u(0, t) = e^{-\pi t} \text{ for } 0 \leq t \leq 1 \\ u(1, t) = -e^{-\pi t} \text{ for } 0 \leq t \leq 1 \end{cases}$

3. Prove that if $f(x)$ is a degree 3 polynomial, then $u(x, t) = f(x) + ctf''(x)$ is a solution of the initial value problem $u_t = cu_{xx}, u(x, 0) = f(x)$.

4. Is the Backward Difference Method unconditionally stable for the heat equation if $c < 0$? Explain.

5. Verify the eigenvector equation (8.13).

6. Show that the nonzero vectors v_j in (8.12), for all integers m, consist of only m distinct vectors, up to change in sign.

8.1 Computer Problems

1. Solve the equation $u_t = 2u_{xx}$ for $0 \leq x \leq 1, 0 \leq t \leq 1$, with the initial and boundary conditions that follow, using the Forward Difference Method with step sizes $h = 0.1$ and $k = 0.002$. Plot the approximate solution, using the MATLAB mesh command. What happens if $k > 0.003$ is used? Compare with the exact solutions from Exercise 1.

   (a) $\begin{cases} u(x, 0) = 2\cosh x \text{ for } 0 \leq x \leq 1 \\ u(0, t) = 2e^{2t} \text{ for } 0 \leq t \leq 1 \\ u(1, t) = (e^2 + 1)e^{2t-1} \text{ for } 0 \leq t \leq 1 \end{cases}$ (b) $\begin{cases} u(x, 0) = e^x \text{ for } 0 \leq x \leq 1 \\ u(0, t) = e^{2t} \text{ for } 0 \leq t \leq 1 \\ u(1, t) = e^{2t+1} \text{ for } 0 \leq t \leq 1 \end{cases}$

2. Consider the equation $\pi u_t = u_{xx}$ for $0 \leq x \leq 1, 0 \leq t \leq 1$ with the initial and boundary conditions that follow. Set step size $h = 0.1$. For what step sizes k is the Forward Difference Method stable? Apply the Forward Difference Method with step sizes $h = 0.1$, $k = 0.01$, and compare with the exact solution from Exercise 2.

   (a) $\begin{cases} u(x, 0) = \sin \pi x \text{ for } 0 \leq x \leq 1 \\ u(0, t) = 0 \text{ for } 0 \leq t \leq 1 \\ u(1, t) = 0 \text{ for } 0 \leq t \leq 1 \end{cases}$ (b) $\begin{cases} u(x, 0) = \cos \pi x \text{ for all } 0 \leq x \leq 1 \\ u(0, t) = e^{-\pi t} \text{ for } 0 \leq t \leq 1 \\ u(1, t) = -e^{-\pi t} \text{ for } 0 \leq t \leq 1 \end{cases}$

3. Use the Backward Difference Method to solve the problems of Computer Problem 1. Make a table of the exact value, the approximate value, and error at $(x, t) = (0.5, 1)$ for step sizes $h = 0.02$ and $k = 0.02, 0.01, 0.005$.

4. Use the Backward Difference Method to solve the problems of Computer Problem 2. Make a table of the exact value, the approximate value, and error at $(x, t) = (0.3, 1)$ for step sizes $h = 0.1$ and $k = 0.02, 0.01, 0.005$.

5. Use the Crank–Nicolson Method to solve the problems of Computer Problem 1. Make a table of the exact value, the approximate value, and error at $(x, t) = (0.5, 1)$ for step sizes $h = k = 0.02, 0.01, 0.005$.

6. Use the Crank–Nicolson Method to solve the problems of Computer Problem 2. Make a table of the exact value, the approximate value, and error at $(x, t) = (0.3, 1)$ for step sizes $h = k = 0.1, 0.05, 0.025$.

7. Set $D = 1$ and find the smallest C for which the population of (8.26), on the patch [0, 10], survives in the long run. Use the Crank–Nicolson Method to approximate the solution, and try to confirm that your results do not depend on the step size choices. Compare your results with the survival rule (8.27).

8. Setting $C = D = 1$ in the population model (8.26), use Crank–Nicolson to find the minimum patch size that allows the population to survive. Compare with the rule (8.27).

8.2 HYPERBOLIC EQUATIONS

Hyperbolic equations put less stringent constraints on explicit methods. In this section, the stability of Finite Difference Methods is explored in the context of a representative hyperbolic equation called the wave equation. The CFL condition will be introduced, which is, in general, a necessary condition for stability of the PDE solver.

8.2.1 The wave equation

Consider the partial differential equation

$$u_{tt} = c^2 u_{xx} \tag{8.28}$$

for $a \leq x \leq b$ and $t \geq 0$. Comparing with the normal form (8.1), we compute $B^2 - 4AC = 4c^2 > 0$, so the equation is hyperbolic. This example is called the **wave equation** with wave speed c. Typical initial and boundary conditions needed to specify a unique solution are

$$\begin{cases} u(x, 0) = f(x) \text{ for all } a \leq x \leq b \\ u_t(x, 0) = g(x) \text{ for all } a \leq x \leq b \\ u(a, t) = l(t) \text{ for all } t \geq 0 \\ u(b, t) = r(t) \text{ for all } t \geq 0 \end{cases} \tag{8.29}$$

Compared with the heat equation example, extra initial data are needed due to the higher-order time derivative in the equation. Intuitively speaking, the wave equation describes the time evolution of a wave propagating along the x-direction. To specify what happens, we need to know the initial shape of the wave and the initial velocity of the wave at each point.

The wave equation models a wide variety of phenomena, from magnetic waves in the sun's atmosphere to the oscillation of a violin string. The equation involves an amplitude u, which for the violin represents the physical displacement of the string. For a sound wave traveling in air, u represents the local air pressure.

We will apply the Finite Difference Method to the wave equation (8.28) and analyze its stability. The Finite Difference Method operates on a grid as in Figure 8.1, just as in the parabolic case. The grid points are (x_i, t_j), where $x_i = a + ih$ and $t_j = jk$, for step sizes h and k. As before, we will represent the approximation to the solution $u(x_i, t_j)$ by w_{ij}.

To discretize the wave equation, the second partial derivatives are replaced by the centered-difference formula (8.4) in both the x and t directions:

$$\frac{w_{i,j+1} - 2w_{ij} + w_{i,j-1}}{k^2} - c^2 \frac{w_{i-1,j} - 2w_{ij} + w_{i+1,j}}{h^2} = 0.$$

Setting $\sigma = ck/h$, we can solve for the solution at the next time step and write the discretized equation as

$$w_{i,j+1} = (2 - 2\sigma^2)w_{ij} + \sigma^2 w_{i-1,j} + \sigma^2 w_{i+1,j} - w_{i,j-1}. \tag{8.30}$$

The formula (8.30) cannot be used for the first time step, since values at two prior times, $j - 1$ and j, are needed. This is similar to the problem with starting multi-step ODE methods. To solve the problem, we can introduce the three-point centered-difference formula to approximate the first time derivative of the solution u:

$$u_t(x_i, t_j) \approx \frac{w_{i,j+1} - w_{i,j-1}}{2k}.$$

Substituting initial data at the first time step (x_i, t_1) yields

$$g(x_i) = u_t(x_i, t_0) \approx \frac{w_{i1} - w_{i,-1}}{2k},$$

or in other words,

$$w_{i,-1} \approx w_{i1} - 2kg(x_i). \tag{8.31}$$

Substituting (8.31) into the finite difference formula (8.30) for $j = 0$ gives

$$w_{i1} = (2 - 2\sigma^2)w_{i0} + \sigma^2 w_{i-1,0} + \sigma^2 w_{i+1,0} - w_{i1} + 2kg(x_i),$$

which can be solved for w_{i1} to yield

$$w_{i1} = (1 - \sigma^2)w_{i0} + kg(x_i) + \frac{\sigma^2}{2}(w_{i-1,0} + w_{i+1,0}). \tag{8.32}$$

Formula (8.32) is used for the first time step. This is the way the initial velocity information g enters the calculation. For all later time steps, formula (8.30) is used. Since second-order formulas have been used for both space and time derivatives, the error of this Finite Difference Method will be $O(h^2) + O(k^2)$ (see Computer Problems 3 and 4).

To write the Finite Difference Method in matrix terms, define

$$A = \begin{bmatrix} 2 - 2\sigma^2 & \sigma^2 & 0 & \cdots & 0 \\ \sigma^2 & 2 - 2\sigma^2 & \sigma^2 & \ddots & \vdots \\ 0 & \sigma^2 & 2 - 2\sigma^2 & \ddots & 0 \\ \vdots & \ddots & \ddots & \ddots & \sigma^2 \\ 0 & \cdots & 0 & \sigma^2 & 2 - 2\sigma^2 \end{bmatrix}. \tag{8.33}$$

The initial equation (8.32) can be written

$$
\begin{bmatrix} w_{11} \\ \vdots \\ w_{m1} \end{bmatrix} = \frac{1}{2}A \begin{bmatrix} w_{10} \\ \vdots \\ w_{m0} \end{bmatrix} + k \begin{bmatrix} g(x_1) \\ \vdots \\ g(x_m) \end{bmatrix} + \frac{1}{2}\sigma^2 \begin{bmatrix} w_{00} \\ 0 \\ \vdots \\ 0 \\ w_{m+1,0} \end{bmatrix},
$$

and the subsequent steps of (8.30) are given by

$$
\begin{bmatrix} w_{1,j+1} \\ \vdots \\ w_{m,j+1} \end{bmatrix} = A \begin{bmatrix} w_{1j} \\ \vdots \\ w_{mj} \end{bmatrix} - \begin{bmatrix} w_{1,j-1} \\ \vdots \\ w_{m,j-1} \end{bmatrix} + \sigma^2 \begin{bmatrix} w_{0j} \\ 0 \\ \vdots \\ 0 \\ w_{m+1,j} \end{bmatrix}.
$$

Inserting the rest of the extra data, the two equations are written

$$
\begin{bmatrix} w_{11} \\ \vdots \\ w_{m1} \end{bmatrix} = \frac{1}{2}A \begin{bmatrix} f(x_1) \\ \vdots \\ f(x_m) \end{bmatrix} + k \begin{bmatrix} g(x_1) \\ \vdots \\ g(x_m) \end{bmatrix} + \frac{1}{2}\sigma^2 \begin{bmatrix} l(t_0) \\ 0 \\ \vdots \\ 0 \\ r(t_0) \end{bmatrix},
$$

and the subsequent steps of (8.30) are given by

$$
\begin{bmatrix} w_{1,j+1} \\ \vdots \\ w_{m,j+1} \end{bmatrix} = A \begin{bmatrix} w_{1j} \\ \vdots \\ w_{mj} \end{bmatrix} - \begin{bmatrix} w_{1,j-1} \\ \vdots \\ w_{m,j-1} \end{bmatrix} + \sigma^2 \begin{bmatrix} l(t_j) \\ 0 \\ \vdots \\ 0 \\ r(t_j) \end{bmatrix}. \tag{8.34}
$$

▶ **EXAMPLE 8.6**   Apply the explicit Finite Difference Method to the wave equation with wave speed $c = 2$ and initial conditions $f(x) = \sin \pi x$ and $g(x) = l(x) = r(x) = 0$.

Figure 8.10 shows approximate solutions of the wave equation with $c = 2$. The explicit Finite Difference Method is conditionally stable; step sizes have to be chosen carefully to avoid instability of the solver. Part (a) of the figure shows a stable choice of $h = 0.05$ and $k = 0.025$, while part (b) shows the unstable choice $h = 0.05$ and $k = 0.032$. The explicit Finite Difference Method applied to the wave equation is unstable when the time step k is too large relative to the space step h.   ◀

8.2.2 The CFL condition

The matrix form allows us to analyze the stability characteristics of the explicit Finite Difference Method applied to the wave equation. The result of the analysis, stated as Theorem 8.5, explains Figure 8.10.

THEOREM 8.5   The Finite Difference Method applied to the wave equation with wave speed $c > 0$ is stable if $\sigma = ck/h \leq 1$.   ∎

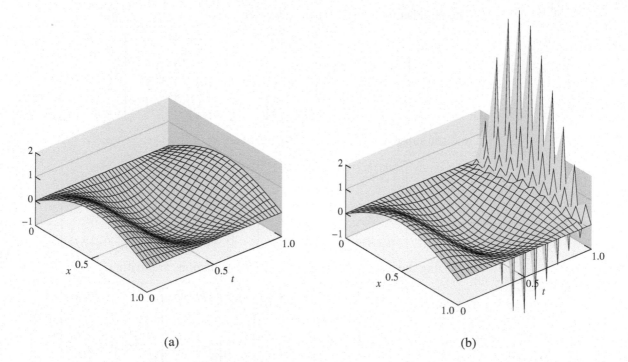

(a)   (b)

Figure 8.10 Wave Equation in Example 8.6 approximated by explicit Finite Difference Method. Space step size is $h = 0.05$. (a) Method is stable for time step $k = 0.025$, (b) unstable for $k = 0.032$.

Proof. Equation (8.34) in vector form is

$$w_{j+1} = Aw_j - w_{j-1} + \sigma^2 s_j, \tag{8.35}$$

where s_j holds the side conditions. Since w_{j+1} depends on both w_j and w_{j-1}, to study error magnification we rewrite (8.35) as

$$\begin{bmatrix} w_{j+1} \\ w_j \end{bmatrix} = \begin{bmatrix} A & -I \\ I & 0 \end{bmatrix} \begin{bmatrix} w_j \\ w_{j-1} \end{bmatrix} + \sigma^2 \begin{bmatrix} s_j \\ 0 \end{bmatrix}, \tag{8.36}$$

to view the method as a one-step recursion. Error will not be magnified as long as the eigenvalues of

$$A' = \begin{bmatrix} A & -I \\ I & 0 \end{bmatrix}$$

are bounded by 1 in absolute value.

Let $\lambda \neq 0, (y, z)^T$ be an eigenvalue/eigenvector pair of A', so that

$$\lambda y = Ay - z$$
$$\lambda z = y,$$

which implies that

$$Ay = \left(\frac{1}{\lambda} + \lambda \right) y,$$

so that $\mu = 1/\lambda + \lambda$ is an eigenvalue of A. The eigenvalues of A lie between $2 - 4\sigma^2$ and 2 (Exercise 5). The assumption that $\sigma \leq 1$ implies that $-2 \leq \mu \leq 2$. To finish, it need only be shown that, for a complex number λ, the fact that $1/\lambda + \lambda$ is real and has magnitude at most 2 implies that $|\lambda| = 1$ (Exercise 6). ❑

The quantity ck/h is called the **CFL number** of the method, after R. Courant, K. Friedrichs, and H. Lewy [1928]. In general, the CFL number must be at most 1 in order for the PDE solver to be stable. Since c is the wave speed, this means that the distance ck traveled by the solution in one time step should not exceed the space step h. Figures 8.10(a) and (b) illustrate CFL numbers of 1 and 1.28, respectively. The constraint $ck \leq h$ is called the **CFL condition** for the wave equation.

Theorem 8.5 states that for the wave equation, the CFL condition implies stability of the Finite Difference Method. For more general hyperbolic equations, the CFL condition is necessary, but not always sufficient for stability. See Morton and Mayers [1996] for further details.

The wave speed parameter c in the wave equation governs the velocity of the propagating wave. Figure 8.11 shows that for $c = 6$, the sine wave initial condition oscillates three times during one time unit, three times as fast as the $c = 2$ case.

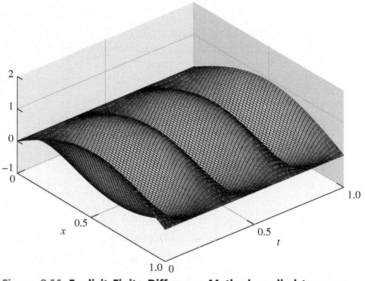

Figure 8.11 Explicit Finite Difference Method applied to wave equation, $c = 6$. The step sizes $h = 0.05, k = 0.008$ satisfy the CFL condition.

▶ **ADDITIONAL EXAMPLES**

1. Show that $u(x, t) = e^{-4(x^2 + 6tx + 9t^2)}$ is a solution of the wave equation $u_{tt} = 9u_{xx}$.

*2. Use the Finite Difference Method with step sizes $h = k = 0.02$ to plot the solution of the wave equation

$$\begin{cases} u_{tt} = \frac{1}{2}u_{xx} \\ u(x, 0) = \sin 3\pi x \\ u_t(x, 0) = 0 \\ u(0, t) = u(1, t) = 0 \end{cases}$$

on $0 \leq x \leq 1, 0 \leq t \leq 2$.

📱 **Solutions** for Additional Examples can be found at bit.ly/2J7pSGr
(* example with video solution)

8.2 Exercises

1. Prove that the functions (a) $u(x,t) = \sin\pi x \cos 4\pi t$, (b) $u(x,t) = e^{-x-2t}$, (c) $u(x,t) = \ln(1 + x + t)$ are solutions of the wave equation with the specified initial-boundary conditions:

   (a) $\begin{cases} u_{tt} = 16u_{xx} \\ u(x,0) = \sin\pi x \text{ for } 0 \le x \le 1 \\ u_t(x,0) = 0 \text{ for } 0 \le x \le 1 \\ u(0,t) = 0 \text{ for } 0 \le t \le 1 \\ u(1,t) = 0 \text{ for } 0 \le t \le 1 \end{cases}$ (b) $\begin{cases} u_{tt} = 4u_{xx} \\ u(x,0) = e^{-x} \text{ for } 0 \le x \le 1 \\ u_t(x,0) = -2e^{-x} \text{ for } 0 \le x \le 1 \\ u(0,t) = e^{-2t} \text{ for } 0 \le t \le 1 \\ u(1,t) = e^{-1-2t} \text{ for } 0 \le t \le 1 \end{cases}$

   (c) $\begin{cases} u_{tt} = u_{xx} \\ u(x,0) = \ln(1+x) \text{ for } 0 \le x \le 1 \\ u_t(x,0) = 1/(1+x) \text{ for } 0 \le x \le 1 \\ u(0,t) = \ln(1+t) \text{ for } 0 \le t \le 1 \\ u(1,t) = \ln(2+t) \text{ for } 0 \le t \le 1 \end{cases}$

2. Prove that the functions (a) $u(x,t) = \sin\pi x \sin 2\pi t$, (b) $u(x,t) = (x + 2t)^5$, (c) $u(x,t) = \sinh x \cosh 2t$ are solutions of the wave equation with the specified initial-boundary conditions:

   (a) $\begin{cases} u_{tt} = 4u_{xx} \\ u(x,0) = 0 \text{ for } 0 \le x \le 1 \\ u_t(x,0) = 2\pi \sin\pi x \text{ for } 0 \le x \le 1 \\ u(0,t) = 0 \text{ for } 0 \le t \le 1 \\ u(1,t) = 0 \text{ for } 0 \le t \le 1 \end{cases}$ (b) $\begin{cases} u_{tt} = 4u_{xx} \\ u(x,0) = x^5 \text{ for } 0 \le x \le 1 \\ u_t(x,0) = 10x^4 \text{ for } 0 \le x \le 1 \\ u(0,t) = 32t^5 \text{ for } 0 \le t \le 1 \\ u(1,t) = (1 + 2t)^5 \text{ for } 0 \le t \le 1 \end{cases}$

   (c) $\begin{cases} u_{tt} = 4u_{xx} \\ u(x,0) = \sinh x \text{ for } 0 \le x \le 1 \\ u_t(x,0) = 0 \text{ for } 0 \le x \le 1 \\ u(0,t) = 0 \text{ for } 0 \le t \le 1 \\ u(1,t) = \frac{1}{2}(e - \frac{1}{e})\cosh 2t \text{ for } 0 \le t \le 1 \end{cases}$

3. Prove that $u_1(x,t) = \sin\alpha x \cos c\alpha t$ and $u_2(x,t) = e^{x+ct}$ are solutions of the wave equation (8.28).

4. Prove that if $s(x)$ is twice differentiable, then $u(x,t) = s(\alpha x + c\alpha t)$ is a solution of the wave equation (8.28).

5. Prove that the eigenvalues of A in (8.33) lie between $2 - 4\sigma^2$ and 2.

6. Let λ be a complex number. (a) Prove that if $\lambda + 1/\lambda$ is a real number, then $|\lambda| = 1$ or λ is real. (b) Prove that if λ is real and $|\lambda + 1/\lambda| \le 2$, then $|\lambda| = 1$.

8.2 Computer Problems

1. Solve the initial-boundary value problems in Exercise 1 on $0 \le x \le 1, 0 \le t \le 1$ by the Finite Difference Method with $h = 0.05, k = h/c$. Use MATLAB's mesh command to plot the solution.

2. Solve the initial-boundary value problems in Exercise 2 on $0 \le x \le 1, 0 \le t \le 1$ by the Finite Difference Method with $h = 0.05$ and k small enough to satisfy the CFL condition. Plot the solution.

3. For the wave equations in Exercise 1, make a table of the approximation and error at $(x,t) = (1/4, 3/4)$ as a function of step sizes $h = ck = 2^{-p}$ for $p = 4, \ldots, 8$.

4. For the wave equations in Exercise 2, make a table of the approximation and error at $(x,t) = (1/4, 3/4)$ as a function of step sizes $h = ck = 2^{-p}$ for $p = 4, \ldots, 8$.

8.3 ELLIPTIC EQUATIONS

The previous sections deal with time-dependent equations. The diffusion equation models the flow of heat as a function of time, and the wave equation follows the motion of a wave. Elliptic equations, the focus of this section, model steady states. For example, the steady-state distribution of heat on a plane region whose boundary is being held at specific temperatures is modeled by an elliptic equation. Since time is usually not a factor in elliptic equations, we will use x and y to denote the independent variables.

DEFINITION 8.6 Let $u(x, y)$ be a twice-differentiable function, and define the **Laplacian** of u as

$$\Delta u = u_{xx} + u_{yy}.$$

For a continuous function $f(x, y)$, the partial differential equation

$$\Delta u(x, y) = f(x, y) \tag{8.37}$$

is called the **Poisson equation**. The Poisson equation with $f(x, y) = 0$ is called the **Laplace equation**. A solution of the Laplace equation is called a **harmonic** function.
❏

Comparing with the normal form (8.1), we compute $B^2 - 4AC < 0$, so the Poisson equation is elliptic. The extra conditions given to pin down a single solution are typically boundary conditions. There are two common types of boundary conditions applied. Dirichlet boundary conditions specify the values of the solution $u(x, y)$ on the boundary ∂R of a region R. Neumann boundary conditions specify values of the directional derivative $\partial u / \partial n$ on the boundary, where n denotes the outward unit normal vector.

▶ **EXAMPLE 8.7** Show that $u(x, y) = x^2 - y^2$ is a solution of the Laplace equation on $[0, 1] \times [0, 1]$ with Dirichlet boundary conditions

$$u(x, 0) = x^2$$
$$u(x, 1) = x^2 - 1$$
$$u(0, y) = -y^2$$
$$u(1, y) = 1 - y^2.$$

The Laplacian is $\Delta u = u_{xx} + u_{yy} = 2 - 2 = 0$. The boundary conditions are listed for the bottom, top, left, and right of the unit square, respectively, and are easily checked by substitution. ◀

The Poisson and Laplace equations are ubiquitous in classical physics because their solutions represent potential energy. For example, an electric field E is the gradient of an electrostatic potential u, or

$$E = -\nabla u.$$

The gradient of the electric field, in turn, is related to the charge density ρ by Maxwell's equation

$$\nabla E = \frac{\rho}{\epsilon},$$

where ϵ is the electrical permittivity. Putting the two equations together yields

$$\Delta u = \nabla(\nabla u) = -\frac{\rho}{\epsilon},$$

the Poisson equation for the potential u. In the special case of zero charge, the potential satisfies the Laplace equation $\Delta u = 0$.

Many other instances of potential energy are modeled by the Poisson equation. The aerodynamics of airfoils at low speeds, known as incompressible irrotational flow, are a solution of the Laplace equation. The gravitational potential u generated by a distribution of mass density ρ satisfies the Poisson equation

$$\Delta u = 4\pi G \rho,$$

where G denotes the gravitational constant. A steady-state heat distribution, such as the limit of a solution of the heat equation as time $t \to \infty$, is modeled by the Poisson equation. In Reality Check 8, a variant of the Poisson equation is used to model the heat distribution on a cooling fin.

We introduce two methods for solving elliptic equations. The first is a Finite Difference Method that closely follows the development for parabolic and hyperbolic equations. The second generalizes the Finite Element Method for solving boundary value problems in Chapter 7. In most of the elliptic equations we consider, the domain is two-dimensional, which will cause a little extra bookkeeping work.

8.3.1 Finite Difference Method for elliptic equations

We will solve the Poisson equation $\Delta u = f$ on a rectangle $[x_l, x_r] \times [y_b, y_t]$ in the plane, with Dirichlet boundary conditions

$$u(x, y_b) = g_1(x)$$
$$u(x, y_t) = g_2(x)$$
$$u(x_l, y) = g_3(y)$$
$$u(x_r, y) = g_4(y)$$

A rectangular mesh of points is shown in Figure 8.12(a), using $M = m - 1$ steps in the horizontal direction and $N = n - 1$ steps in the vertical direction. The mesh sizes in the x and y directions are $h = (x_r - x_l)/M$ and $k = (y_t - y_b)/N$, respectively.

A Finite Difference Method involves approximating derivatives by difference quotients. The centered-difference formula (8.4) can be used for both second derivatives in the Laplacian operator. The Poisson equation $\Delta u = f$ has finite difference form

$$\frac{u(x - h, y) - 2u(x, y) + u(x + h, y)}{h^2} + O(h^2)$$
$$+ \frac{u(x, y - k) - 2u(x, y) + u(x, y + k)}{k^2} + O(k^2) = f(x, y),$$

and in terms of the approximate solution $w_{ij} \approx u(x_i, y_j)$ can be written

$$\frac{w_{i-1,j} - 2w_{ij} + w_{i+1,j}}{h^2} + \frac{w_{i,j-1} - 2w_{i,j} + w_{i,j+1}}{k^2} = f(x_i, y_j) \qquad (8.38)$$

where $x_i = x_l + (i - 1)h$ and $y_j = y_b + (j - 1)k$ for $1 \le i \le m$ and $1 \le j \le n$.

Since the equations in the w_{ij} are linear, we are led to construct a matrix equation to solve for the mn unknowns. This presents a bookkeeping problem: We need to

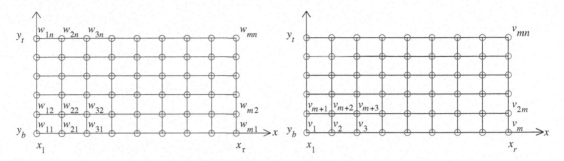

Figure 8.12 Mesh for finite difference solver of Poisson equation with Dirichlet boundary conditions. (a) Original numbering system with double subscripts. (b) Numbering system (8.39) for linear equations, with single subscripts, orders mesh points across rows.

relabel these doubly indexed unknowns into a linear order. Figure 8.12(b) shows an alternative numbering system for the solution values, where we have set

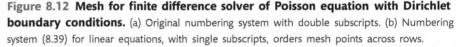

$$v_{i+(j-1)m} = w_{ij}. \tag{8.39}$$

Next, we will construct a matrix A and vector b such that $Av = b$ can be solved for v, and translated back into the solution w on the rectangular grid. Since v is a vector of length mn, A will be an $mn \times mn$ matrix, and each grid point will correspond to its own linear equation.

By definition, the entry A_{pq} is the qth linear coefficient of the pth equation of $Av = b$. For example, (8.38) represents the equation at grid point (i, j), which we call equation number $p = i + (j-1)m$, according to (8.39). The coefficients of the terms $w_{i-1,j}, w_{ij}, \ldots$ in (8.38) are also numbered according to (8.39), which we collect together in Table 8.1.

x	y	Equation number p
i	j	$i + (j-1)m$

x	y	Coefficient number q
i	j	$i + (j-1)m$
$i+1$	j	$i + 1 + (j-1)m$
$i-1$	j	$i - 1 + (j-1)m$
i	$j+1$	$i + jm$
i	$j-1$	$i + (j-2)m$

Table 8.1   Translation table for two-dimensional domains. The equation at grid point (i, j) is numbered p, and its coefficients are A_{pq} for various q, with p and q given in the right column of the table. The table is simply an illustration of (8.39).

According to Table 8.1, labeling by equation number p and coefficient number q, the matrix entries A_{pq} from (8.38) are

$$A_{i+(j-1)m,i+(j-1)m} = -\frac{2}{h^2} - \frac{2}{k^2} \tag{8.40}$$

$$A_{i+(j-1)m,i+1+(j-1)m} = \frac{1}{h^2}$$

$$A_{i+(j-1)m,i-1+(j-1)m} = \frac{1}{h^2}$$

$$A_{i+(j-1)m,i+jm} = \frac{1}{k^2}$$

$$A_{i+(j-1)m,i+(j-2)m} = \frac{1}{k^2}.$$

The right-hand side of the equation corresponding to (i,j) is

$$b_{i+(j-1)m} = f(x_i, y_j).$$

These entries of A and b hold for the interior points $1 < i < m, 1 < j < n$ of the grid in Figure 8.12.

Each boundary point needs an equation as well. Since we assume Dirichlet boundary conditions, they are quite simple:

Bottom	$w_{ij} = g_1(x_i)$ for $j = 1$, $1 \leq i \leq m$
Top side	$w_{ij} = g_2(x_i)$ for $j = n$, $1 \leq i \leq m$
Left side	$w_{ij} = g_3(y_j)$ for $i = 1$, $1 < j < n$
Right side	$w_{ij} = g_4(y_j)$ for $i = m$, $1 < j < n$

The Dirichlet conditions translate via Table 8.1 to

Bottom	$A_{i+(j-1)m,i+(j-1)m} = 1$, $b_{i+(j-1)m} = g_1(x_i)$ for $j = 1$, $1 \leq i \leq m$
Top side	$A_{i+(j-1)m,i+(j-1)m} = 1$, $b_{i+(j-1)m} = g_2(x_i)$ for $j = n$, $1 \leq i \leq m$
Left side	$A_{i+(j-1)m,i+(j-1)m} = 1$, $b_{i+(j-1)m} = g_3(y_j)$ for $i = 1$, $1 < j < n$
Right side	$A_{i+(j-1)m,i+(j-1)m} = 1$, $b_{i+(j-1)m} = g_4(y_j)$ for $i = m$, $1 < j < n$

All other entries of A and b are zero. The linear system $Av = b$ can be solved with appropriate method from Chapter 2. We illustrate this labeling system in the next example.

▶ **EXAMPLE 8.8** Apply the Finite Difference Method with $m = n = 5$ to approximate the solution of the Laplace equation $\Delta u = 0$ on $0, 1 \times 1, 2$ with the following Dirichlet boundary conditions:

$$u(x, 1) = \ln(x^2 + 1)$$
$$u(x, 2) = \ln(x^2 + 4)$$
$$u(0, y) = 2\ln y$$
$$u(1, y) = \ln(y^2 + 1).$$

MATLAB code for the Finite Difference Method follows:

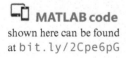

MATLAB code
shown here can be found
at bit.ly/2Cpe6pG

```
% Program 8.5 Finite difference solver for 2D Poisson equation
% with Dirichlet boundary conditions on a rectangle
% Input: rectangle domain [xl,xr]x[yb,yt] with MxN space steps
% Output: matrix w holding solution values
% Example usage: w=poisson(0,1,1,2,4,4)
function w=poisson(xl,xr,yb,yt,M,N)
f=@(x,y) 0; % define input function data
g1=@(x) log(x.^2+1); % define boundary values
g2=@(x) log(x.^2+4); % Example 8.8 is shown
g3=@(y) 2*log(y);
g4=@(y) log(y.^2+1);
m=M+1;n=N+1; mn=m*n;
```

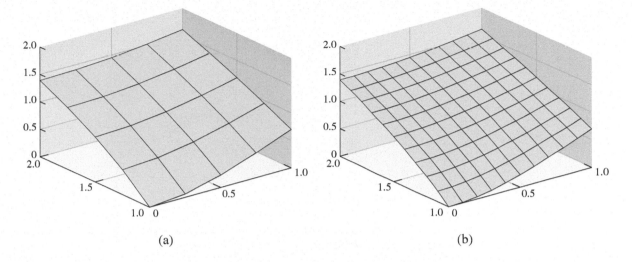

(a)   (b)

Figure 8.13 Finite Difference Method solution for the elliptic PDE in Example 8.8.
(a) $M = N = 4$, mesh sizes $h = k = 0.25$ (b) $M = N = 10$, mesh sizes $h = k = 0.1$.

```
h=(xr-xl)/M;h2=h^2;k=(yt-yb)/N;k2=k^2;
x=xl+(0:M)*h; % set mesh values
y=yb+(0:N)*k:
A=zeros(mn,mn);b=zeros(mn,1);
for i=2:m-1 % interior points
  for j=2:n-1
    A(i+(j-1)*m,i-1+(j-1)*m)=1/h2;A(i+(j-1)*m,i+1+(j-1)*m)=1/h2;
    A(i+(j-1)*m,i+(j-1)*m)=-2/h2-2/k2;
    A(i+(j-1)*m,i+(j-2)*m)=1/k2;A(i+(j-1)*m,i+j*m)=1/k2;
    b(i+(j-1)*m)=f(x(i),y(j));
  end
end
for i=1:m % bottom and top boundary points
  j=1;A(i+(j-1)*m,i+(j-1)*m)=1;b(i+(j-1)*m)~g1(x(i));
  j=n;A(i+(j-1)*m,i+(j-1)*m)=1;b(i+(j-1)*m)=g2(x(i));
end
for j=2:n-1 % left and right boundary points
  i=1;A(i+(j-1)*m,i+(j-1)*m)=1;b(i+(j-1)*m)=g3(y(j));
  i=m;A(i+(j-1)*m,i+(j-1)*m)=1;b(i+(j-1)*m)=g4(y(j));
end
v=A\b; % solve for solution in v labeling
w=reshape(v(1:mn),m,n);  %translate from v to w
mesh(x,y,w')
```

We will use the correct solution $u(x, y) = \ln(x^2 + y^2)$ to compare with the approximation at the nine mesh points in the square. Since $m = n = 5$, the mesh sizes are $h = k = 1/4$.

The solution finds the following nine interior values for u:

$$w_{24} = 1.1390 \quad w_{34} = 1.1974 \quad w_{44} = 1.2878$$
$$w_{23} = 0.8376 \quad w_{33} = 0.9159 \quad w_{43} = 1.0341$$
$$w_{22} = 0.4847 \quad w_{32} = 0.5944 \quad w_{42} = 0.7539$$

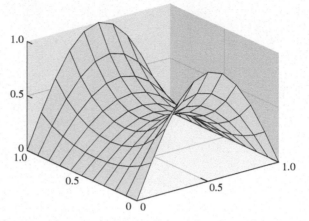

Figure 8.14 Electrostatic potential from the Laplace equation. Boundary conditions set in Example 8.9.

The approximate solution w_{ij} is plotted in Figure 8.13(a). It compares well with the exact solution $u(x, y) = \ln(x^2 + y^2)$ at the same points:

$$u(\tfrac{1}{4}, \tfrac{7}{4}) = 1.1394 \quad u(\tfrac{2}{4}, \tfrac{7}{4}) = 1.1977 \quad u(\tfrac{3}{4}, \tfrac{7}{4}) = 1.2879$$
$$u(\tfrac{1}{4}, \tfrac{6}{4}) = 0.8383 \quad u(\tfrac{2}{4}, \tfrac{6}{4}) = 0.9163 \quad u(\tfrac{3}{4}, \tfrac{6}{4}) = 1.0341$$
$$u(\tfrac{1}{4}, \tfrac{5}{4}) = 0.4855 \quad u(\tfrac{2}{4}, \tfrac{5}{4}) = 0.5947 \quad u(\tfrac{3}{4}, \tfrac{5}{4}) = 0.7538$$

Since second-order finite difference formulas were used, the error of the Finite Difference Method `poisson.m` is second order in h and k. Figure 8.13(b) shows a more accurate approximate solution, for $h = k = 0.1$. The MATLAB code `poisson.m` is written for a rectangular domain, but changes can be made to shift to more general domains. ◄

For another example, we use the Laplace equation to compute a potential.

► **EXAMPLE 8.9**   Find the electrostatic potential on the square $[0, 1] \times [0, 1]$, assuming no charge in the interior and assuming the following boundary conditions:

$$u(x, 0) = \sin \pi x$$
$$u(x, 1) = \sin \pi x$$
$$u(0, y) = 0$$
$$u(1, y) = 0.$$

The potential u satisfies the Laplace equation with Dirichlet boundary conditions. Using mesh size $h = k = 0.1$, or $M = N = 10$ in `poisson.m` results in the plot shown in Figure 8.14. ◄

Reality Check **8** *Heat Distribution on a Cooling Fin*

Heat sinks are used to move excess heat away from the point where it is generated. In this project, the steady-state distribution along a rectangular fin of a heat sink will be modeled. The heat energy will enter the fin along part of one side. The main goal will be to design the dimensions of the fin to keep the temperature within safe tolerances.

The fin shape is a thin rectangular slab, with dimensions $L_x \times L_y$ and width δ cm, where δ is relatively small. Due to the thinness of the slab, we will denote the temperature by $u(x, y)$ and consider it constant along the width dimension.

Heat moves in the following three ways: conduction, convection, and radiation. Conduction refers to the passing of energy between neighboring molecules, perhaps due to the movement of electrons, while in convection the molecules themselves move. Radiation, the movement of energy through photons, will not be considered here.

Conduction proceeds through a conducting material according to Fourier's first law

$$q = -KA\nabla u, \tag{8.41}$$

where q is heat energy per unit time (measured in watts), A is the cross-sectional area of the material, and ∇u is the gradient of the temperature. The constant K is called the **thermal conductivity** of the material. Convection is ruled by Newton's law of cooling,

$$q = -HA(u - u_b), \tag{8.42}$$

where H is a proportionality constant called the **convective heat transfer coefficient** and u_b is the ambient temperature, or **bulk temperature**, of the surrounding fluid (in this case, air).

The fin is a rectangle $[0, L_x] \times [0, L_y]$ by δ cm in the z direction, as illustrated in Figure 8.15(a). Energy equilibrium in a typical $\Delta x \times \Delta y \times \delta$ box interior to the fin, aligned along the x and y axes, says that the energy entering the box per unit time equals the energy leaving. The heat flux into the box through the two $\Delta y \times \delta$ sides and two $\Delta x \times \delta$ sides is by conduction, and through the two $\Delta x \times \Delta y$ sides is by convection, yielding the steady-state equation

$$-K\Delta y\delta u_x(x, y) + K\Delta y\delta u_x(x + \Delta x, y) - K\Delta x\delta u_y(x, y)$$
$$+ K\Delta x\delta u_y(x, y + \Delta y) - 2H\Delta x\Delta y u(x, y) = 0. \tag{8.43}$$

Here, we have set the bulk temperature $u_b = 0$ for convenience; thus, u will denote the difference between the fin temperature and the surroundings.

Dividing through by $\Delta x \Delta y$ gives

$$K\delta\frac{u_x(x + \Delta x, y) - u_x(x, y)}{\Delta x} + K\delta\frac{u_y(x, y + \Delta y) - u_y(x, y)}{\Delta y} = 2Hu(x, y),$$

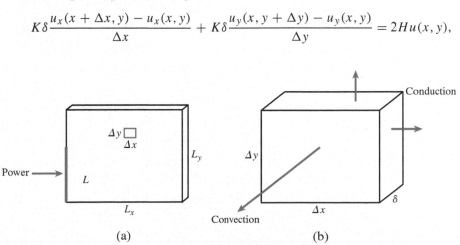

(a) (b)

Figure 8.15 Cooling fin in Reality Check 8. (a) Power input occurs along interval $[0, L]$ on left side of fin. (b) Energy transfer in small interior box is by conduction along the x and y directions, and by convection along the air interface.

and in the limit as $\Delta x, \Delta y \to 0$, the elliptic partial differential equation

$$u_{xx} + u_{yy} = \frac{2H}{K\delta} u \tag{8.44}$$

results.

Similar arguments imply the **convective** boundary condition

$$K u_{\text{normal}} = Hu$$

where u_{normal} is the partial derivative with respect to the outward normal direction $\vec{n}$. The convective boundary condition is known as a **Robin** boundary condition, one that involves both the function value and its derivative. Finally, we will assume that power enters the fin along one side according to Fourier's law,

$$u_{\text{normal}} = \frac{P}{L\delta K},$$

where P is the total power and L is the length of the input.

On a discrete grid with step sizes h and k, respectively, the finite difference approximation (5.8) can be used to approximate the PDE (8.44) as

$$\frac{u_{i+1,j} - 2u_{ij} + u_{i-1,j}}{h^2} + \frac{u_{i,j+1} - 2u_{ij} + u_{i,j-1}}{k^2} = \frac{2H}{K\delta} u_{ij}.$$

This discretization is used for the interior points (x_i, y_j) where $1 < i < m$, $1 < j < n$ for integers m, n. The fin edges obey the Robin conditions using the first derivative approximation

$$f'(x) = \frac{-3f(x) + 4f(x+h) - f(x+2h)}{2h} + O(h^2).$$

To apply this approximation to the fin edges, note that the outward normal direction translates to

$$u_{\text{normal}} = -u_y \text{ on bottom edge}$$
$$u_{\text{normal}} = u_y \text{ on top edge}$$
$$u_{\text{normal}} = -u_x \text{ on left edge}$$
$$u_{\text{normal}} = u_x \text{ on right edge.}$$

Second, note that the second-order first derivative approximation above yields

$$u_y \approx \frac{-3u(x,y) + 4u(x, y+k) - u(x, y+2k)}{2k} \text{ on bottom edge}$$
$$u_y \approx \frac{-3u(x,y) + 4u(x, y-k) - u(x, y-2k)}{-2k} \text{ on top edge}$$
$$u_x \approx \frac{-3u(x,y) + 4u(x+h, y) - u(x+2h, y)}{2h} \text{ on left edge}$$
$$u_x \approx \frac{-3u(x,y) + 4u(x-h, y) - u(x-2h, y)}{-2h} \text{ on right edge.}$$

Putting both together, the Robin boundary condition leads to the difference equations

$$\frac{-3u_{i1} + 4u_{i2} - u_{i3}}{2k} = -\frac{H}{K} u_{i1} \text{ on bottom edge}$$

$$\frac{-3u_{in} + 4u_{i,n-1} - u_{i,n-2}}{2k} = -\frac{H}{K}u_{in} \text{ on top edge}$$

$$\frac{-3u_{1j} + 4u_{2j} - u_{3j}}{2h} = -\frac{H}{K}u_{1j} \text{ on left edge}$$

$$\frac{-3u_{mj} + 4u_{m-1,j} - u_{m-2,j}}{2h} = -\frac{H}{K}u_{mj} \text{ on right edge.}$$

If we assume that the power enters along the left side of the fin, Fourier's law leads to the equation

$$\frac{-3u_{1j} + 4u_{2j} - u_{3j}}{2h} = -\frac{P}{L\delta K}. \tag{8.45}$$

There are mn equations in the mn unknowns u_{ij}, $1 \le i \le m$, $1 \le j \le n$ to solve.

Assume that the fin is composed of aluminum, whose thermal conductivity is $K = 1.68$ W/cm $^\circ$C (watts per centimeter-degree Celsius). Assume that the convective heat transfer coefficient is $H = 0.005$ W/cm^2 $^\circ$C, and that the room temperature is $u_b = 20^\circ$C.

Suggested activities:

1. Begin with a fin of dimensions 2×2 cm, with 1 mm thickness. Assume that 5W of power is input along the entire left edge, as if the fin were attached to dissipate power from a CPU chip with $L = 2$ cm side length. Solve the PDE (8.44) with $M = N = 10$ steps in the x and y directions. Use the mesh command to plot the resulting heat distribution over the xy-plane. What is the maximum temperature of the fin, in $^\circ$C ?

2. Increase the size of the fin to 4×4 cm. Input 5W of power along the interval $[0, 2]$ on the left side of the fin, as in the previous step. Plot the resulting distribution, and find the maximum temperature. Experiment with increased values of M and N. How much does the solution change?

3. Find the maximum power that can be dissipated by a 4×4 cm fin while keeping the maximum temperature less than 80°C. Assume that the bulk temperature is 20°C and the power input is along 2 cm, as in steps 1 and 2.

4. Replace the aluminum fin with a copper fin, with thermal conductivity $K = 3.85$ W/cm $^\circ$C. Find the maximum power that can be dissipated by a 4×4 cm fin with the 2 cm power input in the optimal placement, while keeping the maximum temperature below 80°C.

5. Plot the maximum power that can be dissipated in step 4 (keeping maximum temperature below 80 degrees) as a function of thermal conductivity, for $1 \le K \le 5$ W/cm$^\circ$C.

6. Redo step 4 for a water-cooled fin. Assume that water has a convective heat transfer coefficient of $H = 0.1$ W/cm^2 $^\circ$C, and that the ambient water temperature is maintained at 20°C.

The design of cooling fins for desktop and laptop computers is a fascinating engineering problem. To dissipate ever greater amounts of heat, several fins are needed in a small space, and fans are used to enhance convection near the fin edges. The addition of fans to complicated fin geometry moves the simulation into the realm of computational fluid dynamics, a vital area of modern applied mathematics. 📍

8.3.2 Finite Element Method for elliptic equations

A somewhat more flexible approach to solving partial differential equations arose from the structural engineering community in the mid-20th century. The Finite

Element Method converts the differential equation into a variational equivalent called the weak form of the equation, and uses the powerful idea of orthogonality in function spaces to stabilize its calculations. Moreover, the resulting system of linear equations can have considerable symmetry in its structure matrix, even when the underlying geometry is complicated.

We will apply finite elements by using the Galerkin Method, as introduced in Chapter 7 for ordinary differential equation boundary value problems. The method for PDEs follows the same steps, although the bookkeeping requirements are more extensive. Consider the Dirichlet problem for the elliptic equation

$$\Delta u + r(x, y)u = f(x, y) \quad \text{in } R$$
$$u = g(x, y) \quad \text{on } S \tag{8.46}$$

where the solution $u(x, y)$ is defined on a region R in the plane bounded by a piecewise-smooth closed curve S.

We will use an L^2 function space over the region R, as in Chapter 7. Let

$$L^2(R) = \left\{ \text{functions } \phi(x, y) \text{ on } R \,\middle|\, \int\int_R \phi(x, y)^2 \, dx \, dy \text{ exists and is finite} \right\}.$$

Denote by $L_0^2(R)$ the subspace of $L^2(R)$ consisting of functions that are zero on the boundary S of the region R.

The goal will be to minimize the squared error of the elliptic equation in (8.46) by forcing the residual $\Delta u(x, y) + r(x, y)u(x, y) - f(x, y)$ to be orthogonal to a large subspace of $L^2(R)$. Let $\phi_1(x, y), \dots, \phi_P(x, y)$ be elements of $L^2(R)$. The orthogonality assumption takes the form

$$\int\int_R (\Delta u + ru - f)\phi_p \, dx \, dy = 0,$$

or

$$\int\int_R (\Delta u + ru)\phi_p \, dx \, dy = \int\int_R f\phi_p \, dx \, dy \tag{8.47}$$

for each $1 \le p \le P$. The form (8.47) is called the **weak form** of the elliptic equation (8.46).

The version of integration by parts needed to apply the Galerkin Method is contained in the following fact:

THEOREM 8.7   **Green's First Identity.** Let R be a bounded region with piecewise smooth boundary S. Let u and v be smooth functions, and let n denote the outward unit normal along the boundary. Then

$$\int\int_R v\Delta u = \int_S v\frac{\partial u}{\partial n} dS - \int\int_R \nabla u \cdot \nabla v. \qquad \blacksquare$$

The directional derivative can be calculated as

$$\frac{\partial u}{\partial n} = \nabla u \cdot (n_x, n_y),$$

where (n_x, n_y) denotes the outward normal unit vector on the boundary S of R. Green's identity applied to the weak form (8.47) yields

$$\int_S \phi_p \frac{\partial u}{\partial n} \, dS - \int\int_R (\nabla u \cdot \nabla \phi_p) \, dx \, dy + \int\int_R ru\phi_p \, dx \, dy = \int\int_R f\phi_p \, dx \, dy. \tag{8.48}$$

The essence of the Finite Element Method is to substitute

$$w(x, y) = \sum_{q=1}^{P} v_q \phi_q(x, y) \tag{8.49}$$

for u into the weak form of the partial differential equation, and then determine the unknown constants v_q. Assume for the moment that ϕ_p belongs to $L_0^2(R)$, that is, $\phi_p(S) = 0$. Substituting the form (8.49) into (8.48) results in

$$-\int\int_R \left(\sum_{q=1}^{P} v_q \nabla \phi_q\right) \cdot \nabla \phi_p \, dx\, dy + \int\int_R r \left(\sum_{q=1}^{P} v_q \phi_q\right) \phi_p \, dx\, dy = \int\int_R f \phi_p \, dx\, dy$$

for each ϕ_p in $L_0^2(R)$. Factoring out the constants v_q yields

$$\sum_{q=1}^{P} v_q \left[\int\int_R \nabla \phi_q \cdot \nabla \phi_p \, dx\, dy - \int\int_R r \phi_q \phi_p \, dx\, dy\right] = -\int_R f \phi_p dx\, dy. \tag{8.50}$$

For each ϕ_p belonging to $L_0^2(R)$, we have developed a linear equation in the unknowns $v_1, \ldots, v_P$. In matrix form, the equation is $Av = b$, where the entries of the pth row of A and b are

$$A_{pq} = \int\int_R \nabla \phi_q \cdot \nabla \phi_p \, dx\, dy - \int\int_R r \phi_q \phi_p \, dx\, dy \tag{8.51}$$

and

$$b_p = -\int\int_R f \phi_p \, dx\, dy. \tag{8.52}$$

We are now prepared to choose explicit functions for the finite elements ϕ_p and plan a computation. We follow the lead of Chapter 7 in choosing linear B-splines, piecewise-linear functions of x, y that live on triangles in the plane. For concreteness, let the region R be a rectangle, and form a triangulation with nodes (x_i, y_j) chosen from a rectangular grid. We will reuse the $M \times N$ grid from the previous section, shown in Figure 8.16(a), where we set $m = M + 1$ and $n = N + 1$. As before, we will denote the grid step size in the x and y directions as h and k, respectively. Figure 8.16(b) shows the triangulation of the rectangular region that we will use.

Our choice of finite element functions ϕ_p from $L^2(R)$ will be the $P = mn$ piecewise-linear functions, each of which takes the value 1 at one grid point in Figure 8.16(a) and zero at the other $mn - 1$ grid points. In other words, $\phi_1, \ldots, \phi_{mn}$ are determined by the equality $\phi_{i+(j-1)m}(x_i, y_j) = 1$ and $\phi_{i+(j-1)m}(x_{i'}, y_{j'}) = 0$ for all other grid points $(x_{i'}, y_{j'})$, while being linear on each triangle in Figure 8.16(b). We are once again using the numbering system of Table 8.1, on page 400. Each $\phi_p(x, y)$ is differentiable, except along the triangle edges, and is therefore a Riemann-integrable function belonging to $L^2(R)$. Note that for every nonboundary point (x_i, y_j) of the rectangle R, $\phi_{i+(j-1)m}$ belongs to $L_0^2(R)$. Moreover, due to assumption (8.49), they satisfy

$$w(x_i, y_j) = \sum_{i=1}^{m}\sum_{j=1}^{n} v_{i+(j-1)m} \phi_{i+(j-1)m}(x_i, y_j) = v_{i+(j-1)m}$$

for $i = 1, \ldots, m$, $j = 1, \ldots, n$. Therefore, the approximation w to the correct solution u at (x_i, v_j) will be directly available once the system $Av = b$ is solved. This convenience is the reason B-splines are a good choice for finite element functions.

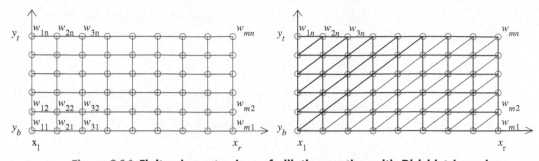

Figure 8.16 Finite element solver of elliptic equation with Dirichlet boundary conditions. (a) Mesh is same as used for finite difference solver. (b) A possible triangulation of the region. Each interior point is a vertex of six different triangles.

It remains to calculate the matrix entries (8.51) and (8.52) and solve $Av = b$. To calculate these entries, we gather a few facts about B-splines in the plane. The integrals of the piecewise-linear functions are easily approximated by the two-dimensional Midpoint Rule. Define the **barycenter** of a region in the plane as the point $(\overline{x}, \overline{y})$ where

$$\overline{x} = \frac{\int \int_R x \, dx \, dy}{\int \int_R 1 \, dx \, dy}, \quad \overline{y} = \frac{\int \int_R y \, dx \, dy}{\int \int_R 1 \, dx \, dy}.$$

If R is a triangle with vertices $(x_1, y_1), (x_2, y_2), (x_3, y_3)$, then the barycenter is (see Exercise 8)

$$\overline{x} = \frac{x_1 + x_2 + x_3}{3}, \quad \overline{y} = \frac{y_1 + y_2 + y_3}{3}.$$

LEMMA 8.8   The average value of a linear function $L(x, y)$ on a plane region R is $L(\overline{x}, \overline{y})$, the value at the barycenter. In other words, $\int \int_R L(x, y) \, dx \, dy = L(\overline{x}, \overline{y}) \cdot$ area (R). ∎

Proof. Let $L(x, y) = a + bx + cy$. Then

$$\int \int_R L(x, y) \, dx \, dy = \int \int_R (a + bx + cy) \, dx \, dy$$

$$= a \int \int_R dx \, dy + b \int \int_R x \, dx \, dy + c \int \int_R y \, dx \, dy$$

$$= \text{area} (R) \cdot (a + b\overline{x} + c\overline{y}).$$

□

Lemma 8.8 leads to a generalization of the Midpoint Rule of Chapter 5 that is useful for approximating the entries of (8.51) and (8.52). Taylor's Theorem for functions of two variables says that

$$f(x, y) = f(\overline{x}, \overline{y}) + \frac{\partial f}{\partial x}(\overline{x}, \overline{y})(x - \overline{x}) + \frac{\partial f}{\partial y}(\overline{x}, \overline{y})(y - \overline{y})$$

$$+ O((x - \overline{x})^2, (x - \overline{x})(y - \overline{y}), (y - \overline{y})^2)$$

$$= L(x, y) + O((x - \overline{x})^2, (x - \overline{x})(y - \overline{y}), (y - \overline{y})^2).$$

Therefore,

$$
\int\int_R f(x,y)\,dx\,dy = \int\int_R L(x,y)\,dx\,dy + \int\int_R O((x-\bar{x})^2, (x-\bar{x})(y-\bar{y}), (y-\bar{y})^2)\,dx\,dy
$$

$$
= \text{area}\,(R) \cdot L(\bar{x},\bar{y}) + O(h^4) = \text{area}\,(R) \cdot f(\bar{x},\bar{y}) + O(h^4),
$$

where h is the **diameter** of R, the largest distance between two points of R, and where we have used Lemma 8.8. This is the Midpoint Rule in two dimensions.

Midpoint Rule in two dimensions

$$
\int\int_R f(x,y)\,dx\,dy = \text{area}\,(R) \cdot f(\bar{x},\bar{y}) + O(h^4), \tag{8.53}
$$

where $(\bar{x},\bar{y})$ is the barycenter of the bounded region R and $h = \text{diam}(R)$.

The Midpoint Rule shows that to apply the Finite Element Method with $O(h^2)$ convergence, we need to only approximate the integrals in (8.51) and (8.52) by evaluating integrands at triangle barycenters. For the B-spline functions ϕ_p, this is particularly easy. Proofs of the next two lemmas are deferred to Exercises 9 and 10.

LEMMA 8.9 Let $\phi(x,y)$ be a linear function on the triangle T with vertices (x_1,y_1), (x_2,y_2), (x_3,y_3), satisfying $\phi(x_1,y_1) = 1, \phi(x_2,y_2) = 0$, and $\phi(x_3,y_3) = 0$. Then $\phi(\bar{x},\bar{y}) = 1/3$. ∎

LEMMA 8.10 Let $\phi_1(x,y)$ and $\phi_2(x,y)$ be the linear functions on the triangle T with vertices (x_1,y_1), (x_2,y_2), and (x_3,y_3), satisfying $\phi_1(x_1,y_1) = 1, \phi_1(x_2,y_2) = 0, \phi_1(x_3,y_3) = 0$, $\phi_2(x_1,y_1) = 0, \phi_2(x_2,y_2) = 1$, and $\phi_2(x_3,y_3) = 0$. Let $f(x,y)$ be a twice-differentiable function. Set

$$
d = \det \begin{bmatrix} 1 & 1 & 1 \\ x_1 & x_2 & x_3 \\ y_1 & y_2 & y_3 \end{bmatrix}.
$$

Then

(a) the triangle T has area $|d|/2$

(b) $\nabla\phi_1(x,y) = \left(\dfrac{y_2 - y_3}{d}, \dfrac{x_3 - x_2}{d} \right)$

(c) $\int\int_T \nabla\phi_1 \cdot \nabla\phi_1 \,dx\,dy = \dfrac{(x_2 - x_3)^2 + (y_2 - y_3)^2}{2|d|}$

(d) $\int\int_T \nabla\phi_1 \cdot \nabla\phi_2 \,dx\,dy = \dfrac{-(x_1 - x_3)(x_2 - x_3) - (y_1 - y_3)(y_2 - y_3)}{2|d|}$

(e) $\int\int_T f\phi_1\phi_2 \,dx\,dy = f(\bar{x},\bar{y})|d|/18 + O(h^4) = \int\int_T f\phi_1^2 \,dx\,dy$

(f) $\int\int_T f\phi_1 \,dx\,dy = f(\bar{x},\bar{y})|d|/6 + O(h^4)$

where $(\bar{x},\bar{y})$ is the barycenter of T and $h = \text{diam}(T)$. ∎

We can now calculate the matrix entries of A. Consider a vertex (x_i, y_j) that is not on the boundary S of the rectangle. Then $\phi_{i+(j-1)m}$ belongs to $L_0^2(R)$, and according to (8.51) with $p = q = i + (j-1)m$, the matrix entry $A_{i+(j-1)m, i+(j-1)m}$ is composed of two integrals. The integrands are zero outside of the six triangles shown

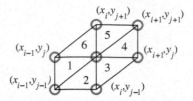

Figure 8.17 Detail of the (i, j) interior point from Figure 8.16(b). Each interior point (x_i, y_j) is surrounded by six triangles, numbered as shown. The B-spline function $\phi_{i+(j-1)m}$ is linear, takes the value 1 at the center, and is zero outside of these six triangles.

in Figure 8.17. The triangles have horizontal and vertical sides h and k, respectively. For the first integral, summing from triangle 1 to triangle 6, respectively, we can use Lemma 8.10(c) to sum the six contributions

$$\frac{k^2}{2hk} + \frac{h^2}{2hk} + \frac{h^2 + k^2}{2hk} + \frac{k^2}{2hk} + \frac{h^2}{2hk} + \frac{h^2 + k^2}{2hk} = \frac{2(h^2 + k^2)}{hk}. \tag{8.54}$$

For the second integral of (8.51), we use Lemma 8.10(e). Again, the integrals are zero except for the six triangles shown. The barycenters of the six triangles are

$$B_1 = (x_i - \frac{2}{3}h, y_j - \frac{1}{3}k)$$

$$B_2 = (x_i - \frac{1}{3}h, y_j - \frac{2}{3}k)$$

$$B_3 = (x_i + \frac{1}{3}h, y_j - \frac{1}{3}k)$$

$$B_4 = (x_i + \frac{2}{3}h, y_j + \frac{1}{3}k)$$

$$B_5 = (x_i + \frac{1}{3}h, y_j + \frac{2}{3}k)$$

$$B_6 = (x_i - \frac{1}{3}h, y_j + \frac{1}{3}k). \tag{8.55}$$

The second integral contributes $-(hk/18)[r(B_1) + r(B_2) + r(B_3) + r(B_4) + r(B_5) + r(B_6)]$, and so summing up (8.54) and (8.55),

$$A_{i+(j-1)m,i+(j-1)m} = \frac{2(h^2 + k^2)}{hk} - \frac{hk}{18}[r(B_1) + r(B_2) + r(B_3)$$
$$+ r(B_4) + r(B_5) + r(B_6)]. \tag{8.56}$$

Similar usage of Lemma 8.10 (see Exercise 12) shows that

$$A_{i+(j-1)m,i-1+(j-1)m} = -\frac{k}{h} - \frac{hk}{18}[r(B_6) + r(B_1)]$$

$$A_{i+(j-1)m,i-1+(j-2)m} = -\frac{hk}{18}[r(B_1) + r(B_2)]$$

$$A_{i+(j-1)m,i+(j-2)m} = -\frac{h}{k} - \frac{hk}{18}[r(B_2) + r(B_3)]$$

$$A_{i+(j-1)m,i+1+(j-1)m} = -\frac{k}{h} - \frac{hk}{18}[r(B_3) + r(B_4)]$$

$$A_{i+(j-1)m,i+1+jm} = -\frac{hk}{18}[r(B_4) + r(B_5)]$$

$$A_{i+(j-1)m,i+jm} = -\frac{h}{k} - \frac{hk}{18}r(B_5) + r(B_6). \tag{8.57}$$

Calculating the entries b_p makes use of Lemma 8.10(f), which implies that for $p = i + (j - 1)m$,

$$b_{i+(j-1)m} = -\frac{hk}{6}f(B_1) + f(B_2) + f(B_3) + f(B_4) + f(B_5) + f(B_6). \tag{8.58}$$

For finite element functions on the boundary, $\phi_{i+(j-1)m}$ does not belong to $L_0^2(R)$, and the equations

$$A_{i+(j-1)m,i+(j-1)m} = 1$$
$$b_{i+(j-1)m} = g(x_i, y_j) \tag{8.59}$$

will be used to guarantee the Dirichlet boundary condition $v_{i+(j-1)m} = g(x_i, y_j)$, where (x_i, y_j) is a boundary point.

With these formulas, it is straightforward to build a MATLAB implementation of the finite element solver on a rectangle with Dirichlet boundary conditions. The program consists of setting up the matrix A and vector b using (8.56) – (8.59), and solving $Av = b$. Although the MATLAB backslash operation is used in the code, for real applications it might be replaced by a sparse solver as in Chapter 2.

MATLAB code

shown here can be found at bit.ly/2EpdrHq

```
% Program 8.6 Finite element solver for 2D PDE
% with Dirichlet boundary conditions on a rectangle
% Input: rectangle domain [xl,xr]x[yb,yt] with MxN space steps
% Output: matrix w holding solution values
% Example usage: w=poissonfem(0,1,1,2,4,4)
function w=poissonfem(xl,xr,yb,yt,M,N)
f=@(x,y) 0; % define input function data
r=@(x,y) 0;
g1=@(x) log(x.^2+1); % define boundary values on bottom
g2=@(x) log(x.^2+4); % top
g3=@(y) 2*log(y);    % left side
g4=@(y) log(y.^2+1); % right side
m=M+1; n=N+1; mn=m*n;
h=(xr-xl)/M; h2=h^2; k=(yt-yb)/N; k2=k^2; hk=h*k;
x=xl+(0:M)*h; % set mesh values
y=yb+(0:N)*k;
A=zeros(mn,mn); b=zeros(mn,1);
for i=2:m-1 % interior points
 for j=2:n-1
  rsum=r(x(i)-2*h/3,y(j)-k/3)+r(x(i)-h/3,y(j)-2*k/3)...
      +r(x(i)+h/3,y(j)-k/3);
  rsum=rsum+r(x(i)+2*h/3,y(j)+k/3)+r(x(i)+h/3,y(j)+2*k/3)...
      +r(x(i)-h/3,y(j)+k/3);
  A(i+(j-1)*m,i+(j-1)*m)=2*(h2+k2)/(hk)-hk*rsum/18;
  A(i+(j-1)*m,i-1+(j-1)*m)=-k/h-hk*(r(x(i)-h/3,y(j)+k/3)...
                  +r(x(i)-2*h/3,y(j)-k/3))/18;
  A(i+(j-1)*m,i-1+(j-2)*m)=-hk*(r(x(i)-2*h/3,y(j)-k/3)...
                  +r(x(i)-h/3,y(j)-2*k/3))/18;
  A(i+(j-1)*m,i+(j-2)*m)=-h/k-hk*(r(x(i)-h/3,y(j)-2*k/3)...
                  +r(x(i)+h/3,y(j)-k/3))/18;
  A(i+(j-1)*m,i+1+(j-1)*m)=-k/h-hk*(r(x(i)+h/3,y(j)-k/3)...
                  +r(x(i)+2*h/3,y(j)+k/3))/18;
```

```
          A(i+(j-1)*m,i+1+j*m)=-hk*(r(x(i)+2*h/3,y(j)+k/3)...
                             +r(x(i)+h/3,y(j)+2*k/3))/18;
          A(i+(j-1)*m,i+j*m)=-h/k-hk*(r(x(i)+h/3,y(j)+2*k/3)...
                             +r(x(i)-h/3,y(j)+k/3))/18;
          fsum=f(x(i)-2*h/3,y(j)-k/3)+f(x(i)-h/3,y(j)-2*k/3)...
              +f(x(i)+h/3,y(j)-k/3);
          fsum=fsum+f(x(i)+2*h/3,y(j)+k/3)+f(x(i)+h/3,y(j)+2*k/3)...
              +f(x(i)-h/3,y(j)+k/3);
          b(i+(j-1)*m)=-h*k*fsum/6;
         end
        end
        for i=1:m % boundary points
          j=1;A(i+(j-1)*m,i+(j-1)*m)=1;b(i+(j-1)*m)=g1(x(i));
          j=n;A(i+(j-1)*m,i+(j-1)*m)=1;b(i+(j-1)*m)=g2(x(i));
        end
        for j=2:n-1
          i=1;A(i+(j-1)*m,i+(j-1)*m)=1;b(i+(j-1)*m)=g3(y(j));
          i=m;A(i+(j-1)*m,i+(j-1)*m)=1;b(i+(j-1)*m)=g4(y(j));
        end
        v=A\b; % solve for solution in v numbering
        w=reshape(v(1:mn),m,n);
        mesh(x,y,w')
```

▶ **EXAMPLE 8.10**   Apply the Finite Element Method with $M = N = 4$ to approximate the solution of the Laplace equation $\Delta u = 0$ on $[0, 1] \times [1, 2]$ with the Dirichlet boundary conditions:

$$u(x, 1) = \ln(x^2 + 1)$$
$$u(x, 2) = \ln(x^2 + 4)$$
$$u(0, y) = 2 \ln y$$
$$u(1, y) = \ln(y^2 + 1).$$

Since $M = N = 4$, there is a $mn \times mn$ linear system to solve. Sixteen of the 25 equations are evaluation of the boundary conditions. Solving $Av = b$ yields

$$\begin{array}{lll}
w_{24} = 1.1390 & w_{34} = 1.1974 & w_{44} = 1.2878 \\
w_{23} = 0.8376 & w_{33} = 0.9159 & w_{43} = 1.0341 \\
w_{22} = 0.4847 & w_{32} = 0.5944 & w_{42} = 0.7539
\end{array}$$

agreeing with the results in Example 8.8.   ◀

▶ **EXAMPLE 8.11**   Apply the Finite Element Method with $M = N = 16$ to approximate the solution of the elliptic Dirichlet problem

$$\begin{cases}
\Delta u + 4\pi^2 u = 2 \sin 2\pi y \\
u(x, 0) = 0 \text{ for } 0 \le x \le 1 \\
u(x, 1) = 0 \text{ for } 0 \le x \le 1 \\
u(0, y) = 0 \text{ for } 0 \le y \le 1 \\
u(1, y) = \sin 2\pi y \text{ for } 0 \le y \le 1.
\end{cases}$$

We define $r(x, y) = 4\pi^2$ and $f(x, y) = 2 \sin 2\pi y$. Since $m = n = 17$, the grid is 17×17, meaning that the matrix A is 289×289. The solution is computed approximately within a maximum error of about 0.023, compared with the correct solution $u(x, y) = x^2 \sin 2\pi y$. The approximate solution w is shown in Figure 8.18.   ◀

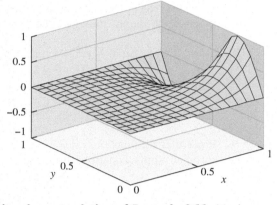

Figure 8.18 Finite element solution of Example 8.11. Maximum error on $0, 1 \times 0, 1$ is 0.023.

▶ **ADDITIONAL EXAMPLES**

1. Prove that $u(x, y) = e^{1-x^2-y^2}$ is the solution on $-1, 1 \times -1.1$ of the boundary value problem

$$
\begin{cases}
\Delta u = 4(x^2 + y^2 - 1)u \\
u(x, -1) = u(x, 1) = e^{-x^2} \\
u(-1, y) = u(1, y) = e^{-y^2}.
\end{cases}
$$

2. Use the Finite Element Method with $h = k = 0.1$ to approximate the solution of the boundary value problem in Additional Exercise 8.3.1. Plot the solution on the square $-1, 1 \times -1, 1$.

⊡ **Solutions** for Additional Examples can be found at `bit.ly/2NQeM9B`

8.3 Exercises

⊡ **Solutions** for Exercises numbered in blue can be found at `bit.ly/2R1lYAC`

1. Show that $u(x, y) - \ln(x^2 + y^2)$ is a solution to the Laplace equation with Dirichlet boundary conditions of Example 8.8.

2. Show that (a) $u(x, y) = x^2 y - 1/3\, y^3$ and (b) $u(x, y) = 1/6\, x^4 - x^2 y^2 + 1/6\, y^4$ are harmonic functions.

3. Prove that the functions (a) $u(x, y) = e^{-\pi y} \sin \pi x$, (b) $u(x, y) = \sinh \pi x \sin \pi y$ are solutions of the Laplace equation with the specified boundary conditions:

(a)
$$
\begin{cases}
u(x, 0) = \sin \pi x \text{ for } 0 \le x \le 1 \\
u(x, 1) = e^{-\pi} \sin \pi x \text{ for } 0 \le x \le 1 \\
u(0, y) = 0 \text{ for } 0 \le y \le 1 \\
u(1, y) = 0 \text{ for } 0 \le y \le 1
\end{cases}
$$

(b)
$$
\begin{cases}
u(x, 0) = 0 \text{ for } 0 \le x \le 1 \\
u(x, 1) = 0 \text{ for } 0 \le x \le 1 \\
u(0, y) = 0 \text{ for } 0 \le y \le 1 \\
u(1, y) = \sinh \pi \sin \pi y \text{ for } 0 \le y \le 1
\end{cases}
$$

4. Prove that the functions (a) $u(x, y) = e^{-xy}$, (b) $u(x, y) = (x^2 + y^2)^{3/2}$ are solutions of the specified Poisson equation with the given boundary conditions:

(a)
$$
\begin{cases}
\Delta u = e^{-xy}(x^2 + y^2) \\
u(x, 0) = 1 \text{ for } 0 \le x \le 1 \\
u(x, 1) = e^{-x} \text{ for } 0 \le x \le 1 \\
u(0, y) = 1 \text{ for } 0 \le y \le 1 \\
u(1, y) = e^{-y} \text{ for } 0 \le y \le 1
\end{cases}
$$

(b)
$$
\begin{cases}
\Delta u = 9\sqrt{x^2 + y^2} \\
u(x, 0) = x^3 \text{ for } 0 \le x \le 1 \\
u(x, 1) = (1 + x^2)^{3/2} \text{ for } 0 \le x \le 1 \\
u(0, y) = y^3 \text{ for } 0 \le y \le 1 \\
u(1, y) = (1 + y^2)^{3/2} \text{ for } 0 \le y \le 1
\end{cases}
$$

5. Prove that the functions (a) $u(x, y) = \sin \frac{\pi}{2} xy$, (b) $u(x, y) = e^{xy}$ are solutions of the specified elliptic equation with the given Dirichlet boundary conditions:

(a) $\begin{cases} \Delta u + \frac{\pi^2}{4}(x^2 + y^2)u = 0 \\ u(x, 0) = 0 \text{ for } 0 \le x \le 1 \\ u(x, 1) = \sin \frac{\pi}{2} x \text{ for } 0 \le x \le 1 \\ u(0, y) = 0 \text{ for } 0 \le y \le 1 \\ u(1, y) = \sin \frac{\pi}{2} y \text{ for } 0 \le y \le 1 \end{cases}$ (b) $\begin{cases} \Delta u = (x^2 + y^2)u \\ u(x, 0) = 1 \text{ for } 0 \le x \le 1 \\ u(x, 1) = e^x \text{ for } 0 \le x \le 1 \\ u(0, y) = 1 \text{ for } 0 \le y \le 1 \\ u(1, y) = e^y \text{ for } 0 \le y \le 1 \end{cases}$

6. Prove that the functions (a) $u(x, y) = e^{x+2y}$, (b) $u(x, y) = y/x$ are solutions of the specified elliptic equation with the given Dirichlet boundary conditions:

(a) $\begin{cases} \Delta u = 5u \\ u(x, 0) = e^x \text{ for } 0 \le x \le 1 \\ u(x, 1) = e^{x+2} \text{ for } 0 \le x \le 1 \\ u(0, y) = e^{2y} \text{ for } 0 \le y \le 1 \\ u(1, y) = e^{2y+1} \text{ for } 0 \le y \le 1 \end{cases}$ (b) $\begin{cases} \Delta u = \dfrac{2u}{x^2} \\ u(x, 0) = 0 \text{ for } 1 \le x \le 2 \\ u(x, 1) = 1/x \text{ for } 1 \le x \le 2 \\ u(1, y) = y \text{ for } 0 \le y \le 1 \\ u(2, y) = y/2 \text{ for } 0 \le y \le 1 \end{cases}$

7. Prove that the functions (a) $u(x, y) = x^2 + y^2$, (b) $u(x, y) = y^2/x$ are solutions of the specified elliptic equation with the given Dirichlet boundary conditions:

(a) $\begin{cases} \Delta u + \dfrac{u}{x^2 + y^2} = 5 \\ u(x, 1) = x^2 + 1 \text{ for } 1 \le x \le 2 \\ u(x, 2) = x^2 + 4 \text{ for } 1 \le x \le 2 \\ u(1, y) = y^2 + 1 \text{ for } 1 \le y \le 2 \\ u(2, y) = y^2 + 4 \text{ for } 1 \le y \le 2 \end{cases}$ (b) $\begin{cases} \Delta u - \dfrac{2u}{x^2} = \dfrac{2}{x} \\ u(x, 0) = 0 \text{ for } 1 \le x \le 2 \\ u(x, 2) = 4/x \text{ for } 1 \le x \le 2 \\ u(1, y) = y^2 \text{ for } 0 \le y \le 2 \\ u(2, y) = y^2/2 \text{ for } 0 \le y \le 2 \end{cases}$

8. Show that the barycenter of a triangle with vertices (x_1, y_1), (x_2, y_2), (x_3, y_3) is $\bar{x} = (x_1 + x_2 + x_3)/3$, $\bar{y} = (y_1 + y_2 + y_3)/3$.

9. Prove Lemma 8.9.

10. Prove Lemma 8.10.

11. Derive the barycenter coordinates of (8.55).

12. Derive the matrix entries in (8.57).

13. Show that the Laplace equation $\Delta T = 0$ on the rectangle $0, L \times 0, H$ with Dirichlet boundary conditions $T = T_0$ on the three sides $x = 0, x = L$, and $y = 0$, and $T = T_1$ on the side $y = H$ has solution

$$T(x, y) = T_0 + \sum_{k=0}^{\infty} C_k \sin \frac{(2k+1)\pi x}{L} \sinh \frac{(2k+1)\pi y}{L},$$

where

$$C_k = \frac{4(T_1 - T_0)}{(2k+1)\pi \sinh \frac{(2k+1)\pi H}{L}}.$$

8.3 Computer Problems

1. Solve the Laplace equation problems in Exercise 3 on $0 \le x \le 1, 0 \le y \le 1$ by the Finite Difference Method with $h = k = 0.1$. Use MATLAB's mesh command to plot the solution.

2. Solve the Poisson equation problems in Exercise 4 on $0 \le x \le 1, 0 \le y \le 1$ by the Finite Difference Method with $h = k = 0.1$. Plot the solution.

3. Use the Finite Difference Method with $h = k = 0.1$ to approximate the electrostatic potential on the square $0 \le x, y \le 1$ from the Laplace equation with the specified boundary conditions. Plot the solution.

(a)
$$\begin{cases} u(x,0) = 0 \text{ for } 0 \le x \le 1 \\ u(x,1) = \sin \pi x \text{ for } 0 \le x \le 1 \\ u(0,y) = 0 \text{ for } 0 \le y \le 1 \\ u(1,y) = 0 \text{ for } 0 \le y \le 1 \end{cases}$$

(b)
$$\begin{cases} u(x,0) = \sin \frac{\pi}{2} x \text{ for } 0 \le x \le 1 \\ u(x,1) = \cos \frac{\pi}{2} x \text{ for } 0 \le x \le 1 \\ u(0,y) = \sin \frac{\pi}{2} y \text{ for } 0 \le y \le 1 \\ u(1,y) = \cos \frac{\pi}{2} y \text{ for } 0 \le y \le 1 \end{cases}$$

4. Use the Finite Difference Method with $h = k = 0.1$ to approximate the electrostatic potential on the square $0 \le x, y \le 1$ from the Laplace equation with the specified boundary conditions. Plot the solution.

(a)
$$\begin{cases} u(x,0) = 0 \text{ for } 0 \le x \le 1 \\ u(x,1) = x^3 \text{ for } 0 \le x \le 1 \\ u(0,y) = 0 \text{ for } 0 \le y \le 1 \\ u(1,y) = y^2 \text{ for } 0 \le y \le 1 \end{cases}$$

(b)
$$\begin{cases} u(x,0) = 0 \text{ for } 0 \le x \le 1 \\ u(x,1) = x \sin \frac{\pi}{2} x \text{ for } 0 \le x \le 1 \\ u(0,y) = 0 \text{ for } 0 \le y \le 1 \\ u(1,y) = y \text{ for } 0 \le y \le 1 \end{cases}$$

5. Hydrostatic pressure can be expressed as the hydraulic head, defined as the equivalent height u of a column of water exerting that pressure. In an underground reservoir, steady-state groundwater flow satisfies the Laplace equation $\Delta u = 0$. Assume that the reservoir has dimensions $2 \text{ km} \times 1 \text{ km}$, and water table heights

$$\begin{cases} u(x,0) = 0.01 \text{ for } 0 \le x \le 2 \\ u(x,1) = 0.01 + 0.003x \text{ for } 0 \le x \le 2 \\ u(0,y) = 0.01 \text{ for } 0 \le y \le 1 \\ u(2,y) = 0.01 + 0.006y^2 \text{ for } 0 \le y \le 1 \end{cases}$$

on the reservoir boundary, in kilometers. Compute the head $u(1, 1/2)$ at the center of the reservoir.

6. The steady-state temperature u on a heated copper plate satisfies the Poisson equation

$$\Delta u = -\frac{D(x,y)}{K},$$

where $D(x, y)$ is the power density at (x, y) and K is the thermal conductivity. Assume that the plate is the shape of the rectangle $[0, 4] \times [0, 2]$ cm whose boundary is kept at a constant 30°C, and that power is generated at the constant rate $D(x, y) = 5$ watts/cm³. The thermal conductivity of copper is $K = 3.85$ watts/cm°C. (a) Plot the temperature distribution on the plate. (b) Find the temperature at the center point $(x, y) = (2, 1)$.

7. For the Laplace equations in Exercise 3, make a table of the finite difference approximation and error at $(x, y) = (1/4, 3/4)$ as a function of step sizes $h = k = 2^{-p}$ for $p = 2, \ldots, 5$.

8. For the Poisson equations in Exercise 4, make a table of the finite difference approximation and error at $(x, y) = (1/4, 3/4)$ as a function of step sizes $h = k = 2^{-p}$ for $p = 2, \ldots, 5$.

9. Solve the Laplace equation problems in Exercise 3 on $0 \le x \le 1, 0 \le y \le 1$ by the Finite Element Method with $h = k = 0.1$. Use MATLAB's mesh command to plot the solution.

10. Solve the Poisson equation problems in Exercise 4 on $0 \le x \le 1, 0 \le y \le 1$ by the Finite Element Method with $h = k = 0.1$. Plot the solution.

11. Solve the elliptic partial differential equations in Exercise 5 by the Finite Element Method with $h = k = 0.1$. Plot the solution.

12. Solve the elliptic partial differential equations in Exercise 6 by the Finite Element Method with $h = k = 1/16$. Plot the solution.

13. Solve the elliptic partial differential equations in Exercise 7 by the Finite Element Method with $h = k = 1/16$. Plot the solution.

14. Solve the elliptic partial differential equations with Dirichlet boundary conditions by the Finite Element Method with $h = k = 0.1$. Plot the solution.

(a) $\begin{cases} \Delta u + \sin \pi xy = (x^2 + y^2)u \\ u(x,0) = 0 \text{ for } 0 \le x \le 1 \\ u(x,1) = 0 \text{ for } 0 \le x \le 1 \\ u(0,y) = 0 \text{ for } 0 \le y \le 1 \\ u(1,y) = 0 \text{ for } 0 \le y \le 1 \end{cases}$
(b) $\begin{cases} \Delta u + (\sin \pi xy)u = e^{2xy} \\ u(x,0) = 0 \text{ for } 0 \le x \le 1 \\ u(x,1) = 0 \text{ for } 0 \le x \le 1 \\ u(0,y) = 0 \text{ for } 0 \le y \le 1 \\ u(1,y) = 0 \text{ for } 0 \le y \le 1 \end{cases}$

15. For the elliptic equations in Exercise 5, make a table of the Finite Element Method approximation and error at $(x,y) = (1/4, 3/4)$ as a function of step sizes $h = k = 2^{-p}$ for $p = 2, \ldots, 5$.

16. For the elliptic equations in Exercise 6, make a log–log plot of the maximum error of the Finite Element Method as a function of step sizes $h = k = 2^{-p}$ for $p = 2, \ldots, 6$.

17. For the elliptic equations in Exercise 7, make a log–log plot of the maximum error of the Finite Element Method as a function of step sizes $h = k = 2^{-p}$ for $p = 2, \ldots, 6$.

18. Solve the Laplace equation with Dirichlet boundary conditions from Exercise 13 on $[0,1] \times [0,1]$ with $T_0 = 0$ and $T_1 = 10$ using (a) a finite difference approximation and (b) the Finite Element Method. Make log–log plots of the error at particular locations in the rectangle as a function of step sizes $h = k = 2^{-p}$ for p as large as possible. Explain any simplifications you are making to evaluate the correct solution at those locations.

8.4 NONLINEAR PARTIAL DIFFERENTIAL EQUATIONS

In the previous sections of this chapter, finite difference and finite element methods have been analyzed and applied to linear PDEs. For the nonlinear case, an extra wrinkle is necessary to make our previous methods appropriate.

To make matters concrete, we will focus on the implicit Backward Difference Method of Section 8.1 and its application to nonlinear diffusion equations. Similar changes can be applied to any of the methods we have studied to make them available for use on nonlinear equations.

8.4.1 Implicit Newton solver

We illustrate the approach with a typical nonlinear example

$$u_t + uu_x = Du_{xx}, \tag{8.60}$$

known as **Burgers' equation**. The equation is nonlinear due to the product term uu_x. This elliptic equation, named after J.M. Burgers (1895–1981), is a simplified model of fluid flow. When the diffusion coefficient $D = 0$, it is called the inviscid Burgers' equation. Setting $D > 0$ corresponds to adding viscosity to the model.

This diffusion equation will be discretized in the same way as the heat equation in Section 8.1. Consider the grid of points as shown in Figure 8.1. We will denote the approximate solution at (x_i, t_j) by w_{ij}. Let M and N be the total number of steps in the x and t directions, and let $h = (b-a)/M$ and $k = T/N$ be the step sizes in the x and t directions. Applying backward differences to u_t and central differences to the other terms yields

$$\frac{w_{ij} - w_{i,j-1}}{k} + w_{ij}\left(\frac{w_{i+1,j} - w_{i-1,j}}{2h}\right) = \frac{D}{h^2}(w_{i+1,j} - 2w_{ij} + w_{i-1,j}),$$

or

$$w_{ij} + \frac{k}{2h}w_{ij}(w_{i+1,j} - w_{i-1,j}) - \sigma(w_{i+1,j} - 2w_{ij} + w_{i-1,j}) - w_{i,j-1} = 0 \quad (8.61)$$

where we have set $\sigma = Dk/h^2$. Note that due to the quadratic terms in the w variables, we cannot directly solve for $w_{i+1,j}, w_{ij}, w_{i-1,j}$, explicitly or implicitly. Therefore, we call on Multivariate Newton's Method from Chapter 2 to do the solving.

To clarify our implementation, denote the unknowns in (8.61) by $z_i = w_{ij}$. At time step j, we are trying to solve the equations

$$F_i(z_1,\ldots,z_m) = z_i + \frac{k}{2h}z_i(z_{i+1} - z_{i-1}) - \sigma(z_{i+1} - 2z_i + z_{i-1}) - w_{i,j-1} = 0$$

(8.62)

for the m unknowns $z_1,\ldots,z_m$. Note that the last term $w_{i,j-1}$ is known from the previous time step, and is treated as a known quantity.

The first and last equations will be replaced by appropriate boundary conditions. For example, in the case of Burgers' equation with Dirichlet boundary conditions

$$\begin{cases} u_t + uu_x = Du_{xx} \\ u(x,0) = f(x) \text{ for } x_l \leq x \leq x_r \\ u(x_l,t) = l(t) \text{ for all } t \geq 0 \\ u(x_r,t) = r(t) \text{ for all } t \geq 0, \end{cases} \quad (8.63)$$

we will add the equations

$$F_1(z_1,\ldots,z_m) = z_1 - l(t_j) = 0$$
$$F_m(z_1,\ldots,z_m) = z_m - r(t_j) = 0. \quad (8.64)$$

Now there are m nonlinear algebraic equations in m unknowns.

To apply Multivariate Newton's Method, we must compute the Jacobian $DF(\vec{z}) = \partial\vec{F}/\partial\vec{z}$, which according to (8.62) and (8.64) will have the tridiagonal form

$$\begin{bmatrix} 1 & 0 & \cdots & & & \\ -\sigma - \frac{kz_2}{2h} & 1 + 2\sigma + \frac{k(z_3 - z_1)}{2h} & -\sigma + \frac{kz_2}{2h} & & & \\ & -\sigma - \frac{kz_3}{2h} & 1 + 2\sigma + \frac{k(z_4 - z_2)}{2h} & -\sigma + \frac{kz_3}{2h} & & \\ & & \ddots & \ddots & \ddots & \\ & & & -\sigma - \frac{kz_{m-1}}{2h} & 1 + 2\sigma + \frac{k(z_m - z_{m-2})}{2h} & -\sigma + \frac{kz_{m-1}}{2h} \\ & & & \cdots & 0 & 1 \end{bmatrix}.$$

The top and bottom rows of DF will in general depend on boundary conditions. Once DF has been constructed, we solve for the $z_i = w_{ij}$ by the Multivariate Newton iteration

$$\vec{z}^{K+1} = \vec{z}^K - DF(\vec{z}^K)^{-1}F(\vec{z}^K). \quad (8.65)$$

▶ **EXAMPLE 8.12**   Use the Backward Difference Equation with Newton iteration to solve Burgers' equation

$$\begin{cases} u_t + u u_x = D u_{xx} \\ u(x,0) = \dfrac{2D\beta\pi\sin\pi x}{\alpha + \beta\cos\pi x} \text{ for } 0 \le x \le 1 \\ u(0,t) = 0 \text{ for all } t \ge 0 \\ u(1,t) = 0 \text{ for all } t \ge 0. \end{cases} \tag{8.66}$$

MATLAB code for the Dirichlet boundary condition version of our Newton solver follows, where we have set $\alpha = 5, \beta = 4$. The program uses three Newton iterations for each time step. For typical problems, this will be sufficient, but more may be needed for difficult cases. Note that Gaussian elimination or equivalent is carried out in the Newton iteration; as usual, no explicit matrix inversion is needed.

⌨ MATLAB code

shown here can be found
at bit.ly/2P6OHYr

```
% Program 8.7  Implicit Newton solver for Burgers equation
% input: space interval [xl,xr], time interval [tb,te],
%         number of space steps M, number of time steps N
% output: solution w
% Example usage: w=burgers(0,1,0,2,20,40)
function w=burgers(xl,xr,tb,te,M,N)
alf=5;bet=4;D=.05;
f=@(x) 2*D*bet*pi*sin(pi*x)./(alf+bet*cos(pi*x));
l=@(t) 0*t;
r=@(t) 0*t;
h=(xr-xl)/M; k=(te-tb)/N; m=M+1; n=N;
sigma=D*k/(h*h);
w(:,1)=f(xl+(0:M)*h)';               % initial conditions
w1=w;
for j=1:n
  for it=1:3                         % Newton iteration
    DF1=zeros(m,m);DF2=zeros(m,m);
    DF1=diag(1+2*sigma*ones(m,1))+diag(-sigma*ones(m-1,1),1);
    DF1=DF1+diag(-sigma*ones(m-1,1),-1);
    DF2=diag([0;k*w1(3:m)/(2*h);0])-diag([0;k*w1(1:(m-2))/(2*h);0]);
    DF2=DF2+diag([0;k*w1(2:m-1)/(2*h)],1)...
        -diag([k*w1(2:m-1)/(2*h);0],-1);
    DF=DF1+DF2;
    F=-w(:,j)+(DF1+DF2/2)*w1;       % Using Lemma 8.11
    DF(1,:)=[1 zeros(1,m-1)];       % Dirichlet conditions for DF
    DF(m,:)=[zeros(1,m-1) 1];
    F(1)=w1(1)-l(j);F(m)=w1(m)-r(j); % Dirichlet conditions for F
    w1=w1-DF\F;
  end
  w(:,j+1)=w1;
end
x=xl+(0:M)*h;t=tb+(0:n)*k;
mesh(x,t,w')                         % 3-D plot of solution w
```

The code is a straightforward implementation of the Newton iteration (8.65), along with a convenient fact about homogeneous polynomials. Consider, for example, the polynomial $P(x_1, x_2, x_3) = x_1 x_2 x_3^2 + x_1^4$, which is called homogeneous of degree 4, since it consists entirely of degree 4 terms in x_1, x_2, x_3. The partial derivatives of P with respect to the three variables are contained in the gradient

$$\nabla P = (x_2 x_3^2 + 4x_1^3, x_1 x_3^2, 2x_1 x_2 x_3).$$

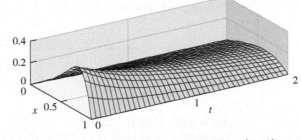

Figure 8.19 Approximate solution to Burgers' equation (8.66). Homogeneous Dirichlet boundary conditions are assumed, with step sizes $h = k = 0.05$.

The remarkable fact is that we can recover P by multiplying the gradient by the vector of variables, with an extra multiple of 4:

$$\nabla P \cdot \begin{bmatrix} x_1 \\ x_2 \\ x_3 \end{bmatrix} = (x_2 x_3^2 + 4x_1^3)x_1 + x_1 x_3^2 x_2 + 2x_1 x_2 x_3 x_3 = 4x_1 x_2 x_3^2 + 4x_1^4 = 4P.$$

In general, define the polynomial $P(x_1, \ldots, x_m)$ to be **homogeneous** of degree d if

$$P(cx_1, \ldots, cx_m) = c^d P(x_1, \ldots, x_m) \tag{8.67}$$

for all c.

LEMMA 8.11 Let $P(x_1, \ldots, x_m)$ be a homogeneous polynomial of degree d. Then

$$\nabla P \cdot \begin{bmatrix} x_1 \\ \vdots \\ x_m \end{bmatrix} = dP.$$

■

Proof. Differentiating (8.67) with respect to c yields

$$x_1 P_{x_1}(cx_1, \ldots, cx_m) + \ldots + x_m P_{x_m}(cx_1, \ldots, cx_m) = dc^{d-1} P(x_1, \ldots, x_m)$$

using the multivariable chain rule. Evaluating at $c = 1$ results in the desired conclusion.
□

Using this fact allows us to write code very compactly for partial differential equations with polynomial terms, as long as we group terms of the same degree together. Note how the matrix DF1 in Program 8.7 collects derivatives of degree 1 terms of F; DF2 collects derivatives of degree 2 terms. Then we can define the Jacobian matrix DF as the sum of derivatives of degree 1 and 2 terms, and essentially for free, define the function F as the sum of degree 0, 1, and 2 terms. Lemma 8.11 is used to identify the degree d terms of F as gradient times variables, divided by d. The added convenience of this simplification will be even more welcome when we proceed to more difficult problems.

For certain boundary conditions, an explicit solution for Burgers' equation is known. The solution to the Dirichlet problem (8.66) is

$$u(x, t) = \frac{2D\beta \pi e^{-D\pi^2 t} \sin \pi x}{\alpha + \beta e^{-D\pi^2 t} \cos \pi x}. \tag{8.68}$$

We can use the exact solution to measure the accuracy of our approximation method, as a function of the step sizes h and k. Using the parameters $\alpha = 5, \beta = 4$, and the diffusion coefficient $D = 0.05$, we find the errors at $x = 1/2$ after one time unit are as follows:

h	k	$u(0.5, 1)$	$w(0.5, 1)$	error
0.01	0.04	0.153435	0.154624	0.001189
0.01	0.02	0.153435	0.154044	0.000609
0.01	0.01	0.153435	0.153749	0.000314

We see the roughly first-order decrease in error as a function of time step size k, as expected with the implicit Backward Difference Method. ◄

Another interesting category of nonlinear PDEs is comprised of **reaction-diffusion equations**. A fundamental example of a nonlinear reaction-diffusion equation is due to the evolutionary biologist and geneticist R.A. Fisher (1890–1962), a successor of Darwin who helped create the foundations of modern statistics. The equation was originally derived to model how genes propagate. The general form of **Fisher's equation** is

$$u_t = Du_{xx} + f(u), \tag{8.69}$$

where $f(u)$ is a polynomial in u. The reaction part of the equation is the function f; the diffusion part is Du_{xx}. If homogeneous Neumann boundary conditions are used, the constant, or equilibrium state $u(x,t) \equiv C$ is a solution whenever $f(C) = 0$. The equilibrium state turns out to be stable if $f'(C) < 0$, meaning that nearby solutions tend toward the equilibrium state.

► **EXAMPLE 8.13** Use the Backward Difference Equation with Newton iteration to solve Fisher's equation with homogeneous Neumann boundary conditions

$$\begin{cases} u_t = Du_{xx} + u(1 - u) \\ u(x, 0) = 0.5 + 0.5\cos \pi x \text{ for } 0 \le x \le 1 \\ u_x(0, t) = 0 \text{ for all } t \ge 0 \\ u_x(1, t) = 0 \text{ for all } t \ge 0. \end{cases} \tag{8.70}$$

Note that $f(u) = u(1 - u)$, implying that $f'(u) = 1 - 2u$. The equilibrium $u = 0$ satisfies $f'(0) = 1$, and the other equilibrium solution $u = 1$ satisfies $f'(1) = -1$. Therefore, solutions are likely to tend toward the equilibrium $u = 1$.

The discretization retraces the derivation that was carried out for Burgers' equation:

$$\frac{w_{ij} - w_{i,j-1}}{k} = \frac{D}{h^2}(w_{i+1,j} - 2w_{ij} + w_{i-1,j}) + w_{ij}(1 - w_{ij}),$$

or

$$(1 + 2\sigma - k(1 - w_{ij}))w_{ij} - \sigma(w_{i+1,j} + w_{i-1,j}) - w_{i,j-1} = 0. \tag{8.71}$$

This results in the nonlinear equations

$$F_i(z_1, \ldots, z_m) = (1 + 2\sigma - k(1 - z_i))z_i - \sigma(z_{i+1} + z_{i-1}) - w_{i,j-1} = 0 \tag{8.72}$$

to solve for the $z_i = w_{ij}$ at the jth time step. The first and last equations will establish the Neumann boundary conditions:

$$F_1(z_1, \ldots, z_m) = (-3z_0 + 4z_1 - z_2)/(2h) = 0$$
$$F_m(z_1, \ldots, z_m) = (-z_{m-2} + 4z_{m-1} - 3z_m)/(-2h) = 0$$

The Jacobian DF has the form

$$
\begin{bmatrix}
-3 & 4 & -1 & & & & \\
-\sigma & 1 + 2\sigma - k + 2kz_2 & -\sigma & & & & \\
 & -\sigma & 1 + 2\sigma - k + 2kz_3 & -\sigma & & & \\
 & & \ddots & \ddots & \ddots & & \\
 & & & & -\sigma & 1 + 2\sigma - k + 2kz_{m-1} & -\sigma \\
 & & & & & -1 & 4 & -3
\end{bmatrix}.
$$

After altering the function F and Jacobian DF, the Newton iteration implemented in Program 8.7 can be used to solve Fisher's equation. Lemma 8.11 can be used to separate the degree 1 and 2 parts of DF. Neumann boundary conditions are also applied, as shown in the code fragment below:

```
DF1=diag(1-k+2*sigma*ones(m,1))+diag(-sigma*ones(m-1,1),1);
DF1=DF1+diag(-sigma*ones(m-1,1),-1);
DF2=diag(2*k*w1);
DF=DF1+DF2;
F=-w(:,j)+(DF1+DF2/2)*w1;
DF(1,:)=[-3 4 -1 zeros(1,m-3)];F(1)=DF(1,:)*w1;
DF(m,:)=[zeros(1,m-3) -1 4 -3];F(m)=DF(m,:)*w1;
```

Figure 8.20 shows approximate solutions of Fisher's equation with $D = 1$ that demonstrate the tendency to relax to the attracting equilibrium $u(x,t) \equiv 1$. Of course, $u(x,t) \equiv 0$ is also a solution of (8.69) with $f(u) = u(1 - u)$, and will be found by the initial data $u(x,0) = 0$. Almost any other initial data, however, will eventually approach $u = 1$ as t increases. ◄

While Example 8.13 covers the original equation considered by Fisher, there are many generalized versions for other choices of the polynomial $f(u)$. See the Computer Problems for more explorations into this reaction-diffusion equation. Next, we will investigate a higher-dimensional version of Fisher's equation.

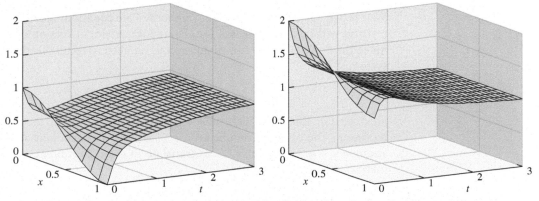

Figure 8.20 Two solutions to Fisher's equation. Both solutions tend toward the equilibrium solution $u(x,t) = 1$ as t increases. (a) Initial condition $u(x,0) = 0.5 + 0.5\cos\pi x$. (b) Initial condition $u(x,0) = 1.5 + 0.5\cos\pi x$. Homogeneous Neumann boundary conditions are assumed, with step sizes $h = k = 0.1$.

8.4.2 Nonlinear equations in two space dimensions

Solving partial differential equations with two-dimensional domains requires us to combine techniques from previous sections. The implicit Backward Difference Method with Newton iteration will handle the nonlinearity, and we will need to apply the accordion-style coordinates of Table 8.1 to do the bookkeeping for the two-dimensional domain.

We begin by extending Fisher's equation from one space dimension to two.

▶ **EXAMPLE 8.14**   Apply the Backward Difference Method with Newton's iteration to Fisher's equation on the unit square $[0, 1] \times [0, 1]$:

$$\begin{cases} u_t = D\Delta u + u(1 - u) \\ u(x, y, 0) = 2 + \cos\pi x \cos\pi y \text{ for } 0 \leq x, y \leq 1 \\ u_{\bar{n}}(x, y, t) = 0 \text{ on rectangle boundary, for all } t \geq 0. \end{cases} \tag{8.73}$$

Here D is the diffusion coefficient, and $u_{\bar{n}}$ denotes the directional derivative in the outward normal direction. We are assuming Neumann, or no-flux, boundary conditions on the rectangle boundary.

In this section, the two discretization subscripts will represent the two space coordinates x and y, and we will use superscripts to denote time steps. Assuming M steps in the x direction and N steps in the y direction, we will define step sizes $h = (x_r - x_l)/M$ and $k = (y_t - y_b)/N$. The discretized equations at nonboundary grid points, for $1 < i < m = M + 1, 1 < j < n = N + 1$, are

$$\frac{w_{ij}^t - w_{ij}^{t-\Delta t}}{\Delta t} = \frac{D}{h^2}(w_{i+1,j}^t - 2w_{ij}^t + w_{i-1,j}^t) + \frac{D}{k^2}(w_{i,j+1}^t - 2w_{ij}^t + w_{i,j-1}^t)$$
$$+ w_{ij}^t(1 - w_{ij}^t), \tag{8.74}$$

which can be rearranged to the form $F_{ij}(w^t) = 0$, or

$$\left(\frac{1}{\Delta t} + \frac{2D}{h^2} + \frac{2D}{k^2} - 1\right)w_{ij}^t - \frac{D}{h^2}w_{i+1,j}^t - \frac{D}{h^2}w_{i-1,j}^t - \frac{D}{k^2}w_{i,j+1}^t - \frac{D}{k^2}w_{i,j-1}^t$$
$$+ (w_{ij}^t)^2 - \frac{w_{ij}^{t-\Delta t}}{\Delta t} = 0. \tag{8.75}$$

We need to solve the F_{ij} equations implicitly. The equations are nonlinear, so Newton's method will be used as it was for the one-dimensional version of Fisher's equation. Since the domain is now two-dimensional, we need to recall the alternative coordinate system (8.39)

$$v_{i+(j-1)m} = w_{ij},$$

illustrated in Table 8.1. There will be mn equations F_{ij}, and in the v coordinates, (8.75) represents the equation numbered $i + (j - 1)m$. The Jacobian matrix DF will have size $mn \times mn$. Using Table 8.1 to translate to the v coordinates, we get the Jacobian matrix entries

$$DF_{i+(j-1)m,i+(j-1)m} = \left(\frac{1}{\Delta t} + \frac{2D}{h^2} + \frac{2D}{k^2} - 1\right) + 2w_{ij}$$

$$DF_{i+(j-1)m,i+1+(j-1)m} = -\frac{D}{h^2}$$

$$DF_{i+(j-1)m,i-1+(j-1)m} = -\frac{D}{h^2}$$

$$DF_{i+(j-1)m,i+jm} = -\frac{D}{k^2}$$

$$DF_{i+(j-1)m,i+(j-2)m} = -\frac{D}{k^2}$$

for the interior points of the grid. The outside points of the grid are governed by the homogenous Neumann boundary conditions

Bottom	$(3w_{ij} - 4w_{i,j+1} + w_{i,j+2})/(2k) = 0$	for $j = 1, 1 \le i \le m$
Top side	$(3w_{ij} - 4w_{i,j-1} + w_{i,j-2})/(2k) = 0$	for $j = n, 1 \le i \le m$
Left side	$(3w_{ij} - 4w_{i+1,j} + w_{i+2,j})/(2h) = 0$	for $i = 1, 1 < j < n$
Right side	$(3w_{ij} - 4w_{i-1,j} + w_{i-2,j})/(2h) = 0$	for $i = m, 1 < j < n$

The Neumann conditions translate via Table 8.1 to

Bottom $\quad DF_{i+(j-1)m,i+(j-1)m} = 3, \ \ DF_{i+(j-1)m,i+jm} = -4, \ \ DF_{i+(j-1)m,i+(j+1)m} = 1,$

$\qquad\qquad b_{i+(j-1)m} = 0$ for $j = 1, 1 \le i \le m$

Top $\quad DF_{i+(j-1)m,i+(j-1)m} = 3, \ \ DF_{i+(j-1)m,i+(j-2)m} = -4, \ \ DF_{i+(j-1)m,i+(j-3)m} = 1,$

$\qquad\qquad b_{i+(j-1)m} = 0$ for $j = n, 1 \le i \le m$

Left $\quad DF_{i+(j-1)m,i+(j-1)m} = 3, \ \ DF_{i+(j-1)m,i+1+(j-1)m} = -4,$

$\quad DF_{i+(j-1)m,i+2+(j-1)m} = 1,$

$\qquad\qquad b_{i+(j-1)m} = 0$ for $i = 1, 1 < j < n$

Right $\quad DF_{i+(j-1)m,i+(j-1)m} = 3, \ \ DF_{i+(j-1)m,i-1+(j-1)m} = -4,$

$\quad DF_{i+(j-1)m,i-2+(j-1)m} = 1,$

$\qquad\qquad b_{i+(j-1)m} = 0$ for $i = m, 1 < j < n$

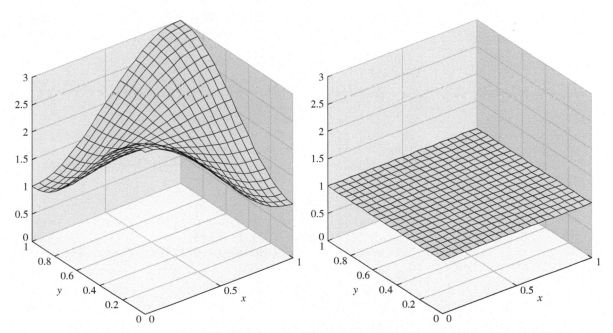

Figure 8.21 Fisher's equation with Neumann boundary conditions on a two-dimensional domain. The solution tends toward the equilibrium solution $u(x, y, t) = 1$ as t increases. (a) The initial condition $u(x, y, 0) = 2 + \cos \pi x \cos \pi y$. (b) Approximate solution after 5 time units. Step sizes $h = k = \Delta t = 0.05$.

The Newton iteration is carried out in the following program. Note that Lemma 8.11 has been used to divide the contributions to DF into degree 1 and degree 2 terms.

☐ **MATLAB code**

shown here can be found
at bit.ly/2yuprlB

```
% Program 8.8 Backward difference method with Newton iteration
%                for Fisher's equation with two-dim domain
% input: space region [xl xr]x[yb yt], time interval [tb te],
% M,N space steps in x and y directions, tsteps time steps
% output: solution mesh [x,y,w]
% Example usage: [x,y,w]=fisher2d(0,1,0,1,0,5,20,20,100);
function [x,y,w]=fisher2d(xl,xr,yb,yt,tb,te,M,N,tsteps)
f=@(x,y) 2+cos(pi*x).*cos(pi*y)
delt=(te-tb)/tsteps;
D=1;
m=M+1;n=N+1;mn=m*n;
h=(xr-xl)/M;k=(yt-yb)/N;
x=linspace(xl,xr,m);y=linspace(yb,yt,n);
for i=1:m            %Define initial u
  for j=1:n
     w(i,j)=f(x(i),y(j));
  end
end
for tstep=1:tsteps
 v=[reshape(w,mn,1)];
 wold=w;
 for it=1:3
   b=zeros(mn,1);DF1=zeros(mn,mn);DF2=zeros(mn,mn);
   for i=2:m-1
     for j=2:n-1
       DF1(i+(j-1)*m,i-1+(j-1)*m)=-D/h^2;
       DF1(i+(j-1)*m,i+1+(j-1)*m)=-D/h^2;
       DF1(i+(j-1)*m,i+(j-1)*m)= 2*D/h^2+2*D/k^2-1+1/(1*delt);
       DF1(i+(j-1)*m,i+(j-2)*m)=-D/k^2;DF1(i+(j-1)*m,i+j*m)=-D/k^2;
       b(i+(j-1)*m)=-wold(i,j)/(1*delt);
       DF2(i+(j-1)*m,i+(j-1)*m)=2*w(i,j);
     end
   end
   for i=1:m      % bottom and top
     j=1; DF1(i+(j-1)*m,i+(j-1)*m)=3;
     DF1(i+(j-1)*m,i+j*m)=-4;DF1(i+(j-1)*m,i+(j+1)*m)=1;
     j=n; DF1(i+(j-1)*m,i+(j-1)*m)=3;
     DF1(i+(j-1)*m,i+(j-2)*m)=-4;DF1(i+(j-1)*m,i+(j-3)*m)=1;
   end
   for j=2:n-1   % left and right
     i=1; DF1(i+(j-1)*m,i+(j-1)*m)=3;
     DF1(i+(j-1)*m,i+1+(j-1)*m)=-4;DF1(i+(j-1)*m,i+2+(j-1)*m)=1;
     i=m; DF1(i+(j-1)*m,i+(j-1)*m)=3;
     DF1(i+(j-1)*m,i-1+(j-1)*m)=-4;DF1(i+(j-1)*m,i-2+(j-1)*m)=1;
   end
   DF=DF1+DF2;
   F=(DF1+DF2/2)*v+b;
   v=v-DF\F;
   w=reshape(v(1:mn),m,n);
 end
 mesh(x,y,w');axis([xl xr yb yt tb te]);
 xlabel('x');ylabel('y');drawnow
end
```

The dynamical behavior of the two-dimensional Fisher's equation is similar to that of the one-dimensional version in Figure 8.20, where we saw convergence to the stable equilibrium solution at $u(x,t) = 1$. Figure 8.21(a) shows the initial data $f(x,y) = 2 + \cos\pi x \cos\pi y$. The solution after $t = 5$ time units is shown in Figure 8.21(b). The solution relaxes quickly toward the stable equilibrium at $u(x,y,t) = 1$. ◄

The mathematician Alan Turing (1912–1954), in a landmark paper (Turing [1952]), proposed a possible explanation for many shapes and structures found in biology. Certain reaction-diffusion equations that model chemical concentrations gave rise to interesting spatial patterns, including stripes and hexagonal shapes. These were seen as a stunning example of emergent order in nature, and are now known as **Turing patterns**.

Turing found that just by adding a diffusive term to a model of a stable chemical reaction, he could cause stable, spatially constant equilibriums, such as the one in Figure 8.21(b), to become unstable. This so-called **Turing instability** causes a transition in which patterns evolve into a new, spatially varying steady-state solution. Of course, this is the opposite of the effect of diffusion we have seen so far, of averaging or smoothing initial conditions over time.

An interesting example of a Turing instability is found in the **Brusselator model**, proposed by the Belgian chemist I. Prigogine in the late 1960's. The model consists of two coupled PDEs, each representing one species of a two-species chemical reaction.

► **EXAMPLE 8.15**   Apply the Backward Difference Method with Newton's iteration to the Brusselator equation with homogeneous Neumann boundary conditions on the square $[0, 40] \times [0, 40]$:

$$
\begin{cases}
p_t = D_p \Delta p + p^2 q + C - (K+1)p \\
q_t = D_q \Delta q - p^2 q + Kp \\
\\
p(x, y, 0) = C + 0.1 \quad \text{for } 0 \le x, y \le 40 \\
q(x, y, 0) = K/C + 0.2 \quad \text{for } 0 \le x, y \le 40 \\
u_{\vec{n}}(x, y, t) = 0 \text{ on rectangle boundary, for all } t \ge 0.
\end{cases}
\tag{8.76}
$$

The system of two coupled equations has variables p, q, two diffusion coefficients $D_p, D_q > 0$, and two other parameters $C, K > 0$. According to Exercise 5, the Brusselator has an equilibrium solution at $p \equiv C, q \equiv K/C$. It is known that the equilibrium is stable for small values of the parameter K, and that a Turing instability is encountered when

$$
K > \left(1 + C\sqrt{\frac{D_p}{D_q}}\right)^2.
\tag{8.77}
$$

The discretized equations at the interior grid points, for $1 < i < m, 1 < j < n$, are

$$
\frac{p_{ij}^t - p_{ij}^{t-\Delta t}}{\Delta t} - \frac{D_p}{h^2}(p_{i+1,j}^t - 2p_{ij}^t + p_{i-1,j}^t) - \frac{D_p}{k^2}(p_{i,j+1}^t - 2p_{ij}^t + p_{i,j-1}^t)
$$
$$
- (p_{ij}^t)^2 q_{ij}^t - C + (K+1)p_{ij}^t = 0
$$
$$
\frac{q_{ij}^t - q_{ij}^{t-\Delta t}}{\Delta t} - \frac{D_q}{h^2}(q_{i+1,j}^t - 2q_{ij}^t + q_{i-1,j}^t) - \frac{D_q}{k^2}(q_{i,j+1}^t - 2q_{ij}^t + q_{i,j-1}^t)
$$
$$
+ (p_{ij}^t)^2 q_{ij}^t - Kp_{ij}^t = 0
$$

This is the first example we have encountered with two coupled variables, p and q. The alternative coordinate vector v will have length $2mn$, and (8.39) will be extended to

$$\begin{aligned} v_{i+(j-1)m} &= p_{ij} && \text{for } 1 \le i \le m, 1 \le j \le n \\ v_{mn+i+(j-1)m} &= q_{ij} && \text{for } 1 \le i \le m, 1 \le j \le n. \end{aligned} \quad (8.78)$$

The Neumann boundary conditions are essentially the same as Example 8.14, now for each variable p and q. Note that there are degree 1 and degree 3 terms to differentiate for the Jacobian DF. Using Table 8.1 expanded in a straightforward way to cover two variables, and Lemma 8.11, we arrive at the following MATLAB code:

⊡ **MATLAB code**

shown here can be found at bit.ly/2Otu09z

```
% Program 8.9 Backward difference method with Newton iteration
%                for the Brusselator
% input: space region [xl,xr]x[yb,yt], time interval [tb,te],
%   M,N space steps in x and y directions, tsteps time steps
% output: solution mesh [x,y,w]
% Example usage: [x,y,p,q]=brusselator(0,40,0,40,0,20,40,40,20);
function [x,y,p,q]=brusselator(xl,xr,yb,yt,tb,te,M,N,tsteps)
Dp=1;Dq=8;C=4.5;K=9;
fp=@(x,y) C+0.1;
fq=@(x,y) K/C+0.2;
delt=(te-tb)/tsteps;
m=M+1;n=N+1;mn=m*n;mn2=2*mn;
h=(xr-xl)/M;k=(yt-yb)/N;
x=linspace(xl,xr,m);y=linspace(yb,yt,n);
for i=1:m              %Define initial conditions
  for j=1:n
    p(i,j)=fp(x(i),y(j));
    q(i,j)=fq(x(i),y(j));
  end
end
for tstep=1:tsteps
  v=[reshape(p,mn,1);reshape(q,mn,1)];
  pold=p;qold=q;
  for it=1:3
    DF1=zeros(mn2,mn2);DF3=zeros(mn2,mn2);
    b=zeros(mn2,1);
    for i=2:m-1
      for j=2:n-1
        DF1(i+(j-1)*m,i-1+(j-1)*m)=-Dp/h^2;
        DF1(i+(j-1)*m,i+(j-1)*m)= Dp*(2/h^2+2/k^2)+K+1+1/(1*delt);
        DF1(i+(j-1)*m,i+1+(j-1)*m)=-Dp/h^2;
        DF1(i+(j-1)*m,i+(j-2)*m)=-Dp/k^2;
        DF1(i+(j-1)*m,i+j*m)=-Dp/k^2;
        b(i+(j-1)*m)=-pold(i,j)/(1*delt)-C;
        DF1(mn+i+(j-1)*m,mn+i-1+(j-1)*m)=-Dq/h^2;
        DF1(mn+i+(j-1)*m,mn+i+(j-1)*m)= Dq*(2/h^2+2/k^2)+1/(1*delt);
        DF1(mn+i+(j-1)*m,mn+i+1+(j-1)*m)=-Dq/h^2;
        DF1(mn+i+(j-1)*m,mn+i+(j-2)*m)=-Dq/k^2;
        DF1(mn+i+(j-1)*m,mn+i+j*m)=-Dq/k^2;
        DF1(mn+i+(j-1)*m,i+(j-1)*m)=-K;
        DF3(i+(j-1)*m,i+(j-1)*m)=-2*p(i,j)*q(i,j);
        DF3(i+(j-1)*m,mn+i+(j-1)*m)=-p(i,j)^2;
        DF3(mn+i+(j-1)*m,i+(j-1)*m)=2*p(i,j)*q(i,j);
```

```
            DF3(mn+i+(j-1)*m,mn+i+(j-1)*m)=p(i,j)^2;
            b(mn+i+(j-1)*m)=-qold(i,j)/(1*delt);
        end
    end
    for i=1:m    % bottom and top Neumann conditions
        j=1;DF1(i+(j-1)*m,i+(j-1)*m)=3;
        DF1(i+(j-1)*m,i+j*m)=-4;
        DF1(i+(j-1)*m,i+(j+1)*m)=1;
        j=n;DF1(i+(j-1)*m,i+(j-1)*m)=3;
        DF1(i+(j-1)*m,i+(j-2)*m)=-4;
        DF1(i+(j-1)*m,i+(j-3)*m)=1;
        j=1;DF1(mn+i+(j-1)*m,mn+i+(j-1)*m)=3;
        DF1(mn+i+(j-1)*m,mn+i+j*m)=-4;
        DF1(mn+i+(j-1)*m,mn+i+(j+1)*m)=1;
        j=n;DF1(mn+i+(j-1)*m,mn+i+(j-1)*m)=3;
        DF1(mn+i+(j-1)*m,mn+i+(j-2)*m)=-4;
        DF1(mn+i+(j-1)*m,mn+i+(j-3)*m)=1;
    end
    for j=2:n-1    %left and right Neumann conditions
        i=1;DF1(i+(j-1)*m,i+(j-1)*m)=3;
        DF1(i+(j-1)*m,i+1+(j-1)*m)=-4;
        DF1(i+(j-1)*m,i+2+(j-1)*m)=1;
        i=m;DF1(i+(j-1)*m,i+(j-1)*m)=3;
        DF1(i+(j-1)*m,i-1+(j-1)*m)=-4;
        DF1(i+(j-1)*m,i-2+(j-1)*m)=1;
        i=1;DF1(mn+i+(j-1)*m,mn+i+(j-1)*m)=3;
        DF1(mn+i+(j-1)*m,mn+i+1+(j-1)*m)=-4;
        DF1(mn+i+(j-1)*m,mn+i+2+(j-1)*m)=1;
        i=m;DF1(mn+i+(j-1)*m,mn+i+(j-1)*m)=3;
        DF1(mn+i+(j-1)*m,mn+i-1+(j-1)*m)=-4;
        DF1(mn+i+(j-1)*m,mn+i-2+(j-1)*m)=1;
    end
    DF=DF1+DF3;
    F=(DF1+DF3/3)*v+b;
    v=v-DF\F;
    p=reshape(v(1:mn),m,n);q=reshape(v(mn+1:mn2),m,n);
    end
    contour(x,y,p');drawnow;
end
```

Figure 8.22 shows contour plots of solutions of the Brusselator. In a contour plot, the closed curves trace level sets of the variable $p(x, y)$. In models, p and q represent chemical concentrations which self-organize into the varied patterns shown in the plots. ◄

Reaction-diffusion equations with a Turing instability are routinely used to model pattern formation in biology, including butterfly wing patterns, animal coat markings, fish and shell pigmentation, and many other examples. Turing patterns have been found experimentally in chemical reactions such as the CIMA (chlorite-iodide-malonic acid) starch reaction. Models for glycolysis and the Gray–Scott equations for chemical reactions are closely related to the Brusselator.

The use of reaction-diffusion equations to study pattern formation is just one direction among several of great contemporary interest. Nonlinear partial differential equations are used to model a variety of temporal and spatial phenomena throughout engineering and the sciences. Another important class of problems is described by the

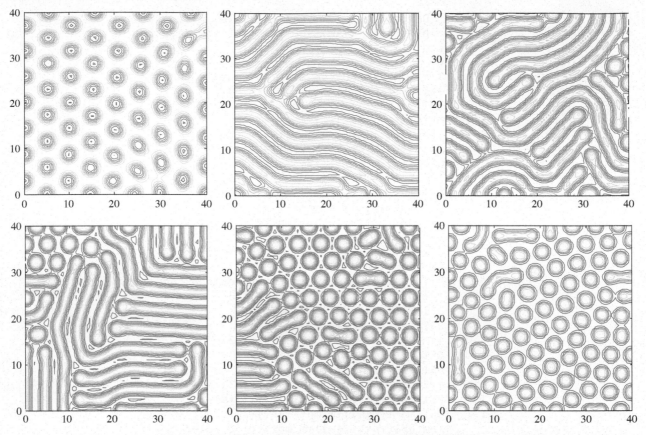

Figure 8.22 Pattern formation in the Brusselator. Contour plots of solutions $p(x, y)$ at $t = 2000$ show Turing patterns. Parameters are $D_p = 1, D_q = 8, C = 4.5$ and (a) $K = 7$ (b) $K = 8$ (c) $K = 9$ (d) $K = 10$ (e) $K = 11$ (f) $K = 12$. Settings for the finite differences are $h = k = 0.5, \Delta t = 1$.

Navier–Stokes equations, which represent incompressible fluid flow. Navier–Stokes is used to model phenomena as diverse as film coatings, lubrication, blood dynamics in arteries, air flow over an airplane wing and the turbulence of stellar gas. Improving finite difference and finite element solvers for linear and nonlinear partial differential equations stands as one of the most active areas of research in computational science.

▶ **ADDITIONAL EXAMPLES**

*1. Find all constant solutions of Fisher's equation $u_t = u_{xx} + 5u^2 - u^3 - 6u$ and check their stability.

2. (a) Adapt the burgers.m code to solve the Fisher's equation with Neumann boundary conditions

$$\begin{cases} u_t = u_{xx} + 5u^2 - u^3 - 6u \\ u_x(0, t) = u_x(1, t) = 0 \\ u(x, 0) = 1 + 3\cos \pi x \end{cases}$$

using step sizes $h = k = 0.05$ on $0 \le x \le 1, 0 \le t \le 2$. (b) How does the solution change for the initial condition $u(x, 0) = 5 + 3\cos \pi x$?

📱 **Solutions** for Additional Examples can be found at bit.ly/2yLmRqw (* example with video solution)

8.4 Exercises

Solutions
for Exercises
numbered in blue
can be found at
bit.ly/2PIxIJ6

1. Show that for any constant c, the function $u(x,t) = c$ is an equilibrium solution of Burgers' equation $u_t + uu_x = Du_{xx}$.

2. Show that over an interval $[x_l, x_r]$ not containing 0, the function $u(x,t) = x^{-1}$ is a time-invariant solution of the Burgers' equation $u_t + uu_x = -\frac{1}{2}u_{xx}$.

3. Show that the function $u(x,t)$ in (8.68) is a solution of the Burgers' equation with Dirichlet boundary conditions (8.66).

4. Find all stable equilibrium solutions of Fisher's equation (8.69) when $f(u) = u(u-1)(2-u)$.

5. Show that the Brusselator has an equilibrium solution at $p \equiv C, q \equiv K/C$.

6. For parameter settings $D_p = 1, D_q = 8, C = 4.5$ of the Brusselator, for what values of K is the equilibrium solution $p \equiv C, q \equiv K/C$ stable? See Computer Problems 5 and 6.

8.4 Computer Problems

Solutions for
Computer Problems
numbered in blue can
be found at
bit.ly/2CT5JDK

1. Solve Burgers' equation (8.63) with $D = 1$ on $[0,1]$ with initial condition $f(x) = \sin 2\pi x$ and boundary conditions $l(t) = r(t) = 0$, using step sizes (a) $h = k = 0.1$ and (b) $h = k = 0.02$. Plot the approximate solutions for $0 \le t \le 1$. Which equilibrium solution does the solution approach as time increases?

2. Solve Burgers' equation on the interval $[0,1]$ with homogeneous Dirichlet boundary conditions and the initial condition given in (8.66) with parameters $\alpha = 4, \beta = 3$, and $D = 0.2$. Plot the approximate solution using step sizes $h = 0.01, k = 1/16$, and make a log–log plot of the approximation error at $x = 1/2, t = 1$ as a function of k for $k = 2^{-p}, p = 4, \ldots, 8$.

3. Solve Fisher's equation (8.69) with $D = 1, f(u) = u(u-1)(2-u)$ and homogeneous Neumann boundary conditions, using initial condition (a) $f(x) = 1/2 + \cos 2\pi x$ (b) $f(x) = 3/2 - \cos 2\pi x$. Equations (8.71) and (8.72) must be redone for the new $f(u)$. Plot the approximate solution for $0 \le t \le 2$ for step sizes $h = k = 0.05$. Which equilibrium solution does the solution approach as time increases? What effect does changing D to 0.1 or 0.01 have on the behavior of the solution?

4. Solve Fisher's equation with $D = 1, f(u) = u(u-1)(2-u)$ on a two-dimensional space domain. Equations (8.74) and (8.75) must be redone for the new $f(u)$. Assume homogeneous Neumann boundary conditions, and the initial conditions of (8.73). Plot the approximate solution for integer times $t = 0, \ldots, 5$ for step sizes $h = k = 0.05$ and $\Delta t = 0.05$. Which equilibrium solution does the solution approach as time increases? What effect does changing D to 0.1 or 0.01 have on the behavior of the solution?

5. Solve the Brusselator equations for $D_p = 1, D_q = 8, C = 4.5$ and (a) $K = 4$ (b) $K = 5$ (c) $K = 6$ (d) $K = 6.5$. Using homogeneous Neumann boundary conditions and initial conditions $p(x, y, 0) = 1 + \cos \pi x \cos \pi y, q(x, y, 0) = 2 + \cos 2\pi x \cos 2\pi y$, estimate the least value T for which $|p(x, y, t) - C| < 0.01$ for all $t > T$.

6. Plot contour plots of solutions $p(x, y, 2000)$ of the Brusselator for $D_p = 1, D_q = 8$, $C = 4.5$ and $K = 7.2, 7.4, 7.6$, and 7.8. Use step sizes $h = k = 0.5, \Delta t = 1$. These plots fill in the range between Figure 8.22.

Software and Further Reading

There is a rich literature on partial differential equations and their applications to science and engineering. Recent textbooks with an applied viewpoint include Haberman [2012], Logan [2015], Evans [2010], Strauss [1992], and Gockenbach [2010]. Many textbooks provide deeper information about numerical methods for PDEs, such as finite difference and finite element methods, including Strikwerda [1989], Lapidus and Pinder [1982], Hall and Porsching [1990], and Morton and Mayers [2006]. Brenner and Scott [2007], Ames [1992], Strang and Fix [2008] are primarily directed toward the Finite Element Method.

MATLAB's PDE toolbox is highly recommended. It has become extremely popular as a companion in PDE and engineering mathematics courses. Maple has an analogous package called PDEtools. Several stand-alone software packages have been developed for numerical PDEs, for general use or targeting special problems. ELLPACK (Rice and Boisvert [1984]) and PLTMG (Bank [1998]) are freely available packages for solving elliptic partial differential equations in general regions of the plane. Both are available at Netlib.

Finite Element Method software includes freeware FEAST (Finite Element and Solution Tools), FreeFEM, and PETSc (Portable Extensible Toolkit for Scientific Computing) and commercial software COMSOL, NASTRAN, and DIFFPACK, among many others.

The NAG library contains several routines for finite difference and finite element methods. The program D03EAF solves the Laplace equation in two dimensions by means of an integral equation method; D03EEF uses a seven-point finite difference formula and handles many types of boundary conditions. The routines D03PCF and D03PFF handle parabolic and hyperbolic equations, respectively.

Random Numbers and Applications

Brownian motion is a model of random behavior, proposed by Robert Brown in 1827. His initial interest was to understand the erratic movement of pollen particles floating on the surface of water, buffeted by nearby molecules. The model's applications have far outgrown the original context.

Financial analysts today think of asset prices in the same way, as fickle entities buffeted by the conflicting momenta of numerous investors. In 1973, Fischer Black and Myron Scholes made a novel use of exponential Brownian motion to provide accurate valuations of stock options. Immediately recognized as an important innovation, the Black–Scholes formula was programmed into some of the first portable calculators designed for use on the trading floors on Wall Street. This work was awarded the Nobel Prize in Economics in 1997 and remains pervasive in financial theory and practice.

Reality Check Reality Check 9 on page 486 explores Monte Carlo simulation and this famous formula.

The previous three chapters concerned deterministic models governed by differential equations. Given proper initial and boundary conditions, the solution is mathematically certain and can be determined with appropriate numerical methods to prescribed accuracy. A stochastic model, on the other hand, includes uncertainty due to noise as part of its definition.

Computational simulation of a stochastic system requires the generation of random numbers to mimic the noise. This chapter begins with some fundamental facts about random numbers and their use in simulation. The second section covers one of the most important uses of random numbers, Monte Carlo simulation, and the third section introduces random walks and Brownian motion. In the last section, the basic ideas of stochastic calculus are covered, including many standard examples of stochastic differential equations (SDEs) that have proved to be useful in physics, biology, and finance. The computational methods for SDEs are based on the ODE solvers developed in Chapter 7, but extended to include noise terms.

Basic concepts of probability are occasionally needed in this chapter. These extra prerequisites, such as expected value, variance, and independence of random variables, are important in Sections 9.2–9.4.

9.1 RANDOM NUMBERS

Everyone has intuition about what random numbers are, but it is surprisingly difficult to define the notion precisely. Nor is it easy to find simple and effective methods of producing them. Of course, with computers working according to prescribed, deterministic rules assigned by the programmer, there is no such thing as a program that produces truly random numbers. We will settle for producing pseudo-random numbers, which is simply a way of saying that we will consider deterministic programs that work the same way every time and that produce strings of numbers that look as random as possible.

The goal of a random number generator is for the output numbers to be independent and identically distributed. By "independent," we mean that each new number x_n should not depend on (be more or less likely due to) the preceding number x_{n-1}, or in fact all preceding numbers $x_{n-1}, x_{n-2}, \ldots$. By "identically distributed," we mean that if the histogram of x_n were plotted over many different repetitions of random number generation, it would look the same as the histogram of x_{n-1}. In other words, independent means that x_n is independent of x_{n-1}, x_{n-2}, etc., and identically distributed means the distribution of x_n is independent of n. The desired histogram, or distribution, may be a uniform distribution of real numbers between 0 and 1, or it may be more sophisticated, such as a normal distribution.

Of course, the independence part of the definition of random numbers is at odds with practical computer-based methods of random number generation, which produce completely predictable and repeatable streams of numbers. In fact, repeatability can be extremely useful for some simulation purposes. The trick is to make the numbers *appear* independent of one another, even though the generation method may be anything but independent. The term **pseudo-random** number is reserved for this situation—deterministically generated numbers that strive to be random in the sense of being independent and identically distributed.

The fact that highly dependent means are used to produce something purporting to be independent explains why there is no perfect software-based, all-purpose random number generator. As John Von Neumann said in 1951, "Anyone who considers arithmetical methods of producing random digits is, of course, in a state of sin." The main hope is that the particular hypothesis the user wants to test by using random numbers is insensitive to the dependencies and deficiencies of the chosen generator.

Random numbers are representatives chosen from a fixed probability distribution. There are many possible choices for the distribution. To keep prerequisites to a minimum, we will restrict our attention to two possibilities: the uniform distribution and the normal distribution.

9.1.1 Pseudo-random numbers

The simplest set of random numbers is the uniform distribution on the interval $[0, 1]$. These numbers correspond to putting on a blindfold and choosing numbers from the interval, with no preference to any particular area of the interval. Each real number in the interval is equally likely to be chosen. How can we produce a string of such numbers with a computer program?

Here is a first try at producing uniform (pseudo-) random numbers in [0, 1]. Pick a starting integer $x_0 \neq 0$, called the **seed**. Then produce the sequence of numbers u_i according to the iteration

$$x_i = 13x_{i-1} \quad (\text{mod } 31)$$
$$u_i = \frac{x_i}{31}, \tag{9.1}$$

that is, multiply the x_{i-1} by 13, evaluate modulo 31, and then divide by 31 to get the next pseudo-random number. The resulting sequence will repeat only after running through all 30 nonzero numbers $1/31, \ldots, 30/31$. In other words, the **period** of this random number generator is 30. There is nothing that appears random about this sequence of numbers. Once the seed is chosen, it cycles through the 30 possible numbers in a predetermined order. The earliest random number generators followed the same logic, although with a larger period.

With $x_0 = 3$ as random seed, here are the first 10 numbers generated by our method:

x	u
8	0.2581
11	0.3548
19	0.6129
30	0.9677
18	0.5806
17	0.5484
4	0.1290
21	0.6774
25	0.8065
15	0.4839

We begin with $3 * 13 = 39 \rightarrow 8 \pmod{31}$, so that the uniform random number is $8/31 \approx 0.2581$. The second random number is $8 * 13 = 104 \rightarrow 11 \pmod{31}$, yielding $11/31 \approx 0.3548$, and so forth, as it runs through the 30 possible random numbers.

This is an example of the most basic type of random number generator.

DEFINITION 9.1   A **linear congruential generator (LCG)** has form

$$x_i = ax_{i-1} + b \quad (\text{mod } m)$$
$$u_i = \frac{x_i}{m}, \tag{9.2}$$

for **multiplier** a, **offset** b, and **modulus** m.   ❏

In the foregoing generator, $a = 13, b = 0$, and $m = 31$. We will keep $b = 0$ in the next two examples. The conventional wisdom is that nonzero b adds little but extra complication to the random number generator.

One application of random numbers is to approximate the average of a function by substituting random numbers from the range of interest. This is the simplest form of the Monte Carlo technique, which we will discuss in more detail in the next section.

▶ **EXAMPLE 9.1**   Approximate the area under the curve $y = x^2$ in [0, 1].

By definition, the mean value of a function on $[a, b]$ is

$$\frac{1}{b-a} \int_a^b f(x)\, dx,$$

so the area in question is exactly the mean value of $f(x) = x^2$ on $[0, 1]$. This mean value can be approximated by averaging the function values at random points in the interval, as shown in Figure 9.1. The function average

$$\frac{1}{10} \sum_{i=1}^{10} f(u_i)$$

for the first 10 uniform random numbers generated by our method is 0.350, not too far from the correct answer, $1/3$. Using all 30 random numbers in the average results in the improved estimate 0.328.

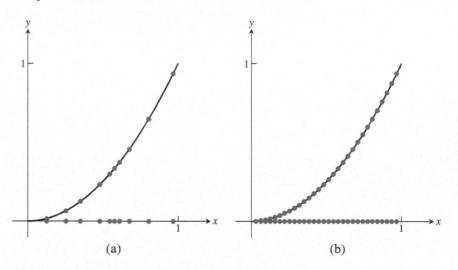

(a)   (b)

Figure 9.1 Averaging a function by using random numbers. (a) The first 10 random numbers from elementary generator (9.1) with seed $x_0 = 3$ give the average 0.350. (b) Using all 30 gives the more accurate average 0.328.

◀

We will call the application in Example 9.1 the Monte Carlo Type 1 problem, since it reduced to a function average. Note that we have exhausted the 30 random numbers that generator (9.1) can provide. If more accuracy is required, more numbers are needed. We can stay with the LCG model, but the multiplier a and modulus m need to be increased.

Park and Miller [1998] proposed a linear congruential generator that is often called the "minimal standard" generator because it is about as good as possible with very simple code. This random number generator was used in MATLAB version 4 in the 1990s.

Minimal standard random number generator

$$x_i = ax_{i-1} \pmod{m}$$
$$u_i = \frac{x_i}{m}, \tag{9.3}$$

where $m = 2^{31} - 1, a = 7^5 = 16807$, and $b = 0$.

An integer of the form $2^p - 1$ that is a prime number, where p is an integer, is called a **Mersenne prime**. Euler discovered this Mersenne prime in 1772. The repetition time of the minimal standard random number generator is the maximum possible $2^{31} - 2$, meaning that it takes on all nonzero integers below the maximum before repeating, as long as the seed is nonzero. This is approximately 2×10^9 numbers, perhaps sufficient for the 20th century, but not generally sufficient now that computers routinely execute that many clock cycles per second.

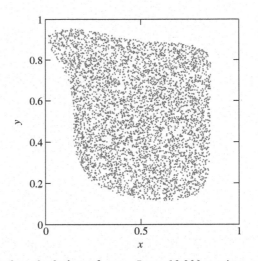

Figure 9.2 Monte Carlo calculation of area. From 10,000 random pairs in $[0, 1] \times [0, 1]$, the ones that satisfy the inequality in Example 9.2 are plotted. The proportion of plotted random pairs is an approximation to the area.

► **EXAMPLE 9.2**   Find the area of the set of points (x, y) that satisfy

$$4(2x - 1)^4 + 8(2y - 1)^8 < 1 + 2(2y - 1)^3(3x - 2)^2.$$

We will call this a Monte Carlo Type 2 problem. There is no clear way to describe this area as the average value of a function of one variable, since we cannot solve for y. However, given a candidate (x, y), we can easily check whether or not it belongs to the set. We will equate the desired area with the probability that a given random pair $(x, y) = (u_i, u_{i+1})$ belongs to the set and try to approximate that probability.

Figure 9.2 shows this idea carried out with 10,000 random pairs generated by the Minimal Standard LCG. The proportion of pairs in the unit square $0 \leq x, y \leq 1$ that satisfy the inequality, and are plotted in the figure, is 0.547, which we will take as an approximation to the area.   ◄

Although we have made a distinction between two types of Monte Carlo problems, there is no firm boundary between them. What they have in common is that they are both computing the average of a function. This is explicit in the previous "type 1" example. In the "type 2" example, we are trying to compute the average of the **characteristic function** of the set, the function that takes the value 1 for points inside the set and 0 for points outside. The main difference here is that unlike the function $f(x) = x^2$ in Example 9.1, the characteristic function of a set is discontinuous—there is an abrupt transition at the boundary of the set. We can also easily imagine combinations of types 1 and 2. (See Computer Problem 8.)

One of the most infamous random number generators is the `randu` generator, used on many early IBM computers and ported from there to many others. Traces of it can be easily found on the Internet with a search engine, so it is apparently still in use.

The `randu` **generator**

$$x_i = ax_{i-1} \quad (\text{mod } m)$$
$$u_i = \frac{x_i}{m}, \tag{9.4}$$

where $a = 65539 = 2^{16} + 3$ and $m = 2^{31}$.

The random seed $x_0 \neq 0$ is chosen arbitrarily. The nonprime modulus was originally selected to make the modulus operation as fast as possible, and the multiplier was selected primarily because its binary representation was simple. The serious problem with this generator is that it flagrantly disobeys the independence postulate for random numbers. Notice that

$$\begin{aligned} a^2 - 6a &= (2^{16} + 3)^2 - 6(2^{16} + 3) \\ &= 2^{32} + 6 \cdot 2^{16} + 9 - 6 \cdot 2^{16} - 18 \\ &= 2^{32} - 9. \end{aligned}$$

Therefore, $a^2 - 6a + 9 = 0 \ (\text{mod } m)$, so

$$\begin{aligned} x_{i+2} - 6x_{i+1} + 9x_i &= a^2 x_i - 6ax_i + 9x_i \quad (\text{mod } m) \\ &= 0 \quad (\text{mod } m). \end{aligned}$$

Dividing by m yields

$$u_{i+2} = 6u_{i+1} - 9u_i \quad (\text{mod } 1). \tag{9.5}$$

The problem is not that u_{i+2} is predictable from the two previous numbers generated. Of course, it will be predictable even from one previous number, because the generator is deterministic. The problem lies with the small coefficients in the relation (9.5), which make the correlation between the random numbers very noticeable. Figure 9.3(a) shows a plot of 10,000 random numbers generated

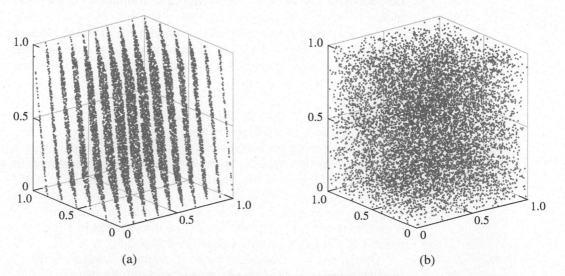

(a)   (b)

Figure 9.3 Comparison of two random number generators. Ten thousand triples (u_i, u_{i+1}, u_{i+2}) are plotted for (a) `randu` and (b) the Minimal Standard generator.

by randu and plotted in triples (u_i, u_{i+1}, u_{i+2}). One consequence of relation (9.5) is that all triples of random numbers will lie on one of 15 planes, as can be seen in the figure. Indeed, $u_{i+2} - 6u_{i+1} + 9u_i$ must be an integer, and the only possibilities are the integers between -5, in case u_{i+1} is relatively large and u_i, u_{i+2} are small, and $+9$, in the opposite case. The planes $u_{i+2} - 6u_{i+1} + 9u_i = k$, for $-5 \leq k \leq 9$, are the 15 planes seen in Figure 9.3. Exercise 5 asks you to analyze another well-known random number generator for a similar deficiency.

The Minimal Standard LCG does not suffer from this problem, at least to the same degree. Since m and a in (9.3) are relatively prime, relations between successive u_i with small coefficients, like the one in (9.5), are much more difficult to come by, and any correlations between three successive random numbers from this generator are much more complicated. This can be seen in Figure 9.3(b), which compares a plot of 10,000 random numbers generated by the Minimal Standard random number generator with a similar plot from randu.

► **EXAMPLE 9.3** Use randu to approximate the volume of the ball of radius 0.04 centered at $(1/3, 1/3, 1/2)$.

Although the ball has a nonzero volume, a straightforward attempt to approximate the volume with randu comes up with 0. The Monte Carlo approach is to randomly generate points in the three-dimensional unit cube and count the proportion of generated points that lie in the ball as the approximate volume.

The point $(1/3, 1/3, 1/2)$ lies midway between the planes $9x - 6y + z = 1$ and $9x - 6y + z = 2$, at a distance of $1/(2\sqrt{118}) \approx 0.046$ from each plane. Therefore, generating the three-dimensional point $(x, y, z) = (u_i, u_{i+1}, u_{i+2})$ from randu can never result in a point contained in the specified ball. Monte Carlo approximations of this innocent problem will be spectacularly unsuccessful because of the choice of random number generator. Surprisingly, difficulties of this type went largely unnoticed during the 1960s and 1970s, when this generator was heavily relied upon for computer simulations. ◄

Random numbers in current versions of MATLAB are no longer generated by LCGs. Starting with MATLAB 5, a lagged Fibonacci generator, developed by G. Marsaglia et al. [1991], has been used in the command rand. All possible floating point numbers between 0 and 1 are used. MATLAB claims that the period of this method is greater than 2^{1400}, which is far more than the total number of steps run by all MATLAB programs since its creation.

Thus far, we have focused on generating pseudo-random numbers for the interval $[0, 1]$. To generate a uniform distribution of random numbers in the general interval $[a, b]$, we need to stretch by $b - a$, the length of the new interval. Thus, each random number r generated in $[0, 1]$ should be replaced by $(b - a)r + a$.

This can be done for each dimension independently. For example, to generate a uniform random point in the rectangle $[1, 3] \times [2, 8]$ in the xy-plane, generate the pair r_1, r_2 of uniform random numbers and then use $(2r_1 + 1, 6r_2 + 2)$ for the random point.

9.1.2 Exponential and normal random numbers

An **exponential** random variable V chooses positive numbers according to the **probability distribution function** $p(x) = ae^{-ax}$ for $a > 0$. In other words, a histogram of exponential random numbers $r_1, \ldots, r_n$ will tend toward $p(x)$ as $n \to \infty$.

Using a uniform random number generator from the previous section, it is fairly easy to generate exponential random numbers. The **cumulative distribution function** is

$$P(x) = \text{Prob}(V \leq x) = \int_0^x p(x)dx = 1 - e^{-ax}.$$

The main idea is to choose the exponential random variable so that the $\text{Prob}(V \leq x)$ is uniform between 0 and 1. Namely, given a uniform random number u, set

$$u = \text{Prob}(V \leq x) = 1 - e^{-ax}$$

and solve for x, yielding

$$x = \frac{-\ln(1-u)}{a}. \tag{9.6}$$

Therefore, formula (9.6) generates exponential random numbers, using uniform random numbers u as inputs.

This idea works in general. Let $P(x)$ be the cumulative distribution function of the random variable that needs to be generated. Let $Q(x) = P^{-1}(x)$ be the inverse function. If $U[0, 1]$ denotes uniform random numbers from $[0, 1]$, then $Q(U[0, 1])$ will generate the required random variables. All that remains is to find ways to make evaluation of Q as efficient as possible.

The standard **normal**, or **Gaussian** random variable $N(0, 1)$ chooses real numbers according to the probability distribution function

$$p(x) = \frac{1}{\sqrt{2\pi}} e^{-\frac{x^2}{2}},$$

the shape of the famous bell curve. The variable $N(0, 1)$ has mean 0 and variance 1. More generally, the normal random variable $N(\mu, \sigma^2) = \mu + \sigma N(0, 1)$ has mean μ and variance σ^2. Since this variable is just a scaled version of the standard normal random variable $N(0, 1)$, we will focus on methods of generating the latter.

Although we could directly apply the inverse of the cumulative distribution function as just outlined, it turns out to be more efficient to generate two normal random numbers at a time. The two-dimensional standard normal distribution has probability distribution function $p(x, y) = (1/2\pi)e^{-(x^2+y^2)/2}$, or $p(r) = (1/2\pi)e^{-r^2/2}$ in polar coordinates. Since $p(r)$ has polar symmetry, we need only generate the radial distance r according to $p(r)$ and then choose an arbitrary angle θ uniform in $[0, 2\pi]$. Since $p(r)$ is an exponential distribution for r^2 with parameter $a = 1/2$, generate r by

$$r^2 = \frac{-\ln(1-u_1)}{1/2}$$

from formula (9.6), where u_1 is a uniform random number. Then

$$n_1 = r\cos 2\pi u_2 = \sqrt{-2\ln(1-u_1)}\cos 2\pi u_2$$
$$n_2 = r\sin 2\pi u_2 = \sqrt{-2\ln(1-u_1)}\sin 2\pi u_2 \tag{9.7}$$

is a pair of independent normal random numbers, where u_2 is a second uniform random number. Note that $1 - u_1$ can be replaced by u_1 in the formula, since the distribution $U[0, 1]$ is unchanged after subtraction from 1. This is the **Box–Muller Method** (Box and Muller [1958]) for generating normal random numbers. Square root, log, cosine, and sine evaluations are required for each pair.

A more efficient version of Box–Muller follows if u_1 is generated in a different way. Choose x_1, x_2 from $U0, 1$ and define $u_1 = x_1^2 + x_2^2$ if the expression is less than 1. If not, throw x_1 and x_2 away and start over. Note that u_1 chosen in this way is $U0, 1$ (see Exercise 6). The advantage is that we can define u_2 as $2\pi u_2 = \arctan x_2/x_1$, the angle made by the line segment connecting the origin to the point (x_1, x_2), making u_2 uniform on $0, 1$. Since $\cos 2\pi u_2 = x_1/\sqrt{u_1}$ and $\sin 2\pi u_2 = x_2/\sqrt{u_1}$, formula (10.7) translates to

$$n_1 = x_1 \sqrt{\frac{-2\ln(u_1)}{u_1}}$$

$$n_2 = x_2 \sqrt{\frac{-2\ln(u_1)}{u_1}}, \tag{10.8}$$

where $u_1 = x_1^2 + x_2^2$, computed without the cosine and sine evaluations of (10.7).

The revised Box–Muller Method is a **rejection method**, since some inputs are not used. Comparing the area of the unit square $-1, 1 \times -1, 1$ to the unit circle, rejection will occur $(4 - \pi)/4 \approx 21\%$ of the time. This is an acceptable price to pay to avoid the sine and cosine evaluations.

There are more sophisticated methods for generating normal random numbers. See Knuth [1997] for more details. MATLAB's randn command, for example, uses the "ziggurat" algorithm of Marsaglia and Tsang [2000], essentially a very efficient way of inverting the cumulative distribution function.

▶ **ADDITIONAL EXAMPLES**

1. Approximate the average value of $f(x) = \frac{1}{1+x^2}$ on the interval $0, 1$ using one period of the linear congruential random number generator with $a = 7, b = 0, m = 11$. Compare with the exact value.

*2. (a) Use calculus to find the area of the region inside the circle of radius 1 centered at $(1, 0)$ but outside the ellipse $x^2 + 4y^2 = 4$. (b) Find a Monte Carlo approximation of the area.

⌐▯ **Solutions** for Additional Examples can be found at bit.ly/2Ai6zHu
(* example with video solution)

9.1 Exercises

⌐▯ **Solutions**
for Exercises
numbered in blue
can be found at
bit.ly/2yPnCPh

1. Find the period of the linear congruential generator defined by (a) $a = 2, b = 0, m = 5$ (b) $a = 4, b = 1, m = 9$.

2. Find the period of the LCG defined by $a = 4, b = 0, m = 9$. Does the period depend on the seed?

3. Approximate the area under the curve $y = x^2$ for $0 \le x \le 1$, using the LCG with (a) $a = 2, b = 0, m = 5$ (b) $a = 4, b = 1, m = 9$.

4. Approximate the area under the curve $y = 1 - x$ for $0 \le x \le 1$, using the LCG with (a) $a = 2, b = 0, m = 5$ (b) $a = 4, b = 1, m = 9$.

5. Consider the RANDNUM-CRAY random number generator, used on the Cray X-MP, one of the first supercomputers. This LCG used $m = 2^{48}, a = 2^{24} + 3$, and $b = 0$. Prove that $u_{i+2} = 6u_{i+1} - 9u_i$ (mod 1). Is this worrisome? See Computer Problems 9 and 10.

6. Prove that $u_1 = x_1^2 + x_2^2$ in the Box–Muller Rejection Method is a uniform random number on $0, 1$. (Hint: Show that for $0 \le y \le 1$, the probability that $u_1 \le y$ is equal to y. To do so, express it as the ratio of the disk area of radius $\sqrt{y}$ to the area of the unit circle.)

9.1 Computer Problems

1. Implement the Minimal Standard random number generator, and find the Monte Carlo approximation of the volume in Example 9.3. Use 10^6 three-dimensional points with seed $x_0 = 1$. How close is your approximation to the correct answer?

2. Implement randu and find the Monte Carlo approximation of the volume in Example 9.3, as in Computer Problem 1. Verify that no point (u_i, u_{i+1}, u_{i+2}) enters the given ball.

3. (a) Using calculus, find the area bounded by the two parabolas $P_1(x) = x^2 - x + 1/2$ and $P_2(x) = -x^2 + x + 1/2$. (b) Estimate the area as a Type 1 Monte Carlo simulation, by finding the average value of $P_2(x) - P_1(x)$ on 0, 1. Find estimates for $n = 10^i$ for $2 \le i \le 6$. (c) Same as (b), but estimate as a Type 2 Monte Carlo problem: Find the proportion of points in the square $0, 1 \times 0, 1$ that lie between the parabolas. Compare the efficiency of the two Monte Carlo approaches.

4. Carry out the steps of Computer Problem 3 for the subset of the first quadrant bounded by the polynomials $P_1(x) = x^3$ and $P_2(x) = 2x - x^2$.

5. Use $n = 10^4$ pseudo-random points to estimate the interior area of the ellipses (a) $13x^2 + 34xy + 25y^2 \le 1$ in $-1 \le x, y \le 1$ and (b) $40x^2 + 25y^2 + y + 9/4 \le 52xy + 14x$ in $0 \le x, y \le 1$. Compare your estimate with the correct areas (a) $\pi/6$ and (b) $\pi/18$, and report the error of the estimate. Repeat with $n = 10^6$ and compare results.

6. Use $n = 10^4$ pseudo-random points to estimate the interior volume of the ellipsoid defined by $2 + 4x^2 + 4z^2 + y^2 \le 4x + 4z + y$, contained in the unit cube $0 \le x, y, z \le 1$. Compare your estimate with the correct volume $\pi/24$, and report the error. Repeat with $n = 10^6$ points.

7. (a) Use calculus to evaluate the integral $\int_0^1 \int_{x^2}^{\sqrt{x}} xy \, dy \, dx$. (b) Use $n = 10^6$ pairs in the unit square $0, 1 \times 0, 1$ to estimate the integral as a Type 1 Monte Carlo problem. (Average the function that is equal to xy if (x, y) is in the integration domain and 0 if not.)

8. Use 10^6 random pairs in the unit square to estimate $\int_A xy \, dx \, dy$, where A is the area described by Example 9.2.

9. Implement the questionable random number generator from Exercise 5, and draw the plot analogous to Figure 10.3.

10. Devise a Monte Carlo approximation problem that completely foils the RANDNUM-CRAY generator of Exercise 5, following the ideas of Example 9.3.

10.2 MONTE CARLO SIMULATION

We have already seen examples of two types of Monte Carlo simulation. In this section, we explore the range of problems that are suited for this technique and discuss some of the refinements that make it work better, including quasi-random numbers. We will need to use the language of random variables and expected values in this section.

10.2.1 Power laws for Monte Carlo estimation

We would like to understand the convergence rate of Monte Carlo simulation. At what rate does the estimation error decrease as the number of points n used in the estimate grows? This is similar to the convergence questions in Chapter 6 for the quadrature methods and in Chapters 7, 8, and 9 for differential equation solvers. In the previous cases, they were posed as questions about error versus step size. Cutting the step size is analogous to adding more random numbers in Monte Carlo simulations.

Think of Type 1 Monte Carlo as the calculation of a function mean using random samples, then multiplying by the volume of the integration region. Calculating a function mean can be viewed as calculating the mean of a probability distribution given by that function. We will use the notation $E(X)$ for the expected value of the random variable X. The **variance** of a random variable X is $E[(X - E(X))^2]$, and the **standard deviation** of X is the square root of its variance. The error expected in estimating the mean will decrease with the number n of random points, in the following way:

Type 1 or Type 2 Monte Carlo with pseudo-random numbers.

$$\text{Error} \propto n^{-\frac{1}{2}} \tag{9.9}$$

To understand this formula, view the integral as the volume of the domain times the mean value A of the function over the domain. Consider the identical random variables X_i corresponding to a function evaluation at a random point. Then the mean value is the expected value of the random variable $Y = (X_1 + \cdots + X_n)/n$, or

$$E\left[\frac{X_1 + \cdots + X_n}{n}\right] = nA/n = A,$$

Convergence   A Monte Carlo Type 1 estimate does something very similar to the Composite Midpoint Method of Chapter 5. We found there that the error is proportional to the step size h, which is roughly equivalent to $1/n$ when the number of function evaluations is taken into account. This is more efficient than the square root power law of Monte Carlo.

However, Monte Carlo comes into its own with problems like Example 9.2. Although convergence to the correct value is still slow, it is not clear how to set up the problem as a Type 1 problem, in order to apply Chapter 5 techniques.

and the variance of Y is

$$E\left[\left(\frac{X_1 + \cdots + X_n}{n} - A\right)^2\right] = \frac{1}{n^2}\sum E[(X_i - A)^2] = \frac{1}{n^2}n\sigma^2 = \frac{\sigma^2}{n},$$

where σ is the original variance of each X_i. Therefore, the standard deviation of Y decreases as $\sigma/\sqrt{n}$. This argument applies to both Type 1 and Type 2 Monte Carlo simulation.

▶ **EXAMPLE 9.4**   Find Type 1 and Type 2 Monte Carlo estimates, using pseudo-random numbers for the area under the curve of $y = x^2$ in $[0, 1]$.

This is an extension of the Type 1 Monte Carlo Example 9.1, where we pay attention to the error as a function of the number n of random points. For each trial, we generate n uniform random numbers x in $[0, 1]$ and find the average value of $y = x^2$. The error is the absolute value of the difference of the average value and the correct answer $1/3$. We average the error over 500 trials for each n and plot the results as the lower curve in Figure 9.4.

For Type 2 Monte Carlo, we generate uniform random pairs (x, y) in the unit square $[0, 1] \times [0, 1]$ and track the proportion that satisfies $y < x^2$. Again, the error is

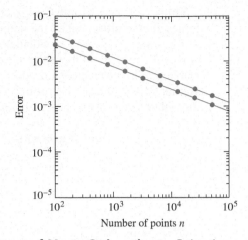

Figure 9.4 Mean error of Monte Carlo estimate. Estimation error in Example 9.4, as Type 1 (lower curve) and Type 2 (upper curve) Monte Carlo problems when pseudo-random numbers are used. The power law dependence has exponent $-1/2$ for both types.

averaged over 500 trials and plotted as the upper curve in Figure 9.4. Although the type 2 error is slightly greater than the type 1 error, both follow the square root power law (9.9). ◄

Is the randomness of the samples really required for a Type 2 Monte Carlo problem? Why not use a rectangular, regular grid of samples to solve a problem like Example 9.2, instead of random numbers? Of course, we would lose the ability to stop after an arbitrary number n of samples, unless there was some random-like way to order them, to avoid huge bias in the estimate. It turns out that there is a middle ground, which keeps the advantages of the regular grid but orders the numbers so as to appear random. This is the topic of the next section.

9.2.2 Quasi-random numbers

The idea of quasi-random numbers is to sacrifice the independence property of random numbers when it is not really essential to the problem being solved. Sacrificing independence means that quasi-random numbers are not only not random, but unlike pseudo-random numbers, they do not pretend to be random. This sacrifice is made in the hope of faster convergence to the correct value in a Monte Carlo setting. Sequences of quasi-random numbers are designed to be self-avoiding rather than independent. That is, the stream of numbers tries to efficiently fill in the gaps left by previous numbers and to avoid clustering. The comparison with pseudo-random numbers is illustrated in Figure 9.5.

There are many ways to produce quasi-random numbers. Perhaps the most popular way goes back to a suggestion of Van der Corput in 1935, called a **base-p low-discrepancy sequence**. We give the implementation due to Halton [1960]. Let p be a prime number, for example, $p = 2$. Write the first n integers in base p arithmetic. Assuming that the ith integer has representation $b_k b_{k-1} \cdots b_2 b_1$, we will assign the ith random number to be $0.b_1 b_2 \cdots b_{k-1} b_k$, again written in base p arithmetic. In other words, write the ith integer in base p, then reverse the digits, and put them on the other side of the decimal point to get the ith uniform random number in $[0, 1]$. Setting $p = 2$ gives the following list for the first eight random numbers:

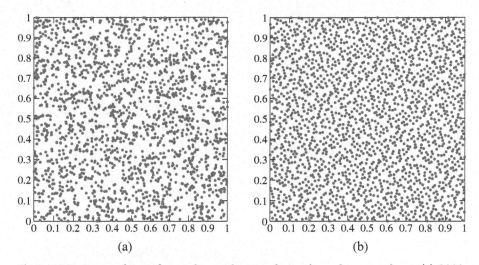

Figure 9.5 Comparison of pseudo-random and quasi-random numbers. (a) 2000 pairs of pseudo-random numbers, produced by MATLAB's rand. (b) 2000 pairs of quasi-random numbers, produced by Halton's low-discrepancy sequences, base 2 in x-coordinate and base 3 in y-coordinate.

i	$(i)_2$	$(u_i)_2$	u_i
1	1	.1	0.5
2	10	.01	0.25
3	11	.11	0.75
4	100	.001	0.125
5	101	.101	0.625
6	110	.011	0.375
7	111	.111	0.875
8	1000	.0001	0.0625

Setting $p = 3$ gives the Halton base-3 sequence:

i	$(i)_3$	$(u_i)_3$	u_i
1	1	.1	$0.\overline{3}$
2	2	.2	$0.\overline{6}$
3	10	.01	$0.\overline{1}$
4	11	.11	$0.\overline{4}$
5	12	.21	$0.\overline{7}$
6	20	.02	$0.\overline{2}$
7	21	.12	$0.\overline{5}$
8	22	.22	$0.\overline{8}$

MATLAB code for the Halton sequence is shown next. It is a simple and straightforward version of the original low-discrepancy idea. For greater efficiency, it can be coded on the bit level.

🖥 **MATLAB code**
shown here can be found
at bit.ly/2yk4n0U

```
% Program 9.1  Quasi-random number generator
% Halton sequence in base p
% Input: prime number p, random numbers required n
% Output: array u of quasi-random numbers in [0,1]
% Example usage: halton(2,100)
```

```
function u=halton(p,n)
b=zeros(ceil(log(n+1)/log(p)),1);    % largest number of digits
for j=1:n
  i=1;
  b(1)=b(1)+1;                        % add one to current integer
  while b(i)>p-1+eps                  % this loop does carrying
    b(i)=0;                           %    in  base p
    i=i+1;
    b(i)=b(i)+1;
  end
  u(j)=0;
  for k=1:length(b(:))               % add up reversed digits
    u(j)=u(j)+b(k)*p^(-k);
  end
end
end
```

For any prime number, the Halton sequence will give a set of quasi-random numbers. To generate a sequence of d-dimensional vectors, we can use a different prime for each coordinate. It is important to remember that quasi-random numbers are not independent; their usefulness lies in their self-avoiding property. For Monte Carlo problems, they are much more efficient than pseudo-random numbers, as we shall see next.

The reason for the use of quasi-random numbers is that they result in faster convergence of Monte Carlo estimates. That means that as a function of n, the number of function evaluations, the error decreases at a rate proportional to a larger negative power of n than the corresponding rate for pseudo-random numbers. The following error formulas should be compared with the corresponding formulas (9.9) for pseudo-random numbers (let d denote the dimension of random numbers being generated):

Type 1 Monte Carlo with quasi-random numbers

$$\text{Error} \propto (\ln n)^d n^{-1} \tag{9.10}$$

Type 2 Monte Carlo with quasi-random numbers

$$\text{Error} \propto n^{-\frac{1}{2}-\frac{1}{2d}} \tag{9.11}$$

The error is dominated by what happens at the discontinuities. In place of a proof, we describe what happens in the case of the Type 2 examples we have encountered, where the function is a characteristic function of a subset of d-dimensional space that has a $(d-1)$-dimensional boundary. In this case, the number of discontinuity points, along the boundary of the set, is proportional to $(n^{1/d})^{d-1}$. This follows from the fact that the boundary is $(d-1)$-dimensional, and there are on the order of $n^{1/d}$ grid points along each of the d dimensions. These points "randomly" take on the values 0 or 1, depending on which side of the boundary they lie on. Since the errors at all other points are much smaller, the variance of the function evaluation is, on average,

$$\frac{n^{\frac{d-1}{d}}}{n} = n^{-\frac{1}{d}},$$

and the standard deviation is the square root $n^{-\frac{1}{2d}}$. By the same argument as in the pseudo-random Monte Carlo case, when we are averaging over n points, the standard deviation is cut by a factor of $\sqrt{n}$, leaving the standard deviation of the quasi-Monte Carlo method to be

$$\frac{n^{-1/2d}}{n^{1/2}} = n^{-\frac{1}{2}-\frac{1}{2d}}.$$

▶ **EXAMPLE 9.5** Find a Monte Carlo estimate by using quasi-random numbers for the area under the curve of $y = x^2$ in [0, 1].

This is a Type 1 Monte Carlo problem, where x-coordinates can be generated in [0, 1] to find the average value of $f(x) = x^2$ as an approximation of the area. We use the Halton sequence with prime number $p = 2$ to generate 10^5 quasi-random numbers. The results, in comparison with the same strategy using pseudo-random numbers, are shown in Figure 9.6. The quasi-random numbers are clearly superior, as previously predicted. ◀

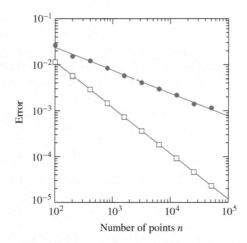

Figure 9.6 Mean error of Type 1 Monte Carlo estimate. Estimate of the integral of Example 9.1. Circles represent error when pseudo-random numbers are used, squares correspond to quasi-random. Note the power law dependence with exponent −1/2 and −1, respectively, for pseudo- and quasi-random numbers.

▶ **EXAMPLE 9.6** Find a quasi-random Monte Carlo estimate for the area in Example 9.2.

For various n, quasi-random samples in the unit square were generated by the Halton sequence. For multidimensional applications, it is convenient to use Halton sequences of different prime numbers p for each coordinate. The area is a subset of a two-dimensional space with a one-dimensional boundary, so $d = 2$. The proportion that satisfied the defining condition in Example 9.2 was determined, and the error was calculated. The error was averaged over 50 trials and plotted in Figure 9.7(a). The exponent of the power law for a Type 2 Monte Carlo problem in dimension two is $-1/2 - 1/(2d) = -1/2 - 1/4 = -3/4$, which is the approximate slope of the lower curve. The same calculation for pseudo-random numbers, with a square root power law, is shown in the figure for comparison. ◀

▶ **EXAMPLE 9.7** Find a quasi-random Monte Carlo estimate for the volume of the three-dimensional ball of radius one in R^3.

We proceed similarly to Example 9.6. Because the type 2 problem occurs in dimension three, the exponent of the power law is $-1/2 - 1/6 = -2/3$, which is approximately the slope of the lower curve in Figure 9.7(b). ◀

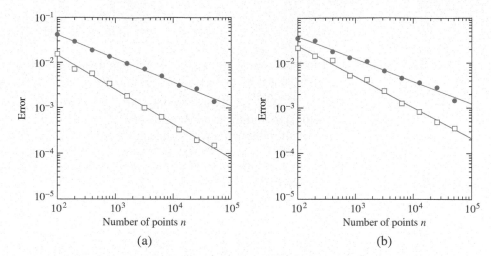

Figure 9.7 Mean error of Monte Carlo Type 2 estimate. Circles represent error when pseudo-random numbers are used, squares for quasi-random. (a) Estimate of the area in Example 9.2, a Type 2 Monte Carlo problem in dimension $d = 2$. The errors follow power laws with exponents $-1/2$ and $-3/4$, respectively, for pseudo- and quasi-random numbers. (b) Estimate of the volume of the three-dimensional ball of diameter 1, a Type 2 Monte Carlo problem in dimension $d = 3$. The errors follow power laws with exponents $-1/2$ and $-2/3$.

▶ **ADDITIONAL EXAMPLES**

1. Consider the region inside the circle of radius 1 centered at $(1, 0)$ but outside the ellipse $x^2 + 4y^2 = 4$. Find a quasi-random Monte Carlo approximation of the area and compare with the pseudo-random approximation in Additional Example 9.1.2.

2. (a) Use calculus to find the average determinant of a 2×2 symmetric matrix with uniform random entries from $[0, 1]$. (b) Carry out a Monte Carlo estimate of the average with pseudo-random numbers.

🖥 Solutions for Additional Examples can be found at `bit.ly/2CTu8J9`

9.2 Computer Problems

1. Carry out the Monte Carlo approximation in Computer Problem 9.1.3 with $n = 10^k$ quasi-random numbers from the Halton sequence for $k = 2, 3, 4$, and 5. For part (c), use `halton(2,n)` and `halton(3,n)` for the x and y coordinates, respectively.

2. Carry out the Monte Carlo approximation in Computer Problem 9.1.4 with quasi-random numbers.

3. Carry out the Monte Carlo approximation in Computer Problem 9.1.5 with $n = 10^4$ and $n = 10^5$ quasi-random points.

4. Carry out the Monte Carlo approximation in Computer Problem 9.1.6 with $n = 10^4$ and $n = 10^5$ quasi-random points.

5. Compute Monte Carlo and quasi-Monte Carlo approximations of the volume of the four-dimensional ball of radius 1 with $n = 10^5$ points. Compare with the exact volume $\pi^2/2$.

6. One of the best-known Monte Carlo problems is the Buffon needle. If a needle is dropped on a floor painted with black and white stripes, each the same width as the length of the needle, then the probability is $2/\pi$ that the needle will straddle both colors. (a) Prove this result analytically. Consider the distance d of the needle's midpoint to the nearest edge,

and its angle θ with the stripes. Express the probability as a simple integral. (b) Design a Monte Carlo Type 2 simulation that approximates the probability, and carry it out with $n = 10^6$ pseudo-random pairs (d, θ).

7. (a) What proportion of 2×2 matrices with entries in the interval $[0, 1]$ have positive determinant? Find the exact value, and approximate with a Monte Carlo simulation. (b) What proportion of symmetric 2×2 matrices with entries in $[0, 1]$ have positive determinant? Find the exact value and approximate with a Monte Carlo simulation.

8. Run a Monte Carlo simulation to approximate the proportion of 2×2 matrices with entries in $[-1, 1]$ whose eigenvalues are both real.

9. What proportion of 4×4 matrices with entries in $[0, 1]$ undergo no row exchanges under partial pivoting? Use a Monte Carlo simulation involving MATLAB's lu command to estimate this probability.

9.3 DISCRETE AND CONTINUOUS BROWNIAN MOTION

Although previous chapters of this book have focused largely on principles that are important for the mathematics of deterministic models, these models are only a part of the arsenal of modern techniques. One of the most important applications of random numbers is to make stochastic modeling possible.

We will begin with one of the simplest stochastic models, the random walk, also called discrete Brownian motion. The basic principles that underlie this discrete model are essentially the same for the more sophisticated models that follow, based on continuous Brownian motion.

9.3.1 Random walks

A **random walk** W_t is defined on the real line by starting at $W_0 = 0$ and moving a step of length s_i at each integer time i, where the s_i are independent and identically distributed random variables. Here, we will assume each s_i is $+1$ or -1 with equal probability $1/2$. **Discrete Brownian motion** is defined to be the random walk given by the sequence of accumulated steps

$$W_t = W_0 + s_1 + s_2 + \cdots + s_t,$$

for $t = 0, 1, 2, \ldots$ Figure 9.8 illustrates a single realization of discrete Brownian motion.

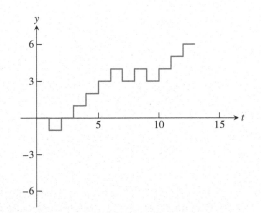

Figure 9.8 A single realization of a random walk. The path hits the boundary of the (vertical) interval $[-3, 6]$ at the 12th step. Random walks escape through the top of this interval one-third of the time, on average.

The following MATLAB code carries out a random walk of 10 steps:

```
t=10;
w=0;
for i=1:t
  if rand>1/2
    w=w+1;
  else
    w=w-1;
  end
end
```

Since a random walk is a probabilistic device, we will need to use some concepts from elementary probability. For each t, the value of W_t is a random variable. Stringing together a number of random variables $\{W_0, W_1, W_2, \ldots\}$ is by definition a **stochastic process**.

The expected value of a single step s_i of the random walk W_t is $(0.5)(1) + (0.5) \times (-1) = 0$, and the variance of s_i is $E[(s_i - 0)^2] = (0.5)(1)^2 + (0.5)(-1)^2 = 1$. The expected value of the random walk after an integer t steps is $E(W_t) = E(s_1 + \cdots + s_t) = E(s_1) + \cdots + E(s_t) = 0$, and the variance is $V(W_t) = V(s_1 + \cdots + s_t) = V(s_1) + \cdots + V(s_t) = t$, because variance is additive over independent random variables.

The mean and variance are statistical quantities that summarize information about a probability distribution. The fact that the mean of W_t is 0 and the variance is t indicates that if we compute n different realizations of the random variable W_t, then the

$$\textbf{sample mean} = E_{\text{sample}}(W_t) = \frac{W_t^1 + \cdots + W_t^n}{n}$$

and

$$\textbf{sample variance} = V_{\text{sample}}(W_t) = \frac{(W_t^1 - E_s)^2 + \cdots + (W_t^n - E_s)^2}{n - 1}$$

should approximate 0 and t, respectively. The **sample standard deviation**, defined to be the square root of the sample variance, is also called the **standard error** of the mean.

Many interesting applications of random walks are based on escape times, also called first passage times. Let a, b be positive integers, and consider the first time the random walk starting at 0 reaches the boundary of the interval $[-b, a]$. This is called the **escape time** of the random walk. It can be shown (Steele [2001]) that the probability that the escape happens at a (rather than $-b$) is exactly $b/(a + b)$.

▶ **EXAMPLE 9.8** Use a Monte Carlo simulation to approximate the probability that the random walk exits the interval $[-3, 6]$ through the top boundary 6.

This should happen 1/3 of the time. We will compute the sample mean and the error of the probability of escaping through $a = 6$ as a Type 2 Monte Carlo problem. We run n random walks until escape, and record the proportion that reach 6 before -3. For various values of n, we find the following table:

n	top exits	prob	error
100	35	0.3500	0.0167
200	72	0.3600	0.0267
400	135	0.3375	0.0042
800	258	0.3225	0.0108
1600	534	0.3306	0.0027
3200	1096	0.3425	0.0092
6400	2213	0.3458	0.0124

The error is the absolute value of the difference between the estimate and the correct probability 1/3. The error decreases gradually as more random walks are used, but irregularly, as the table shows. Figure 9.9 shows this error averaged over 50 trials. With this averaging, the errors show the square root power law decrease that is characteristic of Monte Carlo simulation.   ◄

The expected length of the escape time from $[-b, a]$ is known (Steele [2001]) to be ab. We can use the same simulation to investigate the efficiency of Monte Carlo on this problem.

► **EXAMPLE 9.9**   Use a Monte Carlo simulation to estimate the escape time for a random walk escaping the interval $[-3, 6]$.

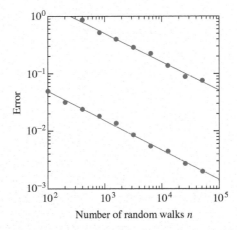

Figure 9.9 Error of Monte Carlo estimation for escape problem. Estimation error versus number of random walks for the probability of escaping $[-3, 6]$ by hitting 6 is shown in the lower curve. The expected value of the probability is 1/3. The upper curve shows estimation error for the escape time of the same problem. The expected value is 18 time steps. The errors were averaged over 50 trials.

The expected value of the escape time is $ab = 18$. A sample calculation shows the following table:

n	average esc. time	error
100	18.84	0.84
200	17.47	0.53
400	19.64	1.64
800	18.53	0.53
1600	18.27	0.27
3200	18.16	0.16
6400	18.05	0.05

Again, the error gradually decreases at an erratic rate. To see the square root power law for the error, we must average over several trials for each n. The result of 50 trials is shown in Figure 9.9. ◀

9.3.2 Continuous Brownian motion

In the previous section, we found that the standard random walk at t time steps has expected value 0 and variance t. Imagine now that double the number of steps are taken per unit time. If a step is taken every $1/2$ time unit, the expected value of the random walk at time t is still 0, but the variance is changed to

$$V(W_t) = V(s_1 + \cdots + s_{2t}) = V(s_1) + \cdots + V(s_{2t}) = 2t,$$

since $2t$ steps have been taken. In order to represent noise in a continuous model such as a differential equation, a continuous version of the random walk is needed. Doubling the number of steps per unit time is a good start, but to keep the variance fixed while we increase the number of steps, we will need to reduce the (vertical) size of each step. If we increase the number of steps by a factor k, we need to change the step height by a factor $1/\sqrt{k}$ to keep the variance the same as before. This is because multiplication of a random variable by a constant changes the variance by the square of the constant.

Therefore, W_t^k is defined to be the random walk that takes a step s_i^k of horizontal length $1/k$, and with step height $\pm 1/\sqrt{k}$ with equal probability. Then the expected value at time t is still

$$E(W_t^k) = \sum_{i=1}^{kt} E(s_i^k) = \sum_{i=1}^{kt} 0 = 0,$$

and the variance is

$$V(W_t^k) = \sum_{i=1}^{kt} V(s_i^k) = \sum_{i=1}^{kt} \left[\left(\frac{1}{\sqrt{k}} \right)^2 (.5) + \left(-\frac{1}{\sqrt{k}} \right)^2 (.5) \right] = kt\frac{1}{k} = t. \qquad (9.12)$$

If we decrease the step size and step height of the random walk in this precise way as k grows, the variance and standard deviation stays constant, independent of the number k of steps per unit time. Figure 9.10(b) shows a realization of W_t^k, where $k = 25$, so 250 individual steps were taken over 10 time units. The mean and variance at $t = 10$ are the same as in Figure 9.10(a).

The limit W_t^∞ of this progression as $k \to \infty$ yields **continuous Brownian motion**. Now time t is a real number, and $B_t \equiv W_t^\infty$ is a random variable for each $t \geq 0$. Continuous Brownian motion B_t has three important properties:

Property 1 For each t, the random variable B_t is normally distributed with mean 0 and variance t.

Property 2 For each $t_1 < t_2$, the normal random variable $B_{t_2} - B_{t_1}$ is independent of the random variable B_{t_1}, and in fact independent of all $B_s, 0 \leq s \leq t_1$.

Property 3 Brownian motion B_t can be represented by continuous paths.

The appearance of the normal distribution is a consequence of the Central Limit Theorem, a deep fact about probability.

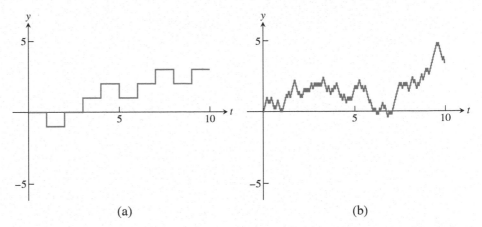

Figure 9.10 Discrete Brownian motion. (a) Random walk W_t of 10 steps. (b) Random walk W_t^{25} using 25 times more steps than (a), but with step height $1/\sqrt{25}$. The mean and variance of the height at time $t = 10$ are identical (0 and 10, respectively) for processes (a) and (b).

Computer simulation of Brownian motion is based on respecting these three properties. Establish a grid of steps

$$0 = t_0 \leq t_1 \leq \cdots \leq t_n$$

on the t-axis, and start with $B_0 = 0$. Property 2 says that the increment $B_{t_1} - B_{t_0}$ is a normal random variable, and its mean and variance are 0 and t_1. Therefore, a realization of the random variable B_{t_1} can be made by choosing from the normal distribution $N(0, t_1) = \sqrt{t_1 - t_0} N(0, 1)$; in other words, by multiplying a standard normal random number by $\sqrt{t_1 - t_0}$. To find B_{t_2}, we proceed similarly. The distribution of $B_{t_2} - B_{t_1}$ is $N(0, t_2 - t_1) = \sqrt{t_2 - t_1} N(0, 1)$, so we choose a standard normal random number, multiply by $\sqrt{t_2 - t_1}$, and add it to B_{t_1} to get B_{t_2}. In general, the increment of Brownian motion is the square root of the time step multiplied by a standard normal random number.

In MATLAB, we can write an approximation to Brownian motion by using the built-in normal random number generator randn. Here we use step size $\Delta t = 1/25$, as in Figure 9.10(b).

```
k=250;
sqdelt=sqrt(1/25);
b=0;
for i=1:k
  b=b+sqdelt*randn;
end
```

Escape time statistics for continuous Brownian motion are identical to those for random walks. Let a, b be positive numbers (not necessarily integers), and consider the first time that continuous Brownian motion starting at 0 reaches the boundary of the interval $[-b, a]$. This is called the escape time of Brownian motion from the interval. It can be shown that the probability that the escape happens at a (rather than $-b$) is exactly $b/(a + b)$. Moreover, the expected value of the escape time is ab. Computer Problem 5 asks the reader to illustrate these facts with Monte Carlo simulations.

▶ **ADDITIONAL EXAMPLES**

1. Carry out a Monte Carlo estimate of the probability of a random walk on the interval $[-7, 6]$ escaping through the top of the interval.

2. Carry out a Monte Carlo estimate of the mean escape time of Brownian motion on the interval $[-7, 6]$.

🖥 **Solutions** for Additional Examples can be found at `bit.ly/2EybkB6`

9.3 Computer Problems

🖥 **Solutions** for Computer Problems numbered in **blue** can be found at `bit.ly/2Aisizf`

1. Design a Monte Carlo simulation to estimate the probability of a random walk reaching the top a of the given interval $[-b, a]$. Carry out $n = 10000$ random walks. Calculate the error by comparing with the correct answer. (a) $[-2, 5]$ (b) $[-5, 3]$ (c) $[-8, 3]$

2. Calculate the mean escape time for the random walks in Computer Problem 1. Carry out $n = 10000$ random walks. Calculate the error by comparing with the correct answer.

3. In a **biased random walk**, the probability of going up one unit is $0 < p < 1$, and the probability of going down one unit is $q = 1 - p$. Design a Monte Carlo simulation with $n = 10000$ to find the probability that the biased random walk with $p = 0.7$ on the interval in Computer Problem 1 reaches the top. Calculate the error by comparing with the correct answer $[(q/p)^b - 1]/[(q/p)^{a+b} - 1]$ for $p \neq q$.

4. Carry out Computer Problem 3 for escape time. The mean escape time for the biased random walk with $p \neq q$ is $[b - (a + b)(1 - (q/p)^b)/(1 - (q/p)^{a+b})]/[q - p]$.

5. Design a Monte Carlo simulation to estimate the probability that Brownian motion escapes through the top of the given interval $[-b, a]$. Use $n = 1000$ Brownian motion paths of step size $\Delta t = 0.01$. Calculate the error by comparing with the correct answer $b/(a + b)$. (a) $[-2, 5]$ (b) $[-2, \pi]$ (c) $[-8/3, 3]$.

6. Calculate the mean escape time for Brownian motion for the intervals in Computer Problem 5. Carry out $n = 1000$ Brownian motion paths of step size $\Delta t = 0.01$. Calculate the error by comparing with the correct answer.

7. The **Arcsine Law** of Brownian motion holds that for $0 \leq t_1 \leq t_2$, the probability that a path does not cross zero in the time interval $[t_1, t_2]$ is $(2/\pi) \arcsin \sqrt{t_1/t_2}$. Carry out a Monte Carlo simulation of this probability by using 10,000 paths with $\Delta t = 0.01$, and compare with the correct probability, for the time intervals: (a) $3 < t < 5$ (b) $2 < t < 10$ (c) $8 < t < 10$.

9.4 STOCHASTIC DIFFERENTIAL EQUATIONS

Ordinary differential equations are deterministic models. Given an ODE and an appropriate initial condition, there is a unique solution, meaning that the future evolution of the solution is completely determined. Such omniscience is not always available to the modeler. For many systems, although some parts may be easily modeled, other parts may appear to move randomly—seemingly independently of the current system state. In such situations, instead of abandoning the idea of a model, it is common to add a noise term to the differential equation to represent the random effects. The result is called a stochastic differential equation (SDE).

In this section, we discuss some elementary stochastic differential equations and explain how to approximate solutions numerically. The solutions will be continuous stochastic processes like Brownian motion. We begin with some necessary definitions and a brief introduction to Ito calculus. For full details, the reader may consult Klebaner [1998], Oksendal [1998], and Steele [2001].

9.4.1 Adding noise to differential equations

Solutions to ordinary differential equations are functions. Solutions to stochastic differential equations, on the other hand, are stochastic processes.

DEFINITION 9.2 A set of random variables x_t indexed by real numbers $t \geq 0$ is called a **continuous-time stochastic process**. ☐

Each instance, or **realization** of the stochastic process is a choice of the random variable x_t for each t, and is therefore a function of t.

Brownian motion B_t is a stochastic process. Any (deterministic) function $f(t)$ can also be trivially considered as a stochastic process, with variance $V(f(t)) = 0$. The solution of the SDE initial value problem

$$\begin{cases} dy = r\,dt + \sigma\,dB_t \\ y(0) = 0 \end{cases}, \qquad (9.13)$$

with constants r and σ, is the stochastic process $y(t) = rt + \sigma B_t$, although we need to define some terms.

Notice that the SDE (9.13) is given in differential form, unlike the derivative form of an ODE. That is because many interesting stochastic processes, like Brownian motion, are continuous, but not differentiable. Therefore, the meaning of the SDE

$$dy = f(t, y)\,dt + g(t, y)\,dB_t$$

is, by definition, the integral equation

$$y(t) = y(0) + \int_0^t f(s, y)\,ds + \int_0^t g(s, y)\,dB_s,$$

where we must still define the meaning of the last integral, called an Ito integral.

Let $a = t_0 < t_1 < \cdots < t_{n-1} < t_n = b$ be a grid of points on the interval $[a, b]$. The Riemann integral is defined as a limit

$$\int_a^b f(t)\,dt = \lim_{\Delta t \to 0} \sum_{i=1}^n f(t_i')\Delta t_i,$$

where $\Delta t_i = t_i - t_{i-1}$ and $t_{i-1} \leq t_i' \leq t_i$. Similarly, the **Ito integral** is the limit

$$\int_a^b f(t)\,dB_t = \lim_{\Delta t \to 0} \sum_{i=1}^n f(t_{i-1})\Delta B_i,$$

where $\Delta B_i = B_{t_i} - B_{t_{i-1}}$, a step of Brownian motion across the interval. While the t_i' in the Riemann integral may be chosen at any point in the interval (t_{i-1}, t_i), the corresponding point for the Ito integral is required to be the left endpoint of that interval.

Because f and B_t are random variables, so is the Ito integral $I = \int_a^b f(t)\,dB_t$. The **differential** dI is a notational convenience; thus,

$$I = \int_a^b f\,dB_t$$

is equivalent by definition to

$$dI = f\,dB_t.$$

The differential dB_t of Brownian motion B_t is called **white noise**.

▶ **EXAMPLE 9.10**   Solve the stochastic differential equation $dy(t) = r\, dt + \sigma\, dB_t$ with initial condition $y(0) = y_0$.

We are assuming that r and σ are constant real numbers. The (deterministic) ordinary differential equation

$$y'(t) = r \tag{9.14}$$

has solution $y(t) = y_0 + rt$, a straight line as a function of time t. If r is positive, the solution moves up with constant slope; if r is negative, the solution moves down.

Adding white noise $\sigma\, dB_t$ for a constant real number σ to the right-hand side yields the stochastic differential equation

$$dy(t) = r\, dt + \sigma\, dB_t. \tag{9.15}$$

Integrating both sides gives

$$y(t) - y(0) = \int_0^t dy = \int_0^t r\, ds + \int_0^t \sigma\, dB_s = rt + \sigma B_t.$$

This confirms that the solution is the stochastic process

$$y(t) = y_0 + rt + \sigma B_t, \tag{9.16}$$

a combination of drift (the rt term) and the diffusion of Brownian motion.

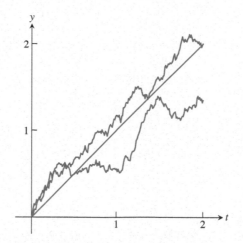

Figure 9.11 Solutions to Example 9.10. A solution $y(t) = rt$ of the ODE $y'(t) = r$ is shown, along with two different realizations of the solution process $y(t) = rt + \sigma B(t)$ for (9.15). The parameters are $r = 1$ and $\sigma = 0.3$.

Figure 9.11 shows two solutions of the SDE (9.15) alongside the unique solution to the ODE (9.14). Strictly speaking, the latter is also a solution to (9.15), representing the realization that goes with all noise inputs $z_i = 0$. This is a possible, but highly unlikely, particular realization of the solution stochastic process.   ◀

To solve SDEs analytically, we need to introduce the basic manipulation rule for stochastic differentials, called the Ito formula.

Ito formula

If $y = f(t, x)$, then

$$dy = \frac{\partial f}{\partial t}(t, x)\, dt + \frac{\partial f}{\partial x}(t, x)\, dx + \frac{1}{2}\frac{\partial^2 f}{\partial x^2}(t, x)\, dx\, dx, \tag{9.17}$$

where the $dx\, dx$ term is interpreted by using the identities $dt\, dt = 0$, $dt\, dB_t = dB_t\, dt = 0$, and $dB_t\, dB_t = dt$.

The Ito formula is the stochastic analogue to the chain rule of conventional calculus. Although it is expressed in differential form for ease of understanding, its meaning is no more and no less than the equality of the Ito integral of both sides of the equation. It is proved by referring the equation back to the definition of Ito integral (Oksendal [1998]).

▶ **EXAMPLE 9.11**   Prove that $y(t) = B_t^2$ is a solution of the SDE $dy = dt + 2B_t\, dB_t$.

To use the Ito formula, write $y = f(t, x)$, where $x = B_t$ and $f(t, x) = x^2$. According to (9.17),

$$dy = f_t\, dt + f_x\, dx + \frac{1}{2}f_{xx}\, dx\, dx$$

$$= 0\, dt + 2x\, dx + \frac{1}{2}2 dx\, dx$$

$$= 2B_t\, dB_t + dB_t\, dB_t$$

$$= 2B_t\, dB_t + dt.$$

◀

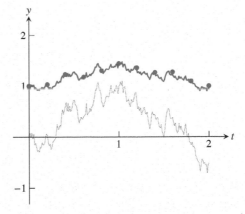

Figure 9.12 Solution to the exponential Brownian motion SDE (9.19).
The solution (9.18) is plotted as a solid curve along with the Euler–Maruyama approximation, plotted as circles. The dotted curve is the Brownian motion path for the corresponding realization. Parameters are set to $r = 0.1$, $\sigma = 0.3$, and $\Delta t = 0.2$.

▶ **EXAMPLE 9.12**   Show that geometric Brownian motion

$$y(t) = y_0 e^{(r - \frac{1}{2}\sigma^2)t + \sigma B_t} \tag{9.18}$$

satisfies the stochastic differential equation

$$dy = ry\, dt + \sigma y\, dB_t. \tag{9.19}$$

Write $y = f(t, x) = y_0 e^x$, where $x = (r - \frac{1}{2}\sigma^2)t + \sigma B_t$. By the Ito formula,

$$dy = y_0 e^x \, dx + \frac{1}{2} y_0 e^x \, dx \, dx,$$

where $dx = (r - 1/2\sigma^2) \, dt + \sigma \, dB_t$. Using the differential identities from the Ito formula, we obtain

$$dx \, dx = \sigma^2 \, dt.$$

Therefore,

$$
\begin{aligned}
dy &= y_0 e^x \left(r - \frac{1}{2}\sigma^2 \right) dt + y_0 e^x \sigma \, dB_t + \frac{1}{2} y_0 \sigma^2 e^x \, dt \\
&= y_0 e^x r \, dt + y_0 e^x \sigma \, dB_t \\
&= ry \, dt + \sigma y \, dB_t.
\end{aligned}
$$

◄

Figure 9.12 shows a realization of geometric Brownian motion with constant **drift coefficient** r and **diffusion coefficient** σ. This model is widely used in financial modeling. In particular, geometric Brownian motion is the underlying model for the Black–Scholes equations that are used to price financial derivatives.

Examples 9.11 and 9.12 are exceptions. Just as in the case of ODEs, relatively few SDEs have closed-form solutions. More often, it is necessary to use numerical approximation techniques.

9.4.2 Numerical methods for SDEs

We can approximate a solution to an SDE in a way that is similar to the Euler Method from Chapter 6. The Euler–Maruyama Method works by discretizing the time axis, just as Euler does. We define the approximate solution path at a grid of points

$$a = t_0 < t_1 < t_2 < \cdots < t_n = b$$

and will assign approximate y-values

$$w_0 < w_1 < w_2 < \cdots < w_n$$

at the respective t points. Given the SDE initial value problem

$$\begin{cases} dy(t) = f(t, y)dt + g(t, y)dB_t \\ y(a) = y_a \end{cases}, \tag{9.20}$$

we compute the solution approximately:

Euler–Maruyama Method

$$
\begin{aligned}
&w_0 = y_0 \\
&\textbf{for } i = 0, 1, 2, \ldots \\
&\quad w_{i+1} = w_i + f(t_i, w_i)(\Delta t_i) + g(t_i, w_i)(\Delta B_i) \\
&\textbf{end}
\end{aligned}
\tag{9.21}
$$

where

$$
\begin{aligned}
\Delta t_i &= t_{i+1} - t_i \\
\Delta B_i &= B_{t_{i+1}} - B_{t_i}.
\end{aligned}
\tag{9.22}
$$

The crucial part is how to model the Brownian motion ΔB_i. Define $N(0, 1)$ to be the standard random variable that is normally distributed with mean 0 and standard deviation 1. Each random number ΔB_i is computed in accordance with the description in Section 9.3.2 as

$$\Delta B_i = z_i \sqrt{\Delta t_i}, \qquad (9.23)$$

where z_i is chosen from $N(0, 1)$. In MATLAB, the z_i can be generated by the randn command. Again, notice the departure from the deterministic ODE case. Each set of $\{w_0, \ldots, w_n\}$ we produce is an approximate realization of the solution stochastic process $y(t)$, which depends on the random numbers z_i that were chosen. Since B_t is a stochastic process, each realization will be different, and so will our approximations.

As a first example, we show how to apply the Euler–Maruyama Method to the exponential Brownian motion SDE (9.19). The Euler–Maruyama Method has form

$$
\begin{aligned}
w_0 &= y_0 \\
w_{i+1} &= w_i + r w_i (\Delta t_i) + \sigma w_i (\Delta B_i),
\end{aligned}
\qquad (9.24)
$$

according to (9.21). A correct realization (generated from the solution (9.18)) and the corresponding Euler–Maruyama approximation are shown in Figure 9.12. By "corresponding," we mean that the approximation used the same Brownian motion realization (also shown in Figure 9.12) as the correct solution. Note the close agreement between the correct solution and the approximating points, plotted as small circles every 0.2 time units.

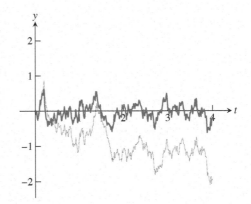

Figure 9.13 Solution to Langevin equation (9.25). The upper path is the solution approximation for parameters $r = 10, \sigma = 1$, computed by the Euler–Maruyama Method. The dotted path is the corresponding Brownian motion realization.

▶ **EXAMPLE 9.13**   Numerically solve the **Langevin equation**

$$dy = -ry \, dt + \sigma \, dB_t, \qquad (9.25)$$

where r and σ are positive constants.

Contrary to the preceding examples, it is not possible to analytically derive the solution to this equation in terms of simple processes. The solution of the Langevin equation is a stochastic process called the **Ornstein–Uhlenbeck process**. Figure 9.13 shows one realization of the approximate solution. It was generated from an Euler–Maruyama approximation, using the steps

$$
\begin{aligned}
w_0 &= y_0 \\
w_{i+1} &= w_i - r w_i (\Delta t_i) + \sigma (\Delta B_i)
\end{aligned}
\qquad (9.26)
$$

for $i = 1, \ldots, n$.

This SDE is used to model systems that tend to revert to a particular state, in this case the state $y = 0$, in the presence of a noisy background. We can think of a bowl containing a ping-pong ball that is in a car being driven over a rough road. The ball's distance $y(t)$ from the center of the bowl might be modeled by the Langevin equation. ◄

Next, we discuss the concept of order for SDE solvers. The idea is the same as for ODE solvers, aside from the differences caused by the fact that a solution to an SDE is a stochastic process, and each computed trajectory is only one realization of that process. Each realization of Brownian motion will force a different realization of the solution $y(t)$. If we fix a point $T > 0$ on the t-axis, each solution path started at $t = 0$ gives us a random value at T—that is, $y(T)$ is a random variable. Also, each computed solution path $w(t)$, using Euler–Maruyama, for example, gives us a random value at T, so that $w(T)$ is a random variable as well. The difference between the values at time T, $e(T) = y(T) - w(T)$, is therefore a random variable. The concept of order quantifies the expected value of the error in a manner similar to that for ODE solvers.

DEFINITION 9.3 An SDE solver has **order** m if the expected value of the error is of mth order in the step size; that is, if for any time T, $E\{|y(T) - w(T)|\} = O((\Delta t)^m)$ as the step size $\Delta t \to 0$. ☐

It is a surprise that unlike the ODE case where the Euler Method has order 1, the Euler–Maruyama Method for SDEs has order $m = 1/2$. To build an order 1 method for SDEs, another term in the "stochastic Taylor series" must be added to the method. Let

$$\begin{cases} dy(t) = f(t, y)\, dt + g(t, y)\, dB_t \\ y(0) = y_0 \end{cases}$$

be the SDE.

Milstein Method

$w_0 = y_0$
for $i = 0, 1, 2, \ldots$

$$\begin{aligned} w_{i+1} = w_i &+ f(t_i, w_i)(\Delta t_i) + g(t_i, w_i)(\Delta B_i) \\ &+ \tfrac{1}{2} g(t_i, w_i) \frac{\partial g}{\partial y}(t_i, w_i)((\Delta B_i)^2 - \Delta t_i) \end{aligned} \qquad (9.27)$$

end

The Milstein Method has order one. Note that the Milstein Method is identical to the Euler–Maruyama Method if there is no y term in the diffusion part $g(y, t)$ of the equation. In case there is, Milstein will converge to the correct stochastic solution process more quickly than Euler–Maruyama as the step size h goes to zero.

► **EXAMPLE 9.14** Apply the Milstein Method to geometric Brownian motion.

The equation is

$$dy = ry\, dt + \sigma y\, dB_t \qquad (9.28)$$

with solution process

$$y = y_0 e^{(r - \frac{1}{2}\sigma^2)t + \sigma B_t}. \qquad (9.29)$$

We discussed the Euler–Maruyama approximation previously. Using constant step size Δt, the Milstein Method becomes

$$w_0 = y_0$$

$$w_{i+1} = w_i + r w_i \Delta t + \sigma w_i \Delta B_i + \frac{1}{2}\sigma^2 w_i ((\Delta B_i)^2 - \Delta t). \qquad (9.30)$$

Applying the Euler–Maruyama Method and the Milstein Method with decreasing step sizes Δt results in successively improved approximations, as the following table shows:

Δt	Euler–Maruyama	Milstein
2^{-1}	0.169369	0.063864
2^{-2}	0.136665	0.035890
2^{-3}	0.086185	0.017960
2^{-4}	0.060615	0.008360
2^{-5}	0.048823	0.004158
2^{-6}	0.035690	0.002058
2^{-7}	0.024277	0.000981
2^{-8}	0.016399	0.000471
2^{-9}	0.011897	0.000242
2^{-10}	0.007913	0.000122

SPOTLIGHT ON

Convergence   The orders of the methods introduced here for SDEs, 1/2 for Euler–Maruyama and 1 for Milstein, would be considered low by ODE standards. Higher-order methods can be developed for SDEs, but are much more complicated as the order grows. Whether higher-order methods are needed in a given application depends on how the resulting approximate solutions are to be used. In the ODE case, the usual assumption is that the initial condition and the equation are known with high accuracy. Then it makes sense to calculate the solution as closely as possible to the same accuracy, and cheap higher-order methods are called for. In many situations, the advantages of higher-order SDE solvers are not so obvious; and if they come with added computational expense, these solvers may not be warranted.

The two columns represent the average, over 100 realizations, of the error $|w(T) - y(T)|$ at $T = 8$. Note that the realizations of $w(t)$ and $y(t)$ share the same Brownian motion increments ΔB_i. The orders 1/2 for Euler–Maruyama and 1 for Milstein are clearly visible in the table. Cutting the step size by a factor of 4 is required to reduce the error by a factor of 2 with the Euler–Maruyama Method. For the Milstein Method, cutting the step size by a factor of 2 achieves the same result. The data in the table is plotted on a log–log scale in Figure 9.14. ◄

A disadvantage of the Milstein Method is that the partial derivative appears in the approximation method, which must be provided by the user. This is analogous to Taylor methods for solving ordinary differential equations. For that reason, Runge–Kutta Methods were developed for ODEs, which trade these extra partial derivatives in the Taylor expansion for extra function evaluations.

In the SDE context, the same trade can be made with the Milstein Method, resulting in a first-order method than requires evaluation of $g(t, y)$ at two places on each step. A heuristic derivation can be carried out by making the replacement

$$\frac{\partial g}{\partial y}(t_i, w_i) \approx \frac{g(t_i, w_i + g(t_i, w_i)\sqrt{\Delta t_i}) - g(t_i, w_i)}{g(t_i, w_i)\sqrt{\Delta t_i}}$$

in the Milstein formula, which leads to the following method.

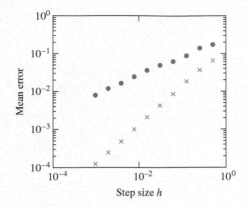

Figure 9.14 Error in the Euler–Maruyama and Milstein Methods. Solution paths are computed for the geometric Brownian motion equation (9.28) and are compared with the correct answer given by (9.29). The absolute difference is plotted versus step size h for the two different methods. The Euler–Maruyama errors are plotted as circles, and the Milstein errors as crosses. Note the slopes 1/2 and 1, respectively, on the log-log plot.

First-Order Stochastic Runge-Kutta Method

$w_0 = y_0$
for $i = 0, 1, 2, \ldots$
$\quad w_{i+1} = w_i + f(t_i, w_i)\Delta t_i + g(t_i, w_i)\Delta B_i$
$$\quad + \frac{1}{2\sqrt{\Delta t_i}} \Big[g(t_i, w_i + g(t_i, w_i)\sqrt{\Delta t_i}) - g(t_i, w_i) \Big] \Big[(\Delta B_i)^2 - \Delta t_i \Big]$$

end

▶ **EXAMPLE 9.15** Use the Euler–Maruyama Method, the Milstein Method, and the First-Order Stochastic Runge-Kutta Method to solve the SDE

$$dy = -2e^{-2y} \, dt + 2e^{-y} \, dB_t. \tag{9.31}$$

This example has an interesting cautionary property that is worth discussing. We can find an explicit solution, but it exists only for a finite time span. Using Ito's formula (9.17), we can show that $y(t) = \ln(2B_t + e^{y_0})$ is a solution, as long as the quantity inside the logarithm is positive. At the first time t when the Brownian motion realization causes $2B_t + e^{y_0}$ to be negative, the solution stops existing.

The Euler–Maruyama Method for this equation is

$$w_0 = y_0$$
$$w_{i+1} = w_i - 2e^{-2w_i}(\Delta t_i) + 2e^{-w_i}(\Delta B_i).$$

The Milstein Method is

$$w_0 = y_0$$
$$w_{i+1} = w_i - 2e^{-2w_i}(\Delta t_i) + 2e^{-w_i}(\Delta B_i) - 2e^{-2w_i}\Big[(\Delta B_i)^2 - \Delta t_i \Big].$$

The First-Order Stochastic Runge–Kutta Method is

$$w_0 = y_0$$
$$w_{i+1} = w_i - 2e^{-2w_i}(\Delta t_i) + 2e^{-w_i}(\Delta B_i)$$
$$+ \frac{1}{2\sqrt{\Delta t_i}} \Big[2e^{-(w_i + 2e^{-w_i}\sqrt{\Delta t_i})} - 2e^{-w_i} \Big] \Big[(\Delta B_i)^2 - \Delta t_i \Big].$$

A Milstein Method solution on the interval $0 \le t \le 4$ is shown in Figure 9.15. ◀

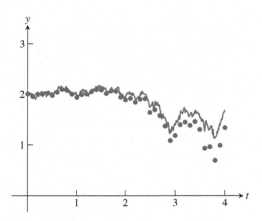

Figure 9.15 Solution to equation (9.31). Correct solution is shown along with Milstein approximation plotted as circles.

The stochastic processes we have seen up to now have had variances that increase with t. The variance of Brownian motion, for example, is $V(B_t) = t$. We finish the section with a remarkable example for which the end of the realization is as predictable as the beginning.

▶ **EXAMPLE 9.16**   Numerically solve the **Brownian bridge** SDE

$$
\begin{cases}
dy = \dfrac{y_1 - y}{t_1 - t}\, dt + dB_t \\
y(t_0) = y_0,
\end{cases}
\tag{9.32}
$$

where y_1 and $t_1 > t_0$ are given.

The solution of the Brownian bridge (9.32) is illustrated in Figure 9.16. Because the target slope adaptively changes as the path is created, all realizations of

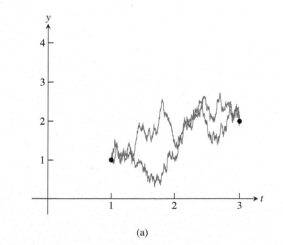

(a)

Figure 9.16 Brownian bridge. Two realizations of the solution of (9.32). The end-points are $(t_0, y_0) = (1, 1)$ and $(t_1, y_1) = (3, 2)$.

the solution process end at the desired point (t_1, y_1). The solution paths can be considered as stochastically generated "bridges" between the two given points (t_0, y_0) and (t_1, y_1). ◄

▶ **ADDITIONAL EXAMPLES**

*1. Show that the stochastic differential equation

$$dy = (y/2 + e^{B_t}) \, dt + (y + e^{B_t}) \, dB_t$$

with initial condition $y(0) = 0$ has solution $y(t) = B_t e^{B_t}$.

2. Use the Euler–Maruyama Method to find an approximate solution to the initial value problem of Additional Example 1. Plot the approximate solution along with the exact solution using the same Brownian motion.

⊡ **Solutions** for Additional Examples can be found at bit.ly/2J7tM24 (* example with video solution)

9.4 Exercises

⊡ **Solutions** for Exercises numbered in blue can be found at bit.ly/2Ct3Psn

1. Use Ito's formula to show that the solutions of the SDE initial value problems

(a) $\begin{cases} dy = B_t \, dt + t \, dB_t \\ y(0) = c \end{cases}$ (b) $\begin{cases} dy = 2B_t \, dB_t \\ y(0) = c \end{cases}$

are (a) $y(t) = t B_t + c$ (b) $y(t) = B_t^2 - t + c$.

2. Use Ito's formula to show that the solutions of the SDE initial value problems

(a) $\begin{cases} dy = (1 - B_t^2)e^{-2y} \, dt + 2B_t e^{-y} \, dB_t \\ y(0) = 0 \end{cases}$ (b) $\begin{cases} dy = B_t \, dt + \sqrt[3]{9y^2} \, dB_t \\ y(0) = 0 \end{cases}$

are (a) $y(t) = \ln(1 + B_t^2)$ (b) $y(t) = \frac{1}{3}B_t^3$.

3. Use Ito's formula to show that the solutions of the SDE initial value problems

(a) $\begin{cases} dy = ty \, dt + e^{t^2/2} \, dB_t \\ y(0) = 1 \end{cases}$ (b) $\begin{cases} dy = 3(B_t^2 - t) \, dB_t \\ y(0) = 0 \end{cases}$

are (a) $y(t) = (1 + B_t)e^{t^2/2}$ (b) $y(t) = B_t^3 - 3t B_t$.

4. Use Ito's formula to show that the solutions of the SDE initial value problems

(a) $\begin{cases} dy = -\frac{1}{2}y \, dt + \sqrt{1 - y^2} \, dB_t \\ y(0) = 0 \end{cases}$ (b) $\begin{cases} dy = y(1 + 2\ln y) \, dt + 2y B_t \, dB_t \\ y(0) = 1 \end{cases}$

are (a) $y(t) = \sin B_t$ and (b) $y(t) = e^{B_t^2}$.

5. Use the Ito formula to show that the solution of equation (9.31) is $\ln(2B_t + e^{y_0})$.

6. (a) Solve the ODE analogue of the Brownian bridge:

$$\begin{cases} y' = \dfrac{y_1 - y}{t_1 - t} \\ y(t_0) = y_0. \end{cases} \tag{9.33}$$

Does the solution reach the point (t_1, y_1) as the Brownian bridge does? Answer the same questions for the variants

(b) $\begin{cases} y' = \dfrac{y_1 - y_0}{t_1 - t_0} \\ y(t_0) = y_0 \end{cases}$ (c) $\begin{cases} dy = \dfrac{y_1 - y_0}{t_1 - t_0} \, dt + dB_t \\ y(t_0) = y_0 \end{cases}$

9.4 Computer Problems

Solutions for Computer Problems numbered in blue can be found at bit.ly/2EzAJdA

1. Use the Euler–Maruyama Method to find approximate solutions to the SDE initial value problems of Exercise 1. Use initial condition $y(0) = 0$. Plot the correct solution (found by keeping track of the Brownian motion B_t, using the same random increments) along with the approximate solution on the interval $[0, 10]$, using step size $h = 0.01$. Plot the error on the interval in a semilog plot.

2. Use the Euler–Maruyama Method to find approximate solutions to the SDE initial value problems of Exercise 2. Use initial condition $y(0) = 1$. Plot the correct solution along with the approximate solution on the interval $[0, 1]$, using step size $h = 0.01$. Plot the error on the interval in a semilog plot.

3. Apply the Euler–Maruyama Method with step size $h = 0.01$ to approximate solutions of Exercise 3 on the interval $[0, 2]$. Plot two realizations of the solution stochastic process.

4. Apply the Euler–Maruyama Method with step size $h = 0.01$ to approximate solutions of Exercise 4 on the interval $[0, 1]$. Plot two realizations of the solution stochastic process.

5. Find Euler–Maruyama approximate solutions to

$$\begin{cases} dy = B_t \, dt + \sqrt[3]{9y^2} \, dB_t \\ y(0) = 0 \end{cases}$$

on the interval $[0, 1]$ for step sizes $h = 0.1, 0.01$, and 0.001. For each step size, run 5000 realizations of the approximate solution, and find the average error at $t = 1$. Make a table of the average error at $t = 1$ versus step size. Does the average error scale according to theory?

6. Use the Euler–Maruyama Method to solve the SDE initial value problem $dy = y \, dt + y \, dB_t$, $y(0) = 1$. Plot the approximate solution and the correct solution $y(t) = e^{\frac{1}{2}t + B_t}$. Use a step size of $h = 0.1$ on the interval $0 \le t \le 2$.

7. Use the Milstein Method to find approximate solutions to the SDE initial value problem of Exercise 2(b). Plot the correct solution along with the approximate solution on the interval $[0, 5]$, using step size $h = 0.1$. Plot the error on the interval, using a semilog plot.

8. Use the Milstein Method to find approximate solutions to the SDE initial value problem of Exercise 4(a). Plot the correct solution along with the approximate solution on the interval $[0, 2]$, using step size $h = 0.1$. Plot the error on the interval, using a semilog plot.

9. Use the First-Order Stochastic Runge–Kutta Method to find approximate solutions to the SDE initial value problem of Exercise 2(b). Plot the correct solution along with the approximate solution on the interval $[0, 5]$, using step size $h = 0.1$. Plot the error on the interval, using a semilog plot.

10. Use the First-Order Stochastic Runge–Kutta Method to find approximate solutions to the SDE initial value problem of Exercise 4(a). Plot the correct solution along with the approximate solution on the interval $[0, 2]$, using step size $h = 0.1$. Plot the error on the interval, using a semilog plot.

11. Find Milstein approximate solutions to

$$\begin{cases} dy = B_t \, dt + \sqrt[3]{9y^2} \, dB_t \\ y(0) = 0 \end{cases}.$$

on the interval $[0, 1]$ for step sizes $h = 0.1, 0.01$, and 0.001. For each step size, run 5000 realizations of the approximate solution, and find the average error at $t = 1$. Make a table of the average error at $t = 1$ versus step size. Does the average error scale according to theory?

12. Perform a Monte Carlo estimate of $y(1)$, where $y(t)$ is the Euler–Maruyama solution of the Langevin equation

$$\begin{cases} dy = -y\,dt + dB_t \\ y(0) = e \end{cases}.$$

Average $n = 1000$ realizations with step size $h = 0.01$. Compare with the expected value of $y(1)$, which is 1.

Reality Check 9 *The Black–Scholes Formula*

Monte Carlo simulation and stochastic differential equation models are heavily used in financial calculations. A **financial derivative** is a financial instrument whose value is derived from the value of another instrument. In particular, an **option** is the right, but not the obligation, to complete a particular financial transaction.

A (European) **call** option is the right to buy one share of a security at a prear-ranged price, called the **strike price**, at a future date, called the **exercise date**. Calls are commonly purchased and sold by corporations to manage risk, and by individuals and mutual funds as part of investment strategies. Our goal is to calculate the value of the call option.

For example, a $15 December call for ABC Corp. represents the right to buy one share for $15 in December. Assume that the price of ABC on June 1 is $12. What is the value of such a right? On the exercise date, the value of a $K call is definite. It is $\max(X - K, 0)$, where X is the current market price of the stock. That is because, if $X > K$, the right to buy ABC at $K is worth $X - K$; and if $X < K$, the right to buy at K is worthless, since we can buy as much as we want at an even lower price. While the value of an option on the exercise date is clear, the difficulty is valuing the call at some time prior to expiration.

In the 1960s, Fisher Black and Myron Scholes explored the hypothesis of geo-metric Brownian motion,

$$dX = mX\,dt + \sigma X\,dB_t, \tag{9.34}$$

as the stock model, where m is the drift, or growth rate, of the stock and σ is the diffu-sion constant, or **volatility**. Both m and σ can be estimated from past stock price data. The insight of Black and Scholes was to develop an arbitrage theory that replicates the option through judicious balancing of stock holding and cash borrowing at the prevailing interest rate r. The result of their argument was that the correct call value, with expiration date T years into the future, is the present value of the expected option value at expiration time, where the underlying stock price $X(t)$ satisfies the SDE

$$dX = rX\,dt + \sigma X\,dB_t. \tag{9.35}$$

That is, for a stock price $X = X_0$ at time $t = 0$, the value of the call with expiration date $t = T$ is the expected value

$$C(X, T) = e^{-rT} E(\max(X(T) - K, 0)), \tag{9.36}$$

where $X(t)$ is given by (9.35). The surprise in their derivation was the replacement of drift m in (9.34) by the interest rate r in (9.35). In fact, the projected growth rate of the stock turns out to be irrelevant to the option price! This follows from the no-arbitrage assumption, a keystone of the Black–Scholes theory, that says that there are no risk-free gains available in an efficient market.

Formula (9.36) depends on the expected value of the random variable $X(T)$, which is only available through simulation. So, in addition to this insight, Black and Scholes [1973] provided a closed-form expression for the call price, namely,

$$C(X, T) = XN(d_1) - Ke^{-rT}N(d_2), \qquad (9.37)$$

where $N(x) = \frac{1}{\sqrt{2\pi}} \int_{-\infty}^{x} e^{-s^2/2} ds$ is the normal cumulative distribution function and

$$d_1 = \frac{\ln(X/K) + (r + \frac{1}{2}\sigma^2)T}{\sigma\sqrt{T}}, \quad d_2 = \frac{\ln(X/K) + (r - \frac{1}{2}\sigma^2)T}{\sigma\sqrt{T}}.$$

Equation (9.37) is known as the **Black–Scholes formula**.

Suggested activities:

Assume that one share of company ABC stock has a price of $12. Consider a European call option with strike price $15 and exercise date six months from today, so that $T = 0.5$ years. Assume that there is a fixed interest rate of $r = 0.05$ and that the volatility of the stock is 0.35 (i.e., 35 percent per year).

1. Perform a Monte Carlo simulation to compute the expected value in (9.36). Use the Euler–Maruyama Method to approximate the solution of (9.35), with a step size of $h = 0.01$ and initial value $X_0 = 12$. Note that SDE (9.34) is not relevant to this calculation. Carry out at least 10000 repetitions.

2. Compare your approximation in step 1 with the correct value from the Black–Scholes formula (9.37). The function $N(x)$ can be computed using the MATLAB error function erf as $N(x) = (1 + \text{erf}(x/\sqrt{2}))/2$.

3. Replace Euler–Maruyama with the Milstein Method, and repeat step 1. Compare the errors of the two methods.

4. A (European) **put** differs from a call in that it represents the right to sell, not buy, at the strike price. The value of a put is

$$P(X, T) = e^{-rT}E(\max(K - X(T), 0)), \qquad (9.38)$$

using $X(T)$ from (9.35). Calculate the value through Monte Carlo simulation for the same data as in step 1, using both Euler–Maruyama and Milstein Methods.

5. Compare your approximation in step 4 with the Black–Scholes formula for a put:

$$P(X, T) = Ke^{-rT}N(-d_2) - XN(-d_1). \qquad (9.39)$$

6. A down-and-out **barrier option** has a payout that is canceled if the stock crosses a given level. Consider the barrier call with strike price $K = \$15$ and barrier $L = \$11$. The payoff is $\max(X - K, 0)$ if $X(t) > L$ for $0 < t < T$, and 0 otherwise. Design and carry out a Monte Carlo simulation, using the geometric Brownian motion (9.35) and with (9.36) modified for the barrier option payout. You may need to make the step size h very small to get sufficient accuracy when barrier crossings are involved. Compare with the exact value

$$V(X, T) = C(X, T) - \left(\frac{X}{L}\right)^{1-2r/\sigma^2} C(L^2/X, T),$$

where $C(X, T)$ is the standard European call value with strike price K. See Wilmott et al. [1995], McDonald [2005], and Hull [2008] for details on more exotic options, their pricing formulas, and the role of Monte Carlo simulation in finance.

Software and Further Reading

The textbook Gentle [2003] is an introduction to the problem of generating random numbers. Other classic sources in the field are Knuth [1997] and Neiderreiter [1992]. Comparison of random number generation methods and a discussion of common evaluation criteria can be found in Hellekalek [1998].

The randu problem is addressed in Marsaglia [1968]. The minimum standard generator was introduced in Park and Miller [1988]. MATLAB's random number generator is based on the subtract-with-borrow methods described by Marsaglia and Zaman [1991]. Comprehensive sources for information on Monte Carlo and its applications include Fishman [1996] and Rubenstein [1981].

Modern textbooks on stochastic differential equations include Oksendal [2010] and Klebaner [2005]. Proper study in this area requires a solid background in basic probability. The computational aspects of SDEs are comprehensively treated in Kloeden and Platen [1992] and the more application-oriented handbook Kloeden et al. [1994]. The article Higham [2001] is a very readable introduction that includes MATLAB software for basic algorithms. Steele [2001] is an introduction to stochastic differential equations illustrated by numerous financial applications.

10

Trigonometric Interpolation and the FFT

The digital signal processing (DSP) chip is the backbone of advanced consumer electronics. Cellular phones, CD and DVD controllers, automobile electronics, personal digital assistants, digital modems, cameras, and televisions all make use of these ubiquitous devices. The hallmark of the DSP chip is its ability to do rapid digital calculations, including the Fast Fourier Transform (FFT).

One of the most basic functions of DSP is to separate desired input information from unwanted noise by filtering. The ability to extract signals from a cluttered background is an important part of the ongoing quest to build reliable speech recognition software. It is also a key element of pattern recognition devices, used by soccer-playing robot dogs to turn sensory inputs into usable data.

Reality Check Reality Check 10 on page 515 describes the Wiener filter, a fundamental building block of noise reduction via DSP.

Not even the most optimistic trigonometry teacher of a half-century ago could have envisioned the impact sines and cosines have had on modern technology. As we learned in Chapter 4, trig functions of multiple frequencies are natural interpolating functions for periodic data. The Fourier transform is almost unreasonably efficient at carrying out the interpolation and is irreplaceable in the data-intensive applications of modern signal processing.

The efficiency of trigonometric interpolation is bound up with the concept of orthogonality. We will see that orthogonal basis functions make interpolation and least squares fitting of data much simpler and more accurate. The Fourier transform exploits this orthogonality and provides an efficient means of interpolation with sines and cosines. The computational breakthrough of Cooley and Tukey called the Fast Fourier Transform (FFT) means that the Discrete Fourier Transform (DFT) can be computed very cheaply.

This chapter covers the basic ideas of the DFT, including a short introduction to complex numbers. The role of the DFT in trigonometric interpolation and least squares approximation is featured and viewed as a special case of approximation by orthogonal basis functions. This is the essence of digital filtering and signal processing.

10.1 THE FOURIER TRANSFORM

The French mathematician Jean Baptiste Joseph Fourier, after escaping the guillotine during the French Revolution and going to war alongside Napoleon, found time to develop a theory of heat conduction. To make the theory work, he needed to expand functions—not in terms of polynomials, as Taylor series, but in a revolutionary way first developed by Euler and Bernoulli—in terms of sine and cosine functions. Although rejected by the leading mathematicians of the time due to a perceived lack of rigor, today Fourier's methods pervade many areas of applied mathematics, physics, and engineering. In this section, we introduce the Discrete Fourier Transform and describe an efficient algorithm to compute it, the Fast Fourier Transform.

10.1.1 Complex arithmetic

The bookkeeping requirements of trigonometric functions can be greatly simplified by adopting the language of complex numbers. Every complex number has form $z = a + bi$, where $i = \sqrt{-1}$. Each z is represented geometrically as a two-dimensional vector of size a along the real (horizontal) axis, and size b along the imaginary (vertical) axis, as shown in Figure 10.1. The **complex magnitude** of the number $z = a + bi$ is defined to be $|z| = \sqrt{a^2 + b^2}$ and is exactly the distance of the complex number from the origin in the complex plane. The **complex conjugate** of a complex number $z = a + bi$ is $\bar{z} = a - bi$.

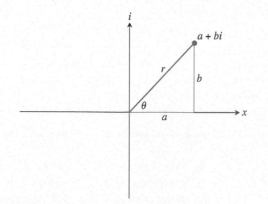

Figure 10.1 Representation of a complex number. The real and imaginary parts are a and bi, respectively. The polar representation is $a + bi = re^{i\theta}$.

The celebrated **Euler formula** for complex arithmetic says $e^{i\theta} = \cos\theta + i\sin\theta$. The complex magnitude of $z = e^{i\theta}$ is 1, so complex numbers of this form lie on the unit circle in the complex plane, as shown in Figure 10.2. Any complex number $a + bi$ can be written in its **polar representation**

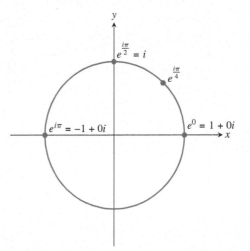

Figure 10.2 Unit circle in the complex plane. Complex numbers of the form $e^{i\theta}$ for some angle θ have magnitude one and lie on the unit circle.

$$z = a + bi = re^{i\theta}, \tag{10.1}$$

where r is the complex magnitude $|z| = \sqrt{a^2 + b^2}$ and $\theta = \arctan b/a$.

The unit circle in the complex plane corresponds to complex numbers of magnitude $r = 1$. To multiply together the two numbers $e^{i\theta}$ and $e^{i\gamma}$ on the unit circle, we could convert to trigonometric functions and then multiply:

$$e^{i\theta}e^{i\gamma} = (\cos\theta + i\sin\theta)(\cos\gamma + i\sin\gamma)$$
$$= \cos\theta\cos\gamma - \sin\theta\sin\gamma + i(\sin\theta\cos\gamma + \sin\gamma\cos\theta).$$

Recognizing the cos addition formula and the sin addition formula, we can rewrite this as

$$\cos(\theta + \gamma) + i\sin(\theta + \gamma) = e^{i(\theta+\gamma)}.$$

Equivalently, just add the exponents:

$$e^{i\theta}e^{i\gamma} = e^{i(\theta+\gamma)}. \tag{10.2}$$

Equation (10.2) shows that the product of two numbers on the unit circle gives a new point on the unit circle whose angle is the sum of the two angles. The Euler formula hides the trigonometry details, like the sine and cosine addition formulas, and makes the bookkeeping much easier. This is the reason we introduce complex arithmetic into the study of trigonometric interpolation. Although it can be done entirely in the real numbers, the Euler formula has a profound simplifying effect.

We single out a special subset of magnitude 1 complex numbers. A complex number z is an nth **root of unity** if $z^n = 1$. On the real number line, there are only two roots of unity, -1 and 1. In the complex plane, however, there are many. For example, i itself is a 4th root of unity, because $i^4 = (-1)^2 = 1$.

An nth root of unity is called **primitive** if it is not a kth root of unity for any $k < n$. By this definition, -1 is a primitive second root of unity and a nonprimitive fourth root of unity. It is easy to check that for any integer n, the complex number $\omega_n = e^{-i2\pi/n}$ is a primitive nth root of unity. The number $e^{i2\pi/n}$ is also a primitive nth root of unity, but we will follow the usual convention of using the former for the basis of the Fourier transform. Figure 10.3 shows a primitive eighth root of unity $\omega_8 = e^{-i2\pi/8}$ and the other seven roots of unity, which are powers of ω_8.

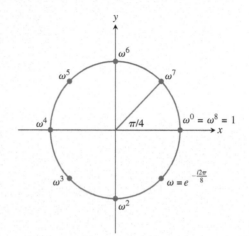

Figure 10.3 Roots of unity. The eight 8th roots of unity are shown. They are generated by $\omega = e^{-2\pi/8}$, meaning that each is ω^k for some integer k. Although ω and ω^3 are primitive 8th roots of unity, ω^2 is not, because it is also a 4th root of unity.

Here is a key identity that we will need later to simplify our computations of the Discrete Fourier Transform. Let ω denote the nth root of unity $\omega = e^{-i2\pi/n}$ where $n > 1$. Then

$$1 + \omega + \omega^2 + \omega^3 + \cdots + \omega^{n-1} = 0. \tag{10.3}$$

The proof of this identity follows from the telescoping sum

$$(1 - \omega)(1 + \omega + \omega^2 + \omega^3 + \cdots + \omega^{n-1}) = 1 - \omega^n = 0. \tag{10.4}$$

Since the first term on the left is not zero, the second must be. A similar method of proof shows that

$$1 + \omega^2 + \omega^4 + \omega^6 + \cdots + \omega^{2(n-1)} = 0,$$
$$1 + \omega^3 + \omega^6 + \omega^9 + \cdots + \omega^{3(n-1)} = 0,$$

$$\vdots$$

$$1 + \omega^{n-1} + \omega^{(n-1)2} + \omega^{(n-1)3} + \cdots + \omega^{(n-1)(n-1)} = 0. \tag{10.5}$$

The next one is different:

$$1 + \omega^n + \omega^{2n} + \omega^{3n} + \cdots + \omega^{n(n-1)} = 1 + 1 + 1 + 1 + \cdots + 1$$
$$= n. \tag{10.6}$$

This information is collected into the following lemma.

LEMMA 10.1 **Primitive roots of unity.** Let ω be a primitive nth root of unity and k be an integer. Then

$$\sum_{j=0}^{n-1} \omega^{jk} = \begin{cases} n & \text{if } k/n \text{ is an integer} \\ 0 & \text{otherwise} \end{cases}.$$

∎

Exercise 6 asks the reader to fill in the details of the proof.

10.1.2 Discrete Fourier Transform

Let $x = [x_0, \ldots, x_{n-1}]^T$ be a (real-valued) n-dimensional vector, and denote $\omega = e^{-i2\pi/n}$. Here is the fundamental definition of this chapter.

DEFINITION 10.2 The **Discrete Fourier Transform** (DFT) of $x = [x_0, \ldots, x_{n-1}]^T$ is the n-dimensional vector $y = [y_0, \ldots, y_{n-1}]$, where $\omega = e^{-i2\pi/n}$ and

$$y_k - \frac{1}{\sqrt{n}} \sum_{j=0}^{n-1} x_j \omega^{jk}. \tag{10.7}$$

❑

For example, Lemma 10.1 shows that the DFT of $x = [1, 1, \ldots, 1]$ is $y = [\sqrt{n}, 0, \ldots, 0]$. In matrix terms, this definition says

$$\begin{bmatrix} y_0 \\ y_1 \\ y_2 \\ \vdots \\ y_{n-1} \end{bmatrix} = \begin{bmatrix} a_0 + ib_0 \\ a_1 + ib_1 \\ a_2 + ib_2 \\ \vdots \\ a_{n-1} + ib_{n-1} \end{bmatrix} = \frac{1}{\sqrt{n}} \begin{bmatrix} \omega^0 & \omega^0 & \omega^0 & \cdots & \omega^0 \\ \omega^0 & \omega^1 & \omega^2 & \cdots & \omega^{n-1} \\ \omega^0 & \omega^2 & \omega^4 & \cdots & \omega^{2(n-1)} \\ \omega^0 & \omega^3 & \omega^6 & \cdots & \omega^{3(n-1)} \\ \vdots & \vdots & \vdots & & \vdots \\ \omega^0 & \omega^{n-1} & \omega^{2(n-1)} & \cdots & \omega^{(n-1)^2} \end{bmatrix} \begin{bmatrix} x_0 \\ x_1 \\ x_2 \\ \vdots \\ x_{n-1} \end{bmatrix}. \tag{10.8}$$

Each $y_k = a_k + ib_k$ is a complex number. The $n \times n$ matrix in (10.8) is called the **Fourier matrix**

$$F_n = \frac{1}{\sqrt{n}} \begin{bmatrix} \omega^0 & \omega^0 & \omega^0 & \cdots & \omega^0 \\ \omega^0 & \omega^1 & \omega^2 & \cdots & \omega^{n-1} \\ \omega^0 & \omega^2 & \omega^4 & \cdots & \omega^{2(n-1)} \\ \omega^0 & \omega^3 & \omega^6 & \cdots & \omega^{3(n-1)} \\ \vdots & \vdots & \vdots & & \vdots \\ \omega^0 & \omega^{n-1} & \omega^{2(n-1)} & \cdots & \omega^{(n-1)^2} \end{bmatrix}. \tag{10.9}$$

Except for the top row, each row of the Fourier matrix adds to zero, and the same is true for the columns, since F_n is a symmetric matrix. The Fourier matrix has an explicit inverse

$$F_n^{-1} = \frac{1}{\sqrt{n}} \begin{bmatrix} \omega^0 & \omega^0 & \omega^0 & \cdots & \omega^0 \\ \omega^0 & \omega^{-1} & \omega^{-2} & \cdots & \omega^{-(n-1)} \\ \omega^0 & \omega^{-2} & \omega^{-4} & \cdots & \omega^{-2(n-1)} \\ \omega^0 & \omega^{-3} & \omega^{-6} & \cdots & \omega^{-3(n-1)} \\ \vdots & \vdots & \vdots & & \vdots \\ \omega^0 & \omega^{-(n-1)} & \omega^{-2(n-1)} & \cdots & \omega^{-(n-1)^2} \end{bmatrix}, \tag{10.10}$$

and the **Inverse Discrete Fourier Transform** of the vector y is $x = F_n^{-1}y$. Checking that (10.10) is the inverse of the matrix F_n requires Lemma 11.1 about nth roots of unity. See Exercise 8.

Let $z = e^{i\theta} = \cos\theta + i\sin\theta$ be a point on the unit circle. Then its reciprocal $e^{-i\theta} = \cos\theta - i\sin\theta$ is its complex conjugate. Therefore, the inverse DFT is the matrix of complex conjugates of the entries of F_n:

$$F_n^{-1} = \overline{F}_n. \tag{10.11}$$

DEFINITION 10.3   The **magnitude** of a complex vector v is the real number $||v|| = \sqrt{\overline{v}^T v}$. A square complex matrix F is **unitary** if $\overline{F}^T F = I$.   ❑

A unitary matrix, like the Fourier matrix, is the complex version of a real orthogonal matrix. If F is unitary, then $||Fv||^2 = \overline{v}^T \overline{F}^T F v = \overline{v}^T v = ||v||^2$. Thus, the magnitude of a vector is unchanged upon multiplication on the left by F—or F^{-1} for that matter.

Applying the Discrete Fourier Transform is a matter of multiplying by the $n \times n$ matrix F_n, and therefore requires $O(n^2)$ operations (specifically n^2 multiplications and $n(n-1)$ additions). The Inverse Discrete Fourier Transform, which is applied by multiplication by F_n^{-1}, is also an $O(n^2)$ process. In Section 10.1.3, we develop a version of the DFT that requires significantly fewer operations, called the Fast Fourier Transform.

▶ **EXAMPLE 10.1**   Find the DFT of the vector $x = [1, 0, -1, 0]^T$.

Let ω be the 4th root of unity, or $\omega = e^{-i\pi/2} = \cos(\pi/2) - i\sin(\pi/2) = -i$. Applying the DFT, we get

$$\begin{bmatrix} y_0 \\ y_1 \\ y_2 \\ y_3 \end{bmatrix} = \frac{1}{\sqrt{4}} \begin{bmatrix} 1 & 1 & 1 & 1 \\ 1 & \omega & \omega^2 & \omega^3 \\ 1 & \omega^2 & \omega^4 & \omega^6 \\ 1 & \omega^3 & \omega^6 & \omega^9 \end{bmatrix} \begin{bmatrix} 1 \\ 0 \\ -1 \\ 0 \end{bmatrix} = \frac{1}{2} \begin{bmatrix} 1 & 1 & 1 & 1 \\ 1 & -i & -1 & i \\ 1 & -1 & 1 & -1 \\ 1 & i & -1 & -i \end{bmatrix} \begin{bmatrix} 1 \\ 0 \\ -1 \\ 0 \end{bmatrix} = \begin{bmatrix} 0 \\ 1 \\ 0 \\ 1 \end{bmatrix}.$$
$$\tag{10.12}$$

◀

The MATLAB command `fft` carries out the DFT with a slightly different normalization, so that $F_n x$ is computed by `fft(x)/sqrt(n)`. The inverse command `ifft` is the inverse of `fft`. Therefore, $F_n^{-1}y$ is computed by the MATLAB command `ifft(y)*sqrt(n)`. In other words, MATLAB's `fft` and `ifft` commands are inverses of one another, although their normalization differs from the definition given here, which has the advantage that F_n and F_n^{-1} are unitary matrices.

Even if the vector x has components that are real numbers, there is no reason for the components of y to be real numbers. But if the x_j are real, the complex numbers y_k have a special property:

LEMMA 10.4   Let $\{y_k\}$ be the DFT of $\{x_j\}$, where the x_j are real numbers. Then (a) y_0 is real, and (b) $y_{n-k} = \overline{y}_k$ for $k = 1, \ldots, n-1$.   ■

Proof. The reason for (a) is clear from (10.7), since y_0 is the sum of the x_j's divided by $\sqrt{n}$. Part (b) follows from the fact that

$$\omega^{n-k} = e^{-i2\pi(n-k)/n} = e^{-i2\pi} e^{i2\pi k/n} = \cos(2\pi k/n) + i\sin(2\pi k/n)$$

while

$$\omega^k = e^{-i2\pi k/n} = \cos(2\pi k/n) - i\sin(2\pi k/n),$$

implying that $\omega^{n-k} = \overline{\omega^k}$. From the definition of Fourier transform,

$$y_{n-k} = \frac{1}{\sqrt{n}} \sum_{j=0}^{n-1} x_j (\omega^{n-k})^j$$

$$= \frac{1}{\sqrt{n}} \sum_{j=0}^{n-1} x_j (\overline{\omega^k})^j$$

$$= \frac{1}{\sqrt{n}} \sum_{j=0}^{n-1} \overline{x_j (\omega^k)^j} = \overline{y_k}.$$

Here we have used the fact that the product of complex conjugates is the conjugate of the product. ☐

Lemma 10.4 has an interesting consequence. Let n be even and the $x_0, \ldots, x_{n-1}$ be real numbers. Then the DFT replaces them with exactly n other real numbers $a_0, a_1, b_1, a_2, b_2, \ldots, a_{n/2}$, the real and imaginary parts of the Fourier transform $y_0, \ldots, y_{n-1}$. For example, the $n = 8$ DFT has the form

$$F_8 \begin{bmatrix} x_0 \\ x_1 \\ x_2 \\ x_3 \\ x_4 \\ x_5 \\ x_6 \\ x_7 \end{bmatrix} = \begin{bmatrix} a_0 \\ a_1 + ib_1 \\ a_2 + ib_2 \\ a_3 + ib_3 \\ a_4 \\ a_3 - ib_3 \\ a_2 - ib_2 \\ a_1 - ib_1 \end{bmatrix} = \begin{bmatrix} y_0 \\ \vdots \\ y_{\frac{n}{2}-1} \\ y_{\frac{n}{2}} \\ \overline{y_{\frac{n}{2}-1}} \\ \vdots \\ \overline{y_1} \end{bmatrix}. \tag{10.13}$$

10.1.3 The Fast Fourier Transform

As mentioned in the last section, the Discrete Fourier Transform applied to an n-vector in the traditional way requires $O(n^2)$ operations. Cooley and Tukey [1965] found a way to accomplish the DFT in $O(n \log n)$ operations in an algorithm called the **Fast Fourier Transform** (FFT). The popularity of the FFT for data analysis followed almost immediately. The field of signal processing converted from primarily analog to digital largely due to this algorithm. We will explain their method and show its superiority to the naive DFT (10.8) through an operation count.

We can write the DFT $F_n x$ as

$$\begin{bmatrix} y_0 \\ \vdots \\ y_{n-1} \end{bmatrix} = \frac{1}{\sqrt{n}} M_n \begin{bmatrix} x_0 \\ \vdots \\ x_{n-1} \end{bmatrix},$$

where

$$M_n = \begin{bmatrix} \omega^0 & \omega^0 & \omega^0 & \cdots & \omega^0 \\ \omega^0 & \omega^1 & \omega^2 & \cdots & \omega^{n-1} \\ \omega^0 & \omega^2 & \omega^4 & \cdots & \omega^{2(n-1)} \\ \omega^0 & \omega^3 & \omega^6 & \cdots & \omega^{3(n-1)} \\ \vdots & \vdots & \vdots & & \vdots \\ \omega^0 & \omega^{n-1} & \omega^{2(n-1)} & \cdots & \omega^{(n-1)^2} \end{bmatrix}.$$

Complexity   The achievement of Cooley and Tukey to reduce the complexity of the DFT from $O(n^2)$ operations to $O(n \log n)$ operations opened up a world of possibilities for Fourier transform methods. A method that scales "almost linearly" with the size of the problem is very valuable. For example, there is a possibility of using it for real-time data, since analysis can occur approximately at the same timescale that data are acquired. The development of the FFT was followed a short time later with specialized circuitry for implementing it, now represented by DSP chips for digital signal processing that are ubiquitous in electronic systems for analysis and control.

We will show how to compute $z = M_n x$ recursively. To complete the DFT requires dividing by $\sqrt{n}$, or $y = F_n x = z/\sqrt{n}$.

We start by showing how the $n = 4$ case works, to get the main idea across. The general case will then be clear. Let $\omega = e^{-i2\pi/4} = -i$. The Discrete Fourier Transform is

$$\begin{bmatrix} z_0 \\ z_1 \\ z_2 \\ z_3 \end{bmatrix} = \begin{bmatrix} \omega^0 & \omega^0 & \omega^0 & \omega^0 \\ \omega^0 & \omega^1 & \omega^2 & \omega^3 \\ \omega^0 & \omega^2 & \omega^4 & \omega^6 \\ \omega^0 & \omega^3 & \omega^6 & \omega^9 \end{bmatrix} \begin{bmatrix} x_0 \\ x_1 \\ x_2 \\ x_3 \end{bmatrix}. \tag{10.14}$$

Write out the matrix product, but rearrange the order of the terms so that the even-numbered terms come first:

$$z_0 = \omega^0 x_0 + \omega^0 x_2 + \omega^0(\omega^0 x_1 + \omega^0 x_3)$$
$$z_1 = \omega^0 x_0 + \omega^2 x_2 + \omega^1(\omega^0 x_1 + \omega^2 x_3)$$
$$z_2 = \omega^0 x_0 + \omega^4 x_2 + \omega^2(\omega^0 x_1 + \omega^4 x_3)$$
$$z_3 = \omega^0 x_0 + \omega^6 x_2 + \omega^3(\omega^0 x_1 + \omega^6 x_3)$$

Using the fact that $\omega^4 = 1$, we can rewrite these equations as

$$z_0 = (\omega^0 x_0 + \omega^0 x_2) + \omega^0(\omega^0 x_1 + \omega^0 x_3)$$
$$z_1 = (\omega^0 x_0 + \omega^2 x_2) + \omega^1(\omega^0 x_1 + \omega^2 x_3)$$
$$z_2 = (\omega^0 x_0 + \omega^0 x_2) + \omega^2(\omega^0 x_1 + \omega^0 x_3)$$
$$z_3 = (\omega^0 x_0 + \omega^2 x_2) + \omega^3(\omega^0 x_1 + \omega^2 x_3)$$

Notice that each term in parentheses in the top two lines is repeated verbatim in the bottom two lines. Define

$$u_0 = \mu^0 x_0 + \mu^0 x_2$$
$$u_1 = \mu^0 x_0 + \mu^1 x_2$$

and

$$v_0 = \mu^0 x_1 + \mu^0 x_3$$
$$v_1 = \mu^0 x_1 + \mu^1 x_3,$$

where $\mu = \omega^2$ is a 2nd root of unity. Both $u = (u_0, u_1)^T$ and $v = (v_0, v_1)^T$ are essentially DFTs with $n = 2$; more precisely,

$$u = M_2 \begin{bmatrix} x_0 \\ x_2 \end{bmatrix}$$

$$v = M_2 \begin{bmatrix} x_1 \\ x_3 \end{bmatrix}.$$

We can write the original M_4x as

$$
\begin{aligned}
z_0 &= u_0 + \omega^0 v_0 \\
z_1 &= u_1 + \omega^1 v_1 \\
z_2 &= u_0 + \omega^2 v_0 \\
z_3 &= u_1 + \omega^3 v_1.
\end{aligned}
$$

In summary, the calculation of the DFT(4) has been reduced to a pair of DFT(2)s plus some extra multiplications and additions.

Ignoring the $1/\sqrt{n}$ for a moment, DFT(n) can be reduced to computing two DFT($n/2$)s plus $2n - 1$ extra operations ($n - 1$ multiplications and n additions). A careful count of the additions and multiplications necessary yields Theorem 10.5.

THEOREM 10.5   **Operation Count for FFT.** Let n be a power of 2. Then the Fast Fourier Transform of size n can be completed in $n(2\log_2 n - 1) + 1$ additions and multiplications, plus a division by $\sqrt{n}$.   ∎

Proof. Ignore the square root, which is applied at the end. The result is equivalent to saying that the DFT(2^m) can be completed in $2^m(2m - 1) + 1$ additions and multiplications. In fact, we saw above how a DFT(n), where n is even, can be reduced to a pair of DFT($n/2$)s. If n is a power of two—say, $n = 2^m$—then we can recursively break down the problem until we get to DFT(1), which is multiplication by the 1×1 identity matrix, taking zero operations. Starting from the bottom up, DFT(1) takes no operations, and DFT(2) requires two additions and a multiplication: $y_0 = u_0 + 1v_0, y_1 = u_0 + \omega v_0$, where u_0 and v_0 are DFT(1)s (that is, $u_0 = y_0$ and $v_0 = y_1$).

DFT(4) requires two DFT(2)s plus $2 * 4 - 1 = 7$ further operations, for a total of $2(3) + 7 = 2^m(2m - 1) + 1$ operations, where $m = 2$. We proceed by induction: Assume that this formula is correct for a given m. Then DFT(2^{m+1}) takes two DFT(2^m)s, which take $2(2^m(2m - 1) + 1)$ operations, plus $2 \cdot 2^{m+1} - 1$ extras (to complete equations similar to (10.15)), for a total of

$$
\begin{aligned}
2(2^m(2m - 1) + 1) + 2^{m+2} - 1 &= 2^{m+1}(2m - 1 + 2) + 2 - 1 \\
&= 2^{m+1}(2(m + 1) - 1) + 1.
\end{aligned}
$$

Therefore, the formula $2^m(2m - 1) + 1$ operations is proved for the fast version of DFT(2^m), from which the result follows.   □

The fast algorithm for the DFT can be exploited to make a fast algorithm for the inverse DFT without further work. The inverse DFT is the complex conjugate matrix $\overline{F}_n$. To carry out the inverse DFT of a complex vector y, just conjugate, apply the FFT, and conjugate the result, because

$$
F_n^{-1} y = \overline{F}_n y = \overline{F_n \overline{y}}. \tag{10.15}
$$

▶ **ADDITIONAL EXAMPLES**

1. (a) Compute the Discrete Fourier Transform of the vector $x = [1\ 2\ 3\ 2]$, and (b) apply the Inverse Discrete Fourier Transform to the result.

*2. (a) Compute the Fast Fourier Transform of $x = [5\ 10\ 5\ 0]$, and (b) apply the Inverse Fast Fourier Transform to the result.

⬛ **Solutions** for Additional Examples can be found at bit.ly/2ypRXED
(* example with video solution)

10.1 Exercises

1. Find the DFT of the following vectors: (a) $[0, 1, 0, -1]$ (b) $[1, 1, 1, 1]$ (c) $[0, -1, 0, 1]$ (d) $[0, 1, 0, -1, 0, 1, 0, -1]$

2. Find the DFT of the following vectors: (a) $[3/4, 1/4, -1/4, 1/4]$ (b) $[9/4, 1/4, -3/4, 1/4]$ (c) $[1, 0, -1/2, 0]$ (d) $[1, 0, -1/2, 0, 1, 0, -1/2, 0]$

3. Find the inverse DFT of the following vectors: (a) $[1, 0, 0, 0]$ (b) $[1, 1, -1, 1]$ (c) $[1, -i, 1, i]$ (d) $[1, 0, 0, 0, 3, 0, 0, 0]$

4. Find the inverse DFT of the following vectors: (a) $[0, -i, 0, i]$ (b) $[2, 0, 0, 0]$ (c) $[1/2, 1/2, 0, 1/2]$ (d) $[1, 3/2, 1/2, 3/2]$

5. (a) Write down all fourth roots of unity and all primitive fourth roots of unity. (b) Write down all primitive seventh roots of unity. (c) How many primitive pth roots of unity exist for a prime number p?

6. Prove Lemma 10.1.

7. Find the real numbers $a_0, a_1, b_1, a_2, b_2, \ldots, a_{n/2}$ as in (10.13) for the Fourier transforms in Exercise 1.

8. Prove that the matrix in (10.10) is the inverse of the Fourier matrix F_n.

10.2 TRIGONOMETRIC INTERPOLATION

What does the Discrete Fourier transform actually do? In this section, we present an interpretation of the output vector y of the Fourier transform as interpolating coefficients for evenly spaced data in order to make its workings more understandable.

10.2.1 The DFT Interpolation Theorem

Let $[c, d]$ be an interval and let n be a positive integer. Define $\Delta t = (d - c)/n$ and $t_j = c + j\Delta t$ for $j = 0, \ldots, n - 1$ to be evenly spaced points in the interval. For a given input vector x to the Fourier transform, we will interpret the component x_j as the jth component of a measured signal. For example, we could think of the components of x as a series of measurements, measured at the discrete, evenly spaced times t_j, as shown in Figure 10.4.

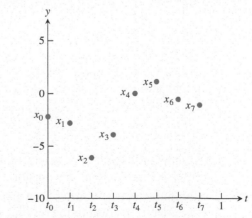

Figure 10.4 The components of x viewed as a time series. The Fourier transform is a way to compute the trigonometric polynomial that interpolates this data.

Let $y = F_n x$ be the DFT of x. Since x is the inverse DFT of y, we can write an explicit formula for the components of x from (10.10), remembering that $\omega = e^{-i2\pi/n}$:

$$x_j = \frac{1}{\sqrt{n}} \sum_{k=0}^{n-1} y_k (\omega^{-k})^j = \frac{1}{\sqrt{n}} \sum_{k=0}^{n-1} y_k e^{i2\pi kj/n} = \sum_{k=0}^{n-1} y_k \frac{e^{\frac{i2\pi k(t_j - c)}{d-c}}}{\sqrt{n}}. \tag{10.16}$$

We can view this as interpolation of the points (t_j, x_j) by trigonometric basis functions where the coefficients are y_k. Theorem 10.6 is a simple restatement of (10.16), saying that data points (t_j, x_j) are interpolated by basis functions $e^{i2\pi k(t-c)/(d-c)}/\sqrt{n}$ for $k = 0, \ldots, n-1$, with interpolation coefficients given by $F_n x$.

THEOREM 10.6 **DFT Interpolation Theorem.** Given an interval $[c, d]$ and positive integer n, let $t_j = c + j(d-c)/n$ for $j = 0, \ldots, n-1$, and let $x = (x_0, \ldots, x_{n-1})$ denote a vector of n numbers. Define $\vec{a} + \vec{b}i = F_n x$, where F_n is the Discrete Fourier Transform matrix. Then the complex function

$$Q(t) = \frac{1}{\sqrt{n}} \sum_{k=0}^{n-1} (a_k + ib_k) e^{i2\pi k(t-c)/(d-c)}$$

satisfies $Q(t_j) = x_j$ for $j = 0, \ldots, n-1$. Furthermore, if the x_j are real, the real function

$$P(t) = \frac{1}{\sqrt{n}} \sum_{k=0}^{n-1} \left(a_k \cos \frac{2\pi k(t-c)}{d-c} - b_k \sin \frac{2\pi k(t-c)}{d-c} \right)$$

satisfies $P(t_j) = x_j$ for $j = 0, \ldots, n-1$. ∎

In other words, the Fourier transform F_n transforms data $\{x_j\}$ into interpolation coefficients.

The explanation for the last part of the theorem is that, using the Euler formula, we can rewrite the interpolation function in (10.16) as

$$Q(t) = \frac{1}{\sqrt{n}} \sum_{k=0}^{n-1} (a_k + ib_k) \left(\cos \frac{2\pi k(t-c)}{d-c} + i \sin \frac{2\pi k(t-c)}{d-c} \right).$$

Separate the interpolating function $Q(t) = P(t) + iI(t)$ into its real and imaginary parts. Since the x_j are real numbers, only the real part of $Q(t)$ is needed to interpolate the x_j. The real part is

$$P(t) = P_n(t) = \frac{1}{\sqrt{n}} \sum_{k=0}^{n-1} \left(a_k \cos \frac{2\pi k(t-c)}{d-c} - b_k \sin \frac{2\pi k(t-c)}{d-c} \right). \tag{10.17}$$

A subscript n identifies the number of terms in the trigonometric model. We will sometimes call P_n an **order n trigonometric function**. Lemma 10.4 and the following Lemma 10.7 can be used to simplify the interpolating function $P_n(t)$ further:

LEMMA 10.7 Let $t = j/n$, where j and n are integers. Let k be an integer. Then

$$\cos 2(n-k)\pi t = \cos 2k\pi t \quad \text{and} \quad \sin 2(n-k)\pi t = -\sin 2k\pi t. \tag{10.18}$$

∎

In fact, the cosine addition formula yields $\cos 2(n - k)\pi j/n = \cos(2\pi j - 2jk\pi/n) = \cos(-2jk\pi/n)$ and similarly for sine.

Lemma 10.7, together with Lemma 10.4, implies that the latter half of the trigonometric expansion (10.17) is redundant. We can interpolate at the t_j's by using only the first half of the terms (except for a change of sign for the sine terms). By Lemma 10.4, the coefficients from the latter half of the expansion are the same as those from the first half (except for a change of sign for the sin terms). Thus, the changes of sign cancel one another out, and we have shown that the simplified version of P_n is

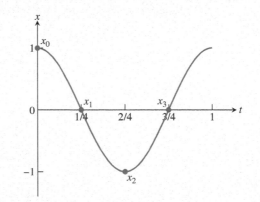

Figure 10.5 Trigonometric interpolation. The input vector x is $[1, 0, -1, 0]^T$. Formula (10.19) gives the interpolating function to be $P_4(t) = \cos 2\pi t$.

$$P_n(t) = \frac{a_0}{\sqrt{n}} + \frac{2}{\sqrt{n}} \sum_{k=1}^{n/2-1} \left(a_k \cos \frac{2k\pi(t - c)}{d - c} - b_k \sin \frac{2k\pi(t - c)}{d - c} \right)$$
$$+ \frac{a_{n/2}}{\sqrt{n}} \cos \frac{n\pi(t - c)}{d - c}.$$

To write this expression, we have assumed that n is even. The formula is slightly different for n odd. See Exercise 5.

COROLLARY 10.8 For an even integer n, let $t_j = c + j(d - c)/n$ for $j = 0, \ldots, n - 1$, and let $x = (x_0, \ldots, x_{n-1})$ denote a vector of n real numbers. Define $\vec{a} + \vec{b}i = F_n x$, where F_n is the Discrete Fourier Transform. Then the function

$$P_n(t) = \frac{a_0}{\sqrt{n}} + \frac{2}{\sqrt{n}} \sum_{k=1}^{n/2-1} \left(a_k \cos \frac{2k\pi(t - c)}{d - c} - b_k \sin \frac{2k\pi(t - c)}{d - c} \right)$$
$$+ \frac{a_{n/2}}{\sqrt{n}} \cos \frac{n\pi(t - c)}{d - c} \tag{10.19}$$

satisfies $P_n(t_j) = x_j$ for $j = 0, \ldots, n - 1$. ∎

▶ **EXAMPLE 10.2** Find the trigonometric interpolant for Example 10.1.

The interval is $[c, d] = [0, 1]$. Let $x = [1, 0, -1, 0]^T$ and compute its DFT to be $y = [0, 1, 0, 1]^T$. The interpolating coefficients are $a_k + ib_k = y_k$. Therefore, $a_0 = a_2 = 0, a_1 = a_3 = 1$, and $b_0 = b_1 = b_2 = b_3 = 0$. According to (10.19), we only need a_0, a_1, a_2, and b_1. A trigonometric interpolating function for x is given by

$$P_4(t) = \frac{a_0}{2} + (a_1\cos 2\pi t - b_1\sin 2\pi t) + \frac{a_2}{2}\cos 4\pi t$$
$$= \cos 2\pi t.$$

The interpolation of the points (t, x), where $t = [0, 1/4, 1/2, 3/4]$ and $x = [1, 0, -1, 0]$, is shown in Figure 10.5. ◀

▶ **EXAMPLE 10.3**   Find the trigonometric interpolant for the temperature data from Example 4.6: $x = [-2.2, -2.8, -6.1, -3.9, 0.0, 1.1, -0.6, -1.1]$ on the interval $[0, 1]$.

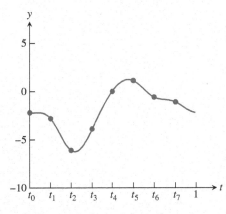

Figure 10.6 Trigonometric interpolation of data from Example 4.6. The data $t = [0, 1/8, 2/8, 3/8, 4/8, 5/8, 6/8, 7/8], x = [-2.2, -2.8, -6.1, -3.9, 0.0, 1.1, -0.6, -1.1]$ are interpolated with the use of the Fourier transform with $n = 8$. The plot is made by Program 10.1 with $p = 100$.

The Fourier transform output, accurate to four decimal places, is

$$y = \begin{bmatrix} -5.5154 \\ -1.0528 + 3.6195i \\ 1.5910 - 1.1667i \\ -0.5028 - 0.2695i \\ -0.7778 \\ -0.5028 + 0.2695i \\ 1.5910 + 1.1667i \\ -1.0528 - 3.6195i \end{bmatrix}.$$

According to formula (10.19), the interpolating function is

$$P_8(t) = \frac{-5.5154}{\sqrt{8}} - \frac{1.0528}{\sqrt{2}}\cos 2\pi t - \frac{3.6195}{\sqrt{2}}\sin 2\pi t$$

$$+ \frac{1.5910}{\sqrt{2}}\cos 4\pi t + \frac{1.1667}{\sqrt{2}}\sin 4\pi t$$

$$- \frac{0.5028}{\sqrt{2}}\cos 6\pi t + \frac{0.2695}{\sqrt{2}}\sin 6\pi t$$

$$- \frac{0.7778}{\sqrt{8}}\cos 8\pi t$$

$$= -1.95 - 0.7445 \cos 2\pi t - 2.5594 \sin 2\pi t$$
$$+ 1.125 \cos 4\pi t + 0.825 \sin 4\pi t$$
$$- 0.3555 \cos 6\pi t + 0.1906 \sin 6\pi t$$
$$- 0.2750 \cos 8\pi t. \tag{10.20}$$

Figure 10.6 shows the data points and the trigonometric interpolating function. ◄

10.2.2 Efficient evaluation of trigonometric functions

Corollary 10.8 is a powerful statement about interpolation. Although it appears complicated at first, there is another way to evaluate and plot the trigonometric interpolating polynomial in Figures 10.5 and 10.6, using the DFT to do all the work instead of plotting the sines and cosines of (10.19). After all, we know from Theorem 10.6 that multiplying the vector x of data points by F_n changes data to interpolation coefficients. Conversely, we can turn interpolation coefficients into data points. Instead of evaluating (10.19), just invert the DFT: Multiply the vector of interpolation coefficients $\{a_k + ib_k\}$ by F_n^{-1}.

Of course, if we follow the operation F_n by its inverse, F_n^{-1}, we just get the original data points back and gain nothing. Instead, we will let $p \geq n$ be a larger number. We plan to view (10.19) as an order p trigonometric function and then invert the Fourier transform to evaluate the curve at the p equally spaced points. We can take p large enough to get a continuous-looking plot.

To view the coefficients of $P_n(t)$ as the coefficients of an order p trigonometric polynomial, notice that we can rewrite (10.19) as

$$P_p(t) = \frac{\sqrt{\frac{p}{n}} a_0}{\sqrt{p}} + \frac{2}{\sqrt{p}} \sum_{k=1}^{p/2-1} \left(\sqrt{\frac{p}{n}} a_k \cos \frac{2k\pi(t-c)}{d-c} - \sqrt{\frac{p}{n}} b_k \sin \frac{2k\pi(t-c)}{d-c} \right)$$
$$+ \frac{\sqrt{\frac{p}{n}} a_{n/2}}{\sqrt{p}} \cos n\pi t, \tag{10.21}$$

where we set $a_k = b_k = 0$ for $k = \frac{n}{2} + 1, \ldots, \frac{p}{2}$. We conclude from (10.21) that the way to produce p points lying on the curve (10.19) at $t_j = c + j(d-c)/n$ for $j = 0, \ldots, n-1$ is to multiply the Fourier coefficients by $\sqrt{p/n}$ and then invert the DFT.

We write MATLAB code to implement this idea. Roughly speaking, we want to implement

$$F_p^{-1} \sqrt{\frac{p}{n}} F_n x$$

using MATLAB's commands `fft` and `ifft`, where

$$F_p^{-1} = \sqrt{p} \cdot \text{ifft} \quad \text{and} \quad F_n = \frac{1}{\sqrt{n}} \cdot \text{fft}.$$

Putting the pieces together, this corresponds to the following operations:

$$\sqrt{p} \cdot \text{ifft}_{[p]} \sqrt{\frac{p}{n}} \frac{1}{\sqrt{n}} \cdot \text{fft}_{[n]} = \frac{p}{n} \cdot \text{ifft}_{[p]} \cdot \text{fft}_{[n]}. \tag{10.22}$$

Of course, F_p^{-1} can only be applied to a length p vector, so we need to place the degree n Fourier coefficients into a length p vector before inverting. The short program `dftinterp.m` carries out these steps.

📟 **MATLAB code**

shown here can be found
at bit.ly/2OttZT3

```
%Program 10.1 Fourier interpolation
%Interpolate n data points on [c,d] with trig function P(t)
%  and plot interpolant at p (>=n) evenly spaced points.
%Input: interval [c,d], data points x, even number of data
%  points n, even number p>=n
%Output: data points of interpolant xp
function xp=dftinterp(inter,x,n,p)
c=inter(1);d=inter(2);t=c+(d-c)*(0:n-1)/n;  tp=c+(d-c)*(0:p-1)/p;
y=fft(x);                      % apply DFT
yp=zeros(p,1);                 % yp will hold coefficients for ifft
yp(1:n/2+1)=y(1:n/2+1);        % move n frequencies from n to p
yp(p-n/2+2:p)=y(n/2+2:n);      % same for upper tier
xp=real(ifft(yp))*(p/n);       % invert fft to recover data
plot(t,x,'o',tp,xp)            % plot data points and interpolant
```

Running the function dftinterp([0,1],[-2.2 -2.8 -6.1 -3.9 0.0 1.1 -0.6 -1.1],8,100), for example, produces the $p = 100$ plotted points in Figure 10.6 without explicitly using sines or cosines. A few comments on the code are in order. The goal is to apply $fft_{[n]}$, followed by $ifft_{[p]}$, and then multiply by p/n. After applying fft to the n values in x, the coefficients in the vector y are moved from the n frequencies in $P_n(t)$ to a vector yp holding p frequencies, where $p \geq n$. There are many higher frequencies among the p frequencies that are not used by P_n, which leads to zero coefficients in those high frequencies, in positions $n/2 + 2$ to $p/2 + 1$. The upper half of the entries in yp gives a recapitulation of the lower half, with complex conjugates and in reverse order, following (10.13). After the DFT is inverted with the ifft command, although theoretically the result is real, computationally there may be a small imaginary part due to rounding. This is removed by applying the real command.

A particularly simple and useful case is $c = 0, d = n$. The data points x_j are collected at the integer interpolation nodes $s_j = j$ for $j = 0, \ldots, n - 1$. The points (j, x_j) are interpolated by the trigonometric function

$$P_n(s) = \frac{a_0}{\sqrt{n}} + \frac{2}{\sqrt{n}} \sum_{k=1}^{n/2-1} \left(a_k \cos \frac{2k\pi}{n} s - b_k \sin \frac{2k\pi}{n} s \right) + \frac{a_{n/2}}{\sqrt{n}} \cos \pi s. \qquad (10.23)$$

In Chapter 11, we will use integer interpolation nodes exclusively, for compatibility with the usual conventions for audio and image data compression algorithms.

▶ **ADDITIONAL EXAMPLES**

1. Use the DFT to find the trigonometric interpolating function for the following data.

t	0	1/4	1/2	3/4
x	5	10	5	0

2. Use dftinterp.m to find the order 8 trigonometric interpolation polynomial for the data and plot along with the data points.

t	0	1/2	1	3/2	2	5/2	3	7/2
x	4	2	1	3	6	2	1	5

📟 **Solutions** for Additional Examples can be found at bit.ly/2AkGyHP

10.2 Exercises

1. Use the DFT and Corollary 10.8 to find the trigonometric interpolating function for the following data:

(a)

t	x
0	0
$\frac{1}{4}$	1
$\frac{1}{2}$	0
$\frac{3}{4}$	-1

(b)

t	x
0	1
$\frac{1}{4}$	1
$\frac{1}{2}$	-1
$\frac{3}{4}$	-1

(c)

t	x
0	-1
$\frac{1}{4}$	1
$\frac{1}{2}$	-1
$\frac{3}{4}$	1

(d)

t	x
0	1
$\frac{1}{4}$	1
$\frac{1}{2}$	1
$\frac{3}{4}$	1

2. Use (10.23) to find the trigonometric interpolating function for the following data:

(a)

t	x
0	0
1	1
2	0
3	-1

(b)

t	x
0	1
1	1
2	-1
3	-1

(c)

t	x
0	1
1	2
2	4
3	1

(d)

t	x
0	1
1	0
2	1
3	0

3. Find the trigonometric interpolating function for the following data:

(a)

t	x
0	0
$\frac{1}{8}$	1
$\frac{1}{4}$	0
$\frac{3}{8}$	-1
$\frac{1}{2}$	0
$\frac{5}{8}$	1
$\frac{3}{4}$	0
$\frac{7}{8}$	-1

(b)

t	x
0	1
$\frac{1}{8}$	2
$\frac{1}{4}$	1
$\frac{3}{8}$	0
$\frac{1}{2}$	1
$\frac{5}{8}$	2
$\frac{3}{4}$	1
$\frac{7}{8}$	0

(c)

t	x
0	1
$\frac{1}{8}$	1
$\frac{1}{4}$	1
$\frac{3}{8}$	1
$\frac{1}{2}$	0
$\frac{5}{8}$	0
$\frac{3}{4}$	0
$\frac{7}{8}$	0

(d)

t	x
0	1
$\frac{1}{8}$	-1
$\frac{1}{4}$	1
$\frac{3}{8}$	-1
$\frac{1}{2}$	1
$\frac{5}{8}$	-1
$\frac{3}{4}$	1
$\frac{7}{8}$	-1

4. Find the trigonometric interpolating function for the following data:

(a)

t	x
0	0
1	1
2	0
3	-1
4	0
5	1
6	0
7	-1

(b)

t	x
0	1
1	2
2	1
3	0
4	1
5	2
6	1
7	0

(c)

t	x
0	1
1	0
2	1
3	0
4	1
5	0
6	1
7	0

(d)

t	x
0	-1
1	0
2	0
3	0
4	1
5	0
6	0
7	0

5. Find a version of (10.19) for the interpolating function in the case where n is odd.

10.2 Computer Problems

1. Find the order 8 trigonometric interpolating function $P_8(t)$ for the following data:

(a)

t	x
0	0
$\frac{1}{8}$	1
$\frac{1}{4}$	2
$\frac{3}{8}$	3
$\frac{1}{2}$	4
$\frac{5}{8}$	5
$\frac{3}{4}$	6
$\frac{7}{8}$	7

(b)

t	x
0	2
$\frac{1}{8}$	-1
$\frac{1}{4}$	0
$\frac{3}{8}$	1
$\frac{1}{2}$	1
$\frac{5}{8}$	3
$\frac{3}{4}$	-1
$\frac{7}{8}$	-1

(c)

t	x
0	3
1	1
2	4
3	2
4	3
5	1
6	4
7	2

(d)

t	x
1	1
2	-2
3	5
4	3
5	-2
6	-3
7	1
8	2

Plot the data points and $P_8(t)$.

2. Find the order 8 trigonometric interpolating function $P_8(t)$ for the following data:

(a)

t	x
0	6
$\frac{1}{8}$	5
$\frac{1}{4}$	4
$\frac{3}{8}$	3
$\frac{1}{2}$	2
$\frac{5}{8}$	1
$\frac{3}{4}$	0
$\frac{7}{8}$	-1

(b)

t	x
0	3
$\frac{1}{8}$	1
$\frac{1}{4}$	2
$\frac{3}{8}$	-1
$\frac{1}{2}$	-1
$\frac{5}{8}$	-2
$\frac{3}{4}$	3
$\frac{7}{8}$	0

(c)

t	x
0	1
2	2
4	4
6	-1
8	0
10	1
12	0
14	2

(d)

t	x
-7	2
-5	1
-3	0
-1	5
1	7
3	2
5	1
7	-4

Plot the data points and $P_8(t)$.

3. Find the order $n = 8$ trigonometric interpolating function for $f(t) = e^t$ at the evenly spaced points $(j/8, f(j/8))$ for $j = 0, \ldots, 7$. Plot $f(t)$, the data points, and the interpolating function.

4. Plot the interpolating function $P_n(t)$ on $[0, 1]$ in Computer Problem 3, along with the data points and $f(t) = e^t$ for (a) $n = 16$ (b) $n = 32$.

5. Find the order 8 trigonometric interpolating function for $f(t) = \ln t$ at the evenly spaced points $(1 + j/8, f(1 + j/8))$ for $j = 0, \ldots, 7$. Plot $f(t)$, the data points, and the interpolating function.

6. Plot the interpolating function $P_n(t)$ on $[0, 1]$ in Computer Problem 5, along with the data points and $f(t) = \ln t$ for (a) $n = 16$ (b) $n = 32$.

10.3 THE FFT AND SIGNAL PROCESSING

The DFT Interpolation Theorem 10.6 is just one application of the Fourier transform. In this section, we look at interpolation from a more general point of view, which will show how to find least squares approximations by using trigonometric functions. These ideas form the basis of modern signal processing. They will make a second appearance in Chapter 11, applied to the Discrete Cosine Transform.

10.3.1 Orthogonality and interpolation

The deceptively simple interpolation result of Theorem 10.6 was made possible by the fact that $F_n^{-1} = \overline{F}_n^T = \overline{F}_n$, making F_n a unitary matrix. We encountered the real version of this definition in Chapter 4, where we called a matrix U orthogonal if $U^{-1} = U^T$. Now we study a particular form for an orthogonal matrix that will translate immediately into a good interpolant.

THEOREM 10.9
Orthogonal Function Interpolation Theorem. Let $f_0(t), \ldots, f_{n-1}(t)$ be functions of t and $t_0, \ldots, t_{n-1}$ be real numbers. Assume that the $n \times n$ matrix

$$A = \begin{bmatrix} f_0(t_0) & f_0(t_1) & \cdots & f_0(t_{n-1}) \\ f_1(t_0) & f_1(t_1) & \cdots & f_1(t_{n-1}) \\ \vdots & \vdots & & \vdots \\ f_{n-1}(t_0) & f_{n-1}(t_1) & \cdots & f_{n-1}(t_{n-1}) \end{bmatrix} \tag{10.24}$$

is a real $n \times n$ orthogonal matrix. If $y = Ax$, the function

$$F(t) = \sum_{k=0}^{n-1} y_k f_k(t)$$

interpolates $(t_0, x_0), \ldots, (t_{n-1}, x_{n-1})$, that is $F(t_j) = x_j$ for $j = 0, \ldots, n-1$. ∎

Proof. The fact $y = Ax$ implies that

$$x = A^{-1}y = A^T y,$$

and it follows that

$$x_j = \sum_{k=0}^{n-1} a_{kj} y_k = \sum_{k=0}^{n-1} y_k f_k(t_j)$$

for $j = 0, \ldots, n-1$, which completes the proof. □

▶ **EXAMPLE 10.4**
Let $[c, d]$ be an interval and let n be an even positive integer. Show that the assumptions of Theorem 10.9 are satisfied for $t_j = c + j(d - c)/n$, $j = 0, \ldots, n-1$, and

$$f_0(t) = \sqrt{\frac{1}{n}}$$

$$f_1(t) = \sqrt{\frac{2}{n}} \cos \frac{2\pi(t - c)}{d - c}$$

$$f_2(t) = \sqrt{\frac{2}{n}} \sin \frac{2\pi(t - c)}{d - c}$$

$$f_3(t) = \sqrt{\frac{2}{n}} \cos \frac{4\pi(t - c)}{d - c}$$

$$f_4(t) = \sqrt{\frac{2}{n}} \sin \frac{4\pi(t - c)}{d - c}$$

$$\vdots$$

$$f_{n-1}(t) = \frac{1}{\sqrt{n}} \cos \frac{n\pi(t - c)}{d - c}.$$

The matrix is

$$
A = \sqrt{\frac{2}{n}}
\begin{bmatrix}
\frac{1}{\sqrt{2}} & \frac{1}{\sqrt{2}} & \cdots & \frac{1}{\sqrt{2}} \\
1 & \cos\frac{2\pi}{n} & \cdots & \cos\frac{2\pi(n-1)}{n} \\
0 & \sin\frac{2\pi}{n} & \cdots & \sin\frac{2\pi(n-1)}{n} \\
\vdots & \vdots & & \vdots \\
\frac{1}{\sqrt{2}} & \frac{1}{\sqrt{2}}\cos\pi & \cdots & \frac{1}{\sqrt{2}}\cos(n-1)\pi
\end{bmatrix}.
\tag{10.25}
$$

Lemma 10.10 shows that the rows of A are pairwise orthogonal. ◄

LEMMA 10.10   Let $n \geq 1$ and k, l be integers. Then

$$
\sum_{j=0}^{n-1} \cos\frac{2\pi jk}{n} \cos\frac{2\pi jl}{n} =
\begin{cases}
n & \text{if both } (k-l)/n \text{ and } (k+l)/n \text{ are integers} \\
\frac{n}{2} & \text{if exactly one of } (k-l)/n \text{ and } (k+l)/n \text{ is an integer} \\
0 & \text{if neither is an integer}
\end{cases}
$$

$$
\sum_{j=0}^{n-1} \cos\frac{2\pi jk}{n} \sin\frac{2\pi jl}{n} = 0
$$

$$
\sum_{j=0}^{n-1} \sin\frac{2\pi jk}{n} \sin\frac{2\pi jl}{n} =
\begin{cases}
0 & \text{if both } (k-l)/n \text{ and } (k+l)/n \text{ are integers} \\
\frac{n}{2} & \text{if } (k-l)/n \text{ is an integer and } (k+l)/n \text{ is not} \\
-\frac{n}{2} & \text{if } (k+l)/n \text{ is an integer and } (k-l)/n \text{ is not} \\
0 & \text{if neither is an integer}
\end{cases}
$$

■

The proof of this lemma follows from Lemma 10.1. See Exercise 5.

Returning to Example 10.4, let $y = Ax$. Theorem 10.9 immediately gives the interpolating function

$$
F(t) = \frac{1}{\sqrt{n}} y_0
$$

$$
+ \sqrt{\frac{2}{n}} y_1 \cos\frac{2\pi(t-c)}{d-c} + \sqrt{\frac{2}{n}} y_2 \sin\frac{2\pi(t-c)}{d-c}
$$

$$
+ \sqrt{\frac{2}{n}} y_3 \cos\frac{4\pi(t-c)}{d-c} + \sqrt{\frac{2}{n}} y_4 \sin\frac{4\pi(t-c)}{d-c}
$$

$$
\vdots
$$

$$
+ \frac{1}{\sqrt{n}} y_{n-1} \cos\frac{n\pi(t-c)}{d-c}
\tag{10.26}
$$

for the points (t_j, x_j), in agreement with (10.19).

► **EXAMPLE 10.5**   Use the basis functions of Example 10.4 to interpolate the data points $x = [-2.2, -2.8, -6.1, -3.9, 0.0, 1.1, -0.6, -1.1]$ from Example 10.3.

Computing the product of the 8×8 matrix A with x yields

$$Ax = \sqrt{\frac{2}{8}} \begin{bmatrix} \frac{1}{\sqrt{2}} & \frac{1}{\sqrt{2}} & \frac{1}{\sqrt{2}} & \cdots & \frac{1}{\sqrt{2}} \\ 1 & \cos 2\pi \frac{1}{8} & \cos 2\pi \frac{2}{8} & \cdots & \cos 2\pi \frac{7}{8} \\ 0 & \sin 2\pi \frac{1}{8} & \sin 2\pi \frac{2}{8} & \cdots & \sin 2\pi \frac{7}{8} \\ 1 & \cos 4\pi \frac{1}{8} & \cos 4\pi \frac{2}{8} & \cdots & \cos 4\pi \frac{7}{8} \\ 0 & \sin 4\pi \frac{1}{8} & \sin 4\pi \frac{2}{8} & \cdots & \sin 4\pi \frac{7}{8} \\ 1 & \cos 6\pi \frac{1}{8} & \cos 6\pi \frac{2}{8} & \cdots & \cos 6\pi \frac{7}{8} \\ 0 & \sin 6\pi \frac{1}{8} & \sin 6\pi \frac{2}{8} & \cdots & \sin 6\pi \frac{7}{8} \\ \frac{1}{\sqrt{2}} & \frac{1}{\sqrt{2}} \cos\pi & \frac{1}{\sqrt{2}} \cos 2\pi & \cdots & \frac{1}{\sqrt{2}} \cos 7\pi \end{bmatrix} \begin{bmatrix} -2.2 \\ -2.8 \\ -6.1 \\ -3.9 \\ 0.0 \\ 1.1 \\ -0.6 \\ -1.1 \end{bmatrix} = \begin{bmatrix} -5.5154 \\ -1.4889 \\ -5.1188 \\ 2.2500 \\ 1.6500 \\ -0.7111 \\ 0.3812 \\ -0.7778 \end{bmatrix}.$$

The formula (10.26) gives the interpolating function,

$$\begin{aligned} P(t) = -1.95 &- 0.7445 \cos 2\pi t - 2.5594 \sin 2\pi t \\ &+ 1.125 \cos 4\pi t + 0.825 \sin 4\pi t \\ &- 0.3555 \cos 6\pi t + 0.1906 \sin 6\pi t \\ &- 0.2750 \cos 8\pi t, \end{aligned}$$

in agreement with Example 10.3. ◀

10.3.2 Least squares fitting with trigonometric functions

Corollary 10.8 showed how the DFT makes it easy to interpolate n evenly spaced data points on [0, 1] by a trigonometric function of form

$$P_n(t) = \frac{a_0}{\sqrt{n}} + \frac{2}{\sqrt{n}} \sum_{k=1}^{n/2-1} (a_k \cos 2k\pi t - b_k \sin 2k\pi t) + \frac{a_{n/2}}{\sqrt{n}} \cos n\pi t. \quad (10.27)$$

Note that the number of terms is n, equal to the number of data points. (As usual in this chapter, we assume that n is even.) The more data points there are, the more cosines and sines are added to help with the interpolation.

SPOTLIGHT ON

Orthogonality   In Chapter 4, we established the normal equations $A^T A \bar{x} = A^T b$ for solving least squares approximation to data by basis functions. The point of Theorem 10.9 is to find special cases that make the normal equations trivial, greatly simplifying the least squares procedure. This leads to an extremely useful theory of so-called orthogonal functions. Major examples include the Fourier transform in this chapter and the cosine transform in Chapter 11.

As we found in Chapter 3, when the number of data points n is large, it becomes less common to fit a model function exactly. In fact, a common application of a model is to forget a few details (lossy compression) in order to simplify matters. A second reason to move away from interpolation, discussed in Chapter 4, is the case where the data points themselves are assumed to be inexact, so that rigorous enforcement of an interpolating function is inappropriate.

In either of these situations, we are motivated to do a least squares fit with a function of type (10.27). Since the coefficients a_k and b_k occur linearly in the model, we can proceed with the same program described in Chapter 4, using the normal equations to solve for the best coefficients. When we try this, we find a surprising result, which will send us right back to the DFT.

Return to Theorem 10.9. Let n denote the number of data points x_j, which we think of as occurring at evenly spaced times $t_j = j/n$ in $[0, 1]$, for simplicity. We will introduce the even positive integer m to denote the number of basis functions to use in the least squares fit. That is, we will fit to the first m of the basis functions, $f_0(t), \ldots, f_{m-1}(t)$. The function used to fit the n data points will be

$$P_m(t) = \sum_{k=0}^{m-1} c_k f_k(t), \tag{10.28}$$

where the c_k are to be determined. When $m = n$, the problem is still interpolation. When $m < n$, we have changed to the compression problem. In this case, we expect to match the data points using P_m with minimum squared error.

The least squares problem is to find coefficients $c_0, \ldots, c_{m-1}$ such that the equality

$$\sum_{k=0}^{m-1} c_k f_k(t_j) = x_j$$

is met with as little error as possible. In matrix terms,

$$A_m^T c = x, \tag{10.29}$$

where A_m is the matrix of the first m rows of A. Under the assumptions of Theorem 10.9, A_m^T has pairwise orthonormal columns. When we set up the normal equations

$$A_m A_m^T c = A_m x$$

for c, $A_m A_m^T$ is the identity matrix. Therefore, the least squares solution,

$$c = A_m x, \tag{10.30}$$

is easy to calculate. We have proved the following useful result, which extends Theorem 10.9:

THEOREM 10.11   **Orthogonal Function Least Squares Approximation Theorem.** Let $m \leq n$ be integers, and assume that data $(t_0, x_0), \ldots, (t_{n-1}, x_{n-1})$ are given. Set $y = Ax$, where A is an orthogonal matrix of form (10.24). Then the interpolating polynomial for basis functions $f_0(t), \ldots, f_{n-1}(t)$ is

$$F_n(t) = \sum_{k=0}^{n-1} y_k f_k(t), \tag{10.31}$$

and the best least squares approximation, using only the functions $f_0, \ldots, f_{m-1}$, is

$$F_m(t) = \sum_{k=0}^{m-1} y_k f_k(t). \tag{10.32}$$

∎

This is a beautiful and useful fact. It says that, given n data points, to find the best least squares trigonometric function with $m < n$ terms fitting the data, it suffices to compute the actual interpolant with n terms and keep only the desired first m terms.

In other words, the interpolating coefficients Ax for x degrade as gracefully as possible when terms are dropped from the highest frequencies. Keeping the m lowest terms in the n-term expansion guarantees the best fit possible with m lowest frequency terms. This property reflects the "orthogonality" of the basis functions.

The reasoning preceding Theorem 10.11 is easily adapted to prove something more general. We showed how to find the least squares solution for the first m basis functions, but in truth, the order was not relevant; we could have specified any subset of the basis functions. The least squares solution is found simply by dropping all terms in (10.31) that are not included in the subset. The version (10.32) is a "low-pass" filter, assuming that the lower index functions go with lower "frequencies"; but by changing the subset of basis functions kept, we can pass any frequencies of interest simply by dropping the undesired coefficients.

Now we return to the trigonometric polynomial (10.27) and demonstrate how to fit an order m version to n data points, where $m < n$. The basis functions used are the functions of Example 10.4, which satisfy the assumptions of Theorem 10.9. Theorem 10.11 shows that, whatever the interpolating coefficients, the coefficients of the best least squares approximation of order m are found by dropping all terms above order m. We have arrived at the following application:

COROLLARY 10.12 Let $[c, d]$ be an interval, let $m < n$ be even positive integers, $x = (x_0, \ldots, x_{n-1})$ a vector of n real numbers, and let $t_j = c + j(d - c)/n$ for $j = 0, \ldots, n - 1$. Let $\{a_0, a_1, b_1, a_2, b_2, \ldots, a_{n/2-1}, b_{n/2-1}, a_{n/2}\} = F_n x$ be the interpolating coefficients for x so that

$$x_j = P_n(t_j) = \frac{a_0}{\sqrt{n}} + \frac{2}{\sqrt{n}} \sum_{k=1}^{\frac{n}{2}-1} \left(a_k \cos \frac{2k\pi(t_j - c)}{d - c} - b_k \sin \frac{2k\pi(t_j - c)}{d - c} \right)$$
$$+ \frac{a_{\frac{n}{2}}}{\sqrt{n}} \cos \frac{n\pi(t_j - c)}{d - c}$$

for $j = 0, \ldots, n - 1$. Then

$$P_m(t) = \frac{a_0}{\sqrt{n}} + \frac{2}{\sqrt{n}} \sum_{k=1}^{\frac{m}{2}-1} \left(a_k \cos \frac{2k\pi(t - c)}{d - c} - b_k \sin \frac{2k\pi(t - c)}{d - c} \right) + \frac{2a_{\frac{m}{2}}}{\sqrt{n}} \cos \frac{n\pi(t - c)}{d - c}$$

is the best least squares fit of order m to the data (t_j, x_j) for $j = 0, \ldots, n - 1$. ∎

Another way of appreciating the power of Theorem 10.11 is to compare it with the monomial basis functions we have used previously for least squares models. The best least squares parabola fit to the points $(0, 3), (1, 3), (2, 5)$ is $y = x^2 - x + 3$. In other words, the best coefficients for the model $y = a + bx + cx^2$ for this data are $a = 3, b = -1$, and $c = 1$ (in this case because the squared error is zero—this is the interpolating parabola). Now let's fit to a subset of the basis functions—say, change the model to $y = a + bx$. We calculate the best line fit to be $a = 8/3, b = 1$. Note that the coefficients for the degree 1 fit have no apparent relation to their corresponding coefficients for the degree 2 fit. This is exactly what *doesn't* happen for trigonometric basis functions. An interpolating fit, or any least squares fit to the form (10.28), explicitly contains all the information about lower order least squares fits.

Because of the extremely simple answer DFT has for least squares, it is especially simple to write a computer program to carry out the steps. Let $m < n < p$ be integers, where n is the number of data points, m is the order of the least squares trigonometric model, and p governs the resolution of the plot of the best model. We can think of least squares as "filtering out" the highest frequency contributions of the order n interpolant and leaving only the lowest m frequency contributions. That explains the name of the following MATLAB function:

MATLAB code shown here can be found at bit.ly/2J4yeyv

```
% Program 10.2 Least squares trigonometric fit
% Least squares fit of n data points on [0,1] with trig function
%    where 2 <=m <=n. Plot best fit at p (>=n) points.
% Input: interval [c,d], data points x, even number m,
%    even number of data points n, even number p>=n
% Output: filtered points xp
function xp=dftfilter(inter,x,m,n,p)
c=inter(1); d=inter(2);
t=c+(d-c)*(0:n-1)/n;          % time points for data (n)
tp=c+(d-c)*(0:p-1)/p          % time points for interpolant (p)
y=fft(x);                     % compute interpolation coefficients
yp=zeros(p,1);                % will hold coefficients for ifft
yp(1:m/2)=y(1:m/2);           % keep only first m frequencies
yp(m/2+1)=real(y(m/2+1));     % since m is even, keep cos term only
if(m<n)                       % unless at the maximum frequency,
   yp(p-m/2+1)=yp(m/2+1);     %    add complex conjugate to
end                          %    corresponding place in upper tier
yp(p-m/2+2:p)=y(n-m/2+2:n);  % more conjugates for upper tier
xp=real(ifft(yp))*(p/n);     % invert fft to recover data
plot(t,x,'o',tp,xp)          % plot data and least square approx
```

► **EXAMPLE 10.6**   Fit the temperature data from Example 10.3 by least squares trigonometric functions of orders 4 and 6.

The point of Corollary 10.12 is that we can just interpolate the data points by applying F_n and then chop off terms at will to get the least squares fit of lower orders. The result from Example 10.3 was that

$$
\begin{aligned}
P_8(t) = &-1.95 - 0.7445\cos 2\pi t - 2.5594\sin 2\pi t \\
&+1.125\cos 4\pi t + 0.825\sin 4\pi t \\
&-0.3555\cos 6\pi t + 0.1906\sin 6\pi t \\
&-0.2750\cos 8\pi t
\end{aligned}
\tag{10.33}
$$

Therefore, the least squares models of orders 4 and 6 are

$$
P_4(t) = -1.95 - 0.7445\cos 2\pi t - 2.5594\sin 2\pi t + 1.125\cos 4\pi t
$$
$$
\begin{aligned}
P_6(t) = &-1.95 - 0.7445\cos 2\pi t - 2.5594\sin 2\pi t \\
&+1.125\cos 4\pi t + 0.825\sin 4\pi t - 0.3555\cos 6\pi t.
\end{aligned}
$$

Figure 10.7 shows both least squares fits, generated by

```
dftfilter([0,1],[-2.2,-2.8,-6.1,-3.9,0.0,1.1,-0.6,-1.1],m,8,200)
```

for $m = 4$ and 6, respectively. The $m = 4$ fit matches the explicit least squares fit to the basis functions $1, \cos 2\pi t, \sin 2\pi t, \cos 4\pi t$ carried out in Example 4.6 and plotted in Figure 4.5(b).   ◄

The program dftfilter.m can be made more efficient. It computes the order n interpolant and then ignores $n - m$ coefficients. Of course, one look at the Fourier matrix F_n shows that if we want to know only the first m Fourier coefficients of n data points, we can multiply x by only the top m rows of F_n and leave it at that. In other words, it would suffice to replace the $n \times n$ matrix F_n by an $m \times n$ submatrix. An improved version of dftfilter.m would make use of this fact.

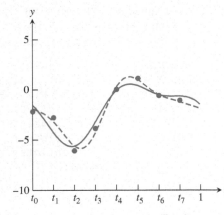

Figure 10.7 Least squares trigonometric fits for Example 10.6. Fits for $m=4$ (solid curve) and 6 (dashed curve) are shown. The input vector x is $[-2.2, -2.8, -6.1, -3.9, 0.0, 1.1, -0.6, -1.1]^T$. The fit for $m=8$ is trigonometric interpolation, shown in Figure 10.6.

10.3.3 Sound, noise, and filtering

The `dftfilter.m` code of the last section is an introduction to the vast area of digital signal processing. We are using the Fourier transform as a way to transfer the information of a signal $\{x_0, \dots, x_{n-1}\}$ from the "time domain" to the "frequency domain," where it is easier to manipulate. When we finish changing what we want to change, we send the signal back to the time domain by an inverse FFT.

If x represents an audio signal, this is helpful because of the way our hearing system is constructed. The human ear contains structures that respond to frequencies, and so the building blocks in the frequency domain are directly meaningful. We will illustrate this by introducing some basic concepts of audio and signal processing and a few convenient MATLAB commands.

An audio signal consists of a set of real numbers indexed by time. Each real number represents a sound intensity. When an audio signal is played back, the speaker head is made to vibrate so that the amplitude matches the signal, causing the surrounding air to vibrate at the same frequencies. When the sound waves reach your ear, you perceive sound.

MATLAB provides an audio signal of the first 9 seconds of Handel's *Hallelujah Chorus* for us to sample. The curve in Figure 10.8 shows the first $2^8 = 256$ values of the file, which consists of sound intensities. The sampling rate of the music is $2^{13} = 8192$ Hz, meaning that intensities are represented at the rate of 2^{13} per second, evenly spaced. To access the signal, type

```
>> load handel
```

which puts the variables Fs and y in the workspace. The former variable is the sampling rate $Fs = 8192$. The variable y is a length 73113 vector containing the sound signal. The MATLAB command

```
>> sound(y,Fs)
```

plays the signal on your computer speakers, if available, at the correct sampling rate Fs.

The *Hallelujah Chorus* data can be used to implement the filtering of Corollary 10.12. Using `dftfilter.m` with the first $n = 256$ samples of the signal, and $m = 64$

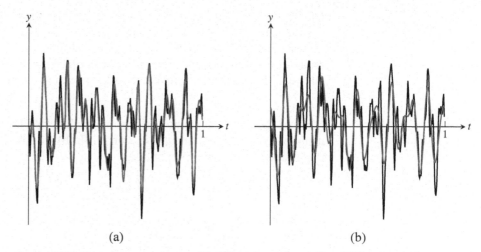

(a)   (b)

Figure 10.8 Sound curve along with filtered versions. First 1/32 second of *Hallelujah Chorus* (256 points on black curve) along with filtered version (blue curve) with (a) 64 basis functions, a 4:1 compression ratio and (b) 32 basis functions, an 8:1 compression ratio.

and 32 basis functions, results in the blue curves of Figure 10.8. The reader may want to explore filtering with other audio files.

One common audio file format is the `.wav` format. A stereo `.wav` file carries two paired signals to be played from two different speakers. For example, using the MATLAB command

```
>> [y,Fs]=wavread('castanets')
```

will extract the stereo signal from the file `castanets.wav` and load it into MATLAB as an $n \times 2$ matrix y, each column a separate sound signal. (The file `castanets.wav` is a common audio test file and can be easily found by a Web search.) The MATLAB command `wavwrite` reverses the process, creating a `.wav` file from simple sound signals.

Filtering is used in two ways. It can be used to match the original sound wave as closely as possible with a simpler function. This is a form of compression. Instead of using 256 numbers to store the wave, we could instead just store the lowest m frequency components and then reconstruct the wave when needed by using Corollary 10.12. In Figure 10.8(a), we used $m = 64$ real numbers in place of the original 256, a 4:1 compression ratio. Note that the compression is lossy, in that the original wave has not been reproduced exactly.

SPOTLIGHT ON

> **Compression**   Filtering is a form of lossy compression. In the case of an audio signal, the goal is to reduce the amount of data required to store or transmit the sound without compromising the musical effect or spoken information the signal is designed to represent. This is best done in the frequency domain, which means applying the DFT, manipulating the frequency components, and then inverting the DFT.

The second major application of filtering is to remove noise. Given a music file where the music or speech was corrupted by high-frequency noise (or hiss), eliminating the higher frequency contributions may be important to enhancing the sound.

Of course, so-called low-pass filters are blunt hammers—a high-frequency part of the desired sound, possibly in overtones not even obvious to the listener, may be deleted as well. The topic of filtering is part of a vast literature on signal processing, and the reader is referred to Oppenheim and Schafer [2009] for further study. In Reality Check 10, we investigate a filter of widespread application called the Wiener filter.

▶ **ADDITIONAL EXAMPLES**

*1 (a) Find the best least squares trigonometric function approximation to the data using basis functions 1 and $\cos \frac{1}{2}\pi t$ on $[0, 4]$. (b) Add basis function $\sin \frac{1}{2}\pi t$ and find the best least squares approximation.

t	0	1	2	3
x	1	4	2	3

2. Use `dftfilter.m` to plot the order 4, 6, and 8 least squares trigonometric approximation functions for the following data.

t	0	1/2	1	3/2	2	5/2	3	7/2
x	4	2	1	3	6	2	1	5

▭ **Solutions** for Additional Examples can be found at `bit.ly/2CVSfY3`
(* example with video solution)

10.3 Exercises

▭ **Solutions** for Exercises numbered in blue can be found at `bit.ly/2PJ1jSw`

1. Find the best order 2 least squares approximation to the data in Exercise 10.2.1, using the basis functions 1 and $\cos 2\pi t$.

2. Find the best order 3 least squares approximation to the data in Exercise 10.2.1, using the basis functions 1, $\cos 2\pi t$, and $\sin 2\pi t$.

3. Find the best order 4 least squares approximation to the data in Exercise 10.2.3, using the basis functions 1, $\cos 2\pi t$, $\sin 2\pi t$, and $\cos 4\pi t$.

4. Find the best order 4 least squares approximation to the data in Exercise 10.2.4, using the basis functions 1, $\cos \frac{\pi}{4} t$, $\sin \frac{\pi}{4} t$, and $\cos \frac{\pi}{2} t$.

5. Prove Lemma 10.10. (*Hint:* Express $\cos 2\pi jk/n$ as $(e^{i2\pi jk/n} + e^{-i2\pi jk/n})/2$, and write everything in terms of $\omega = e^{-i2\pi/n}$, so that Lemma 10.1 can be applied.)

10.3 Computer Problems

▭ **Solutions** for Computer Problems numbered in blue can be found at `bit.ly/2yq6q3r`

1. Find the least squares trigonometric approximating functions of orders $m = 2$ and 4 for the following data points:

(a)
t	y
0	3
$\frac{1}{4}$	0
$\frac{1}{2}$	-3
$\frac{3}{4}$	0

(b)
t	y
0	2
$\frac{1}{4}$	0
$\frac{1}{2}$	5
$\frac{3}{4}$	1

(c)
t	y
0	5
1	2
2	6
3	1

(d)
t	y
1	-1
2	1
3	4
4	3
5	3
6	2

Using `dftfilter.m`, plot the data points and the approximating functions, as in Figure 10.7.

2. Find the least squares trigonometric approximating functions of orders 4, 6, and 8 for the following data points:

(a)

t	y
0	3
$\frac{1}{8}$	0
$\frac{1}{4}$	-3
$\frac{3}{8}$	0
$\frac{1}{2}$	3
$\frac{5}{8}$	0
$\frac{3}{4}$	-6
$\frac{7}{8}$	0

(b)

t	y
0	1
$\frac{1}{8}$	0
$\frac{1}{4}$	-2
$\frac{3}{8}$	1
$\frac{1}{2}$	3
$\frac{5}{8}$	0
$\frac{3}{4}$	-2
$\frac{7}{8}$	1

(c)

t	y
0	1
$\frac{1}{8}$	2
$\frac{1}{4}$	3
$\frac{3}{8}$	1
$\frac{1}{2}$	-1
$\frac{5}{8}$	-1
$\frac{3}{4}$	-3
$\frac{7}{8}$	0

(d)

t	y
0	4.2
$\frac{1}{8}$	5.0
$\frac{1}{4}$	3.8
$\frac{3}{8}$	1.6
$\frac{1}{2}$	-2.0
$\frac{5}{8}$	-1.4
$\frac{3}{4}$	0.0
$\frac{7}{8}$	1.0

Plot the data points and the approximating functions, as in Figure 10.7.

3. Plot the least squares trigonometric approximation function of orders $m = n/2, n/4$, and $n/8$, along with the vector x containing the first 2^{14} sound intensity values from MATLAB's handel sound file. (This covers about 2 seconds of audio. The MATLAB code dftfilter can be used with $p = n$. Make three separate plots.) Use the MATLAB sound command to compare the original with the approximation. What has been lost?

4. Download castanets.wav from an appropriate website, and form a vector containing the signal at the first 2^{14} sample times. Carry out the steps of Computer Problem 3 for each stereo channel separately.

5. Gather 24 consecutive hourly temperature readings from a newspaper or website. Plot the data points along with (a) the trigonometric interpolating function and least squares approximating functions of order (b) $m = 6$ and (c) $m = 12$.

Reality Check 10 The Wiener Filter

Let c be a clean audio signal, and add a vector r of the same length to c. Is the resulting signal $x = c + r$ noisy? If $r = c$, we would not consider r noise, since the result would be a louder, but still clean, version of c. By definition, noise is uncorrelated with the signal. In other words, if r is noise, the expected value of the inner product $c^T r$ is zero. We will exploit this lack of correlation next.

In a typical application, we are presented with a noisy signal x and asked to find c. The signal c might be the value of an important system variable, being monitored in a noisy environment. Or, as in our example below, c might be an audio sample that we want to bring out of noise. In the middle of the 20th century, Norbert Wiener suggested looking for the optimal filter for removing the noise from x, in the sense of least squares error. He suggested finding a real, diagonal matrix Φ such that the Euclidean norm of

$$F^{-1}\Phi F x - c$$

is as small as possible, where F denotes the Discrete Fourier Transform. The idea is to clean up the signal x by applying the Fourier transform, operating on the frequency components by multiplying by Φ, and then inverting the Fourier transform. This is called filtering in the frequency domain, since we are changing the Fourier-transformed version of x rather than x itself.

To find the best diagonal matrix Φ, note that

$$
\begin{aligned}
\|F^{-1}\Phi Fx - c\|_2 &= \|\Phi Fx - Fc\|_2 \\
&= \|\Phi F(c+r) - Fc\|_2 \\
&= \|(\Phi - I)C + \Phi R\|_2,
\end{aligned}
\tag{10.34}
$$

where we set $C = Fc$ and $R = Fr$ to be the Fourier transforms. Note also that the definition of noise implies

$$
\overline{C}^T R = \overline{Fc}^T Fr = c^T \overline{F}^T Fr = c^T r = 0.
$$

We will use this as motivation to ignore the cross-terms in the norm, so that the squared magnitude reduces to

$$
\begin{aligned}
\left(\overline{(\Phi - I)C + \Phi R}\right)^T ((\Phi - I)C + \Phi R) &= \left(\overline{C}^T(\Phi - I) + \overline{R}^T\Phi\right)((\Phi - I)C + \Phi R) \\
&\approx \overline{C}^T(\Phi - I)^2 C + \overline{R}^T\Phi^2 R \\
&= \sum_{i=1}^{n}(\phi_i - 1)^2|C_i|^2 + \phi_i^2|R_i|^2.
\end{aligned}
\tag{10.35}
$$

To find the diagonal entries ϕ_i that minimize this expression, differentiate with respect to each ϕ_i separately to obtain

$$
2(\phi_i - 1)|C_i|^2 + 2\phi_i|R_i|^2 = 0
$$

for each i, or, solving for ϕ_i,

$$
\phi_i = \frac{|C_i|^2}{|C_i|^2 + |R_i|^2}.
\tag{10.36}
$$

This formula gives Wiener's values for the entries of the diagonal matrix Φ, to minimize the difference between the filtered version $F^{-1}\Phi Fx$ and the clean signal c. The only problem is that in typical cases, we don't know C or R and must make some approximations to apply the formula.

Your job is to investigate ways of putting together an approximation. Let $X = Fx$ be the Fourier transform. Again using the uncorrelatedness of signal and noise, approximate

$$
|X_i|^2 \approx |C_i|^2 + |R_i|^2.
$$

Then we can write the optimal choice as

$$
\phi_i \approx \frac{|X_i|^2 - |R_i|^2}{|X_i|^2}
\tag{10.37}
$$

and use our best knowledge of the noise level. For example, if the noise is uncorrelated Gaussian noise (modeled by adding a normal random number independently to each sample of the clean signal), we could replace $|R_i|^2$ in (10.37) with the constant $(p\sigma)^2$, where σ is the standard deviation of the noise and p is a parameter near one to be chosen. Note that

$$
\sum_{i=1}^{n}|R_i|^2 = \overline{R}^T R = r\overline{F}^T Fr = r^T r = \sum_{i=1}^{n} r_i^2.
$$

In the following code, we add 50 percent noise to the Handel signal, and use $p = 1.3$ standard deviations to approximate R_i:

MATLAB code
shown here can be found
at bit.ly/2COMmM4

```
load handel                           % y is clean signal
c=y(1:40000);                         % work with first 40K samples
p=1.3;                                % parameter for cutoff
noise=std(c)*.50;                     % 50 percent noise
n=length(c);                          % n is length of signal
r=noise*randn(n,1);                   % pure noise
x=c+r;                                % noisy signal
fx=fft(x);sfx=conj(fx).*fx;           % take fft of signal, and
sfcapprox=max(sfx-n*(p*noise)^2,0);   % apply cutoff
phi=sfcapprox./sfx;                   % define phi as derived
xout=real(ifft(phi.*fx));             % invert the fft
% then compare sound(x) and sound(xout)
```

Suggested activities:

1. Run the code to form the filtered signal yf, and use MATLAB's sound command to compare the input and output signals.

2. Compute the mean squared error (MSE) of the input (ys) and output (yf) by comparing with the clean signal (yc).

3. Find the best value of the parameter p for 50 percent noise. Compare the value that minimizes MSE to the one that sounds best to the ear.

4. Change the noise level to 10 percent, 25 percent, 100 percent, 200 percent, and repeat Step 3. Summarize your conclusions.

5. Design a fair comparison of the Wiener filter with the low-pass filter described in Section 10.2, and carry out the comparison.

6. Download a .wav file of your choice, add noise, and carry out the aforementioned steps.

Software and Further Reading

Good sources for further reading on the Discrete Fourier Transform include Briggs [1995], Brigham [1988], and Briggs and Henson [1995]. The original breakthrough of Cooley and Tukey appeared in Cooley and Tukey [1965], and computational improvements that have continued as the central place of the Fast Fourier Transform in modern signal processing have been acknowledged (Winograd [1978], Van Loan [1992], and Chu and George [1999]). The FFT is an important algorithm in its own right and, additionally, is used as a building block in other algorithms because of its efficient implementation. For example, it is used by MATLAB to compute the Discrete Cosine Transform, defined in Chapter 11. Interestingly, the divide-and-conquer strategy used by Cooley and Tukey was later successfully applied to many other computational problems.

MATLAB's fft command is based on the "Fastest Fourier Transform in the West" (FFTW), developed in the 1990s at MIT (Frigo and Johnson [1998]). In case the size n is not a power of two, the program breaks down the problem, using the prime factors of n, into smaller "codelets" optimized for particular fixed sizes. More information on the FFTW, including downloadable code, is available at http://www.fftw.org. Netlib's FFTPACK (Swarztrauber [1982]) is a package of Fortran subprograms for the Fast Fourier Transform, optimized for use in parallel implementations.

11

Compression

The increasingly rapid movement of information around the world relies on ingenious methods of data representation, which are in turn made possible by orthogonal transformations. The JPEG format for image representation is based on the Discrete Cosine Transform developed in this chapter. The MPEG-1 and MPEG-2 formats for TV and video data and the H.263 format for video phones are also based on the DCT, but with extra emphasis on compressing in the time dimension.

Sound files can be compressed into a variety of different formats, including MP3, Advanced Audio Coding (used by Apple's iTunes and XM satellite radio), Microsoft's Windows Media Audio (WMA), and other state-of-the-art methods. What these formats have in common is that the core compression is done by a variant of the DCT called the Modified Discrete Cosine Transform.

Reality Check Reality Check 11 on page 552 explores implementation of the MDCT into a simple, working algorithm to compress audio.

In Chapters 4 and 10, we observed the usefulness of orthogonality to represent and compress data. Here, we introduce the Discrete Cosine Transform (DCT), a variant of the Fourier transform that can be computed in real arithmetic. It is currently the method of choice for compression of sound and image files.

The simplicity of the Fourier transform stems from orthogonality, due to its representation as a complex unitary matrix. The Discrete Cosine Transform has a representation as a real orthogonal matrix, and so the same orthogonality properties make it simple to apply and easy to invert. Its similarity to the Discrete Fourier Transform (DFT) is close enough that fast versions of the DCT exist, in analogy to the Fast Fourier Transform (FFT).

In this chapter, the basic properties of the DCT are explained, and the links to working compression formats are investigated. The well-known JPEG format, for example, applies the two-dimensional DCT to 8×8 pixel blocks of an image, and stores the results using Huffman coding. The details of JPEG compression are investigated as a case study in Sections 11.2–11.3.

A modified version of the Discrete Cosine Transform, called the Modified Discrete Cosine Transform (MDCT), is the basis of most modern audio compression formats. The MDCT is the current gold standard for compression of sound files. We will introduce MDCT and investigate its application for coding and decoding, which provides the core technology of file formats such as MP3 and AAC (Advanced Audio Coding).

11.1 THE DISCRETE COSINE TRANSFORM

In this section, we introduce the Discrete Cosine Transform. This transform interpolates data, using basis functions that are all cosine functions, and involves only real computations. Its orthogonality characteristics make least squares approximations simple, as in the case of the Discrete Fourier Transform.

11.1.1 One-dimensional DCT

Let n be a positive integer. The one-dimensional Discrete Cosine Transform of order n is defined by the $n \times n$ matrix C whose entries are

$$C_{ij} = \frac{\sqrt{2}}{\sqrt{n}} a_i \cos \frac{i(2j+1)\pi}{2n} \tag{11.1}$$

for $i, j = 0, \ldots, n-1$, where

$$a_i \equiv \begin{cases} 1/\sqrt{2} & \text{if } i = 0, \\ 1 & \text{if } i = 1, \ldots, n-1 \end{cases}$$

or

$$C = \sqrt{\frac{2}{n}} \begin{bmatrix} \frac{1}{\sqrt{2}} & \frac{1}{\sqrt{2}} & \cdots & \frac{1}{\sqrt{2}} \\ \cos \frac{\pi}{2n} & \cos \frac{3\pi}{2n} & \cdots & \cos \frac{(2n-1)\pi}{2n} \\ \cos \frac{2\pi}{2n} & \cos \frac{6\pi}{2n} & \cdots & \cos \frac{2(2n-1)\pi}{2n} \\ \vdots & \vdots & & \vdots \\ \cos \frac{(n-1)\pi}{2n} & \cos \frac{(n-1)3\pi}{2n} & \cdots & \cos \frac{(n-1)(2n-1)\pi}{2n} \end{bmatrix}. \tag{11.2}$$

With two-dimensional images, the convention is to begin with 0 instead of 1. The notation will be easier if we extend this convention to matrix numbering, as we have done in (11.1). *In this chapter, subscripts for $n \times n$ matrices will go from 0 to $n-1$.* For simplicity, we will treat only the case where n is even in the following discussion.

DEFINITION 11.1   Let C be the matrix defined in (11.2). The **Discrete Cosine Transform** (DCT) of $x = [x_0, \ldots, x_{n-1}]^T$ is the n-dimensional vector $y = [y_0, \ldots, y_{n-1}]^T$, where

$$y = Cx. \tag{11.3}$$

❑

Note that C is a real orthogonal matrix, meaning that its transpose is its inverse:

$$C^{-1} = C^T = \sqrt{\frac{2}{n}} \begin{bmatrix} \frac{1}{\sqrt{2}} & \cos \frac{\pi}{2n} & \cdots & \cos \frac{(n-1)\pi}{2n} \\ \frac{1}{\sqrt{2}} & \cos \frac{3\pi}{2n} & \cdots & \cos \frac{(n-1)3\pi}{2n} \\ \vdots & \vdots & & \vdots \\ \frac{1}{\sqrt{2}} & \cos \frac{(2n-1)\pi}{2n} & \cdots & \cos \frac{(n-1)(2n-1)\pi}{2n} \end{bmatrix}. \tag{11.4}$$

The rows of an orthogonal matrix are pairwise orthogonal unit vectors. The orthogonality of C follows from the fact that the columns of C^T are the unit eigenvectors of the real symmetric $n \times n$ matrix

$$
\begin{bmatrix}
1 & -1 & & & & \\
-1 & 2 & -1 & & & \\
& -1 & 2 & -1 & & \\
& & \ddots & \ddots & \ddots & \\
& & & -1 & 2 & -1 \\
& & & & -1 & 1
\end{bmatrix}. \tag{11.5}
$$

Exercise 6 asks the reader to verify this fact.

The fact that C is a real orthogonal matrix is what makes the DCT useful. The Orthogonal Function Interpolation Theorem 10.9 applied to the matrix C implies Theorem 11.2.

THEOREM 11.2   **DCT Interpolation Theorem.** Let $x = [x_0, \ldots, x_{n-1}]^T$ be a vector of n real numbers. Define $y = [y_0, \ldots, y_{n-1}]^T = Cx$, where C is the Discrete Cosine Transform matrix of order n. Then the real function

$$
P_n(t) = \frac{1}{\sqrt{n}} y_0 + \frac{\sqrt{2}}{\sqrt{n}} \sum_{k=1}^{n-1} y_k \cos \frac{k(2t+1)\pi}{2n}
$$

satisfies $P_n(j) = x_j$ for $j = 0, \ldots, n-1$.   ∎

Proof. Follows directly from Theorem 10.9.   □

Theorem 11.2 shows that the $n \times n$ matrix C transforms n data points into n interpolation coefficients. Like the Discrete Fourier Transform, the Discrete Cosine Transform gives coefficients for a trigonometric interpolation function. Unlike the DFT, the DCT uses cosine terms only and is defined entirely in terms of real arithmetic.

▶ **EXAMPLE 11.1**   Use the DCT to interpolate the points $(0, 1), (1, 0), (2, -1), (3, 0)$.

It is helpful to notice, using elementary trigonometry, that the 4×4 DCT matrix can be viewed as

$$
C = \frac{1}{\sqrt{2}}
\begin{bmatrix}
\frac{1}{\sqrt{2}} & \frac{1}{\sqrt{2}} & \frac{1}{\sqrt{2}} & \frac{1}{\sqrt{2}} \\
\cos \frac{\pi}{8} & \cos \frac{3\pi}{8} & \cos \frac{5\pi}{8} & \cos \frac{7\pi}{8} \\
\cos \frac{2\pi}{8} & \cos \frac{6\pi}{8} & \cos \frac{10\pi}{8} & \cos \frac{14\pi}{8} \\
\cos \frac{3\pi}{8} & \cos \frac{9\pi}{8} & \cos \frac{15\pi}{8} & \cos \frac{21\pi}{8}
\end{bmatrix}
=
\begin{bmatrix}
a & a & a & a \\
b & c & -c & -b \\
a & -a & -a & a \\
c & -b & b & -c
\end{bmatrix}, \tag{11.6}
$$

where

$$
a = \frac{1}{2}, b = \frac{1}{\sqrt{2}} \cos \frac{\pi}{8} = \frac{\sqrt{2 + \sqrt{2}}}{2\sqrt{2}}, c = \frac{1}{\sqrt{2}} \cos \frac{3\pi}{8} = \frac{\sqrt{2 - \sqrt{2}}}{2\sqrt{2}}. \tag{11.7}
$$

The order-4 DCT multiplied by the data $x = (1, 0, -1, 0)^T$ is

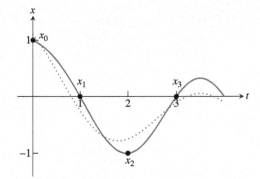

Figure 11.1 DCT interpolation and least squares approximation. The data points are (j, x_j), where $x = [1, 0, -1, 0]$. The DCT interpolating function $P_4(t)$ of (11.8) is shown as a solid curve, along with the least squares DCT approximation function $P_3(t)$ of (11.9) as a dotted curve.

$$
\begin{bmatrix} a & a & a & a \\ b & c & -c & -b \\ a & -a & -a & a \\ c & -b & b & -c \end{bmatrix}
\begin{bmatrix} 1 \\ 0 \\ -1 \\ 0 \end{bmatrix}
=
\begin{bmatrix} 0 \\ c+b \\ 2a \\ c-b \end{bmatrix}
=
\begin{bmatrix} 0 \\ \dfrac{\sqrt{2-\sqrt{2}}+\sqrt{2+\sqrt{2}}}{2\sqrt{2}} \\ 1 \\ \dfrac{\sqrt{2-\sqrt{2}}-\sqrt{2+\sqrt{2}}}{2\sqrt{2}} \end{bmatrix}
\approx
\begin{bmatrix} 0.0000 \\ 0.9239 \\ 1.0000 \\ -0.3827 \end{bmatrix}.
$$

According to Theorem 11.2 with $n = 4$, the function

$$
P_4(t) = \frac{1}{\sqrt{2}} \left[0.9239 \cos \frac{(2t+1)\pi}{8} + \cos \frac{2(2t+1)\pi}{8} - 0.3827 \cos \frac{3(2t+1)\pi}{8} \right] \quad (11.8)
$$

interpolates the four data points. The function $P_4(t)$ is plotted as the solid curve in Figure 11.1. ◄

11.1.2 The DCT and least squares approximation

Just as the DCT Interpolation Theorem 11.2 is an immediate consequence of Theorem 10.9, the least squares result Theorem 10.11 shows how to find a DCT least squares approximation of the data, using only part of the basis functions. Because of the orthogonality of the basis functions, this can be accomplished by simply dropping the higher frequency terms.

SPOTLIGHT ON

Orthogonality The idea behind least squares approximation is that finding the shortest distance from a point to a plane (or subspace in general) means constructing the perpendicular from the point to the plane. This construction is carried out by the normal equations, as we saw in Chapter 4. In Chapters 10 and 11, this concept is applied to approximate data as closely as possible with a relatively small set of basis functions, resulting in compression. The basic message is to choose the basis functions to be orthogonal, as reflected in the rows of the DCT matrix. Then the normal equations become computationally very simple (see Theorem 10.11).

THEOREM 11.3   **DCT Least Squares Approximation Theorem.** Let $x = [x_0, \ldots, x_{n-1}]^T$ be a vector of n real numbers. Define $y = [y_0, \ldots, y_{n-1}]^T = Cx$, where C is the Discrete Cosine Transform matrix. Then, for any positive integer $m \leq n$, the choice of coefficients $y_0, \ldots, y_{m-1}$ in

$$P_m(t) = \frac{1}{\sqrt{n}} y_0 + \frac{\sqrt{2}}{\sqrt{n}} \sum_{k=1}^{m-1} y_k \cos \frac{k(2t+1)\pi}{2n}$$

minimizes the squared approximation error $\sum_{j=0}^{n-1}(P_m(j) - x_j)^2$ of the n data points. ∎

Proof. Follows directly from Theorem 10.11.   □

Referring to Example 11.1, if we require the best least squares approximation to the same four data points, but use the three basis functions

$$1, \cos \frac{(2t+1)\pi}{8}, \cos \frac{2(2t+1)\pi}{8}$$

only, the solution is

$$P_3(t) = \frac{1}{2} \cdot 0 + \frac{1}{\sqrt{2}}\left[0.9239 \cos \frac{(2t+1)\pi}{8} + \cos \frac{2(2t+1)\pi}{8} \right]. \tag{11.9}$$

Figure 11.1 compares the least squares solution P_3 with the interpolating function P_4.

▶ **EXAMPLE 11.2**   Use the DCT and Theorem 11.3 to find least squares fits to the data $t = 0, \ldots, 7$ and $x = [-2.2, -2.8, -6.1, -3.9, 0.0, 1.1, -0.6, -1.1]^T$ for $m = 4, 6$, and 8.

Setting $n = 8$, we find that the DCT of the data is

$$y = Cx = \begin{bmatrix} -5.5154 \\ -3.8345 \\ 0.5833 \\ 4.3715 \\ 0.4243 \\ -1.5504 \\ -0.6243 \\ -0.5769 \end{bmatrix}.$$

According to Theorem 11.2, the discrete cosine interpolant of the eight data points is

$$P_8(t) = \frac{1}{\sqrt{8}}(-5.5154) + \frac{1}{2}\left[-3.8345 \cos \frac{(2t+1)\pi}{16} + 0.5833 \cos \frac{2(2t+1)\pi}{16} \right.$$

$$+ 4.3715 \cos \frac{3(2t+1)\pi}{16} + 0.4243 \cos \frac{4(2t+1)\pi}{16}$$

$$- 1.5504 \cos \frac{5(2t+1)\pi}{16} - 0.6243 \cos \frac{6(2t+1)\pi}{16}$$

$$\left. - 0.5769 \cos \frac{7(2t+1)\pi}{16} \right].$$

The interpolant P_8 is plotted in Figure 11.2, along with the least squares fits P_6 and P_4. The latter are obtained, according to Theorem 11.3, by keeping the first six, or first four terms, respectively, of P_8.   ◀

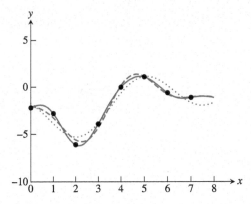

Figure 11.2 DCT interpolation and least squares approximation. The solid curve is the DCT interpolant of the data points in Example 11.2. The dashed curve is the least squares fit from the first six terms only, and the dotted curve represents four terms.

▶ **ADDITIONAL EXAMPLES**

1. Find the DCT of the data vector x and find the corresponding interpolating function $P_4(t)$.

t	0	1	2	3
x	5	10	5	0

*2. Plot the order 4, 6, and 8 DCT interpolating functions for the data

t	0	1	2	3	4	5	6	7
x	4	2	1	3	6	2	1	5

 Solutions for Additional Examples can be found at bit.ly/2CTq8bL (* example with video solution)

11.1 Exercises

1. Use the 2×2 DCT matrix and Theorem 11.2 to find the DCT interpolating function for the data points.

(a)
t	x
0	3
1	3

(b)
t	x
0	2
1	-2

(c)
t	x
0	3
1	1

(d)
t	x
0	4
1	-1

2. Describe the $m = 1$ least squares DCT approximation in terms of the input data $(0, x_0), (1, x_1)$.

3. Find the DCT of the following data vectors x, and find the corresponding interpolating function $P_n(t)$ for the data points $(i, x_i), i = 0, \ldots, n - 1$ (you may state your answers in terms of the b and c defined in (11.7)):

(a)
t	x
0	1
1	0
2	1
3	0

(b)
t	x
0	1
1	1
2	1
3	1

(c)
t	x
0	1
1	0
2	0
3	0

(d)
t	x
0	1
1	2
2	3
3	4

4. Find the DCT least squares approximation with $m = 2$ terms for the data in Exercise 3.

5. Carry out the trigonometry needed to establish equations (11.6) and (11.7).

6. (a) Prove the trigonometric formula $\cos(x + y) + \cos(x - y) = 2\cos x \cos y$ for any x, y. (b) Show that the columns of C^T are eigenvectors of the matrix T in (11.5), and identify the eigenvalues. (c) Show that the columns of C^T are unit vectors.

7. Extend the DCT Interpolation Theorem 11.2 to the interval $[c, d]$ as follows. Let n be a positive integer and set $\Delta_t = (d - c)/n$. Use the DCT to produce a polynomial $P_n(t)$ that satisfies $P_n(c + j\Delta_t) = x_j$ for $j = 0, \ldots, n - 1$.

11.1 Computer Problems

Solutions for Computer Problems numbered in blue can be found at
bit.ly/2ypa262

1. Plot the data from Exercise 3, along with the DCT interpolant and the DCT least squares approximation with $m = 2$ terms.

2. Plot the data along with the $m = 4, 6,$ and 8 DCT least squares approximations.

	t	x		t	x		t	x		t	x
	0	3		0	4		0	3		0	4
	1	5		1	1		1	−1		1	2
	2	−1		2	−3		2	−1		2	−4
(a)	3	3	(b)	3	0	(c)	3	3	(d)	3	2
	4	1		4	0		4	3		4	4
	5	3		5	2		5	−1		5	2
	6	−2		6	−4		6	−1		6	−4
	7	4		7	0		7	3		7	2

3. Plot the function $f(t)$, the data points $(j, f(j))$, $j = 0, \ldots, 7$, and the DCT interpolation function. (a) $f(t) = e^{-t/4}$ (b) $f(t) = \cos \frac{\pi}{2} t$.

11.2 TWO-DIMENSIONAL DCT AND IMAGE COMPRESSION

The two-dimensional Discrete Cosine Transform is often used to compress small blocks of an image, as small as 8×8 pixels. The compression is lossy, meaning that some information from the block is ignored. The key feature of the DCT is that it helps organize the information so that the part that is ignored is the part that the human eye is least sensitive to. More precisely, the DCT will show us how to interpolate the data with a set of basis functions that are in descending order of importance as far as the human visual system is concerned. The less important interpolation terms can be dropped if desired, just as a newspaper editor cuts a long story on deadline.

Later, we will apply what we have learned about the DCT to compress images. Using the added tools of quantization and Huffman coding, each 8×8 block of an image can be reduced to a bit stream that is stored with bit streams from the other blocks of the image. The complete bit stream is decoded, when the image needs to be uncompressed and displayed, by reversing the encoding process. We will describe this approach, called Baseline JPEG, the default method for storing JPEG images.

11.2.1 Two-dimensional DCT

The two-dimensional Discrete Cosine Transform is simply the one-dimensional DCT applied in two dimensions, one after the other. It can be used to interpolate or approximate data given on a two-dimensional grid, in a straightforward analogy to the one-dimensional case. In the context of image processing, the two-dimensional grid represents a block of pixel values—say, grayscale intensities or color intensities.

In this chapter only, we will list the vertical coordinate first and the horizontal coordinate second when referring to a two-dimensional point, as shown in Figure 11.3. The goal is to be consistent with the usual matrix convention, where the i index of entry x_{ij} changes along the vertical direction, and j along the horizontal. A major application of this section is to pixel files representing images, which are most naturally viewed as matrices of numbers.

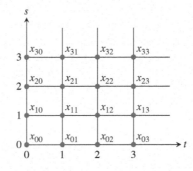

Figure 11.3 Two-dimensional grid of data points. The 2D-DCT can be used to interpolate function values on a square grid, such as pixel values of an image.

Figure 11.3 shows a grid of (s, t) points in the two-dimensional plane with assigned values x_{ij} at each rectangular grid point (s_i, t_j). For concreteness, we will use the integer grid $s_i = \{0, 1, \ldots, n - 1\}$ (remember, along the vertical axis) and $t_j = \{0, 1, \ldots, n - 1\}$ along the horizontal axis. The purpose of the two-dimensional DCT is to construct an interpolating function $F(s, t)$ that fits the n^2 points (s_i, t_j, x_{ij}) for $i, j = 0, \ldots, n - 1$. The 2D-DCT accomplishes this in an optimal way from the point of view of least squares, meaning that the fit degrades gracefully as basis functions are dropped from the interpolating function.

The 2D-DCT is the one-dimensional DCT applied successively to both horizontal and vertical directions. Consider the matrix X consisting of the values x_{ij}, as in Figure 11.3. To apply the 1D-DCT in the horizontal s-direction, we first need to transpose X, then multiply by C. The resulting columns are the 1D-DCT's of the rows of X. Each column of CX^T corresponds to a fixed t_i. To do a 1D-DCT in the t-direction means moving across the rows; so, again, transposing and multiplying by C yields

$$C(CX^T)^T = CXC^T. \tag{11.10}$$

DEFINITION 11.4 The **two-dimensional Discrete Cosine Transform** (2D-DCT) of the $n \times n$ matrix X is the matrix $Y = CXC^T$, where C is defined in (11.1). ☐

▶ **EXAMPLE 11.3** Find the 2D Discrete Cosine Transform of the data in Figure 11.4(a).

From the definition and (11.6), the 2D-DCT is the matrix

$$Y = CXC^T = \begin{bmatrix} a & a & a & a \\ b & c & -c & -b \\ a & -a & -a & a \\ c & -b & b & -c \end{bmatrix} \begin{bmatrix} 1 & 1 & 1 & 1 \\ 1 & 0 & 0 & 1 \\ 1 & 0 & 0 & 1 \\ 1 & 1 & 1 & 1 \end{bmatrix} \begin{bmatrix} a & b & a & c \\ a & c & -a & -b \\ a & -c & -a & b \\ a & -b & a & -c \end{bmatrix}$$

$$= \begin{bmatrix} 3 & 0 & 1 & 0 \\ 0 & 0 & 0 & 0 \\ 1 & 0 & -1 & 0 \\ 0 & 0 & 0 & 0 \end{bmatrix}. \tag{11.11}$$

◀

The inverse of the 2D-DCT is easy to express in terms of the DCT matrix C. Since $Y = CXC^T$ and C is orthogonal, the X is recovered as $X = C^T YC$.

DEFINITION 11.5   The **inverse two-dimensional Discrete Cosine Transform** of the $n \times n$ matrix Y is the matrix $X = C^T YC$.   ◻

As we have seen, there is a close connection between inverting an orthogonal transform (like the 2D-DCT) and interpolation. The goal of interpolation is to recover the original data points from functions that are constructed with the interpolating coefficients that came out of the transform. Since C is an orthogonal matrix, $C^{-1} = C^T$. The inversion of the 2D-DCT can be written as a fact about interpolation, $X = C^T YC$, since in this equation the x_{ij} are being expressed in terms of products of cosines.

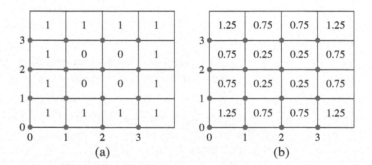

Figure 11.4 Two-dimensional data for Example 11.3. (a) The 16 data points (i, j, x_{ij}). (b) Values of the least squares approximation (11.14) at the grid points.

To write a useful expression for the interpolating function, recall the definition of C in (11.1),

$$C_{ij} = \frac{\sqrt{2}}{\sqrt{n}} a_i \cos \frac{i(2j+1)\pi}{2n} \tag{11.12}$$

for $i, j = 0, \ldots, n-1$, where

$$a_i \equiv \begin{cases} 1/\sqrt{2} & \text{if } i = 0, \\ 1 & \text{if } i = 1, \ldots, n-1 \end{cases}.$$

According to the rules of matrix multiplication, the equation $X = C^T YC$ translates to

$$
\begin{aligned}
x_{ij} &= \sum_{k=0}^{n-1} \sum_{l=0}^{n-1} C_{ik}^T y_{kl} C_{lj} \\
&= \sum_{k=0}^{n-1} \sum_{l=0}^{n-1} C_{ki} y_{kl} C_{lj} \\
&= \frac{2}{n} \sum_{k=0}^{n-1} \sum_{l=0}^{n-1} y_{kl} a_k a_l \cos \frac{k(2i+1)\pi}{2n} \cos \frac{l(2j+1)\pi}{2n}.
\end{aligned} \tag{11.13}
$$

This is exactly the interpolation statement we were looking for.

THEOREM 11.6   **2D-DCT Interpolation Theorem.** Let $X = (x_{ij})$ be a matrix of n^2 real numbers. Let $Y = (y_{kl})$ be the two-dimensional Discrete Cosine Transform of X. Define $a_0 = 1/\sqrt{2}$ and $a_k = 1$ for $k > 0$. Then the real function

$$P_n(s, t) = \frac{2}{n} \sum_{k=0}^{n-1} \sum_{l=0}^{n-1} y_{kl} a_k a_l \cos \frac{k(2s+1)\pi}{2n} \cos \frac{l(2t+1)\pi}{2n}$$

satisfies $P_n(i, j) = x_{ij}$ for $i, j = 0, \dots, n-1$.   ∎

Returning to Example 11.3, the only nonzero interpolation coefficients are $y_{00} = 3$, $y_{02} = y_{20} = 1$, and $y_{22} = -1$. Writing out the interpolation function in the Theorem 11.6 yields

$$\begin{aligned} P_4(s, t) &= \frac{2}{4} \left[\frac{1}{2} y_{00} + \frac{1}{\sqrt{2}} y_{02} \cos \frac{2(2t+1)\pi}{8} + \frac{1}{\sqrt{2}} y_{20} \cos \frac{2(2s+1)\pi}{8} \right. \\ &\quad \left. + y_{22} \cos \frac{2(2s+1)\pi}{8} \cos \frac{2(2t+1)\pi}{8} \right] \\ &= \frac{1}{2} \left[\frac{1}{2}(3) + \frac{1}{\sqrt{2}}(1) \cos \frac{2(2t+1)\pi}{8} + \frac{1}{\sqrt{2}}(1) \cos \frac{2(2s+1)\pi}{8} \right. \\ &\quad \left. + (-1) \cos \frac{2(2s+1)\pi}{8} \cos \frac{2(2t+1)\pi}{8} \right] \\ &= \frac{3}{4} + \frac{1}{2\sqrt{2}} \cos \frac{(2t+1)\pi}{4} + \frac{1}{2\sqrt{2}} \cos \frac{(2s+1)\pi}{4} \\ &\quad - \frac{1}{2} \cos \frac{(2s+1)\pi}{4} \cos \frac{(2t+1)\pi}{4}. \end{aligned}$$

Checking the interpolation, we get, for example,

$$P_4(0, 0) = \frac{3}{4} + \frac{1}{4} + \frac{1}{4} - \frac{1}{4} = 1$$

and

$$P_4(1, 2) = \frac{3}{4} - \frac{1}{4} - \frac{1}{4} - \frac{1}{4} = 0,$$

agreeing with the data in Figure 11.4. The constant term y_{00}/n of the interpolation function is called the "DC" component of the expansion (for "direct current"). It is the simple average of the data; the nonconstant terms contain the fluctuations of the data about this average value. In this example, the average of the 12 ones and 4 zeros is $y_{00}/4 = 3/4$.

Least squares approximations with the 2D-DCT are done in the same way as with the 1D-DCT. For example, implementing a low-pass filter would mean simply deleting the "high-frequency" components, those whose coefficients have larger indices, from the interpolating function. In Example 11.3, the best least squares fit to the basis functions

$$\cos \frac{i(2s+1)\pi}{8} \cos \frac{j(2t+1)\pi}{8}$$

for $i + j \leq 2$ is given by dropping all terms that do not satisfy $i + j \leq 2$. In this case, the only nonzero "high-frequency" term is the $i = j = 2$ term, leaving

$$P_2(s, t) = \frac{3}{4} + \frac{1}{2\sqrt{2}} \cos \frac{(2t+1)\pi}{4} + \frac{1}{2\sqrt{2}} \cos \frac{(2s+1)\pi}{4}. \tag{11.14}$$

This least squares approximation is shown in Figure 11.4(b).

Defining the DCT matrix C in MATLAB can be done through the code fragment

```
for i=1:n
  for j=1:n
    C(i,j)=cos((i-1)*(2*j-1)*pi/(2*n));
  end
end
C=sqrt(2/n)*C;
C(1,:)=C(1,:)/sqrt(2);
```

Alternatively, if MATLAB's Signal Processing Toolbox is available, the one-dimensional DCT of a vector x can be computed as

```
>> y=dct(x);
```

To carry out the 2D-DCT of a matrix X, we fall back on equation (11.10), or

```
>> Y=C*X*C'
```

If MATLAB's dct is available, the command

```
>> Y=dct(dct(X')')
```

computes the 2D-DCT with two applications of the 1D-DCT.

11.2.2 Image compression

The concept of orthogonality, as represented in the Discrete Cosine Transform, is crucial to performing image compression. Images consist of pixels, each represented by a number (or three numbers, for color images). The convenient way that methods like the DCT can carry out least squares approximation makes it easy to reduce the number of bits needed to represent the pixel values, while degrading the picture only slightly, and perhaps imperceptibly to human viewers.

Figure 11.5(a) shows a grayscale rendering of a 256×256 array of pixels. The grayness of each pixel is represented by one byte, a string of 8 bits representing $0 = 00000000$ (black) to $255 = 11111111$ (white). We can think of the information shown in the figure as a 256×256 array of integers. Represented in this way, the picture holds $(256)^2 = 2^{16} = 64K$ bytes of information.

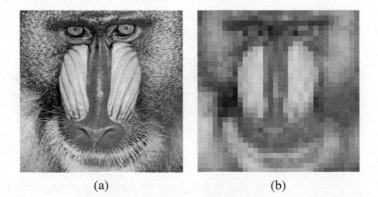

(a)   (b)

Figure 11.5 Grayscale image. (a) Each pixel in the 256×256 grid is represented by an integer between 0 and 255. (b) Crude compression—each 8×8 square of pixels is colored by its average grayscale value.

MATLAB imports grayscale or RGB (Red-Green-Blue) values of images from standard image formats. For example, given a grayscale image file `picture.jpg`, the command

```
>> x = imread('picture.jpg');
```

puts the matrix of grayscale values into the double precision variable x. If the JPEG file is a color image, the array variable will have a third dimension to index the three colors. We will restrict attention to gray scale to begin our discussion; extension to color is straightforward.

An $m \times n$ matrix of grayscale values can be rendered by MATLAB with the commands

```
>> imagesc(x);colormap(gray)
```

while an $m \times n \times 3$ matrix of RGB color is rendered with the `imagesc(x)` command alone. A common formula for converting a color RGB image to gray scale is

$$X_{\text{gray}} = 0.2126R + 0.7152G + 0.0722B, \tag{11.15}$$

or in MATLAB code,

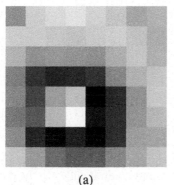

110	168	176	182	170	159	134	145
166	168	164	161	165	171	159	141
146	118	124	122	119	145	162	144
102	34	22	25	38	111	146	159
107	49	130	159	2	29	117	164
95	71	153	207	15	30	122	150
112	21	0	19	0	30	132	135
163	129	83	67	69	107	139	159

-18	40	48	54	42	31	6	17
38	40	36	33	37	43	31	13
18	-10	-4	-6	-9	17	34	16
-26	-94	-106	-103	-90	-17	18	31
-21	-79	2	31	-126	-99	-11	36
-33	-57	25	79	-113	-98	-6	22
-16	107	-128	-109	-128	-98	4	7
35	1	-45	-61	-59	-21	11	31

(a)   (b)   (c)

Figure 11.6 Example of 8×8 **block.** (a) Grayscale view (b) Grayscale pixel values (c) Grayscale pixel values minus 128.

```
>> x=double(x);
>> r=x(:,:,1);g=x(:,:,2);b=x(:,:,3);
>> xgray=0.2126*r+0.7152*g+0.0722*b;
>> xgray=uint8(xgray);
>> imagesc(xgray);colormap(gray)
```

Note that we have converted the default MATLAB data type `uint8`, or unsigned integers, to double precision reals before we do the computation. It is best to convert back to `uint8` type before rendering the picture with `imagesc`.

Figure 11.5(b) shows a crude method of compression, where each 8×8 pixel block is replaced by its average pixel value. The amount of data compression is considerable—there are only $(32)^2 = 2^{10}$ blocks, each now represented by a single integer—but the resulting image quality is poor. Our goal is to compress less harshly, by replacing each 8×8 block with a few integers that better carry the information of the original image.

To begin, we simplify the problem to a single 8×8 block of pixels, as shown in Figure 11.6(a). The block was taken from the center of the subject's left eye in Figure 11.5. Figure 11.6(b) shows the one-byte integers that represent the grayscale intensities of the 64 pixels. In Figure 11.6(c), we have subtracted $256/2 = 128$ from the pixel numbers to make them approximately centered around zero. This step is not essential, but better use of the 2D-DCT will result because of this centering.

To compress the 8×8 pixel block shown, we will transform the matrix of grayscale pixel values

$$X = \begin{bmatrix} -18 & 40 & 48 & 54 & 42 & 31 & 6 & 17 \\ 38 & 40 & 36 & 33 & 37 & 43 & 31 & 13 \\ 18 & -10 & -4 & -6 & -9 & 17 & 34 & 16 \\ -26 & -94 & -106 & -103 & -90 & -17 & 18 & 31 \\ -21 & -79 & 2 & 31 & -126 & -99 & -11 & 36 \\ -33 & -57 & 25 & 79 & -113 & -98 & -6 & 22 \\ -16 & -107 & -128 & -109 & -128 & -98 & 4 & 7 \\ 35 & 1 & -45 & -61 & -59 & -21 & 11 & 31 \end{bmatrix} \tag{11.16}$$

and rely on the 2D-DCT's ability to sort information according to its importance to the human visual system. We calculate the 2D-DCT of X to be

$$Y = C_8 X C_8^T = \begin{bmatrix} -121 & -66 & 127 & -65 & 27 & 98 & 7 & -25 \\ 200 & 22 & -124 & 34 & -36 & -62 & 5 & 6 \\ 113 & 43 & -32 & 55 & -25 & -75 & -21 & 12 \\ -10 & 35 & -69 & -131 & 28 & 54 & -4 & -24 \\ -14 & -18 & 16 & 1 & -5 & -27 & 14 & -6 \\ -124 & -74 & 47 & 60 & -1 & -16 & -8 & 13 \\ 81 & 35 & -57 & -54 & -7 & 6 & 1 & -16 \\ -16 & 11 & 5 & -15 & 11 & 12 & -1 & 9 \end{bmatrix}, \tag{11.17}$$

after rounding to the nearest integer for simplicity. This rounding adds a small amount of extra error and is not strictly necessary, but again it will help the compression. Note that due to the larger amplitudes, there is a tendency for more of the information to be stored in the top left part of the transform matrix Y, compared with the lower right. The lower right represents higher frequency basis functions that are often less important to the visual system. Nevertheless, because the 2D-DCT is an invertible transform, the information in Y can be used to completely reconstruct the original image, up to the rounding.

The first compression strategy we try will be a form of low-pass filtering. As discussed in the last section, least squares approximation with the 2D-DCT is just a matter of dropping terms from the interpolation function $P_8(s,t)$. For example, we can cut off the contribution of functions with relatively high spatial frequency by setting all $y_{kl} = 0$ for $k + l \geq 7$ (recall that we continue to number matrix entries as $0 \leq k, l \leq 7$). After low-pass filtering, the transform coefficients are

$$Y_{\text{low}} = \begin{bmatrix} -121 & -66 & 127 & -65 & 27 & 98 & 7 & 0 \\ 200 & 22 & -124 & 34 & -36 & -62 & 0 & 0 \\ 113 & 43 & -32 & 55 & -25 & 0 & 0 & 0 \\ -10 & 35 & -69 & -131 & 0 & 0 & 0 & 0 \\ -14 & -18 & 16 & 0 & 0 & 0 & 0 & 0 \\ -124 & -74 & 0 & 0 & 0 & 0 & 0 & 0 \\ 81 & 0 & 0 & 0 & 0 & 0 & 0 & 0 \\ 0 & 0 & 0 & 0 & 0 & 0 & 0 & 0 \end{bmatrix}. \tag{11.18}$$

To reconstruct the image, we apply the inverse 2D-DCT as $C_8^T Y_{\text{low}} C_8$ and get the grayscale pixel values shown in Figure 11.7. The image in part (a) is similar to the original in Figure 11.6(a), but different in detail.

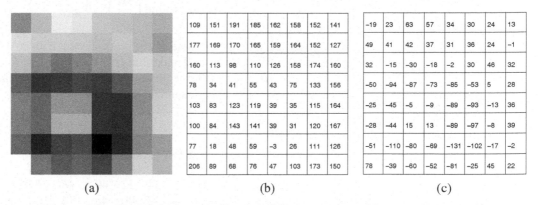

| (a) | (b) | (c) |

Figure 11.7 Result of low-pass filtering. (a) Filtered image (b) Grayscale pixel values, after transforming and adding 128 (c) Inverse transformed data.

How much have we compressed the information from the 8×8 block? The original picture can be reconstructed (losslessly, except for the integer rounding) by inverse transforming the 2D-DCT (11.17) and adding back the 128. In doing the low-pass filtering with matrix (11.17), we have cut the storage requirements approximately in half, while retaining most of the qualitative visual aspects of the block.

11.2.3 Quantization

The idea of quantization will allow the effects of low-pass filtering to be achieved in a more selective way. Instead of completely ignoring coefficients, we will retain low-accuracy versions of some coefficients at a lower storage cost. This idea exploits the same aspects of the human visual system—that it is less sensitive to higher spatial frequencies. The main idea is to assign fewer bits to store information about the lower right corner of the transform matrix Y, instead of throwing it away.

Quantization modulo q

Quantization: $z = \text{round}\left(\dfrac{y}{q}\right)$

Dequantization: $\overline{y} = qz$ (11.19)

Here, "round" means "to the nearest integer." The **quantization error** is the difference between the input y and the output $\overline{y}$ after quantizing and dequantizing. The maximum error of quantization modulo q is $q/2$.

► **EXAMPLE 11.4** Quantize the numbers $-10, 3$, and 65 modulo 8.

The quantized values are $-1, 0$, and 8. Upon dequantizing, the results are $-8, 0$, and 64. The errors are $|-2|, |3|$, and $|1|$, respectively, each less than $q/2 = 4$. ◄

Returning to the image example, the number of bits allowed for each frequency can be chosen arbitrarily. Let Q be an 8×8 matrix called the **quantization matrix**.

The entries $q_{kl}, 0 \le k, l \le 7$ will regulate how many bits we assign to each entry of the transform matrix Y. Replace Y by the compressed matrix

$$Y_Q = \left[\text{round} \left(\frac{y_{kl}}{q_{kl}} \right) \right], 0 \le k, l \le 7. \tag{11.20}$$

The matrix Y is divided entrywise by the quantization matrix. The subsequent rounding is where the loss occurs, and makes this method a form of lossy compression. Note that the larger the entry of Q, the more is potentially lost to quantization.

As a first example, **linear quantization** is defined by the matrix

$$q_{kl} = 8p(k + l + 1) \text{ for } 0 \le k, l \le 7 \tag{11.21}$$

for some constant p, called the **loss parameter**. Thus,

$$Q = p \begin{bmatrix}
8 & 16 & 24 & 32 & 40 & 48 & 56 & 64 \\
16 & 24 & 32 & 40 & 48 & 56 & 64 & 72 \\
24 & 32 & 40 & 48 & 56 & 64 & 72 & 80 \\
32 & 40 & 48 & 56 & 64 & 72 & 80 & 88 \\
40 & 48 & 56 & 64 & 72 & 80 & 88 & 96 \\
48 & 56 & 64 & 72 & 80 & 88 & 96 & 104 \\
56 & 64 & 72 & 80 & 88 & 96 & 104 & 112 \\
64 & 72 & 80 & 88 & 96 & 104 & 112 & 120
\end{bmatrix}.$$

In MATLAB, the linear quantization matrix can be defined by Q=p*8./hilb(8);

The loss parameter p is a knob that can be turned to trade bits for visual accuracy. The smaller the loss parameter, the better the reconstruction will be. The resulting set of numbers in the matrix Y_Q represents the new quantized version of the image.

To decompress the file, the Y_Q matrix is dequantized by reversing the process, which is entrywise multiplication by Q. This is the lossy part of image coding. Replacing the entries y_{kl} by dividing by q_{kl} and rounding, and then reconstructing by multiplying by q_{kl}, one has potentially added error of size $q_{kl}/2$ to y_{kl}. This is the quantization error. The larger the q_{kl}, the larger the potential error in reconstructing the image. On the other hand, the larger the q_{kl}, the smaller the integer entries of Y_Q, and the fewer bits will be needed to store them. This is the trade-off between image accuracy and file size.

In fact, quantization accomplishes two things: Many small contributions from higher frequencies are immediately set to zero by (11.20), and the contributions that remain nonzero are reduced in size, so that they can be transmitted or stored by using fewer bits. The resulting set of numbers are converted to a bit stream with the use of Huffman coding, discussed in the next section.

Next, we demonstrate the complete series of steps for compression of a matrix of pixel values in MATLAB. The output of MATLAB's imread command is an $m \times n$ matrix of 8-bit integers for a grayscale photo, or three such matrices for a color photo. (The three matrices carry information for red, green, and blue, respectively; we discuss color in more detail below.) An 8-bit integer is called a uint8, to distinguish it from a double, as studied in Chapter 0, which requires 64 bits of storage. The command double(x) converts the uint8 number x into the double format, and the command uint8(x) does the reverse by rounding x to the nearest integer between 0 and 255.

The following four commands carry out the conversion, centering, transforming, and quantization of a square $n \times n$ matrix X of uint8 numbers, such as the 8×8 pixel matrices considered above. Denote by C the $n \times n$ DCT matrix.

```
>> Xd=double(X);
>> Xc=Xd-128;
>> Y=C*Xc*C';
>> Yq=round(Y./Q);
```

At this point the resulting Yq is stored or transmitted. To recover the image requires undoing the four steps in reverse order:

```
>> Ydq=Yq.*Q;
>> Xdq=C'*Ydq*C;
>> Xe=Xdq+128;
>> Xf=uint8(Xe);
```

After dequantization, the inverse DCT transform is applied, the offset 128 is added back, and the double format is converted back to a matrix Xf of uint8 integers.

When linear quantization is applied to (11.17) with $p = 1$, the resulting coefficients are

$$Y_Q = \begin{bmatrix} -15 & -4 & 5 & -2 & 1 & 2 & 0 & 0 \\ 13 & 1 & -4 & 1 & -1 & -1 & 0 & 0 \\ 5 & 1 & -1 & 1 & 0 & -1 & 0 & 0 \\ 0 & 1 & -1 & -2 & 0 & 1 & 0 & 0 \\ 0 & 0 & 0 & 0 & 0 & 0 & 0 & 0 \\ -3 & -1 & 1 & 1 & 0 & 0 & 0 & 0 \\ 1 & 1 & -1 & -1 & 0 & 0 & 0 & 0 \\ 0 & 0 & 0 & 0 & 0 & 0 & 0 & 0 \end{bmatrix}. \qquad (11.22)$$

The reconstructed image block, formed by dequantizing and inverse-transforming Y_Q, is shown in Figure 11.8(a). Small differences can be seen in comparison with the original block, but it is more faithful than the low-pass filtering reconstruction.

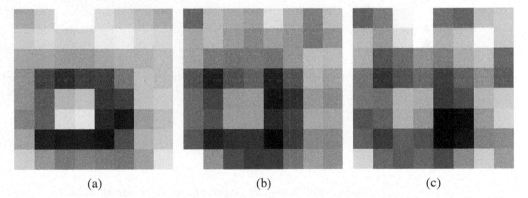

(a)   (b)   (c)

Figure 11.8 Result of linear quantization. Loss parameter is (a) $p = 1$ (b) $p = 2$ (c) $p = 4$.

After linear quantization with $p = 2$, the quantized transform coefficients are

$$Y_Q = \begin{bmatrix} -8 & -2 & 3 & -1 & 0 & 1 & 0 & 0 \\ 6 & 0 & -2 & 0 & 0 & -1 & 0 & 0 \\ 2 & 1 & 0 & 1 & 0 & -1 & 0 & 0 \\ 0 & 0 & -1 & -1 & 0 & 0 & 0 & 0 \\ 0 & 0 & 0 & 0 & 0 & 0 & 0 & 0 \\ -1 & -1 & 0 & 0 & 0 & 0 & 0 & 0 \\ 1 & 0 & 0 & 0 & 0 & 0 & 0 & 0 \\ 0 & 0 & 0 & 0 & 0 & 0 & 0 & 0 \end{bmatrix}, \qquad (11.23)$$

and after linear quantization with $p = 4$, the quantized transform coefficients are

$$Y_Q = \begin{bmatrix} -4 & -1 & 1 & -1 & 0 & 1 & 0 & 0 \\ 3 & 0 & -1 & 0 & 0 & 0 & 0 & 0 \\ 1 & 0 & 0 & 0 & 0 & 0 & 0 & 0 \\ 0 & 0 & 0 & -1 & 0 & 0 & 0 & 0 \\ 0 & 0 & 0 & 0 & 0 & 0 & 0 & 0 \\ -1 & 0 & 0 & 0 & 0 & 0 & 0 & 0 \\ 0 & 0 & 0 & 0 & 0 & 0 & 0 & 0 \\ 0 & 0 & 0 & 0 & 0 & 0 & 0 & 0 \end{bmatrix}. \tag{11.24}$$

Figure 11.8 shows the result of linear quantization for the three different values of loss parameter p. Notice that the larger the value of the loss parameter p, the more entries of the matrix Y_Q are zeroed by the quantization procedure, the smaller are the data requirements for representing the pixels, and the less faithfully the original image has been reconstructed.

Next, we quantize all $32 \times 32 = 1024$ blocks of the image in Figure 11.5. That is, we carry out 1024 independent versions of the previous example. The results for loss parameter $p = 1, 2$, and 4 are shown in Figure 11.9. The image has begun to deteriorate significantly by $p = 4$.

We can make a rough calculation to quantify the amount of image compression due to quantization. The original image uses a pixel value from 0 to 255, which is one byte, or 8 bits. For each 8×8 block, the total number of bits needed without compression is $8(8)^2 = 512$ bits.

Now, assume that linear quantization is used with loss parameter $p = 1$. Assume that the maximum entry of the transform Y is 255. Then the largest possible entries of Y_Q, after quantization by Q, are

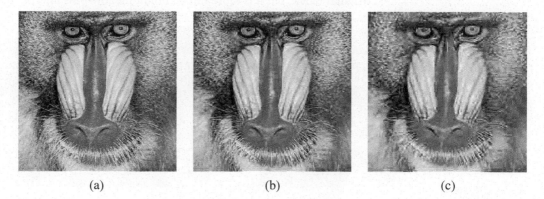

(a)   (b)   (c)

Figure 11.9 Result of linear quantization for all 1024 8×8 blocks. Loss parameters are (a) $p = 1$ (b) $p = 2$ (c) $p = 4$.

$$\begin{bmatrix} 32 & 16 & 11 & 8 & 6 & 5 & 5 & 4 \\ 16 & 11 & 8 & 6 & 5 & 5 & 4 & 4 \\ 11 & 8 & 6 & 5 & 5 & 4 & 4 & 3 \\ 8 & 6 & 5 & 5 & 4 & 4 & 3 & 3 \\ 6 & 5 & 5 & 4 & 4 & 3 & 3 & 3 \\ 5 & 5 & 4 & 4 & 3 & 3 & 3 & 2 \\ 5 & 4 & 4 & 3 & 3 & 3 & 2 & 2 \\ 4 & 4 & 3 & 3 & 3 & 2 & 2 & 2 \end{bmatrix}.$$

Since both positive and negative entries are possible, the number of bits necessary to store each entry is

$$
\begin{bmatrix}
7 & 6 & 5 & 5 & 4 & 4 & 4 & 4 \\
6 & 5 & 5 & 4 & 4 & 4 & 4 & 4 \\
5 & 5 & 4 & 4 & 4 & 4 & 4 & 3 \\
5 & 4 & 4 & 4 & 4 & 4 & 3 & 3 \\
4 & 4 & 4 & 4 & 4 & 3 & 3 & 3 \\
4 & 4 & 4 & 4 & 3 & 3 & 3 & 3 \\
4 & 4 & 4 & 3 & 3 & 3 & 3 & 3 \\
4 & 4 & 3 & 3 & 3 & 3 & 3 & 3
\end{bmatrix} .
$$

The sum of these 64 numbers is 249, or $249/64 \approx 3.89$ bits/pixel, which is less than one-half the number of bits (512, or 8 bits/pixel) needed to store the original pixel values of the 8×8 image matrix. The corresponding statistics for other values of p are shown in the following table:

p	total bits	bits/pixel
1	249	3.89
2	191	2.98
4	147	2.30

As seen in the table, the number of bits necessary to represent the image is reduced by a factor of 2 when $p = 1$, with little recognizable change in the image. This compression is due to quantization. In order to compress further, we can take advantage of the fact that many of the high-frequency terms in the transform are zero after quantization. This is most efficiently done by using Huffman and run-length coding, introduced in the next section.

Linear quantization with $p = 1$ is close to the default JPEG quantization. The quantization matrix that provides the most compression with the least image degradation has been the subject of much research and discussion. The JPEG standard includes an appendix called "Annex K: Examples and Guidelines," which contains a Q based on experiments with the human visual system. The matrix

$$
Q_Y = p
\begin{bmatrix}
16 & 11 & 10 & 16 & 24 & 40 & 51 & 61 \\
12 & 12 & 14 & 19 & 26 & 58 & 60 & 55 \\
14 & 13 & 16 & 24 & 40 & 57 & 69 & 56 \\
14 & 17 & 22 & 29 & 51 & 87 & 80 & 62 \\
18 & 22 & 37 & 56 & 68 & 109 & 103 & 77 \\
24 & 35 & 55 & 64 & 81 & 104 & 113 & 92 \\
49 & 64 & 78 & 87 & 103 & 121 & 120 & 101 \\
72 & 92 & 95 & 98 & 112 & 100 & 103 & 99
\end{bmatrix}
\tag{11.25}
$$

is widely used in currently distributed JPEG encoders. Setting the loss parameter $p = 1$ should give virtually perfect reconstruction as far as the human visual system is concerned, while $p = 4$ usually introduces noticeable defects. To some extent, the visual quality depends on the pixel size: If the pixels are small, some errors may go unnoticed.

So far, we have discussed grayscale images only. It is fairly easy to extend application to color images, which can be expressed in the RGB color system. Each pixel is assigned three integers, one each for red, green, and blue intensity. One approach to image compression is to repeat the preceding processing independently for each of the three colors, treating each as if it were gray scale, and then to reconstitute the image from its three colors at the end.

Although the JPEG standard does not take a position on how to treat color, the method often referred to as Baseline JPEG uses a more delicate approach. Define the **luminance** $Y = 0.299R + 0.587G + 0.114B$ and the **color differences** $U = B - Y$ and $V = R - Y$. This transforms the RGB color data to the YUV system. This is a completely reversible transform, since the RGB values can be found as $B = U + Y$, $R = V + Y$, and $G = (Y - 0.299R - 0.114B)/(0.587)$. Baseline JPEG applies the DCT filtering previously discussed independently to Y, U, and V, using the quantization matrix Q_Y from Annex K for the luminance variable Y and the quantization matrix

$$Q_C = \begin{bmatrix} 17 & 18 & 24 & 47 & 99 & 99 & 99 & 99 \\ 18 & 21 & 26 & 66 & 99 & 99 & 99 & 99 \\ 24 & 26 & 56 & 99 & 99 & 99 & 99 & 99 \\ 47 & 66 & 99 & 99 & 99 & 99 & 99 & 99 \\ 99 & 99 & 99 & 99 & 99 & 99 & 99 & 99 \\ 99 & 99 & 99 & 99 & 99 & 99 & 99 & 99 \\ 99 & 99 & 99 & 99 & 99 & 99 & 99 & 99 \\ 99 & 99 & 99 & 99 & 99 & 99 & 99 & 99 \end{bmatrix} \tag{11.26}$$

for the color differences U and V. After reconstructing Y, U, and V, they are put back together and converted back to RGB to reconstitute the image.

Because of the less important roles of U and V in the human visual system, more aggressive quantization is allowed for them, as seen in (11.26). Further compression can be derived from an array of additional ad hoc tricks—for example, by averaging the color differences and treating them on a less fine grid.

► **ADDITIONAL EXAMPLES**

1. Find the 2D-DCT of the following matrix X and write the DCT interpolating function $P_4(s, t)$.

$$X = \begin{bmatrix} 1 & 0 & 0 & 1 \\ 0 & 1 & 1 & 0 \\ 0 & 1 & 1 & 0 \\ 1 & 0 & 0 & 1 \end{bmatrix}$$

2. (a) Find the 2D-DCT of the matrix X. (b) Find the least-squares low-pass filtered approximation to X by setting $Y_{kl} = 0$ for $k + l \geq 4$. (c) Same as part (b), but set $Y_{kl} = 0$ for $k + l \geq 3$.

$$X = \begin{bmatrix} 1 & 2 & 3 & 4 \\ 2 & 2 & 2 & 2 \\ 4 & 3 & 2 & 1 \\ 6 & 5 & 4 & 3 \end{bmatrix}$$

▭ **Solutions** for Additional Examples can be found at bit.ly/2P4ulzd

11.2 Exercises

▭ **Solutions** for Exercises numbered in blue can be found at bit.ly/2Pfk9nC

1. Find the 2D-DCT of the following data matrices X, and find the corresponding interpolating function $P_2(s, t)$ for the data points $(i, j, x_{ij}), i, j = 0, 1$:

(a) $\begin{bmatrix} 1 & 0 \\ 0 & 0 \end{bmatrix}$ (b) $\begin{bmatrix} 1 & 0 \\ 1 & 0 \end{bmatrix}$ (c) $\begin{bmatrix} 1 & 1 \\ 1 & 1 \end{bmatrix}$ (d) $\begin{bmatrix} 1 & 0 \\ 0 & 1 \end{bmatrix}$

2. Find the 2D-DCT of the data matrix X, and find the corresponding interpolating function $P_n(s, t)$ for the data points $(i, j, x_{ij}), i, j = 0, \ldots, n - 1$.

(a) $\begin{bmatrix} 1 & 0 & -1 & 0 \\ 1 & 0 & -1 & 0 \\ 1 & 0 & -1 & 0 \\ 1 & 0 & -1 & 0 \end{bmatrix}$ (b) $\begin{bmatrix} 1 & 0 & 0 & 0 \\ 0 & 1 & 0 & 0 \\ 0 & 0 & 1 & 0 \\ 0 & 0 & 0 & 1 \end{bmatrix}$

(c) $\begin{bmatrix} 0 & 0 & 0 & 0 \\ 0 & 1 & 1 & 0 \\ 0 & 1 & 1 & 0 \\ 0 & 0 & 0 & 0 \end{bmatrix}$ (d) $\begin{bmatrix} 3 & 3 & 3 & 3 \\ 3 & -1 & -1 & 3 \\ 3 & 3 & 3 & 3 \\ 3 & -1 & -1 & 3 \end{bmatrix}$

3. Find the least squares approximation, using the basis functions $1, \cos\frac{(2s+1)\pi}{8}, \cos\frac{(2t+1)\pi}{8}$ for the data in Exercise 2.

4. Use the quantization matrix $Q = \begin{bmatrix} 10 & 20 \\ 20 & 100 \end{bmatrix}$ to quantize the matrices that follow. State the quantized matrix, the (lossy) dequantized matrix, and the matrix of quantization errors.

(a) $\begin{bmatrix} 24 & 24 \\ 24 & 24 \end{bmatrix}$ (b) $\begin{bmatrix} 32 & 28 \\ 28 & 45 \end{bmatrix}$ (c) $\begin{bmatrix} 54 & 54 \\ 54 & 54 \end{bmatrix}$

11.2 Computer Problems

Solutions for Computer Problems numbered in blue can be found at bit.ly/2CrRftA

1. Find the 2D-DCT of the data matrix X.

(a) $\begin{bmatrix} -1 & 1 & -1 & 1 \\ -2 & 2 & -2 & 2 \\ -3 & 3 & -3 & 3 \\ -4 & 4 & -4 & 4 \end{bmatrix}$ (b) $\begin{bmatrix} 1 & 2 & -1 & -2 \\ -1 & -2 & 1 & 2 \\ 1 & 2 & -1 & -2 \\ -1 & -2 & 1 & 2 \end{bmatrix}$

(c) $\begin{bmatrix} 1 & 3 & 1 & -1 \\ 2 & 1 & 0 & 1 \\ 1 & -1 & 2 & 3 \\ 3 & 2 & 1 & 0 \end{bmatrix}$ (d) $\begin{bmatrix} -3 & -2 & -1 & 0 \\ -2 & -1 & 0 & 1 \\ -1 & 0 & 1 & 2 \\ 0 & 1 & 2 & 3 \end{bmatrix}$

2. Using the 2D-DCT from Computer Problem 1, find the least squares low-pass filtered approximation to X by setting all transform values $Y_{kl} = 0$ for $k + l \geq 4$.

3. Obtain a grayscale image file of your choice, and use the imread command to import into MATLAB. Crop the resulting matrix so that each dimension is a multiple of 8. If necessary, converting a color RGB image to gray scale can be accomplished by the standard formula (11.15).

   (a) Extract an 8×8 pixel block, for example, by using the MATLAB command xb=x(81:88,81:88). Display the block with the imagesc command.

   (b) Apply the 2D-DCT.

   (c) Quantize by using linear quantization with $p = 1, 2$, and 4. Print out each Y_Q.

   (d) Reconstruct the block by using the inverse 2D-DCT, and compare with the original. Use MATLAB commands colormap(gray) and imagesc(X,[0 255]).

   (e) Carry out (a)–(d) for all 8×8 blocks, and reconstitute the image in each case.

4. Carry out the steps of Computer Problem 3, but quantize by the JPEG-suggested matrix (11.25) with $p = 1$.

5. Obtain a color image file of your choice. Carry out the steps of Computer Problem 3 for colors R, G, and B separately, using linear quantization, and recombine as a color image.

6. Obtain a color image, and transform the RGB values to luminance/color difference coordinates. Carry out the steps of Computer Problem 3 for Y, U, and V separately by using JPEG quantization, and recombine as a color image.

11.3 HUFFMAN CODING

Lossy compression for images requires making a trade of accuracy for file size. If the reductions in accuracy are small enough to be unnoticeable for the intended purpose of the image, the trade may be worthwhile. The loss of accuracy occurs at the quantization step, after transforming to separate the image into its spatial frequencies. Lossless compression refers to further compression that may be applied without losing any more accuracy, simply due to efficient coding of the DCT-transformed, quantized image.

In this section, we discuss lossless compression. As a relevant application, there are simple, efficient methods for turning the quantized DCT transform matrix from the last section into a JPEG bit stream. Finding out how to do this will take us on a short tour of basic information theory.

11.3.1 Information theory and coding

Consider a message consisting of a string of symbols. The symbols are arbitrary; let us assume that they come from a finite set. In this section, we consider efficient ways to encode such a string in binary digits, or bits. The shorter the string of bits, the easier and cheaper it will be to store or transmit the message.

▶ **EXAMPLE 11.5** Encode the message ABAACDAB as a binary string.

Since there are four symbols, a convenient binary coding might associate two bits with each letter. For example, we could choose the correspondence

A	00
B	01
C	10
D	11

Then the message would be coded as

$$(00)(01)(00)(00)(10)(11)(00)(01).$$

With this code, a total of 16 bits is required to store or transmit the message. ◀

It turns out that there are more efficient coding methods. To understand them, we first have to introduce the idea of information. Assume that there are k different symbols, and denote by p_i the probability of the appearance of symbol i at any point in the string. The probability might be known a priori, or it may be estimated empirically by dividing the number of appearances of symbol i in the string by the length of the string.

DEFINITION 11.7 The **Shannon information**, or **Shannon entropy** of the string is $I = -\sum_{i=1}^{k} p_i \log_2 p_i$. ❑

The definition is named after C. Shannon of Bell Laboratories, who did seminal work on information theory in the middle of the 20th century. The Shannon information of a string is considered an average of the number of bits per symbol that is needed, at minimum, to code the message. The logic is as follows: On average, if a symbol appears p_i of the time, then one expects to need $-\log_2 p_i$ bits to represent it. For example, a symbol that appears 1/8 of the time could be represented by one of the $-\log_2(1/8) = 3$-bit symbols $000, 001, \ldots, 111$, of which there are 8. To find the average bits per symbol over all symbols, we should weight the bits per symbol i by its probability p_i. This means that the average number of bits/symbol for the entire message is the sum I in the definition.

▶ **EXAMPLE 11.6**   Find the Shannon information of the string ABAACDAB.

The empirical probabilities of appearance of the symbols A, B, C, D are $p_1 = 4/8 = 2^{-1}, p_2 = 2/8 = 2^{-2}, p_3 = 1/8 = 2^{-3}, p_4 = 2^{-3}$, respectively. The Shannon information is

$$-\sum_{i=1}^{4} p_i \log_2 p_i = \frac{1}{2}1 + \frac{1}{4}2 + \frac{1}{8}3 + \frac{1}{8}3 = \frac{7}{4}.$$ ◀

Thus, Shannon information estimates that at least 1.75 bits/symbol are needed to code the string. Since the string has length 8, the optimal total number of bits should be $(1.75)(8) = 14$, not 16, as we coded the string earlier.

In fact, the message can be sent in the predicted 14 bits, using the method known as **Huffman coding**. The goal is to assign a unique binary code to each symbol that reflects the probability of encountering the symbol, with more common symbols receiving shorter codes.

The algorithm works by building a tree from which the binary code can be read. Begin with two symbols with the smallest probability, and consider the "combined" symbol, assigning to it the combined probability. The two symbols form one branching of the tree. Then repeat this step, combining symbols and working up the branches of the tree, until there is only one symbol group left, which corresponds to the top of the tree. Here, we first combined the least probable symbols C and D into a symbol CD with probability 1/4. The remaining probabilities are A (1/2), B (1/4), and CD (1/4). Again, we combine the two least likely symbols to get A (1/2), BCD (1/2). Finally, combining the remaining two gives ABCD (1). Each combination forms a branch of the Huffman tree:

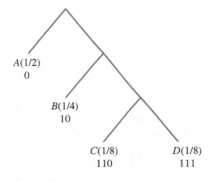

Once the tree is completed, the Huffman code for each symbol can be read by traversing the tree from the top, writing a 0 for a branch to the left and a 1 for a branch to the right, as shown above. For example, A is represented by 0, and C is represented by two rights and a left, 110. Now the string of letters ABAACDAB can be translated to a bit stream of length 14:

$$(0)(10)(0)(0)(110)(111)(0)(10).$$

The Shannon information of the message provides a lower bound for the bits/symbol of the binary coding. In this case, the Huffman code has achieved the Shannon information bound of $14/8 = 1.75$ bits/symbol. Unfortunately, this is not always possible, as the next example shows.

▶ **EXAMPLE 11.7** Find the Shannon information and a Huffman coding of the message ABRA CADABRA.

The empirical probabilities of the six symbols are

A	5/12
B	2/12
R	2/12
C	1/12
D	1/12
_	1/12

Note that the space has been treated as a symbol. The Shannon information is

$$-\sum_{i=1}^{6} p_i \log_2 p_i = -\frac{5}{12}\log_2 \frac{5}{12} - 2\frac{1}{6}\log_2 \frac{1}{6} - 3\frac{1}{12}\log_2 \frac{1}{12} \approx 2.28 \text{ bits/symbol.}$$

This is the theoretical minimum for the average bits/symbol for coding the message ABRA CADABRA. To find the Huffman coding, proceed as already described. We begin by combining the symbols D and __, although any two of the three with probability $1/12$ could have been chosen for the lowest branch. The symbol A comes in last, since it has highest probability. One Huffman coding is displayed in the diagram.

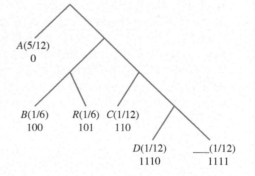

Note that A has a short code, due to the fact that it is a popular symbol in the message. The coded binary sequence for ABRA CADABRA is

$$(0)(100)(101)(0)(1111)(110)(0)(1110)(0)(100)(101)(0),$$

which has length 28 bits. The average for this coding is $28/12 = 2\frac{1}{3}$ bits/symbol, slightly larger than the theoretical minimum previously calculated. Huffman codes cannot always match the Shannon information, but they often come very close. ◀

The secret of a Huffman code is the following: Since each symbol occurs only at the end of a tree branch, no complete symbol code can be the beginning of another symbol code. Therefore, there is no ambiguity when translating the code back into symbols.

11.3.2 Huffman coding for the JPEG format

This section is devoted to an extended example of Huffman coding in practice. The JPEG image compression format is ubiquitous in modern digital photography. It makes a fascinating case study due to the juxtaposition of theoretical mathematics and engineering considerations.

The binary coding of transform coefficients for a JPEG image file uses Huffman coding in two different ways, one for the DC component (the $(0, 0)$ entry of the transform matrix) and another for the other 63 entries of the 8×8 matrix, the so-called AC components.

DEFINITION 11.8   Let y be an integer. The **size** of y is defined to be

$$L = \begin{cases} \text{floor}(\log_2 |y|) + 1 & \text{if } y \neq 0 \\ 0 & \text{if } y = 0 \end{cases}.$$ ⬜

Huffman coding for JPEG has three ingredients: a Huffman tree for the DC components, another Huffman tree for the AC components, and an integer identifier table. The first part of the coding for the entry $y = y_{00}$ is the binary coding for the size of y, from the following Huffman tree for DC components, called the **DPCM tree**, for Differential Pulse Code Modulation.

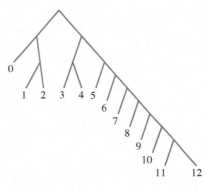

Again, the tree is to be interpreted by coding a 0 or 1 when going down a branch to the left or right, respectively. The first part is followed by a binary string from the following integer identifier table:

L	entry	binary
0	0	- -
1	$-1,1$	0,1
2	$-3,-2,2,3$	00,01,10,11
3	$-7,-6,-5,-4,4,5,6,7$	000,001,010,011,100,101,110,111
4	$-15,-14,\ldots,-8,8,\ldots,14,15$	0000,0001,......0111,1000,......,1110,1111
5	$-31,-30,\ldots,-16,16,\ldots,30,31$	00000,00001,......,01111,10000......,11110,11111
6	$-63,-62,\ldots,-32,32,\ldots,62,63$	000000,000001,...,011111,100000,...,111110,111111
⋮	⋮	⋮

As an example, the entry $y_{00} = 13$ would have size $L = 4$. According to the DPCM tree, the Huffman code for 4 is (101). The table shows that the extra digits for 13 are (1101), so the concatenation of the two parts, 1011101, would be stored for the DC component.

Since there are often correlations between the DC components of nearby 8×8 blocks, only the differences from block to block are stored after the first block. The differences are stored, moving from left to right, using the DPCM tree.

For the remaining 63 AC components of the 8×8 block, **Run Length Encoding (RLE)** is used as a way to efficiently store long runs of zeros. The conventional order for storing the 63 components is the zigzag pattern

$$\begin{bmatrix} 0 & 1 & 5 & 6 & 14 & 15 & 27 & 28 \\ 2 & 4 & 7 & 13 & 16 & 26 & 29 & 42 \\ 3 & 8 & 12 & 17 & 25 & 30 & 41 & 43 \\ 9 & 11 & 18 & 24 & 31 & 40 & 44 & 53 \\ 10 & 19 & 23 & 32 & 39 & 45 & 52 & 54 \\ 20 & 22 & 33 & 38 & 46 & 51 & 55 & 60 \\ 21 & 34 & 37 & 47 & 50 & 56 & 59 & 61 \\ 35 & 36 & 48 & 49 & 57 & 58 & 62 & 63 \end{bmatrix}. \tag{11.27}$$

Instead of coding the 63 numbers themselves, a zero run–length pair (n, L) is coded, where n denotes the length of a run of zeros, and L represents the size of the next nonzero entry. The most common codes encountered in typical JPEG images, and their default codings according to the JPEG standard, are shown in the Huffman tree for AC components.

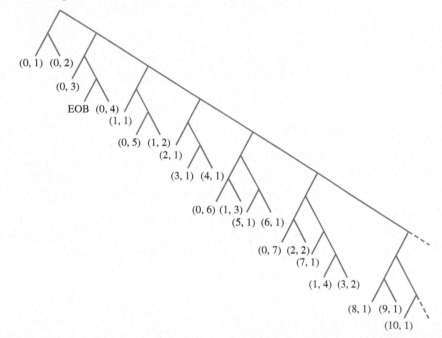

In the bit stream, the Huffman code from the tree (which only identifies the size of the entry) is immediately followed by the binary code identifying the integer, from the previous table. For example, the sequence of entries $-5, 0, 0, 0, 2$ would be represented as $(0, 3) -5 (3, 2) 2$, where $(0, 3)$ means no zeros followed by a size 3 number, and $(3, 2)$ represents 3 zeros followed by a size 2 number. From the Huffman tree, we find that $(0, 3)$ codes as (100), and $(3, 2)$ as (111110111). The identifier for -5 is (010) and for 2 is (10), from the integer identifier table. Therefore, the bit stream used to code $-5, 0, 0, 0, 2$ is $(100)(010)(111110111)(10)$.

The preceding Huffman tree shows only the most commonly occurring JPEG run-length codes. Other useful codes are $(11, 1) = 1111111001$, $(12, 1) = 1111111010$, and $(13, 1) = 11111111000$.

▶ **EXAMPLE 11.8**   Code the quantized DCT transform matrix in (11.24) for a JPEG image file.

The DC entry $y_{00} = -4$ has size 3, coded as (100) by the DPCM tree, and extra bits (011) from the integer identifier table. Next, we consider the AC coefficient string. According to (11.27), the AC coefficients are ordered as $-1, 3, 1, 0, 1, -1, -1$, seven zeros, 1, four zeros, -1, three zeros, -1, and the remainder all zeros. The run-length encoding begins with -1, which has size 1 and so contributes $(0, 1)$ from the run-length code. The next number 3 has size 2 and contributes $(0, 2)$. The zero run-length pairs are

$$(0, 1) -1 \ (0, 2) \ 3 \ (0, 1) \ 1 \ (1, 1) \ 1 \ (0, 1) -1 \ (0, 1) -1$$
$$(7, 1) \ 1 \ (4, 1) -1 \ (3, 1) -1 \ \text{EOB}.$$

Here, EOB stands for "end-of-block" and means that the remainder of the entries consists of zeros. Next, we read the bit representatives from the Huffman tree on page 542 and the integer identifier table. The bit stream that stores the 8×8 block from the photo in Figure 11.8(c) is listed below, where the parentheses are included only for human readability:

$$(100)(011)$$
$$(00)(0)(01)(11)(00)(1)(1100)(1)(00)(0)(00)(0)$$
$$(11111010)(1)(111011)(0)(111010)(0)(1010)$$

The pixel block in Figure 11.8(c), which is a reasonable approximation of the original Figure 11.6(a), is exactly represented by these 54 bits. On a per-pixel basis, this works out to $54/64 \approx 0.84$ bits/pixel. Note the superiority of this coding to the bits/pixel achieved by low-pass filtering and quantization alone. Given that the pixels started out as 8-bit integers, the 8×8 image has been compressed by more than a factor of 9:1. ◄

Decompressing a JPEG file consists of reversing the compression steps. The JPEG reader decodes the bit stream to run-length symbols, which form 8×8 DCT transform blocks that in turn are finally converted back to pixel blocks with the use of the inverse DCT.

▶ **ADDITIONAL EXAMPLES**

1. For the phrase NUMERICAL ANALYSIS, find the probability of each symbol and the Shannon information of the phrase.

2. Find the binary code for the following quantized DCT matrix using the JPEG format for an image file.

$$\begin{bmatrix} 5 & 1 & 1 & 0 & 0 & 0 & 0 & 0 \\ 2 & 0 & 0 & 0 & -1 & 0 & 0 & 0 \\ -1 & 0 & 1 & 0 & 0 & 0 & 0 & 0 \\ 0 & -1 & 0 & 0 & 0 & 0 & 0 & 0 \\ 0 & 0 & 0 & 0 & 0 & 0 & 0 & 0 \\ 1 & 0 & 0 & 0 & 0 & 0 & 0 & 0 \\ 0 & 0 & 0 & 0 & 0 & 0 & 0 & 0 \\ 0 & 0 & 0 & 0 & 0 & 0 & 0 & 0 \end{bmatrix}$$

 Solutions for Additional Examples can be found at bit.ly/2PKaAda

11.3 Exercises

⊑⊓ **Solutions**
for Exercises
numbered in blue
can be found at
bit.ly/2q2hdw0

1. Find the probability of each symbol and the Shannon information for the messages.
   (a) BABBCABB (b) ABCACCAB (c) ABABCABA

2. Draw a Huffman tree and use it to code the messages in Exercise 1. Compare the Shannon information with the average number of bits needed per symbol.

3. Draw a Huffman tree and convert the message, including spaces and punctuation marks, to a bit stream by using Huffman coding. Compare the Shannon information with the average number of bits needed per symbol. (a) AY CARUMBA! (b) COMPRESS THIS MESSAGE (c) SHE SELLS SEASHELLS BY THE SEASHORE

4. Translate the transformed, quantized image components (a) (11.22) and (b) (11.23) to bit streams, using JPEG Huffman coding.

11.4 MODIFIED DCT AND AUDIO COMPRESSION

We return to the problem of one-dimensional signals and discuss state-of-the-art approaches to audio compression. Although one might think that one dimension is easier to handle than two, the challenge is that the human auditory system is very sensitive in the frequency domain, and unwanted artifacts introduced by compression and decompression are even more readily detected. For that reason, it is common for sound compression methods to make use of sophisticated tricks designed to hide the fact that compression has occurred.

First we introduce DCT4, a new version of the Discrete Cosine Transform, and the so-called Modified Discrete Cosine Transform (MDCT). The MDCT is represented by a matrix that is not square and so, unlike the DCT and DCT4, is not invertible. However, when applied on overlapping windows, it can be used to completely reconstruct the original data stream. More importantly, it can be combined with quantization to carry out lossy compression with minimal degradation of sound quality. The MDCT is at the core of most of the current widely supported sound compression formats, such as MP3, AAC, and WMA.

11.4.1 Modified Discrete Cosine Transform

We begin with a slightly different form of the DCT introduced earlier. There are four different versions of the DCT that are commonly used—we used version DCT1 for image compression in the previous section. Version DCT4 is most popular for sound compression.

DEFINITION 11.9   The **Discrete Cosine Transform (version 4)** (DCT4) of $x = (x_0, \ldots, x_{n-1})^T$ is the n-dimensional vector

$$y = Ex,$$

where E is the $n \times n$ matrix

$$E_{ij} = \sqrt{\frac{2}{n}} \cos \frac{(i + \frac{1}{2})(j + \frac{1}{2})\pi}{n}. \qquad (11.28)$$

□

Just as in the DCT1, the matrix E in DCT4 is a real orthogonal matrix: It is square and its columns are pairwise orthogonal unit vectors. The latter follows from the fact that the columns of E are the unit eigenvectors of the real symmetric $n \times n$ matrix

$$\begin{bmatrix} 1 & -1 & & & & \\ -1 & 2 & -1 & & & \\ & -1 & 2 & -1 & & \\ & & \ddots & \ddots & \ddots & \\ & & & -1 & 2 & -1 \\ & & & & -1 & 3 \end{bmatrix}. \tag{11.29}$$

Exercise 6 asks the reader to verify this fact.

Next, we note two important facts about the columns of the DCT4 matrix. Treat n as fixed, and consider not only the n columns in DCT4, but the column vectors defined by (11.28) for all positive and negative integers j.

LEMMA 11.10   Denote by c_j the jth column of the (extended) DCT4 matrix (11.28). Then (a) $c_j = c_{-1-j}$ for all integers j (the columns are symmetric around $j = -\frac{1}{2}$), and (b) $c_j = -c_{2n-1-j}$ for all integers j (the columns are antisymmetric around $j = n - \frac{1}{2}$). ■

Proof.   To prove part (a) of the lemma, write $j = -\frac{1}{2} + (j + \frac{1}{2})$ and $-1 - j = -\frac{1}{2} - (j + \frac{1}{2})$. Using equation (11.28) yields

$$c_j = c_{-\frac{1}{2}+(j+\frac{1}{2})} = \sqrt{\frac{2}{n}} \cos \frac{(i+\frac{1}{2})(j+\frac{1}{2})\pi}{n} = \sqrt{\frac{2}{n}} \cos \frac{(i+\frac{1}{2})(-j-\frac{1}{2})\pi}{n}$$

$$= c_{-\frac{1}{2}-(j+\frac{1}{2})} = c_{-1-j}$$

for $i = 0, \ldots, n - 1$.

For the proof of (b), set $r = n - \frac{1}{2} - j$. Then $j = n - \frac{1}{2} - r$ and $2n - 1 - j = n - \frac{1}{2} + r$, and we must show that $c_{n-\frac{1}{2}-r} + c_{n-\frac{1}{2}+r} = 0$. By the cosine addition formula,

$$c_{n-\frac{1}{2}-r} = \sqrt{\frac{2}{n}} \cos \frac{(2i+1)(n-r)\pi}{2n}$$

$$= \sqrt{\frac{2}{n}} \cos \frac{2i+1}{2}\pi \cos \frac{(2i+1)r\pi}{2n} + \sqrt{\frac{2}{n}} \sin \frac{2i+1}{2}\pi \sin \frac{(2i+1)r\pi}{2n}$$

$$c_{n-\frac{1}{2}+r} = \sqrt{\frac{2}{n}} \cos \frac{(2i+1)(n+r)\pi}{2n}$$

$$= \sqrt{\frac{2}{n}} \cos \frac{2i+1}{2}\pi \cos \frac{(2i+1)r\pi}{2n} - \sqrt{\frac{2}{n}} \sin \frac{2i+1}{2}\pi \sin \frac{(2i+1)r\pi}{2n}$$

for $i = 0, \ldots, n - 1$. Since $\cos \frac{1}{2}(2i+1)\pi = 0$ for all integers i, the sum $c_{n-\frac{1}{2}-r} + c_{n-\frac{1}{2}+r} = 0$, as claimed. □

We will use the DCT4 matrix E to build the Modified Discrete Cosine Transform. Assume that n is even. We are going to create a new matrix, using the columns $c_{\frac{n}{2}}, \ldots, c_{\frac{5}{2}n-1}$. Lemma 11.10 shows that for any integer j, the column c_j can be expressed as one of the columns of DCT4—that is, one of the c_j for $0 \leq i \leq n - 1$, as shown in Figure 11.10, up to a possible sign change

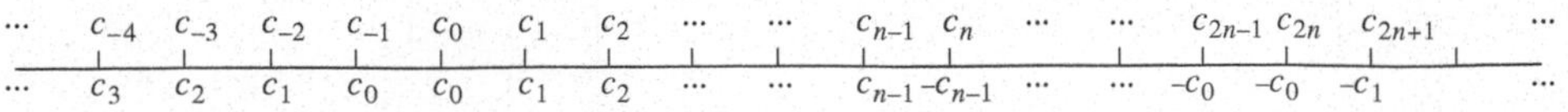

$$\begin{array}{ccccccccccccccc}
\cdots & c_{-4} & c_{-3} & c_{-2} & c_{-1} & c_0 & c_1 & c_2 & \cdots & \cdots & c_{n-1} & c_n & \cdots & \cdots & c_{2n-1} \ c_{2n} \ c_{2n+1} & \cdots \\
\cdots & c_3 & c_2 & c_1 & c_0 & c_0 & c_1 & c_2 & \cdots & \cdots & c_{n-1} \ -c_{n-1} & \cdots & \cdots & -c_0 & -c_0 & -c_1 & \cdots
\end{array}$$

Figure 11.10 Illustration of Lemma 11.10. The columns $c_0, \ldots, c_{n-1}$ make up the $n \times n$ DCT4 matrix. For integers j outside that range, the column defined by c_j in equation (11.28) still corresponds to one of the n columns of DCT4, shown directly below it in the Figure. This illustrates Lemma 11.10.

DEFINITION 11.11 Let n be an even positive integer. The **Modified Discrete Cosine Transform** (MDCT) of $x = (x_0, \ldots, x_{2n-1})^T$ is the n-dimensional vector

$$y = Mx, \tag{11.30}$$

where M is the $n \times 2n$ matrix

$$M_{ij} = \sqrt{\frac{2}{n}} \cos \frac{(i + \frac{1}{2})(j + \frac{n}{2} + \frac{1}{2})\pi}{n} \tag{11.31}$$

for $0 \le i \le n - 1$ and $0 \le j \le 2n - 1$. ☐

Note the major difference from the previous forms of the DCT: The MDCT of a length $2n$ vector is a length n vector. For this reason, the MDCT is not directly invertible, but we will see later that the same effect will be achieved by overlapping the length $2n$ vectors.

Comparing with Definition 11.9 allows us to write the MDCT matrix M in terms of the DCT4 columns and then simplify, using Lemma 11.10:

$$\begin{aligned}
M &= \left[c_{\frac{n}{2}} \cdots c_{\frac{5}{2}n-1} \right] \\
&= \left[c_{\frac{n}{2}} \cdots c_{n-1} | c_n \cdots c_{\frac{3}{2}n-1} | c_{\frac{3}{2}n} \cdots c_{2n-1} | c_{2n} \cdots c_{\frac{5}{2}n-1} \right] \\
&= \left[c_{\frac{n}{2}} \cdots c_{n-1} | -c_{n-1} \cdots -c_{\frac{n}{2}} | -c_{\frac{n}{2}-1} \cdots -c_0 | -c_0 \cdots -c_{\frac{n}{2}-1} \right]. \tag{11.32}
\end{aligned}$$

For example, the $n = 4$ MDCT matrix is

$$M = [c_2 c_3 | c_4 c_5 | c_6 c_7 | c_8 c_9] = [c_2 c_3 | -c_3 -c_2 | -c_1 -c_0 | -c_0 -c_1].$$

To simplify notation, let A and B denote the left and right halves of the DCT4 matrix, so that $E = [A|B]$. Define the permutation matrix formed by reversing the columns of the identity matrix, left for right:

$$R = \begin{bmatrix} & & 1 \\ & \iddots & \\ 1 & & \end{bmatrix}.$$

The permutation matrix R reverses columns right for left when multiplying a matrix on the right. When multiplying on the left, it reverses rows top to bottom. Note that R is a symmetric orthogonal matrix, since $R^{-1} = R^T = R$. Now (11.32) can be written more simply as

$$M = (B| - BR| - AR| - A), \tag{11.33}$$

where AR and BR are versions of A and B in which the order of the columns has been reversed, left for right.

The action of MDCT can be expressed in terms of DCT4. Let

$$
x = \begin{bmatrix} x_1 \\ x_2 \\ x_3 \\ x_4 \end{bmatrix}
$$

be a $2n$-vector, where each x_i is a length $n/2$ vector (remember that n is even). Then, by the characterization of M in (11.33),

$$
\begin{aligned}
Mx &= Bx_1 - BRx_2 - ARx_3 - Ax_4 \\
&= [A|B]\begin{bmatrix} -Rx_3 - x_4 \\ x_1 - Rx_2 \end{bmatrix} = E\begin{bmatrix} -Rx_3 - x_4 \\ x_1 - Rx_2 \end{bmatrix},
\end{aligned} \tag{11.34}
$$

where E is the $n \times n$ DCT4 matrix and Rx_2 and Rx_3 represent x_2 and x_3 with their entries reversed top to bottom. This is very helpful—we can express the output of M in terms of an orthogonal matrix E.

Since the $n \times 2n$ matrix M of the MDCT is not a square matrix, it is not invertible. However, two adjacent MDCT's can have rank $2n$ in total, and working together, can reconstruct the input x-values perfectly, as we now show.

The "inverse" MDCT is represented by the $2n \times n$ matrix $N = M^T$, which has transposed entries

$$
N_{ij} = \sqrt{\frac{2}{n}} \cos \frac{(j + \frac{1}{2})(i + \frac{n}{2} + \frac{1}{2})\pi}{n}. \tag{11.35}
$$

It is not an actual inverse, although it is as close as it can be for a rectangular matrix. By transposing (11.33), we have

$$
N = \begin{bmatrix} B^T \\ -RB^T \\ -RA^T \\ -A^T \end{bmatrix}, \tag{11.36}
$$

using our earlier notation $E = [A|B]$ for the DCT4. We know that since E is an orthogonal matrix,

$$
\begin{aligned}
A^T A &= I \\
B^T B &= I \\
A^T B = B^T A &= 0,
\end{aligned}
$$

where I denotes the $n \times n$ identity matrix.

Now we are ready to calculate NM, to see in what sense N inverts the MDCT matrix M. Let x be partitioned into four parts, as before. According to (11.34) and (11.36), the orthogonality of A and B, and the fact that $R^2 = I$, we have

$$
\begin{aligned}
NM \begin{bmatrix} x_1 \\ x_2 \\ x_3 \\ x_4 \end{bmatrix} &= \begin{bmatrix} B^T \\ -RB^T \\ -RA^T \\ -A^T \end{bmatrix} [A(-Rx_3 - x_4) + B(x_1 - Rx_2)] \\
&= \begin{bmatrix} x_1 - Rx_2 \\ -Rx_1 + x_2 \\ x_3 + Rx_4 \\ Rx_3 + x_4 \end{bmatrix}.
\end{aligned} \tag{11.37}
$$

In audio compression algorithms, MDCT is applied to vectors of data that overlap. The reason is that any artifacts due to the ends of the vectors will occur with a fixed frequency, because of the constant vector length. The auditory system is even more sensitive to periodic errors than the visual system; after all, an error of fixed frequency is a tone of that frequency, which the ear is designed to pick up. Assume that the data will be presented in overlapped fashion. Let

$$
Z_1 = \begin{bmatrix} x_1 \\ x_2 \\ x_3 \\ x_4 \end{bmatrix} \text{ and } Z_2 = \begin{bmatrix} x_3 \\ x_4 \\ x_5 \\ x_6 \end{bmatrix}
$$

be two $2n$-vectors for an even integer n, where each x_i is a length $n/2$ vector. The vectors Z_1 and Z_2 overlap by half of their length. Since (11.37) shows that

$$
NMZ_1 = \begin{bmatrix} x_1 - Rx_2 \\ -Rx_1 + x_2 \\ x_3 + Rx_4 \\ Rx_3 + x_4 \end{bmatrix} \text{ and } NMZ_2 = \begin{bmatrix} x_3 - Rx_4 \\ -Rx_3 + x_4 \\ x_5 + Rx_6 \\ Rx_5 + x_6 \end{bmatrix}, \tag{11.38}
$$

we can reconstruct the n-vector $[x_3, x_4]$ exactly by averaging the bottom half of NMZ_1 and the top half of NMZ_2:

$$
\begin{bmatrix} x_3 \\ x_4 \end{bmatrix} = \frac{1}{2}(NMZ_1)_{n,\dots,2n-1} + \frac{1}{2}(NMZ_2)_{0,\dots,n-1}. \tag{11.39}
$$

This equality is how N is used to decode the signal after being coded by M.

This result is summarized in Theorem 11.12.

THEOREM 11.12   **Inversion of MDCT through overlapping.** Let M be the $n \times 2n$ MDCT matrix, and $N = M^T$. Let u_1, u_2, u_3 be n-vectors, and set

$$
v_1 = M \begin{bmatrix} u_1 \\ u_2 \end{bmatrix} \text{ and } v_2 = M \begin{bmatrix} u_2 \\ u_3 \end{bmatrix}.
$$

Then the n-vectors w_1, w_2, w_3, w_4 defined by

$$
\begin{bmatrix} w_1 \\ w_2 \end{bmatrix} = Nv_1 \text{ and } \begin{bmatrix} w_3 \\ w_4 \end{bmatrix} = Nv_2
$$

satisfy $u_2 = \frac{1}{2}(w_2 + w_3)$. ■

This is exact reconstruction. Theorem 11.12 is customarily used with a long signal of concatenated n-vectors $[u_1, u_2, \dots, u_m]$. The MDCT is applied to adjacent pairs to get a transformed signal $(v_1, v_2, \dots, v_{m-1})$. Now the lossy compression comes in. The v_i are frequency components, so we can choose to keep certain frequencies and de-emphasize others. We will take up this direction in the next section.

After shrinking the content of the v_i by quantization or other means, $(u_2, \dots, u_{m-1})$ can be decompressed by Theorem 11.12. Note that we cannot recover u_1 and u_m; they should either be unimportant parts of the signal or padding that is added beforehand.

▶ **EXAMPLE 11.9**   Use the overlapped MDCT to transform the signal $x = [1, 2, 3, 4, 5, 6]$. Then invert the transform to reconstruct the middle section $[3, 4]$.

We will overlap the vectors $[1, 2, 3, 4]$ and $[3, 4, 5, 6]$. Let $n = 2$ and set

$$E_2 = \begin{bmatrix} \cos\frac{\pi}{8} & \cos\frac{3\pi}{8} \\ \cos\frac{3\pi}{8} & \cos\frac{9\pi}{8} \end{bmatrix} = \begin{bmatrix} b & c \\ c & -b \end{bmatrix}.$$

Note that our definitions of b and c have changed slightly from (11.7) to be compatible with the MDCT. Applying the 2×4 MDCT gives

$$v_1 = M \begin{bmatrix} 1 \\ 2 \\ 3 \\ 4 \end{bmatrix} = E_2 \begin{bmatrix} -R(3) - 4 \\ 1 - R(2) \end{bmatrix} = E_2 \begin{bmatrix} -7 \\ -1 \end{bmatrix} = \begin{bmatrix} -7b - c \\ b - 7c \end{bmatrix} = \begin{bmatrix} -6.8498 \\ -1.7549 \end{bmatrix}$$

$$v_2 = M \begin{bmatrix} 3 \\ 4 \\ 5 \\ 6 \end{bmatrix} = E_2 \begin{bmatrix} -R(5) - 6 \\ 3 - R(4) \end{bmatrix} = E_2 \begin{bmatrix} -11 \\ -1 \end{bmatrix} = \begin{bmatrix} -11b - c \\ b - 11c \end{bmatrix} = \begin{bmatrix} -10.5454 \\ -3.2856 \end{bmatrix}.$$

The transformed signal is represented by

$$[v_1|v_2] = \begin{bmatrix} -6.8498 & -10.5454 \\ -1.7549 & -3.2856 \end{bmatrix}.$$

To invert the MDCT, define A and B by

$$E_2 = \begin{bmatrix} A & B \end{bmatrix} = \begin{bmatrix} b & c \\ c & -b \end{bmatrix}$$

and calculate

$$\begin{bmatrix} w_1 \\ w_2 \end{bmatrix} = Nv_1 = \begin{bmatrix} B^T v_1 \\ -RB^T v_1 \\ R\Lambda^T v_1 \\ -A^T v_1 \end{bmatrix} = \begin{bmatrix} c & -b \\ -c & b \\ -b & c \\ -b & -c \end{bmatrix} \begin{bmatrix} -7b - c \\ b - 7c \end{bmatrix} = \begin{bmatrix} -1 \\ 1 \\ 7 \\ 7 \end{bmatrix}$$

$$\begin{bmatrix} w_3 \\ w_4 \end{bmatrix} = Nv_2 = \begin{bmatrix} B^T v_2 \\ -RB^T v_2 \\ -RA^T v_2 \\ -A^T v_2 \end{bmatrix} = \begin{bmatrix} c & -b \\ -c & b \\ -b & -c \\ -b & -c \end{bmatrix} \begin{bmatrix} -11b - c \\ b - 11c \end{bmatrix} = \begin{bmatrix} -1 \\ 1 \\ 11 \\ 11 \end{bmatrix},$$

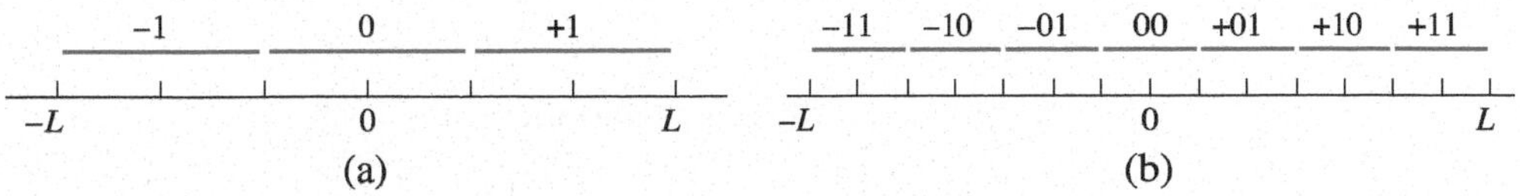

Figure 11.11 Bit quantization. Illustration of (11.39). (a) 2 bits (b) 3 bits.

where we have used the fact $b^2 + c^2 = 1$. The result of Theorem 11.12 is that we can recover the overlap $[3, 4]$ by

$$u_2 = \frac{1}{2}(w_2 + w_3) = \frac{1}{2}\left(\begin{bmatrix} 7 \\ 7 \end{bmatrix} + \begin{bmatrix} -1 \\ 1 \end{bmatrix} \right) = \begin{bmatrix} 3 \\ 4 \end{bmatrix}. \quad ◀$$

The definition and use of MDCT is less direct than the use of the DCT, discussed earlier in the chapter. Its advantage is that it allows overlapping of adjacent vectors in an efficient way. The effect is to average contributions from two vectors, reducing artifacts from abrupt transitions seen at boundaries. As in the case of DCT, we can filter or quantize the transform coefficients before reconstructing the signal in order to improve or compress the signal. Next, we show how the MDCT can be used for compression by adding a quantization step.

11.4.2 Bit quantization

Lossy compression of audio signals is achieved by quantizing the output of a signal's MDCT. In this section, we will expand on the quantization used for image compression, to allow more control over the number of bits used to represent the lossy version of the signal.

Start with the open interval of real numbers $(-L, L)$. Assume that the goal is to represent a number in $(-L, L)$ by b bits, and that we are willing to live with a little error. We will use one bit for the sign and quantize to a binary integer of $b - 1$ bits. The formula follows:

b-**bit quantization of** $(-L, L)$

Quantization: $z = \text{round}\left(\dfrac{y}{q}\right)$, where $q = \dfrac{2L}{2^b - 1}$

Dequantization: $\overline{y} = qz$ (11.40)

As an example, we show how to represent the numbers in the interval $(-1, 1)$ by 4 bits. Set $q = 2(1)/(2^4 - 1) = 2/15$, and quantize by q. The number $y = -0.3$ is represented by

$$\frac{-0.3}{2/15} = -\frac{9}{4} \longrightarrow -2 \longrightarrow -010,$$

and the number $y = 0.9$ is represented by

$$\frac{0.9}{2/15} = \frac{27}{4} = 6.75 \longrightarrow 7 \longrightarrow +111.$$

Dequantization reverses the process. The quantized version of -0.3 is dequantized as

$$(-2)q = (-2)(2/15) = -4/15 \approx -0.2667$$

and the quantized version of 0.9 as

$$(7)q = (7)(2/15) = 14/15 \approx 0.9333.$$

In both cases, the quantization error is 1/30.

▶ **EXAMPLE 11.10**   Quantize the MDCT output of Example 11.9 to 4-bit integers. Then dequantize, invert the MDCT, and find the quantization error.

All transform entries lie in the interval $(-12, 12)$. Using $L = 12$, four-bit quantization requires $q = 2(12)/(2^4 - 1) = 1.6$. Then

$$v_1 = \begin{bmatrix} -6.8498 \\ -1.7549 \end{bmatrix} \longrightarrow \begin{bmatrix} \text{round}(\frac{-6.8948}{1.6}) \\ \text{round}(\frac{-1.7549}{1.6}) \end{bmatrix} \longrightarrow \begin{bmatrix} -4 \\ -1 \end{bmatrix} \longrightarrow \begin{matrix} -100 \\ -001 \end{matrix}$$

and

$$v_2 = \begin{bmatrix} -10.5454 \\ -3.2856 \end{bmatrix} \longrightarrow \begin{bmatrix} \text{round}(\frac{-10.5454}{1.6}) \\ \text{round}(\frac{-3.2856}{1.6}) \end{bmatrix} \longrightarrow \begin{bmatrix} -7 \\ -2 \end{bmatrix} \longrightarrow \begin{matrix} -111 \\ -010 \end{matrix}.$$

The transform variables v_1, v_2 can be stored as four 4-bit integers, for a total of 16 bits. Dequantization with $q = 1.6$ is

$$\begin{bmatrix} -4 \\ -1 \end{bmatrix} \longrightarrow \begin{bmatrix} -6.4 \\ -1.6 \end{bmatrix} = \bar{v}_1$$

and

$$\begin{bmatrix} -7 \\ -2 \end{bmatrix} \longrightarrow \begin{bmatrix} -11.2 \\ -3.2 \end{bmatrix} = \bar{v}_2.$$

Applying the inverse MDCT yields

$$\begin{bmatrix} w_1 \\ w_2 \end{bmatrix} = N\bar{v}_1 = \begin{bmatrix} -0.9710 \\ 0.9710 \\ 6.5251 \\ 6.5251 \end{bmatrix},$$

$$\begin{bmatrix} w_3 \\ w_4 \end{bmatrix} = N\bar{v}_2 = \begin{bmatrix} -1.3296 \\ 1.3296 \\ 11.5720 \\ 11.5720 \end{bmatrix},$$

and the reconstructed signal

$$u_2 = \frac{1}{2}(w_2 + w_3) = \frac{1}{2}\left(\begin{bmatrix} 6.5251 \\ 6.5251 \end{bmatrix} + \begin{bmatrix} -1.3296 \\ 1.3296 \end{bmatrix} \right) = \begin{bmatrix} 2.5977 \\ 3.9274 \end{bmatrix}.$$

The quantization error is the difference between the original and reconstructed signals:

$$\left| \begin{bmatrix} 2.5977 \\ 3.9274 \end{bmatrix} - \begin{bmatrix} 3 \\ 4 \end{bmatrix} \right| = \begin{bmatrix} 0.4023 \\ 0.0726 \end{bmatrix}. \qquad \blacktriangleleft$$

Coding of audio files is usually done by using a preset allocation of bits for prescribed frequency ranges. Reality Check 11 guides the reader through construction of a complete **codec**, or code–decode protocol, that uses the MDCT along with bit quantization.

▶ **ADDITIONAL EXAMPLES**

*1. Apply 3-bit quantization to the numbers $-3, -1, \pi$, and 4 in the interval $[-L, L] = [-5, 5]$. Then dequantize and evaluate the quantization errors.

2. Apply the MDCT to the signal $x_i = \sin \frac{\pi i}{3} + \cos \frac{\pi i}{7}, i = 0, \dots, 47$. (a) Calculate $v_1 = M[x_0 \dots x_{31}]^T, v_2 = M[x_{16} \dots x_{47}]^T, y = Nv_1, z = Nv_2$, and check that $([y_{16} \dots y_{31}] + [z_0 \dots z_{15}])/2$ reproduces $[x_{16} \dots x_{31}]$. (b) Quantize and dequantize the v_i on the interval $[-L, L] = [-3, 3]$ with 2, 3, and 4 bits, respectively, to compare the reconstructions of $[x_{16} \dots x_{31}]$.

🖳 **Solutions** for Additional Examples can be found at bit.ly/2yRqWt5
(* example with video solution)

11.4 Exercises

1. Find the MDCT of the input. Express the answer in terms of $b = \cos \pi/8$ and $c = \cos 3\pi/8$. (a) $[1, 3, 5, 7]$ (b) $[-2, -1, 1, 2]$ (c) $[4, -1, 3, 5]$

2. Find the MDCT of the two overlapping length 4 windows of the given input, as in Example 11.9. Then reconstruct the middle section, using the inverse MDCT. (a) $[-3, -2, -1, 1, 2, 3]$ (b) $[1, -2, 2, -1, 3, 0]$ (c) $[4, 1, -2, -3, 0, 3]$

3. Quantize each real number in $(-1, 1)$ to 4 bits, and then dequantize and compute the quantization error. (a) $2/3$ (b) 0.6 (c) $3/7$

4. Repeat Exercise 3, but quantize to 8 bits.

5. Quantize each real number in $(-4, 4)$ to 8 bits, and then dequantize and compute the quantization error. (a) $3/2$ (b) $-7/5$ (c) 2.9 (d) π

6. Show that the DCT4 $n \times n$ matrix is an orthogonal matrix for each even integer n.

7. Reconstruct the middle section of the data in Exercise 2 after quantizing to 4 bits in $(-6, 6)$. Compare with the correct middle section.

8. Reconstruct the middle section of the data in Exercise 2 after quantizing to 6 bits in $(-6, 6)$. Compare with the correct middle section.

9. Explain why the n-dimensional column vector c_k defined by (11.28) for any integer k can be expressed in terms of a column $c_{k'}$ for $0 \le k' \le n - 1$. Express c_{5n} and c_{6n} in this way.

10. Find an upper bound for the quantization error (the error caused by quantization, followed by dequantization) when converting a real number to a b-bit integer in the interval $(-L, L)$.

11.4 Computer Problems

1. Write a MATLAB program to accept as input a vector, apply the MDCT to each of the length $2n$ windows, and reconstruct the overlapped length n sections, as in Example 11.9. Demonstrate that it works on the following input signals.
   (a) $n = 4, x = [1\ 2\ 3\ 4\ 5\ 6\ 7\ 8\ 9\ 10\ 11\ 12]$ (b) $n = 4, x_i = \cos(i\pi/6)$ for $i = 0, \ldots, 11$
   (c) $n = 8, x_i = \cos(i\pi/10)$ for $i = 0, \ldots, 63$

2. Adapt your program from Computer Problem 1 to apply b-bit quantization before reconstructing the overlaps. Then reconstruct the examples from that problem, and compute the reconstruction errors by comparing with the original input.

Reality Check 📍 **11** *A Simple Audio Codec*

Efficient transmission and storage of audio files is a key part of modern communications, and the part played by compression is crucial. In this Reality Check, you will put together a bare-bones compression–decompression protocol based on the ability of the MDCT to split the audio signal into its frequency components and the bit quantization method of Section 11.4.2.

The MDCT is applied to an input window of $2n$ signal values and provides an output of n frequency components that approximate the data (and together with the next window, interpolates the latter n input points). The compression part of the algorithm consists of coding the frequency components after quantization to save space, as demonstrated in Example 11.10.

In common audio storage formats, the way the bits are allocated to the various frequency components during quantization is based on **psychoacoustics**, the science of human sound perception. Techniques such as **frequency masking**, the empirical fact that the ear can handle only one dominant sound in each frequency range at a given time, are used to decide which frequency components are most and least important to preserve. More quantization bits are allocated to more important components. Most competitive methods are based on the MDCT and differ on how the psychoacoustic factors are treated. In our description, we will take a simplified approach that ignores most psychoacoustic factors and relies simply on **importance filtering**, the tendency to apportion more bits to frequency components of greater magnitude.

We begin with the reconstruction of a pure tone. Frequencies perceptible to the human ear range from around 100 Hz (cycles per second) to a few thousand Hz. The MDCT, using $n = 32$, catalogues frequencies starting at 64 Hz. A pure 64 Hz tone is expressed mathematically as $x(t) = \cos 2\pi (64) t$, where t is measured in seconds. Let F_s denote the sampling frequency, or number of samples per second. That means that $t = 1/F_s, 2/F_s, \ldots, F_s/F_s$ represent one second worth of points in time. Sampling $x(t)$ at these times yields

$$x_j = \cos(2\pi 64 j / F_s).$$

The rows of the $n \times 2n$ MDCT matrix M represent the length $2n$ fundamental interpolation functions. Note from (11.31) that the ith row, for $0 \le i \le n - 1$, is

$$\sqrt{\frac{2}{n}} \cos \frac{(2i + 1)(j + \frac{n+1}{2})\pi}{2n}$$

for $j = 0. \ldots, 2n - 1$. Thus the first row traverses a length π interval, one-half period of cosine; the second row traverses 3/2 periods, etc.

A signal of frequency 64 Hz, given a sampling rate of $F_s = 2^{13} = 8192$, corresponds to the vector $\cos(2\pi 64 j / 2^{13})$. In the first length $2n$ window, j is moving from 1 to $2n$, so the argument of cosine moves from about 0 to $2\pi (64) 2n / 2^{13} = 2\pi (64) 2(32)/2^{13} = \pi$, i.e., one-half period of cosine.

The MATLAB commands

```
Fs=2^(13);
x=cos(2*pi*110*(1:Fs)/Fs);
sound(x,Fs);
```

play one second of a 110 Hz tone. The sampling frequency Fs of $2^{13} = 8192$ bytes/sec is quite common, corresponding to $2^{16} = 65536$ bits/sec, referred to as a 64Kb/sec sampling rate for an audio file. Higher quality files are often sampled at two or three times this rate, at 128 or 192 Kbs.

The MATLAB program below applies the MDCT and quantizes, followed by dequantization and inverse MDCT on the overlapped segments, as described in Section 11.4. In this way, the effect of the quanitization error that accompanies lossy compression can be examined. Note that to avoid distortion in the MATLAB sound command, it is helpful to scale signal amplitudes to no more than about 0.3 in absolute value.

MATLAB code
shown here can be found at bit.ly/2EscAFV

```
% Program 11.1 Audio codec
% input: column vector x of input signal
% output: column vector out of output signal
% Example usage: out=simplecodec((cos((1:2^(13))*2*pi*440/2^(13)))');
%                example signal is 1 sec. pure tone of frequency f=440Hz
function out=simplecodec(x)
len=numel(x);              % length of signal
n=2^5;                     % length of processing window
```

```
nw=floor(len/n);                    % number of length n windows in x
x=x(1:n*nw);                        % cut x to integer number of n windows
Fs=2^(13);                          % Fs = sampling rate
b=4; L=1;                           % b = quantiz. bits, [-L,L] amplitude range
q=2*L/(2^b-1);                      % q used for b bits on interval [-L, L]
for i=1:n                           % form the MDCT matrix
  for j=1:2*n
    M(i,j)= cos((i-1+1/2)*(j-1+1/2+n/2)*pi/n);
  end
end
M=sqrt(2/n)*M;
N=M';                               % inverse MDCT
sound(0.3*x/max(abs(x)),Fs)         % play the input signal (scale to max = 0.3)
out=[];
for k=1:nw-1                         % loop over length 2n windows
  x0=x(1+(k-1)*n:2*n+(k-1)*n);      % column vector of signal in current window
  y0=M*x0;                          % apply MDCT
  y1=round(y0/q);                   % quantize transform components
% Storage/transmission of file occurs here
  y2=y1*q;                          % dequantize transform components
  w(:,k)=N*y2;                      % invert the MDCT
  if(k>1)
      w2=w(n+1:2*n,k-1);w3=w(1:n,k);
      out=[out;(w2+w3)/2];          % collect the reconstructed signal
  end                               % (out has length 2n less than length of x)
end
pause(2)
sound(0.3*out/max(abs(out)),Fs)% play the reconstructed signal
```

Suggested activities:

1. Investigate the ability of MDCT to represent pure tones. Begin with $b = 4$ bits per window of size $n = 32$. Pick a tone of frequency between 100 Hz and 1000 Hz, and calculate the difference (as RMSE) between the original signal and the signal after encoding/decoding. You should cut the original signal to `xshort = x(n+1: end-n);` for comparison with the output signal, since the latter lacks n entries at the left and right ends. Plot a short section of the original and decoded signal.

2. Build chords and evaluate the RMSE as in Step 1. Simple intervals can be constructed by a simple addition of multiple pure tones. Rational ratios of frequencies with low numerators and denominators are pleasing to the ear: A 2 : 1 ratio of frequencies gives an octave, 1.25 : 1 ratio gives a third, a 1.5 : 1 gives a fifth, and so forth. How does the RMSE depend on the number of bits used in the coder?

3. A "windowing function" is often used to reduce codec error, due to the fact that the function being represented is not periodic over the window, but is being represented by periodic functions. The windowing function scales the input signal x smoothly to zero at each end of the window, partially mitigating this problem. A common choice is to replace x_j with $x_j h_j$, where

$$h_j = \sqrt{2}\sin\frac{(j-\frac{1}{2})\pi}{2n}$$

for a length $2n$ window, where $j = 1, \ldots, 2n$. To undo the windowing function, multiply the inverse MDCT output w componentwise by the same h_j. This results in multiplying w_2 componentwise by the second half of the h_j, $j = n + 1, \ldots, 2n$, and w_3

by the first half h_j, $j = 1, \ldots, n$ before combining into the decoded signal. Compare RMSE, plots, and audible sound as in Steps 1 and 2.

4. Explain the method for undoing the windowing that is suggested in Step 3. In other words, assume that if Z_1 and Z_2 are each multiplied componentwise by the entire windowing function h, and NMZ_1 and NMZ_2 in equation (11.38) are each multiplied componentwise by h, that equation (11.39) still holds.

5. Import a .wav file with the MATLAB audioread command, or download an audio file of your choice. (Alternatively, load handel can be used. If you download a stereo file, you will need to work with each channel separately.) Reproduce the file (or a segment of it) using various values of b and with and without windowing. Compute RMSE for your choices of parameters and exhibit the results using the sound command.

6. Introduce importance sampling. Make a new test tone that is a combination of pure tones. Modify the code so that each of the 32 frequency components of y has its own number b_k of bits for quantization. Propose a method that makes b_k larger if the contributions $|y_k|$ are larger, on average. Count the number of bits required to hold the signal, and refine your proposal.

7. Build two separate subprograms, a coder and a decoder. The coder should write a file (or MATLAB variable) of bits representing the quantized output of the MDCT and print the number of bits used. The decoder should load the file written by the coder and reconstruct the signal.

Software and Further Reading

For good practical introductions to data compression, see Nelson and Gailly [1995], Storer [1988], and Sayood [1996]. General references on image and sound compression are Bhaskaran and Konstandtinides [1995]. Rao and Yip [1990] is a good source for information on the Discrete Cosine Transform. The seminal article on Huffman coding is Huffman [1952].

We have introduced the baseline JPEG standard (Wallace [1991]) for image compression. The full standard is available in Pennebaker and Mitchell [1993]. The recently introduced JPEG-2000 standard (Taubman and Marcellin [2002]) allows wavelet compression in place of DCT.

Most protocols for sound compression are based on the Modified Discrete Cosine Transform (Wang and Vilermo [2003], Malvar [1992]). More specific information can be found on the individual formats like MP3 (shorthand for MPEG audio layer 3, see Hacker [2000]), AAC (Advanced Audio Coding, used in Apple iTunes and QuickTime video, and XM satellite radio), and the open-source audio format Ogg Vorbis.

12

Eigenvalues and Singular Values

The World Wide Web makes vast amounts of information easily accessible to the casual user—so vast, in fact, that navigation with a powerful search engine is essential. Technology has also provided miniaturization and low-cost sensors, making great quantities of data available to researchers. How can access to large amounts of information be exploited in an efficient way?

Many aspects of search technology, and knowledge discovery in general, benefit from treatment as an eigenvalue or singular value problem. Numerical methods to solve these high-dimensional problems generate projections to distinguished lower dimensional subspaces. This is exactly the simplification that complex data environments most need.

Reality Check Reality Check 12 on page 575 explores what has been called the largest ongoing eigenvalue computation in the world, used by one of the well-known Web search providers.

Computational methods for locating eigenvalues are based on the fundamental idea of Power Iteration, a type of fixed-point iteration for eigenspaces. A sophisticated version of the idea, called the QR algorithm, is the standard algorithm for determining all eigenvalues of typical matrices.

The singular value decomposition reveals the basic structure of a matrix and is heavily used in statistical applications to find relations between data. In this chapter, we survey methods for finding the eigenvalues and eigenvectors of a square matrix, and the singular values and singular vectors of a general matrix.

12.1 POWER ITERATION METHODS

There is no direct method for computing eigenvalues. The situation is analogous to root-finding, in that all feasible methods depend on some type of iteration. To begin the section, we consider whether the problem might be reducible to root-finding.

Appendix A shows a method for calculating eigenvalues and eigenvectors of an $m \times m$ matrix. This approach, based on finding the roots of the degree m characteristic polynomial, works well for 2×2 matrices. For larger matrices, the procedure requires a rootfinder of the type studied in Chapter 1.

The difficulty of this approach to finding eigenvalues becomes clear if we recall the example of the Wilkinson polynomial of Chapter 1. There we found that very small changes in the coefficients of a polynomial can change the roots of the polynomial by arbitrarily large amounts. In other words, the condition number of the input/output problem taking coefficients to roots can be extremely large. Because our calculation of the coefficients of the characteristic polynomial will be subject to errors on the order of machine roundoff or larger, calculation of eigenvalues by this approach is susceptible to large errors. This difficulty is serious enough to warrant eliminating the method of finding roots of the characteristic polynomial as a pathway to the accurate calculation of eigenvalues.

A simple example of poor accuracy for this method follows from the existence of the Wilkinson polynomial. If we are trying to find the eigenvalues of the matrix

$$A = \begin{bmatrix} 1 & 0 & \cdots & 0 \\ 0 & 2 & & \vdots \\ \vdots & & \ddots & \vdots \\ 0 & 0 & \cdots & 20 \end{bmatrix},$$ (12.1)

we will calculate the coefficients of the characteristic polynomial $P(x) = (x - 1)(x - 2)\cdots(x - 20)$ and use a rootfinder to find the roots. However, as shown in Chapter 1, some of the roots of the machine version of $P(x)$ are far from the roots of the true version of $P(x)$, which are the eigenvalues of A.

This section introduces methods based on multiplying high powers of the matrix times a vector, which usually will turn into an eigenvector as the power is raised. We will refine the idea later, but it is the main thrust of the most sophisticated methods.

12.1.1 Power Iteration

The motivation behind Power Iteration is that multiplication by a matrix tends to move vectors toward the dominant eigenvector direction.

SPOTLIGHT ON

> **Conditioning**   The large errors that the "characteristic polynomial method" are subject to are not the fault of the rootfinder. A perfectly accurate rootfinder would fare no better. When the polynomial is multiplied out to determine its coefficients for entry into the rootfinder, the coefficients will, in general, be subject to errors on the order of machine epsilon. The rootfinder will then be asked to find the roots of the slightly wrong polynomial, which, as we have seen, can have disastrous consequences. There is no general fix to this problem. The only way to fight the problem would be to increase the size of the mantissa representing floating point numbers, which would have the effect of lowering machine epsilon. If machine epsilon could be made lower than $1/\text{cond}(P)$, then accuracy could be assured for the eigenvalues. Of course, this is not really a solution, but just another step in an unwinnable arms race. If higher precision computing is used, we can always extend the Wilkinson polynomial to a higher degree to find an even higher condition number.

DEFINITION 12.1 Let A be an $m \times m$ matrix. A **dominant eigenvalue** of A is an eigenvalue λ whose magnitude is greater than all other eigenvalues of A. If it exists, an eigenvector associated to λ is called a **dominant eigenvector**. ❑

The matrix

$$A = \begin{bmatrix} 1 & 3 \\ 2 & 2 \end{bmatrix}$$

has a dominant eigenvalue of 4 with eigenvector $[1, 1]^T$, and an eigenvector that is smaller in magnitude, -1, with associated eigenvector $[-3, 2]^T$. Let us observe the result of multiplying the matrix A times a "random" vector, say $[-5, 5]^T$:

$$x_1 = Ax_0 = \begin{bmatrix} 1 & 3 \\ 2 & 2 \end{bmatrix}\begin{bmatrix} -5 \\ 5 \end{bmatrix} = \begin{bmatrix} 10 \\ 0 \end{bmatrix}$$

$$x_2 = A^2x_0 = \begin{bmatrix} 1 & 3 \\ 2 & 2 \end{bmatrix}\begin{bmatrix} 10 \\ 0 \end{bmatrix} = \begin{bmatrix} 10 \\ 20 \end{bmatrix}$$

$$x_3 = A^3x_0 = \begin{bmatrix} 1 & 3 \\ 2 & 2 \end{bmatrix}\begin{bmatrix} 10 \\ 20 \end{bmatrix} = \begin{bmatrix} 70 \\ 60 \end{bmatrix}$$

$$x_4 = A^4x_0 = \begin{bmatrix} 1 & 3 \\ 2 & 2 \end{bmatrix}\begin{bmatrix} 70 \\ 60 \end{bmatrix} = \begin{bmatrix} 250 \\ 260 \end{bmatrix} = 260 \begin{bmatrix} \frac{25}{26} \\ 1 \end{bmatrix}.$$

Multiplying a random starting vector repeatedly by the matrix A has resulted in moving the vector very close to the dominant eigenvector of A. This is no coincidence, as can be seen by expressing x_0 as a linear combination of the eigenvectors

$$x_0 = 1\begin{bmatrix} 1 \\ 1 \end{bmatrix} + 2\begin{bmatrix} -3 \\ 2 \end{bmatrix}$$

and reviewing the calculation in this light:

$$x_1 = Ax_0 = 4\begin{bmatrix} 1 \\ 1 \end{bmatrix} - 2\begin{bmatrix} -3 \\ 2 \end{bmatrix}$$

$$x_2 = A^2x_0 = 4^2\begin{bmatrix} 1 \\ 1 \end{bmatrix} + 2\begin{bmatrix} -3 \\ 2 \end{bmatrix}$$

$$x_3 = A^3x_0 = 4^3\begin{bmatrix} 1 \\ 1 \end{bmatrix} - 2\begin{bmatrix} -3 \\ 2 \end{bmatrix}$$

$$x_4 = A^4x_0 = 4^4\begin{bmatrix} 1 \\ 1 \end{bmatrix} + 2\begin{bmatrix} -3 \\ 2 \end{bmatrix}$$

$$= 256\begin{bmatrix} 1 \\ 1 \end{bmatrix} + 2\begin{bmatrix} -3 \\ 2 \end{bmatrix}.$$

The point is that the eigenvector corresponding to the eigenvalue that is largest in magnitude will dominate the calculation after several steps. In this case, the eigenvalue 4 is largest, and so the calculation moves closer and closer to an eigenvector in its direction $[1, 1]^T$.

To keep the numbers from getting out of hand, it is necessary to normalize the vector at each step. One way to do this is to divide the current vector by its length prior to each step. The two operations, normalization and multiplication by A constitute the method of Power Iteration.

As the steps deliver improved approximate eigenvectors, how do we find approximate eigenvalues? To pose the question more generally, assume that a matrix A and an approximate eigenvector are known. What is the best guess for the associated eigenvalue?

Convergence   Power Iteration is essentially a fixed-point iteration with normalization at each step. Like FPI, it converges linearly, meaning that during convergence, the error decreases by a constant factor on each iteration step. Later in this section, we will encounter a quadratically convergent variant of Power Iteration called Rayleigh Quotient Iteration.

We will appeal to least squares. Consider the eigenvalue equation $x\lambda = Ax$, where x is an approximate eigenvector and λ is unknown. Looked at this way, the coefficient matrix is the $n \times 1$ matrix x. The normal equations say that the least squares answer is the solution of $x^T x\lambda = x^T Ax$, or

$$\lambda = \frac{x^T Ax}{x^T x}, \tag{12.2}$$

known as the **Rayleigh quotient**. Given an approximate eigenvector, the Rayleigh quotient is the best approximate eigenvalue. Applying the Rayleigh quotient to the normalized eigenvector adds an eigenvalue approximation to Power Iteration.

Power Iteration

Given initial vector x_0.

for $j = 1, 2, 3, \ldots$
$\quad u_{j-1} = x_{j-1}/\|x_{j-1}\|_2$
$\quad x_j = Au_{j-1}$
$\quad \lambda_j = u_{j-1}^T Au_{j-1}$
end
$u_j = x_j/\|x_j\|_2$

To find the dominant eigenvector of the matrix A, begin with an initial vector. Each iteration consists of normalizing the current vector and multiplying by A. The Rayleigh quotient is used to approximate the eigenvalue. The MATLAB norm command makes this simple to implement, as shown in the following code:

MATLAB code
shown here can be found
at bit.ly/2yKx5HK

```
% Program 12.1 Power Iteration
% Computes dominant eigenvector of square matrix
% Input: matrix A, initial (nonzero) vector x, number of steps k
% Output: dominant eigenvalue lam, eigenvector u
function [lam,u]=powerit(A,x,k)
for j=1:k
    u=x/norm(x);             % normalize vector
    x=A*u;                   % power step
    lam=u'*x;                % Rayleigh quotient
end
u=x/norm(x);
```

12.1.2 Convergence of Power Iteration

We will prove the convergence of Power Iteration under certain conditions on the eigenvalues. Although these conditions are not completely general, they serve to show why the method succeeds in the clearest possible case. Later, we will assemble successively more sophisticated eigenvalue methods, built on the basic concept of Power Iteration, that cover more general matrices.

THEOREM 12.2 Let A be an $m \times m$ matrix with real eigenvalues $\lambda_1, \ldots, \lambda_m$ satisfying $|\lambda_1| > |\lambda_2| \geq |\lambda_3| \geq \cdots \geq |\lambda_m|$. Assume that the eigenvectors of A span R^m. For almost every initial vector, Power Iteration converges linearly to an eigenvector associated to λ_1 with convergence rate constant $S = |\lambda_2/\lambda_1|$. ∎

Proof. Let $v_1, \ldots, v_n$ be the eigenvectors that form a basis of R^n, with corresponding eigenvalues $\lambda_1, \ldots, \lambda_n$, respectively. Express the initial vector in this basis as $x_0 = c_1 v_1 + \cdots + c_n v_n$ for some coefficients c_i. The phrase "for almost every initial vector" means we can assume that $c_1, c_2 \neq 0$. Applying Power Iteration yields

$$Ax_0 = c_1 \lambda_1 v_1 + c_2 \lambda_2 v_2 + \cdots + c_n \lambda_n v_n$$
$$A^2 x_0 = c_1 \lambda_1^2 v_1 + c_2 \lambda_2^2 v_2 + \cdots + c_n \lambda_n^2 v_n$$
$$A^3 x_0 = c_1 \lambda_1^3 v_1 + c_2 \lambda_2^3 v_2 + \cdots + c_n \lambda_n^3 v_n$$
$$\vdots$$

with normalization at each step. As the number of steps $k \to \infty$, the first term on the right-hand side will dominate, no matter how the normalization is done, because

$$\frac{A^k x_0}{\lambda_1^k} = c_1 v_1 + c_2 \left(\frac{\lambda_2}{\lambda_1}\right)^k v_2 + \cdots + c_n \left(\frac{\lambda_n}{\lambda_1}\right)^k v_n.$$

The assumption that $|\lambda_1| > |\lambda_i|$ for $i > 1$ implies that all but the first term on the right will converge to zero with convergence rate $S \leq |\lambda_2/\lambda_1|$, and exactly that rate, as long as $c_2 \neq 0$. As a result, the method converges to a multiple of the dominant eigenvector v_1, with eigenvalue λ_1. □

The term "almost every" in the theorem's conclusion means that the set of initial vectors x_0 for which the iteration fails is a set of lower dimension in R^m. Specifically, the iteration will succeed at the specified rate if x_0 is not contained in the union of the dimension $m - 1$ planes spanned by $\{v_1, v_3, \ldots, v_m\}$ and $\{v_2, v_3, \ldots, v_m\}$.

12.1.3 Inverse Power Iteration

Power Iteration is limited to locating the eigenvalue of largest magnitude (absolute value). If Power Iteration is applied to the inverse of the matrix, the smallest eigenvalue can be found.

LEMMA 12.3 Let the eigenvalues of the $m \times m$ matrix A be denoted by $\lambda_1, \lambda_2, \ldots, \lambda_m$. (a) The eigenvalues of the inverse matrix A^{-1} are $\lambda_1^{-1}, \lambda_2^{-1}, \ldots, \lambda_m^{-1}$, assuming that the inverse exists. The eigenvectors are the same as those of A. (b) The eigenvalues of the shifted matrix $A - sI$ are $\lambda_1 - s, \lambda_2 - s, \ldots, \lambda_m - s$ and the eigenvectors are the same as those of A. ∎

Proof. (a) $Av = \lambda v$ implies that $v = \lambda A^{-1} v$, and therefore, $A^{-1} v = (1/\lambda) v$. Note that the eigenvector is unchanged. (b) Subtract sIv from both sides of $Av = \lambda v$. Then $(A - sI)v = (\lambda - s)v$ is the definition of eigenvalue for $(A - sI)$, and again the same eigenvector can be used. □

According to Lemma 12.3, the largest magnitude eigenvalue of the matrix A^{-1} is the reciprocal of the smallest magnitude eigenvalue of A. Applying Power Iteration to the inverse matrix, followed by inverting the resulting eigenvalue of A^{-1}, gives the smallest magnitude eigenvalue of A.

To avoid explicit calculation of the inverse of A, we rewrite the application of Power Iteration to A^{-1}, namely,

$$x_{k+1} = A^{-1}x_k \tag{12.3}$$

as the equivalent

$$Ax_{k+1} = x_k, \tag{12.4}$$

which is then solved for x_{k+1} by Gaussian elimination.

Now we know how to find the largest and smallest eigenvalues of a matrix. In other words, for a 100×100 matrix, we are 2 percent finished. How do we find the other 98 percent?

One approach is suggested by Lemma 12.3(b). We can make any of the other eigenvalues small by shifting A by a value close to the eigenvalue. If we happen to know that there is an eigenvalue near 10 (say, 10.05), then $A - 10I$ has an eigenvalue $\lambda = 0.05$. If it is the smallest magnitude eigenvalue of $A - 10I$, then the Inverse Power Iteration $x_{k+1} = (A - 10I)^{-1}x_k$ will locate it. That is, the Inverse Power Iteration will converge to the reciprocal $1/(0.05) = 20$, after which we invert to 0.05 and add the shift back to get 10.05. This trick will locate the eigenvalue that is smallest after the shift—which is another way of saying the eigenvalue nearest to the shift. To summarize, we write

Inverse Power Iteration

Given initial vector x_0 and shift s

for $j = 1, 2, 3, \ldots$
  $u_{j-1} = x_{j-1}/\|x_{j-1}\|_2$
  Solve $(A - sI)x_j = u_{j-1}$
  $\lambda_j = u_{j-1}^T x_j$
end
$u_j = x_j/\|x_j\|_2$

To find the eigenvalue of A nearest to the real number s, apply Power Iteration to $(A - sI)^{-1}$ to get the largest magnitude eigenvalue b of $(A - sI)^{-1}$. The Power Iterations should be done by Gaussian elimination on $(A - sI)y_{k+1} = x_k$. Then $\lambda = b^{-1} + s$ is the eigenvalue of A nearest to s. The eigenvector associated to λ is given directly from the calculation.

```
% Program 12.2 Inverse Power Iteration
% Computes eigenvalue of square matrix nearest to input s
% Input: matrix A, (nonzero) vector x, shift s, steps k
% Output: eigenvalue lam, eigenvector of inv(A-sI)
function [lam,u]=invpowerit(A,x,s,k)
As=A-s*eye(size(A));
for j=1:k
  u=x/norm(x);                    % normalize vector
  x=As\u;                         % power step
  lam=u'*x;                       % Rayleigh Quotient
end
lam=1/lam+s; u=x/norm(x);
```

▶ **EXAMPLE 12.1** Assume that A is a 5×5 matrix with eigenvalues $-5, -2, 1/2, 3/2, 4$. Find the eigenvalue and convergence rate expected when applying (a) Power Iteration (b) Inverse Power Iteration with shift $s = 0$ (c) Inverse Power Iteration with shift $s = 2$.

(a) Power Iteration with a random initial vector will converge to the largest magnitude eigenvalue -5, with convergence rate $S = |\lambda_2|/|\lambda_1| = 4/5$. (b) Inverse Power Iteration (with no shift) will converge to the smallest, $1/2$, because its reciprocal 2 is larger than the other reciprocals $-1/5, -1/2, 2/3$, and $1/4$. The convergence rate will be the ratio of the two largest eigenvalues of the inverse matrix, $S = (2/3)/2 = 1/3$. (c) The Inverse Power Iteration with shift $s = 2$ will locate the eigenvalue nearest to 2, which is $3/2$. The reason is that, after shifting the eigenvalues to $-7, -4, -3/2, -1/2$, and 2, the largest of the reciprocals is -2. After inverting to get $-1/2$ and adding back the shift $s = 2$, we get $3/2$. The convergence rate is again the ratio $(2/3)/2 = 1/3$. ◀

12.1.4 Rayleigh Quotient Iteration

The Rayleigh quotient can be used in conjunction with Inverse Power Iteration. We know that it converges to the eigenvector associated to the eigenvalue with the smallest distance to the shift s, and that convergence is fast if this distance is small. If at any step along the way an approximate eigenvalue were known, it could be used as the shift s, to speed convergence.

Using the Rayleigh quotient as the updated shift in Inverse Power Iteration leads to Rayleigh Quotient Iteration (RQI).

Rayleigh Quotient Iteration

Given initial vector x_0.

for $\quad j = 1, 2, 3, \ldots$
$\qquad u_{j-1} = x_{j-1}/\|x_{j-1}\|$
$\qquad \lambda_{j-1} = u_{j-1}^T A u_{j-1}$
$\qquad$ Solve $(A - \lambda_{j-1} I) x_j = u_{j-1}$
end
$u_j = x_j / \|x_j\|_2$

MATLAB code
shown here can be found
at bit.ly/2J005hp

```
% Program 12.3 Rayleigh Quotient Iteration
% Input: matrix A, initial (nonzero) vector x, number of steps k
% Output: eigenvalue lam and eigenvector u
function [lam,u]=rqi(A,x,k)
for j=1:k
  u=x/norm(x);                   % normalize
  lam=u'*A*u;                    % Rayleigh quotient
  x=(A-lam*eye(size(A)))\u;      % inverse power iteration
end
u=x/norm(x);
lam=u'*A*u;                      % Rayleigh quotient
```

While Inverse Power Iteration converges linearly, Rayleigh Quotient Iteration is quadratically convergent for simple (nonrepeated) eigenvalues and will converge cubically if the matrix is symmetric. This means that very few steps are needed to converge to machine precision for this method. After convergence, the matrix $A - \lambda_{j-1} I$ is singular and no more steps can be performed. As a result, trial and error should be used with Program 12.3 to stop the iteration just before this occurs. Note that the complexity has grown for RQI. Inverse Power Iteration requires only one LU factorization; but for RQI, each step requires a new factorization, since the shift has changed. Even so, Rayleigh Quotient Iteration is the fastest converging method we have presented in this section on finding one eigenvalue at a time. In the next section, we discuss ways to find all eigenvalues of a matrix in the same calculation. The basic engine will remain Power Iteration—it is only the organizational details that will become more sophisticated.

▶ **ADDITIONAL EXAMPLES**

*1 Let $A = \begin{bmatrix} -5 & 4 \\ -8 & 7 \end{bmatrix}$. (a) Find all eigenvalues and eigenvectors of A using the characteristic equation. (b) Apply three steps of Power Iteration with initial vector $[1, 0]$. At each step, approximate the eigenvalue by the Rayleigh quotient.

2. Let $A = \begin{bmatrix} 5 & 2 & -2 \\ -12 & -19 & 12 \\ -12 & -22 & 15 \end{bmatrix}$. (a) Apply 10 steps of the Power Method with initial vector $[1, 1, 1]$ to estimate the dominant eigenvalue of A. (b) Apply 10 steps of the Inverse Power Method with shift 0 and initial vector $[1, 1, 1]$ to estimate the eigenvalue closest to zero.

Solutions for Additional Examples can be found at bit.ly/2EAAb7r
(* example with video solution)

12.1 Exercises

Solutions for Exercises numbered in blue can be found at bit.ly/2OvZOux

1. Find the characteristic polynomial and the eigenvalues and eigenvectors of the following symmetric matrices:

(a) $\begin{bmatrix} 3.5 & -1.5 \\ -1.5 & 3.5 \end{bmatrix}$ (b) $\begin{bmatrix} 0 & 2 \\ 2 & 0 \end{bmatrix}$ (c) $\begin{bmatrix} -0.2 & -2.4 \\ -2.4 & 1.2 \end{bmatrix}$ (d) $\begin{bmatrix} 136 & -48 \\ -48 & 164 \end{bmatrix}$

2. Find the characteristic polynomial and the eigenvalues and eigenvectors of the following matrices:

(a) $\begin{bmatrix} 7 & 9 \\ -6 & -8 \end{bmatrix}$ (b) $\begin{bmatrix} 2 & 6 \\ 1 & 3 \end{bmatrix}$ (c) $\begin{bmatrix} 2.2 & 0.6 \\ -0.4 & 0.8 \end{bmatrix}$ (d) $\begin{bmatrix} 32 & 45 \\ -18 & -25 \end{bmatrix}$

3. Find the characteristic polynomial and the eigenvalues and eigenvectors of the following matrices:

(a) $\begin{bmatrix} 1 & 0 & 1 \\ 0 & 3 & -2 \\ 0 & 0 & 2 \end{bmatrix}$ (b) $\begin{bmatrix} 1 & 0 & -\frac{1}{3} \\ 0 & 1 & \frac{2}{3} \\ -1 & 1 & 1 \end{bmatrix}$ (c) $\begin{bmatrix} -\frac{1}{2} & -\frac{1}{2} & -\frac{1}{6} \\ -1 & 0 & \frac{1}{3} \\ -\frac{1}{2} & \frac{1}{2} & \frac{1}{2} \end{bmatrix}$

4. Prove that a square matrix and its transpose have the same characteristic polynomial, and therefore the same set of eigenvalues.

5. Assume that A is a 3×3 matrix with the given eigenvalues. Decide to which eigenvalue Power Iteration will converge, and determine the convergence rate constant S. (a) $\{3, 1, 4\}$ (b) $\{3, 1, -4\}$ (c) $\{-1, 2, 4\}$ (d) $\{1, 9, 10\}$

6. Assume that A is a 3×3 matrix with the given eigenvalues. Decide to which eigenvalue Power Iteration will converge, and determine the convergence rate constant S. (a) $\{1, 2, 7\}$ (b) $\{1, 1, -4\}$ (c) $\{0, -2, 5\}$ (d) $\{8, -9, 10\}$

7. Assume that A is a 3×3 matrix with the given eigenvalues. Decide to which eigenvalue Inverse Power Iteration with the given shift s will converge, and determine the convergence rate constant S. (a) $\{3, 1, 4\}, s = 0$ (b) $\{3, 1, -4\}, s = 0$ (c) $\{-1, 2, 4\}, s = 0$ (d) $\{1, 9, 10\}, s = 6$

8. Assume that A is a 3×3 matrix with the given eigenvalues. Decide to which eigenvalue Inverse Power Iteration with the given shift s will converge, and determine the convergence rate constant S. (a) $\{3, 1, 4\}, s = 5$ (b) $\{3, 1, -4\}, s = 4$ (c) $\{-1, 2, 4\}, s = 1$ (d) $\{1, 9, 10\}, s = 8$

9. Let $A = \begin{bmatrix} 1 & 2 \\ 4 & 3 \end{bmatrix}$. (a) Find all eigenvalues and eigenvectors of A. (b) Apply three steps of Power Iteration with initial vector $x_0 = (1, 0)$. At each step, approximate the eigenvalue by the current Rayleigh quotient. (c) Predict the result of applying Inverse Power Iteration with shift $s = 0$ (d) with shift $s = 3$.

10. Let $A = \begin{bmatrix} -2 & 1 \\ 3 & 0 \end{bmatrix}$. Carry out the steps of Exercise 9 for this matrix.

11. If A is a 6×6 matrix with eigenvalues $-6, -3, 1, 2, 5, 7$, which eigenvalue of A will the following algorithms find? (a) Power Iteration (b) Inverse Power Iteration with shift $s = 4$ (c) Find the linear convergence rates of the two computations. Which converges faster?

12.1 Computer Problems

Solutions for Computer Problems numbered in blue can be found at bit.ly/2J7DKk0

1. Using the supplied code (or code of your own) for the Power Iteration Method, find the dominant eigenvector of A, and estimate the dominant eigenvalue by calculating a Rayleigh quotient. Compare your conclusions with the corresponding part of Exercise 5.

(a) $\begin{bmatrix} 10 & -12 & -6 \\ 5 & -5 & -4 \\ -1 & 0 & 3 \end{bmatrix}$ (b) $\begin{bmatrix} -14 & 20 & 10 \\ -19 & 27 & 12 \\ 23 & -32 & -13 \end{bmatrix}$

(c) $\begin{bmatrix} 8 & -8 & -4 \\ 12 & -15 & -7 \\ -18 & 26 & 12 \end{bmatrix}$ (d) $\begin{bmatrix} 12 & -4 & -2 \\ 19 & -19 & -10 \\ -35 & 52 & 27 \end{bmatrix}$

2. Using the supplied code (or code of your own) for the Inverse Power Iteration Method, verify your conclusions from Exercise 7, using the appropriate matrix from Computer Problem 1.

3. For the Inverse Power Iteration Method, verify your conclusions from Exercise 8, using the appropriate matrix from Computer Problem 1.

4. Apply Rayleigh Quotient Iteration to the matrices in Computer Problem 1. Try different starting vectors until all three eigenvalues are found.

12.2 QR ALGORITHM

The goal of this section is to develop methods for finding all eigenvalues at once. We begin with a method that works for symmetric matrices, and later supplement it to work in general. Symmetric matrices are easiest to handle because their eigenvalues are real and their unit eigenvectors form an orthonormal basis of R^m (see Appendix A). This motivates applying Power Iteration with m vectors in parallel, where we actively work at keeping the vectors orthogonal to one another.

12.2.1 Simultaneous Iteration

Assume that we begin with m pairwise orthogonal initial vectors $v_1, \ldots, v_m$. After one step of Power Iteration applied to each vector, $Av_1, \ldots, Av_m$ are no longer guaranteed to be orthogonal to one another. In fact, under further multiplications by A, they all would prefer to converge to the dominant eigenvector, according to Theorem 12.2.

To avoid this, we re-orthogonalize the set of m vectors at each step. The simultaneous multiplication by A of the m vectors is efficiently written as the matrix product

$$A[v_1|\cdots|v_m].$$

As we found in Chapter 4, the orthogonalization step can be viewed as factoring the resulting product as QR. If the elementary basis vectors are used as initial vectors, then the first step of Power Iteration followed by re-orthogonalization is $AI = \overline{Q}_1 R_1$, or

$$\left[A\begin{bmatrix}1\\0\\\vdots\\0\end{bmatrix} \middle| A\begin{bmatrix}0\\1\\\vdots\\0\end{bmatrix} \middle| \cdots \middle| A\begin{bmatrix}0\\0\\\vdots\\1\end{bmatrix} \right] = \left[\overline{q}_1^1|\cdots|\overline{q}_m^1\right] \begin{bmatrix} r_{11}^1 & r_{12}^1 & \cdots & r_{1m}^1 \\ & r_{22}^1 & & \vdots \\ & & \ddots & \vdots \\ & & & r_{mm}^1 \end{bmatrix}. \quad (12.5)$$

The $\overline{q}_i^1$ for $i = 1, \ldots, m$ are the new orthogonal set of unit vectors in the Power Iteration process. Next, we repeat the step:

$$A\overline{Q}_1 = \left[A\overline{q}_1^1 | A\overline{q}_2^1 | \cdots | A\overline{q}_m^1 \right]$$

$$= \left[\overline{q}_1^2|\overline{q}_2^2|\cdots|\overline{q}_m^2\right] \begin{bmatrix} r_{11}^2 & r_{12}^2 & \cdots & r_{1m}^2 \\ & r_{22}^2 & & \vdots \\ & & \ddots & \vdots \\ & & & r_{mm}^2 \end{bmatrix}$$

$$= \overline{Q}_2 R_2. \quad (12.6)$$

In other words, we have developed a matrix form of Power Iteration that searches for all m eigenvectors of a symmetric matrix simultaneously.

Normalized Simultaneous Iteration

Set $\quad \overline{Q}_0 = I$
for $\quad j = 1, 2, 3, \ldots$
$\quad\quad A\overline{Q}_j = \overline{Q}_{j+1} R_{j+1}$
end

At the jth step, the columns of $\overline{Q}_j$ are approximations to the eigenvectors of A, and the diagonal elements $r_{11}^j, \ldots, r_{mm}^j$ are approximations to the eigenvalues. In MATLAB code, this algorithm, which we will call Normalized Simultaneous Iteration (NSI), can be written very compactly.

MATLAB code shown here can be found at bit.ly/2P3VpOL

```
% Program 12.4 Normalized Simultaneous Iteration
% Computes eigenvalues/vectors of symmetric matrix
% Input: matrix A, number of steps k
% Output: eigenvalues lam and eigenvector matrix Q
function [lam,Q]=nsi(A,k)
```

```
[m,n]=size(A);
Q=eye(m,m);
for j=1:k
    [Q,R]=qr(A*Q);                  % QR factorization
end
lam=diag(Q'*A*Q);                   % Rayleigh quotient
```

An even more compact way to implement Normalized Simultaneous Iteration is available. Set $\overline{Q}_0 = I$. Then NSI proceeds as follows:

$$
\begin{aligned}
A\overline{Q}_0 &= \overline{Q}_1 R_1 \\
A\overline{Q}_1 &= \overline{Q}_2 R_2 \\
A\overline{Q}_2 &= \overline{Q}_3 R_3
\end{aligned}
$$
$$\vdots$$
$$\tag{12.7}$$

Consider the similar iteration $Q_0 = I$, and

$$
\begin{aligned}
A_0 &\equiv A Q_0 = Q_1 R_1' \\
A_1 &\equiv R_1' Q_1 = Q_2 R_2' \\
A_2 &\equiv R_2' Q_2 = Q_3 R_3'
\end{aligned}
$$
$$\vdots$$
$$\tag{12.8}$$

which we will call the **unshifted QR algorithm**. The only difference is that A is not needed after the first step; it is replaced by the current R_k. Comparing (12.7) and (12.8) shows that we could choose $Q_1 = \overline{Q}_1$ and $R_1 = R_1'$ in (12.7). Furthermore, since

$$\overline{Q}_2 R_2 = A\overline{Q}_1 = Q_1 R_1' \overline{Q}_1 = Q_1 R_1' Q_1 = Q_1 Q_2 R_2', \tag{12.9}$$

we could choose $\overline{Q}_2 = Q_1 Q_2$ and $R_2 = R_2'$ in (12.7). In fact, if we have chosen $\overline{Q}_{k-1} = Q_1 \cdots Q_{k-1}$ and $R_{j-1} = R_{j-1}'$, then

$$
\begin{aligned}
\overline{Q}_j R_j &= A\overline{Q}_{j-1} = A Q_1 \cdots Q_{j-1} \\
&= \overline{Q}_2 R_2 Q_2 \cdots Q_{j-1} \\
&= \overline{Q}_2 Q_3 R_3 Q_3 \cdots Q_{j-1} \\
&= Q_1 Q_2 Q_3 Q_4 R_4 Q_4 \cdots Q_{j-1} \\
&= \cdots = Q_1 \cdots Q_j R_j,
\end{aligned}
\tag{12.10}
$$

and we may define $\overline{Q}_j = Q_1 \cdots Q_j$ and $R_j = R_j'$ in (12.7).

Therefore, the unshifted QR algorithm does the same calculations as Normalized Simultaneous Iteration, with slightly different notation. Note also that

$$A_{j-1} = Q_j R_j = Q_j R_j Q_j Q_j^T = Q_j A_j Q_j^T, \tag{12.11}$$

so that all A_j are similar matrices and have the same set of eigenvalues.

MATLAB code
shown here can be found
at bit.ly/2NMnMfG

```
% Program 12.5 Unshifted QR Algorithm
% Computes eigenvalues/vectors of symmetric matrix
% Input: matrix A, number of steps k
% Output: eigenvalues lam and eigenvector matrix Qbar
function [lam,Qbar]=unshiftedqr(A,k)
```

```
[m,n]=size(A);
Q=eye(m,m);
Qbar=Q; R=A;
for j=1:k
    [Q,R]=qr(R*Q);          % QR factorization
    Qbar=Qbar*Q;            % accumulate Q's
end
lam=diag(R*Q);              % diagonal converges to eigenvalues
```

THEOREM 12.4 Assume that A is a symmetric $m \times m$ matrix with eigenvalues λ_i satisfying $|\lambda_1| > |\lambda_2| > \cdots > |\lambda_m|$. The unshifted QR algorithm converges linearly to the eigenvectors and eigenvalues of A. As $j \to \infty$, A_j converges to a diagonal matrix containing the eigenvalues on the main diagonal and $\overline{Q}_j = Q_1 \cdots Q_j$ converges to an orthogonal matrix whose columns are the eigenvectors. ■

A proof of Theorem 12.4 can be found in Golub and Van Loan [2012]. Normalized Simultaneous Iteration, essentially the same algorithm, converges under the same conditions. Note that the unshifted QR algorithm may fail even for symmetric matrices if the hypotheses of the theorem are not met. See Exercise 5.

Although unshifted QR is an improved version of Power Iteration, the conditions required by Theorem 12.4 are strict, and a couple of improvements are needed to make this eigenvalue finder work more generally—for example, in the case of nonsymmetric matrices. One problem, which also occurs for symmetric matrices, is that unshifted QR is not guaranteed to work in the case of a tie for dominant eigenvector. An example of this is

$$A = \begin{bmatrix} 0 & 1 \\ 1 & 0 \end{bmatrix},$$

which has eigenvalues 1 and -1. Another form of "tie" occurs when the eigenvalues are complex. The eigenvalues of the nonsymmetric matrix

$$A - \begin{bmatrix} 0 & 1 \\ -1 & 0 \end{bmatrix}$$

are i and $-i$, both of complex magnitude 1. Nothing in the definition of the unshifted QR algorithm allows for the computation of complex eigenvalues. Furthermore, unshifted QR does not make use of the trick of Inverse Power Iteration. We found that Power Iteration could be sped up considerably with this trick, and we want to find a way to apply the idea to our new implementation. These refinements are applied next, after introducing the goal of the QR algorithm, which is to reduce the matrix A to its real Schur form.

12.2.2 Real Schur form and the QR algorithm

The way the QR algorithm finds eigenvalues of a matrix A is to locate a similar matrix whose eigenvalues are obvious. An example of the latter is real Schur form.

DEFINITION 12.5 A matrix T has **real Schur form** if it is upper triangular, except possibly for 2×2 blocks on the main diagonal. ❏

For example, a matrix of the form

$$
\begin{bmatrix}
x & x & x & x & x \\
  & x & x & x & x \\
  &   & x & x & x \\
  &   & x & x & x \\
  &   &   &   & x
\end{bmatrix}
$$

has real Schur form. According to Exercise 6, the eigenvalues of a matrix in this form are the eigenvalues of the diagonal block—diagonal entries when the block is 1×1, or the eigenvalues of the 2×2 block in that case. Either way, the eigenvalues of the matrix are quickly calculated.

The value of the definition is that every square matrix with real entries is similar to one of this form. This is the conclusion of the following theorem, proved in Golub and Van Loan [1996]:

THEOREM 12.6   Let A be a square matrix with real entries. Then there exists an orthogonal matrix Q and a matrix T in real Schur form such that $A = Q^T T Q$. ■

The so-called Schur factorization of the matrix A is an "eigenvalue-revealing factorization," meaning that if we can perform it, we will know the eigenvalues and eigenvectors.

The full QR algorithm iteratively moves an arbitrary matrix A toward its Schur factorization by a series of similarity transformations. We will proceed in two stages. First we will install the Inverse Power Iteration idea with shifts and add the idea of deflation to develop the shifted QR algorithm. Then we will develop an improved version that allows for complex eigenvalues.

The shifted version is straightforward to write. Each step consists of applying the shift, completing a QR factorization, and then taking the shift back. In symbols,

$$
\begin{aligned}
A_0 - sI &= Q_1 R_1 \\
A_1 &= R_1 Q_1 + sI.
\end{aligned}
\tag{12.12}
$$

Note that

$$
\begin{aligned}
A_1 - sI &= R_1 Q_1 \\
&= Q_1^T (A_0 - sI) Q_1 \\
&= Q_1^T A_0 Q_1 - sI
\end{aligned}
$$

implies that A_1 is similar to A_0 and so has the same eigenvalues. We repeat this step, generating a sequence A_k of matrices, all similar to $A = A_0$.

What are good choices for the shift s? This leads us to the concept of **deflation** for eigenvalue calculations. We will choose the shift to be the bottom right entry of the matrix A_k. This will cause the iteration, as it converges to real Schur form, to move the bottom row to a row of zeros, except for the bottom right entry. After this entry has converged to an eigenvalue, we deflate the matrix by eliminating the last row and column. Then we proceed to find the rest of the eigenvalues.

A first try at the shifted QR algorithm is given in the MATLAB code shown in Program 12.6. At each step, we apply a shifted QR step, and then check the bottom row. If all entries are small except the diagonal entry a_{nn}, we declare that entry to be an eigenvalue and deflate by ignoring the last row and last column for the rest of the computation. This program will succeed under the hypotheses of Theorem 12.4. Complex eigenvalues, or real eigenvalues of equal magnitude, will cause problems, which we will solve in a more sophisticated version later. Exercise 7 illustrates the shortcomings of this preliminary version of the QR algorithm.

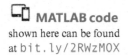

MATLAB code

shown here can be found at `bit.ly/2ylMyyA`

```
% Program 12.6 Shifted QR Algorithm, preliminary version
% Computes eigenvalues of matrices without equal size eigenvalues
% Input: matrix a
% Output: eigenvalues lam
function lam=shiftedqr0(a)
tol=1e-14;
m=size(a,1);lam=zeros(m,1);
n=m;
while n>1
    while max(abs(a(n,1:n-1)))>tol
        mu=a(n,n);         % define shift mu
        [q,r]=qr(a-mu*eye(n));
        a=r*q+mu*eye(n);
    end
    lam(n)=a(n,n);         % declare eigenvalue
    n=n-1;                 % decrement n
    a=a(1:n,1:n);          % deflate
end
lam(1)=a(1,1);             % 1x1 matrix remains
```

Finally, to allow for the calculation of complex eigenvalues, we must allow for the existence of 2×2 blocks on the diagonal of the real Schur form. The improved version of the shifted QR algorithm given in Program 12.7 tries to iterate the matrix to a 1×1 diagonal block in the bottom right corner; if it fails (after a user-specified number of tries), it declares a 2×2 block, finds the pair of eigenvalues, and then deflates by 2. This improved version will converge to real Schur form for most, but not all, input matrices. To round up a final few holdouts, as well as make the algorithm more efficient, we will develop upper Hessenberg form in the next section.

MATLAB code

shown here can be found at `bit.ly/2RWzMOX`

```
% Program 12.7 Shifted QR Algorithm, general version
% Computes real and complex eigenvalues of square matrix
% Input: matrix a
% Output: eigenvalues lam
function lam=shiftedqr(a)
tol=1e-14;kounttol=500;
m=size(a,1);lam=zeros(m,1);
n=m;
while n>1
    kount=0;
    while max(abs(a(n,1:n-1)))>tol & kount<kounttol
        kount=kount+1;     % keep track of number of qr's
        mu=a(n,n);         % shift is mu
        [q,r]=qr(a-mu*eye(n));
        a=r*q+mu*eye(n);
    end
    if kount<kounttol      % have isolated 1x1 block
        lam(n)=a(n,n);     % declare eigenvalue
        n=n-1;
        a=a(1:n,1:n);      % deflate by 1
    else                   % have isolated 2x2 block
        disc=(a(n-1,n-1)-a(n,n))^2+4*a(n,n-1)*a(n-1,n);
        lam(n)=(a(n-1,n-1)+a(n,n)+sqrt(disc))/2;
        lam(n-1)=(a(n-1,n-1)+a(n,n)-sqrt(disc))/2;
        n=n-2;
```

```
            a=a(1:n,1:n);        % deflate by 2
      end
end
if n>0;lam(1)=a(1,1);end    % only a 1x1 block remains
```

Even in its general form, the shifted QR algorithm fails for the following example:

$$A = \begin{bmatrix} 0 & 0 & 0 & 1 \\ 0 & 0 & -1 & 0 \\ 0 & 1 & 0 & 0 \\ -1 & 0 & 0 & 0 \end{bmatrix} \tag{12.13}$$

Matrices like this one, with a repeated complex eigenvalue, may not be moved into real Schur form by shifted QR. The extra assistance needed for these more difficult examples is to replace A by a similar matrix in upper Hessenberg form, which is the focus of the next section.

12.2.3 Upper Hessenberg form

Efficiency of the QR algorithm increases considerably if we first put A into upper Hessenberg form. The idea is to apply similarity transformations, before beginning the QR iteration, that put as many zeros into A as possible while preserving all eigenvalues. In addition, upper Hessenberg form will eliminate the final difficulty with the version of QR algorithm we have developed—convergence to multiple complex eigenvalues—by ensuring that the QR iteration will always proceed to 1×1 or 2×2 blocks.

DEFINITION 12.7   The $m \times n$ matrix A is in **upper Hessenberg form** if $a_{ij} = 0$ for $i > j + 1$.   ◻

A matrix of the form

$$\begin{bmatrix} x & x & x & x & x \\ x & x & x & x & x \\ & x & x & x & x \\ & & x & x & x \\ & & & x & x \end{bmatrix}$$

is upper Hessenberg. There is a finite algorithm for putting matrices in upper Hessenberg form by similarity transformations.

THEOREM 12.8   Let A be a square matrix. There exists an orthogonal matrix Q such that $A = QBQ^T$ and B is in upper Hessenberg form.   ■

We will construct B by using the Householder reflectors of Section 4.3.3, where they were used to construct the QR factorization. However, there is a major difference: Now we care about multiplication by the reflector H on the left *and* right of the matrix, since we want to end up with a similar matrix with identical eigenvalues. Because of this, we must be less aggressive about the zeros we can install into A.

Define x to be the $n - 1$ vector consisting of all but the first entry of the first column of A. Let $\hat{H}_1$ be the Householder reflector that moves x to $(\pm||x||, 0, \ldots, 0)$. (As noted in Chapter 4, we should choose the sign as $-\text{sign}(x_1)$ to avoid cancellation problems in practice, but the theory holds for either choice.) Let H_1 be the orthogonal matrix formed by inserting $\hat{H}_1$ into the bottom $(n - 1) \times (n - 1)$ corner of the $n \times n$ identity matrix. Then we have

$$H_1 A = \begin{bmatrix} 1 & 0 & 0 & 0 & 0 \\ 0 & & & & \\ 0 & & \hat{H}_1 & & \\ 0 & & & & \\ 0 & & & & \end{bmatrix} \begin{bmatrix} x & x & x & x & x \\ x & x & x & x & x \\ x & x & x & x & x \\ x & x & x & x & x \\ x & x & x & x & x \end{bmatrix} = \begin{bmatrix} x & x & x & x & x \\ x & x & x & x & x \\ 0 & x & x & x & x \\ 0 & x & x & x & x \\ 0 & x & x & x & x \end{bmatrix}.$$

Before we can evaluate our success in putting zeros in the matrix, we need to finish the similarity transformation by multiplying by H_1^{-1} on the right. Recall that Householder reflectors are symmetric orthogonal matrices, so that $H_1^{-1} = H_1^T = H_1$. Thus,

$$H_1 A H_1 = \begin{bmatrix} x & x & x & x & x \\ x & x & x & x & x \\ 0 & x & x & x & x \\ 0 & x & x & x & x \\ 0 & x & x & x & x \end{bmatrix} \begin{bmatrix} 1 & 0 & 0 & 0 & 0 \\ 0 & & & & \\ 0 & & \hat{H}_1 & & \\ 0 & & & & \\ 0 & & & & \end{bmatrix} = \begin{bmatrix} x & x & x & x & x \\ x & x & x & x & x \\ 0 & x & x & x & x \\ 0 & x & x & x & x \\ 0 & x & x & x & x \end{bmatrix}.$$

The zeros made in $H_1 A$ are not changed in the matrix $H_1 A H_1$. However, note that if we would have tried to eliminate all but one nonzero in the first column, as we did in the QR factorization of the last section, we would have failed to keep the zeros when multiplying on the right. In fact, there is no finite algorithm that computes a similarity transformation between an arbitrary matrix and an upper triangular matrix. If there were, this chapter would be much shorter, since we could read off the eigenvalues of the arbitrary matrix from the diagonal of the similar, upper triangular matrix.

The next step in achieving upper Hessenberg form is to repeat the previous step, using for x the $(n-2)$-dimensional vector consisting of the lower $n-2$ entries of the second column. Let $\hat{H}_2$ be the $(n-2) \times (n-2)$ Householder reflector for the new x, and define H_2 to be the identity matrix with $\hat{H}_2$ in the bottom corner. Then

$$H_2(H_1 A H_1) = \begin{bmatrix} 1 & 0 & 0 & 0 & 0 \\ 0 & 1 & 0 & 0 & 0 \\ 0 & 0 & & & \\ 0 & 0 & & \hat{H}_2 & \\ 0 & 0 & & & \end{bmatrix} \begin{bmatrix} x & x & x & x & x \\ x & x & x & x & x \\ 0 & x & x & x & x \\ 0 & x & x & x & x \\ 0 & x & x & x & x \end{bmatrix} = \begin{bmatrix} x & x & x & x & x \\ x & x & x & x & x \\ 0 & x & x & x & x \\ 0 & 0 & x & x & x \\ 0 & 0 & x & x & x \end{bmatrix},$$

and further, check that like H_1, multiplication on the right by H_2 does not adversely affect the zeros already obtained. If $n = 5$, then after one more step, we obtain the 5×5 matrix

$$H_3 H_2 H_1 A H_1^T H_2^T H_3^T = H_3 H_2 H_1 A (H_3 H_2 H_1)^T = Q A Q^T$$

in upper Hessenberg form. Since the matrix is similar to A, it has the same eigenvalues and multiplicities as A. In general, for an $n \times n$ matrix A, $n - 2$ Householder steps are needed to put A into upper Hessenberg form.

► **EXAMPLE 12.2** Put $\begin{bmatrix} 2 & 1 & 0 \\ 3 & 5 & -5 \\ 4 & 0 & 0 \end{bmatrix}$ into upper Hessenberg form.

Let $x = [3, 4]$. Earlier, we found the Householder reflector

$$\hat{H}_1 x = \begin{bmatrix} 0.6 & 0.8 \\ 0.8 & -0.6 \end{bmatrix} \begin{bmatrix} 3 \\ 4 \end{bmatrix} = \begin{bmatrix} 5 \\ 0 \end{bmatrix}.$$

Therefore,

$$H_1 A = \begin{bmatrix} 1 & 0 & 0 \\ 0 & 0.6 & 0.8 \\ 0 & 0.8 & -0.6 \end{bmatrix} \begin{bmatrix} 2 & 1 & 0 \\ 3 & 5 & -5 \\ 4 & 0 & 0 \end{bmatrix} = \begin{bmatrix} 2 & 1 & 0 \\ 5 & 3 & -3 \\ 0 & 4 & -4 \end{bmatrix}$$

and

$$A' \equiv H_1 A H_1 = \begin{bmatrix} 2 & 1 & 0 \\ 5 & 3 & -3 \\ 0 & 4 & -4 \end{bmatrix} \begin{bmatrix} 1 & 0 & 0 \\ 0 & 0.6 & 0.8 \\ 0 & 0.8 & -0.6 \end{bmatrix} = \begin{bmatrix} 2.0 & 0.6 & 0.8 \\ 5.0 & -0.6 & 4.2 \\ 0.0 & -0.8 & 5.6 \end{bmatrix}.$$

The result is a matrix A' that is in upper Hessenberg form and is similar to A. ◀

Next we implement the preceding strategy and build an algorithm for finding Q, using Householder reflections:

MATLAB code
shown here can be found
at bit.ly/2ymYSi1

```
% Program 12.8 Upper Hessenberg form
% Input: matrix a
% Output: Hessenberg form matrix a and reflectors v
% Usage: [a,v]=hessen(a) yields similar matrix a of
%   Hessenberg form and a matrix v whose columns hold
%   the v's defining the Householder reflectors.
function [a,v]=hessen(a)
[m,n]=size(a);
v=zeros(m,m);
for k=1:m-2
  x=a(k+1:m,k);
  v(1:m-k,k)=-sign(x(1)+eps)*norm(x)*eye(m-k,1)-x;
  v(1:m-k,k)=v(1:m-k,k)/norm(v(1:m-k,k));
  a(k+1:m,k:m)=a(k+1:m,k:m)-2*v(1:m-k,k)*v(1:m-k,k)'*a(k+1:m,k:m);
  a(1:m,k+1:m)=a(1:m,k+1:m)-2*a(:,k+1:m)*v(1:m-k,k)*v(1:m-k,k)';
end
```

One advantage of upper Hessenberg form for eigenvalue computations is that only 2×2 blocks can occur along the diagonal during the QR algorithm, eliminating the difficulty caused by repeated complex eigenvalues of the previous section.

▶ **EXAMPLE 12.3**   Find the eigenvalues of the matrix (12.13).

For

$$A = \begin{bmatrix} 0 & 0 & 0 & 1 \\ 0 & 0 & -1 & 0 \\ 0 & 1 & 0 & 0 \\ -1 & 0 & 0 & 0 \end{bmatrix},$$

the similar matrix with upper Hessenberg form given by Householder reflectors is

$$A' = \begin{bmatrix} 0 & 1 & 0 & 0 \\ -1 & 0 & 0 & 0 \\ 0 & 0 & 0 & -1 \\ 0 & 0 & 1 & 0 \end{bmatrix},$$

where $A' = QAQ^T$ and

$$Q = \begin{bmatrix} 1 & 0 & 0 & 0 \\ 0 & 0 & 0 & 1 \\ 0 & 0 & -1 & 0 \\ 0 & 1 & 0 & 0 \end{bmatrix}.$$

The matrix A' is already in real Schur form. Its eigenvalues are the eigenvalues of the two 2×2 matrices along the main diagonal, which are repeated pairs of $\{i, -i\}$. ◄

Thus, we finally have a complete method for finding all eigenvalues of an arbitrary square matrix A. The matrix is first put into upper Hessenberg form with the use of a similarity transformation (Program 12.8), and then the shifted QR algorithm is applied (Program 12.7). The MATLAB `eig` command provides accurate eigenvalues based on this progression of calculations.

There are a few rare matrices that cause the shifted QR algorithm to fail, and some extra enhancements are needed in production code. These matrices have the property that at least three eigenvalues share the same magnitude. See Computer Problem 7 for an example.

There are many alternative techniques to accelerate convergence of the QR algorithm that are not covered here. The QR algorithm is designed for full matrices. For large sparse systems, alternative methods will usually be more efficient; see Saad [2003].

► **ADDITIONAL EXAMPLES**

1. Put the matrix $A = \begin{bmatrix} 4 & 0 & 2 \\ 8 & 5 & 10 \\ 6 & 10 & -5 \end{bmatrix}$ into upper Hessenberg form.

2. Use MATLAB code to put the matrix $A = \begin{bmatrix} 5 & 2 & -2 \\ -12 & -19 & 12 \\ -12 & -22 & 15 \end{bmatrix}$ into upper Hessenberg form and use the shifted QR algorithm to solve for all eigenvalues.

🖵 **Solutions** for Additional Examples can be found at `bit.ly/2AijOYW`

12.2 Exercises

🖵 **Solutions** for Exercises numbered in blue can be found at `bit.ly/2OvpxDg`

1. Put the following matrices in upper Hessenberg form:

(a) $\begin{bmatrix} 1 & 0 & 1 \\ 1 & 1 & 0 \\ 1 & 0 & 0 \end{bmatrix}$ (b) $\begin{bmatrix} 0 & 0 & 1 \\ 0 & 1 & 0 \\ 1 & 0 & 0 \end{bmatrix}$ (c) $\begin{bmatrix} 2 & 1 & 0 \\ 4 & 1 & 1 \\ 3 & 0 & 1 \end{bmatrix}$ (d) $\begin{bmatrix} 1 & 1 & 0 \\ 2 & 3 & 1 \\ 2 & 1 & 0 \end{bmatrix}$

2. Put the matrix $\begin{bmatrix} 1 & 0 & 2 & 3 \\ -1 & 0 & 5 & 2 \\ 2 & -2 & 0 & 0 \\ 2 & -1 & 2 & 0 \end{bmatrix}$ into upper Hessenberg form.

3. Show that a symmetric matrix in Hessenberg form is tridiagonal.

4. Call a square matrix **stochastic** if the entries of each column add to one. Prove that a stochastic matrix (a) has an eigenvalue equal to one, and (b) all eigenvalues are, at most, one in absolute value.

5. Carry out Normalized Simultaneous Iteration with the following matrices, and explain how it fails:

(a) $\begin{bmatrix} 0 & 1 \\ 1 & 0 \end{bmatrix}$ (b) $\begin{bmatrix} 0 & 1 \\ -1 & 0 \end{bmatrix}$

6. (a) Show that the determinant of a matrix in real Schur form is the product of the determinants of the 1×1 and 2×2 blocks on the main diagonal. (b) Show that the

eigenvalues of a matrix in real Schur form are the eigenvalues of the 1×1 and 2×2 blocks on the main diagonal.

7. Decide whether the preliminary version of the QR algorithm finds the correct eigenvalues, both before and after changing to Hessenberg form.

$$\text{(a)} \begin{bmatrix} 1 & 0 & 0 \\ 0 & 0 & 1 \\ 0 & 1 & 0 \end{bmatrix} \quad \text{(b)} \begin{bmatrix} 0 & 0 & 1 \\ 0 & 1 & 0 \\ 1 & 0 & 0 \end{bmatrix}$$

8. Decide whether the general version of the QR algorithm finds the correct eigenvalues, both before and after changing to Hessenberg form, for the matrices in Exercise 7.

12.2 Computer Problems

1. Apply the shifted QR algorithm (preliminary version `shiftedqr0`) with tolerance 10^{-14} directly to the following matrices:

$$\text{(a)} \begin{bmatrix} -3 & 3 & 5 \\ 1 & -5 & -5 \\ 6 & 6 & 4 \end{bmatrix} \quad \text{(b)} \begin{bmatrix} 3 & 1 & 2 \\ 1 & 3 & -2 \\ 2 & 2 & 6 \end{bmatrix}$$

$$\text{(c)} \begin{bmatrix} 17 & 1 & 2 \\ 1 & 17 & -2 \\ 2 & 2 & 20 \end{bmatrix} \quad \text{(d)} \begin{bmatrix} -7 & -8 & 1 \\ 17 & 18 & -1 \\ -8 & -8 & 2 \end{bmatrix}$$

2. Apply the shifted QR algorithm method directly to find all eigenvalues of the following matrices:

$$\text{(a)} \begin{bmatrix} 3 & 1 & -2 \\ 4 & 1 & 1 \\ -3 & 0 & 3 \end{bmatrix} \quad \text{(b)} \begin{bmatrix} 1 & 5 & 4 \\ 2 & -4 & -3 \\ 0 & -2 & 4 \end{bmatrix}$$

$$\text{(c)} \begin{bmatrix} 1 & 1 & -2 \\ 4 & 2 & -3 \\ 0 & -2 & 2 \end{bmatrix} \quad \text{(d)} \begin{bmatrix} 5 & -1 & 3 \\ 0 & 6 & 1 \\ 3 & 3 & -3 \end{bmatrix}$$

3. Apply the shifted QR algorithm method directly to find all eigenvalues of the following matrices:

$$\text{(a)} \begin{bmatrix} -1 & 1 & 3 \\ 3 & 3 & -2 \\ -5 & 2 & 7 \end{bmatrix} \quad \text{(b)} \begin{bmatrix} 7 & -33 & -15 \\ 2 & 26 & 7 \\ -4 & -50 & -13 \end{bmatrix}$$

$$\text{(c)} \begin{bmatrix} 8 & 0 & 5 \\ -5 & 3 & -5 \\ 10 & 0 & 13 \end{bmatrix} \quad \text{(d)} \begin{bmatrix} -3 & -1 & 1 \\ 5 & 3 & -1 \\ -2 & -2 & 0 \end{bmatrix}$$

4. Repeat Computer Problem 3, but precede the application of the QR iteration with reduction to upper Hessenberg form. Print the Hessenberg form and the eigenvalues.

5. Apply the shifted QR algorithm directly to find all real and complex eigenvalues of the following matrices:

$$\text{(a)} \begin{bmatrix} 4 & 3 & 1 \\ -5 & -3 & 0 \\ 3 & 2 & 1 \end{bmatrix} \quad \text{(b)} \begin{bmatrix} 3 & 2 & 0 \\ -4 & -2 & 1 \\ 2 & 1 & 0 \end{bmatrix}$$

$$\text{(c)} \begin{bmatrix} 7 & 2 & -4 \\ -8 & 0 & 7 \\ 2 & -1 & -2 \end{bmatrix} \quad \text{(d)} \begin{bmatrix} 11 & 4 & -2 \\ -10 & 0 & 5 \\ 4 & 1 & 2 \end{bmatrix}$$

6. Use the shifted QR algorithm to find the eigenvalues. In each matrix, all eigenvalues have equal magnitude, so Hessenberg may be needed. Compare the results of QR algorithm before and after reduction to Hessenberg form.

$$
\text{(a)} \quad
\begin{bmatrix}
-5 & -10 & -10 & 5 \\
4 & 16 & 11 & -8 \\
12 & 13 & 8 & -4 \\
22 & 48 & 28 & -19
\end{bmatrix}
\qquad
\text{(b)} \quad
\begin{bmatrix}
7 & 6 & 6 & -3 \\
-26 & -20 & -19 & 10 \\
0 & -1 & 0 & 0 \\
-36 & -28 & -24 & 13
\end{bmatrix}
$$

$$
\text{(c)} \quad
\begin{bmatrix}
13 & 10 & 10 & -5 \\
-20 & -16 & -15 & 8 \\
-12 & -9 & -8 & 4 \\
-30 & -24 & -20 & 11
\end{bmatrix}
$$

7. There are a few remaining matrices for which the shifted QR algorithm will not converge to the exact eigenvalues without extra help. An unassuming but notorious example is the matrix in Hessenberg form

$$
A =
\begin{bmatrix}
0 & 0 & 1 \\
1 & 0 & 0 \\
0 & 1 & 0
\end{bmatrix}.
$$

(a) Find the exact eigenvalues. (b) What is the result of applying `shiftedqr.m` to this matrix? (c) Add uniform random numbers of size ϵ_{mach} to A and repeat part (b). Compare with the exact eigenvalues.

Reality Check **12** *How Search Engines Rate Page Quality*

Web search engines such as `Google.com` distinguish themselves by the quality of their returns to search queries. We will discuss a rough approximation of Google's method for judging the quality of Web pages by using knowledge of the network of links that exists on the Web.

When a Web search is initiated, there is a rather complex series of tasks that are carried out by the search engine. One obvious task is word-matching, to find pages that contain the query words, in the title or body of the page. Another key task is to rate the pages that are identified by the first task, to help the user wade through the possibly large set of choices. For very specific queries, there may be only a few text matches, all of which can be returned to the user. (In the early days of the Web, there was a game to try to discover search queries that resulted in exactly one hit.) In the case of very specific queries, the quality of the returned pages is not so important, since no sorting may be necessary. The need for a quality ranking becomes apparent for more general queries. For example, the Google query "new automobile" returns several million pages, beginning with automobile buying services, a reasonably useful outcome. How is the ranking determined?

The answer to this question is that `Google.com` assigns a nonnegative real number, called the **page rank**, to each Web page that it indexes. The page rank is computed by Google in what is one of the world's largest ongoing Power Iterations for determining eigenvectors. Consider a graph where each of n nodes represents a Web page, and a directed edge from node i to node j means that page i contains a Web link to page j, so that you can move from page i to page j with one click. Figure 12.1 shows two examples of such graphs (note in some cases there are double arrows):

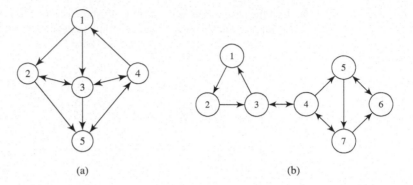

(a)   (b)

Figure 12.1 Two directed graphs of Web pages and links. Each circle represents a Web page. Each directed edge from one page to another means that the first page contains at least one link to the second.

The information in the graph can be expressed as a matrix. Let A denote the **adjacency matrix**, an $n \times n$ matrix whose ijth entry is 1 if there is a link from node j to node i, and 0 otherwise. For the graph in Figure 12.1(a), the adjacency matrix A is the 5×5 matrix

$$\text{FROM}$$

$$\text{TO} \quad \begin{bmatrix} 0 & 0 & 0 & 1 & 0 \\ 1 & 0 & 1 & 0 & 0 \\ 1 & 1 & 0 & 1 & 0 \\ 0 & 0 & 1 & 0 & 1 \\ 0 & 1 & 1 & 1 & 0 \end{bmatrix}.$$

In other words, row i represents the incoming arrows *to* node i, while column j represents the outgoing arrows *from* node j.

The breakthrough that the founders of Google made is very easy to understand, in retrospect. Suppose you start on an arbitrary node of the network (an internet page) and move to another node by following an arrow. If there is more than one arrow, say three arrows to choose from, then pick an arrow at random (equal probability for each) and move to the new node. Continue this forever. If you keep track of the proportion of time that you spend at each node, that is the "page rank" of the node. It turns out that page rank gives a very intuitive value of the importance of the Web page. This allowed Google search to list pages from highest to lowest page rank. Of course, unlike the above example, the real network has billions of nodes, and an even larger number of arrows.

Eigenvectors come into the story as a shortcut to compute the page rank without wandering forever around the graph. Suppose

$$\begin{bmatrix} p_1 \\ p_2 \\ p_3 \\ p_4 \\ p_5 \end{bmatrix}$$

is the vector of probabilities of being at nodes $1, 2, 3, 4$ and 5 during the wandering. Alternatively, think of the p_i as the proportion of time spent by the surfer at node i. Let's also divide each column of the adjacency matrix A by its column sum, and call the result the **google matrix** G. For the example above,

$$G = \begin{bmatrix} 0 & 0 & 0 & \frac{1}{3} & 0 \\ \frac{1}{2} & 0 & \frac{1}{3} & 0 & 0 \\ \frac{1}{2} & \frac{1}{2} & 0 & \frac{1}{3} & 0 \\ 0 & 0 & \frac{1}{3} & 0 & 1 \\ 0 & \frac{1}{2} & \frac{1}{3} & \frac{1}{3} & 0 \end{bmatrix}.$$

A simple way to construct G is to define D to be the diagonal matrix whose entries are the column sums

$$D = \begin{bmatrix} 2 & 0 & 0 & 0 & 0 \\ 0 & 2 & 0 & 0 & 0 \\ 0 & 0 & 3 & 0 & 0 \\ 0 & 0 & 0 & 3 & 0 \\ 0 & 0 & 0 & 0 & 1 \end{bmatrix}$$

and then set

$$G = AD^{-1} = \begin{bmatrix} 0 & 0 & 0 & 1 & 0 \\ 1 & 0 & 1 & 0 & 0 \\ 1 & 1 & 0 & 1 & 0 \\ 0 & 0 & 1 & 0 & 1 \\ 0 & 1 & 1 & 1 & 0 \end{bmatrix} \begin{bmatrix} \frac{1}{2} & 0 & 0 & 0 & 0 \\ 0 & \frac{1}{2} & 0 & 0 & 0 \\ 0 & 0 & \frac{1}{3} & 0 & 0 \\ 0 & 0 & 0 & \frac{1}{3} & 0 \\ 0 & 0 & 0 & 0 & 1 \end{bmatrix} = \begin{bmatrix} 0 & 0 & 0 & \frac{1}{3} & 0 \\ \frac{1}{2} & 0 & \frac{1}{3} & 0 & 0 \\ \frac{1}{2} & \frac{1}{2} & 0 & \frac{1}{3} & 0 \\ 0 & 0 & \frac{1}{3} & 0 & 1 \\ 0 & \frac{1}{2} & \frac{1}{3} & \frac{1}{3} & 0 \end{bmatrix}.$$

Here we are using the fact that multiplying a matrix on the right by a diagonal matrix multiplies the ith column by the ith diagonal entry.

The key fact is the following connection between the google matrix and the vector $\vec{p}$ of probabilities:

$$G \begin{bmatrix} p_1 \\ p_2 \\ p_3 \\ p_4 \\ p_5 \end{bmatrix} = \begin{bmatrix} p_1 \\ p_2 \\ p_3 \\ p_4 \\ p_5 \end{bmatrix}, \tag{12.14}$$

where G is the google matrix. In the above example, it means that

$$\begin{bmatrix} 0 & 0 & 0 & \frac{1}{3} & 0 \\ \frac{1}{2} & 0 & \frac{1}{3} & 0 & 0 \\ \frac{1}{2} & \frac{1}{2} & 0 & \frac{1}{3} & 0 \\ 0 & 0 & \frac{1}{3} & 0 & 1 \\ 0 & \frac{1}{2} & \frac{1}{3} & \frac{1}{3} & 0 \end{bmatrix} \begin{bmatrix} p_1 \\ p_2 \\ p_3 \\ p_4 \\ p_5 \end{bmatrix} = \begin{bmatrix} p_1 \\ p_2 \\ p_3 \\ p_4 \\ p_5 \end{bmatrix}.$$

Looked at as a shipping problem, node 2 is being shipped 1/2 of node 1's probability and 1/3 of node 3's probability, and in a steady state, that should match p_2, the probability that node 2 is shipping out.

We recognize (12.14) as the eigenvalue equation $G\vec{p} = 1 \cdot \vec{p}$ with eigenvalue 1 and eigenvector $\vec{p}$. So instead of needing a computer simulation to wander through the graph, all we have to do is compute the eigenvector corresponding to eigenvalue 1. The five entries in the eigenvector will be the page ranks of the five nodes of the graph. The eigenvector of G corresponding to eigenvalue 1 is

$$\vec{p} = \begin{bmatrix} 0.1042 \\ 0.1250 \\ 0.2187 \\ 0.3125 \\ 0.2396 \end{bmatrix}.$$

These are the page ranks of pages 1 through 5. Therefore, node 4 is the most influential Web page, followed by 5, 3, 2, and 1 in that order. Here we have "normalized" the eigenvector by finding the scalar multiple that makes the sum of the entries equal to 1, so that they are interpretable as probabilities. (Any scalar multiple of an eigenvector is also an eigenvector corresponding to the same eigenvalue.)

A matrix like G, whose columns each add up to 1, is called a **stochastic matrix**. According to Exercise 12.2.4, a stochastic matrix always has an eigenvalue equal to 1, and all other eigenvalues are smaller in absolute value. The idea to measure influence by steady-state probabilities goes back at least to Pinski and Narin [1976].

Suggested activities:

1. Verify the page rank eigenvector p for Figure 12.1(a).

2. Find the adjacency matrix, google matrix, and page rank eigenvector for Figure 12.1(b).

3. An innovation of Brin and Page [1998], the originators of Google, was the *jump probability q*, a number between 0 and 1 that represents the probability that the surfer moves to a random page on the Web, instead of clicking on a link on the current page. Explain why this means we should replace the adjacency matrix A in the above reasoning by the matrix $A' = qI + (1 - q)A$, where I is the $n \times n$ identity matrix. Prove that $G = A'D^{-1}$ is a stochastic matrix for any q. (Note that D still consists of the columns sums of A.)

4. Find the page rank eigenvectors from both graphs in Figure 12.1 with jump probability (a) $q = 0.15$ and (b) $q = 0.5$. Describe the resulting changes in page rank, quantitatively and qualitatively.

5. Set $q = 0.15$. Suppose that Page 2 in the Figure 12.1(a) network attempts to improve its page rank by persuading Pages 1 and 3 to more prominently display its links to Page 2. Model this by replacing A_{21} and A_{23} by 2 in the adjacency matrix. Does this strategy succeed? What other changes in relative page ranks do you see?

6. Study the effect of removing Page 5 from the Figure 12.1(b) network. (All links to and from Page 5 are deleted.) Which page ranks increase, and which decrease?

7. Design your own network, compute page ranks, and analyze according to the preceding questions.

12.3 SINGULAR VALUE DECOMPOSITION

The image of the unit sphere in $\mathbb{R}^n$ under an $m \times n$ matrix is an ellipsoid in $\mathbb{R}^m$. This interesting fact underlies the singular value decomposition, which has many applications in matrix analysis in general and especially for compression purposes. In Figure 12.2, think of taking the vector v corresponding to each point on the unit circle, multiplying by A, and then plotting the endpoint of the resulting vector Av. The result is the ellipse shown. In order to describe the ellipse, it helps to use an orthonormal set of vectors to define the basis of a coordinate system.

12.3.1 Geometry of the SVD

Figure 12.2 is an illustration of the ellipse that corresponds to the matrix

$$A = \begin{bmatrix} 3 & 0 \\ 0 & \frac{1}{2} \end{bmatrix}.$$

(12.15)

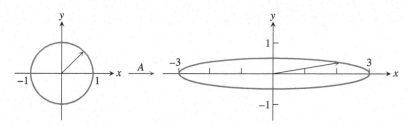

Figure 12.2 The image of the unit circle under a 2×2 **matrix.** The unit circle in R^2 is mapped to the ellipse with semimajor axes $(3,0)$ and $(0,1/2)$ by matrix A in (12.15).

We first consider the case $m \geq n$ of tall, thin matrices. All facts concerning the opposite case $m \leq n$ will be derived by transposing the corresponding facts about this case.

We will see in Theorem 12.11 that for every $m \times n$ matrix A, there are orthonormal sets $\{u_1, \ldots, u_m\}$ and $\{v_1, \ldots, v_n\}$, together with nonnegative numbers $s_1 \geq \cdots \geq s_n \geq 0$, satisfying

$$Av_1 = s_1 u_1$$
$$Av_2 = s_2 u_2$$
$$\vdots$$
$$Av_n = s_n u_n. \tag{12.16}$$

The vectors are visualized in Figure 12.3. The v_i are called the **right singular vectors** of the matrix A, the u_i are the **left singular vectors** of A, and the s_i are the **singular values** of A. (The reason for this terminology will become clear shortly.)

This useful fact immediately explains why a 2×2 matrix maps the unit circle into an ellipse. We can think of the v_i's as the basis of a rectangular coordinate system on which A acts in a simple way: It produces the basis vectors of a new coordinate system, the u_i's, with some stretching quantified by the scalars s_i. The stretched basis vectors $s_i u_i$ are the semimajor axes of the ellipse, as shown in Figure 12.3.

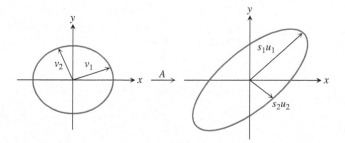

Figure 12.3 The ellipse associated to a matrix. Every 2×2 matrix A can be viewed in the following simple way: There is a coordinate system $\{v_1, v_2\}$ for which A sends $\mathbf{v_1} \to s_1 \mathbf{u_1}$ and $\mathbf{v_1} \to s_2 \mathbf{u_1}$, where $\{u_1, u_2\}$ is another coordinate system and s_1, s_2 are nonnegative numbers. This picture extends to a transformation from $\mathbb{R}^n$ to $\mathbb{R}^m$ for an $m \times n$ matrix.

▶ **EXAMPLE 12.4** Find the singular values and singular vectors for the matrix (12.15) represented in Figure 12.2.

Clearly, the matrix stretches by 3 in the x-direction and shrinks by a factor of $1/2$ in the y-direction. The singular vectors and values of A are

$$A \begin{bmatrix} 1 \\ 0 \end{bmatrix} = 3 \begin{bmatrix} 1 \\ 0 \end{bmatrix}$$

$$A \begin{bmatrix} 0 \\ 1 \end{bmatrix} = \frac{1}{2} \begin{bmatrix} 0 \\ 1 \end{bmatrix}. \tag{12.17}$$

The vectors $3(1,0)$ and $\frac{1}{2}(0,1)$ form the semimajor axes of the ellipse. The right singular vectors are $[1,0], [0,1]$, and the left singular vectors are $[1,0], [0,1]$. The singular values are 3 and $1/2$. ◀

▶ **EXAMPLE 12.5** Find the singular values and singular vectors of

$$A = \begin{bmatrix} 0 & -\frac{1}{2} \\ 3 & 0 \\ 0 & 0 \end{bmatrix}. \tag{12.18}$$

This is a slight variation on Example 12.4. The matrix exchanges the x- and y-axes, with some changing of scale, and adds a z-axis, along which nothing happens. The singular vectors and values of A are

$$Av_1 = A \begin{bmatrix} 1 \\ 0 \end{bmatrix} = 3 \begin{bmatrix} 0 \\ 1 \\ 0 \end{bmatrix} = s_1 u_1$$

$$Av_2 = A \begin{bmatrix} 0 \\ 1 \end{bmatrix} = \frac{1}{2} \begin{bmatrix} -1 \\ 0 \\ 0 \end{bmatrix} = s_2 u_2. \tag{12.19}$$

The right singular vectors are $[1,0], [0,1]$, and the left singular vectors are $[0,1,0], [-1,0,0]$. The singular values are $3, 1/2$. Notice that we always require the s_i to be a nonnegative number, and any necessary negative signs are absorbed in the u_i and v_i. ◀

There is a standard way to keep track of this information, in a matrix factorization of the $m \times n$ matrix A. Form an $m \times m$ matrix U whose columns are the left singular vectors u_i, an $n \times n$ matrix V whose columns are the right singular vectors v_i, and a diagonal $m \times n$ matrix S whose diagonal entries are the singular values s_i. Then the **singular value decomposition** (SVD) of the $m \times n$ matrix A is

$$A = USV^T. \tag{12.20}$$

Example 12.5 has the SVD representation

$$\begin{bmatrix} 0 & -\frac{1}{2} \\ 3 & 0 \\ 0 & 0 \end{bmatrix} = \begin{bmatrix} 0 & -1 & 0 \\ 1 & 0 & 0 \\ 0 & 0 & 1 \end{bmatrix} \begin{bmatrix} 3 & 0 \\ 0 & \frac{1}{2} \\ 0 & 0 \end{bmatrix} \begin{bmatrix} 1 & 0 \\ 0 & 1 \end{bmatrix}. \tag{12.21}$$

Since U and V are square matrices with orthonormal columns, they are orthogonal matrices. Note that we had to add a third column u_3 to U to complete the basis of R^3. Finally, the terminology can be explained. The u_i (v_i) are the left (right) singular vectors because they appear on that side in the matrix representation (12.20).

12.3.2 Finding the SVD in general

We have shown two simple examples of the SVD. To show that the SVD exists for a general matrix A, we need the following lemma:

LEMMA 12.10   Let A be an $m \times n$ matrix. The eigenvalues of $A^T A$ are nonnegative.   ■

> **Proof.** Let v be a unit eigenvector of $A^T A$, and $A^T A v = \lambda v$. Then
>
> $$0 \le ||Av||^2 = v^T A^T A v = \lambda v^T v = \lambda.$$
>   ❑

For an $m \times n$ matrix A, the $n \times n$ matrix $A^T A$ is symmetric, so its eigenvectors are orthogonal and its eigenvalues are real. Lemma 12.10 shows that the eigenvalues are nonnegative real numbers and so should be expressed as $s_1^2 \ge \cdots \ge s_n^2$, where the corresponding orthonormal set of eigenvectors is $\{v_1, \ldots, v_n\}$. This already gives us two-thirds of the SVD. Use the following directions to find the u_i for $1 \le i \le m$:

If $s_i \ne 0$, define u_i by the equation $s_i u_i = A v_i$.
Choose each remaining u_i as an arbitrary unit vector subject to being orthogonal to $u_1, \ldots, u_{i-1}$.

The reader should check that this choice implies that $u_1, \ldots, u_m$ are pairwise orthogonal unit vectors, and therefore another orthonormal basis of R^m. In fact, $u_1, \ldots, u_m$ forms an orthonormal set of eigenvectors of $A A^T$. (See Exercise 4.) Summarizing, we have proved the following Theorem:

THEOREM 12.11   Let A be an $m \times n$ matrix where $m \ge n$. Then there exist two orthonormal bases $\{v_1, \ldots, v_n\}$ of R^n, and $\{u_1, \ldots, u_m\}$ of R^m, and real numbers $s_1 \ge \cdots \ge s_n \ge 0$ such that $A v_i = s_i u_i$ for $1 \le i \le n$. The columns of $V = [v_1 | \ldots | v_n]$, the right singular vectors, are the set of orthonormal eigenvectors of $A^T A$; and the columns of $U = [u_1 | \ldots | u_m]$, the left singular vectors, are the set of orthonormal eigenvectors of $A A^T$.   ■

The SVD is not unique for a given matrix A. In the defining equation $A v_1 = s_1 u_1$, for example, replacing v_1 by $-v_1$ and u_1 by $-u_1$ does not change the equality, but changes the matrices U and V.

We conclude from this theorem that the image of the unit sphere of vectors is an ellipsoid of vectors, centered at the origin, with semimajor axes $s_i u_i$. Figure 12.3 shows that the unit circle of vectors is mapped into an ellipse with axes $\{s_1 u_1, s_2 u_2\}$. To find where Ax goes for a vector x, we can write $x = a_1 v_1 + a_2 v_2$ (where $a_1 v_1$ ($a_2 v_2$) is the projection of x onto the direction v_1 (v_2)), and then $Ax = a_1 s_1 u_1 + a_2 s_2 u_2$.

The matrix representation (12.20) follows directly from Theorem 12.11. Define S to be an $m \times n$ diagonal matrix whose entries are $s_1 \ge \cdots \ge s_{\min\{m,n\}} \ge 0$. Define U to be the matrix whose columns are $u_1, \ldots, u_m$, and V to be the matrix whose columns are $v_1, \ldots, v_n$. Notice that $U S V^T v_i = s_i u_i$ for $i = 1, \ldots, m$. Since the matrices A and $U S V^T$ agree on the basis $v_1, \ldots, v_n$, they are identical $m \times n$ matrices.

▶ **EXAMPLE 12.6**   Find the singular value decomposition of the 4×2 matrix

$$A = \begin{bmatrix} 3 & 3 \\ -3 & -3 \\ -1 & 1 \\ 1 & -1 \end{bmatrix}.$$

The eigenvectors and eigenvalues of

$$A^T A = \begin{bmatrix} 20 & 16 \\ 16 & 20 \end{bmatrix}, \tag{12.22}$$

arranged in decreasing size of eigenvalue, are $v_1 = [1/\sqrt{2}, 1/\sqrt{2}]^T$, $s_1^2 = 36$; and $v_2 = [1/\sqrt{2}, -1/\sqrt{2}]^T$, $s_1^2 = 4$. The singular values are $s_1 = 6$ and $s_2 = 2$. According to the previous directions, u_1 and u_2 are defined by

$$6u_1 = Av_1 = \begin{bmatrix} 3\sqrt{2} \\ -3\sqrt{2} \\ 0 \\ 0 \end{bmatrix} \qquad 2u_2 = Av_2 = \begin{bmatrix} 0 \\ 0 \\ -\sqrt{2} \\ \sqrt{2} \end{bmatrix} \tag{12.23}$$

yielding

$$u_1 = \begin{bmatrix} \frac{1}{\sqrt{2}} \\ -\frac{1}{\sqrt{2}} \\ 0 \\ 0 \end{bmatrix} \qquad u_2 = \begin{bmatrix} 0 \\ 0 \\ -\frac{1}{\sqrt{2}} \\ \frac{1}{\sqrt{2}} \end{bmatrix}.$$

We choose

$$u_3 = \begin{bmatrix} \frac{1}{\sqrt{2}} \\ \frac{1}{\sqrt{2}} \\ 0 \\ 0 \end{bmatrix} \qquad u_4 = \begin{bmatrix} 0 \\ 0 \\ \frac{1}{\sqrt{2}} \\ \frac{1}{\sqrt{2}} \end{bmatrix}$$

to complete the orthonormal basis of $\mathbb{R}^4$. We have determined the SVD to be

$$A = \begin{bmatrix} 3 & 3 \\ -3 & -3 \\ -1 & 1 \\ 1 & -1 \end{bmatrix} = USV^T = \begin{bmatrix} \frac{1}{\sqrt{2}} & 0 & \frac{1}{\sqrt{2}} & 0 \\ -\frac{1}{\sqrt{2}} & 0 & \frac{1}{\sqrt{2}} & 0 \\ 0 & -\frac{1}{\sqrt{2}} & 0 & \frac{1}{\sqrt{2}} \\ 0 & \frac{1}{\sqrt{2}} & 0 & \frac{1}{\sqrt{2}} \end{bmatrix} \begin{bmatrix} 6 & 0 \\ 0 & 2 \\ 0 & 0 \\ 0 & 0 \end{bmatrix} \begin{bmatrix} \frac{1}{\sqrt{2}} & \frac{1}{\sqrt{2}} \\ \frac{1}{\sqrt{2}} & -\frac{1}{\sqrt{2}} \end{bmatrix}.$$

$$\tag{12.24}$$

A good way to remember the matrix shapes involved in the SVD is that S has the same shape as A; in this case, 4×2. However, notice that because of the zeros in S, the third and fourth columns of U do not materially participate in constructing A. For rectangular matrices, there is an alternative version of the SVD called the **reduced SVD**, which for this matrix is

$$A = \begin{bmatrix} 3 & 3 \\ -3 & -3 \\ -1 & 1 \\ 1 & -1 \end{bmatrix} = U_0 S_0 V^T = \begin{bmatrix} \frac{1}{\sqrt{2}} & 0 \\ -\frac{1}{\sqrt{2}} & 0 \\ 0 & -\frac{1}{\sqrt{2}} \\ 0 & \frac{1}{\sqrt{2}} \end{bmatrix} \begin{bmatrix} 6 & 0 \\ 0 & 2 \end{bmatrix} \begin{bmatrix} \frac{1}{\sqrt{2}} & \frac{1}{\sqrt{2}} \\ \frac{1}{\sqrt{2}} & -\frac{1}{\sqrt{2}} \end{bmatrix}.$$

$$\tag{12.25}$$

In the reduced SVD, either U_0 (if $m \geq n$) or V_0 (if $m \leq n$) have the same shape as A. The reduced SVD is analogous to the reduced QR factorization of Chapter 4. ◀

To find the SVD of a matrix with $m \leq n$, apply what we have done above to A^T to get $A^T = USV^T$. Then $A = (USV^T)^T = VS^T U^T$ is the SVD of A.

▶ **EXAMPLE 12.7** Find the singular value decomposition of the 2×3 matrix

$$A = \begin{bmatrix} -1 & 3 & 7 \\ 7 & 4 & 1 \end{bmatrix}.$$

Since $m \le n$, we find the SVD of A^T and then transpose the result. The eigenvectors and eigenvalues of

$$AA^T = \begin{bmatrix} 59 & 12 \\ 12 & 66 \end{bmatrix}, \tag{12.26}$$

arranged in decreasing size of eigenvalue, are $v_1 = [3/5, 4/5]^T, s_1^2 = 75$; and $v_2 = [-4/5, 3/5]^T, s_2^2 = 50$. The singular values are $s_1 = \sqrt{75} = 5\sqrt{3}$ and $s_2 = \sqrt{50} = 5\sqrt{2}$. This implies u_1 and u_2 are defined by

$$5\sqrt{3}u_1 = A^T v_1 = \begin{bmatrix} 5 \\ 5 \\ 5 \end{bmatrix} \qquad 5\sqrt{2}u_2 = A^T v_2 = \begin{bmatrix} 5 \\ 0 \\ -5 \end{bmatrix}$$

yielding

$$u_1 = \begin{bmatrix} \frac{1}{\sqrt{3}} \\ \frac{1}{\sqrt{3}} \\ \frac{1}{\sqrt{3}} \end{bmatrix} \qquad u_2 = \begin{bmatrix} \frac{1}{\sqrt{2}} \\ 0 \\ -\frac{1}{\sqrt{2}} \end{bmatrix}.$$

Choosing $u_3 = \left[\frac{1}{\sqrt{6}}, -\frac{2}{\sqrt{6}}, \frac{1}{\sqrt{6}}\right]^T$ to complete the basis of $\mathbb{R}^3$, we find the SVD of A^T to be

$$A^T = \begin{bmatrix} -1 & 7 \\ 3 & 4 \\ 7 & 1 \end{bmatrix} = USV^T = \begin{bmatrix} \frac{1}{\sqrt{3}} & \frac{1}{\sqrt{2}} & \frac{1}{\sqrt{6}} \\ \frac{1}{\sqrt{3}} & 0 & -\frac{2}{\sqrt{6}} \\ \frac{1}{\sqrt{3}} & -\frac{1}{\sqrt{2}} & \frac{1}{\sqrt{6}} \end{bmatrix} \begin{bmatrix} 5\sqrt{3} & 0 \\ 0 & 5\sqrt{2} \\ 0 & 0 \end{bmatrix} \begin{bmatrix} 3/5 & 4/5 \\ -4/5 & 3/5 \end{bmatrix}, \tag{12.27}$$

and the SVD of A is

$$A = \begin{bmatrix} -1 & 3 & 7 \\ 7 & 4 & 1 \end{bmatrix} = \begin{bmatrix} 3/5 & 4/5 \\ -4/5 & 3/5 \end{bmatrix} \begin{bmatrix} 5\sqrt{3} & 0 & 0 \\ 0 & 5\sqrt{2} & 0 \end{bmatrix} \begin{bmatrix} \frac{1}{\sqrt{3}} & \frac{1}{\sqrt{3}} & \frac{1}{\sqrt{3}} \\ \frac{1}{\sqrt{2}} & 0 & -\frac{1}{\sqrt{2}} \\ \frac{1}{\sqrt{6}} & -\frac{2}{\sqrt{6}} & \frac{1}{\sqrt{6}} \end{bmatrix}. \tag{12.28}$$

The reduced SVD of A is

$$A = \begin{bmatrix} -1 & 3 & 7 \\ 7 & 4 & 1 \end{bmatrix} = \begin{bmatrix} 3/5 & 4/5 \\ -4/5 & 3/5 \end{bmatrix} \begin{bmatrix} 5\sqrt{3} & 0 \\ 0 & 5\sqrt{2} \end{bmatrix} \begin{bmatrix} \frac{1}{\sqrt{3}} & \frac{1}{\sqrt{3}} & \frac{1}{\sqrt{3}} \\ \frac{1}{\sqrt{2}} & 0 & -\frac{1}{\sqrt{2}} \end{bmatrix}. \tag{12.29}$$

◀

The MATLAB command for the singular value decomposition is svd, and

[u,s,v]=svd(A)

will return all three matrices of the factorization. The command

[u,s,v]=svd(A,0)

will return the reduced SVD if $m \ge n$.

▶ **ADDITIONAL EXAMPLES**

*1. Find the singular values and singular vectors of the symmetric matrix
$$A = \begin{bmatrix} 0 & 1 \\ 0 & -1 \end{bmatrix}.$$

2. Find the reduced singular value decomposition of the matrix
$$A = \begin{bmatrix} 6 & 2 & 10 & -4 & 6 & 4 \\ 8 & 11 & 5 & 3 & 8 & -3 \end{bmatrix}.$$

Solutions for Additional Examples can be found at bit.ly/2ys0LtQ
(* example with video solution)

12.3 Exercises

Solutions for Exercises numbered in blue can be found at bit.ly/2CRnh3f

1. Find the SVD of the following symmetric matrices by hand calculation, and describe geometrically the action of the matrix on the unit circle:

(a) $\begin{bmatrix} -3 & 0 \\ 0 & 2 \end{bmatrix}$ (b) $\begin{bmatrix} 0 & 0 \\ 0 & 3 \end{bmatrix}$ (c) $\begin{bmatrix} \frac{3}{2} & -\frac{1}{2} \\ -\frac{1}{2} & \frac{3}{2} \end{bmatrix}$

(d) $\begin{bmatrix} -\frac{3}{2} & \frac{1}{2} \\ \frac{1}{2} & -\frac{3}{2} \end{bmatrix}$ (e) $\begin{bmatrix} 0.75 & 1.25 \\ 1.25 & 0.75 \end{bmatrix}$

2. Find the SVD of the following matrices by hand calculation:

(a) $\begin{bmatrix} 3 & 0 \\ 4 & 0 \end{bmatrix}$ (b) $\begin{bmatrix} 6 & -2 \\ 8 & \frac{3}{2} \end{bmatrix}$ (c) $\begin{bmatrix} 0 & 1 \\ 0 & 0 \end{bmatrix}$

(d) $\begin{bmatrix} -4 & -12 \\ 12 & 11 \end{bmatrix}$ (e) $\begin{bmatrix} 0 & -2 \\ -1 & 0 \end{bmatrix}$

3. The SVD is not unique. How many different SVDs exist for Example 12.4? List them.

4. (a) Prove that the u_i as defined in Theorem 12.11 are eigenvectors of AA^T. (b) Prove that the u_i are unit vectors. (c) Prove that they form an orthonormal basis of R^m.

5. Prove that for any constants a and b, the nonzero singular values of the 4×2 matrices are 2a and 2b. Find the reduced SVD in terms of a and b.

(a) $\begin{bmatrix} a & a \\ -a & -a \\ b & -b \\ -b & b \end{bmatrix}$ (b) $\begin{bmatrix} -a & a \\ a & -a \\ b & b \\ b & b \end{bmatrix}$

6. Prove that for any constants a and b, the nonzero singular values of the 4×2 matrices are 25a and 25b. Find the reduced SVD in terms of a and b.

(a) $\begin{bmatrix} 9a & 12a \\ -12a & -16a \\ -16b & 12b \\ 12b & -9b \end{bmatrix}$ (b) $\begin{bmatrix} -12a & 9a \\ 16a & -12a \\ 9b & 12b \\ 12b & 16b \end{bmatrix}$

7. Prove that for any constants a and b, the nonzero singular values of the 4×2 matrices are 10a and 10b. Find the reduced SVD in terms of a and b.

$$(a) \begin{bmatrix} 3a+4b & 4a-3b \\ -3a-4b & -4a+3b \\ 3a-4b & 4a+3b \\ -3a+4b & -4a-3b \end{bmatrix} \qquad (b) \begin{bmatrix} 4a-3b & 3a+4b \\ -4a+3b & -3a-4b \\ 4a+3b & 3a-4b \\ -4a-3b & -3a+4b \end{bmatrix}$$

8. Prove that for any constants a and b, the nonzero singular values of the 4×2 matrices are $25\sqrt{2}a$ and $25\sqrt{2}b$. Find the reduced SVD in terms of a and b.

$$(a) \begin{bmatrix} 16a-9b & 12a+12b \\ -12a+12b & -9a-16b \\ 16a+9b & 12a-12b \\ -12a-12b & -9a+16b \end{bmatrix} \qquad (b) \begin{bmatrix} -16a+9b & -12a-12b \\ 16a+9b & 12a-12b \\ -12a-12b & -9a+16b \\ -12a+12b & -9a-16b \end{bmatrix}$$

12.4 APPLICATIONS OF THE SVD

In this section, we gather some useful properties of the SVD and indicate some of their widespread uses. For example, the SVD turns out to be the best means of finding the rank of a matrix. The determinant and inverse of a square matrix, if it exists, can be found from the SVD. Perhaps the most useful applications of the SVD follow from the low rank approximation property.

12.4.1 Properties of the SVD

Assume in the following that $A = USV^T$ is the singular value decomposition. The **rank** of an $m \times n$ matrix A is the number of linearly independent rows (or equivalently, columns).

Property 1 The rank of the matrix $A = USV^T$ is the number of nonzero entries in S.

Proof. Since U and V^T are invertible matrices, rank(A) = rank(S), and the latter is the number of nonzero diagonal entries. ☐

Property 2 If A is an $n \times n$ matrix, $|\det(A)| = s_1 \cdots s_n$.

Proof. Since $U^T U = I$ and $V^T V = I$, the determinants of U and V^T are 1 or -1, due to the fact that the determinant of a product equals the product of the determinants. Property 2 follows from the factorization $A = USV^T$. ☐

Property 3 If A is an invertible $m \times m$ matrix, then $A^{-1} = VS^{-1}U^T$.

Proof. By Property 1, S is invertible, meaning all $s_i > 0$. Now Property 3 follows from the fact that if $A_1, A_2,$ and A_3 are invertible matrices, then $(A_1 A_2 A_3)^{-1} = A_3^{-1} A_2^{-1} A_1^{-1}$. ☐

For example, the SVD

$$\begin{bmatrix} 0 & 1 \\ 1 & \frac{3}{2} \end{bmatrix} = \begin{bmatrix} \frac{1}{\sqrt{5}} & -\frac{2}{\sqrt{5}} \\ \frac{2}{\sqrt{5}} & \frac{1}{\sqrt{5}} \end{bmatrix} \begin{bmatrix} 2 & 0 \\ 0 & \frac{1}{2} \end{bmatrix} \begin{bmatrix} \frac{1}{\sqrt{5}} & \frac{2}{\sqrt{5}} \\ \frac{2}{\sqrt{5}} & -\frac{1}{\sqrt{5}} \end{bmatrix}$$

from (12.29) shows that the inverse matrix is

$$\begin{bmatrix} 0 & 1 \\ 1 & \frac{3}{2} \end{bmatrix}^{-1} = \begin{bmatrix} \frac{1}{\sqrt{5}} & \frac{2}{\sqrt{5}} \\ \frac{2}{\sqrt{5}} & -\frac{1}{\sqrt{5}} \end{bmatrix} \begin{bmatrix} \frac{1}{2} & 0 \\ 0 & 2 \end{bmatrix} \begin{bmatrix} \frac{1}{\sqrt{5}} & \frac{2}{\sqrt{5}} \\ -\frac{2}{\sqrt{5}} & \frac{1}{\sqrt{5}} \end{bmatrix} = \begin{bmatrix} -\frac{3}{2} & 1 \\ 1 & 0 \end{bmatrix}. \quad (12.30)$$

Property 4 The $m \times n$ matrix A can be written as the sum of rank-one matrices

$$A = \sum_{i=1}^{r} s_i u_i v_i^T, \quad (12.31)$$

where r is the rank of A, and u_i and v_i are the ith columns of U and V, respectively.

Proof.

$$A = USV^T = U \begin{bmatrix} s_1 & & \\ & \ddots & \\ & & s_r \end{bmatrix} V^T$$

$$= U \left(\begin{bmatrix} s_1 & & \\ & & \\ & & \end{bmatrix} + \begin{bmatrix} & & \\ & s_2 & \\ & & \end{bmatrix} + \cdots + \begin{bmatrix} & & \\ & & \\ & & s_r \end{bmatrix} \right) V^T$$

$$= s_1 u_1 v_1^T + s_2 u_2 v_2^T + \cdots + s_r u_r v_r^T$$

$\square$

Property 4 is the low rank approximation property of the SVD. The best least squares approximation to A of rank $p \le r$ is provided by retaining the first p terms of (12.31).

▶ **EXAMPLE 12.8** Find the best rank-one approximation of the matrix $\begin{bmatrix} 0 & 1 \\ 1 & \frac{3}{2} \end{bmatrix}$.

Writing out (12.31) yields

$$\begin{bmatrix} 0 & 1 \\ 1 & \frac{3}{2} \end{bmatrix} = \begin{bmatrix} \frac{1}{\sqrt{5}} & -\frac{2}{\sqrt{5}} \\ \frac{2}{\sqrt{5}} & \frac{1}{\sqrt{5}} \end{bmatrix} \begin{bmatrix} 2 & 0 \\ 0 & \frac{1}{2} \end{bmatrix} \begin{bmatrix} \frac{1}{\sqrt{5}} & \frac{2}{\sqrt{5}} \\ \frac{2}{\sqrt{5}} & -\frac{1}{\sqrt{5}} \end{bmatrix}$$

$$= \begin{bmatrix} \frac{1}{\sqrt{5}} & -\frac{2}{\sqrt{5}} \\ \frac{2}{\sqrt{5}} & \frac{1}{\sqrt{5}} \end{bmatrix} \left(\begin{bmatrix} 2 & 0 \\ 0 & 0 \end{bmatrix} + \begin{bmatrix} 0 & 0 \\ 0 & \frac{1}{2} \end{bmatrix} \right) \begin{bmatrix} \frac{1}{\sqrt{5}} & \frac{2}{\sqrt{5}} \\ \frac{2}{\sqrt{5}} & -\frac{1}{\sqrt{5}} \end{bmatrix}$$

$$= 2 \begin{bmatrix} \frac{1}{\sqrt{5}} \\ \frac{2}{\sqrt{5}} \end{bmatrix} \begin{bmatrix} \frac{1}{\sqrt{5}} & \frac{2}{\sqrt{5}} \end{bmatrix} + \frac{1}{2} \begin{bmatrix} -\frac{2}{\sqrt{5}} \\ \frac{1}{\sqrt{5}} \end{bmatrix} \begin{bmatrix} \frac{2}{\sqrt{5}} & -\frac{1}{\sqrt{5}} \end{bmatrix}$$

$$= \begin{bmatrix} \frac{2}{5} & \frac{4}{5} \\ \frac{4}{5} & \frac{8}{5} \end{bmatrix} + \begin{bmatrix} -\frac{2}{5} & \frac{1}{5} \\ \frac{1}{5} & -\frac{1}{10} \end{bmatrix}. \quad (12.32)$$

Notice how the original matrix is separated into a larger contribution plus a smaller contribution, because of the different sizes of the singular values. The best rank-one approximation of the matrix is given by the first rank-one matrix

$$\begin{bmatrix} \frac{2}{5} & \frac{4}{5} \\ \frac{4}{5} & \frac{8}{5} \end{bmatrix},$$

while the second matrix provides small corrections. This is the main idea behind the dimension reduction and compression applications of the SVD. ◀

The next two sections introduce two closely related uses for the SVD. In dimension reduction, the focus is on the approximation of a large collection of multidimensional vectors by a collection of vectors spanning fewer dimensions. The other application is lossy compression, reducing the amount of information needed to approximately represent a matrix. Both applications rely on Property 4 concerning low rank approximation.

12.4.2 Dimension reduction

The idea is to project data into a lower dimension. Assume that $a_1, \ldots, a_n$ comprise a collection of m-dimensional vectors. In data-rich applications, m is far less than n. The goal of dimension reduction is to replace $a_1, \ldots, a_n$ with n vectors that span $p < m$ dimensions, while minimizing the error associated with doing so. Usually we begin with set of vectors with mean zero. If not, we can subtract the mean to achieve this and add it back later.

The SVD gives a straightforward way to carry out the dimension reduction. Consider the data vectors as columns of an $m \times n$ matrix $A = [a_1 | \cdots | a_n]$, and calculate the singular value decomposition $A = USV^T$. Let e_j denote the jth elementary basis vector (all zeros except for jth entry 1). Then $Ae_j = a_j$. Using the rank-p approximation

$$A \approx A_p = \sum_{i=1}^{p} s_i u_i v_i^T$$

of Property 4, we can project a_j into the p-dimensional space spanned by the columns $u_1, \ldots, u_p$ of U by

$$a_j = Ae_j \approx A_p e_j. \tag{12.33}$$

Since multiplying a matrix times e_j just picks out the jth column, we can more efficiently describe our finding as the following:

The space $\langle u_1, \ldots, u_p \rangle$ spanned by the left singular vectors $u_1, \ldots, u_p$ is the best-approximating dimension-p subspace to $a_1, \ldots, a_n$ in the sense of least squares, and the orthogonal projections of the columns a_i of A into this space are the columns of A_p. In other words, the projection of a collection of vectors $a_1, \ldots, a_n$ to their best least squares p-dimensional subspace is precisely the best rank-p approximation matrix A_p. The vectors u_i are often called the *principal components* of the data set.

▶ **EXAMPLE 12.9**   Find the best one-dimensional subspace fitting the data vectors $[-4, -4.5]$, $[0.8, 1.9]$, $[2.6, -0.7]$, $[0.6, 3.3]$.

The four data vectors are shown as points in Figure 12.4. We want to find the subspace that minimizes the sum of squared errors from projecting the vectors into that subspace, and then find the projected vectors.

Use the data vectors as columns of the data matrix

$$A = \begin{bmatrix} -4 & 0.8 & 2.6 & 0.6 \\ -4.5 & 1.9 & -0.7 & 3.3 \end{bmatrix}$$

and find its reduced SVD, which is

$$USV^T = \begin{bmatrix} 0.6 & -0.8 \\ 0.8 & 0.6 \end{bmatrix} \begin{bmatrix} 5\sqrt{2} & 0 \\ 0 & 3 \end{bmatrix} \begin{bmatrix} -0.6\sqrt{2} & 0.2\sqrt{2} & 0.1\sqrt{2} & 0.3\sqrt{2} \\ 1/6 & 1/6 & -5/6 & 1/2 \end{bmatrix}.$$

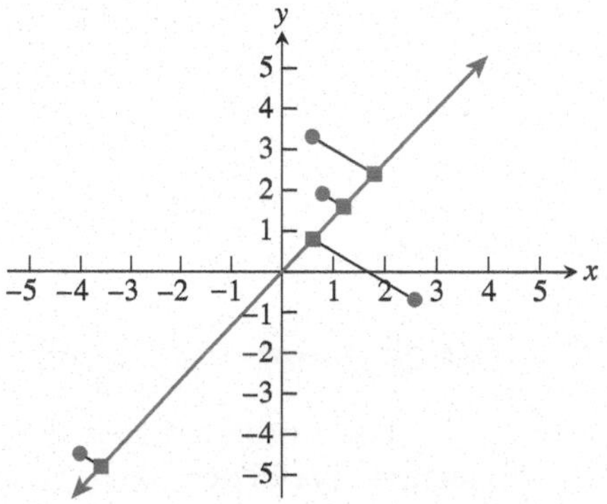

Figure 12.4 Dimension reduction by SVD. Four data points are projected orthogonally to the best one-dimensional subspace. The square symbols show the projections of the four data points onto the subspace.

The best one-dimensional subspace, shown as a line in Figure 12.4, is spanned by $u_1 = [0.6, 0.8]^T$. Dimension reduction of the data set to a subspace of dimension $p = 1$ is equivalent to setting $s_2 = 0$ and reconstituting the matrix. In other words, $A_1 = US_1V^T$, where

$$S_1 = \begin{bmatrix} 5\sqrt{2} & 0 \\ 0 & 0 \end{bmatrix}.$$

Thus, the columns of

$$\begin{aligned} A_1 &= \begin{bmatrix} 0.6 & -0.8 \\ 0.8 & 0.6 \end{bmatrix} \begin{bmatrix} 5\sqrt{2} & 0 \\ 0 & 0 \end{bmatrix} \begin{bmatrix} -0.6\sqrt{2} & 0.2\sqrt{2} & 0.1\sqrt{2} & 0.3\sqrt{2} \\ 1/6 & 1/6 & -5/6 & 1/2 \end{bmatrix} \\ &= \begin{bmatrix} -3.6 & 1.2 & 0.6 & 1.8 \\ -4.8 & 1.6 & 0.8 & 2.4 \end{bmatrix} \end{aligned} \tag{12.34}$$

are the four projected vectors in $\mathbb{R}^1$ corresponding to the four original data vectors. They are shown as square symbols in Figure 12.4. The line shown is the line that simultaneously maximizes the sum of squares of the projections of the data onto the line, and minimizes the sum of squares of the orthogonal distances from the points to the line. ◀

12.4.3 Compression

Property 4 can also be used to compress the information in a matrix. Note that each term in the rank-one expansion of Property 4 is specified by using two vectors u_i, v_i and one more number s_i. If A is an $n \times n$ matrix, we can attempt lossy compression of A by throwing away the terms at the end of the sum in Property 4, the ones with smaller s_i. Each term in the expansion requires $2n + 1$ numbers to store or transmit.

For example, if $n = 8$, the matrix is specified by 64 numbers, but we could transmit or store the first term in the expansion by using only $2n + 1 = 17$ numbers. If most of the information is captured by the first term—for example, if the first singular value is much larger than the rest—there may be a 75 percent savings in space by working this way.

As an example, return to the 8×8 pixel block shown in Figure 11.6. After subtracting 128 to center the pixel values around 0, the matrix is given in equation (11.16). The singular values of this 8×8 matrix are as follows:

$$
\begin{array}{c}
387.78 \\
216.74 \\
83.77 \\
62.69 \\
34.75 \\
21.47 \\
10.50 \\
4.35
\end{array}
$$

The original block is shown in Figure 12.5(c), along with the compressed versions in (a) and (b). Figure 12.5(a) corresponds to replacing the matrix with the first term in the expansion of Property 4, the best rank-one approximation of the pixel value matrix. As remarked previously, this achieves approximately 4:1 compression. In Figure 12.5(b), two terms are used, for an approximate compression ratio of 2:1. (Of course, we are simplifying the discussion here by going without quantization tricks. It would help to carry the coefficients corresponding to smaller singular values with less precision, as done in Chapter 11.)

The grayscale photo in Figure 11.5 is a 256×256 pixel image. We can also apply Property 4 to the entire matrix, after subtracting 128 from each pixel entry. The 256 singular values of the matrix vary in size from 8108 to 0.46. Figure 12.6 shows the reconstructed image that results from keeping p of the terms of the rank-one expansion in Property 4. For $p = 8$, only $8(2(256) + 1) = 4104$ numbers need to be stored, compared with $(256)^2 = 65536$ original pixel values, about a 16:1 compression ratio. In Figure 12.6(c), where 32 terms are kept, the compression ratio is approximately 4:1.

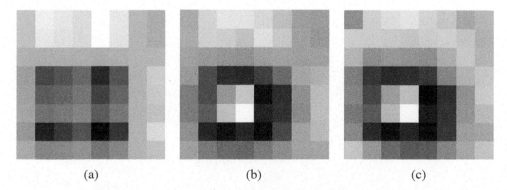

| (a) | (b) | (c) |

Figure 12.5 Result of compression and decompression by SVD. Number of singular values retained: (a) $p = 1$ (b) $p = 2$ (c) all.

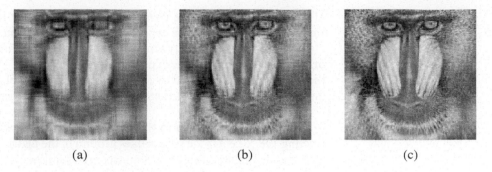

(a)   (b)   (c)

Figure 12.6 Result of compression and decompression by SVD. Number of singular values retained: (a) $p = 8$ (b) $p = 16$ (c) $p = 32$.

12.4.4 Calculating the SVD

If A is a real symmetric matrix, the SVD reduces to the eigenvalue computation discussed earlier in the chapter. In this case, the unit eigenvectors form an orthogonal basis. If we define a matrix V to hold the unit eigenvectors as columns, then $AV = US$ expresses the eigenvector equation, where S is a diagonal matrix holding the absolute values of the eigenvalues and U is the same as V, but with the sign of column switched if the eigenvalue is negative, as discussed in (12.26). Since U and V are orthogonal matrices,

$$A = USV^T$$

is a singular value decomposition of A.

For a general, nonsymmetric $m \times n$ matrix A, there are two distinct computational approaches for determining the SVD. The first and most obvious method is to form $A^T A$ and to find its eigenvalues. According to Theorem 12.11, this reveals the columns v_i of V, and by normalizing the vectors $Av_i = s_i u_i$, we get both the singular values and the columns of U.

This method is not recommended, however, for all but simple examples. If the condition number of A is large, then the condition number of $A^T A$, often of magnitude the square of the condition number of A, may become prohibitively large, and digits of accuracy may be lost.

Fortunately, there is an alternative method of finding the eigenvectors of $A^T A$ that avoids forming the matrix product. Consider the matrix

$$B = \begin{bmatrix} 0 & A^T \\ A & 0 \end{bmatrix}. \tag{12.34}$$

Notice that B is a symmetric $(m + n) \times (m + n)$ matrix (check its transpose). Therefore, it has real eigenvalues and a basis of eigenvectors. Let $[v, w]$ denote a $(m + n)$-vector that is an eigenvector of B. Then

$$\begin{bmatrix} A^T w \\ Av \end{bmatrix} = \begin{bmatrix} 0 & A^T \\ A & 0 \end{bmatrix} \begin{bmatrix} v \\ w \end{bmatrix} = \lambda \begin{bmatrix} v \\ w \end{bmatrix},$$

or $Av = \lambda w$. Multiplying on the left by A^T yields

$$A^T Av = \lambda A^T w = \lambda^2 v, \tag{12.35}$$

showing that w is an eigenvector of $A^T A$ with corresponding eigenvalue λ^2. Note that we can determine the eigenvalues and eigenvectors of $A^T A$ in this way without ever forming the matrix $A^T A$.

Therefore, the second and preferred method for computing singular values and singular vectors begins with putting the symmetric matrix B into upper Hessenberg form. Because of the symmetry, upper Hessenberg is equivalent to tridiagonal. Then methods like the shifted QR algorithm can be applied to find the eigenvalues, which are the squares of the singular values, and the eigenvectors, whose n top entries are the singular vectors v_i. Although this approach seems to double the size of the matrix, it avoids increasing the condition number unnecessarily, and there are more efficient ways to implement this idea (which we will not pursue here) that avoid the need for extra storage.

▶ **ADDITIONAL EXAMPLES**

1. Use MATLAB's svd command to find the best rank-two approximation to the 5×5 Hilbert matrix.

2. Project the vectors

$$\begin{bmatrix} 1 \\ -3 \\ -1 \\ 4 \\ 1 \end{bmatrix}, \quad \begin{bmatrix} 8 \\ -14 \\ -1 \\ 11 \\ 1 \end{bmatrix}, \quad \begin{bmatrix} 2 \\ -1 \\ 1 \\ -2 \\ 3 \end{bmatrix}, \quad \begin{bmatrix} -1 \\ -2 \\ -6 \\ 13 \\ 5 \end{bmatrix}$$

onto the best least squares plane in $\mathbb{R}^5$.

⊡ **Solutions** for Additional Examples can be found at bit.ly/2P9C8eP

12.4 Computer Problems

⊡ **Solutions** for Computer Problems numbered in blue can be found at bit.ly/2yPo7Jh

1. Use MATLAB's svd command to find the best rank-one approximation of the following matrices:

(a) $\begin{bmatrix} 1 & 2 \\ 2 & 3 \end{bmatrix}$ (b) $\begin{bmatrix} 1 & 4 \\ 2 & 3 \end{bmatrix}$ (c) $\begin{bmatrix} 1 & 2 & 4 \\ 1 & 3 & 3 \\ 0 & 0 & 1 \end{bmatrix}$ (d) $\begin{bmatrix} 1 & 5 & 3 \\ 2 & -3 & 2 \\ -3 & 1 & 1 \end{bmatrix}$

2. Find the best rank-two approximation to the following matrices:

(a) $\begin{bmatrix} 1 & 2 & 4 \\ 1 & 3 & 3 \\ 0 & 0 & 1 \end{bmatrix}$ (b) $\begin{bmatrix} 2 & -2 & 4 \\ 1 & -1 & 2 \\ -3 & 3 & -6 \end{bmatrix}$ (c) $\begin{bmatrix} 1 & 5 & 3 \\ 2 & -3 & 2 \\ -3 & 1 & 1 \end{bmatrix}$

3. Find the best least squares approximating line for the following vectors, and the projections of the vectors onto the one-dimensional subspace:

(a) $\begin{bmatrix} 1 \\ 4 \end{bmatrix}, \begin{bmatrix} 1 \\ 5 \end{bmatrix}, \begin{bmatrix} 2 \\ 4 \end{bmatrix}$ (b) $\begin{bmatrix} 2 \\ 0 \end{bmatrix}, \begin{bmatrix} 4 \\ 1 \end{bmatrix}, \begin{bmatrix} 3 \\ 2 \end{bmatrix}$ (c) $\begin{bmatrix} 1 \\ 2 \\ 4 \end{bmatrix}, \begin{bmatrix} 2 \\ 3 \\ 5 \end{bmatrix}, \begin{bmatrix} 2 \\ 1 \\ 6 \end{bmatrix}, \begin{bmatrix} 1 \\ 1 \\ 3 \end{bmatrix}$

4. Find the best least squares approximating plane for the following three-dimensional vectors, and the projections of the vectors onto the subspace:

(a) $\begin{bmatrix} 1 \\ 2 \\ 4 \end{bmatrix}, \begin{bmatrix} 2 \\ 3 \\ 5 \end{bmatrix}, \begin{bmatrix} 2 \\ 1 \\ 6 \end{bmatrix}, \begin{bmatrix} 1 \\ 1 \\ 3 \end{bmatrix}$ (b) $\begin{bmatrix} 2 \\ 3 \\ 1 \end{bmatrix}, \begin{bmatrix} -1 \\ 4 \\ 0 \end{bmatrix}, \begin{bmatrix} 7 \\ -2 \\ 1 \end{bmatrix}, \begin{bmatrix} 1 \\ 1 \\ 0 \end{bmatrix}$

5. Write a MATLAB program that uses the matrix of (12.34) to compute the singular values of a matrix. Use the upper Hessenberg code given earlier, and use shifted QR to solve the resulting eigenvalue problem. Apply your method to find the singular values of the following matrices:

(a) $\begin{bmatrix} 3 & 0 \\ 4 & 0 \end{bmatrix}$ (b) $\begin{bmatrix} 6 & -2 \\ 8 & \frac{3}{2} \end{bmatrix}$ (c) $\begin{bmatrix} 0 & 1 \\ 0 & 0 \end{bmatrix}$ (d) $\begin{bmatrix} -4 & -12 \\ 12 & 11 \end{bmatrix}$ (e) $\begin{bmatrix} 0 & -2 \\ -1 & 0 \end{bmatrix}$

6. Continuing Computer Problem 5, add code to find the full SVD of the matrices.

7. Use the code developed in Computer Problem 6 to find the full SVD of the following matrices, and compare your results with MATLAB's svd command (your answer should agree up to the choice of minus signs in u_i, v_i):

(a) $\begin{bmatrix} 1 & 3 & 0 \\ 4 & 5 & 0 \\ 2 & 5 & 3 \end{bmatrix}$ (b) $\begin{bmatrix} 1 & 0 & 2 & 4 \\ 1 & 1 & 1 & 3 \end{bmatrix}$ (c) $\begin{bmatrix} 0 & 1 & 3 \\ 1 & 3 & 1 \\ 2 & -1 & 3 \\ 0 & 1 & -1 \end{bmatrix}$ (d) $\begin{bmatrix} 0 & 1 & 3 & 1 \\ -1 & 1 & 1 & 0 \\ 0 & 1 & 3 & -1 \\ 2 & -1 & -1 & 2 \end{bmatrix}$

8. Import a photo, using MATLAB's imread command. Use the SVD to create 8:1, 4:1, and 2:1 compressed versions of the photo. If the photo is in color, compress each of the RGB colors separately.

Software and Further Reading

The modern era of eigenvalue calculation was initiated by Wilkinson [1965], and the QR algorithm and upper Hessenberg form were already present in Wilkinson and Reinsch [1971]. Other influential references on eigenvalue calculations are Stewart [1973], Parlett [1998], Golub and Van Loan [1996], and the revealing articles Parlett [2000] and Watkins [1982].

Lapack (Anderson et al. [1990]) provides routines for reductions to upper Hessenberg form and for the symmetric and nonsymmetric eigenvalue problem. These routines are descended from the Eispack package (Smith et al. [1970]) developed in the 1960s. Netlib's DGEHRD reduces a real matrix to upper Hessenberg form by using Householder reflectors, and DHSEQR implements the QR algorithm for calculating eigenvalues and the Schur form for a real upper Hessenberg matrix. NAG provides F08NEF and F08PEF, respectively, for the same two operations. There are analogous programs for complex matrices.

Saad [2003] and Bai et al. [2000] consider state-of-the-art methods for large eigenvalue problems. Cuppen [1981] introduced the divide-and-conquer method for the tridiagonal symmetric eigenvalue problem. Arpack is a suite for Arnoldi iteration for large sparse problems, and Parpack is an extension for parallel processors.

Algorithms for the singular value decomposition include Lapack's original DGESVD, and the divide-and-conquer method DGESDD that is preferable for large matrices. Complex versions are also available.

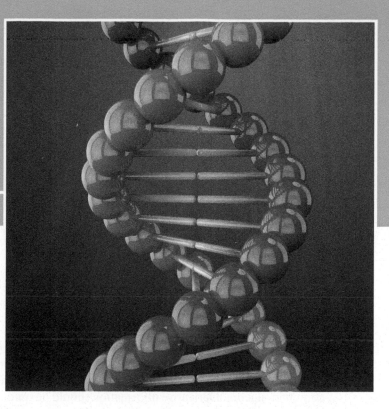

13

Optimization

The discovery of the double helix structure of DNA in 1953 has led, a half century later, to a nearly complete sequencing of the human genome. The sequence holds instructions for folding strings of amino acids into individual proteins that perform the activities of life, but written in a coded language. This information now awaits translation, so that it can be directed toward a detailed understanding of physiological function. A host of potential applications, including gene therapy and rational drug design, may promote the early prevention, diagnosis, and cure of disease.

The folding of amino acids into functional proteins depends crucially on Van der Waals forces, the microscopic attraction and repulsion between unbound atoms. Atomic cluster models, where these forces are modeled by the Lennard-Jones potential, are studied for minimum energy configurations, bringing the problem into the realm of optimization.

Reality Check Reality Check 13 on page 609 applies the optimization techniques of the chapter to solve this energy minimization problem.

Optimization refers to finding the maximum or minimum of a real-valued function, called the **objective function**. Since locating the maximum of a function $f(x)$ is equivalent to locating the minimum of $-f(x)$, it suffices to consider minimization alone in developing computational methods.

Some optimization problems call for a minimum of the objective function subject to several equality and inequality constraints. For example, although x_1 is the global minimum of the function in Figure 13.1, x_2 would be the minimum subject to the constraint $x \geq 0$. In particular, the field of linear programming considers problems where the objective function and constraints are linear. In this chapter, we will keep things simple and consider unconstrained optimization only.

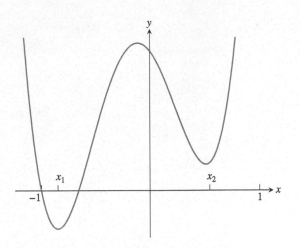

Figure 13.1 The minimization problem for $f(x) = 5x^4 + 3x^3 - 4x^2 - x +$ **2.** The solution of the unconstrained minimization problem $\min_x f(x)$ is x_1.

Methods for unconstrained optimization fall into two groups, depending on whether derivatives of the objective function $f(x)$ are used. If an algebraic function is known for $f(x)$, in most cases the derivatives can be easily determined by hand or computer algebra. Derivative information should be used if possible, but there are several reasons why it might not be available. In particular, the objective function may be too complicated, too high dimensional, or not known in a form that may be differentiated.

13.1 UNCONSTRAINED OPTIMIZATION WITHOUT DERIVATIVES

In this section, the assumption is made that the objective function $f(x)$ can be evaluated for any input x, but that the derivative $f'(x)$ (or partial derivatives if f is a function of several variables) is not available. We will discuss three methods for optimizing without derivatives: Golden Section Search, Successive Parabolic Interpolation, and the Nelder–Mead Method. The first two apply only to functions $f(x)$ of one scalar variable, while Nelder–Mead can search through several dimensions.

13.1.1 Golden Section Search

Golden Section Search is an efficient method for finding a minimum of a function $f(x)$ of one variable, once a bracketing interval is known.

DEFINITION 13.1   The continuous function $f(x)$ is called **unimodal** on the interval $[a, b]$ if there is exactly one relative minimum or maximum on $[a, b]$, and f is strictly decreasing or increasing at all other points.   ❏

A unimodal function either increases to a relative maximum in $[a, b]$ and then decreases as x moves from a to b, or decreases to a relative minimum and then increases.

Assume that f is unimodal and has a relative minimum on $[a, b]$. Choose two points x_1 and x_2 inside the interval, so that $a < x_1 < x_2 < b$, as shown in Figure 13.2 for the case $[a, b] = [0, 1]$. We will replace the original interval by a new, smaller inter-

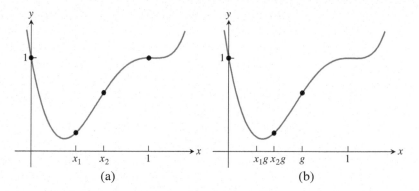

Figure 13.2 Golden Section Search. (a) Evaluate the objective function at two points x_1, x_2 within the current interval $[0, 1]$. If $f(x_1) \le f(x_2)$, then the new interval will be $[0, x_2]$. (b) In the next step, set $g = x_2$ and repeat the same comparison with $x_1 g$ and $x_2 g$.

val that continues to bracket a relative minimum, according to the following rule: If $f(x_1) \le f(x_2)$, then retain the interval $[a, x_2]$ at the next step. If $f(x_1) > f(x_2)$, retain the interval $[x_1, b]$.

Note that in either case the new interval contains a relative minimum of the unimodal function f. For example, if $f(x_1) < f(x_2)$, as shown in Figure 13.2, then because of the unimodal assumption, the minimum must be to the left of x_2. This is because f must decrease to the left of the minimum, so $f(x_1) < f(x_2)$ means that x_2 must be to the right of the minimum. Likewise, $f(x_1) > f(x_2)$ implies that $[x_1, b]$ contains the minimum. Since the new interval is smaller than the previous interval $[a, b]$, progress has been made toward locating the minimum. This basic step is then repeated until the interval containing the minimum is as small as desired. The method is reminiscent of the Bisection Method for locating roots.

Next we discuss how x_1 and x_2 should be placed in the interval $[a, b]$. In each step, we would like to reduce the length of the interval as much as possible, using as little work as possible. The way of doing this is shown in Figure 13.3 for the interval $[a, b] = [0, 1]$. Accept two criteria for the choice of x_1 and x_2: (a) Make them symmetric with respect to the interval (since we have no information about which side of the interval the minimum lies in), and (b) choose them such that no matter which choice

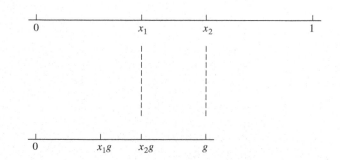

Figure 13.3 Choice of proportions in Golden Section Search. The ratio of the top segment to the bottom segment is $1/g = (1 + \sqrt{5})/2$, the **golden section**. The points x_1 and x_2 are chosen exactly, so that no matter whether the new interval is $[0, x_2]$ or $[x_1, 1]$, one point can be reused as a new interior point, reducing the number of new objective function evaluations to one per step.

is made for the new interval, both x_1 and x_2 are used in the next step. That is, require (a) $x_1 = 1 - x_2$, and (b) $x_1 = x_2^2$. As shown in Figure 13.3, if the new interval is $[0, x_2]$, criterion (b) ensures that the original x_1 will be the "x_2" for the next interval; therefore, only one new function evaluation, namely, $f(x_1 g)$, will be necessary. Likewise, if the new interval is $[x_1, 1]$, then x_2 will become the new "x_1." This ability to reuse function evaluations means that after the first step, only a single evaluation of the objective function is needed per step.

Criteria (a) and (b) together imply that $x_2^2 + x_2 - 1 = 0$. The positive solution of this quadratic equation is $x_2 = g = (\sqrt{5} - 1)/2$. To start the method, the objective function f must be known to be unimodal on $[a, b]$, and then f is evaluated at the interior points x_1 and x_2, where $a < x_1 = a + (1 - g)(b - a) < x_2 = a + g(b - a) < b$. Note that x_1 and x_2 are set at exactly $1 - g$ and g of the way between a and b. The new interval is chosen as has been shown and this basic step is repeated. The new interval has length g times the previous interval, so after k steps the current interval has length $g^k(b - a)$. The midpoint of the final interval is correct within an uncertainty of one-half of the length of the final interval, $g^k(b - a)/2$. We have proved the following theorem:

THEOREM 13.2   After k steps of Golden Section Search with starting interval $[a, b]$, the midpoint of the final interval is within $g^k(b - a)/2$ of the minimum, where $g = (\sqrt{5} - 1)/2 \approx 0.618$. ∎

Golden Section Search

Given f unimodal with minimum in $[a, b]$

for $i = 1, 2, 3, \ldots$
  $g = (\sqrt{5} - 1)/2$
  if $f(a + (1 - g)(b - a)) < f(a + g(b - a))$
    $b = a + g(b - a)$
  else
    $a = a + (1 - g)(b - a)$
  end
end

The final interval $[a,b]$ contains a minimum.

MATLAB code for Golden Section Search requires one function evaluation per step after step one, as mentioned before.

```
% Program 13.1 Golden Section Search for minimum of f(x)
% Start with unimodal f(x) and minimum in [a,b]
% Input: function f, interval [a,b], number of steps k
% Output: approximate minimum y
function y=gss(f,a,b,k)
g=(sqrt(5)-1)/2;
x1 = a+(1-g)*(b-a);
x2 = a+g*(b-a);
f1=f(x1);f2=f(x2);
for i=1:k
   if f1 < f2              % if f(x1) < f(x2), replace b with x2
     b=x2; x2=x1; x1=a+(1-g)*(b-a);
     f2=f1; f1=f(x1);  % single function evaluation
   else                   % otherwise, replace a with x1
     a=x1; x1=x2; x2=a+g*(b-a);
```

```
      f1=f2; f2=f(x2);  % single function evaluation
   end
end
y=(a+b)/2;
```

SPOTLIGHT ON

Convergence   According to Theorem 13.2, Golden Section Search converges linearly to the minimum with linear convergence rate $g \approx 0.618$. It is interesting to notice the many similarities of this method to the Bisection Method of Chapter 1 for finding roots. Although they solve different problems, both are globally convergent, meaning that if started with the right conditions (unimodality on $[a, b]$ for Golden Section Search, and $f(a)f(b) < 0$ for bisection), they are both guaranteed to converge to a solution. Neither requires derivative information. Both require one function evaluation per step and both are linearly convergent. Bisection is slightly faster, with linear convergence rate $K = 0.5 < g = 0.618$. They both belong to the valuable category of "slow, but sure" methods.

▶ **EXAMPLE 13.1**   Use Golden Section Search to find the minimum of $f(x) = x^6 - 11x^3 + 17x^2 - 7x + 1$ on the interval $[0, 1]$.

Figure 13.2 shows the first two steps of the method. On the first step, $x_1 = 1 - g$ and $x_2 = g$, where $g = (\sqrt{5} - 1)/2$. Since $f(x_1) < f(x_2)$, the interval $[0, 1]$ is replaced with $[0, g]$. The new x_1, x_2 are the previous x_1g, x_2g, respectively. On the second step, again $f(x_1) < f(x_2)$, so the interval $[0, g]$ is replaced with $[0, x_2]$. The first 15 steps are shown in the following table:

step	a	x_1	x_2	b
0	0.0000	0.3820	0.6180	1.0000
1	0.0000	0.2361	0.3820	0.6180
2	0.0000	0.1459	0.2361	0.3820
3	0.1459	0.2361	0.2918	0.3820
4	0.2361	0.2918	0.3262	0.3820
5	0.2361	0.2705	0.2918	0.3262
6	0.2705	0.2918	0.3050	0.3262
7	0.2705	0.2837	0.2918	0.3050
8	0.2705	0.2786	0.2837	0.2918
9	0.2786	0.2837	0.2868	0.2918
10	0.2786	0.2817	0.2837	0.2868
11	0.2817	0.2837	0.2849	0.2868
12	0.2817	0.2829	0.2837	0.2849
13	0.2829	0.2837	0.2841	0.2849
14	0.2829	0.2834	0.2837	0.2841
15	0.2834	0.2837	0.2838	0.2841

After 15 steps, we can say that the minimum is between 0.2834 and 0.2838.   ◀

13.1.2 Successive Parabolic Interpolation

In Golden Section Search, no use is made of the function evaluations $f(x_1)$ and $f(x_2)$, except to compare them. A decision is made on how to proceed, no matter how much larger one is than the other. In this section, we describe a new method that is less wasteful of the function values; it uses them to build a local model of the function f.

The local model chosen is a parabola, which we know from Chapter 3 is uniquely determined by three points. Begin with three points r, s, t in the vicinity of the minimum, as shown in Figure 13.4. Evaluate the objective function f at the three points and draw the parabola through them. Divided differences give

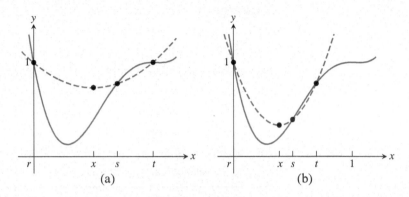

Figure 13.4 Successive Parabolic Interpolation. (a) A parabola is drawn through the three current points r, s, t, and the minimum x of the parabola is used to replace the current s. (b) The step is repeated with the new r, s, t.

$$
\begin{array}{c|ccc}
r & f(r) & & \\
 & & d_1 & \\
s & f(s) & & d_3 \\
 & & d_2 & \\
t & f(t) & &
\end{array}
$$

where $d_1 = (f(s) - f(r))/(s - r), d_2 = (f(t) - f(s))/(t - s)$, and $d_3 = (d_2 - d_1)/(t - r)$. Therefore, we can express the parabola as

$$P(x) = f(r) + d_1(x - r) + d_3(x - r)(x - s). \tag{13.1}$$

Setting the derivative of $P(x) = 0$ to find the minimum of the parabola yields the formula

$$x = \frac{r + s}{2} - \frac{(f(s) - f(r))(t - r)(t - s)}{2[(s - r)(f(t) - f(s)) - (f(s) - f(r))(t - s)]} \tag{13.2}$$

for the new approximation for the minimum. In SPI, the new x may replace the least recent or least optimal of r, s, t, and the step is repeated as needed. There is no guarantee of convergence for SPI, unlike the case of Golden Section Search. However, it is usually faster when it does converge, because it uses the function evaluation information more wisely.

Successive Parabolic Interpolation

Start with approximate minima r, s, t

for $i = 1, 2, 3, \ldots$

$\qquad x = \dfrac{r + s}{2} - \dfrac{(f(s) - f(r))(t - r)(t - s)}{2[(s - r)(f(t) - f(s)) - (f(s) - f(r))(t - s)]}$

$\qquad t = s$

$\qquad s = r$

$\qquad r = x$

end

In the following MATLAB code, the minimum of the parabola replaces the least recent of the three current points:

MATLAB code
shown here can be found
at bit.ly/2pWIss3

```
% Program 13.2 Successive Parabolic Interpolation
% Input: inline function f, initial guesses r,s,t, steps k
% Output: approximate minimum x
function x=spi(f,r,s,t,k)
x(1)=r;x(2)=s;x(3)=t;
fr=f(r);fs=f(s);ft=f(t);
for i=4:k+3
 x(i)=(r+s)/2-(fs-fr)*(t-r)*(t-s)/(2*((s-r)*(ft-fs)...
       -(fs-fr)*(t-s)));
 t=s;s=r;r=x(i);
 ft=fs;fs=fr;fr=f(r);            % single function evaluation
end
```

▶ **EXAMPLE 13.2**   Use Successive Parabolic Interpolation to find the minimum of $f(x) = x^6 - 11x^3 + 17x^2 - 7x + 1$ on the interval $[0, 1]$.

Using starting points $r = 0, s = 0.7, t = 1$, we compute the following steps:

step	x	$f(x)$
0	1.00000000000000	1.00000000000000
0	0.70000000000000	0.77464900000000
0	0.00000000000000	1.00000000000000
1	0.50000000000000	0.39062500000000
2	0.38589683548538	0.20147287814500
3	0.33175129602524	0.14844165724673
4	0.23735573316721	0.14933737764402
5	0.28526617269372	0.13172660338164
6	0.28516942161639	0.13172426136234
7	0.28374069464218	0.13170646451792
8	0.28364647631123	0.13170639859035
9	0.28364826437569	0.13170639856301
10	0.28364835832962	0.13170639856295
11	0.28364835808377	0.13170639856295
12	0.28364833218729	0.13170639856295

We conclude that the minimum is near $x_{\min} = 0.2836483$. Note that after 12 steps we have far outdone the accuracy of Golden Section Search with fewer function evaluations. We have used no derivative information about the objective function, although we have used the knowledge of the precise values of f, while GSS needed only to know comparisons between values.

Note also from the table a curiosity near the end. As discussed in Chapter 1, functions are very flat near relative maxima and minima. Since numbers within 10^{-7} of $x_{\min}$ give the same minimum function value, we cannot go beyond this accuracy while using IEEE double precision, no matter how many steps we can afford to run. Since minima typically occur where derivatives of the function are zero, this difficulty is not the fault of the optimization method, but endemic to floating point computation.

The progression from GSS to SPI is similar to that from the Bisection Method to the Secant Method and Inverse Quadratic Interpolation. Building a local model for the function and acting as if it were the objective function helps to speed convergence.   ◀

13.1.3 Nelder–Mead search

For a function of more than one variable, the methods become more sophisticated. Nelder–Mead search tries to roll a polyhedron downhill to the lowest possible level. For this reason, it is also called the **downhill simplex method**. It uses no derivative information about the objective function.

Assume that the function to be minimized is a function of n variables f. The method begins with $n + 1$ initial guess vectors $x_1, \ldots, x_{n+1}$ belonging to R^n that together form the vertices of an n-dimensional simplex. For example, if $n = 2$, the three initial guesses form the vertices of a triangle in the plane.

The vertices of the simplex are tested and put into ascending order according to their function values $y_1 < y_2 < \cdots < y_{n+1} = y_h$. The simplex vector $x_h = x_{n+1}$ that is least optimal is replaced according to the flowchart shown in Figure 13.5. First we define the centroid $\overline{x}$ of the face of the simplex that omits x_h. Then we test the function value $y_r = f(x_r)$ of the reflection point $x_r = 2\overline{x} - x_h$, as shown in Figure 13.5(a). If the new value y_r lies in the range $y_1 < y_r < y_n$, we replace the worst point x_n with x_r, sort the vertices by their function values, and repeat the step.

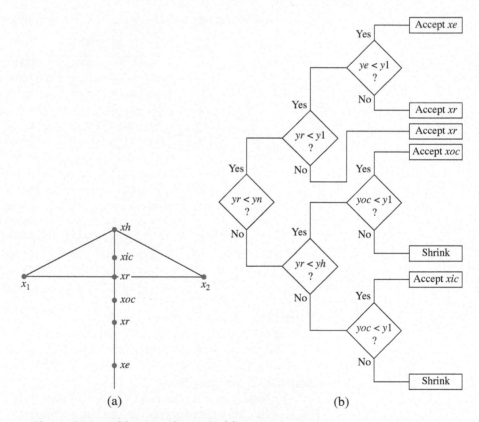

(a)   (b)

Figure 13.5 Nelder-Mead search. (a) Points along the line connecting the highest function point x_h and the centroid $\overline{x}$ are tested. (b) A flowchart describing one step of the method.

In case y_r is lower than the current minimum y_1, an extrapolation attempt is made, using $x_e = 3\overline{x} - 2x_h$, to see whether we should move even further in this direction. The better of x_e and x_r is accepted for the step. On the other hand, in case y_r is greater than y_n (the current maximum once x_{n+1} is ignored), a further test is made, either at the outside contraction point $x_{oc} = 1.5\overline{x} - 0.5x_h$ or at the inside contraction point

$x_{ic} = 0.5\overline{x} + 0.5x_h$, as shown in the figure. Failure to show improvement at either one of these points means that no progress is being made by branching out and that the method should look more locally for the optimum. It accomplishes this by shrinking the simplex by a factor of 2 in the direction of the current minimum x_1 before going to the next step. The MATLAB code follows. The function f should be defined in the variables $x(1), x(2), \ldots, x(n)$.

MATLAB code

shown here can be found
at bit.ly/2PEYfHl

```
% Program 13.3 Nelder-Mead Search
% Input:  function f, best guess xbar (column vector),
%     initial search radius rad and number of steps k
% Output: matrix x whose columns are vertices of simplex,
%     function values y of those vertices
function [x,y]=neldermead(f,xbar,rad,k)
n=length(xbar);
x(:,1)=xbar;              % each column of x is a simplex vertex
x(:,2:n+1)=xbar*ones(1,n)+rad*eye(n,n);
for j=1:n+1
  y(j)=f(x(:,j));        % evaluate obj function f at each vertex
end
[y,r]=sort(y);           % sort the function values in ascending order
x=x(:,r);                %     and rank the vertices the same way
for i=1:k
  xbar=mean(x(:,1:n)')'; % xbar is the centroid of the face
  xh=x(:,n+1);           %     omitting the worst vertex xh
  xr = 2*xbar - xh; yr = f(xr);
  if yr < y(n)
    if yr < y(1)         % try expansion xe
      xe = 3*xbar - 2*xh; ye = f(xe);
      if ye < yr         % accept expansion
        x(:,n+1) = xe; y(n+1) = f(xe);
      else               % accept reflection
        x(:,n+1) = xr; y(n+1) = f(xr);
      end
    else                 % xr is middle of pack, accept reflection
      x(:,n+1) = xr; y(n+1) = f(xr);
    end
  else                   % xr is still the worst vertex, contract
    if yr < y(n+1)       % try outside contraction xoc
      xoc = 1.5*xbar - 0.5*xh; yoc = f(xoc);
      if yoc < yr        % accept outside contraction
        x(:,n+1) = xoc; y(n+1) = f(xoc);
      else               % shrink simplex toward best point
        for j=2:n+1
          x(:,j) = 0.5*x(:,1)+0.5*x(:,j); y(j) = f(x(:,j));
        end
      end
    else                 % xr is even worse than the previous worst
      xic = 0.5*xbar+0.5*xh; yic = f(xic);
      if yic < y(n+1) % accept inside contraction
        x(:,n+1) = xic; y(n+1) = f(xic);
      else               % shrink simplex toward best point
        for j=2:n+1
          x(:,j) = 0.5*x(:,1)+0.5*x(:,j); y(j) = f(x(:,j));
        end
      end
    end
  end
```

```
    end
    [y,r] = sort(y);      % resort the obj function values
    x=x(:,r);             %    and rank the vertices the same way
end
```

The code implements the flowchart in Figure 13.5(b). The number of iteration steps is required as an input. Computer Problem 8 asks the reader to rewrite the code with a stopping criterion based on a user-given error tolerance. A common stopping criterion is to require both that the simplex has reduced in size to within a small distance tolerance and that the maximum spread of the function values at the vertices is within a small tolerance. MATLAB implements the Nelder–Mead Method in its fmin-search command.

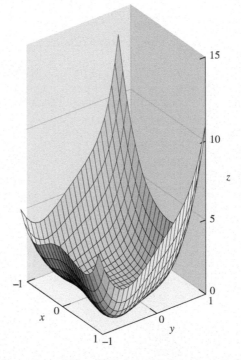

Figure 13.6 Surface plot of two-dimensional function. Graph of $z = 5x^4 + 4x^2y - xy^3 + 4y^4 - x$. Minimum is found by the Nelder–Mead Method to occur at $\approx (0.4923, -0.3643)$.

▶ **EXAMPLE 13.3**   Locate the minimum of the function $f(x, y) = 5x^4 + 4x^2y - xy^3 + 4y^4 - x$, using the Nelder–Mead Method.

The function is shown in Figure 13.6. We define the function f of two variables by

```
>> f=@(x) 5*x(1)^4+4*x(1)^2*x(2)-x(1)*x(2)^3+4*x(2)^4-x(1)
```

and run 60 steps of the Nelder–Mead Method in Program 13.3 with the command

```
>> [x,y]=neldermead(f,[1;1],1,60)

x =

    0.492307778751573   0.492307773822840   0.492307807617628
```

$$-0.364285558245531 \quad -0.364285542189284 \quad -0.364285562179872$$

$$y =$$

$$-0.457521622634071 \quad -0.457521622634070 \quad -0.457521622634069$$

We used the vector $[x, y] = [1, 1]$ as the starting guess and an initial radius of 1, but a wide range of choices will work. After 60 steps the simplex has shrunk to a triangle whose vertices are the three columns in the output vector x. To four correct decimal places, the minimum of -0.4575 occurs at the point $[x, y] = [0.4923, -0.3643]$. ◄

▶ **ADDITIONAL EXAMPLES**

1. Compare the Golden Section Search and Successive Parabolic Interpolation Methods for locating the minimum of the function $f(x) = x^2 + \sin x$ accurate to 8 decimal places.

*2. Use the Nelder–Mead Method to find the distance between the ellipsoids $(x - 3)^2 + 2(y - 1)^2 + (z - 6)^2 = 9$ and $x^2 + 2y^2 + z^2 = 9$.

Solutions for Additional Examples can be found at bit.ly/2q3314U (* example with video solution)

13.1 Exercises

Solutions for Exercises numbered in blue can be found at bit.ly/2EARGUZ

1. Prove that the functions are unimodal on some intervals and find the absolute minimum and where it occurs. (a) $f(x) = e^x + e^{-x}$ (b) $f(x) = x^6$ (c) $f(x) = 2x^4 + x$ (d) $f(x) = x - \ln x$

2. Find the absolute minimum in the given intervals and at which x it occurs.
(a) $f(x) = \cos x, [3, 4]$ (b) $f(x) = 2x^3 + 3x^2 - 12x + 3, [0, 2]$
(c) $f(x) = x^3 + 6x^2 + 5, [-5, 5]$ (d) $f(x) = 2x + e^{-x}, [-5, 5]$

13.1 Computer Problems

Solutions for Computer Problems numbered in blue can be found at bit.ly/2Pbq255

1. Plot the function $y = f(x)$, and find a length-one starting interval on which f is unimodal around each relative minimum. Then apply Golden Section Search to locate each of the function's relative minima to within five correct digits.
(a) $f(x) = 2x^4 + 3x^2 - 4x + 5$ (b) $f(x) = 3x^4 + 4x^3 - 12x^2 + 5$
(c) $f(x) = x^6 + 3x^4 - 2x^3 + x^2 - x - 7$ (d) $f(x) = x^6 + 3x^4 - 12x^3 + x^2 - x - 7$

2. Apply Successive Parabolic Interpolation to the functions in Computer Problem 1. Locate the minima to within five correct digits.

3. Find the point on the hyperbola $y = 1/x$ closest to the point $(2, 3)$ in two different ways: (a) by Newton's Method applied to find a critical point (b) by Golden Section Search on the square of the distance between a point on the conic and $(2, 3)$.

4. Find the point on the ellipse $4x^2 + 9y^2 = 4$ farthest from $(1, 5)$, using methods (a) and (b) of Computer Problem 3.

5. Use the Nelder–Mead Method to find the minimum of $f(x, y) = e^{-x^2 y^2} + (x - 1)^2 + (y - 1)^2$. Try various initial conditions, and compare answers. How many correct digits can you obtain by using this method?

6. Apply the Nelder–Mead Method to find the minima of the following functions to six correct decimal places (each function has two minima):

(a) $f(x, y) = x^4 + y^4 + 2x^2y^2 + 6xy - 4x - 4y + 1$
(b) $f(x, y) = x^6 + y^6 + 3x^2y^2 - x^2 - y^2 - 2xy$

7. Apply the Nelder–Mead Method to find the minimum of the Rosenbrock function $f(x, y) = 100(y - x^2)^2 + (x - 1)^2$.

8. Rewrite Program 13.3 to accommodate a stopping criterion for Nelder–Mead based on a user-specified error tolerance. Demonstrate by finding the minima of the objective functions in Computer Problem 6 to six correct decimal places.

13.2 UNCONSTRAINED OPTIMIZATION WITH DERIVATIVES

Derivatives contain information about the rates of increase and decrease of a function, and in the case of partial derivatives, also the directions of fastest increase and decrease. If such information is available about the objective function, then it can be exploited to find the optimum more efficiently.

13.2.1 Newton's Method

If the function is continuously differentiable and the derivative can be evaluated, then the optimization problem can be expressed as a root-finding problem. Let us begin in one dimension, where the translation is simplest.

At a minimum x^* of a continuously differentiable function $f(x)$, the first derivative must be zero. The methods of Chapter 1 can be used to solve the resulting equation $f'(x) = 0$. If the objective function is unimodal and has a minimum on an interval, then starting Newton's Method with an initial guess close to the minimum x^* will result in convergence to x^*. Newton's Method applied to $f'(x) = 0$ becomes the iteration

$$x_{k+1} = x_k - \frac{f'(x_k)}{f''(x_k)}. \tag{13.3}$$

While Newton's Method (13.3) will find points at which $f'(x) = 0$, in general, such points need not be minima. It is important to have a reasonably close initial guess for the optimum and to check the points for their optimality once located.

For optimization of a function $f(x_1, \ldots, x_n)$ by this method, Newton's Method in several variables is used. As in the one-dimensional case, we want to set the derivative to zero and solve. We thus have

$$\nabla f = 0, \tag{13.4}$$

where

$$\nabla f = \left[\frac{\partial f}{\partial x_1}(x_1, \ldots, x_n), \ldots, \frac{\partial f}{\partial x_n}(x_1, \ldots, x_n) \right]$$

denotes the gradient of f.

Newton's Method for vector-valued functions from Chapter 2 allows (13.4) to be solved. Setting $F(x) = \nabla f(x)$, the iterative step of Newton's Method will set $x_{k+1} = x_k + v$, where v is the solution of $DF(x_k)v = -F(x_k)$. The Jacobian matrix DF of the gradient is

$$H_f = DF = \begin{bmatrix} \frac{\partial^2 f}{\partial x_1 \partial x_1} & \cdots & \frac{\partial^2 f}{\partial x_1 \partial x_n} \\ \vdots & & \vdots \\ \frac{\partial^2 f}{\partial x_n \partial x_1} & \cdots & \frac{\partial^2 f}{\partial x_n \partial x_n} \end{bmatrix}, \tag{13.5}$$

which is the Hessian matrix of f. The Newton step is therefore

$$\begin{cases} H_f(x_k)v = -\nabla f(x_k) \\ x_{k+1} = x_k + v \end{cases}. \tag{13.6}$$

▶ **EXAMPLE 13.4**   Locate the minimum of the function $f(x, y) = 5x^4 + 4x^2y - xy^3 + 4y^4 - x$, using Newton's Method.

The function is shown in Figure 13.6. The gradient is $\nabla f = (20x^3 + 8xy - y^3 - 1, 4x^2 - 3xy^2 + 16y^3)$, and the Hessian is

$$H_f(x, y) = \begin{bmatrix} 60x^2 + 8y & 8x - 3y^2 \\ 8x - 3y^2 & -6xy + 48y^2 \end{bmatrix}.$$

Applying 10 steps of Newton's Method (13.6) gives the results:

step	x	y	$f(x, y)$
0	1.00000000000000	1.00000000000000	11.00000000000000
1	0.64429530201342	0.63758389261745	1.77001867827422
2	0.43064034542956	0.39233298702231	0.10112006537534
3	0.33877971433352	0.19857714160717	−0.17818585977225
4	0.50009733696780	−0.44771929519763	−0.42964065053918
5	0.49737350571430	−0.37972645728644	−0.45673719664708
6	0.49255000651877	−0.36497753746514	−0.45752009007757
7	0.49230831759106	−0.36428704569173	−0.45752162262701
8	0.49230778672681	−0.36428555993321	−0.45752162263407
9	0.49230778672434	−0.36428555992634	−0.45752162263407
10	0.49230778672434	−0.36428555992634	−0.45752162263407

Newton's Method has converged within computer accuracy to the minimum value near −0.4575. Note another feature of minimization using Newton's Method: We have achieved machine accuracy in the solution, unlike the one-dimensional case of Successive Parabolic Interpolation. The reason is that we are no longer working with the objective function, but have recast the problem solely as a root-finding problem involving the gradient. Since ∇f has a simple root at the optimum, there is no difficulty getting forward error close to machine epsilon.   ◀

Newton's Method is often the method of choice if it is possible to compute the Hessian. In two-dimensional problems, the Hessian is commonly available. In high dimension n, it may be just feasible to compute the gradient, an n-dimensional vector, at each point, but infeasible to construct the $n \times n$ Hessian. The next two methods are usually slower than Newton's Method, but require only the gradient to be computed at various points.

13.2.2 Steepest Descent

The fundamental idea behind **Steepest Descent**, also called **Gradient Search**, is to search for a minimum of the function by moving in the direction of steepest decline from the current point. Since the gradient ∇f points in the direction of steepest growth of f, the opposite direction $-\nabla f$ is the line of steepest descent. How far should we go along this direction? Now that we have reduced the problem to minimizing along a line, let one of the one-dimensional methods decide how far to go. After the new minimum along the line of steepest descent is located, repeat the process, starting at that point. That is, find the gradient at the new point, and do a one-dimensional minimization in the new direction.

The Steepest Descent algorithm is an iterative loop.

Steepest Descent

for $i = 0, 1, 2, \ldots$
   $v = \nabla f(x_i)$
   Minimize $f(x - sv)$ for scalar $s = s^*$
   $x_{i+1} = x_i - s^* v$
end

We will apply Steepest Descent to the objective function of Example 13.3.

▶ **EXAMPLE 13.5**   Locate the minimum of the function $f(x, y) = 5x^4 + 4x^2 y - xy^3 + 4y^4 - x$, using Steepest Descent.

We follow the preceding steps, using Successive Parabolic Interpolation as the one-dimensional minimizer. The results for 25 steps are as follows:

step	x	y	$f(x, y)$
0	1.00000000000000	−1.00000000000000	11.00000000000000
5	0.40314579518113	−0.27992088271756	−0.41964888830651
10	0.49196895085112	−0.36216404374206	−0.45750680523754
15	0.49228284433776	−0.36426635686172	−0.45752161934016
20	0.49230786417532	−0.36428539567277	−0.45752162263389
25	0.49230778262142	−0.36428556578033	−0.45752162263407

Convergence is slower compared with the Newton's Method, for a good reason. Newton's Method is solving an equation and is using the first and second derivatives (including the Hessian). Steepest Descent is actually minimizing by following the downhill direction and is using only first derivative information.   ◀

13.2.3 Conjugate Gradient Search

In Chapter 2, the Conjugate Gradient Method was used to solve symmetric positive–definite matrix equations. Now we will return to the method, viewed from a different direction.

Solving $Ax = w$ when A is symmetric and positive-definite is equivalent to finding the minimum of a paraboloid. In two dimensions, for example, the solution of the linear system

$$\begin{bmatrix} a & b \\ b & c \end{bmatrix} \begin{bmatrix} x_1 \\ x_2 \end{bmatrix} = \begin{bmatrix} e \\ f \end{bmatrix} \tag{13.7}$$

is the minimum of the paraboloid

$$f(x_1, x_2) = \frac{1}{2}ax_1^2 + bx_1x_2 + \frac{1}{2}cx_2^2 - ex_1 - fx_2. \tag{13.8}$$

The reason is that the gradient of f is

$$\nabla f = [ax_1 + bx_2 - e, bx_1 + cx_2 - f].$$

The gradient is zero at the minimum, which gives the previous matrix equation. Positive-definiteness means the paraboloid is concave up.

The key observation is that the residual $r = w - Ax$ of the linear system (13.7) is $-\nabla f(x)$, the direction of steepest descent of the function f at the point x. Suppose we have chosen a search direction, denoted by the vector d. To minimize f in (13.8) along that direction is to find the α that minimizes the function $h(\alpha) = f(x + \alpha d)$. We will set the derivative to zero to find the minimum:

$$
\begin{aligned}
0 &= \nabla f \cdot d \\
&= (A(x + \alpha d) - (e, f)^T) \cdot d \\
&= (\alpha A d - r)^T d.
\end{aligned}
$$

This implies that

$$
\alpha = \frac{r^T d}{d^T A d} = \frac{r^T r}{d^T A d},
$$

where the last equality follows from Theorem 2.16 on the Conjugate Gradient Method.

We conclude from this calculation that we could alternatively solve for the minimum of a paraboloid by using the Conjugate Gradient Method, but replacing

$$
r_i = -\nabla f
$$

and

$$
\alpha_i = \alpha \text{ that minimizes } f(x_{i-1} + \alpha d_{i-1}).
$$

In fact, in looking at it this way, notice that we have expressed conjugate gradient completely in terms of f. No mention of the matrix A remains. We can run the algorithm in this form for general f. Near regions where f has a parabolic shape, the method will move toward the bottom very quickly. The new algorithm has the following steps:

Conjugate Gradient Search

Let x_0 be the initial guess and set $d_0 = r_0 = -\nabla f$.

for $i = 1, 2, 3, \ldots$
   $\alpha_i = \alpha$ that minimizes $f(x_{i-1} + \alpha d_{i-1})$
   $x_i = x_{i-1} + \alpha_i d_{i-1}$
   $r_i = -\nabla f(x_i)$
   $\beta_i = \dfrac{r_i^T r_i}{r_{i-1}^T r_{i-1}}$
   $d_i = r_i + \beta_i d_{i-1}$
end

We will try out the new method on a familiar example.

▶ **EXAMPLE 13.6**   Locate the minimum of the function $f(x, y) = 5x^4 + 4x^2 y - xy^3 + 4y^4 - x$, using Conjugate Gradient Search.

We follow the preceding steps, using Successive Parabolic Interpolation as the one-dimensional minimizer. The results for 20 steps are as follows:

step	x	y	$f(x, y)$
0	1.00000000000000	-1.00000000000000	.11000000000000
5	0.46038657599935	-0.38316114029860	-0.44849953420621
10	0.49048892807181	-0.36106561127830	-0.45748477171484
15	0.49243714956128	-0.36421661473526	-0.45752147604312
20	0.49231477751583	-0.36429817275371	-0.45752162206984

The subject of unconstrained optimization is vast, and the methods of this chapter represent only the tip of the iceberg. **Trust region methods** form local models, as Successive Parabolic Interpolation or Conjugate Gradient Search do, but allow use of them only within a specified region that narrows as the search progresses. The routine `fminunc` of the MATLAB Optimization Toolbox is an example of a trust region method. **Simulated annealing** is a stochastic method that attempts to progress lower on the objective function, but will accept an upward step with a small, positive probability, in order to avoid convergence to a nonoptimal local minima. **Genetic algorithms** and evolutionary computation in general propose entirely new approaches to optimization and are still being actively explored.

Constrained optimization takes as a goal the minimization of an objective function subject to a set of constraints. The most common subset of these problems, linear programming, has been solved by the simplex method since its development in the mid-20th century, although new and often faster algorithms based on interior point methods have emerged fairly recently. Quadratic and nonlinear programming problems require more sophisticated methods. Consult the references for entry points into this literature. ◄

▶ **ADDITIONAL EXAMPLES**

*1. Use Newton's Method to find the minimum of $f(x, y) = x^2 + y^2 + \sin(x + 3y)$.

2. Apply Conjugate Gradient Search to find the minimum of $f(x, y, z) = x^2 + x + y^2 + 2y + z^2 + 3z - e^{-x^2-y^2-z^2}$.

⌨ **Solutions** for Additional Examples can be found at `bit.ly/2PK9e1V` (* example with video solution)

13.2 Computer Problems

⌨ **Solutions** for Computer Problems numbered in blue can be found at `bit.ly/2ypjSVr`

1. Use Newton's Method to find the minimum of $f(x, y) = e^{-x^2 y^2} + (x - 1)^2 + (y - 1)^2$. Try various initial conditions, and compare answers. How many correct digits can you obtain with this method?

2. Apply Newton's Method to find the minima of the following functions to six correct decimal places (each function has two minima):
   (a) $f(x, y) = x^4 + y^4 + 2x^2 y^2 + 6xy - 4x - 4y + 1$
   (b) $f(x, y) = x^6 + y^6 + 3x^2 y^2 - x^2 - y^2 - 2xy$

3. Find the minimum of the Rosenbrock function $f(x, y) = 100(y - x^2)^2 + (x - 1)^2$ by (a) Newton's Method and (b) Steepest Descent. Use starting guess $(2, 2)$. After how many steps does the solution stop improving? Explain the difference in accuracy that is achieved.

4. Use the Steepest Descent to find the minima of the functions in Computer Problem 2.

5. Use Conjugate Gradient Search to find the minima of the functions in Computer Problem 2.

6. Find the minima to five correct digits by Conjugate Gradient Search:
   (a) $f(x, y) = x^4 + 2y^4 + 3x^2y^2 + 6x^2y - 3xy^2 + 4x - 2y$
   (b) $f(x, y) = x^6 + x^2y^4 + y^6 + 3x + 2y$

Reality Check **13** *Molecular Conformation and Numerical Optimization*

The function of a protein follows its form: The knobs and creases of the molecular shapes enable the bindings and blockings that are integral to their roles. The forces that govern the **conformation**, or folding, of amino acids into proteins are due to bonds between individual atoms and to weaker intermolecular interactions between unbound atoms such as electrostatic and Van der Waals forces. For densely packed molecules such as proteins, the latter are especially important.

One current approach to predicting the conformations of the proteins is to find the minimum potential energy of the total configuration of amino acids. The Van der Waals forces are modeled by the Lennard-Jones potential

$$U(r) = \frac{1}{r^{12}} - \frac{2}{r^6},$$

where r denotes the distance between two atoms. Figure 13.7 shows the energy well that is defined by the potential. The force is attractive for distances $r > 1$, but turns strongly repulsive when atoms try to come closer than $r = 1$. For a cluster of atoms with positions $(x_1, y_1, z_1), \ldots, (x_n, y_n, z_n)$, the objective function to be minimized is the sum of the pairwise Lennard-Jones potentials

$$U = \sum_{i<j} \left(\frac{1}{r_{ij}^{12}} - \frac{2}{r_{ij}^6} \right)$$

over all pairs of atoms, where

$$r_{ij} = \sqrt{(x_i - x_j)^2 + (y_i - y_j)^2 + (z_i - z_j)^2}$$

denotes the distance between atoms i and j. The variables in the optimization problem are the rectangular coordinates of the atoms.

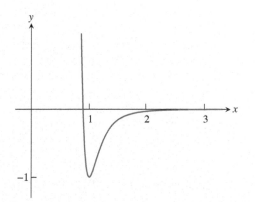

Figure 13.7 **The Lennard-Jones potential** $U(r) = r^{-12} - 2r^{-6}$. The energy minimum is -1, achieved at $r = 1$.

There are translational and rotational symmetries to consider: The total energy does not change if the cluster is moved in a straight line or rotated. To deal with the symmetries, we will limit the possible configurations by fixing the first atom at the origin $v_1 = (0, 0, 0)$ and requiring the second atom to lie on the z-axis at $v_2 = (0, 0, z_2)$. The remaining position variables $(x_3, y_3, z_3), \ldots, (x_n, y_n, z_n)$ are left to be arranged in a configuration that minimizes the potential energy U.

With the help of Figure 13.7, it is simple to arrange four or fewer atoms at the lowest possible Lennard-Jones energy. Note that the minimum of the single potential has the value -1 at $r = 1$. Thus, two atoms can sit exactly one unit from another, so that the energy is exactly at the bottom of the trough. Three atoms can sit in a triangle whose side is the same common distance, and a fourth atom can be placed at the same distance from the three vertices, say, above the triangle, creating an equilateral tetrahedron. The total energy U for the $n = 2, 3$, and 4 cases is -1 times the number of interactions, or $-1, -3$, and -6, respectively.

Placement of the fifth atom, however, is not so obvious. There is no point equidistant from the tetrahedron vertices of the $n = 4$ case, and a new technique is needed—numerical optimization.

Suggested activities:

1. Write a function file that returns the potential energy. Apply Nelder–Mead to find the minimal energy for $n = 5$. Try several initial guesses until you are convinced you have the absolute minimum. How many steps are required?

2. Use the MATLAB command `plot3` to plot the five atoms in the minimum energy configuration as circles, and connect all circles with line segments to view the conformed molecule.

3. Extend the function in Step 1 so that it returns f and the gradient vector ∇f. Apply Gradient Search for the $n = 5$ case. Find the minimum energy as before.

4. If the MATLAB Optimization Toolbox is available, apply the command `fminunc`, using only the objective function f.

5. Apply `fminunc`, using f and ∇f.

6. Apply the previous methods to $n = 6$. Rank the methods according to reliability and efficiency.

7. Determine and plot minimum-energy conformations for larger n. Information on minimum-energy Lennard-Jones clusters for n up to several hundred is posted at several Internet sites, so your answers can be readily checked.

The protein folding problem has become a hotbed of multidisciplinary optimization research. Simulated annealing and powerful quasi-Newton's Methods are often used to predict conformation of complicated molecules, with ever more realistic modeling of the intermolecular forces. The Protein Data Bank `http://www.rcsb.org/pdb` is a useful worldwide archive of structural data on biological macromolecules. Extensive lists of experimentally measured atom positions are available there for use in testing and validation of hypotheses concerning the forces and energy minimization.

Software and Further Reading

Introductory texts on optimization include Dennis and Schnabel [1987], Nocedal and Wright [1999], and Griva et al. [2008]. The useful guide Moré and Wright [1987] contains references to many software packages designed particularly for optimization.

A large set of test problems of varying types are found in Floudas et al. [1999]. The Optimization Technology Center run by Northwestern University and the Argonne National Lab `http://www.ece.northwestern.edu/OTC` has many links to available software.

The `opt` directory of Netlib contains a number of freely available optimization routines, including: `hooke` (derivative-free unconstrained optimization, via Hooke and Jeeves Method), `praxis` (unconstrained optimization, without requiring derivatives), and `tn` (Newton's Method for unconstrained or simple-bound optimization). WNLIB by Chapman and Naylor includes routines for unconstrained and constrained nonlinear optimization based on conjugate-gradient and conjugate-directions algorithms (as well as a general simulated annealing routine).

The MATLAB Optimization Toolbox includes routines for a variety of constrained and unconstrained nonlinear optimization problems. The TOMLAB Optimization Environment offers a broad variety of nonlinear optimization tools based on MATLAB toolboxes. It has a unified input–output format, an optional GUI, and automatic handling of derivatives. The optimization listings at `mathtools.net` include many solvers written in MATLAB and other languages.

Appendix A: Matrix Algebra

We begin with a short review of the basic definitions in matrix algebra.

A.1 MATRIX FUNDAMENTALS

A **vector** is an array of numbers

$$u = \begin{bmatrix} u_1 \\ u_2 \\ \vdots \\ u_n \end{bmatrix}.$$

If the list contains n numbers, it is called an n-dimensional vector. We will often make a distinction between the foregoing vertically arranged array, or **column vector**, and a horizontally arranged array

$$u = [u_1, \ldots, u_n]$$

called a **row vector**. An $m \times n$ **matrix** is an $m \times n$ array of numbers having the form

$$A = \begin{bmatrix} a_{11} & \cdots & a_{1n} \\ \vdots & & \vdots \\ a_{m1} & \cdots & a_{mn} \end{bmatrix}.$$

Each (horizontal) row of A can be considered as a row vector of A, and each (vertical) column as a column vector.

Matrix–vector multiplication makes a vector out of a matrix and a vector. The matrix–vector product is defined as

$$Au = \begin{bmatrix} a_{11} & \cdots & a_{1n} \\ \vdots & & \vdots \\ a_{m1} & \cdots & a_{mn} \end{bmatrix} \begin{bmatrix} u_1 \\ u_2 \\ \vdots \\ u_n \end{bmatrix} = \begin{bmatrix} a_{11}u_1 + a_{12}u_2 + \cdots + a_{1n}u_n \\ \vdots \\ a_{m1}u_1 + a_{m2}u_2 + \cdots + a_{mn}u_n \end{bmatrix}. \quad \text{(A.1)}$$

Note that in order to multiply an $m \times n$ matrix by a d-dimensional vector, it is required that $n = d$.

In matrix–matrix multiplication, an $m \times n$ matrix is multiplied by an $n \times p$ matrix to yield an $m \times p$ matrix as the product. Multiplying matrices can be expressed in terms of matrix–vector multiplication. Let C be an $n \times p$ matrix written in terms of its column vectors

$$C = \begin{bmatrix} c_1| & \cdots & |c_p \end{bmatrix}.$$

Then the matrix–matrix product of A and C is

$$AC = A\begin{bmatrix} c_1| & \cdots & |c_p \end{bmatrix} = \begin{bmatrix} Ac_1| & \cdots & |Ac_p \end{bmatrix}.$$

A system of m linear equations in n unknowns can be written in matrix form as

$$\begin{bmatrix} a_{11} & \cdots & a_{1n} \\ \vdots & & \vdots \\ a_{m1} & \cdots & a_{mn} \end{bmatrix} \begin{bmatrix} x_1 \\ x_2 \\ \vdots \\ x_n \end{bmatrix} = \begin{bmatrix} b_1 \\ b_2 \\ \vdots \\ b_n \end{bmatrix},$$

which we call a **matrix equation**.

The $n \times n$ **identity matrix** I_n is the matrix with $I_{ii} = 1$ for $1 \le i \le n$ and $I_{ij} = 0$ for $i \ne j$. The identity matrix serves as the identity for the operation of matrix multiplication, as $AI_n = I_n A = A$ for each $n \times n$ matrix A. For an $n \times n$ matrix A, the **inverse** A^{-1} of A is an $n \times n$ matrix satisfying $AA^{-1} = A^{-1}A = I_n$. If A has an inverse, it is called **invertible**; a noninvertible matrix is called **singular**.

The **transpose** of an $m \times n$ matrix A is the matrix A^T whose entries are $A^T_{ij} = A_{ji}$. The rule for the transpose of a product is $(AB)^T = B^T A^T$.

There are two important ways to multiply two vectors together. Let

$$
u = \begin{bmatrix} u_1 \\ \vdots \\ u_n \end{bmatrix} \text{ and } v = \begin{bmatrix} v_1 \\ \vdots \\ v_n \end{bmatrix}.
$$

The **inner product** $u^T v$ transposes u to a row vector; then ordinary matrix multiplication gives

$$
u^T v = u_1 v_1 + \cdots + u_n v_n.
$$

Thus the product of $1 \times n$ by $n \times 1$ yields a 1×1 matrix, or real number, as the result. Two column vectors are **orthogonal** if $u^T v = 0$. The **outer product** uv^T multiplies an $n \times 1$ column by a $1 \times n$ row. Ordinary matrix multiplication gives an $n \times n$ matrix result

$$
uv^T = \begin{bmatrix} u_1 v_1 & u_1 v_2 & \cdots & u_1 v_n \\ u_2 v_1 & u_2 v_2 & \cdots & u_2 v_n \\ \vdots & & & \vdots \\ u_n v_1 & \cdots & \cdots & u_n v_n \end{bmatrix}.
$$

An outer product is a rank-one matrix.

Each matrix product AB can be represented as the sum of outer products of the columns of A with the rows of B. More precisely,

Outer product sum rule

Let A and B be $m \times p$ and $p \times n$ matrices, respectively. Then

$$
AB = \sum_{i=1}^{p} a_i b_i^T
$$

where a_i is the ith column of A and b_i^T is the ith row of B.

The case $n = 1$ is sometimes called the "alternate form of matrix–vector multiplication." For example,

$$
\begin{bmatrix} 1 & 2 & 3 \\ 4 & 5 & 6 \\ 7 & 8 & 9 \end{bmatrix} \begin{bmatrix} -3 \\ 1 \\ 2 \end{bmatrix} = \begin{bmatrix} 1 \\ 4 \\ 7 \end{bmatrix} \begin{bmatrix} -3 \end{bmatrix} + \begin{bmatrix} 2 \\ 5 \\ 8 \end{bmatrix} \begin{bmatrix} 1 \end{bmatrix} + \begin{bmatrix} 3 \\ 6 \\ 9 \end{bmatrix} \begin{bmatrix} 2 \end{bmatrix}
$$

illustrates why, when viewed as a linear transformation, the range of a matrix is equivalent to its column space.

Because of the high computational complexity of computing the matrix inverse, it is avoided or minimized whenever possible. One trick that helps is the Sherman–Morrison formula. Assume that the inverse of an $n \times n$ matrix A is already known, and that the inverse of the modified matrix $A + uv^T$ is needed, where u and v are n-vectors.

THEOREM A.1   (Sherman–Morrison formula) If $v^T A^{-1} u \neq -1$, then $A + uv^T$ is invertible and

$$(A + uv^T)^{-1} = A^{-1} - \frac{A^{-1} uv^T A^{-1}}{1 + v^T A^{-1} u}. \qquad \blacksquare$$

The Sherman–Morrison formula is proved by multiplying $A + uv^T$ times the expression in the formula. The matrix $A + uv^T$ is called a **rank-one update** of A, since uv^T is a rank-one matrix. (See the discussion of Broyden's Method in Chapter 2 for an important application of the Sherman–Morrison formula. Elementary facts about matrices can be found in linear algebra texts such as Strang [2006] and Lay [2015].)

A.2 SYSTEMS OF LINEAR EQUATIONS

A system of linear equations is efficiently represented as a matrix equation. For example, the system

$$x_1 - x_2 + 2x_3 = -3$$
$$x_2 - x_3 = 3$$
$$2x_1 + x_2 + 3x_3 = 1$$

is equivalent to the matrix equation $Ax = b$:

$$\begin{bmatrix} 1 & -1 & 2 \\ 0 & 1 & -1 \\ 2 & 1 & 3 \end{bmatrix} \begin{bmatrix} x_1 \\ x_2 \\ x_3 \end{bmatrix} = \begin{bmatrix} -3 \\ 3 \\ 1 \end{bmatrix}.$$

The solution of the system is $[x_1, x_2, x_3] = [1, 2, -1]$. A system of linear equations is called **consistent** if a solution exists. A system of linear equations can have no solutions, one solution, or infinitely many solutions.

If the $n \times n$ matrix A is invertible, then $Ax = b$ is consistent, because $x = A^{-1}b$. If A is not invertible, A is singular, and the system $Ax = b$ may or may not be consistent, depending on b. For example, the matrix equation

$$\begin{bmatrix} 1 & 1 & 0 \\ 0 & 1 & 1 \\ 1 & 1 & 0 \end{bmatrix} \begin{bmatrix} x_1 \\ x_2 \\ x_3 \end{bmatrix} = \begin{bmatrix} 3 \\ 5 \\ 3 \end{bmatrix}$$

is consistent with solution $[x_1, x_2, x_3] = [1, 2, 3]$, while

$$\begin{bmatrix} 1 & 1 & 0 \\ 0 & 1 & 1 \\ 1 & 1 & 0 \end{bmatrix} \begin{bmatrix} x_1 \\ x_2 \\ x_3 \end{bmatrix} = \begin{bmatrix} 2 \\ 5 \\ 3 \end{bmatrix}$$

is clearly inconsistent, since $x_1 + x_2$ cannot be simultaneously 2 and 3. The inconsistency is possible because A is a singular matrix.

An elementary fact from linear algebra holds that an $n \times n$ matrix A is singular if and only if there is a nontrivial (nonzero) solution to the homogeneous equation $Ax = \vec{0}$. For example,

$$\begin{bmatrix} 1 & 1 & 0 \\ 0 & 1 & 1 \\ 1 & 1 & 0 \end{bmatrix} \begin{bmatrix} 1 \\ -1 \\ 1 \end{bmatrix} = \begin{bmatrix} 0 \\ 0 \\ 0 \end{bmatrix}.$$

Any scalar multiple of $v = [x_1, x_2, x_3] = [1, -1, 1]$ is also a solution of the homogeneous equation, so that $Ax = 0$ has infinitely many solutions. Moreover, for a singular matrix A, if $Ax = b$ is consistent, then there are infinitely many solutions, since

$$A(x + cv) = Ax + cAv = b + \vec{0} = b$$

for all scalars c.

To summarize, if the $n \times n$ matrix A is invertible, then $Ax = b$ has exactly one solution for each b. If A is singular, then $Ax = b$ will either have no solution, or infinitely many solutions, depending on b.

A.3 BLOCK MULTIPLICATION

Matrix multiplication can be done blockwise, a fact that will be very helpful in Chapter 12. If two matrices are divided into blocks whose sizes are compatible with matrix multiplication, then the matrix product can be carried out by matrix multiplication of the blocks. For example, the product of two 3×3 matrices can be carried out in the following blocks:

$$AB = \begin{bmatrix} x & x & x \\ x & x & x \\ x & x & x \end{bmatrix} \begin{bmatrix} x & x & x \\ x & x & x \\ x & x & x \end{bmatrix} = \begin{bmatrix} A_{11} & A_{12} \\ A_{21} & A_{22} \end{bmatrix} \begin{bmatrix} B_{11} & B_{12} \\ B_{21} & B_{22} \end{bmatrix}$$

$$= \begin{bmatrix} A_{11}B_{11} + A_{12}B_{21} & A_{11}B_{12} + A_{12}B_{22} \\ A_{21}B_{11} + A_{22}B_{21} & A_{21}B_{12} + A_{22}B_{22} \end{bmatrix}$$

Here A_{11} and B_{11} are 1×1 matrices, A_{12} and B_{12} are 1×2 matrices, and so forth. For example,

$$\begin{bmatrix} 1 & 2 & 3 \\ 0 & 1 & 3 \\ 2 & 2 & 4 \end{bmatrix} \begin{bmatrix} 2 & 4 & 1 \\ 1 & 0 & 1 \\ 3 & 1 & 2 \end{bmatrix} = \begin{bmatrix} 1 \cdot 2 + [2\ 3]\begin{bmatrix}1\\3\end{bmatrix} & 1[4\ 1] + [2\ 3]\begin{bmatrix}0 & 1\\1 & 2\end{bmatrix} \\ \begin{bmatrix}0\\2\end{bmatrix}2 + \begin{bmatrix}1 & 3\\2 & 4\end{bmatrix}\begin{bmatrix}1\\3\end{bmatrix} & \begin{bmatrix}0\\2\end{bmatrix}[4\ 1] + \begin{bmatrix}1 & 3\\2 & 4\end{bmatrix}\begin{bmatrix}0 & 1\\1 & 2\end{bmatrix} \end{bmatrix}$$

$$= \begin{bmatrix} 13 & 7 & 9 \\ 10 & 3 & 7 \\ 18 & 12 & 12 \end{bmatrix}.$$

Doing the multiplication blockwise gives the same result as doing it without blocks. This alternative way of looking at matrix multiplication is not meant to reduce computations, but to assist with bookkeeping, especially with eigenvalue computations in Chapter 12.

The only necessary compatibility required of the blocks is that the column groupings of A must exactly match the row groupings of B. In the preceding example, the first column of A is in one group, and the last two columns are in another. For matrix B, the first *row* is in one group and the last two *rows* are in another. As another example, we can multiply the 3×5 matrix A and the 5×2 matrix B in the following blocks:

$$
\begin{bmatrix} x & x & x & x & x \\ x & x & x & x & x \\ \hline x & x & x & x & x \end{bmatrix}
\begin{bmatrix} x & x \\ \hline x & x \\ x & x \\ \hline x & x \\ x & x \end{bmatrix}
$$

$$
= \begin{bmatrix} A_{11} & A_{12} & A_{13} \\ \hline A_{21} & A_{22} & A_{23} \end{bmatrix}
\begin{bmatrix} B_{11} & B_{12} \\ \hline B_{21} & B_{22} \\ \hline B_{31} & B_{32} \end{bmatrix}
$$

$$
= \begin{bmatrix} A_{11}B_{11} + A12B_{21} + A_{13}B_{31} & A_{11}B_{12} + A_{12}B_{22} + A_{13}B_{32} \\ \hline A_{21}B_{11} + A_{22}B_{21} + A_{23}B_{31} & A_{21}B_{12} + A_{22}B_{22} + A_{23}B_{32} \end{bmatrix}
$$

In this case, the three groups of columns of A are matched by the three groups of rows of B. On the other hand, the groupings of rows of A and columns of B do not need to match; they may be done arbitrarily.

A.4 EIGENVALUES AND EIGENVECTORS

We begin with a short review of the basic concepts of eigenvalues and eigenvectors.

DEFINITION A.2 Let A be an $m \times m$ matrix and x a nonzero m-dimensional real or complex vector. If $Ax = \lambda x$ for some real or complex number λ, then λ is called an **eigenvalue** of A and x is the corresponding **eigenvector**. ❒

For example, the matrix $A = \begin{bmatrix} 1 & 3 \\ 2 & 2 \end{bmatrix}$ has an eigenvector $\begin{bmatrix} 1 \\ 1 \end{bmatrix}$, and corresponding eigenvalue 4.

Eigenvalues are the roots λ of the **characteristic polynomial** $\det(A - \lambda I)$. If λ is an eigenvalue of A, then any nonzero vector in the nullspace of $A - \lambda I$ is an eigenvector corresponding to λ. For this example,

$$
\det(A - \lambda I) = \det \begin{bmatrix} 1 - \lambda & 3 \\ 2 & 2 - \lambda \end{bmatrix} = (\lambda - 1)(\lambda - 2) - 6 = (\lambda - 4)(\lambda + 1),
$$
(A.2)

so the eigenvalues are $\lambda = 4, -1$. The eigenvectors corresponding to $\lambda = 4$ are found in the nullspace of

$$
A - 4I = \begin{bmatrix} -3 & 3 \\ 2 & -2 \end{bmatrix}
$$
(A.3)

and so consist of all nonzero multiples of $\begin{bmatrix} 1 \\ 1 \end{bmatrix}$. Similarly, the eigenvectors corresponding to $\lambda = -1$ are all nonzero multiples of $\begin{bmatrix} 3 \\ -2 \end{bmatrix}$.

DEFINITION A.3 The $m \times m$ matrices A_1 and A_2 are **similar**, denoted $A_1 \sim A_2$, if there exists an invertible $m \times m$ matrix S such that $A_1 = SA_2S^{-1}$. ❒

Similar matrices have identical eigenvalues, because their characteristic polynomials are identical:

$$A_1 - \lambda I = SA_2S^{-1} - \lambda I = S(A_2 - \lambda I)S^{-1} \tag{A.4}$$

implies that

$$\det(A_1 - \lambda I) = (\det S)\det(A_2 - \lambda I)\det S^{-1} = \det(A_2 - \lambda I). \tag{A.5}$$

If a matrix A has eigenvectors that form a basis for R^m, then A is similar to a diagonal matrix, and A is called **diagonalizable**. In fact, assume that $Ax_i = \lambda_i x_i$ for $i = 1, \ldots, m$, and define the matrix

$$S = [\ x_1 \ \cdots \ x_m\ ].$$

Then you can check that the matrix equation

$$AS = S \begin{bmatrix} \lambda_1 & & \\ & \ddots & \\ & & \lambda_m \end{bmatrix} \tag{A.6}$$

holds. The matrix S is invertible because its columns span R^m. Therefore, A is similar to the diagonal matrix containing its eigenvalues.

Not all matrices are diagonalizable, even in the 2×2 case. In fact, all 2×2 matrices are similar to one of the following three types:

$$A_1 = \begin{bmatrix} a & 0 \\ 0 & b \end{bmatrix}$$

$$A_2 = \begin{bmatrix} a & 1 \\ 0 & a \end{bmatrix}$$

$$A_3 = \begin{bmatrix} a & -b \\ b & a \end{bmatrix}.$$

Remember that eigenvalues are identical for similar matrices. A matrix is similar to a matrix of form A_1 if there are two eigenvectors that span R^2; a matrix is similar to a matrix of form A_2 if there is a repeated eigenvalue with only one dimensional space of eigenvectors; and to A_3 if it has a complex pair of eigenvalues.

A.5 SYMMETRIC MATRICES

For a symmetric matrix, all eigenvectors are orthogonal to one another, and together they span the underlying space. In other words, symmetric matrices always have an orthonormal basis of eigenvectors.

DEFINITION A.4   A set of vectors is **orthonormal** if the elements of the set are unit vectors that are pairwise orthogonal.   ❏

In terms of dot products, orthonormality of the set $\{w_1, \ldots, w_m\}$ means $w_i^T w_j = 0$ if $i \neq j$, and $w_i^T w_i = 1$, for $1 \leq i, j \leq m$. For example, the sets $\{(1, 0, 0), (0, 1, 0), (0, 0, 1)\}$ and $\{(\sqrt{2}/2, \sqrt{2}/2), (\sqrt{2}/2, -\sqrt{2}/2)\}$ are orthonormal sets.

THEOREM A.5   Assume that A is a symmetric $m \times m$ matrix with real entries. Then the eigenvalues are real numbers, and the set of unit eigenvectors of A is an orthonormal set $\{w_1, \ldots, w_m\}$ that forms a basis of R^m. ∎

▶ **EXAMPLE A.1**   Find the eigenvalues and eigenvectors of

$$A = \begin{bmatrix} 0 & 1 \\ 1 & \frac{3}{2} \end{bmatrix}. \tag{A.7}$$

Calculating as before, the eigenvalue/eigenvector pairs are $2, (1, 2)^T$ and $-1/2, (-2, 1)^T$. Note that as the theorem promises, the eigenvectors are orthogonal. The corresponding orthonormal basis of unit eigenvectors is

$$\left\{ \begin{bmatrix} \frac{1}{\sqrt{5}} \\ \frac{2}{\sqrt{5}} \end{bmatrix}, \begin{bmatrix} -\frac{2}{\sqrt{5}} \\ \frac{1}{\sqrt{5}} \end{bmatrix} \right\}.$$

◀

The following theorem will be helpful for studying iterative methods in Chapter 2:

DEFINITION A.6   The **spectral radius** $\rho(A)$ of a square matrix A is the maximum magnitude of its eigenvalues. ☐

THEOREM A.7   If the $n \times n$ matrix A has spectral radius $\rho(A) < 1$, and b is arbitrary, then, for any vector x_0, the iteration $x_{k+1} = Ax_k + b$ converges. In fact, there exists a unique x_* such that $\lim_{k \to \infty} x_k = x_*$ and $x_* = Ax_* + b$. ∎

Moreover, if $b = 0$, then x_* is either the zero vector or an eigenvector of A with eigenvalue 1. The latter is ruled out because of the spectral radius, leading to the following fact that is useful in Chapter 8:

COROLLARY A.8   If the $n \times n$ matrix A has spectral radius $\rho(A) < 1$, then, for any initial vector x_0, the iteration $x_{k+1} = Ax_k$ converges to 0. ∎

A.6 VECTOR CALCULUS

In this section, the derivatives of scalar-valued and vector-valued functions are defined, and the product rules involving them are collected for later use.

Let $f(x_1, \ldots, x_n)$ be a scalar-valued function of n variables. The **gradient** of f is the vector-valued function

$$\nabla f(x_1, \ldots, x_n) = [f_{x_1}, \ldots, f_{x_n}],$$

where the subscripts denote partial derivatives of f with respect to that variable.

Let

$$F(x_1, \ldots, x_n) = \begin{bmatrix} f_1(x_1, \ldots, x_n) \\ \vdots \\ f_n(x_1, \ldots, x_n) \end{bmatrix}$$

be a vector-valued function of n variables. The **Jacobian** of F is the matrix

$$DF(x_1, \ldots, x_n) = \begin{bmatrix} \nabla f_1 \\ \vdots \\ \nabla f_n \end{bmatrix}.$$

Now we can state the product rules for two typical products in matrix algebra. Both have straightforward proofs when they are written in components and the single-variable product rule is applied. Let $u(x_1, \ldots, x_n)$ and $v(x_1, \ldots, x_n)$ be vector-valued functions, and let $A(x_1, \ldots, x_n)$ be an $n \times n$ matrix function. The dot product $u^T v$ is a scalar function. The first formula shows how to take its gradient. The matrix vector product Av is a vector whose Jacobian is expressed in the second rule.

Vector dot product rule

$$\nabla(u^T v) = v^T Du + u^T Dv$$

Matrix/vector product rule

$$D(Av) = A \cdot Dv + \sum_{i=1}^{n} v_i Da_i,$$

where a_i denotes the ith column of A.

Appendix B: Introduction to MATLAB

MATLAB is a general-purpose computing environment that is ideally suited for implementing mathematical and numerical methods. It is used as a high-powered calculator for small problems and as a full-featured programming language for large problems. A helpful feature of MATLAB is its long list of high-quality library functions that can make complicated calculations short, precise, and easy to write in high-level code.

This section contains a brief introduction to MATLAB's commands and features. Much more detailed accounts can be found in MATLAB's help facilities, the MATLAB *User's Guide*, in books such as Sigmon [2002], Hahn [2002], and on websites devoted to the package.

B.1 STARTING MATLAB

On PC-based systems, MATLAB is started by clicking the appropriate icon and ended by clicking on File/Exit. On Unix-based systems, type MATLAB at the system prompt:

```
$ matlab
```

Then type

```
>> exit
```

to exit.
Type the command

```
>> a=5
```

followed by the return key. MATLAB will echo the information back to you. Type the additional commands

```
>> b=3
>> c=a+b
>> c=a*b
>> d=log(c)
>> who
```

to get an idea of how MATLAB works. You may include a semicolon after a statement to suppress echoing of the value. The who command gives a list of all variables you have defined.

MATLAB has an extensive online help facility. Type help log for information on the log command. The PC version of MATLAB has a Help menu that contains descriptions and usage suggestions on all commands.

To erase the value of the variable a, type clear a. Typing clear will erase all previously defined variables. To recover a previous command, use the up cursor key. If you run out of room on the current command line, end the line with three periods and a return; then resume typing on the next line.

To save values of variables for your next login, type save, then load on your next login to MATLAB. For a transcript of part or all of the MATLAB session, type diary filename to start logging, and diary off to end. Use a filename of your choice for filename. This is helpful for submitting your work for an assignment. The diary command produces a file that can be viewed or printed once your MATLAB session is over.

MATLAB normally performs all computations in IEEE double precision, about 16 decimal digits of accuracy. The numeric display format can be changed with the format statement. Typing format long will change the way numbers are displayed until further notice. For example, the number 1/3 will be displayed differently depending on the current format:

```
format short      0.3333
format short e    3.3333E-001
format long       0.33333333333333
format long e     3.333333333333333E-001
format bank       0.33
format hex        3fd5555555555555
```

More control over formatting output is given by the fprintf command. The commands

```
>> x=0:0.1:1;
>> y=x.^2;
>> fprintf('%8.5f %8.5f \n',[x;y])
```

print the table

```
0.00000   0.00000
0.10000   0.01000
0.20000   0.04000
0.30000   0.09000
0.40000   0.16000
0.50000   0.25000
0.60000   0.36000
0.70000   0.49000
0.80000   0.64000
0.90000   0.81000
1.00000   1.00000
```

B.2 GRAPHICS

To plot data, express the data as vectors in the X and Y directions. For example, the commands

```
>> a=[0.0 0.4 0.8 1.2 1.6 2.0];
>> b=sin(a);
>> plot(a,b)
```

will draw a piecewise-linear approximation to the graph of $y = \sin x$ on $0 \leq x \leq 2$, as shown in Figure B.1(a). In this case, a and b are 6-dimensional vectors, or 6-element arrays. The font of the axis numbers can be set to 16-point, for example, by the command set(gca,'FontSize',16). A shorter way to define the vector a is the command

```
>> a=0:0.4:2;
```

This command defines a to be a vector whose entries begin at 0, increment by 0.4, and end at 2, identical to the previous longer definition. A more accurate version of one entire cycle of the sine curve results from

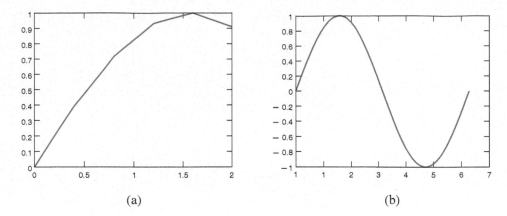

Figure B.1 MATLAB **figures.** (a) Piecewise-linear plot of $f(x) = \sin x$, with x increment of 0.4. (b) Another piecewise plot looks smooth because the x increment is 0.02.

```
>> a=0:0.02:2*pi;
>> b=sin(a);
>> plot(a,b)
```

and is shown in Figure B.1(b).

To draw the graph of $y = x^2$ on $0 \le x \le 2$, one could use

```
>> a=0:0.02:2;
>> b=a.^2;
>> plot(a,b)
```

The "." character preceding the power operator may be unexpected. It causes the power operator to be vectorized, that is, to square each entry of the vector a. As we will see in the next section, MATLAB treats every variable as a matrix. Omitting the period in this instance would mean multiplying the 101×1 matrix a by itself, under the rules of matrix multiplication, which is impossible. If you ask MATLAB to do this, it will complain. In general, MATLAB interprets an operation preceded by a period to mean that the operation should be applied entry-wise, not as matrix multiplication.

There are more advanced techniques for plotting graphs. MATLAB will choose axis scaling automatically if it is not specified, as in Figure B.1. To choose the axis scaling manually, use the `axis` command. For example, following a plot with the command

```
>> v=[-1 1 0 10]; axis(v)
```

sets the graphing window to $[-1, 1] \times [0, 10]$. The `grid` command draws a grid behind the plot.

Use the command `plot(x1,y1,x2,y2,x3,y3)` to plot three curves in the same graph window, where `xi, yi` are pairs of vectors of the same lengths. Type `help plot` to see the choices of solid, dotted, and dashed line types and various symbol types (circles, dots, triangles, squares, etc.) for plots. Semilog plots are available through the `semilogy` and `semilogx` commands.

The subplot command splits the graph window into multiple parts. The statement `subplot(abc)` breaks the window into an $a \times b$ grid and uses the c box for the plot. For example,

```
>> subplot(121),plot(x,y)
>> subplot(122),plot(x,z)
```

plots the first graph on the left side of the screen and the second on the right. The figure command opens up new plot windows and moves among them, if you need to view several different plots at once.

Three-dimensional surface plots are drawn with the command mesh. For example, the function $z = \sin(x^2 + y^2)$ on the domain $[-1, 1] \times [-2, 2]$ can be graphed by

```
>> [x,y]=meshgrid(-1:0.1:1,-2:0.1:2);
>> z=sin(x.^2+y.^2);
>> mesh(x,y,z)
```

The vector x created by meshgrid is 41 rows of the 21-vector -1:0.1:1, and similarly, y is 21 columns of the column vector -2:0.1:2. The graph produced by this code is shown in Figure B.2. Replacing mesh with surf plots a colored surface over the mesh.

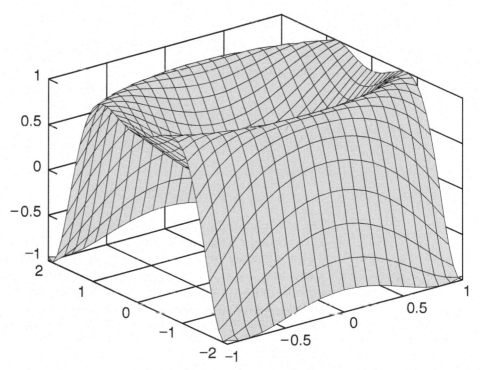

Figure B.2 Three-dimensional MATLAB plot. The mesh command is used to plot surfaces.

B.3 PROGRAMMING IN MATLAB

More sophisticated results can be achieved by writing programs in the MATLAB language. A **script file** is a file containing a list of MATLAB commands. The filename of a script file has a suffix of .m, so such files are sometimes called **m-files**. For example, you might use your favorite editor, or the MATLAB editor if available, to create the file cubrt.m, containing the following lines:

```
% The program cubrt.m finds a cube root by iteration
y=1;
n=15;
z=input('Enter z:');
```

```
for i = 1:n
  y = 2*y/3 + z/(3*y^2)
end
```

To run the program, type `cubrt` at the MATLAB prompt. The reason that this code converges to the cube root will become evident from our study of Newton's Method in Chapter 1. Notice that the semicolon was dropped from the line that defines the new y by iteration. This allows you to see the progression of approximants as they approach the cube root.

With the graphics ability of MATLAB, we can analyze the data from the cube root algorithm. Consider the program `cubrt1.m`:

```
% The program cubrt1.m finds cube roots and displays its progress
y(1)=1;
n=15;
z=input('Enter z:');
for i = 1:n-1
  y(i+1) = 2*y(i)/3 + z/(3*y(i)^2);
end
plot(1:n,y)
title('Iterative method for cube roots')
xlabel('Iteration number')
ylabel('Approximate cube root')
```

Run the foregoing program with $z = 64$. When finished, type the commands

```
>> e=y-4;
>> plot(1:n,e)
>> semilogy(1:n,e)
```

The first command subtracts the correct cube root 4 from each entry of the vector y. This remainder is the error e at each step of the iteration. The second command plots the error, and the third plots the error in a semilog plot, using logarithmic units in the y-direction.

Creating a script file to hold MATLAB code is preferred if the calculation will take more than a few lines. A script file can call other script files, including itself. (Typing ⟨cntl⟩-C will usually abort runaway MATLAB processes.)

B.4 FLOW CONTROL

The `for` loop was introduced in the previous cube root program. MATLAB has a number of commands to control the flow of a program. A number of these, including `while` loops and `if` and `break` statements, will be familiar to anyone with knowledge of a high-level programming language. For example,

```
n=5;
 for i=1:n
  for j=1:n
    a(i,j)=1/(i+j-1);
  end
 end
 a
```

creates and displays the 5×5 Hilbert matrix. The semicolon avoids repeated printing of partial results, and the final a displays the final result. Note that each `for` must be matched with an end. It is a good idea, though not required by MATLAB, to indent loops for greater readability.

The while command works similarly:

```
n=5;i=1;
while i<=n
  j=1;
  while j<=n
    a(i,j)=1/(i+j-1);
    j=j+1;
  end
  i=i+1;
end
a
```

This produces the same result as the double for loop.

The if statement is used to make flow decisions, and the break command provides an exit jump out of the next inner loop. Both are illustrated as follows:

```
% To compute the nth derivative of sin(x) at x=0
n=input('Enter n, negative number to quit:)
if n<=0,break,end
r=rem(n,4)  % rem is the remainder function
if r==0
  y=0
elseif r==1
    y=1
elseif r==2
    y=0
else
    y=-1
end
y
```

The logical operators & and | stand for AND, OR, respectively. The error command stops execution of the m-file and reports information to the user.

B.5 FUNCTIONS

In addition to built-in library functions like sin and exp, MATLAB allows the definition of user-defined functions. The command

```
>> f=@(x) exp(sin(2*x))
```

creates a function with input x and output $f(x) = e^{\sin 2x}$. After defining f as above, the command

```
>> f(0)
```

returns the correct result $e^{\sin 2(0)} = 1$. Moreover, the definition with @ assigns a **function handle** to f that can be passed to another function. If we create another function

```
>> firstderiv=@(f,x,h)  (f(x+h)-f(x-h))/(2*h)
```

with three inputs f,x,h, the command

```
>> firstderiv(f,0,0.0001)
```

returns an approximation to the derivative at 0. Here, we have used the user-defined function handle f as an input to the user-defined MATLAB function firstderiv.

A MATLAB function may have several inputs and several outputs. An example of a vector-valued function of several variables having three inputs and three outputs is the following function that converts rectangular to spherical coordinates:

```
>> rec2sph=@(x,y,z) [sqrt(x^2+y^2+z^2) acos(z/sqrt(x^2+y^2+z^2))...
    atan2(y,x)]
```

This method of defining functions is useful when the function can be defined on one line. For more complicated examples, MATLAB allows a second way to define a function, through a special m-file. The syntax of the first line must be adhered to, as in the following example, where the filename is cubrtf.m:

```
function y=cubrtf(x)
% Approximates the cube root of x
% Input real number x, output its cube root
y=1;
n=15;
for i = 1:n
  y = 2*y/3 + x/(3*y^2)
end
```

Here, we have transferred the script-file version of the cube root approximator to a MATLAB function. The function can be evaluated by

```
>> c=cubrtf(8)
```

Note that a MATLAB function differs from a script m-file in the first line. The filename, with the .m omitted, should agree with the function name in the first line. Variables in a function file are local by default, but can be made global with the global command.

Combining the two above approaches, a previously defined MATLAB function, such as an m-file function, can be assigned a function handle by prefixing with the @ sign. The function handle can then be passed into another function. For example,

```
>> firstderiv(@cubrtf,1,0.0001)
```

returns the approximation 0.3333 for the derivative of $x^{1/3}$ at $x = 1$.

A more complicated function can use several variables as inputs and several as outputs. For example, here is a function that calls the existing MATLAB functions mean and std and collects both in an array:

```
function [m,sigma]=stat(x)
% Returns sample mean and standard deviation of input vector x
m=mean(x);
sigma=std(x);
```

If this file stat.m resides in your MATLAB path, typing stat(x), where x is a vector, will return the mean and standard deviation of the entries of the vector.

The nargin command provides the number of input arguments to a function. With this command, the work of a function can change, depending on how many arguments are presented to it. An example of nargin is given in Program 0.1 on nested multiplication.

An example of a piecewise-defined function is

$$h(x) = \begin{cases} x+2 & \text{for } x \leq -1 \\ 1 & \text{for } -1 < x \leq 0 \\ \cos x & \text{for } x > 0. \end{cases}$$

The function $h(x)$ can be represented by the creating the MATLAB function file h.m containing

```
function y=h(x)
p1=(x<=-1);
p2=(x>-1).*(x<=0);
p3=(x>0);
y=p1.*(x+2)+p2.*1+p3.*cos(x);
```

Here we are making use of Boolean evaluation of the conditional expressions as 1 if true and 0 if false. We are also using the period preceding arithmetic operations to vectorize them, allowing the input x to be a vector of numbers. Now h can be passed to other MATLAB functions via its function handle @h. For example,

```
>> ezplot(@h,[-3 3])
```

plots the piecewise function h, and

```
>> fzero(@h,1)
```

finds a root of $h(x)$ near 1. Should the result of

```
>> firstderiv(@h,-1,0.0001)
```

be trusted?

B.6 MATRIX OPERATIONS

The key to MATLAB's power and versatility is the sophistication of its variables' data structure. Each variable in MATLAB is an $m \times n$ matrix of double precision floating point numbers. A scalar is simply the special case of a 1×1 matrix. The syntax

```
>> A=[1 2 3
4 5 6]
```

or

```
>> A=[1 2 3;4 5 6]
```

defines a 2×3 matrix A. The command B=A' creates a 3×2 matrix B that is the transpose of A. Matrices of the same size can be added and subtracted with the + and − operators. The command size(A) returns the dimensions of the matrix A, and length(A) returns the maximum of the two dimensions.

MATLAB provides many commands that allow matrices to be easily built. For example, zeros(m,n) produces a matrix full of zeros of size $m \times n$. If A is a matrix, then zeros(size(A)) produces a matrix of zeros of the same size as A. The commands ones(m,n) and eye(m,n) (for the identity matrix) work essentially the same way. For example,

```
>> A=[eye(2) zeros(2,2);zeros(2,2) eye(2)]
```

is a convoluted, but accurate way to construct the 4×4 identity matrix.

The colon operator can be used to extract a submatrix from a matrix. For example,

```
>> b=A(1:3,2)
```

assigns to b the first three entries of the second column of A. The command

```
>> b=A(:,2)
```

assigns to b the entire second column of A, and

```
>> B=A(:,1:3)
```

assigns to B the submatrix consisting of the first three columns of A.

The $m \times n$ matrix A and the $n \times p$ matrix B can be multiplied by the command C=A*B. If the matrices have inappropriate sizes, MATLAB will refuse to do the operation and return an error message.

B.7 ANIMATION AND MOVIES

The field of differential equations includes the study of dynamic systems, or "things that move." MATLAB makes animation easy, and these aspects are exploited in Chapter 6 to follow solutions that are changing with time.

The sample MATLAB program bounce.m given next shows a tennis ball bouncing from wall to wall in a unit square. The first set command sets up parameters of the current figure (gca), including the axis limits $0 \leq x, y \leq 1$. The cla command clears the figure window, and axis square equalizes the units in the x and y directions.

Next, the animatedline command is used to define a line object called ball, along with its properties. The four if statements in the while loop cause the ball to reverse velocity when it hits one of the four walls. The loop also contains an addpoints command that updates the current x and y coordinates of the line object ball, by setting its x and y attributes, respectively. The drawnow command draws all defined objects to the current figure window. The speed of the moving ball can be adjusted with the pause command and through the step sizes hx0 and hy0. The while loop is infinite and can be interrupted by ⟨cntl⟩-C. Here is the program in its entirety:

MATLAB code shown here can be found at bit.ly/2CNXXLn

```
% bounce.m
% Illustrates Matlab animation using the drawnow command
% Usage: Save this file as bounce.m, then type "bounce"
set(gca,'XLim',[0 1],'YLim',[0 1]);
cla
axis square
ball = animatedline('color','r','Marker','o','MarkerSize',12,...
    'MarkerFaceColor','r');
hx0=.05;hy0=.05/sqrt(2);hx=hx0;hy=hy0;
xl=.02;xr=.98;yb=xl;yt=xr;x=.1;y=.1;
while 1 == 1
    if x < xl
        hx= hx0;
    end
    if x > xr
        hx = -hx0;
    end
    if y < yb
        hy = hy0;
    end
    if y > yt
        hy = -hy0;
    end
    x=x+hx;y=y+hy;
    clearpoints(ball);addpoints(ball,x,y);
    drawnow;pause(0.01)
end
```

Using the MATLAB VideoWriter command, it is straightforward to make mp4 videos. Each frame of the video will be a single MATLAB figure. For example, the code segment

```
v=VideoWriter('filename.mp4','MPEG-4');
open(v)
for i=1:n
  (plot a figure)
  frame=getframe;writeVideo(v,frame);
end
close(v)
```

will capture the n still figures and place them into a video file named filename.mp4.

Answers to Selected Exercises

CHAPTER 0

0.1 Exercises

1. (a) $P(x) = 1 + x(1 + x(5 + x(1 + x(6))))$, $P(1/3) = 2$.

   (b) $P(x) = 1 + x(-5 + x(5 + x(4 + x(-3))))$, $P(1/3) = 0$

   (c) $P(x) = 1 + x(0 + x(-1 + x(1 + x(2))))$, $P(1/3) = 77/81$

3. $P(x) = 1 + x^2(2 + x^2(-4 + x^2(1)))$, $P(1/2) = 81/64$

5. (a) 5 (b) 41/4

7. n multiplications and $2n$ additions

0.1 Computer Problems

1. Correct answer from Q is 51.01275208275, error $= 4.76 \times 10^{-12}$

0.2 Exercises

1. (a) 1000000 (b) 10001 (c) 1001111
   (d) 11100011

3. (a) 1010.1 (b) $0.\overline{01}$ (c) $0.\overline{101}$ (d) $1100.\overline{1100}$ (e) $110111.\overline{0110}$ (f) $0.\overline{00011}$

5. 11.0010010000111

7. (a) 85 (b) 93/8 (c) 70/3 (d) 20/3
   (e) 20/7 (f) 48/7 (g) 283/120 (h) 8

0.3 Exercises

1. (a) $1.0000\ldots0000 \times 2^{-2}$ (b) $1.0101\ldots0101 \times 2^{-2}$
   (c) $1.0101\ldots0101 \times 2^{-1}$ (d) $1.11001100\ldots 11001101 \times 2^{-1}$

3. $1 \le k \le 50$

5. (a) $2\epsilon_{mach}$ (b) $4\epsilon_{mach}$

7. (a) 4020000000000000 (b) 4035000000000000
   (c) 3fc0000000000000 (d) 3fd5555555555555
   (e) 3fe5555555555555 (f) 3fb999999999999a
   (g) bfb999999999999a (h) bfc999999999999a

9. (a) Note that $(7/3 - 4/3) - 1 = \epsilon_{mach}$ in double precision. (b) No, $(4/3 - 1/3) - 1 = 0$.

11. No, associative law fails.

13. (a) 2, represented by $010\ldots0$ (b) 2^{-511}, represented by $0010\ldots0$ (c) 0, represented by $10\ldots0$

15. (a) 2^{-50} (b) 0 (c) 2^{-50}

0.4 Exercises

1. (a) Loss of significance near $x = 2\pi n$, n integer. Rewrite as $-1/(1 + \sec x)$ (b) Loss of significance near $x = 0$. Rewrite as $3 - 3x + x^2$ (c) Loss of significance near $x = 0$. Rewrite as $2x/(x^2 - 1)$

3. $x_1 = -(b + \sqrt{b^2 + 4 \times 10^{-12}})/2$, $x_2 = (2 \times 10^{-12})/(b + \sqrt{b^2 + 4 \times 10^{-12}})$

5. -0.125

0.4 Computer Problems

1. (a)

x	original	revised
0.10000000000000	−0.49874791371143	−0.49874791371143
0.01000000000000	−0.49998749979096	−0.49998749979166
0.00100000000000	−0.49999987501429	−0.49999987499998
0.00010000000000	−0.49999999362793	−0.49999999875000
0.00001000000000	−0.50000004133685	−0.49999999998750
0.00000100000000	−0.50004445029084	−0.49999999999987
0.00000010000000	−0.51070259132757	−0.50000000000000
0.00000001000000	0	−0.50000000000000
0.00000000100000	0	−0.50000000000000
0.00000000010000	0	−0.50000000000000
0.00000000001000	0	−0.50000000000000
0.00000000000100	0	−0.50000000000000
0.00000000000010	0	−0.50000000000000
0.00000000000001	0	−0.50000000000000

(b)

x	original	revised
0.10000000000000	2.71000000000000	2.71000000000000
0.01000000000000	2.97010000000001	2.97010000000000
0.00100000000000	2.99700100000000	2.99700100000000
0.00010000000000	2.99970000999905	2.99970001000000
0.00001000000000	2.99997000008379	2.99997000010000
0.00000100000000	2.99999700015263	2.99999700000100
0.00000010000000	2.99999969866072	2.99999970000001
0.00000001000000	2.99999998176759	2.99999997000000
0.00000000100000	2.99999991515421	2.99999999700000
0.00000000010000	3.00000024822111	2.99999999970000

x	original	revised
0.00000000001000	3.00000024822111	2.99999999997000
0.00000000000100	2.99993363483964	2.99999999999700
0.00000000000010	3.00093283556180	2.99999999999970
0.00000000000001	2.99760216648792	2.99999999999997

3. 6.127×10^{-13}

5. 2.23322×10^{-10}

0.5 Exercises

1. (a) $f(0)f(1) = -2 < 0$ implies $f(c) = 0$ for some c in $(0, 1)$ by the Intermediate Value Theorem.
   (b) $f(0)f(1) = -9 < 0$ implies $f(c) = 0$ for some c in $(0, 1)$ (c) $f(0)f(1/2) = -1/2 < 0$ implies $f(c) = 0$ for some c in $(0, 1/2)$.

3. (a) $c = 2/3$ (b) $c = 1/\sqrt{2}$ (c) $c = 1/(e - 1)$

5. (a) $P(x) = 1 + x^2 + 1/2x^4$ (b) $P(x) = 1 - 2x^2 + 2/3x^4$ (c) $P(x) = x - x^2/2 + x^3/3 - x^4/4 + x^5/5$
   (d) $P(x) = x^2 - x^4/3$

7. (a) $P(x) = (x - 1) - (x - 1)^2/2 + (x - 1)^3/3 - (x - 1)^4/4$ (b) $P(0.9) = -0.105358\overline{3}$, $P(1.1) = 0.095308\overline{3}$ (c) error bound = 0.000003387 for $x = 0.9$, 0.000002 for $x = 1.1$ (d) Actual error ≈ 0.00000218 at $x = 0.9$, 0.00000185 at $x = 1.1$

9. $\sqrt{1 + x} = 1 + x/2 \pm x^2/8$. For $x = 1.02$, $\sqrt{1.02} \approx 1.01\pm 0.00005$. Actual value is $\sqrt{1.02} = 1.0099505$, error $= 0.0000495$

CHAPTER 1

1.1 Exercises

1. (a) $[2, 3]$ (b) $[1, 2]$ (c) $[6, 7]$
3. (a) 2.125 (b) 1.125 (c) 6.875
5. (a) $[2, 3]$ (b) 33 steps

1.1 Computer Problems

1. (a) 2.080084 (b) 1.169726 (c) 6.776092
3. (a) Intervals $[-2, -1], [-1, 0], [1, 2]$, roots $-1.641784, -0.168254, 1.810038$ (b) Intervals $[-2, -1], [-0.5, 0.5], [0.5, 1.5]$, roots $-1.023482, 0.163822, 0.788941$ (c) Intervals $[-1.7, -0.7], [-0.7, 0.3], [0.3, 1.3]$, roots $-0.818094, 0, 0.506308$
5. (a) $[1, 2]$, 27 steps, 1.25992105 (b) $[1, 2]$, 27 steps, 1.44224957 (c) $[1, 2]$, 27 steps, 1.70997595
7. first root -17.188498, determinant correct to 2 places; second root 9.708299, determinant correct to 3 places.
9. $H = 635.5$ mm

1.2 Exercises

1. (a) $-\sqrt{3}, \sqrt{3}$ (b) $1, 2$ (c) $(5 \pm \sqrt{17})/2$
3. Check by substitution.
5. B, D
7. (a) loc. convergent (b) divergent (c) divergent
9. (a) 0 is locally convergent, 1 is divergent (b) $1/2$ is locally convergent, $3/4$ is divergent
11. (a) For example, $x = x^3 + e^x$, $x = (x - e^x)^{1/3}$, and $x = \ln(x - x^3)$; (b) For example, $x = 9x^2 + 3/x^3$, $x = 1/9 - 1/(3x^4)$, and $x = (x^5 - 9x^6)/3$
13. (a) $0.3, -1.3$ (b) 0.3 (c) slower
15. All converge to $\sqrt{5}$. From faster to slowest: (B), (C), (A).
17. $g(x) = \sqrt{(1 - x)/2}$ is locally convergent to $1/2$, and $g(x) = -\sqrt{(1 - x)/2}$ is locally convergent to -1.
19. $g(x) = (x + A/x^2)/2$ converges to $A^{1/3}$.
21. (a) Substitute and check (b) $|g'(r)| > 1$ for all three fixed points r
23. $g'(r_2) > 1$
27. (a) $x = x - x^3$ implies $x = 0$ (b) If $0 < x_i < 1$, then $x_{i+1} = x_i - x_i^3 = x_i (1 - x_i^2) < x_i$, and $0 < x_{i+1} < x_i < 1$ (c) The bounded monotonic sequence x_i converges to a limit L, which must be a fixed point. Therefore $L = 0$.
29. (a) $c < -2$ (b) $c = -4$
31. The open interval $(-5/4, 5/4)$ of initial guesses converge to the fixed point $1/4$; the two initial guesses $-5/4, 5/4$ lead to $-5/4$.
33. (a) Choose $a = 0$ and $|b| < 1$, c arbitrary. (b) Choose $a = 0$ and $|b| > 1$, c arbitrary.

1.2 Computer Problems

1. (a) 1.76929235 (b) 1.67282170 (c) 1.12998050
3. (a) 1.73205081 (b) 2.23606798
5. fixed point is $r = 0.641714$ and $S = |g'(r)| \approx 0.959$
7. (a) $0 < x_0 < 1$ (b) $1 < x_0 < 2$ (c) $x_0 > 2.2$, for example

1.3 Exercises

1. (a) $FE = 0.01$, $BE = 0.04$ (b) $FE = 0.01$ $BE = 0.0016$ (c) $FE = 0.01$, $BE = 0.000064$ (d) $FE = 0.01$, $BE = 0.342$
3. (a) 2 (b) $FE = 0.0001$, $BE = 5 \times 10^{-9}$
5. $BE = |a|$ FE
7. (b) $(-1)^j (j - 1)!(20 - j)!$

1.3 Computer Problems

1. (a) $m = 3$ (b) $x_a = -2.0735 \times 10^{-8}$, $FE = 2.0735 \times 10^{-8}$, $BE = 0$
3. (a) $x_a = FE = 0.000169$, $BE = 0$ (b) Terminates after 13 steps, $x_a = -0.00006103$
5. Predicted root $= r + \Delta r = 4 + 4^6 10^{-6}/6 = 4.000682\bar{6}$, actual root $= 4.0006825$

1.4 Exercises

1. (a) $x_1 = 2, x_2 = 18/13$ (b) $x_1 = 1, x_2 = 1$ (c) $x_1 = -1, x_2 = -2/3$
3. (a) $r = -1, e_{i+1} = \frac{5}{2}e_i^2; r = 0, e_{i+1} = 2e_i^2; r = 1, e_{i+1} = \frac{2}{3}e_i$ (b) $r = -1/2, e_{i+1} = 2e_i^2; r = 1, e_{i+1} = 2/3e_i$
5. $r = 0$, Newton's Method; $r = 1/2$, Bisection Method
7. No, $2/3$
9. $x_{i+1} = (x_i + A/x_i)/2$
11. $x_{i+1} = (n - 1)x_i/n + A/(nx_i^{n-1})$
13. (a) 0.75×10^{-12} (b) 0.5×10^{-18}

1.4 Computer Problems

1. (a) 1.76929235 (b) 1.67282170 (c) 1.12998050
3. (a) $r = -2/3, m = 3$ (b) $r = 1/6, m = 2$
5. $r = 3.2362$ m
7. -1.197624, quadratic conv.; 0, linear conv., $m = 4$; 1.530134, quadratic conv.
9. 0.857143, quadratic conv., $M = 2.414$; 2, linear conv., $m = 3, S = 2/3$
11. initial guess $= 1.75$, solution $V = 1.70$ L
13. (a) $3/4$ (c) $f(x)$ fails to be differentiable at $x = 3/4$.
15. 635.5 mm
17. (a) 0.866% per year (b) 0.576% per year

1.5 Exercises

1. (a) $x_2 = 8/5, x_3 = 1.742268$
   (b) $x_2 = 1.578707, x_3 = 1.66016$
   (c) $x_2 = 1.092907, x_3 = 1.119357$
3. (a) $x_3 = -1/5, x_4 = -0.11996018$
   (b) $x_3 = 1.757713, x_4 = 1.662531$
   (c) $x_3 = 1.139481, x_4 = 1.129272$
7. From fastest to slowest, (B), (D), (A), and (C), which does not converge (b) Newton's Method will converge faster.

1.5 Computer Problems

1. (a) 1.76929235 (b) 1.67282170 (c) 1.12998050
3. (a) 1.76929235 (b) 1.67282170 (c) 1.12998050
5. `fzero` converges to the non-root zero, same as Bisection Method

CHAPTER 2

2.1 Exercises

1. (a) $[4, 2]$ (b) $[5, -3]$ (c) $[1, 3]$
3. (a) $[1/3, 1, 1]$ (b) $[2, -1/2, -1]$
5. Approximately 27 times longer.
7. Approximately 61 seconds.

2.1 Computer Problems

1. (a) $[1, 1, 2]$ (b) $[1, 1, 1]$ (c) $[-1, 3, 2]$

2.2 Exercises

1. (a) $\begin{bmatrix} 1 & 0 \\ 3 & 1 \end{bmatrix} \begin{bmatrix} 1 & 2 \\ 0 & -2 \end{bmatrix}$

(b) $\begin{bmatrix} 1 & 0 \\ 2 & 1 \end{bmatrix} \begin{bmatrix} 1 & 3 \\ 0 & -4 \end{bmatrix}$

(c) $\begin{bmatrix} 1 & 0 \\ -5/3 & 1 \end{bmatrix} \begin{bmatrix} 3 & -4 \\ 0 & -14/3 \end{bmatrix}$

3. (a) $[-2, 1]$ (b) $[-1, 1]$
5. $[1, -1, 1, -1]$
7. 5 min, 33 sec
9. 300

2.3 Exercises

1. (a) 7 (b) 8
3. (a) FE $= 2$, BE $= 0.0002$, EMF $= 20001$
   (b) FE $= 1$, BE $= 0.0001$, EMF $= 20001$ (c) FE $= 1$, BE $= 2.0001$, EMF $= 1$ (d) FE $= 3$, BE $= 0.0003$, EMF $= 20001$ (e) FE $= 3.0001$, BE $= 0.0002$, EMF $= 30002.5$
5. (a) RFE $= 3$, RBE $= 3/7$, EMF $= 7$ (b) RFE $= 3$, RBE $= 1/7$, EMF $= 21$ (c) RFE $= 1$, RBE $= 1/7$, EMF $= 7$ (d) RFE $= 2$, RBE $= 6/7$, EMF $= 7/3$
   (e) 21
7. 137/60
9. (a) 1 (b) $\max |d_i| / \min |d_i|$.

15. (a) $\begin{bmatrix} 1 \\ 1 \end{bmatrix}$ (b) $\begin{bmatrix} 1 \\ -1 \\ 1 \end{bmatrix}$

17. LU $= \begin{bmatrix} 1 & 0 & 0 \\ 0.1 & 1 & 0 \\ 0 & -5000 & 1 \end{bmatrix} \begin{bmatrix} 10 & 20 & 1 \\ 0 & -0.01 & 5.9 \\ 0 & 0 & 29501 \end{bmatrix}$,
   largest multiplier $= 5000$

2.3 Computer Problems

Answers given to Computer Problems in this section are illustrative only; results will vary slightly with implementation details.

1.

	n	FE	EMF	cond(A)
(a)	6	5.35×10^{-10}	3.69×10^6	7.03×10^7
(b)	10	1.10×10^{-3}	9.05×10^{12}	1.31×10^{14}

3.

n	FE	EMF	cond(A)
100	4.62×10^{-12}	3590	9900
200	4.21×10^{-11}	23010	39800
300	7.37×10^{-11}	50447	89700
400	1.20×10^{-10}	55019	159600
500	2.56×10^{-10}	91495	249500

5. $n \geq 13$

2.4 Exercises

1. (a) $\begin{bmatrix} 0 & 1 \\ 1 & 0 \end{bmatrix} \begin{bmatrix} 1 & 3 \\ 2 & 3 \end{bmatrix} = \begin{bmatrix} 1 & 0 \\ \frac{1}{2} & 1 \end{bmatrix} \begin{bmatrix} 2 & 3 \\ 0 & \frac{3}{2} \end{bmatrix}$

(b) $\begin{bmatrix} 1 & 0 \\ 0 & 1 \end{bmatrix} \begin{bmatrix} 2 & 4 \\ 1 & 3 \end{bmatrix} = \begin{bmatrix} 1 & 0 \\ \frac{1}{2} & 1 \end{bmatrix} \begin{bmatrix} 2 & 4 \\ 0 & 1 \end{bmatrix}$

(c) $\begin{bmatrix} 0 & 1 \\ 1 & 0 \end{bmatrix} \begin{bmatrix} 1 & 5 \\ 5 & 12 \end{bmatrix} = \begin{bmatrix} 1 & 0 \\ \frac{1}{5} & 1 \end{bmatrix} \begin{bmatrix} 5 & 12 \\ 0 & \frac{13}{5} \end{bmatrix}$

(d) $\begin{bmatrix} 0 & 1 \\ 1 & 0 \end{bmatrix} \begin{bmatrix} 0 & 1 \\ 1 & 0 \end{bmatrix} = \begin{bmatrix} 1 & 0 \\ 0 & 1 \end{bmatrix} \begin{bmatrix} 1 & 0 \\ 0 & 1 \end{bmatrix}$

3. (a) $[-2, 1]$ (b) $[-1, 1, 1]$

5. $\begin{bmatrix} 1 & 0 & 0 & 0 & 0 \\ 0 & 0 & 0 & 0 & 1 \\ 0 & 0 & 1 & 0 & 0 \\ 0 & 0 & 0 & 1 & 0 \\ 0 & 1 & 0 & 0 & 0 \end{bmatrix}$

7. $\begin{bmatrix} 0 & 0 & 1 & 0 \\ 0 & 1 & 0 & 0 \\ 0 & 0 & 0 & 1 \\ 1 & 0 & 0 & 0 \end{bmatrix}$

9. (a) $\begin{bmatrix} 1 & 0 & 0 & 0 \\ 0 & 1 & 0 & 0 \\ 0 & 0 & 1 & 0 \\ 0 & 0 & 0 & 1 \end{bmatrix} \begin{bmatrix} 1 & 0 & 0 & 1 \\ -1 & 1 & 0 & 1 \\ -1 & -1 & 1 & 1 \\ -1 & -1 & -1 & 1 \end{bmatrix}$
$= \begin{bmatrix} 1 & 0 & 0 & 0 \\ -1 & 1 & 0 & 0 \\ -1 & -1 & 1 & 0 \\ -1 & -1 & -1 & 1 \end{bmatrix} \begin{bmatrix} 1 & 0 & 0 & 1 \\ 0 & 1 & 0 & 2 \\ 0 & 0 & 1 & 4 \\ 0 & 0 & 0 & 8 \end{bmatrix}$

(b) $P = I$, L is lower triangular with all non-diagonal entries -1, the nonzero entries of U are $u_{ii} = 1$ for $1 \leq i \leq n - 1$, and $u_{in} = 2^{i-1}$ for $1 \leq i \leq n$.

2.5 Exercises

1. (a) Jacobi $[u_2, v_2] = [7/3, 17/6]$ Gauss-Seidel $[u_2, v_2] = [47/18, 119/36]$ (b) Jacobi $[u_2, v_2, w_2] = [1/2, 1, 1/2]$ Gauss-Seidel $[u_2, v_2, w_2] = [1/2, 3/2, 3/4]$ (c) Jacobi $[u_2, v_2, w_2] = [10/9, -2/9, 2/3]$ Gauss-Seidel $[u_2, v_2, w_2] = [43/27, 14/81, 262/243]$

3. (a) $[u_2, v_2] = [59/16, 213/64]$
   (b) $[u_2, v_2, w_2] = [9/8, 39/16, 81/64]$
   (c) $[u_2, v_2, w_2] = [1, 1/2, 5/4]$

2.5 Computer Problems

1. $n = 100$, 36 steps, BE $= 4.58 \times 10^{-7}$; $n = 100000$, 48 steps, BE $= 2.70 \times 10^{-6}$

5. (a) 21 steps, BE $= 4.78 \times 10^{-7}$ (b) 16 steps, BE $= 1.55 \times 10^{-6}$

2.6 Exercises

1. (a) $x^T A x = x_1^2 + 3x_2^2 > 0$ for $x \neq 0$
   (b) $x^T A x = (x_1 + 3x_2)^2 + x_2^2 > 0$ for $x \neq 0$
   (c) $x_1^2 + 2x_2^2 + 3x_3^2 > 0$ for $x \neq 0$

3. (a) $R = \begin{bmatrix} 1 & 0 \\ 0 & \sqrt{3} \end{bmatrix}$ (b) $R = \begin{bmatrix} 1 & 3 \\ 0 & 1 \end{bmatrix}$
   (c) $R = \begin{bmatrix} 1 & 0 & 0 \\ 0 & \sqrt{2} & 0 \\ 0 & 0 & \sqrt{3} \end{bmatrix}$

5. (a) $R = \begin{bmatrix} 1 & 2 \\ 0 & 2 \end{bmatrix}$ (b) $R = \begin{bmatrix} 2 & -1 \\ 0 & 1/2 \end{bmatrix}$
   (c) $R = \begin{bmatrix} 5 & 1 \\ 0 & 5 \end{bmatrix}$ (d) $R = \begin{bmatrix} 1 & -2 \\ 0 & 1 \end{bmatrix}$

7. (a) $[2, -1]$ (b) $[3, 1]$

9. $x^T A x = (x_1 + 2x_2)^2 + (d - 4)x_2^2$. If $d > 4$, the expressions can be 0 only if $0 = x_2 = x_1 + 2x_2$, which implies $x_1 = x_2 = 0$.

11. $d > 1$

13. (a) $[3, -1]$ (b) $[-1, 1]$

15. $\alpha_1 = 1/A, x_1 = b/A, r_1 = b - Ab/A = 0$

2.6 Computer Problems

1. (a) $[2, 2]$ (b) $[3, -1]$

3. (a) $[-4, 60, -180, 140]$ (b) $[-8, 504, -7560, 46200, -138600, 216216, -168168, 51480]$

2.7 Exercises

1. (a) $\begin{bmatrix} 3u^2 & 0 \\ v^3 & 3uv^2 \end{bmatrix}$ (b) $\begin{bmatrix} v\cos uv & u\cos uv \\ ve^{uv} & ue^{uv} \end{bmatrix}$
   (c) $\begin{bmatrix} 2u & 2v \\ 2(u-1) & 2v \end{bmatrix}$
   (d) $\begin{bmatrix} 2u & 1 & -2w \\ vw\cos uvw & uw\cos uvw & uv\cos uvw \\ vw^4 & uw^4 & 4uvw^3 \end{bmatrix}$

3. (a) $(1/2, \pm\sqrt{3}/2)$ (b) $(\pm 2/\sqrt{5}, \pm 2/\sqrt{5})$
   (c) $(4(1 + \sqrt{6})/5, \pm\sqrt{3 + 8\sqrt{6}}/5)$

5. (a) $x_1 = [0, 1], x_2 = [0, 0]$
   (b) $x_1 = [0, 0], x_2 = [0.8, 0.8]$
   (c) $x_1 = [8, 4], x_2 = [9.0892, -12.6103]$

2.7 Computer Problems

1. (a) $(1/2, \pm\sqrt{3}/2)$ (b) $(\pm 2/\sqrt{5}, \pm 2/\sqrt{5})$
   (c) $(4(1 + \sqrt{6})/5, \pm\sqrt{3 + 8\sqrt{6}}/5)$

3. $\pm[0.50799200040795, 0.86136178666199]$

5. (a) $[1, 1, 1], [1/3, 1/3, 1/3]$ (b) $[1, 2, 3], [17/9, 22/9, 19/9]$

7. (a) 11 steps give the root $(1/2, \sqrt{3}/2)$ to 15 places
   (b) 13 steps give the root $(2/\sqrt{5}, 2/\sqrt{5})$ to 15 places
   (c) 14 steps give the root $(4(1 + \sqrt{6})/5, \sqrt{3 + 8\sqrt{6}}/5)$ to 15 places

9. Same answers as Computer Problem 5

11. Same answers as Computer Problem 5

CHAPTER 3

3.1 Exercises

1. (a) $P(x) = \dfrac{(x-2)(x-3)}{(0-2)(0-3)} + 3\dfrac{x(x-3)}{(2-0)(2-3)}$
   (b) $P(x) = \dfrac{(x+1)(x-3)(x-5)}{(2+1)(2-3)(2-5)}$
   $+\dfrac{(x+1)(x-2)(x-5)}{(3+1)(3-2)(3-5)}$
   $+2\dfrac{(x+1)(x-2)(x-3)}{(5+1)(5-2)(5-3)}$
   (c) $P(x) = -2\dfrac{(x-2)(x-4)}{(0-2)(0-4)} + \dfrac{x(x-4)}{(2-0)(2-4)}$
   $+4\dfrac{x(x-2)}{4(4-2)}$

3. (a) One, $P(x) = 3 + (x + 1)(x - 2)$ (b) None
   (c) Infinitely many, for example $P(x) = 3 + (x + 1)$ $(x - 2) + C(x + 1)(x - 1)(x - 2)(x - 3)^3$, where C is a nonzero constant

5. (a) $P(x) = 4 - 2x$ (b) $P(x) = 4 - 2x + A(x + 2)$ $x(x - 1)(x - 3)$ for $A \neq 0$

7. 4

9. (a) $P(x) = 10(x - 1) \cdots (x - 6)/6!$ (b) Same as (a)

11. None

13. 412

15. $P(x) = -x - (x - 1)(x - 2) \cdots (x - 25)/24!$

17. (a) 316 (b) 465

3.1 Computer Problems

1. (a) 4494564854 (b) 4454831984 (c) 4472888288

3.2 Exercises

1. (a) $P_2(x) = \dfrac{2}{\pi}x - \dfrac{4}{\pi^2}x(x - \pi/2)$ (b) $P_2(\pi/4) = 3/4$ (c) $\pi^3/128 \approx 0.242$ (d) $|\sqrt{2}/2 - 3/4| \approx 0.043$

3. (a) 7.06×10^{-11} (b) at least 9 decimal places, since $7.06 \times 10^{-11} < 0.5 \times 10^{-9}$

5. Expect errors at $x = 0.35$ to be smaller; approximately $5/21$ the size of the error at $x = 0.55$.

3.2 Computer Problems

1. (a) $P_4(x) = 1.433329 + (x - 0.6)(1.98987 + (x - 0.7)(3.2589 + (x - 0.8)(3.680667 + (x - 0.9)(4.000417))))$ (b) $P_4(0.82) = 1.95891, P_4(0.98) = 2.612848$ (c) Upper bound for error at $x = 0.82$ is 0.0000537, actual error is 0.0000234. Upper bound for error at $x = 0.98$ is 0.000217, actual error is 0.000107.

3. -1.952×10^{12} bbl/day. The estimate is nonsensical, due to the Runge phenomenon.

3.3 Exercises

1. (a) $\cos \pi/12, \cos \pi/4, \cos 5\pi/12, \cos 7\pi/12, \cos 3\pi/4,$
$\cos 11\pi/12$
   (b) $2\cos \pi/8, 2\cos 3\pi/8, 2\cos 5\pi/8, 2\cos 7\pi/8$
   (c) $8 + 4\cos \pi/12, 8 + 4\cos \pi/4, 8 + 4\cos 5\pi/12, 8 +$
$4\cos 7\pi/12, 8 + 4\cos 3\pi/4, 8 + 4\cos 11\pi/12$
   (d) $1/5 + 1/2\cos \pi/10, 1/5 + 1/2\cos 3\pi/10, 1/5, 1/5 +$
$1/2\cos 7\pi/10, 1/5 + 1/2\cos 9\pi/10$
3. 0.000118, 3 correct digits
5. 0.00521
7. $d = 14$
9. (a) -1 (b) 1 (c) 0 (d) 1 (e) 1 (f) $-1/2$

3.4 Exercises

1. (a) not a cubic spline (b) cubic spline
3. (a) $c = 9/4$, natural (b) $c = 4$, parabolically terminated and not-a-knot (c) $c = 5/2$, not-a-knot
5. One, $S_1(x) = S_2(x) = x$
7. (a) $\begin{cases} \frac{1}{2}x + \frac{1}{2}x^3 & \text{on } [0,1] \\ 1 + 2(x-1) + \frac{3}{2}(x-1)^2 - \frac{1}{2}(x-1)^3 & \text{on } [1,2] \end{cases}$

   (b) $\begin{cases} 1 - (x+1) + \frac{1}{4}(x+1)^3 & \text{on } [-1,1] \\ 1 + 2(x-1) + \frac{3}{2}(x-1)^2 - \frac{1}{2}(x-1)^3 & \text{on } [1,2] \end{cases}$
9. $-3, -12$
11. (a) One, $S_1(x) = S_2(x) = 2 - 4x + 2x^2$
    (b) Infinitely many, $S_1(x) = S_2(x) = 2 - 4x$
$+2x^2 + cx(x-1)(x-2)$ for arbitrary c.
13. (a) $b_1 = 1, c_3 = -8/9$ (b) No. (c) The clamps are
$S'(0) = 1$ and $S'(3) = -1/3$.
15. Yes. The leftmost and rightmost sections of the spline must be linear.
17. $S_2(x) = 1 + dx^3$ for arbitrary d
19. There are infinitely many parabolas through two arbitrary points with $x_1 \neq x_2$; each is a parabolically-terminated cubic spline.
21. (a) infinitely many (b) $S_1(x) = S_2(x) = x^2 +$
$dx(x-1)(x-2)$ where $d \neq 0$.

3.4 Computer Problems

1. (a)
$$S(x) = \begin{cases} 3 + \frac{8}{3}x - \frac{2}{3}x^3 & \text{on } [0,1] \\ 5 + \frac{2}{3}(x-1) - 2(x-1)^2 + \frac{1}{3}(x-1)^3 & \text{on } [1,2] \\ 4 - \frac{7}{3}(x-2) - (x-2)^2 + \frac{1}{3}(x-2)^3 & \text{on } [2,3] \end{cases}$$
   (b) $S(x)$
$$= \begin{cases} 3 + 2.5629(x+1) - 0.5629(x+1)^3 & \text{on } [-1,0] \\ 5 + 0.8742x - 1.6887x^2 + 0.3176x^3 & \text{on } [0,3] \\ 1 - 0.6824(x-3) + 1.1698(x-3)^2 - 0.4874(x-3)^3 & \text{on } [3,4] \\ 1 + 0.1950(x-4) - 0.2925(x-4)^2 + 0.0975(x-4)^3 & \text{on } [4,5] \end{cases}$$

3. $S(x)$
$$= \begin{cases} 1 + \frac{149}{56}x - \frac{37}{56}x^3 & \text{on } [0,1] \\ 3 + \frac{19}{28}(x-1) - \frac{111}{56}(x-1)^2 + \frac{73}{56}(x-1)^3 & \text{on } [1,2] \\ 3 + \frac{5}{8}(x-2) + \frac{27}{14}(x-2)^2 - \frac{87}{56}(x-2)^3 & \text{on } [2,3] \\ 4 - \frac{5}{28}(x-3) - \frac{153}{56}(x-3)^2 + \frac{51}{56}(x-3)^3 & \text{on } [4,5] \end{cases}$$

1. $S(x)$
$$= \begin{cases} 1 + 1.8006x + \frac{3}{2}x^2 - 1.3006x^3 & \text{on } [0,1] \\ 3 + 0.8988(x-1) - 2.4018(x-1)^2 + 1.5030(x-1)^3 & \text{on } [1,2] \\ 3 + 0.6042(x-2) + 2.1071(x-2)^2 - 1.7113(x-2)^3 & \text{on } [2,3] \\ 4 - 0.3155(x-3) - 3.0268(x-3)^2 + 1.3423(x-3)^3 & \text{on } [4,5] \end{cases}$$

3. $S(x)$
$$= \begin{cases} 1 - 2x + \frac{57}{7}x^2 - \frac{29}{7}x^3 & \text{on } [0,1] \\ 3 + \frac{13}{7}(x-1) - \frac{30}{7}(x-1)^2 + \frac{17}{7}(x-1)^3 & \text{on } [1,2] \\ 3 + \frac{4}{7}(x-2) + 3(x-2)^2 - \frac{18}{7}(x-2)^3 & \text{on } [2,3] \\ 4 - \frac{8}{7}(x-3) - \frac{33}{7}(x-3)^2 + \frac{27}{7}(x-3)^3 & \text{on } [4,5] \end{cases}$$

5. $S(x) =$
$$\begin{cases} x - 0.0006x^2 - 0.1639x^3 & \text{on } [0, \frac{\pi}{8}] \\ \sin \frac{\pi}{8} + 0.9237(x - \frac{\pi}{8}) - 0.1937(x - \frac{\pi}{8})^2 - 0.1396(x - \frac{\pi}{8})^3 & \text{on } [\frac{\pi}{8}, \frac{\pi}{4}] \\ \frac{\sqrt{2}}{2} + 0.7070(x - \frac{\pi}{4}) - 0.3582(x - \frac{\pi}{4})^2 - 0.0931(x - \frac{\pi}{4})^3 & \text{on } [\frac{\pi}{4}, \frac{3\pi}{8}] \\ \sin \frac{3\pi}{8} + 0.3826(x - \frac{3\pi}{8}) - 0.4679(x - \frac{3\pi}{8})^2 - 0.0327(x - \frac{3\pi}{8})^3 & \text{on } [\frac{3\pi}{8}, \frac{\pi}{2}] \end{cases}$$

7. $n = 48$
9. (a) 322.6 (b) 318.8 (c) not-a-knot spline is identical to solution of Exercise 3.1.13

3.5 Exercises

1. (a) $\begin{cases} x(t) = 6t^2 - 5t^3 \\ y(t) = 6t - 12t^2 + 6t^3 \end{cases}$

   (b) $\begin{cases} x(t) = 1 - 3t - 3t^2 + 3t^3 \\ y(t) = 1 - 3t + 3t^2 \end{cases}$

   (c) $\begin{cases} x(t) = 1 + 3t^2 - 2t^3 \\ y(t) = 2 + 3t - 3t^2 \end{cases}$

3. $\begin{cases} x(t) = 1 + 6t^2 - 4t^3 \\ y(t) = 2 + 6t^2 - 4t^3 \end{cases}$ $\begin{cases} x(t) = 3 + 6t^2 - 4t^3 \\ y(t) = 4 - 9t^2 + 6t^3 \end{cases}$
$\begin{cases} x(t) = 5 - 12t^2 + 8t^3 \\ y(t) = 1 + 3t^2 - 2t^3 \end{cases}$

5. The number 3.
7. $\begin{cases} x(t) = -1 + 6t^2 - 4t^3 \\ y(t) = 4t - 4t^2 \end{cases}$

9. (a) $\begin{cases} x(t) = 1 + 3t - 9t^2 + 5t^3 \\ y(t) = 6t^2 - 5t^3 \\ z(t) = 3t^2 - 3t^3 \end{cases}$

   (b) $\begin{cases} x(t) = 1 - 6t^2 + 6t^3 \\ y(t) = 1 + 3t - 9t^2 + 6t^3 \\ z(t) = 2 + 3t - 12t^2 + 8t^3 \end{cases}$

   (c) $\begin{cases} x(t) = 2 + 3t - 12t^2 + 10t^3 \\ y(t) = 1 \\ z(t) = 1 + 6t^2 - 4t^3 \end{cases}$

CHAPTER 4

4.1 Exercises

1. (a) $\bar{x} = [-1/7, 10/7]$, $\|e\|_2 = \sqrt{14}/7$
   (b) $\bar{x} = [-1/2, 2]$, $\|e\|_2 = \sqrt{6}/2$
   (c) $\bar{x} = [16/19, 16/19]$, $\|e\|_2 = 2.013$
3. $\bar{x} = [4, x_2]$ for arbitrary x_2
7. (a) $y = 1/5 - 6/5t$, RMSE $= \sqrt{2/5} \approx 0.6325$
   (b) $y = 6/5 + 1/2t$, RMSE $= \sqrt{26}/10 \approx 0.5099$
9. (a) $y = 0.3481 + 1.9475t - 0.1657t^2$, RMSE
   $= 0.5519$ (b) $y = 2.9615 - 1.0128t + 0.1667t^2$,
   RMSE $= 0.4160$ (c) $y = 4.8 - 1.2t$, RMSE $= 0.4472$
11. $h(t) = 0.475 + 141.525t - 4.905t^2$, max height
   $= 1021.3$ m, landing time $= 28.86$ sec

4.1 Computer Problems

1. (a) $\bar{x} = [2.5246, 0.6616, 2.0934]$, $\|e\|_2 = 2.4135$
   (b) $\bar{x} = [1.2739, 0.6885, 1.2124, 1.7497]$, $\|e\|_2 = 0.8256$
3. (a) $2, 996, 236, 899 + 76, 542, 140(t - 1960)$, RMSE
   $= 36, 751, 088$ (b) $3, 028, 751, 748 + 67, 871, 514(t - 1960) + 216, 766(t - 1960)^2$, RMSE $= 17, 129, 714$;
   1980 estimates: (a) $4, 527, 079, 702$
   (b) $4, 472, 888, 288$; Parabola gives better estimate.
5. (a) $c_1 = 9510.1$, $c_2 = -8314.36$, RMSE $= 518.3$
   (b) selling price $= 68.7$ cents maximizes profit.
7. (a) $y = 0.0769$, RMSE $= 0.2665$
   (b) $y = 0.1748 - 0.02797t^2$, RMSE $= 0.2519$
9. (a) 4 correct decimal places,
   $P_5(t) = 1.000009 + 0.999983t + 1.000012t^2 + 0.999996t^3 + 1.000000t^4 + 1.000000t^5$;
   $\text{cond}(A^T A) = 2.72 \times 10^{13}$ (b) 1 correct decimal
   place, $P_6(t) = 0.99 + 1.02t + 0.98t^2 + 1.01t^3 + t^4 + t^5 + t^6$; $\text{cond}(A^T A) = 2.55 \times 10^{16}$ (c) $P_8(t)$ has no
   correct places, $\text{cond}(A^T A) = 1.41 \times 10^{19}$

4.2 Exercises

1. (a) $y = 3/2 - 1/2 \cos 2\pi t + 3/2 \sin 2\pi t$, $\|e\|_2 = 0$,
   RMSE $= 0$ (b) $y = 7/4 - 1/2 \cos 2\pi t + \sin 2\pi t$,
   $\|e\|_2 = 1/2$, RMSE $= 1/4$ (c) $y = 9/4 + 3/4 \cos 2\pi t$, $\|e\|_2 = 1/\sqrt{2}$, RMSE $= 1/(2\sqrt{2})$
3. (a) $y = 1.932e^{0.3615t}$, $\|e\|_2 = 1.2825$
   (b) $y = 2^{t-1/4}$, $\|e\|_2 = 0.9982$
5. (a) $y = 5.5618t^{-1.3778}$, RMSE $= 0.2707$
   (b) $y = 2.8256t^{0.7614}$, RMSE $= 0.7099$

4.2 Computer Problems

1. $y = 5.5837 + 0.7541 \cos 2\pi t + 0.1220 \sin 2\pi t + 0.1935 \cos 4\pi t$ M bbls/day, RMSE $= 0.1836$
3. $P(t) = 3, 079, 440, 361e^{0.0174(t-1960)}$, 1980 estimate is
   $P(20) = 4, 361, 485, 000$, estimation error ≈ 91 million
5. (a) $t_{max} = -1/c_2$ (b) half-life ≈ 7.81 hrs

4.3 Exercises

1. (a) $\begin{bmatrix} 0.8 & -0.6 \\ 0.6 & 0.8 \end{bmatrix} \begin{bmatrix} 5 & 0.6 \\ 0 & 0.8 \end{bmatrix}$

   (b) $\dfrac{1}{\sqrt{2}} \begin{bmatrix} 1 & 1 \\ 1 & -1 \end{bmatrix} \begin{bmatrix} \sqrt{2} & \frac{3\sqrt{2}}{2} \\ 0 & \frac{\sqrt{2}}{2} \end{bmatrix}$

(c) $\begin{bmatrix} \frac{2}{3} & \frac{\sqrt{2}}{6} & \frac{\sqrt{2}}{2} \\ \frac{1}{3} & -\frac{2\sqrt{2}}{3} & 0 \\ \frac{2}{3} & \frac{\sqrt{2}}{6} & -\frac{\sqrt{2}}{2} \end{bmatrix} \begin{bmatrix} 3 & 1 \\ 0 & \sqrt{2} \\ 0 & 0 \end{bmatrix}$

(d) $\begin{bmatrix} \frac{4}{5} & 0 & -\frac{3}{5} \\ 0 & 1 & 0 \\ \frac{3}{5} & 0 & \frac{4}{5} \end{bmatrix} \begin{bmatrix} 5 & 10 & 5 \\ 0 & 2 & -2 \\ 0 & 0 & 5 \end{bmatrix}$

3. (a)–(d) same as Exercise 1
5. (a)–(d) same as Exercise 1
7. (a) $\bar{x} = [4, -1]$ (b) $\bar{x} = [-11/18, 4/9]$

4.3 Computer Problems

5. (a) $\bar{x} = [1.6154, 1.6615]$, $\|e\|_2 = 0.3038$
   (b) $\bar{x} = [2.0588, 2.3725, 1.5784]$, $\|e\|_2 = 0.2214$
7. (a) $\bar{x} = [1, \ldots, 1]$ to 10 correct decimal places
   (b) $\bar{x} = [1, \ldots, 1]$ to 6 correct decimal places

4.4 Exercises

1. (a) $x_1 = [1/2, 0, 0]$, $x_2 = [1, 0, -1]$
   (b) $x_1 = [1/2, 0, 0]$, $x_2 = [1/2, 1/2, 0]$
   (c) $x_1 = [0, 0, 0]$, $x_2 = [0, 0, 0]$, $x_3 = [0, 0, 1]$

4.5 Exercises

1. (a) $(x_1, y_1) = (2 - \sqrt{2}, 0)$
   (b) $(x_1, y_1) = (1 - \sqrt{2}/2, 0)$

5. (a) $\begin{bmatrix} t_1^{c_2} & c_1 t_1^{c_2} \ln t_1 \\ t_2^{c_2} & c_1 t_2^{c_2} \ln t_2 \\ t_3^{c_2} & c_1 t_3^{c_2} \ln t_3 \end{bmatrix}$

   (b) $\begin{bmatrix} t_1 e^{c_2 t_1} & c_1 t_1^2 e^{c_2 t_1} \\ t_2 e^{c_2 t_2} & c_1 t_2^2 e^{c_2 t_2} \\ t_3 e^{c_2 t_3} & c_1 t_3^2 e^{c_2 t_3} \end{bmatrix}$

4.5 Computer Problems

1. (a) $(\bar{x}, \bar{y}) = (0.410623, 0.055501)$
   (b) $(\bar{x}, \bar{y}) = (0.275549, 0)$
3. (a) $(x, y) = (0, -0.586187)$, $K = 0.329572$
   (b) $(x, y) = (0.556853, 0)$, $K = 1.288037$
5. $c_1 = 15.9$, $c_2 = 2.53$, RMSE $= 0.755$
7. Same as Computer Problem 5.
9. (a) $c_1 = 11.993468$, $c_2 = 0.279608$, $c_3 = 1.802342$,
   RMSE $= 0.441305$
   (b) $c_1 = 12.702778$, $c_2 = 0.159591$, $c_3 = 5.682764$,
   RMSE $= 0.802834$
11. (a) $c_1 = 8.670956$, $c_2 = 0.274184$, $c_3 = 0.981070$,
   $c_4 = 1.232813$, RMSE $= 0.102660$
   (b) $c_1 = 8.683823$, $c_2 = 0.131945$, $c_3 = 0.620292$,
   $c_4 = -1.921257$, RMSE $= 0.199789$

CHAPTER 5

5.1 Exercises

1. (a) 0.9531, error $= 0.0469$ (b) 0.9950,
   error $= 0.0050$ (c) 0.9995, error $= 0.0005$

3. (a) 0.455902, error = 0.044098; error must satisfy 0.0433 ≤ error ≤ 0.0456 (b) 0.495662, error = 0.004338; error must satisfy 0.004330 ≤ error ≤ 0.004355 (c) 0.499567, error = 0.000433; error must satisfy 0.0004330 ≤ error ≤ 0.0004333
5. (a) 2.02020202, error = 0.02020202
(b) 2.00020002, error = 0.00020002
(c) 2.00000200, error = 0.00000200
7. $f'(x) = [(f(x) - f(x - h)]/h + hf''(c)/2$
9. $f'(x) = [3f(x) - 4f(x - h) + f(x - 2h)]/(2h) + O(h^2)$
11. $f'(x) \approx [4f(x + h/2) - 3f(x) - f(x + h)]/h$
13. $f'(x) = [f(x + 3h) + 8f(x) - 9f(x - h)]/(12h) - h^2 f'''(c)/2$, where $x - h < c < x + 3h$
15. $f''(x) = [f(x + 3h) - 4f(x) + 3f(x - h)]/(6h^2) - 2hf'''(c)/3$, where $x - h < c < x + 3h$
17. $f'(x) = [4f(x + 3h) + 5f(x) - 9f(x - 2h)]/(30h) - h^2 f'''(c)$, where $x - 2h < c < x + 3h$

5.1 Computer Problems

1. minimum error at $h = 10^{-5} \approx \epsilon_{mach}^{1/3}$
3. minimum error at $h = 10^{-8} \approx \epsilon_{mach}^{1/2}$
5. (a) minimum error at $h = 10^{-4} \approx \epsilon_{mach}^{1/4}$ (b) same as (a)

5.2 Exercises

1. (a) $m = 1 : 0.500000$, err = 0.166667; $m = 2 : 0.375000$, err = 0.041667; $m = 4 : 0.343750$, err = 0.010417
(b) $m = 1 : 0.785398$, err = 0.214602; $m = 2 : 0.948059$, err = 0.051941; $m = 4 : 0.987116$, err = 0.012884
(c) $m = 1 : 1.859141$, err = 0.140859; $m = 2 : 1.753931$, err = 0.035649; $m = 4 : 1.727222$, err = 0.008940
3. (a) $m = 1 : 1/3$, err = 0; $m = 2 : 1/3$, err = 0; $m = 4 : 1/3$, err = 0 (b) $m = 1 : 1.002280$, err = 0.002280; $m = 2 : 1.000135$, err = 0.000135; $m = 4 : 1.000008$, err = 0.000008 (c) $m = 1 : 1.718861$, err = 0.000579; $m = 2 : 1.718319$, err = 0.000037; $m = 4 : 1.718284$, err = 0.000002
5. (a) $m = 1 : 1.414214$, err = 0.585786; $m = 2 : 1.577350$, err = 0.422650; $m = 4 : 1.698844$; err = 0.301156
(b) $m = 1 : 1.259921$, err = 0.240079; $m = 2 : 1.344022$, err = 0.155978; $m = 4 : 1.400461$, err = 0.099539
(c) $m = 1 : 2.000000$, err = 0.828427; $m = 2 : 2.230710$, err = 0.597717; $m = 4 : 2.402528$, err = 0.425899
7. (a) 1.631729, err = 0.368271
(b) 1.372055, err = 0.127945
(c) 2.307614, err = 0.520814
11. (a) 1 (b) 1 (c) 3
13. $\frac{4h}{3} \sum_{i=1}^{m} [2f(u_i) + 2f(v_i) - f(w_i)] + \frac{7(b - a)h^4}{90} f^{(iv)}(c)$
15. 5

5.2 Computer Problems

1. (a) exact = 2; $m = 16$ approx = 1.998638, err = 1.36×10^{-3}; $m = 32$ approx = 1.999660, err = 3.40×10^{-4} (b) exact = $1/2(1 - \ln 2)$; $m = 16$ approx = 0.153752, err = 3.26×10^{-4}; $m = 32$ approx = 0.153508, err = 8.14×10^{-5} (c) exact = 1; $m = 16$ approx = 1.001444, err = 1.44×10^{-3}; $m = 32$ approx = 1.000361, err = 3.61×10^{-4} (d) exact = $9\ln 3 - 26/9$; $m = 16$ approx = 7.009809, err = 1.12×10^{-2}; $m = 32$ approx = 7.001419, err = 2.80×10^{-3} (e) exact = $\pi^2 - 4$; $m = 16$ approx = 5.837900, err = 3.17×10^{-2}; $m = 32$ approx = 5.861678, err = 7.93×10^{-3} (f) exact = $2\sqrt{5} - \sqrt{15}/2$; $m = 16$ approx = 2.535672, err = 2.80×10^{-5}; $m = 32$ approx = 2.535651, err = 7.00×10^{-6} (g) exact = $\ln(\sqrt{3} + 2)$; $m = 16$ approx = 1.316746, err = 2.11×10^{-4}; $m = 32$ approx = 1.316905, err = 5.29×10^{-5} (h) exact = $\ln(\sqrt{2} + 1)/2$; $m = 16$ approx = 0.440361, err = 3.26×10^{-4}; $m = 32$ approx = 0.440605, err = 8.14×10^{-5}
3. (a) $m = 16$ approx = 1.464420; $m = 32$ approx = 1.463094 (b) $m = 16$ approx = 0.891197; $m = 32$ approx = 0.893925 (c) $m = 16$ approx = 3.977463; $m = 32$ approx = 3.977463 (d) $m = 16$ approx = 0.264269; $m = 32$ approx = 0.264025 (e) $m = 16$ approx = 0.160686; $m = 32$ approx = 0.160936 (f) $m = 16$ approx = -0.278013; $m = 32$ approx = -0.356790 (g) $m = 16$ approx = 0.785276; $m = 32$ approx = 0.783951 (h) $m = 16$ approx = 0.369964; $m = 32$ approx = 0.371168
5. (a) $m = 10 : 1.808922$, err = 0.191078; $m = 100 : 1.939512$, err = 0.060488; $m = 1000 : 1.980871$, err = 0.019129
(b) $m = 10 : 1.445632$, err = 0.054368; $m = 100 : 1.488258$, err = 0.011742; $m = 1000 : 1.497470$, err = 0.002530
(c) $m = 10 : 2.558203$, err = 0.270225; $m = 100 : 2.742884$, err = 0.085543; $m = 1000 : 2.801375$, err = 0.027052
7. (a) $m = 16$ approx = 1.8315299; $m = 32$ approx = 1.83183081 (b) $m = 16$ approx = 2.99986658; $m = 32$ approx = 3.00116293 (c) $m = 16$ approx = 0.91601205; $m = 32$ approx = 0.91597721

5.3 Exercises

1. (a) 1/3 (b) 0.99999157 (c) 1.71828269

5.3 Computer Problems

1. (a) correct = 2, approx = 2.00000010, err = 1.0×10^{-7}
(b) correct $1/2(1 - \ln 2)$, approx = 0.15342640, err = 1.23×10^{-8} (c) correct 1, approx = 1.00000000, err = 3.5×10^{-13} (d) correct $9\ln 3 - 26/9$, approx = 6.99862171, err = 3.00×10^{-9} (e) correct $\pi^2 - 4$, approx = 5.86960486, err = 4.56×10^{-7}
(f) correct $2\sqrt{5} - \sqrt{15}/2$, approx = 2.53564428,

err $= 1.21 \times 10^{-10}$ (g) correct $\ln(\sqrt{3} + 2)$, approx $= 1.31695765$, err $= 2.46 \times 10^{-7}$ (h) correct $\ln(\sqrt{2} + 1)/2$, approx $= 0.44068686$, err $= 6.98 \times 10^{-8}$

5.4 Exercises

1. (a) 0.3750, error $= 0.0417$ (b) 0.9871, error $= 0.0129$ (c) 1.7539, error $= 0.0356$
3. Use same tolerance test as Adaptive Quadrature with Trapezoid Rule, replace Trapezoid Rule with Midpoint Rule.

5.4 Computer Problems

1. (a) 2.00000000, 12606 subintervals (b) 0.15342641, 6204 subintervals (c) 1.00000000, 12424 subintervals (d) 6.99862171, 32768 subintervals (e) 5.86960440, 73322 subintervals (f) 2.53564428, 1568 subintervals (g) 1.31695790, 7146 subintervals (h) 0.44068679, 5308 subintervals
3. first eight decimal places identical to Computer Problem 1 (a) 56 subintervals (b) 46 subintervals (c) 40 subintervals (d) 56 subintervals (e) 206 subintervals (f) 22 subintervals (g) 54 subintervals (h) 52 subintervals
5. first eight decimal places identical to Computer Problem 1 (a) 50 subintervals (b) 44 subintervals (c) 36 subintervals (d) 54 subintervals (e) 198 subintervals (f) 22 subintervals (g) 50 subintervals (h) 52 subintervals
7. Same as Computer Problem 5
9. $\text{erf}(1) = 0.84270079$, $\text{erf}(3) = 0.99997791$

5.5 Exercises

1. (a) 0, error $= 0$ (b) 0.222222, error $= 0.1777778$ (c) 2.342696, error $= 0.007706$ (d) -0.481237, error $= 0.481237$
3. (a) 0, error $- 0$ (b) 0.4, error $- 0$ (c) 2.350402, error $= 2.95 \times 10^{-7}$ (d) -0.002136, error $= 0.002136$
5. (a) 1.999825 (b) 0.15340700 (c) 0.99999463 (d) 6.99867782

CHAPTER 6

6.1 Exercises

3. (a) $y(t) = 1 + t^2/2$ (b) $y(t) = e^{t^3/3}$
(c) $y(t) = e^{t^2 + 2t}$ (d) $y = e^{t^5}$ (e) $y(t) = (3t + 1)^{1/3}$ (f) $y(t) = (3t^4/4 + 1)^{1/3}$
5. (a) $w = [1.0000, 1.0000, 1.0625, 1.1875, 1.3750]$, error $= 0.1250$
(b) $w = [1.0000, 1.0000, 1.0156, 1.0791, 1.2309]$, error $= 0.1648$
(c) $w = [1.0000, 1.5000, 2.4375, 4.2656, 7.9980]$, error $= 12.0875$
(d) $w = [1.0000, 1.0000, 1.0049, 1.0834, 1.5119]$, error $= 1.2064$
(e) $w = [1.0000, 1.2500, 1.4100, 1.5357, 1.6417]$, error $= 0.0543$

(f) $w = [1.0000, 1.0000, 1.0039, 1.0349, 1.1334]$, error $= 0.0717$
7. (b) $c = \arctan y_0$
9. (a) $L = 0$, has unique solution (b) $L = 1$, has unique solution (a) $L = 1$, has unique solution (d) No Lipschitz constant
11. (a) Solutions are $Y(t) = t^2/2$ and $Z(t) = t^2/2 + 1$. $|Y(t) - Z(t)| = 1 \le e^0|1| = 1$ (b) Solutions are $Y(t) = 0$ and $Z(t) = e^t$. $|Y(t) - Z(t)| = e^t \le e^{1(t-0)}|1|$ (c) Solutions are $Y(t) = 0$ and $Z(t) = e^{-t}$. $|Y(t) - Z(t)| = e^{-t} \le e^{1(t-0)}|1| = 1$ (d) Lipschitz condition not satisfied
13. $y(t) = 1/(1 - t)$
15. (a) $[a, b]$
17. (c) $\sqrt{3}$

6.1 Computer Problems

1.

(a)

t_i	w_i	error
0.0	1.0000	0.0000
0.1	1.0000	0.0050
0.2	1.0100	0.0100
0.3	1.0300	0.0150
0.4	1.0600	0.0200
0.5	1.1000	0.0250
0.6	1.1500	0.0300
0.7	1.2100	0.0350
0.8	1.2800	0.0400
0.9	1.3600	0.0450
1.0	1.4500	0.0500

(b)

t_i	w_i	error
0.0	1.0000	0.0000
0.1	1.0000	0.0003
0.2	1.0010	0.0017
0.3	1.0050	0.0040
0.4	1.0140	0.0075
0.5	1.0303	0.0123
0.6	1.0560	0.0186
0.7	1.0940	0.0271
0.8	1.1477	0.0384
0.9	1.2211	0.0540
1.0	1.3200	0.0756

(c)

t_i	w_i	error
0.0	1.0000	0.0000
0.1	1.2000	0.0337
0.2	1.4640	0.0887
0.3	1.8154	0.1784
0.4	2.2874	0.3243
0.5	2.9278	0.5625
0.6	3.8062	0.9527
0.7	5.0241	1.5952
0.8	6.7323	2.6610
0.9	9.1560	4.4431
1.0	12.6352	7.4503

(d)

t_i	w_i	error
0.0	1.0000	0.0000
0.1	1.0000	0.0000
0.2	1.0001	0.0003
0.3	1.0009	0.0016
0.4	1.0049	0.0054
0.5	1.0178	0.0140
0.6	1.0496	0.0313
0.7	1.1176	0.0654
0.8	1.2517	0.1360
0.9	1.5081	0.2968
1.0	2.0028	0.7154

(e)

t_i	w_i	error
0.0	1.0000	0.0000
0.1	1.1000	0.0086
0.2	1.1826	0.0130
0.3	1.2541	0.0156
0.4	1.3177	0.0171
0.5	1.3753	0.0181
0.6	1.4282	0.0187
0.7	1.4772	0.0191
0.8	1.5230	0.0193
0.9	1.5661	0.0195
1.0	1.6069	0.0195

(f)

t_i	w_i	error
0.0	1.0000	0.0000
0.1	1.0000	0.0000
0.2	1.0001	0.0003
0.3	1.0009	0.0011
0.4	1.0036	0.0028
0.5	1.0099	0.0054
0.6	1.0222	0.0092
0.7	1.0429	0.0139
0.8	1.0744	0.0190
0.9	1.1188	0.0239
1.0	1.1770	0.0281

6.2 Exercises

1. (a) $w = [1.0000, 1.0313, 1.1250, 1.2813, 1.5000]$, error $= 0$ (b) $w = [1.0000, 1.0078, 1.0477, 1.1587, 1.4054]$, error $= 0.0097$ (c) $w = [1.0000, 1.7188, 3.3032, 7.0710, 16.7935]$, error $= 3.2920$ (d) $w = [1.0000, 1.0024, 1.0442, 1.3077, 2.7068]$, error $= 0.0115$ (e) $w = [1.0000, 1.2050, 1.3570, 1.4810, 1.5871]$, error $= 0.0003$ (f) $w = [1.0000, 1.0020, 1.0193, 1.0823, 1.2182]$, error $= 0.0132$

3. (a) $w_{i+1} = w_i + ht_iw_i + 1/2h^2(w_i + t_i^2w_i)$
   (b) $w_{i+1} = w_i + h(t_iw_i^2 + w_i^3) + 1/2h^2(w_i^2 + (2t_iw_i + 3w_i^2)(t_iw_i^2 + w_i^3))$
   (c) $w_{i+1} = w_i + hw_i\sin w_i + 1/2h^2(\sin w_i + w_i\cos w_i)w_i\sin w_i$
   (d) $w_{i+1} = w_i + he^{w_it_i^2} + 1/2h^2e^{w_it_i^2}(2t_iw_i + t_i^2e^{w_it_i^2})$

6.2 Computer Problems

1.

(a)

t_i	w_i	error
0.0	1.0000	0
0.1	1.0050	0
0.2	1.0200	0
0.3	1.0450	0
0.4	1.0800	0
0.5	1.1250	0
0.6	1.1800	0
0.7	1.2450	0
0.8	1.3200	0
0.9	1.4050	0
1.0	1.5000	0

(b)

t_i	w_i	error
0.0	1.0000	0.0000
0.1	1.0005	0.0002
0.2	1.0030	0.0003
0.3	1.0095	0.0005
0.4	1.0222	0.0007
0.5	1.0434	0.0008
0.6	1.0757	0.0010
0.7	1.1224	0.0012
0.8	1.1875	0.0014
0.9	1.2767	0.0016
1.0	1.3974	0.0018

(c)

t_i	w_i	error
0.0	1.0000	0.0000
0.1	1.2320	0.0017
0.2	1.5479	0.0048
0.3	1.9832	0.0106
0.4	2.5908	0.0209
0.5	3.4509	0.0394
0.6	4.6864	0.0725
0.7	6.4878	0.1316
0.8	9.1556	0.2378
0.9	13.1694	0.4297
1.0	19.3063	0.7792

(d)

t_i	w_i	error
0.0	1.0000	0.0000
0.1	1.0000	0.0000
0.2	1.0005	0.0001
0.3	1.0029	0.0004
0.4	1.0114	0.0011
0.5	1.0338	0.0021
0.6	1.0845	0.0037
0.7	1.1890	0.0060
0.8	1.3967	0.0090
0.9	1.8158	0.0109
1.0	2.7164	0.0018

(e)

t_i	w_i	error
0.0	1.0000	0.0000
0.1	1.0913	0.0001
0.2	1.1695	0.0001
0.3	1.2384	0.0001
0.4	1.3005	0.0001
0.5	1.3571	0.0001
0.6	1.4093	0.0001
0.7	1.4580	0.0001
0.8	1.5036	0.0001
0.9	1.5466	0.0001
1.0	1.5873	0.0001

(f)

t_i	w_i	error
0.0	1.0000	0.0000
0.1	1.0001	0.0000
0.2	1.0005	0.0001
0.3	1.0022	0.0002
0.4	1.0068	0.0004
0.5	1.0160	0.0006
0.6	1.0323	0.0009
0.7	1.0579	0.0011
0.8	1.0948	0.0014
0.9	1.1443	0.0017
1.0	1.2069	0.0018

6.3 Exercises

1. (a) $\begin{bmatrix} w_1 \\ w_2 \end{bmatrix} = \begin{bmatrix} 1 & 1.25 & 1.5 & 1.7188 & 1.8594 \\ 0 & -0.25 & -0.625 & -1.1563 & -1.875 \end{bmatrix}$
   error $= \begin{bmatrix} 0.3907 \\ 0.4124 \end{bmatrix}$

   (b) $\begin{bmatrix} w_1 \\ w_2 \end{bmatrix} = \begin{bmatrix} 1 & 0.7500 & 0.5000 & 0.2813 & 0.1094 \\ 0 & 0.2500 & 0.3750 & 0.4063 & 0.3750 \end{bmatrix}$
   error $= \begin{bmatrix} 0.0894 \\ 0.0654 \end{bmatrix}$

   (c) $\begin{bmatrix} w_1 \\ w_2 \end{bmatrix} = \begin{bmatrix} 1 & 1.0000 & 0.9375 & 0.8125 & 0.6289 \\ 0 & 0.2500 & 0.5000 & 0.7344 & 0.9375 \end{bmatrix}$
   error $= \begin{bmatrix} 0.0886 \\ 0.0960 \end{bmatrix}$

   (d) $\begin{bmatrix} w_1 \\ w_2 \end{bmatrix} = \begin{bmatrix} 5 & 6.2500 & 9.6875 & 17.2656 & 32.9492 \\ 0 & 2.5000 & 6.8750 & 15.1563 & 31.3672 \end{bmatrix}$
   error $= \begin{bmatrix} 77.3507 \\ 77.0934 \end{bmatrix}$

3. (a) $\begin{cases} y_1' = y_2 \\ y_2' = ty_1 \end{cases}$ (b) $\begin{cases} y_1' = y_2 \\ y_2' = 2ty_2 - 2y_1 \end{cases}$
   (c) $\begin{cases} y_1' = y_2 \\ y_2' = ty_2 + y_1 \end{cases}$

5. (b) $\begin{cases} y_1' = y_2 \\ y_2' = y_3 \\ y_3' = t + y_2 \end{cases}$ (c) $w = [0, 0, 0, 0, 1/256]$ (d) 0.0392

6.3 Computer Problems

1. errors in $[y_1, y_2]$: (a) $[0.1973, 0.1592]$ for $h = 0.1$, $[0.0226, 0.0149]$ for $h = 0.01$ (b) $[0.0328, 0.0219]$ for $h = 0.1$, $[0.0031, 0.0020]$ for $h = 0.01$ (c) $[0.0305, 0.0410]$ for $h = 0.1$, $[0.0027, 0.0042]$ for $h = 0.01$ (d) $[51.4030, 51.3070]$ for $h = 0.1$, $[8.1919, 8.1827]$ for $h = 0.01$. Note that the errors decline roughly by a factor of 10 for a first-order method.

5. (a) Roughly speaking, periodic trajectory consisting of $3\frac{1}{2}$ revolutions clockwise, $2\frac{1}{2}$ revolutions counterclockwise, $3\frac{1}{2}$ revolutions clockwise, $2\frac{1}{2}$ revolutions counterclockwise. The other periodic trajectory is the same with clockwise replaced by counterclockwise.

6.4 Exercises

1. (a) $w = [1.0000, 1.0313, 1.1250, 1.2813, 1.5000]$, error $= 0$ (b) $w = [1.0000, 1.0039, 1.0395, 1.1442, 1.3786]$, error $= 0.0171$ (c) $w = [1.0000, 1.7031, 3.2399, 6.8595, 16.1038]$, error $= 3.9817$ (d) $w = [1.0000, 1.0003, 1.0251, 1.2283, 2.3062]$, error $= 0.4121$ (e) $w = [1.0000, 1.1975, 1.3490, 1.4734, 1.5801]$, error $= 0.0073$ (f) $w = [1.0000, 1.0005, 1.0136, 1.0713, 1.2055]$, error $= 0.0004$

3. (a) $w = [1, 1.0313, 1.1250, 1.2813, 1.5000]$, error $= 0$ (b) $w = [1, 1.0052, 1.0425, 1.1510, 1.3956]$, error $= 1.2476 \times 10^{-5}$ (c) $w = [1, 1.7545, 3.4865, 7.8448, 19.975]$, error $= 0.11007$ (d) $w = [1, 1.001, 1.0318, 1.2678, 2.7103]$, error $= 7.9505 \times 10^{-3}$

(e) $w = [1, 1.2051, 1.3573, 1.4813, 1.5874]$, error $= 4.1996 \times 10^{-5}$ (f) $w = [1, 1.0010, 1.0154, 1.0736, 1.2051]$, error $= 6.0464 \times 10^{-5}$

6.4 Computer Problems

1.

(a)

t_i	w_i	error
0.0	1.0000	0
0.1	1.0050	0
0.2	1.0200	0
0.3	1.0450	0
0.4	1.0800	0
0.5	1.1250	0
0.6	1.1800	0
0.7	1.2450	0
0.8	1.3200	0
0.9	1.4050	0
1.0	1.5000	0

(b)

t_i	w_i	error
0.0	1.0000	0.0000
0.1	1.0003	0.0001
0.2	1.0025	0.0002
0.3	1.0088	0.0003
0.4	1.0212	0.0004
0.5	1.0420	0.0005
0.6	1.0740	0.0007
0.7	1.1201	0.0010
0.8	1.1847	0.0014
0.9	1.2730	0.0020
1.0	1.3926	0.0030

(c)

t_i	w_i	error
0.0	1.0000	0.0000
0.1	1.2310	0.0027
0.2	1.5453	0.0074
0.3	1.9780	0.0158
0.4	2.5814	0.0303
0.5	3.4348	0.0555
0.6	4.6594	0.0995
0.7	6.4430	0.1764
0.8	9.0814	0.3120
0.9	13.0463	0.5528
1.0	19.1011	0.9845

(d)

t_i	w_i	error
0.0	1.0000	0.0000
0.1	1.0000	0.0000
0.2	1.0003	0.0001
0.3	1.0022	0.0002
0.4	1.0097	0.0005
0.5	1.0306	0.0012
0.6	1.0785	0.0024
0.7	1.1778	0.0052
0.8	1.3754	0.0124
0.9	1.7711	0.0338
1.0	2.6107	0.1076

(e)

t_i	w_i	error
0.0	1.0000	0.0000
0.1	1.0907	0.0007
0.2	1.1686	0.0010
0.3	1.2375	0.0011
0.4	1.2995	0.0011
0.5	1.3561	0.0011
0.6	1.4083	0.0011
0.7	1.4570	0.0011
0.8	1.5026	0.0011
0.9	1.5456	0.0010
1.0	1.5864	0.0010

(f)

t_i	w_i	error
0.0	1.0000	0.0000
0.1	1.0000	0.0000
0.2	1.0003	0.0000
0.3	1.0019	0.0001
0.4	1.0062	0.0002
0.5	1.0151	0.0003
0.6	1.0311	0.0003
0.7	1.0564	0.0003
0.8	1.0931	0.0003
0.9	1.1426	0.0001
1.0	1.2051	0.0001

6.6 Exercises

1. (a) $w = [0, 0.0833, 0.2778, 0.6204, 1.1605]$, error $= 0.4422$

   (b) $w = [0, 0.0500, 0.1400, 0.2620, 0.4096]$, error $= 0.0417$

   (c) $w = [0, 0.1667, 0.4444, 0.7963, 1.1975]$, error $= 0.0622$

6.6 Computer Problems

1. (a) $y = 1$, Euler step size ≤ 1.8 (b) $y = 1$, Euler step size $\leq 1/3$

6.7 Exercises

1. (a) $w = [1.0000, 1.0313, 1.1250, 1.2813, 1.5000]$, error $= 0$

   (b) $w = [1.0000, 1.0078, 1.0314, 1.1203, 1.3243]$, error $= 0.0713$

   (c) $w = [1.0000, 1.7188, 3.0801, 6.0081, 12.7386]$, error $= 7.3469$

(d) $w = [1.0000, 1.0024, 1.0098, 1.1257, 1.7540]$, error $= 0.9642$

(e) $w = [1.0000, 1.2050, 1.3383, 1.4616, 1.5673]$, error $= 0.0201$

(f) $w = [1.0000, 1.0020, 1.0078, 1.0520, 1.1796]$, error $= 0.0255$

3. $w_{i+1} = -4w_i + 5w_{i-1} + h[4f_i + 2f_{i-1}]$; No.

7. (a) $0 < a_1 < 2$ (b) $a_1 = 0$

9. (a) second order unstable (b) second order strongly stable (c) third order strongly stable (d) third order unstable (e) third order unstable

11. For example, $a_1 = 0, a_2 = 1, b_1 = 2 - 2b_0, b_2 = b_0$, where $b_0 \neq 0$ is arbitrary.

13. $w_{i+1} = w_{i-1} + h[\frac{7}{3}f_i - \frac{2}{3}f_{i-1} + \frac{1}{3}f_{i-2}]$

15. (a) $a_1 + a_2 + a_3 = 1, -a_2 - 2a_3 + b_0 + b_1 + b_2 + b_3 = 1, a_2 + 4a_3 + 2b_0 - 2b_2 - 4b_3 = 1,$ $-a_2 - 8a_3 + 3b_0 + 3b_2 + 12b_3 = 1, a_2 + 16a_3 + 4b_0 - 4b_2 - 32b_3 = 1$ (c) $P(x) = x^3 - x^2 = x^2(x - 1)$ has simple root at 1.

6.7 Computer Problems

1.

(a)

t_i	w_i	error
0.0	1.0000	0
0.1	1.0050	0
0.2	1.0200	0
0.3	1.0450	0
0.4	1.0800	0
0.5	1.1250	0
0.6	1.1800	0
0.7	1.2450	0
0.8	1.3200	0
0.9	1.4050	0
1.0	1.5000	0

(b)

t_i	w_i	error
0.0	1.0000	0.0000
0.1	1.0005	0.0002
0.2	1.0020	0.0007
0.3	1.0075	0.0015
0.4	1.0191	0.0025
0.5	1.0390	0.0035
0.6	1.0698	0.0048
0.7	1.1146	0.0065
0.8	1.1773	0.0088
0.9	1.2630	0.0121
1.0	1.3788	0.0168

(c)

t_i	w_i	error
0.0	1.0000	0.0000
0.1	1.2320	0.0017
0.2	1.5386	0.0141
0.3	1.9569	0.0368
0.4	2.5355	0.0762
0.5	3.3460	0.1443
0.6	4.4967	0.2621
0.7	6.1533	0.4661
0.8	8.5720	0.8214
0.9	12.1548	1.4443
1.0	17.5400	2.5455

(d)

t_i	w_i	error
0.0	1.0000	0.0000
0.1	1.0000	0.0000
0.2	1.0001	0.0002
0.3	1.0013	0.0012
0.4	1.0070	0.0033
0.5	1.0243	0.0075
0.6	1.0658	0.0150
0.7	1.1534	0.0296
0.8	1.3266	0.0611
0.9	1.6649	0.1400
1.0	2.3483	0.3700

(e)

t_i	w_i	error
0.0	1.0000	0.0000
0.1	1.0913	0.0001
0.2	1.1673	0.0023
0.3	1.2354	0.0032
0.4	1.2970	0.0036
0.5	1.3534	0.0038
0.6	1.4055	0.0039
0.7	1.4542	0.0039
0.8	1.4998	0.0039
0.9	1.5428	0.0038
1.0	1.5836	0.0038

(f)

t_i	w_i	error
0.0	1.0000	0.0000
0.1	1.0001	0.0000
0.2	1.0002	0.0002
0.3	1.0013	0.0007
0.4	1.0050	0.0014
0.5	1.0131	0.0022
0.6	1.0282	0.0032
0.7	1.0528	0.0039
0.8	1.0890	0.0044
0.9	1.1383	0.0044
1.0	1.2011	0.0040

3.

(a)

t_i	w_i	error
0.0	0.0000	0.0000
0.1	0.0050	0.0002
0.2	0.0213	0.0002
0.3	0.0493	0.0005
0.4	0.0916	0.0002
0.5	0.1474	0.0013
0.6	0.2222	0.0001
0.7	0.3105	0.0032
0.8	0.4276	0.0020
0.9	0.5510	0.0086
1.0	0.7283	0.0100

(b)

t_i	w_i	error
0.0	0.0000	0.0000
0.1	0.0050	0.0002
0.2	0.0187	0.0000
0.3	0.0413	0.0005
0.4	0.0699	0.0004
0.5	0.1082	0.0016
0.6	0.1462	0.0027
0.7	0.2032	0.0066
0.8	0.2360	0.0134
0.9	0.3363	0.0297
1.0	0.3048	0.0631

(c)

t_i	w_i	error
0.0	0.0000	0.0000
0.1	0.0200	0.0013
0.2	0.0700	0.0003
0.3	0.1530	0.0042
0.4	0.2435	0.0058
0.5	0.3855	0.0176
0.6	0.4645	0.0367
0.7	0.7356	0.0890
0.8	0.5990	0.2029
0.9	1.4392	0.4739
1.0	0.0394	1.0959

CHAPTER 7

7.1 Exercises

1. Check the hypotheses of Thm. 7.1: (a) $f_y = 1 > 0$, $f_z = 0$
   (b) $f_y = 2 + 4t^2 > 0$, $f_z = 0$ (c) $f_y = 4 > 0$, $f_z = 0$
   (d) $f_y = 1 + 6(y - e^t)^2 > 0$, $f_z = 0$

3. (a) Follows from Thm. 7.1
   (b) $y(t) = \frac{e^{-b\sqrt{c}}y_a - e^{-a\sqrt{c}}y_b}{e^{(a-b)\sqrt{c}} - e^{(b-a)\sqrt{c}}}e^{\sqrt{c}t} + \frac{e^{a\sqrt{c}}y_b - e^{b\sqrt{c}}y_a}{e^{(a-b)\sqrt{c}} - e^{(b-a)\sqrt{c}}}e^{-\sqrt{c}t}$

5. (a) Since $f_y = 2e^{2d-2y}$, $f_z = 0$, existence and uniqueness follow from Thm 7.1.

7. All requirements are easily checked.

9. (a) $\sin 2t, \cos 2t$ (b) $y_a - y_b = 0$ (c) $y_a + y_b = 0$
   (d) no condition, solution always exists

7.1 Computer Problems

1. (a) $y(t) = 1/3te^t$ (b) $y(t) = e^{t^2}$
3. (a) $y(t) = 1/(3t^2)$ (b) $y(t) = \ln(t^2 + 1)$
5. (a) $s = y_2(0) = 1$, exact solution is
   $y_1(t) = \arctan t, y_2 = t^2 + 1$ (b) $s = y_2(0) = 1/3$,
   exact solution is $y_1(t) = e^{t^3}, y_2(t) = 1/3 - t^2$

7.2 Computer Problems

5. (a) $y(t) = \dfrac{e^{1+t} - e^{1-t}}{e^2 - 1}$

(c)

n	h	error
3	1/4	0.00026473
7	1/8	0.00006657
15	1/16	0.00001667
31	1/32	0.00000417
63	1/64	0.00000104
127	1/128	0.00000026

7. Extrapolate by $N_2(h) = (4N(h/2) - N(h))/3$ and
   $N_3(h) = (16N_2(h/2) - N_2(h))/15$ to arrive at estimate
   $y(1/2) \approx 0.443409442296$, error $\approx 3.11 \times 10^{-10}$.

11. 11.786

CHAPTER 8

8.1 Computer Problems

1. Approximate solution at representative points:

(a)

	$x = 0.2$	$x = 0.5$	$x = 0.8$
$t = 0.2$	3.0432	3.3640	3.9901
$t = 0.5$	5.5451	6.1296	7.2705
$t = 0.8$	10.1039	11.1688	13.2477

(b)

	$x = 0.2$	$x = 0.5$	$x = 0.8$
$t = 0.2$	1.8219	2.4593	3.3199
$t = 0.5$	3.3198	4.4811	6.0492
$t = 0.8$	6.0490	8.1651	11.0224

Forward Difference Method is unstable on both parts for $h = 0.1, K > 0.003$.

3.

(a)

h	k	$u(0.5, 1)$	$w(0.5, 1)$	error
0.02	0.02	16.6642	16.7023	0.0381
0.02	0.01	16.6642	16.6834	0.0192
0.02	0.005	16.6642	16.6738	0.0097

(b)

h	k	$u(0.5, 1)$	$w(0.5, 1)$	error
0.02	0.02	12.1825	12.2104	0.0279
0.02	0.01	12.1825	12.1965	0.0140
0.02	0.005	12.1825	12.1896	0.0071

5.

(a)

h	k	$u(0.5, 1)$	$w(0.5, 1)$	error
0.02	0.02	16.664183	16.664504	0.000321
0.01	0.01	16.664183	16.664263	0.000080
0.005	0.005	16.664183	16.664203	0.000020

(b)

h	k	$u(0.5, 1)$	$w(0.5, 1)$	error
0.02	0.02	12.182494	12.182728	0.000235
0.01	0.01	12.182494	12.182553	0.000059
0.005	0.005	12.182494	12.182509	0.000015

7. $C = \pi^2/100$

8.2 Computer Problems

1. Approximate solution at representative points:

(a)

	$x = 0.2$	$x = 0.5$	$x = 0.8$
$t = 0.2$	−0.4755	−0.8090	−0.4755
$t = 0.5$	0.5878	1.0000	0.5878
$t = 0.8$	−0.4755	−0.8090	−0.4755

(b)

	$x = 0.2$	$x = 0.5$	$x = 0.8$
$t = 0.2$	0.5489	0.4067	0.3012
$t = 0.5$	0.3012	0.2231	0.1653
$t = 0.8$	0.1652	0.1224	0.0907

(c)

	$x = 0.2$	$x = 0.5$	$x = 0.8$
$t = 0.2$	0.3364	0.5306	0.6931
$t = 0.5$	0.5306	0.6930	0.8329
$t = 0.8$	0.6931	0.8329	0.9554

3.

(a)

h	k	$w(1/4, 3/4)$	error
2^{-4}	2^{-6}	-0.70710678	0.0
2^{-5}	2^{-7}	-0.70710678	0.0
2^{-6}	2^{-8}	-0.70710678	0.0
2^{-7}	2^{-9}	-0.70710678	0.0
2^{-8}	2^{-10}	-0.70710678	0.0

(b)

h	k	$w(1/4, 3/4)$	error
2^{-4}	2^{-5}	0.17367424	0.00009971
2^{-5}	2^{-6}	0.17374901	0.00002493
2^{-6}	2^{-7}	0.17376771	0.00000623
2^{-7}	2^{-8}	0.17377238	0.00000156
2^{-8}	2^{-9}	0.17377355	0.00000039

(c)

h	k	$w(1/4, 3/4)$	error
2^{-4}	2^{-4}	0.69308400	0.00006318
2^{-5}	2^{-5}	0.69313136	0.00001582
2^{-6}	2^{-6}	0.69314323	0.00000396
2^{-7}	2^{-7}	0.69314619	0.00000099
2^{-8}	2^{-8}	0.69314693	0.00000025

8.3 Computer Problems

1. Approximate solution at representative points:

(a)

	$x = 0.2$	$x = 0.5$	$x = 0.8$
$y = 0.2$	0.3151	0.5362	0.3151
$y = 0.5$	0.1236	0.2103	0.1236
$y = 0.8$	0.0482	0.0821	0.0482

(b)

	$x = 0.2$	$x = 0.5$	$x = 0.8$
$y = 0.2$	0.4006	1.3686	3.6222
$y = 0.5$	0.6816	2.3284	6.1624
$y = 0.8$	0.4006	1.3686	3.6222

3. Approximate solution at representative points:

(a)

	$x = 0.2$	$x = 0.5$	$x = 0.8$
$y = 0.2$	0.0347	0.0590	0.0347
$y = 0.5$	0.1185	0.2016	0.1185
$y = 0.8$	0.3136	0.5336	0.3136

(b)

	$x = 0.2$	$x = 0.5$	$x = 0.8$
$y = 0.2$	0.4579	0.6752	0.8417
$y = 0.5$	0.6752	0.6708	0.6752
$y = 0.8$	0.8417	0.6752	0.4579

5. 11.4 m

7.

(a)

h	k	$w(1/4, 3/4)$	error
2^{-2}	2^{-2}	0.072692	0.005672
2^{-3}	2^{-3}	0.068477	0.001457
2^{-4}	2^{-4}	0.067387	0.000367
2^{-5}	2^{-5}	0.067112	0.000092

(b)

h	k	$w(1/4, 3/4)$	error
2^{-2}	2^{-2}	0.673903	0.059660
2^{-3}	2^{-3}	0.629543	0.015300
2^{-4}	2^{-4}	0.618094	0.003851
2^{-5}	2^{-5}	0.615207	0.000964

11. Approximate solution at representative points:

(a)

	$x = 0.2$	$x = 0.5$	$x = 0.8$
$y = 0.2$	0.0631	0.1571	0.2493
$y = 0.5$	0.1571	0.3839	0.5887
$y = 0.8$	0.2493	0.5887	0.8448

(b)

	$x = 0.2$	$x = 0.5$	$x = 0.8$
$y = 0.2$	1.0405	1.1046	1.1731
$y = 0.5$	1.1046	1.2830	1.4910
$y = 0.8$	1.1731	1.4910	1.8956

13. Approximate solution at representative points:

(a)

	$x = 1.25$	$x = 1.50$	$x = 1.75$
$y = 1.25$	3.1250	3.8125	4.6250
$y = 1.50$	3.8125	4.5000	5.3125
$y = 1.75$	4.6250	5.3125	6.1250

(b)

	$x = 1.25$	$x = 1.50$	$x = 1.75$
$y = 0.50$	0.1999	0.1666	0.1428
$y = 1.00$	0.7999	0.6666	0.5714
$y = 1.50$	1.7999	1.4999	1.2857

15.

(a)

h	k	$w(1/4, 3/4)$	error
2^{-2}	2^{-2}	0.294813	0.004528
2^{-3}	2^{-3}	0.291504	0.001219
2^{-4}	2^{-4}	0.290596	0.000311
2^{-5}	2^{-5}	0.290363	0.000078

(b)

h	k	$w(1/4, 3/4)$	error
2^{-2}	2^{-2}	1.202628	0.003602
2^{-3}	2^{-3}	1.205310	0.000920
2^{-4}	2^{-4}	1.205999	0.000231
2^{-5}	2^{-5}	1.206172	0.000058

8.4 Computer Problems

1. Solution approaches $u = 0$.
3. (a) Solution approaches $u = 0$ (b) Solution approaches $u = 2$

CHAPTER 9

9.1 Exercises

1. (a) 4 (b) 9
3. (a) 0.3 (b) 0.28

9.1 Computer Problems

1. 0.000273, compared with correct volume ≈ 0.000268.
3. (The minimal standard LCG with seed 1 is used in the following answers:)

(a) 1/3 (b)

n	Type 1 estimate	error
10^2	0.327290	0.006043
10^3	0.342494	0.009161
10^4	0.332705	0.000628
10^5	0.333610	0.000277
10^6	0.333505	0.000172

(c)

n	Type 2 estimate	error
10^2	0.28	0.053333
10^3	0.354	0.020667
10^4	0.3406	0.007267
10^5	0.33382	0.000487
10^6	0.333989	0.000656

5. (a) $n = 10^4$: 0.5128, error = 0.010799; $n = 10^6$: 0.524980, error = 0.001381 (b) $n = 10^4$: 0.1744, error = 0.000133; $n = 10^6$: 0.174851, error = 0.000318
7. (a) 1/12 (b) 0.083566, error = 0.000232

9.2 Computer Problems

1. (a) 1/3 (b)

n	Type 1 estimate	error
10^2	0.335414	0.002080
10^3	0.333514	0.000181
10^4	0.333339	0.000006
10^5	0.333334	0.000001

(c)

n	Type 2 estimate	error
10^2	0.35	0.016667
10^3	0.333	0.000333
10^4	0.3339	0.000567
10^5	0.33338	0.000047

3. (a) $n = 10^4$: 0.5232, error = 0.000399; $n = 10^5$: 0.52396, error = 0.000361 (b) $n = 10^4$: 0.1743, error = 0.000233; $n = 10^5$: 0.17455, error = 0.000017
5. Typical results: Monte Carlo estimate 4.9656, error = 0.030798; quasi-Monte Carlo estimate 4.92928, error = 0.005522.
7. (a) exact value = 1/2; $n = 10^6$ Monte Carlo estimate 0.500313 (b) exact value 4/9; $n = 10^6$ Monte Carlo estimate 0.444486
9. $1/24 \approx 4.167\%$

9.3 Computer Problems

Answers in this section use the minimal standard LCG.
1. (a) Monte Carlo = 0.2907, error = 0.0050
   (b) 0.6323, error 0.0073 (c) 0.7322, error 0.0049
3. (a) 0.8199, error = 0.0014 (b) 0.9871, error = 0.0004 (c) 0.9984, error = 0.0006
5. (a) 0.2969, error = 0.0112 (b) 0.3939, error = 0.0049 (c) 0.4600, error = 0.0106
7. (a) 0.5848, error = 0.0207 (b) 0.3106, error = 0.0154 (c) 0.7155, error = 0.0107

9.4 Computer Problems

5. Typical results:

Δt	avg. error
10^{-1}	0.2657
10^{-2}	0.0925
10^{-3}	0.0256

The results show approximate order 1/2.

11.

Δt	avg. error
10^{-1}	0.1394
10^{-2}	0.0202
10^{-3}	0.0026

The results show approximate order 1.

CHAPTER 10

10.1 Exercises

1. (a) $[0, -i, 0, i]$ (b) $[2, 0, 0, 0]$ (c) $[0, i, 0, -i]$
   (d) $[0, 0, -\sqrt{2}i, 0, 0, 0, \sqrt{2}i, 0]$
3. (a) $[1/2, 1/2, 1/2, 1/2]$ (b) $[1, 1, -1, 1]$
   (c) $[1, 1, 1, -1]$ (d) $[2, -1, 2, -1, 2, -1, 2, -1]/\sqrt{2}$
5. (a) 4th roots of unity: $-i, -1, i, 1$; primitive: $-i, i$
   (b) $\omega, \omega^2, \omega^3, \omega^4, \omega^5, \omega^6$ where $\omega = e^{-2\pi i/7}$
   (c) $p - 1$
7. (a) $a_0 = a_1 = a_2 = 0, b_1 = -1$
   (b) $a_0 = 2, a_1 = a_2 = 0, b_1 = 0$
   (c) $a_0 = a_1 = a_2 = 0, b_1 = 1$
   (d) $b_2 = -\sqrt{2}, a_0 = a_1 = a_2 = a_3 = a_4 = b_1 = b_3 = 0$

10.2 Exercises

1. (a) $P_4(t) = \sin 2\pi t$ (b) $P_4(t) = \cos 2\pi t + \sin 2\pi t$
   (c) $P_4(t) = -\cos 4\pi t$ (d) $P_4(t) = 1$
3. (a) $P_8(t) = \sin 4\pi t$ (b) $P_8(t) = 1 + \sin 4\pi t$
   (c) $P_8(t) = \frac{1}{2} + \frac{1}{4}\cos 2\pi t + \frac{\sqrt{2}+1}{4}\sin 2\pi t + \frac{1}{4}\cos 6\pi t + \frac{\sqrt{2}-1}{4}\sin 6\pi t$
   (d) $P_8(t) = \cos 8\pi t$

10.2 Computer Problems

1. (a) $P_8(t) = \frac{7}{2} - \cos 2\pi t - (1 + \sqrt{2})\sin 2\pi t - \cos 4\pi t - \sin 4\pi t - \cos 6\pi t + (1 - \sqrt{2})\sin 6\pi t - \frac{1}{2}\cos 8\pi t$ (b) $P_8(t) = \frac{1}{2} - 0.8107\cos 2\pi t - 0.1036\sin 2\pi t + \cos 4\pi t + \frac{1}{2}\sin 4\pi t + 1.3107\cos 6\pi t - 0.6036\sin 6\pi t$ (c) $P_8(t) = \frac{5}{2} - \frac{1}{2}\cos \frac{\pi}{2}t - \frac{1}{2}\sin \frac{\pi}{2}t + \cos \pi t$ (d) $P_8(t) = \frac{5}{8} + \frac{3}{4}\cos \frac{\pi}{4}(t - 1) +$

$1.3536 \sin \frac{\pi}{4}(t-1) - \frac{7}{4}\cos\frac{\pi}{2}(t-1) - \frac{5}{2}\sin\frac{\pi}{2}(t-1) + \frac{3}{4}\cos\frac{3\pi}{4}(t-1) - 0.6464\sin\frac{3\pi}{4}(t-1) + \frac{5}{8}\cos\pi(t-1)$

3. $P_8(t) = 1.6131 - 0.1253\cos 2\pi t - 0.5050\sin 2\pi t - 0.1881\cos 4\pi t - 0.2131\sin 4\pi t - 0.1991\cos 6\pi t - 0.0886\sin 6\pi t - 0.1007\cos 8\pi t$

5. $P_8(t) = 0.3423 - 0.1115\cos 2\pi(t-1) - 0.2040\sin 2\pi(t-1) - 0.0943\cos 4\pi(t-1) - 0.0859\sin 4\pi(t-1) - 0.0912\cos 6\pi(t-1) - 0.0357\sin 6\pi(t-1) - 0.0453\cos 8\pi(t-1)$

10.3 Exercises

1. (a) $F_2(t) = 0$ (b) $F_2(t) = \cos 2\pi t$ (c) $F_2(t) = 0$
   (d) $F_2(t) = 1$
3. (a) $F_4(t) = 0$ (b) $F_4(t) = 1$ (c) $F_4(t) = \frac{1}{2}$
   $+ \frac{1}{4}\cos 2\pi t + \frac{\sqrt{2}+1}{4}\sin 2\pi t$ (d) $F_4(t) = 0$

10.3 Computer Problems

1. (a) $F_2(t) = F_4(t) = 3\cos 2\pi t$
   (b) $F_2(t) = 2 - \frac{3}{2}\cos 2\pi t$, $F_4(t) = 2 - \frac{3}{2}\cos 2\pi t - \frac{1}{2}\sin 2\pi t + \frac{3}{2}\cos 4\pi t$
   (c) $F_2(t) = \frac{7}{2} - \frac{1}{2}\cos\frac{\pi}{2}t$, $F_4(t) = \frac{7}{2} - \frac{1}{2}\cos\frac{\pi}{2}t + \frac{1}{2}\sin\frac{\pi}{2}t + 2\cos\pi t$
   (d) $F_2(t) = 2 - 2\cos\frac{\pi}{3}(t-1)$, $F_4(t) = 2 - 2\cos\frac{\pi}{3}(t-1) - \cos\frac{2\pi}{3}(t-1)$

CHAPTER 11

11.1 Exercises

1. The DCT matrix is $C = \frac{1}{\sqrt{2}}\begin{bmatrix} 1 & 1 \\ 1 & -1 \end{bmatrix}$, and
   $P_2(t) = \frac{1}{\sqrt{2}}y_0 + y_1\cos\frac{(2t+1)\pi}{4}$
   (a) $y = [3\sqrt{2}, 0]$, $P_2(t) = 3$ (b) $y = [0, 2\sqrt{2}]$,
   $P_2(t) = 2\sqrt{2}\cos\frac{(2t+1)\pi}{4}$ (c) $y = [2\sqrt{2}, \sqrt{2}]$, $P_2(t)$
   $= 2 + \sqrt{2}\cos\frac{(2t+1)\pi}{4}$ (d) $y = [3\sqrt{2}/2, 5\sqrt{2}/2]$,
   $P_2(t) = 3/2 + (5\sqrt{2}/2)\cos\frac{(2t+1)\pi}{4}$.
3. (a) $y = [1, b-c, 0, b+c]$, $P_4(t) =$
   $\frac{1}{2} + \left((b-c)/\sqrt{2}\right)\cos\frac{(2t+1)\pi}{8} + \left((b+c)/\sqrt{2}\right)\cos\frac{3(2t+1)\pi}{8}$
   (b) $y = [2, 0, 0, 0]$, $P_4(t) = 1$ (c) $y =$
   $[1/2, b, 1/2, c]$, $P_4(t) = 1/2 + \left(b/\sqrt{2}\right)\cos\frac{(2t+1)\pi}{8} + (1/2\sqrt{2})\cos\frac{2(2t+1)\pi}{8} + \left(c/\sqrt{2}\right)\cos\frac{3(2t+1)\pi}{8}$
   (d) $y = [5, -(c+3b), 0, (b-3c)]$, $P_4(t) = \frac{5}{2} - \left((c+3b)/\sqrt{2}\right)\cos\frac{(2t+1)\pi}{8} + \left((b-3c)/\sqrt{2}\right)\cos\frac{3(2t+1)\pi}{8}$

11.2 Exercises

1. (a) $Y = \begin{bmatrix} 1/2 & 1/2 \\ 1/2 & 1/2 \end{bmatrix}$, $P_2(s,t) =$
   $\frac{1}{4} + \frac{1}{2\sqrt{2}}\cos\frac{(2s+1)\pi}{4} + \frac{1}{2\sqrt{2}}\cos\frac{(2t+1)\pi}{4} + \frac{1}{2}\cos\frac{(2s+1)\pi}{4}\cos\frac{(2t+1)\pi}{4}$
   (b) $Y = \begin{bmatrix} 1 & 1 \\ 0 & 0 \end{bmatrix}$, $P_2(s,t) = \frac{1}{2} + \frac{1}{\sqrt{2}}\cos\frac{(2t+1)\pi}{4}$
   (c) $Y = \begin{bmatrix} 2 & 0 \\ 0 & 0 \end{bmatrix}$, $P_2(s,t) = 1$
   (d) $Y = \begin{bmatrix} 1 & 0 \\ 0 & 1 \end{bmatrix}$, $P_2(s,t) = \frac{1}{2} + \cos\frac{(2s+1)\pi}{4}\cos\frac{(2t+1)\pi}{4}$

3. (a) $P(t) = \left((b+c)/\sqrt{2}\right)\cos\frac{(2t+1)\pi}{8}$
   (b) $P(t) = 1/4$ (c) $P(t) = 1/4$
   (d) $P(t) = 2 + \sqrt{2}(b-c)\cos\frac{(2s+1)\pi}{8}$

11.2 Computer Problems

1. (a) $\begin{bmatrix} 0 & -3.8268 & 0 & -9.2388 \\ 0 & 1.7071 & 0 & 4.1213 \\ 0 & 0 & 0 & 0 \\ 0 & 0.1213 & 0 & 0.2929 \end{bmatrix}$

   (b) $\begin{bmatrix} 0 & 0 & 0 & 0 \\ 0 & 2.1213 & -0.7654 & -0.8787 \\ 0 & 0 & 0 & 0 \\ 0 & 5.1213 & -1.8478 & -2.1213 \end{bmatrix}$

   (c) $\begin{bmatrix} 4.7500 & 1.4419 & 0.2500 & 0.2146 \\ -0.7886 & 0.5732 & -1.4419 & -1.0910 \\ 0.2500 & 2.6363 & -2.2500 & -0.8214 \\ 0.0560 & -2.0910 & -0.2146 & 0.9268 \end{bmatrix}$

   (d) $\begin{bmatrix} 0 & -4.4609 & 0 & -0.3170 \\ -4.4609 & 0 & 0 & 0 \\ 0 & 0 & 0 & 0 \\ -0.3170 & 0 & 0 & 0 \end{bmatrix}$

11.3 Exercises

1. (a) $P(A) = 1/4$, $P(B) = 5/8$, $P(C) = 1/8$, 1.30
   (b) $P(A) = 3/8$, $P(B) = 1/4$, $P(C) = 3/8$, 1.56
   (c) $P(A) = 1/2$, $P(B) = 3/8$, $P(C) = 1/8$, 1.41
3. (a) 34 bits needed, $34/11 = 3.09$ bits/symbol $> 3.03 =$ Shannon inf. (b) 73 bits needed, $73/21 = 3.48$ bits/symbol $> 3.42 =$ Shannon inf. (c) 108 bits needed, $108/35 = 3.09$ bits/symbol $> 3.04 =$ Shannon inf.

11.4 Exercises

1. (a) $[-12b - 2c, 2b - 12c]$ (b) $[-3b - c, b - 3c]$
   (c) $[-8b + 5c, -5b - 8c]$

3. (a) +101., error = 0   (b)   +101., error = 1/15
   (c)   +011., error = 1/35
5. (a) +0110000., error = 1/170   (b)   −0101101., error
   = 1/85   (c)   +1011100., error = 7/510
   (d)   +1100100., error ≈ 0.0043
7. (a) $\frac{1}{2}(w_2 + w_3) = [-1.2246, 0.9184] \approx [-1, 1]$
   (b) $\frac{1}{2}(w_2 + w_3) = [2.1539, -0.9293] \approx [2, -1]$
   (c) $\frac{1}{2}(w_2 + w_3) = [-1.7844, -3.0832] \approx [-2, -3]$
9. $c_{5n} = -c_{n-1}, c_{6n} = -c_0$

CHAPTER 12

12.1 Exercises

1. (a) $P(\lambda) = (\lambda - 5)(\lambda - 2)$, 2 and $[1, 1]$, 5 and
   $[1, -1]$ (b) $P(\lambda) = (\lambda + 2)(\lambda - 2)$, -2 and $[1, -1]$, 2
   and $[1, 1]$ (c) $P(\lambda) = (\lambda - 3)(\lambda + 2)$, 3 and $[-3, 4]$, -2
   and $[4, 3]$ (d) $P(\lambda) = (\lambda - 100)(\lambda - 200)$, 200 and
   $[-3, 4]$, 100 and $[4, 3]$
3. (a) $P(\lambda) = -(\lambda - 1)(\lambda - 2)(\lambda - 3)$, 3 and $[0, 1, 0]$, 2
   and $[1, 2, 1]$, 1 and $[1, 0, 0]$ (b) $P(\lambda) = -\lambda(\lambda - 1)$
   $(\lambda - 2)$, 2 and $[-1, 2, 3]$, 1 and $[1, 1, 0]$, 0 and $[1, -2, 3]$
   (c) $P(\lambda) = -\lambda(\lambda - 1)(\lambda + 1)$, 1 and $[1, -2, -3]$, 0 and
   $[1, -2, 3]$, -1 and $[1, 1, 0]$
5. (a) $\lambda = 4, S = 3/4$ (b) $\lambda = -4, S = 3/4$
   (c) $\lambda = 4, S = 1/2$ (d) $\lambda = 10, S = 9/10$
7. (a) $\lambda = 1, S = 1/3$ (b) $\lambda = 1, S = 1/3$
   (c) $\lambda = -1, S = 1/2$ (d) $\lambda = 9, S = 3/4$
9. (a)   5 and $[1, 2]$, -1 and $[-1, 1]$ (b)
   $u_1 = [1/\sqrt{17}, 4/\sqrt{17}]$, RQ = 1; $u_2 = [0.4903, 0.8716]$,
   RQ = 4.29; $u_3 = [0.4386, 0.8987]$, RQ = 5.08   (c)   IPI
   converges to $\lambda = -1$.   (d)   IPI converges to $\lambda = 5$.
11. (a)   7   (b)   5   (c) $S = 6/7, S = 1/2$; IPI with $s = 4$
   is faster.

12.1 Computer Problems

1. (a)   converges to 4 and $[1, 1, -1]$ (b)   converges to -4
   and $[1, 1, -1]$ (c)   converges to 4 and $[1, 1, -1]$
   (d)   converges to 10 and $[1, 1, -1]$
3. (a) $\lambda = 4$ (b) $\lambda = 3$ (c) $\lambda = 2$ (d) $\lambda = 9$

12.2 Exercises

1. (a) $\begin{bmatrix} 1 & -\frac{1}{\sqrt{2}} & \frac{1}{\sqrt{2}} \\ -\sqrt{2} & \frac{1}{2} & \frac{1}{2} \\ 0 & \frac{1}{2} & \frac{1}{2} \end{bmatrix}$ (b) $\begin{bmatrix} 1 & 0 & 0 \\ 0 & 0 & -1 \\ 0 & -1 & 0 \end{bmatrix}$

(c) $\begin{bmatrix} 2 & -\frac{4}{5} & -\frac{3}{5} \\ -5 & \frac{37}{25} & -\frac{16}{25} \\ 0 & \frac{9}{25} & \frac{13}{25} \end{bmatrix}$ (d) $\begin{bmatrix} 1 & -\frac{1}{\sqrt{2}} & -\frac{1}{\sqrt{2}} \\ -\sqrt{8} & \frac{5}{2} & \frac{3}{2} \\ 0 & \frac{3}{2} & \frac{1}{2} \end{bmatrix}$

5. (a) NSI fails: $\overline{Q}_k$ does not converge, alternates with
   period of 2. (b) NSI fails: $\overline{Q}_k$ does not converge,
   alternates with period of 2.

7. (a) before: does not converge; after: same (already in
   Hessenberg form) (b) before: does not converge; after:
   does not converge

12.2 Computer Problems

1. (a) $\{-6, 4, -2\}$ (b) $\{6, 4, 2\}$ (c) $\{20, 18, 16\}$
   (d) $\{10, 2, 1\}$
3. (a) $\{3, 3, 3\}$ (b) $\{1, 9, 10\}$ (c) $\{3, 3, 18\}$
   (d) $\{-2, 2, 0\}$
5. (a) $\{2, i, -i\}$ (b) $\{1, i, -i\}$ (c) $\{2 + 3i, 2 - 3i, 1\}$
   (d) $\{5, 4 + 3i, 4 - 3i\}$

12.3 Exercises

1. (a) $\begin{bmatrix} -3 & 0 \\ 0 & 2 \end{bmatrix} = \begin{bmatrix} 1 & 0 \\ 0 & 1 \end{bmatrix} \begin{bmatrix} 3 & 0 \\ 0 & 2 \end{bmatrix} \begin{bmatrix} -1 & 0 \\ 0 & 1 \end{bmatrix}$

Expands by factor of 3 and flips along x-axis,
expands by factor of 2 along y-axis.

(b) $\begin{bmatrix} 0 & 0 \\ 0 & 3 \end{bmatrix} = \begin{bmatrix} 0 & 1 \\ 1 & 0 \end{bmatrix} \begin{bmatrix} 3 & 0 \\ 0 & 0 \end{bmatrix} \begin{bmatrix} 0 & 1 \\ 1 & 0 \end{bmatrix}$

Projects onto y axis and expands by 3 in y-direction.

(c) $\begin{bmatrix} \frac{3}{2} & -\frac{1}{2} \\ -\frac{1}{2} & \frac{3}{2} \end{bmatrix} = \begin{bmatrix} -\frac{1}{\sqrt{2}} & \frac{1}{\sqrt{2}} \\ \frac{1}{\sqrt{2}} & \frac{1}{\sqrt{2}} \end{bmatrix}$
$\begin{bmatrix} 2 & 0 \\ 0 & 1 \end{bmatrix} \begin{bmatrix} -\frac{1}{\sqrt{2}} & \frac{1}{\sqrt{2}} \\ \frac{1}{\sqrt{2}} & \frac{1}{\sqrt{2}} \end{bmatrix}$

Expands into ellipse with major axis of length 4
along the line $y = -x$.

(d) $\begin{bmatrix} -\frac{3}{2} & \frac{1}{2} \\ \frac{1}{2} & -\frac{3}{2} \end{bmatrix} = \begin{bmatrix} -\frac{1}{\sqrt{2}} & \frac{1}{\sqrt{2}} \\ \frac{1}{\sqrt{2}} & \frac{1}{\sqrt{2}} \end{bmatrix}$
$\begin{bmatrix} 2 & 0 \\ 0 & 1 \end{bmatrix} \begin{bmatrix} \frac{1}{\sqrt{2}} & -\frac{1}{\sqrt{2}} \\ -\frac{1}{\sqrt{2}} & -\frac{1}{\sqrt{2}} \end{bmatrix}$

Same as (c), but rotated 180°.

(e) $\begin{bmatrix} \frac{3}{4} & \frac{5}{4} \\ \frac{5}{4} & \frac{3}{4} \end{bmatrix} = \begin{bmatrix} -\frac{1}{\sqrt{2}} & \frac{1}{\sqrt{2}} \\ -\frac{1}{\sqrt{2}} & -\frac{1}{\sqrt{2}} \end{bmatrix}$
$\begin{bmatrix} 2 & 0 \\ 0 & \frac{1}{2} \end{bmatrix} \begin{bmatrix} -\frac{1}{\sqrt{2}} & -\frac{1}{\sqrt{2}} \\ -\frac{1}{\sqrt{2}} & \frac{1}{\sqrt{2}} \end{bmatrix}$

Expands by factor of 2 along line $y = x$ and contracts by
factor of 2 along line $y = -x$, and flips the points on the
circle.

3. Four: $\begin{bmatrix} 3 & 0 \\ 0 & \frac{1}{2} \end{bmatrix} = \begin{bmatrix} 1 & 0 \\ 0 & 1 \end{bmatrix} \begin{bmatrix} 3 & 0 \\ 0 & \frac{1}{2} \end{bmatrix} \begin{bmatrix} 1 & 0 \\ 0 & 1 \end{bmatrix} =$
$\begin{bmatrix} -1 & 0 \\ 0 & 1 \end{bmatrix} \begin{bmatrix} 3 & 0 \\ 0 & \frac{1}{2} \end{bmatrix} \begin{bmatrix} -1 & 0 \\ 0 & 1 \end{bmatrix}$
$= \begin{bmatrix} 1 & 0 \\ 0 & -1 \end{bmatrix} \begin{bmatrix} 3 & 0 \\ 0 & \frac{1}{2} \end{bmatrix} \begin{bmatrix} 1 & 0 \\ 0 & -1 \end{bmatrix}$
$= \begin{bmatrix} -1 & 0 \\ 0 & -1 \end{bmatrix} \begin{bmatrix} 3 & 0 \\ 0 & \frac{1}{2} \end{bmatrix} \begin{bmatrix} -1 & 0 \\ 0 & -1 \end{bmatrix}$

5. (a)

$$\begin{bmatrix} 1/\sqrt{2} & 0 \\ -1/\sqrt{2} & 0 \\ 0 & 1/\sqrt{2} \\ 0 & -1/\sqrt{2} \end{bmatrix} \begin{bmatrix} 2a & 0 \\ 0 & 2b \end{bmatrix} \begin{bmatrix} 1/\sqrt{2} & 1/\sqrt{2} \\ 1/\sqrt{2} & -1/\sqrt{2} \end{bmatrix}$$

(b)

$$\begin{bmatrix} 1/\sqrt{2} & 0 \\ -1/\sqrt{2} & 0 \\ 0 & 1/\sqrt{2} \\ 0 & 1/\sqrt{2} \end{bmatrix} \begin{bmatrix} 2a & 0 \\ 0 & 2b \end{bmatrix} \begin{bmatrix} -1/\sqrt{2} & 1/\sqrt{2} \\ 1/\sqrt{2} & 1/\sqrt{2} \end{bmatrix}$$

7. (a)

$$\begin{bmatrix} 1/2 & 1/2 \\ -1/2 & -1/2 \\ 1/2 & -1/2 \\ -1/2 & 1/2 \end{bmatrix} \begin{bmatrix} 10a & 0 \\ 0 & 10b \end{bmatrix} \begin{bmatrix} 0.6 & 0.8 \\ 0.8 & -0.6 \end{bmatrix}$$

(b)

$$\begin{bmatrix} 1/2 & 1/2 \\ -1/2 & -1/2 \\ 1/2 & -1/2 \\ -1/2 & 1/2 \end{bmatrix} \begin{bmatrix} 10a & 0 \\ 0 & 10b \end{bmatrix} \begin{bmatrix} 0.8 & 0.6 \\ -0.6 & 0.8 \end{bmatrix}$$

12.4 Computer Problems

1. (a) $\begin{bmatrix} 1.1708 & 1.8944 \\ 1.8944 & 3.0652 \end{bmatrix}$ (b) $\begin{bmatrix} 1.5607 & 3.7678 \\ 1.3536 & 3.2678 \end{bmatrix}$

(c) $\begin{bmatrix} 1.0107 & 2.5125 & 3.6436 \\ 0.9552 & 2.3746 & 3.4436 \\ 0.1787 & 0.4442 & 0.6441 \end{bmatrix}$

(d) $\begin{bmatrix} -0.5141 & 5.2343 & 1.9952 \\ 0.2070 & -2.1076 & -0.8033 \\ -0.1425 & 1.4510 & 0.5531 \end{bmatrix}$

3. (a) Best line $y = 3.3028x$; projections are $\begin{bmatrix} 1.1934 \\ 3.9415 \end{bmatrix}$, $\begin{bmatrix} 1.4707 \\ 4.8575 \end{bmatrix}$, $\begin{bmatrix} 1.2774 \\ 4.2188 \end{bmatrix}$.

(b) Best line $y = 0.3620x$; projections are $\begin{bmatrix} 1.7682 \\ 0.6402 \end{bmatrix}$, $\begin{bmatrix} 3.8565 \\ 1.3963 \end{bmatrix}$, $\begin{bmatrix} 3.2925 \\ 1.1921 \end{bmatrix}$.

(c) Best line $(x(t), y(t), z(t)) = [0.3015, 0.3416, 0.8902]t$; projections are $\begin{bmatrix} 1.3702 \\ 1.5527 \\ 4.0463 \end{bmatrix}$, $\begin{bmatrix} 1.8325 \\ 2.0764 \\ 5.4111 \end{bmatrix}$, $\begin{bmatrix} 1.8949 \\ 2.1471 \\ 5.5954 \end{bmatrix}$ $\begin{bmatrix} 0.9989 \\ 1.1319 \\ 2.9498 \end{bmatrix}$.

5. See Exercise 12.3.2 answers.

CHAPTER 13

13.1 Exercises

1. (a) $(0, 1)$ (b) $(0, 0)$ (c) $(-1/2, -3/8)$ (d) $(1, 1)$

13.1 Computer Problems

1. (a) $1/2$ (b) $-2, 1$ (c) 0.47033 (d) 1.43791
3. (a), (b): $(0.358555, 2.788973)$
5. $(1.20881759, 1.20881759)$, about 8 correct places
7. $(1, 1)$

13.2 Computer Problems

1. Minimum is $(1.2088176, 1.2088176)$. Different initial conditions will yield answers that differ by about $\epsilon^{1/2}$.
3. $(1, 1)$. Newton's Method will be accurate to machine precision, since it is finding a simple root. Steepest Descent will have error of size $\approx \epsilon^{1/2}$.
5. Same as Computer Problem 2.

Bibliography

Y. Achdou and O. Pironneau [2005] *Computational Methods for Options Pricing*. SIAM, Philadelphia, PA.

A. Ackleh, E. J. Allen, R. B. Kearfott, and P. Seshaiyer [2009] *Classical and Modern Numerical Analysis: Theory, Methods, and Practice*. Chapman and Hall, New York.

M. Agoston [2005] *Computer Graphics and Geometric Modeling*. Springer, New York.

K. Alligood, T. Sauer, and J. A. Yorke [1996] *Chaos: An Introduction to Dynamical Systems*. Springer, New York.

W. F. Ames [1992] *Numerical Methods for Partial Differential Equations*, 3rd ed. Academic Press, Boston.

E. Anderson, Z. Bai, C. Bischof, J. W. Demmel, J. J. Dongarra, J. Du Croz, A. Greenbaum, S. Hammarling, A. McKenney, and D. Sorensen [1990] "LAPACK: A Portable Linear Algebra Library for High-performance Computers," Computer Science Dept. Technical Report CS-90–105, University of Tennessee, Knoxville.

U. M. Ascher, R. M. Mattheij, and R. B. Russell [1995] *Numerical Solution of Boundary Value Problems for Ordinary Differential Equations*. SIAM, Philadelphia, PA.

U. M. Ascher and L. Petzold [1998] *Computer Methods for Ordinary Differential Equations and Differential-algebraic Equations*. SIAM, Philadelphia, PA.

R. Ashino, M. Nagase, and R. Vaillancourt [2000] "Behind and Beyond the MATLAB ODE Suite." *Computers and Mathematics with Application* **40**, 491–572.

R. Aster, B. Borchers, and C. Thurber [2005] *Parameter Estimation and Inverse Problems*. Academic Press, New York.

O. Axelsson [1994] *Iterative Solution Methods*. Cambridge University Press, New York.

O. Axelsson and V. A. Barker [1984] *Finite Element Solution of Boundary Value Problems for Ordinary Differential Equations*. Academic Press, Orlando, FL.

Z. Bai, J. Demmel, J. Dongarra, A. Ruhe, and H. Van der Vorst [2000] *Templates for the Solution of Algebraic Eigenvalue Problems: A Practical Guide*. SIAM, Philadelphia, PA.

P. B. Bailey, L. F. Shampine, and P. E. Waltman [1968] *Nonlinear Two-Point Boundary-Value Problems*. Academic Press, New York.

R. Bank [1998] "PLTMG, A Software Package for Solving Elliptic Partial Differential Equations", *Users' Guide 8.0*. SIAM, Philadelphia, PA.

R. Barrett, M. Berry, T. Chan, J. Demmel, J. Donato, J. Dongarra, V. Eijkhout, R. Pozo, C. Romine, and H. van der Vorst [1987] *Templates for the Solution of Linear Systems: Building Blocks for Iterative Methods*. SIAM, Philadelphia, PA.

V. Bhaskaran and K. Konstandtinides [1995] *Image and Video Compression Standards: Algorithms and Architectures*. Kluwer Academic Publishers, Boston, MA.

G. Birkhoff and R. Lynch [1984] *Numerical Solution of Elliptic Problems*. SIAM, Philadelphia, PA.

G. Birkhoff and G. Rota [1989] *Ordinary Differential Equations*, 4th ed. John Wiley & Sons, New York.

F. Black and M. Scholes [1973] "The Pricing of Options and Corporate Liabilities." *Journal of Political Economy* **81**, 637–654.

P. Blanchard, R. Devaney, and G. R. Hall [2011] *Differential Equations*, 4th ed. Brooks-Cole, Pacitic Grove, CA.

F. Bornemann, D. Laurie, S. Wagon, and J. Waldvogel [2004] *The SIAM 100-Digit Challenge: A Study in High-Accuracy Numerical Computing*. SIAM, Philadelphia, PA.

W. E. Boyce and R. C. DiPrima [2012] *Elementary Differential Equations and Boundary Value Problems*, 10th ed. John Wiley & Sons, New York.

G. E. P. Box and M. Muller [1958] "A Note on the Generation of Random Normal Deviates." *The Annals Mathematical Statistics* **29**, 610–611.

R. Bracewell [2000] *The Fourier Transform and Its Application*, 3rd ed. McGraw-Hill, New York.

J. H. Bramble [1993] *Multigrid Methods*. John Wiley & Sons, New York.

K. Brandenburg and M. Bosi [1997] "Overview of MPEG Audio: Current and Future Standards for Low Bit Rate Audio Coding." *Journal of the Audio Engineering Society* **45**, 4–21.

M. Braun [1993] *Differential Equations and Their Applications*, 4th ed. Springer-Verlag, New York.

S. Brenner and L. R. Scott [2007] *The Mathematical Theory of Finite Element Methods*, 3rd ed. Springer-Verlag, New York.

R. P. Brent [1973] *Algorithms for Minimization without Derivatives*. Prentice Hall, Englewood Cliffs, NJ.

W. Briggs [1987] *A Multigrid Tutorial*. SIAM, Philadelphia, PA.

W. Briggs and V. E. Henson [1995] *The DFT: An Owner's Manual for the Discrete Fourier Transform*. SIAM, Philadelphia, PA.

E. O. Brigham [1988] *The Fast Fourier Transform and Its Applications*. Prentice-Hall, Englewood Cliffs, NJ.

S. Brin and L. Page [1998] "The Anatomy of a Large-Scale Hypertextual Web Search Engine." *Computer Networks and ISDN systems* **30**, 107–117.

C. G. Broyden [1965] "A Class of Methods for Solving Nonlinear Simultaneous Equations." *Mathematics of Computation* **19**, 577–593.

C. G. Broyden, J. E. Dennis, Jr., and J. J. Moré [1973] "On the Local and Superlinear Convergence of Quasi-Newton Methods." *IMA Journal of Applied Mathematics* **12**, 223–245.

K. Burrage [1995] *Parallel and Sequential Methods for Ordinary Differential Equations.* Oxford University Press, New York.

J. C. Butcher [1987] *Numerical Analysis of Ordinary Differential Equations.* Wiley, London.

E. Cheney [1966] *Introduction to Approximation Theory.* McGraw-Hill, New York.

E. Chu and A. George [1999] *Inside the FFT Black Box.* CRC Press, Boca Raton, FL.

P. G. Ciarlet [1978] *The Finite Element Method for Elliptic Problems.* North-Holland, Amsterdam.

CODEE [1999] *ODE Architect Companion.* John Wiley & Sons, New York.

T. F. Coleman and C. van Loan [1988] *Handbook for Matrix Computations.* SIAM, Philadelphia, PA.

R. D. Cook [1995] *Finite Element Modeling for Stress Analysis.* Wiley, New York.

J. W. Cooley and J. W. Tukey [1965] "An Algorithm for the Machine Calculation of Complex Fourier Series." *Mathematics of Computation* **19**, 297–301.

T. Cormen, C. Leiserson, R. Rivest and C. Stein [2009] *Introduction to Algorithms*, 3rd ed. MIT Press, Cambridge, MA.

R. Courant, K. O. Friedrichs and H. Lewy [1928] "Über die Partiellen Differenzengleichungen der Mathematischen Physik." *Mathematischen Annalen* **100**, 32–74.

J. Crank and P. Nicolson [1947] "A Practical Method for Numerical Evaluation of Solutions of Partial Differential Equations of the Heat Conduction Type." *Proceedings of the Cambridge Philosophical Society* **43**, 1–67.

J. Cuppen [1981] "A Divide and Conquer Method for the Symmetric Tridiagonal Eigenproblem." *Numerische Mathematik* **36**, 177–195.

B. Datta [2010] *Numerical Linear Algebra and Applications*, 2nd ed. SIAM, Philadelphia, PA.

A. Davies and P. Samuels [1996] *An Introduction to Computational Geometry for Curves and Surfaces.* Oxford University Press, Oxford.

P. J. Davis [1975] *Interpolation and Approximation.* Dover, New York.

P. Davis and P. Rabinowitz [1984] *Methods of Numerical Integration*, 2nd ed. Academic Press, New York.

T. Davis [2006] *Direct Methods for Sparse Linear Systems.* SIAM, Philadelphia, PA.

C. de Boor [2001] *A Practical Guide to Splines*, 2nd ed. Springer-Verlag, New York.

J. W. Demmel [1997] *Applied Numerical Linear Algebra.* Society for Industrial and Applied Mathematics, Philadelphia, PA.

J. E. Dennis and Jr., R. B. Schnabel [1987] *Numerical Methods for Unconstrained Optimization and Nonlinear Equations.* SIAM, Philadelphia, PA.

C. S. Desai and T. Kundu [2001] *Introductory Finite Element Method.* CRC Press, Boca Raton, FL.

P. Dierckx [1995] *Curve and Surface Fitting with Splines.* Oxford University Press, New York.

J. R. Dormand [1996] *Numerical Methods for Differential Equations.* CRC Press: Boca Raton, FL.

N. Draper and H. Smith [2001] *Applied Regression Analysis*, 3rd ed. John Wiley and Sons, New York.

T. Driscoll [2009] *Learning MATLAB.* SIAM, Philadelphia, PA.

P. Duhamel and M. Vetterli [1990] "Fast Fourier Transforms: A Tutorial Review and a State of the Art." *Signal Processing* **19**, 259–299.

C. Edwards and D. Penny [2014] *Differential Equations and Boundary Value Problems*, 5th ed. Prentice Hall, Upper Saddle River, NJ.

H. Elman, D. J. Silvester and A. Wathen [2004] *Finite Elements and Fast Iterative Solvers.* Oxford University Press, Oxford, UK.

H. Engels [1980] *Numerical Quadrature and Cubature.* Academic Press, New York.

G. Evans [1993] *Practical Numerical Integration.* John Wiley and Sons, New York.

L. C. Evans [2010] *Partial Differential Equations*, 2nd ed. AMS Publications, Providence, RI.

G. Farin [1990] *Curves and Surfaces for Computer-aided Geometric Design*, 2nd ed. Academic Press, New York.

G. S. Fishman [1996] *Monte Carlo: Concepts, Algorithms, and Applications.* Springer-Verlag, New York.

C. A. Floudas, P. M. Pardalos, C. Adjiman, W. R. Esposito, Z. H. Gms, S. T. Harding, J. L. Klepeis, C. A. Meyer, and C. A. Schweiger [1999] *Handbook of Test Problems in Local and Global Optimization*, Vol. 33, Series titled Nonconvex Optimization and its Applications, Springer, Berlin, Germany.

B. Fornberg [1998] *A Practical Guide to Pseudospectral Methods.* Cambridge University. Press, Cambridge, UK.

J. Fox [1997] *Applied Regression Analysis, Linear Models, and Related Methods.* Sage Publishing, New York.

M. Frigo and S. G. Johnson [1998] "FFTW: An Adaptive Software Architecture for the FFT." *Proceedings ICASSP* **3**, 1381–1384.

C. W. Gear [1971] *Numerical Initial Value Problems in Ordinary Differential Equations*. Prentice-Hall, Englewood Cliffs, NJ.

J. E. Gentle [2003] *Random Number Generation and Monte Carlo Methods*, 2nd ed. Springer-Verlag, New York.

A. George and J. W. Liu [1981] *Computer Solution of Large Sparse Positive Definite Systems*. Prentice Hall, Englewood Cliff, NJ.

M. Gockenbach [2006] *Understanding and Implementing the Finite Element Method*. SIAM, Philadelphia, PA.

M. Gockenbach [2010] *Partial Differential Equations: Analytical and Numerical Methods*, 2nd ed. SIAM, Philadelphia, PA.

D. Goldberg [1991] "What Every Computer Scientist Should Know about Floating Point Arithmetic." *ACM Computing Surveys* **23**, 5–48.

G. H. Golub and C. F. Van Loan [2012] *Matrix Computations*, 4th ed. Johns Hopkins University Press, Baltimore, MD.

D. Gottlieb and S. Orszag [1977] *Numerical Analysis of Spectral Methods: Theory and Applications*. SIAM, Philadelphia, PA.

T. Gowers, J. Barrow-Green, and I. Leader [2008] *The Princeton Companion to Mathematics*. Princeton University Press, Princeton, NJ.

I. Griva, S. Nash, and A. Sofer [2008] *Linear and Nonlinear Programming*, 2nd ed. SIAM, Philadelphia, PA.

C. Grossmann, H. Roos, and M. Stynes [2007] *Numerical Treatment of Partial Differential Equations*. Springer, Berlin, Germany.

B. Guenter and R. Parent [1990] "Motion Control: Computing the Arc Length of Parametric Curves." *IEEE Computer Graphics and Applications* **10**, 72–78.

S. Haber [1970] "Numerical Evaluation of Multiple Integrals." *SIAM Review* **12**, 481–526.

R. Haberman [2012] *Applied Partial Differential Equations with Fourier Series and Boundary Value Problems*, 5th ed. Prentice Hall, Upper Saddle River, NJ.

W. Hackbush [1994] *Iterative Solution of Large Sparse Systems of Equations*. Springer-Verlag, New York.

S. Hacker [2000] *MP3: The Definitive Guide*. O'Reilly Publishing, Sebastopol, CA.

B. Hahn [2002] *Essential MATLAB for Scientists and Engineers*, 3rd ed. Elsevier, Amsterdam.

E. Hairer, S. P. Norsett, and G. Wanner [2011] *Solving Ordinary Differential Equations I: Nonstiff Problems*, 2nd ed., Springer-Verlag, Berlin.

E. Hairer and G. Wanner [2004] *Solving Ordinary Differential Equations II: Stiff and Differential-algebraic Problems*, 2nd ed., Springer-Verlag, Berlin.

C. Hall and T. Porsching [1990] *Numerical Analysis of Partial Differential Equations*. Prentice Hall, Englewood Cliffs, NJ.

J. H. Halton [1960] "On the Efficiency of Certain Quasi-Random Sequences of Points in Evaluating Multi-Dimensional Integrals." *Numerische Mathematik* **2**, 84–90.

M. Heath [2002] *Scientific Computing*, 2nd ed. McGraw-Hill, New York.

P. Hellekalek [1998] "Good Random Number Generators Are (Not So) Easy to Find." *Mathematics and Computers in Simulation* **46**, 485–505.

P. Henrici [1962] *Discrete Variable Methods in Ordinary Differential Equations*. New York, John Wiley & Sons, New York.

M. R. Hestenes and E. Steifel [1952] "Methods of Conjugate Gradients for Solving Linear Systems." *Journal of Research National Bureau of Standards* **49**, 409–436.

R. C. Hibbeler [2014] *Structural Analysis*, 9th ed. Prentice Hall, Englewood Cliffs, NJ.

D. J. Higham [2001] "An Algorithmic Introduction to Numerical Simulation of Stochastic Differential Equations." *SIAM Review* **43**, 525–546.

D. J. Higham and N. J. Higham [2006] *MATLAB Guide*, 2nd ed. SIAM, Philadelphia, PA.

N. J. Higham [2002] *Accuracy and Stability of Numerical Algorithms*, 2nd ed. SIAM, Philadelphia, PA.

B. Hoffmann-Wellenhof, H. Lichtenegger, and J. Collins [2001] *Global Positioning System: Theory and Practice*, 5th ed. Springer-Verlag, New York.

J. Hoffman [2001] *Numerical Methods for Engineers and Scientists*, 2nd ed. CRC Press, New York.

K. Höllig [2003] *Finite Element Methods with B-Splines*. SIAM, Philadelphia, PA.

M. Holmes [2006] *Introduction to Numerical Methods in Differential Equations*. Springer, New York.

M. Holmes [2009] *Introduction to the Foundations of Applied Mathematics*. Springer, New York.

A. S. Householder [1970] *The Numerical Treatment of a Single Nonlinear Equation*. McGraw-Hill, New York.

J. V. Huddleston [2000] *Extensibility and Compressibility in One-dimensional Structures*, 2nd ed. ECS Publishing, Buffalo, NY.

D. A. Huffman [1952] "A Method for the Construction of Minimum-Redundancy Codes." *Proceedings of the IRE* **40**, 1098–1101.

J. C. Hull [2014] *Options, Futures, and Other Derivatives*, 9th ed. Prentice Hall, Upper Saddle River, NJ.

IEEE [1985] Standard for Binary Floating Point Arithmetic, IEEE Std. 754-1985, IEEE, New York.

I. Ipsen [2009] *Numerical Matrix Analysis: Linear Systems and Least Squares*. SIAM, Philadelphia, PA.

A. Iserles [1996] *A First Course in the Numerical Analysis of Differential Equations*, Cambridge University Press, Cambridge, UK.

C. Johnson [2009] *Numerical Solution of Partial Differential Equations by the Finite Element Method.* Dover Publications, New York.

P. Kattan [2007] *MATLAB Guide to Finite Elements*, 2nd ed. Springer, New York.

H. B. Keller [1968] *Numerical Methods of Two-Point Boundary-Value Problems.* Blaisdell, Waltham, MA.

C. T. Kelley [1987] *Iterative Methods for Linear and Nonlinear Problems.* SIAM, Philadelphia, PA.

J. Kepner [2009] *Parallel MATLAB for Multicore and Multinode Computers.* SIAM, Philadelphia, PA.

F. Klebaner [2005] *Introduction to Stochastic Calculus with Applications*, 2nd ed. Imperial College Press, London.

P. Kloeden and E. Platen [1992] *Numerical Solution of Stochastic Differential Equations.* Springer-Verlag, Berlin, Germany.

P. Kloeden, E. Platen, and H. Schurz [1994] *Numerical Solution of SDE through Computer Experiments.* Springer-Verlag, Berlin, Germany.

P. Knaber, and L. Angerman [2003] *Numerical Methods for Elliptic and Parabolic Partial Differential Equations.* Springer, Berlin, Germany.

D. Knuth [1981] *The Art of Computer Programming.* Addison-Wesley, Reading, MA.

D. Knuth [1997] *The Art of Computer Programming, Vol. 2: Seminumerical Algorithms*, 3rd ed. Addison-Wesley, Reading, MA.

E. Kostelich and D. Armbruster [1997] *Introductory Differential Equations: From Linearity to Chaos.* Addison Wesley, Boston, MA.

A. Krommer and C. Ueberhuber [1998] *Computational Integration.* SIAM, Philadelphia, PA.

M. Kutner, C. Nachtsheim, J. Neter, and W. Li [2004] *Applied Linear Statistical Models*, 5th ed. McGraw-Hill, New York.

J. C. Lagarias, J. A. Reeds, M. H. Wright, and P. E. Wright [1998] "Convergence Properties of the Nelder-Mead Simplex Method in Low Dimensions." *SIAM Journal of Optimization* **9**, 112–147.

J. D. Lambert [1991] *Numerical Methods for Ordinary Differential Systems*, John Wiley & Sons, New York.

L. Lapidus and G. F. Pinder [1982] *Numerical Solution of Partial Differential Equations in Science and Engineering.* Wiley-Interscience, New York.

S. Larsson and V. Thomee [2008] *Partial Differential Equations with Numerical Methods.* Springer, Berlin, Germany.

C. L. Lawson and R. J. Hanson [1995] *Solving Least Squares Problems.* SIAM, Philadelphia, PA.

D. Lay [2015] *Linear Algebra and Its Applications*, 5th ed. Pearson Education, Boston, MA.

K. Levenberg [1944] "A Method for the Solution of Certain Nonlinear Problems in Least Squares." *The Quarterly of Applied Mathematics* **2**,164–168.

R. Leveque [2007] *Finite Difference Methods for Ordinary and Partial Differential Equations.* SIAM, Philadelphia, PA.

J. D. Logan [2015] *Applied Partial Differential Equations*, 3rd ed. Springer, New York.

D. L. Logan [2011] *A First Course in the Finite Element Method*, 5th ed. CL-Engineering, New York.

H. S. Malvar [1992] *Signal Processing with Lapped Transforms.* Artech House, Norwood, MA.

D. Marquardt [1963] "An Algorithm for Least-Squares Estimation of Nonlinear Parameters." *SIAM J. on Applied Mathematics* **11**, 431–441.

G. Marsaglia [1968] "Random Numbers Fall Mainly in the Planes." *Proceedings of the National Academy of Sciences* **61**, 25.

G. Marsaglia and A. Zaman [1991] "A New Class of Random Number Generators." *Annals of Applied Probability* **1**, 462–480.

G. Marsaglia and W. W. Tsang [2000] "The Ziggurat Method for Generating Random Variables," *Journal of Statistical Software* **5**, 1–7.

R. McDonald [2006] *Derivatives Markets*, 2nd ed. Pearson Education, Boston, MA.

P. J. McKenna and C. Tuama [2001] "Large Torsional Oscillations in Suspension Bridges Visited Again: Vertical Forcing Creates Torsional Response." *American Mathematical Monthly* **108**, 738–745.

J.-P. Merlet [2000] *Parallel Robots.* Kluwer Academic Publishers, London.

A. R. Mitchell and D. F. Griffiths [1980] *The Finite Difference Method in Partial Differential Equations.* Wiley, New York.

C. Moler [2004] *Numerical Computing with MATLAB.* SIAM, Philadelphia, PA.

J. Moré and S. Wright [1987] *Optimization Software Guide.* SIAM, Philadelphia, PA.

K. W. Morton and D. F. Mayers [2006] *Numerical Solution of Partial Differential Equations*, 2nd ed. Cambridge University Press, Cambridge, UK.

J. A. Nelder and R. Mead [1965] "A Simplex Method for Function Minimization." *Computer Journal* **7**, 308–313.

M. Nelson and J. Gailly [1995] *The Data Compression Book*, 2nd ed. M&T Books, Redwood City, CA.

H. Niederreiter [1992] *Random Number Generation and Quasi-Monte Carlo Methods.* SIAM, Philadelphia, PA.

J. Nocedal and S. Wright [2006] *Numerical Optimization,* 2nd ed. Springer Series in Operations Research. Springer, New York.

B. Oksendal [2010] *Stochastic Differential Equations: An Introduction with Applications*, 6th ed. Springer-Verlag, Berlin, Germany.

A. Oppenheim and R. Schafer [2009] *Discrete-time Signal Processing*, 3rd ed. Prentice Hall, Upper Saddle River, NJ.

J. M. Ortega [1972] *Numerical Analysis: A Second Course*. Academic Press, New York.

A. M. Ostrowski [1966] *Solution of Equations and Systems of Equations*, 2nd ed. Academic Press, New York.

M. Overton [2001] *Numerical Computing with IEEE Floating Point Arithmetic*. SIAM, Philadelphia, PA.

S. Park and K. Miller [1988] "Random Number Generators: Good Ones Are Hard to Find." *Communications of the ACM* **31**, 1192–1201.

B. Parlett [1998] *The Symmetric Eigenvalue Problem*. SIAM, Philadelphia, PA.

B. Parlett [2000] "The QR Algorithm." *Computing in Science and Engineering* **2**, 38–42.

W. Pennebaker and J. Mitchell [1993] *JPEG Still Image Data Compression Standard*. Van Nostrand Reinhold, New York.

R. Piessens, E. de Doncker-Kapenga, C. Ueberhuber, and D. Kahaner [1983] *QUADPACK: A Subroutine Package for Automatic Integration*, Springer, New York.

G. Pinski and F. Narin [1976] "Citation Influence for Journal Aggregates of Scientific Publications: Theory, with Application to the Literature of Physics." *Information Processing and Management* **12**, 297–312.

J. Polking [2003] *Ordinary Differential Equations Using MATLAB*, 3rd ed. Pearson Education, Boston, MA.

H. Prautzsch, W. Boehm, and M. Paluszny [2002] *Bézier and B-Spline Techniques*. Springer, Berlin, Germany.

A. Quarteroni, R. Sacco, and F. Saleri [2000] *Numerical Mathematics*. Springer, Berlin, Germany.

K. R. Rao and J. J. Hwang [1996] *Techniques and Standards for Image, Video, and Audio Coding*. Prentice Hall, Upper Saddle River, NJ.

K. R. Rao and P. Yip [1990] *Discrete Cosine Transform: Algorithms, Advantages, Applications*. Academic Press, Boston, MA.

J. R. Rice and R. F. Boisvert [1984] *Solving Elliptic Problems Using ELLPACK*. Springer-Verlag, New York.

T. J. Rivlin [1981] *An Introduction to the Approximation of Functions*, 2nd ed. Dover, New York.

T. J. Rivlin [1990] *Chebyshev Polynomials*, 2nd ed. John Wliey and Sons, New York.

S. Roberts and J. Shipman [1972] *Two-Point Boundary Value Problems: Shooting Methods*. Elsevier, New York.

R. Y. Rubinstein [1981] *Simulation and the Monte Carlo Method*. John Wiley, New York.

T. Ryan [1997] *Modern Regression Methods*. John Wiley and Sons.

Y. Saad [2003] *Iterative Methods for Sparse Linear Systems*, 2nd ed. SIAM, Philadelphia, PA.

D. Salomon [2005] *Curves and Surfaces for Computer Graphics*. Springer, New York.

K. Sayood [1996] *Introduction to Data Compression*. Morgan Kaufmann Publishers, San Francisco.

M. H. Schultz [1973] *Spline Analysis*. Prentice Hall, Englewood Cliffs, NJ.

L. L. Schumaker [1981] *Spline Functions: Basic Theory*. John Wiley, New York.

L. F. Shampine [1994] *Numerical Solution of Ordinary Differential Equations*. Chapman & Hall, New York.

L. F. Shampine, I. Gladwell, and S. Thompson [2003] *Solving ODEs with MATLAB*. Cambridge University Press, Cambridge, UK.

L. F. Shampine and M. W. Reichelt [1997] "The MATLAB ODE Suite." *SIAM Journal on Scientific Computing* **18**, 1–22.

K. Sigmon and T. Davis [2002] *MATLAB Primer*, 6th ed. CRC Press, Boca Raton, FL.

S. Skiena [2008] *The Algorithm Design Manual*, 2nd ed. Springer, New York.

I. Smith and D. Griffiths [2004] *Programming the Finite Element Method*. John Wiley, New York.

B. T. Smith, J. M. Boyle, Y. Ikebe, V. Klema, and C. B. Moler [1970] *Matrix Eigensystem Routines: EISPACK Guide*, 2nd ed. Springer-Verlag, New York.

W. Stallings [2003] *Computer Organization and Architecture*, 6th ed. Prentice Hall, Upper Saddle River, NJ.

J. M. Steele [2001] *Stochastic Calculus and Financial Applications*. Springer-Verlag, New York.

G. W. Stewart [1973] *Introduction to Matrix Computations*. Academic Press, New York.

G. W. Stewart [1998] *Afternotes on Numerical Analysis: Afternotes Goes to Graduate School*. SIAM, Philadelphia, PA.

J. Stoer and R. Bulirsch [2002] *Introduction to Numerical Analysis*, 3rd ed. Springer-Verlag, New York.

J. A. Storer [1988] *Data Compression: Methods and Theory*. Computer Science Press, Rockville, MD.

G. Strang [2006] *Linear Algebra and Its Applications*, 4th ed. Saunders, Philadelphia.

G. Strang [2007] *Computational Science and Engineering*. Wellesley Cambridge Press, Cambridge, MA.

G. Strang and K. Borre [1997] *Linear Algebra, Geodesy, and GPS*. Wellesley Cambridge Press, Cambridge, MA.

G. Strang and G. J. Fix [2008] *An Analysis of the Finite Element Method*, 2nd ed. Prentice-Hall, Englewood Cliffs, NJ.

J. C. Strikwerda [1989] *Finite Difference Schemes and Partial Differential Equations.* Wadsworth and Brooks-Cole, Pacific Grove, CA.

W. A. Strauss [1992] *Partial Differential Equations: An Introduction.* John Wiley and Sons, New York.

A. Stroud and D. Secrest [1966] *Gaussian Quadrature Formulas*, Prentice Hall, Englewood Cliffs, NJ.

P. N. Swarztrauber [1982] "Vectorizing the FFTs." In: *Parallel Computations*, ed. G. Rodrigue, pp. 51–83. Academic Press, New York.

D. S. Taubman and M. W. Marcellin [2002] *JPEG 2000: Image Compression Fundamentals, Standards and Practice.* Kluwer, Boston, MA.

J. Traub [1964] *Iterative Methods for the Solution of Equations.* Prentice-Hall, Englewood Cliffs, NJ.

N. Trefethen [2001] *Spectral Methods in MATLAB.* SIAM, Philadelphia, PA.

N. Trefethen and D. Bau [1997] *Numerical Linear Algebra.* SIAM, Philadelphia, PA.

A. Turing [1952] "The Chemical Basis of Morphogenesis." *Philosophical Transactions Royal of the Society Lond.* B **237**, 3772.

C. Van Loan [1992] *Computational Frameworks for the Fast Fourier Transform.* SIAM, Philadelphia, PA.

C. Van Loan and K. Fan [2010] *Insight Through Computing: A MATLAB Introduction to Computational Science and Engineering.* SIAM, Philadelphia, PA.

R. S. Varga [2000] *Matrix Iterative Analysis*, 2nd ed. Springer-Verlag, New York.

J. Volder [1959] "The CORDIC Trigonometric Computing Technique." *IRE Transactions on Electronic Computing* **8**, 330–334.

G. K. Wallace [1991] "The JPEG Still Picture Compression Standard." *Communications of the ACM* **34**, 30–44.

H. Wang, J. Kearney, and K. Atkinson [2003] "Arc-length Parameterized Spline Curves for Real-time Simulation." In: *Curve and Surface Design*: Saint Malo 2002, eds. T. Lyche, M. Mazure, and L. Schumaker. Nashboro Press, Brentwood, TN.

Y. Wang and M. Vilermo [2003] "The Modified Discrete Cosine Transform: Its Implications for Audio Coding and Error Concealment." *Journal of the Audio Engineering Society* **51**, 52–62.

D. S. Watkins [1982] "Understanding the QR Algorithm." *SIAM Review* **24**, 427–440.

D. S. Watkins [2007] *The Matrix Eigenvalue Problem: GR and Krylow Subspace Methods.* SIAM, Philadelphia, PA.

J. Wilkinson [1965] *The Algebraic Eigenvalue Problem.* Clarendon Press, Oxford.

J. Wilkinson [1984] "The Perfidious Polynomial." In: Studies in Numerical Analysis, Ed: G. Golub. MAA, Washington, DC.

J. Wilkinson [1994] *Rounding Errors in Algebraic Processes.* Dover, New York.

J. Wilkinson and C. Reinsch [1971] *Handbook for Automatic Computation, Vol. 2: Linear Algebra.* Springer-Verlag, New York.

P. Wilmott, S. Howison, and J. Dewynne [1995] *The Mathematics of Financial Derivatives.* Cambridge University Press, Oxford and New York.

S. Winograd [1978] "On Computing the Discrete Fourier Transform." *Mathematics of Computation* **32**, 175–199.

F. Yamaguchi [1988] *Curves and Surfaces in Computer-aided Geometric Design.* Springer-Verlag, New York.

D. M. Young [1971] *Iterative Solution of Large Linear Systems.* Academic Press, New York.

Index